Molecular Biology of Selenium in Health and Disease

Molecular Biology of Selenium in Health and Disease

Editors

Petra A. Tsuji
Dolph L. Hatfield

MDPI • Basel • Beijing • Wuhan • Barcelona • Belgrade • Manchester • Tokyo • Cluj • Tianjin

Editors
Petra A. Tsuji
Department of Biological
Sciences, Towson University
USA

Dolph L. Hatfield
Scientist Emeritus, Mouse Cancer Genetics Program,
Center for Cancer Research,
National Cancer Institute (NCI)
USA

Editorial Office
MDPI
St. Alban-Anlage 66
4052 Basel, Switzerland

This is a reprint of articles from the Special Issue published online in the open access journal *International Journal of Molecular Sciences* (ISSN 1422-0067) (available at: https://www.mdpi.com/journal/ijms/special_issues/Selenium_Health_Disease).

For citation purposes, cite each article independently as indicated on the article page online and as indicated below:

LastName, A.A.; LastName, B.B.; LastName, C.C. Article Title. *Journal Name* **Year**, *Volume Number*, Page Range.

ISBN 978-3-0365-3307-0 (Hbk)
ISBN 978-3-0365-3308-7 (PDF)

© 2022 by the authors. Articles in this book are Open Access and distributed under the Creative Commons Attribution (CC BY) license, which allows users to download, copy and build upon published articles, as long as the author and publisher are properly credited, which ensures maximum dissemination and a wider impact of our publications.

The book as a whole is distributed by MDPI under the terms and conditions of the Creative Commons license CC BY-NC-ND.

Contents

About the Editors ix

Petra A. Tsuji and Dolph L. Hatfield
Editorial to Special Issue Molecular Biology of Selenium in Health and Disease
Reprinted from: *Int. J. Mol. Sci.* 2022, 23, 808, doi:10.3390/ijms23020808 1

Olivia M. Guillin, Caroline Vindry, Théophile Ohlmann and Laurent Chavatte
Interplay between Selenium, Selenoproteins and HIV-1 Replication in Human CD4 T-Lymphocytes
Reprinted from: *Int. J. Mol. Sci.* 2022, 23, 1394, doi:10.3390/ijms23031394 5

Qian Sun, Sebastian Mehl, Kostja Renko, Petra Seemann, Christian L. Görlich, Julian Hackler, Waldemar B. Minich, George J. Kahaly and Lutz Schomburg
Natural Autoimmunity to Selenoprotein P Impairs Selenium Transport in Hashimoto's Thyroiditis
Reprinted from: *Int. J. Mol. Sci.* 2021, 22, 13088, doi:10.3390/ijms222313088 29

Lenny K. Hong, Shrinidhi Kadkol, Maria Sverdlov, Irida Kastrati, Mostafa Elhodaky, Ryan Deaton, Karen S. Sfanos, Heidi Wang, Li Liu and Alan M. Diamond
Loss of SELENOF Induces the Transformed Phenotype in Human Immortalized Prostate Epithelial Cells
Reprinted from: *Int. J. Mol. Sci.* 2021, 22, 12040, doi:10.3390/ijms222112040 43

Jeyoung Bang, Minguk Han, Tack-Jin Yoo, Lu Qiao, Jisu Jung, Jiwoon Na, Bradley A. Carlson, Vadim N. Gladyshev, Dolph L. Hatfield, Jin-Hong Kim, Lark Kyun Kim and Byeong Jae Lee
Identification of Signaling Pathways for Early Embryonic Lethality and Developmental Retardation in *Sephs1*$^{-/-}$ Mice
Reprinted from: *Int. J. Mol. Sci.* 2021, 22, 11647, doi:10.3390/ijms222111647 59

Jisu Jung, Yoomin Kim, Jiwoon Na, Lu Qiao, Jeyoung Bang, Dongin Kwon, Tack-Jin Yoo, Donghyun Kang, Lark Kyun Kim, Bradley A. Carlson, Dolph L. Hatfield, Jin-Hong Kim and Byeong Jae Lee
Constitutive Oxidative Stress by SEPHS1 Deficiency Induces Endothelial Cell Dysfunction
Reprinted from: *Int. J. Mol. Sci.* 2021, 22, 11646, doi:10.3390/ijms222111646 81

Noelia Fradejas-Villar, Simon Bohleber, Wenchao Zhao, Uschi Reuter, Annika Kotter, Mark Helm, Rainer Knoll, Robert McFarland, Robert W. Taylor, Yufeng Mo, Kenjyo Miyauchi, Yuriko Sakaguchi, Tsutomu Suzuki and Ulrich Schweizer
The Effect of tRNA$^{[Ser]Sec}$ Isopentenylation on Selenoprotein Expression
Reprinted from: *Int. J. Mol. Sci.* 2021, 22, 11454, doi:10.3390/ijms222111454 99

Zhan-Ling Liang, Heng Wee Tan, Jia-Yi Wu, Xu-Li Chen, Xiu-Yun Wang, Yan-Ming Xu and Andy T. Y. Lau
The Impact of ZIP8 Disease-Associated Variants G38R, C113S, G204C, and S335T on Selenium and Cadmium Accumulations: The First Characterization
Reprinted from: *Int. J. Mol. Sci.* 2021, 22, 11399, doi:10.3390/ijms222111399 119

Theresa Wolfram, Leonie M. Weidenbach, Johanna Adolf, Maria Schwarz, Patrick Schädel, André Gollowitzer, Oliver Werz, Andreas Koeberle, Anna P. Kipp and Solveigh C. Koeberle
The Trace Element Selenium Is Important for Redox Signaling in Phorbol Ester-Differentiated THP-1 Macrophages
Reprinted from: *Int. J. Mol. Sci.* **2021**, *22*, 11060, doi:10.3390/ijms222011060 . . . . **139**

Daniel J. Torres, Matthew W. Pitts, Lucia A. Seale, Ann C. Hashimoto, Katlyn J. An, Ashley N. Hanato, Katherine W. Hui, Stella Maris A. Remigio, Bradley A. Carlson, Dolph L. Hatfield and Marla J. Berry
Female Mice with Selenocysteine tRNA Deletion in Agrp Neurons Maintain Leptin Sensitivity and Resist Weight Gain While on a High-Fat Diet
Reprinted from: *Int. J. Mol. Sci.* **2021**, *22*, 11010, doi:10.3390/ijms222011010 . . . . **159**

Atsuki Shimizu, Ryuta Tobe, Riku Aono, Masao Inoue, Satoru Hagita, Kaito Kiriyama, Yosuke Toyotake, Takuya Ogawa, Tatsuo Kurihara, Kei Goto, N. Tejo Prakash and Hisaaki Mihara
Initial Step of Selenite Reduction via Thioredoxin for Bacterial Selenoprotein Biosynthesis
Reprinted from: *Int. J. Mol. Sci.* **2021**, *22*, 10965, doi:10.3390/ijms222010965 . . . . **177**

Jessica A. Canter, Sarah E. Ernst, Kristin M. Peters, Bradley A. Carlson, Noelle R. J. Thielman, Lara Grysczyk, Precious Udofe, Yunkai Yu, Liang Cao, Cindy D. Davis, Vadim N. Gladyshev, Dolph L. Hatfield and Petra A. Tsuji
Selenium and the 15 kDa Selenoprotein Impact Colorectal Tumorigenesis by Modulating Intestinal Barrier Integrity
Reprinted from: *Int. J. Mol. Sci.* **2021**, *22*, 10651, doi:10.3390/ijms221910651 . . . . **187**

Hyunwoo Kang, Yeong Ha Jeon, Minju Ham, Kwanyoung Ko and Ick Young Kim
Compensatory Protection of Thioredoxin-Deficient Cells from Etoposide-Induced Cell Death by Selenoprotein W via Interaction with 14-3-3
Reprinted from: *Int. J. Mol. Sci.* **2021**, *22*, 10338, doi:10.3390/ijms221910338 . . . . **207**

Jordan Sonet, Anne-Laure Bulteau, Zahia Touat-Hamici, Maurine Mosca, Katarzyna Bierla, Sandra Mounicou, Ryszard Lobinski and Laurent Chavatte
Selenoproteome Expression Studied by Non-Radioactive Isotopic Selenium-Labeling in Human Cell Lines
Reprinted from: *Int. J. Mol. Sci.* **2021**, *22*, 7308, doi:10.3390/ijms22147308 . . . . **219**

Sabrina Sales Martinez, Yongjun Huang, Leonardo Acuna, Eduardo Laverde, David Trujillo, Manuel A. Barbieri, Javier Tamargo, Adriana Campa and Marianna K. Baum
Role of Selenium in Viral Infections with a Major Focus on SARS-CoV-2
Reprinted from: *Int. J. Mol. Sci.* **2022**, *23*, 280, doi:10.3390/ijms23010280 . . . . **239**

Petra A. Tsuji, Didac Santesmasses, Byeong J. Lee, Vadim N. Gladyshev and Dolph L. Hatfield
Historical Roles of Selenium and Selenoproteins in Health and Development: The Good, the Bad and the Ugly
Reprinted from: *Int. J. Mol. Sci.* **2022**, *23*, 5, doi:10.3390/ijms23010005 . . . . **259**

Paul R. Copeland and Michael T. Howard
Ribosome Fate during Decoding of UGA-Sec Codons
Reprinted from: *Int. J. Mol. Sci.* **2021**, *22*, 13204, doi:10.3390/ijms222413204 . . . . **279**

Erik Schoenmakers and Krishna Chatterjee
Human Genetic Disorders Resulting in Systemic Selenoprotein Deficiency
Reprinted from: *Int. J. Mol. Sci.* **2021**, *22*, 12927, doi:10.3390/ijms222312927 . . . . **291**

Didac Santesmasses and Vadim N. Gladyshev
Pathogenic Variants in Selenoproteins and Selenocysteine Biosynthesis Machinery
Reprinted from: *Int. J. Mol. Sci.* **2021**, *22*, 11593, doi:10.3390/ijms222111593 . . . . **305**

Chi Ma, Verena Martinez-Rodriguez and Peter R. Hoffmann
Roles for Selenoprotein I and Ethanolamine Phospholipid Synthesis in T Cell Activation
Reprinted from: *Int. J. Mol. Sci.* **2021**, *22*, 11174, doi:10.3390/ijms222011174 . . . . **321**

Briana K. Shimada, Naghum Alfulaij and Lucia A. Seale
The Impact of Selenium Deficiency on Cardiovascular Function
Reprinted from: *Int. J. Mol. Sci.* **2021**, *22*, 10713, doi:10.3390/ijms221910713 . . . . **333**

Lutz Schomburg
Selenium Deficiency Due to Diet, Pregnancy, Severe Illness, or COVID-19—A Preventable Trigger for Autoimmune Disease
Reprinted from: *Int. J. Mol. Sci.* **2021**, *22*, 8532, doi:10.3390/ijms22168532 . . . . **349**

About the Editors

Petra A. Tsuji has been broadly trained in Zoology and Marine Biology and subsequently in the Biomedical Sciences and Public Health. In 2011, she moved from the National Cancer Institute's Cancer Prevention Fellowship Program, where she was a post-doc in Dr. Dolph Hatfield's Molecular Biology of Selenium lab, to Towson University, Towson, MD. Currently, she is an Associate Professor in the Department of Biological Sciences. Together with her students, she investigates molecular strategies to inhibit colon carcinogenesis with dietary nutrients, including the trace mineral selenium. Her passion is to mentor and empower young women in STEM, and she is co-director of Towson University's annual Women in Science Forum, co-faculty advisor to the Women in Science Club, and co-faculty leader of the Hill-Lopes Scholars Program, where she provides inclusive and intentional mentoring to foster the academic and professional skills young women need to succeed in today's STEM workforce.

Dolph L. Hatfield is a Scientist Emeritus in the Mouse Cancer Genetics Program, Center for Cancer Research, National Cancer Institute (NCI), National Institutes of Health (NIH), Bethesda, MD, USA. He has been active in the study of the molecular biology of selenium since the mid-1980s. He joined NCI in 1967 and recently retired as the Chief of the Section on the Molecular Biology of Selenium, Laboratory of Cancer Prevention. Over his career he has won various awards including the NIH Graduate School Student Community Outstanding Mentor Award, the NCI Mentor of Merit Award for Excellence in Mentoring and Guiding the Careers of Trainees in Cancer Research, and the Klaus Schwartz Commemorative Medal. He was also promoted to a RS 0401 Research Biologist in the Senior Biomedical Research Service NCI, and was elected as an AAAS Fellow. He has been active in civil rights issues since 1968 and the homeless in Washington, DC, from the late 1980s. In addition, he has written numerous articles since the 1990s on the abuse of African and Native Americans.

Editorial

Editorial to Special Issue Molecular Biology of Selenium in Health and Disease

Petra A. Tsuji [1],* and Dolph L. Hatfield [2]

1. Department of Biological Sciences, Towson University, 8000 York Road, Towson, MD 21252, USA
2. Scientist Emeritus, Mouse Cancer Genetics Program, Center for Cancer Research, National Cancer Institute, National Institutes of Health, Bethesda, MD 20892, USA; hatfielddolph@gmail.com
* Correspondence: ptsuji@towson.edu; Tel.: +1-410-704-4117

Citation: Tsuji, P.A.; Hatfield, D.L. Editorial to Special Issue Molecular Biology of Selenium in Health and Disease. *Int. J. Mol. Sci.* **2022**, *23*, 808. https://doi.org/10.3390/ijms23020808

Academic Editor: Claudio Santi

Received: 24 December 2021
Accepted: 5 January 2022
Published: 12 January 2022

Publisher's Note: MDPI stays neutral with regard to jurisdictional claims in published maps and institutional affiliations.

Copyright: © 2022 by the authors. Licensee MDPI, Basel, Switzerland. This article is an open access article distributed under the terms and conditions of the Creative Commons Attribution (CC BY) license (https://creativecommons.org/licenses/by/4.0/).

The selenium field expanded at a rapid rate for about 45 years, from the mid-1970's until about 2015 (see [1] for a summary of these 45 years). Then, the pace of major discoveries began to decline. However, we were fortunate enough to obtain many of the major players in the selenium field to write novel, up-to-date research articles and/or reviews on various topics of this fascinating element. Selenium is still regarded as one of the most interesting and health-beneficial elements. It will be fascinating to bear witness to additional discoveries in the field of selenium and selenoproteins in health and disease in the coming years.

There are 21 published manuscripts, including 13 research articles and 8 reviews, with over 120 different contributors, in this Special Issue, entitled "Molecular Biology of Selenium in Health and Disease". Many of the most important subjects in the selenium field are covered. These include a historical perspective of the roles of selenium and selenoproteins in health and development, including its beneficial and detrimental aspects [2], the role of selenium in redox signaling in macrophages [3], as well as in bacterial and human selenoprotein biosynthesis [4], and its proposed cellular transport via a metal cation symporter [5]. Furthermore, selenium deficiency is discussed in the setting of cardiovascular function [6] and as a trigger for autoimmune diseases [7]. This Special Issue also provides insights into selenoprotein mechanisms, including SELENOF in prostate [8] and colon cancers [9], SELENOI in immune response [10], SELENOW's interaction with thioredoxins [11], and SELENOP-mediated selenium transport in Hashimoto's thyroiditis [12]. Furthermore, an overview of the pathogenic variants in selenoproteins genes from a population genomics perspective [13], a review on the human genetic disorders resulting in systemic selenoprotein deficiency [14], and a new strategy to assess the selenoproteome by non-radioactive isotopic labeling are presented [15]. This Special Issue also addresses the mechanistic aspects of selenoprotein synthesis, including the role of selenophosphate synthetase in endothelial cells [16] and mice [17], what is known about the mechanisms of UGA recoding and the fate of ribosomes that fail to incorporate selenocysteine [18], tRNA$^{[Ser]Sec}$ isopentylation and its role in hypothalamic neurons [19], as well as the impact of a conditional tRNA$^{[Ser]Sec}$ knockout on leptin sensitivity and weight gain in agouti-related, peptide-positive hypothalamic neurons [20]. Lastly, a review of selenium and viral infections, including HIV and also the ongoing COVID-19 pandemic [21], and a research article on the interactions between selenium, selenoproteins and HIV-1 replication in human CD4 T-lymphocytes [22], provide important considerations regarding this trace element's impact on viral infections threatening human health.

When we were in the early stages of organizing this Special Issue on Molecular Biology of Selenium in Health and Disease, with the assistance of our internal editor at the *International Journal of Molecular Sciences*, Ms. Lachelle Fang, we asked her if we could dedicate this Special Issue to Professor (Prof.) Dr. Leopold Flohé, M.D., who made many important discoveries to the growth and expansion in the field of this unique element. We were delighted that Editor Fang informed us that we could. In addition, Ms. Fang was most helpful throughout the organization of our Special Issue by handling so much of the correspondence with authors, and by obtaining reviewers for each of the papers. We are certainly indebted to her.

This Special Issue on 'Molecular Biology of Selenium in Health and Disease' within the various *International Journal of Molecular Sciences* Special Issues, is dedicated to Prof. Dr. Leopold Flohé, M.D. (Figure 1) for his many contributions to the selenium field. His seminal discoveries have played important roles in developing the molecular biological aspects of selenium, particularly regarding the essential roles of selenium in health and the development of humans and many other higher lifeforms. These fundamental discoveries included the identity of the first selenium-containing protein, glutathione peroxidase in 1973 [23], a tetrameric selenoenzyme, designated glutathione peroxidase 1 (GPx1). The significance of this finding is regarded as providing the foundation of the selenium field [24], and it linked selenium to the underlying metabolism in humans and other mammals. Prof. Dr. Leopold Flohé's work was largely responsible for the Food and Drug Administration recognizing selenium as a daily supplement for domestic animals in 1979 and for humans in 1981 [25]. Interestingly, almost 10 years passed before another selenoprotein was found in animals. Prof. Dr. Flohé, M.D. and his group had a major hand in identifying the second selenoenzyme in mammals, a monomeric selenoenzme, gluthathione peroxidase 4 (GPx4) [26,27]. Subsequently, GPx4 was found to be an essential selenoprotein in mammalian cell development, while GPx1 was found to be a stress-related or non-essential selenoprotein in mammalian cell development [28].

Figure 1. Photograph of Prof. Dr. Leopold Flohé, M.D (taken from his LinkedIn profile at https://www.linkedin.com/in/leopold-floh%C3%A9-66b6239b, accessed on 22 December 2021).

Prof. Dr. Flohé's efforts were not limited to glutathione peroxidases. In the late 1990's, he and his research team identified additional novel enzymes, tryparedoxin and tryparedoxin peroxidase, which are part of the hydroperoxide metabolism in trypanosomatids [29]. Dr. Flohé's PhD student E. Nogoceke was awarded the 'Paper of the Year' award in 1998 for this work, leading to the discovery of the peroxidase system in trypanosomatids. Interestingly, tryparedoxin peroxidase is not only a homolog of the yeast thiol-dependent antioxidant protein discovered by Earl Stadtman in 1988, and was later found to be trypanosomatid's thioredoxin peroxidase [30]. Dr. Flohé continued his research efforts in the field of redoxbiology, focusing on mitochondria as an important source of superoxides [31,32], and hydroperoxide metabolism in mycobacteria [24,33].

Prof. Dr. Flohé, M.D. has also won numerous prestigious awards and been presented with numerous distinguished honors. He was awarded the Fellowship of the Studienstiftung des Deutschen Volkes (German Academic Scholarship Foundation funded by the Federal Ministry of Education and Research) to support his education from 1962 to 1968, which is Germany's oldest, largest, and most prestigious scholarship foundation. In 1973, he won the Award of the Anna-Monika-Foundation for the basic research he carried out on endogenous depression, and in 1985, he was presented the Claudius-Galenus Award for the production of urokinase by recombinant technology. Prof. Dr. Flohé, M.D. was presented with two superb awards in 1997, an Honorary Degree from the University of Buenos Aires, Argentina for his major achievements in parasitology, and the Klaus Schwarz Commemorative Medal for his pioneering work in research on the trace element selenium.

In 1998, Dr. Flohé was awarded the 'Science and Humanity Prize' for lifetime achievements by the Oxygen Club of California. A few years later, in 2006, he was awarded the Trevor Frank Slater Award and Gold Medal, which is the highest academic and research prize in the redox biology and medicine/free radical field conferred by the Society for Free Radical Research International, for lifetime achievements in science.

Furthermore, in 2010, he was recognized as a Redox Pioneer, an award given to authors whose publications on redox biology have been cited more than 1,000 times, and who have over 20 articles that have been cited more than 100 times [24]. A few years ago, in his honor, the Leopold Flohé Redox Pioneer Young Investigator Award was created by the Society for Free Radical Research International, which selects a scientist below the age of 45 years with "outstanding novel findings in the field of biological redox processes, working already independently, and leading an own group of young scientists having published high quality papers" (https://www.sfrr-europe.org/index.php/awards/leopold-flohe-award, accessed on 11 January 2022). Lastly, in 2019, he was nominated as an honorary member of the international Scientific Network "Selenium Sulfur Redox and Catalysis" (SeSRedCat). It is indeed a pleasure and an honor to dedicate this Special Issue of the *IJMS* to Prof. Dr. Leopold Flohé, M.D.

Author Contributions: All authors have read and agreed to the published version of the manuscript.

Funding: The APC was graciously waived by IJMS.

Conflicts of Interest: The authors declare no conflict of interest.

References

1. Hatfield, D.L.; Schweizer, U.; Tsuji, P.A.; Gladyshev, V.N. *Selenium—Its Molecular Biology and Role in Human Health*, 4th ed.; Hatfield, D.L., Schweizer, U., Tsuji, P.A., Gladyshev, V.N., Eds.; Springer Science+Business Media: New York, NY, USA, 2016.
2. Tsuji, P.A.; Santesmasses, D.; Lee, B.J.; Gladyshev, V.N.; Hatfield, D.L. Historical Roles of Selenium and Selenoproteins in Health and Development: The Good, the Bad and the Ugly. *Int. J. Mol. Sci.* **2022**, *23*, 5. [CrossRef]
3. Wolfram, T.; Weidenbach, L.M.; Adolf, J.; Schwarz, M.; Schädel, P.; Gollowitzer, A.; Werz, O.; Koeberle, A.; Kipp, A.P.; Koeberle, S.C. The Trace Element Selenium Is Important for Redox Signaling in Phorbol Ester-Differentiated THP-1 Macrophages. *Int. J. Mol. Sci.* **2021**, *22*, 11060. [CrossRef] [PubMed]
4. Shimizu, A.; Tobe, R.; Aono, R.; Inoue, M.; Hagita, S.; Kiriyama, K.; Toyotake, Y.; Ogawa, T.; Kurihara, T.; Goto, K.; et al. Initial Step of Selenite Reduction via Thioredoxin for Bacterial Selenoprotein Biosynthesis. *Int. J. Mol. Sci.* **2021**, *22*, 10965. [CrossRef]
5. Liang, Z.-L.; Tan, H.W.; Wu, J.-Y.; Chen, X.-L.; Wang, X.-Y.; Xu, Y.-M.; Lau, A.T.Y. The Impact of ZIP8 Disease-Associated Variants G38R, C113S, G204C, and S335T on Selenium and Cadmium Accumulations: The First Characterization. *Int. J. Mol. Sci.* **2021**, *22*, 11399. [CrossRef]
6. Shimada, B.K.; Alfulaij, N.; Seale, L.A. The Impact of Selenium Deficiency on Cardiovascular Function. *Int. J. Mol. Sci.* **2021**, *22*, 10713. [CrossRef]
7. Schomburg, L. Selenium Deficiency Due to Diet, Pregnancy, Severe Illness, or COVID-19—A Preventable Trigger for Autoimmune Disease. *Int. J. Mol. Sci.* **2021**, *22*, 8532. [CrossRef] [PubMed]
8. Hong, L.K.; Kadkol, S.; Sverdlov, M.; Kastrati, I.; Elhodaky, M.; Deaton, R.; Sfanos, K.S.; Wang, H.; Liu, L.; Diamond, A.M. Loss of SELENOF Induces the Transformed Phenotype in Human Immortalized Prostate Epithelial Cells. *Int. J. Mol. Sci.* **2021**, *22*, 12040. [CrossRef]
9. Canter, J.A.; Ernst, S.E.; Peters, K.M.; Carlson, B.A.; Thielman, N.R.J.; Grysczyk, L.; Udofe, P.; Yu, Y.; Cao, L.; Davis, C.D.; et al. Selenium and the 15kDa Selenoprotein Impact Colorectal Tumorigenesis by Modulating Intestinal Barrier Integrity. *Int. J. Mol. Sci.* **2021**, *22*, 10651. [CrossRef]
10. Ma, C.; Martinez-Rodriguez, V.; Hoffmann, P.R. Roles for Selenoprotein I and Ethanolamine Phospholipid Synthesis in T Cell Activation. *Int. J. Mol. Sci.* **2021**, *22*, 11174. [CrossRef] [PubMed]
11. Kang, H.; Jeon, Y.H.; Ham, M.; Ko, K.; Kim, I.Y. Compensatory Protection of Thioredoxin-Deficient Cells from Etoposide-Induced Cell Death by Selenoprotein W via Interaction with 14-3-3. *Int. J. Mol. Sci.* **2021**, *22*, 10338. [CrossRef]

12. Sun, Q.; Mehl, S.; Renko, K.; Seemann, P.; Görlich, C.L.; Hackler, J.; Minich, W.B.; Kahaly, G.J.; Schomburg, L. Natural Autoimmunity to Selenoprotein P Impairs Selenium Transport in Hashimoto's Thyroiditis. *Int. J. Mol. Sci.* **2021**, *22*, 13088. [CrossRef] [PubMed]
13. Santesmasses, D.; Gladyshev, V.N. Pathogenic Variants in Selenoproteins and Selenocysteine Biosynthesis Machinery. *Int. J. Mol. Sci.* **2021**, *22*, 11593. [CrossRef] [PubMed]
14. Schoenmakers, E.; Chatterjee, K. Human Genetic Disorders Resulting in Systemic Selenoprotein Deficiency. *Int. J. Mol. Sci.* **2021**, *22*, 12927. [CrossRef]
15. Sonet, J.; Bulteau, A.-L.; Touat-Hamici, Z.; Mosca, M.; Bierla, K.; Mounicou, S.; Lobinski, R.; Chavatte, L. Selenoproteome Expression Studied by Non-Radioactive Isotopic Selenium-Labeling in Human Cell Lines. *Int. J. Mol. Sci.* **2021**, *22*, 7308. [CrossRef] [PubMed]
16. Jung, J.; Kim, Y.; Na, J.; Qiao, L.; Bang, J.; Kwon, D.; Yoo, T.-J.; Kang, D.; Kim, L.K.; Carlson, B.A.; et al. Constitutive Oxidative Stress by SEPHS1 Deficiency Induces Endothelial Cell Dysfunction. *Int. J. Mol. Sci.* **2021**, *22*, 11646. [CrossRef]
17. Bang, J.; Han, M.; Yoo, T.-J.; Qiao, L.; Jung, J.; Na, J.; Carlson, B.A.; Gladyshev, V.N.; Hatfield, D.L.; Kim, J.-H.; et al. Identification of Signaling Pathways for Early Embryonic Lethality and Developmental Retardation in $Sephs1^{-/-}$ Mice. *Int. J. Mol. Sci.* **2021**, *22*, 11647. [CrossRef]
18. Copeland, P.R.; Howard, M.T. Ribosome Fate during Decoding of UGA-Sec Codons. *Int. J. Mol. Sci.* **2021**, *22*, 13204. [CrossRef]
19. Fradejas-Villar, N.; Bohleber, S.; Zhao, W.; Reuter, U.; Kotter, A.; Helm, M.; Knoll, R.; McFarland, R.; Taylor, R.W.; Mo, Y.; et al. The Effect of tRNA$^{[Ser]Sec}$ Isopentenylation on Selenoprotein Expression. *Int. J. Mol. Sci.* **2021**, *22*, 11454. [CrossRef]
20. Torres, D.J.; Pitts, M.W.; Seale, L.A.; Hashimoto, A.C.; An, K.J.; Hanato, A.N.; Hui, K.W.; Remigio, S.M.A.; Carlson, B.A.; Hatfield, D.L.; et al. Female Mice with Selenocysteine tRNA Deletion in Agrp Neurons Maintain Leptin Sensitivity and Resist Weight Gain While on a High-Fat Diet. *Int. J. Mol. Sci.* **2021**, *22*, 11010. [CrossRef]
21. Martinez, S.S.; Huang, Y.; Acuna, L.; Laverde, E.; Trujillo, D.; Barbieri, M.A.; Tamargo, J.; Campa, A.; Baum, M.K. Role of Selenium in Viral Infections with a Major Focus on SARS-CoV-2. *Int. J. Mol. Sci.* **2022**, *23*, 280. [CrossRef]
22. Guillin, O.M.; Vindry, C.; Ohlmann, T.; Chavatte, L. Interplay between selenium, selenoproteins and HIV-1 replication in human CD4 T-lymphocytes. *Int. J. Mol. Sci.* **2022**, in press.
23. Flohé, L.; Günzler, W.A.; Schock, H.H. Glutathione peroxidase: A selenoenzyme. *FEBS Lett.* **1973**, *32*, 132–134. [CrossRef]
24. Ursini, F.; Maiorino, M. Redox pioneer: Professor Leopold Flohé. *Antioxid. Redox Signal.* **2010**, *13*, 1617–1622. [CrossRef]
25. National Research Council. *Selenium in Nutrition*, revised ed.; Board of Agriculture, Subcommittee on Selenium, Ed.; The National Academies Press: Washington, DC, USA, 1983. [CrossRef]
26. Brigelius-Flohé, R.; Aumann, K.D.; Blöcker, H.; Gross, G.; Kiess, M.; Klöppel, K.D.; Maiorino, M.; Roveri, A.; Schuckelt, R.; Usani, F.; et al. Phospholipid-hydroperoxide glutathione peroxidase. Genomic DNA, cDNA, and deduced amino acid sequence. *J. Biol. Chem.* **1994**, *269*, 7342–7348. [CrossRef]
27. Schuckelt, R.; Brigelius-Flohé, R.; Maiorino, M.; Roveri, A.; Reumkens, J.; Strassburger, W.; Ursini, F.; Wolf, B.; Flohé, L. Phospholipid hydroperoxide glutathione peroxidase is a selenoenzyme distinct from the classical glutathione peroxidase as evident from cDNA and amino acid sequencing. *Free Radic. Res. Commun.* **1991**, *14*, 343–361. [CrossRef] [PubMed]
28. Carlson, B.A.; Moustafa, M.E.; Sengupta, A.; Schweizer, U.; Shrimali, R.; Rao, M.; Zhong, N.; Wang, S.; Feigenbaum, L.; Lee, B.J.; et al. Selective restoration of the selenoprotein population in a mouse hepatocyte selenoproteinless background with different mutant selenocysteine tRNAs lacking Um34. *J. Biol. Chem.* **2007**, *28*, 32591–32602. [CrossRef]
29. Nogoceke, E.; Gommel, D.U.; Kiess, M.; Kalisz, H.M.; Flohé, L. A unique cascade of oxidoreductases catalyses trypanothione-mediated peroxide metabolism in Crithidia fasciculata. *Biol. Chem.* **1997**, *378*, 827–836. [CrossRef] [PubMed]
30. Kim, K.; Kim, I.H.; Lee, K.Y.; Rhee, S.G.; Stadtman, E.R. The isolation and purification of a specific "protector" protein which inhibits enzyme inactivation by a thiol/Fe(III)/O_2 mixed-function oxidation system. *J. Biol. Chem.* **1988**, *263*, 4704–4711. [CrossRef]
31. Loschen, G.; Azzi, A.; Richter, C.; Flohé, L. Superoxide radicals as precursors of mitochondrial hydrogen peroxide. *FEBS Lett.* **1974**, *42*, 68–72. [CrossRef]
32. Loschen, G.; Flohé, L.; Chance, B. Respiratory chain linked H_2O_2 production in pigeon heart mitochondria. *FEBS Lett.* **1971**, *18*, 261–264. [CrossRef]
33. Jaeger, T.; Budde, H.; Flohé, L.; Menge, U.; Singh, M.; Trujillo, M.; Radi, R. Multiple thioredoxin-mediated routes to detoxify hydroperoxides in *Mycobacterium tuberculosis*. *Arch. Biochem. Biophys.* **2004**, *423*, 182–191. [CrossRef] [PubMed]

Article

Interplay between Selenium, Selenoproteins and HIV-1 Replication in Human CD4 T-Lymphocytes

Olivia M. Guillin [1,2,3,4,5], Caroline Vindry [1,2,3,4,5], Théophile Ohlmann [1,2,3,4,5,*] and Laurent Chavatte [1,2,3,4,5,*]

1. Centre International de Recherche en Infectiologie (CIRI), 69007 Lyon, France; olivia.guillin@ens-lyon.fr (O.M.G.); caroline.vindry@ens-lyon.fr (C.V.)
2. Institut National de la Santé et de la Recherche Médicale (INSERM) Unité U1111, 69007 Lyon, France
3. Ecole Normale Supérieure de Lyon (ENS), 69007 Lyon, France
4. Université Claude Bernard Lyon 1 (UCBL1), 69622 Lyon, France
5. Centre National de la Recherche Scientifique (CNRS), Unité Mixte de Recherche 5308 (UMR5308), 69007 Lyon, France
* Correspondence: theophile.ohlmann@ens-lyon.fr (T.O.); laurent.chavatte@ens-lyon.fr (L.C.); Tel.: +33-4-72-72-89-53 (T.O.); +33-4-72-72-86-74 (L.C.)

Citation: Guillin, O.M.; Vindry, C.; Ohlmann, T.; Chavatte, L. Interplay between Selenium, Selenoproteins and HIV-1 Replication in Human CD4 T-Lymphocytes. *Int. J. Mol. Sci.* **2022**, *23*, 1394. https://doi.org/10.3390/ijms23031394

Academic Editors: Petra A. Tsuji and Dolph L. Hatfield

Received: 10 December 2021
Accepted: 22 January 2022
Published: 26 January 2022

Publisher's Note: MDPI stays neutral with regard to jurisdictional claims in published maps and institutional affiliations.

Copyright: © 2022 by the authors. Licensee MDPI, Basel, Switzerland. This article is an open access article distributed under the terms and conditions of the Creative Commons Attribution (CC BY) license (https://creativecommons.org/licenses/by/4.0/).

Abstract: The infection of CD4 T-lymphocytes with human immunodeficiency virus (HIV), the etiological agent of acquired immunodeficiency syndrome (AIDS), disrupts cellular homeostasis, increases oxidative stress and interferes with micronutrient metabolism. Viral replication simultaneously increases the demand for micronutrients and causes their loss, as for selenium (Se). In HIV-infected patients, selenium deficiency was associated with a lower CD4 T-cell count and a shorter life expectancy. Selenium has an important role in antioxidant defense, redox signaling and redox homeostasis, and most of these biological activities are mediated by its incorporation in an essential family of redox enzymes, namely the selenoproteins. Here, we have investigated how selenium and selenoproteins interplay with HIV infection in different cellular models of human CD4 T lymphocytes derived from established cell lines (Jurkat and SupT1) and isolated primary CD4 T cells. First, we characterized the expression of the selenoproteome in various human T-cell models and found it tightly regulated by the selenium level of the culture media, which was in agreement with reports from non-immune cells. Then, we showed that selenium had no significant effect on HIV-1 protein production nor on infectivity, but slightly reduced the percentage of infected cells in a Jurkat cell line and isolated primary CD4 T cells. Finally, in response to HIV-1 infection, the selenoproteome was slightly altered.

Keywords: selenoproteome; HIV-1; viral infection; glutathione peroxidase; thioredoxin reductase; SELENOS; SELENOO; primary T cells; Jurkat; SupT1; translational control

1. Introduction

The human immunodeficiency virus (HIV) is an enveloped, linear, positive-sense single-stranded RNA virus that belongs to the family of *Retroviridae* (Group VI), genus *Lentivirus*. HIV is the etiological agent of acquired immunodeficiency syndrome (AIDS) and is responsible for a weakened immune system, as it infects immune cells [1]. There are two types of HIV (HIV-1 and HIV-2) that differ in their epidemiological and pathological properties [1]. In contrast to HIV-2 that is mostly confined to West Africa, HIV-1 has spread worldwide, due to its improved infectivity and virulence [2]. There are currently more than 37 million people infected with human immunodeficiency virus (HIV), causing about 1.5 million deaths every year (http://www.who.int/hiv/en/, accessed on 1 December 2021). HIV infection is now considered a chronic disease that requires intensive treatment and can present a variable clinical course. HIV-1 infects immune cells that harbor the CD4 receptor and a co-receptor belonging to the chemokine receptor family (CCR5 and CXCR4) [1]. Therefore, cells infected by HIV-1 are CD4 T lymphocytes, monocytes, macrophages and dendritic cells. A decline in CD4 T cells is characteristic of the progression of HIV infection and is often used as a prognostic marker. No vaccine is currently available, but an effective combination of drugs, called antiretroviral therapy (ART), has been developed to lower the viral load and increase the CD4 T-cell count [3,4]. Worldwide, about a quarter of HIV-infected people are still not receiving ART. In addition, the emergence of drug resistant viruses remains a threat to the effectiveness of ART. The replication but also the latency of the virus is extremely variable from one cell type to another. Lentiviruses are characterized by a long incubation period after the primo infection, which is highly variable from one patient to another. During this time, humans infected with HIV are under chronic oxidative stress and present nutritional deficiencies, particularly with regard to selenium [5]. In HIV-infected people, lower selenium levels have been associated with lower CD4 T-cell counts, faster progression of AIDS, and a 20% increase in the risk of death [6,7]. The beneficial effects of selenium nutritional supplementation were reported in a few studies by a decrease in viral burden, an increase in the time of progression to AIDS and an indirect increase in the number of CD4 T cells [8–10]. These findings remained significant even after correction for the ART regimen [10]. Interestingly, a long-time treatment with ART (more than 2 years) improved selenium levels as compared with HIV-infected patients with a shorter time of exposure to ART or those who are not receiving any treatment [11]. Since the discovery of HIV in the 1980s, several established T-cell lines (including Jurkat, SupT1 and CEM) were found permissive to this virus and were instrumental for investigating the molecular mechanisms of viral replication and virus–host interactions that occur in the course of infection. In contrast to these established cultured cell lines, resting primary CD4 T cells isolated from donors represent a more physiological model relevant to HIV infection, but these cells are more difficult to grow, infect, and they must be activated to increase their susceptibility to HIV infection. Even when stimulated, the infection efficiency of primary cells rarely reaches the levels observed in established T-cell lines. In the present study, we used both cellular models: cultured T cells, SupT1 and Jurkat, and primary CD4 T cells that were isolated from different healthy donors.

Selenium is an essential trace element implicated in many facets of human health and disease. This element is often evoked in the context of immune function and infectious diseases caused by viruses or bacteria [5,12]. The current example of the COVID-19 pandemic, due to the SARS-CoV-2 viral infection, confirms this finding with an association between low levels of selenium in body fluids and poor prognosis of infected patients hospitalized in intensive care units, as reviewed in [13]. Several studies showed that nutritionally deficient humans or animals were more susceptible to a wide variety of infections [5]. The mechanism is probably more complex than expected in the sense that nutritional deficiency does not impact only the host immune response, but also the viral pathogen itself. For instance, dietary selenium deficiency that induces oxidative stress in the host can alter the genome of a virus so that a normally benign or mildly pathogenic virus becomes highly virulent. This phenomenon has been particularly well described in animal models for influenza and coxsackie viruses [14–17], but has not been investigated for other viral models yet. Consequently, it was hypothesized that selenium deficiency could participate in the genetic evolution of a wide variety of viruses and their pathogenicity by a mechanism that remains to be elucidated.

Most of selenium's beneficial effects are attributed to its presence as selenocysteine, a rare amino acid, in a small, but vital group of redox enzymes that constitute the selenoproteins. They are implicated in antioxidant defense, redox homeostasis, redox signaling and possibly other cellular processes [18–20]. Among the selenoproteins, glutathione peroxidases (GPXs) and thioredoxin reductases (TXNRDs) are well characterized [21,22]. While the GPXs are part of the antioxidant defenses that reduce a wide variety of peroxides using glutathione (GSH) as a cofactor [23], the TXNRDs are NADPH-dependent reductases that control the redox balance of a broad range of substrates, including proteins (thioredoxins or glutaredoxin 2) but also small selenium- and sulfur-containing molecules (lipoic acid, DTNB, selenite, selenocystine, etc.) [24]. In addition, a third of the selenoproteome is localized at the endoplasmic reticulum (ER), where they participate in protein folding, calcium homeostasis and ER-stress response [25–29]. These ER-resident selenoproteins include SELENOF, SELENOI, SELENOK, SELENOM, SELENON, SELENOS, SELENOT and DIO2.

Therefore, as a family of redox enzymes, selenoproteins are thought to play important functions in the immune system, as reviewed in [30], although experimental data are still awaited. The selenoproteome is encoded by 25 genes and is primarily controlled by selenium bioavailability, which induces prioritization of selenoprotein biosynthesis [18,31,32]. The hierarchical regulation of the selenoproteome by other exogenous stimuli, cellular sensors or pathophysiological conditions is poorly understood [33–38]. This hierarchy of response induced by variations of the selenium concentration is very specific to each tissue or cell line and relies primarily on a translational control strategy. Indeed, the insertion of selenocysteine in selenoproteins is based on a non-conventional translational mechanism that is unique in many aspects. Indeed, selenocysteine is encoded by UGA, which is normally a stop codon. As the first addition to the genetic code in 1991, selenocysteine is, therefore, often referred to as the 21st amino acid [39,40]. Thus, the cell has developed a singular strategy to recode UGA as a selenocysteine in selenoprotein mRNAs, while it continues to be used as a stop codon for all other cellular mRNAs. This efficiency of UGA recoding as selenocysteine is rather low (between 1% and 5%), even under optimal conditions of selenium levels, and most often results in a truncated protein, with the UGA codon being read as a stop codon [31]. In theory, this poorly efficient UGA codon can be seen as a premature stop codon by the dedicated nonsense mediated decay (NMD) surveillance machinery that safeguards the quality of transcripts in eukaryotic cells [41]. However, even though several selenoprotein mRNAs follow the prerequisites of NMD rules, only a few mRNA transcripts are targeted by this mechanism, and this only occurs upon selenium deprivation and in specific cell lines [34,42]. Additionally, recent experiments using ribosome profiling, a method that allows precise location of the ribosome position on mRNAs, showed that UGA-selenocysteine recoding event is a limiting step, which is very sensitive to selenium level variation [43–45].

Although low serum selenium levels are associated with HIV progression in many epidemiological studies, cellular and molecular evidence is still lacking to clarify the role of selenium and selenoproteins in viral infection. Moreover, very little information is available on the expression and regulation of the selenoproteome in CD4 T cells, either originating from lymphoma or from healthy donors. To date, the only relevant study was performed in vitro and reported a modification of the pattern of selenoprotein expression in response to HIV infection in lymphocytes, as revealed by ^{75}Se radioactive isotope labeling [46]. Based on the migration on SDS-PAGE and radioactive signal intensity, it was suggested that TXNRD1, GPX4 and GPX1 were downregulated in Jurkat cells after HIV-1 infection in favor of one or more low-molecular-weight selenocompounds. However, these experiments were done at a time when only a few selenoproteins were characterized. As the mammalian selenoproteome is now complete [21,47,48] and the physiology of selenium is better understood, this prompted us to investigate further the role of selenium and selenoproteins during HIV infection at the molecular level. Here, we performed a detailed analysis of selenoproteome expression at the mRNA and protein levels in response to varying selenium concentrations both in primary and cultured T cells. We also investigated the replication and pathogenicity of HIV-1 produced in control or selenium-supplemented media. Finally, we studied how the selenoproteins expression in CD4 T cells isolated from healthy donors were affected by HIV infection.

2. Results

2.1. Comparison of Selenoprotein mRNA Expression Pattern in Established and Primary CD4 T-Cells

Among the selenoprotein family, in mammals, several members, such as GPX1, GPX4 and TXNRD1, are ubiquitously and abundantly expressed, while others are highly tissue specific, such as GPX2, GPX3, GPX6, SELENOV and, to a lesser extent, deiodinases. Although it is now well established that the selenoproteome is highly regulated by selenium availability in several cellular models, very little is known about the expression pattern of these selenoproteins and their response to selenium supplementation in human immune cells, and particularly T lymphocytes. Therefore, as a prerequisite, we used RT-qPCR to analyze the levels of selenoprotein mRNAs in established human cell lines (Jurkat and SupT1) and in primary CD4 T cells purified from four different donors as shown in Figures 1–3, respectively. The cells were grown in culture medium supplemented or not with 100 nM of sodium selenite, and total RNA was extracted after three days. Cellular extracts were then referred to as 100 nM and Ctrl, respectively. Of note, the Ctrl media contained an endogenous concentration of 11 nM of selenium, which was essentially provided by the fetal calf serum; thus, after supplementation, the final concentration was 111 nM of total selenium. Interestingly, this selenium concentration of the Ctrl media is rather low and is often considered selenium deficient [49]. In our experiments, the majority of the 25 selenoprotein genes were detected in the various T-lymphocyte extracts. Three classes of mRNAs (SELENOV, GPX3 and GPX6) were below the threshold of detection in all cell types, and three more transcripts (DIO1, DIO2 and GPX2) could not be detected in primary CD4 T cells.

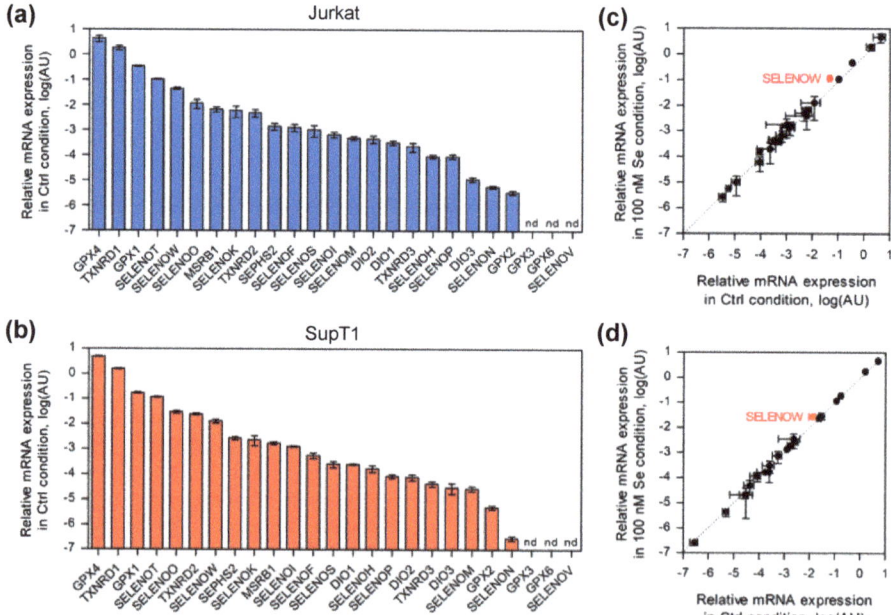

Figure 1. Analysis of selenoprotein mRNA levels in Jurkat and SupT1 in control conditions and in response to 100 nM selenium supplementation of culture media (three days). The geometrical mean of four housekeeping genes (HPCB, RPS13, HRPT, and GAPDH) was used to normalize mRNA abundance. In all panels, mRNA levels are represented in logarithmic scales. The values are given in Table S1. Selenoprotein mRNA levels in control medium (Ctrl) are represented for Jurkat (**a**) and SupT1 (**b**), from most to least abundant (from left to right). To evaluate the impact of selenium supplementation on steady state levels of selenoprotein mRNAs, the values obtained in selenium supplemented conditions (100 nM) were plotted as a function of values obtained with unsupplemented ones (Ctrl) for Jurkat (**c**) SupT1 (**d**) cells (±standard deviation). The experiments were done in biological triplicate and in technical triplicate. The selenoprotein genes with significant changes are labeled in red and the statistical analyses are given in Table S1.

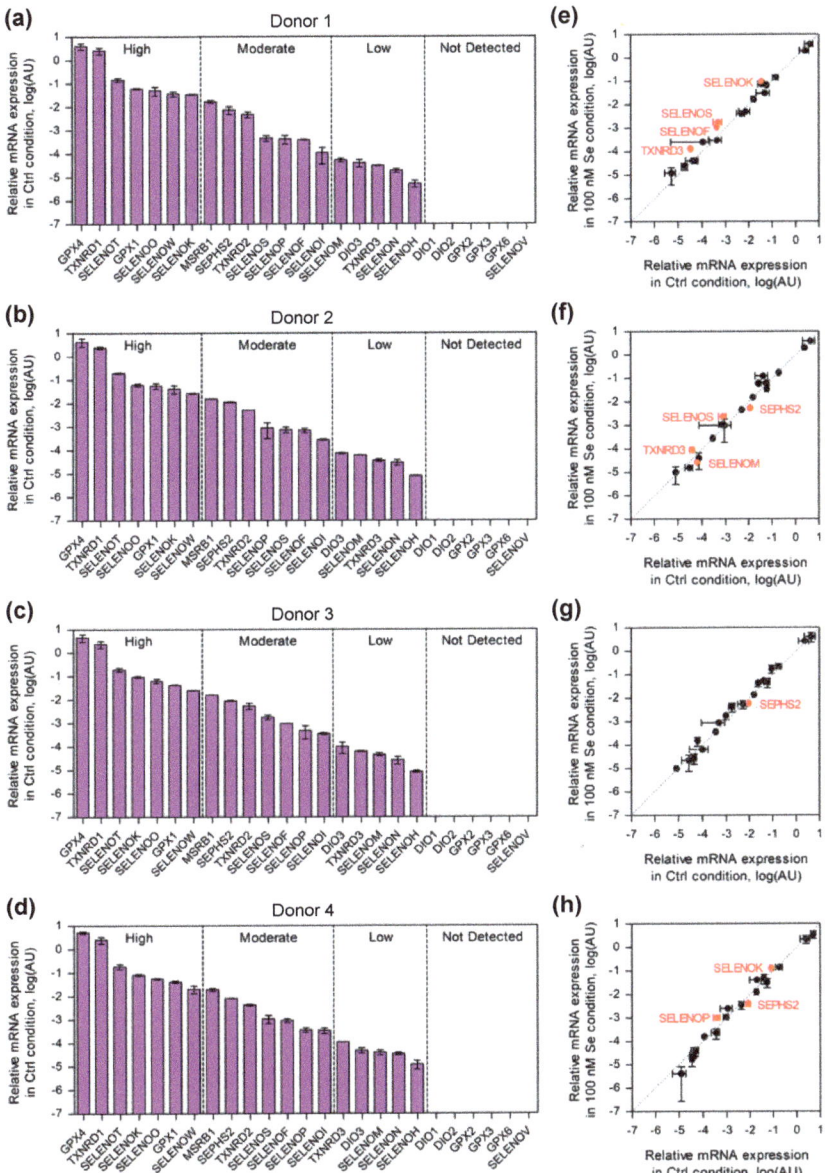

Figure 2. Analysis of selenoprotein mRNA levels in CD4-T cells isolated from donors in control conditions (Ctrl) and selenium-supplemented (100 nM) conditions after three days. The levels are represented in logarithmic scales. The transcripts from four different donors were measured and normalized similarly to what has been done for Jurkat and SupT1 extracts. The mRNAs expressed in control conditions were represented as histograms (**a**–**d**) and arbitrarily separated in high, medium and low abundance by dashed lines. Additionally, the values obtained in selenium-supplemented conditions (100 nM) were plotted as a function of values obtained with unsupplemented ones for every donor (±standard deviation) (**e**–**h**). The values are given in Table S2. The experiments were done in biological duplicates and in technical triplicates. The selenoprotein genes with significant changes are labeled in red and the statistical analyses are given in Table S2.

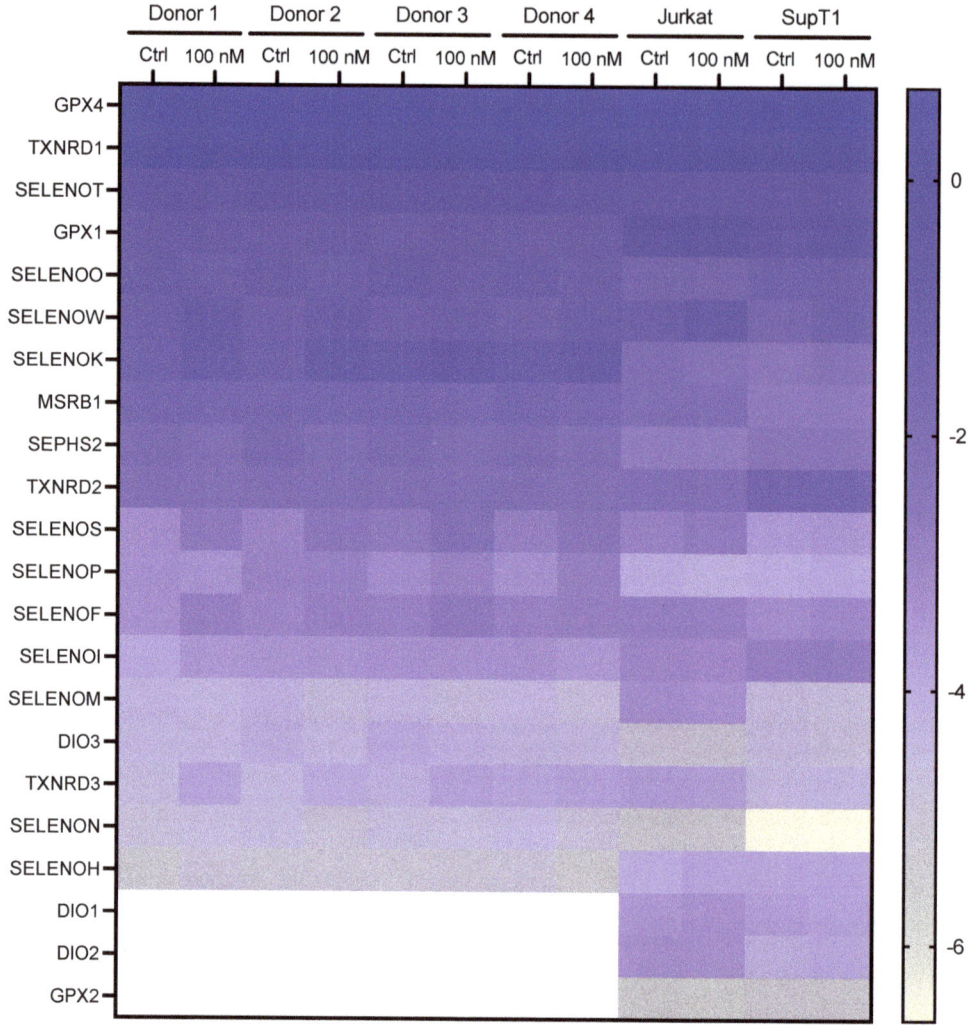

Figure 3. Heatmap representation of selenoprotein mRNA levels (in logarithmic scales) as a function of T-cell types (donors 1 to 4, Jurkat and SupT1) and growth conditions (Ctrl or supplemented with 100 nM of selenium). The values are given in Tables S1 and S2.

Figures 1 and 2 show the distribution pattern of the selenoproteins in Ctrl conditions for the two cell lines and primary cells used in this study. Interestingly, a comparison of the data obtained in the two lymphoma-derived T cells (Figure 1) shows numerous similarities. Firstly, the four most abundant selenoprotein mRNAs present in Jurkat and SupT1 were identical: GPX4, TXNRD1, GPX1 and SELENOT. While the first three proteins are often expressed at high level in most tissues, this is not the case for SELENOT, which is generally restricted to endocrine organs and embryogenesis. Overall, the expression of selenoprotein mRNAs was comparable between these two cellular lymphoma-derived T-cell lines, with two notable exceptions for the TXNRD2 mRNA, which was five times more expressed in SupT1 than in Jurkat cells and for SELENOM mRNA, which was, conversely, 18 times less expressed in SupT1 than in Jurkat (Figure 1a,b and Table S1). We then extended our analysis to primary CD4 T cells isolated from four different healthy donors and activated with magnetic beads coupled to anti-CD3 and anti-CD28 antibodies. As illustrated in Figure 2 and Table S2, a very similar expression pattern of selenoprotein gene expression was observed in the Ctrl conditions, in the four different extracts, although with subtle variations depending on the donor. To simplify the comparison between the four different donors, we arbitrarily separated the transcripts in high, moderate and low abundance categories (see Figure 2a–d). For these three classes of transcripts, we observed variation between the four donor samples. As shown in Figure 2, the most abundant transcripts, which included GPX4, TXNRD1, SELENOT, GPX1, SELENOO, SELENOW, and SELENOK, were similar in all four donors, but with significant donor-specific variation in their rankings, particularly for GPX1 and SELENOK mRNAs (Figure 2a–d). This contrasts with the expression of mRNA transcripts within the moderate and low abundance categories that were very similar between the four different donors. Finally, when comparing these donor-derived CD4 T-cells with lymphoma-derived cell lines, common features emerged, most notably the high expression of GPX4, TXNRD1 and SELENOT transcripts. The high expression of SELENOK mRNA appeared to be more specific to primary cells, although its ranking was variable among donors.

Then, we analyzed the effect of selenium addition on the expression of selenoprotein transcripts. Indeed, the addition of selenium in the growth media is a well-characterized enhancer of selenoprotein expression in many cell lines, although with, generally, a limited impact at the transcriptional level [31,42]. As illustrated in Figure 1c,d and Figure 2e–h, a limited number of selenoprotein transcripts were sensitive to selenium supplementation. As previously observed in non-immune cell lines, only SELENOW transcript was found to be sensitive to selenium supplementation in T cells, although this can be considered statistically significant ($p < 0.05$) only in Jurkat and SupT1 (see Tables S1 and S2 for the quantitative values). Moreover, it appeared that selenium-specific regulation of transcripts was more pronounced in primary CD4 T cells than in lymphoma-derived cell lines, although with a subtle donor-specific pattern. Thus, among the selenoprotein genes that were selectively altered in response to selenium variation in one or several primary cells, we found SELENOK, SEPHS2, SELENOS, SELENOF, SELENOM, SELENOP and TXNRD3.

A heatmap representation of selenoprotein mRNAs levels obtained in various cell types and selenium levels is shown in Figure 3. Although our data show a significant degree of homogeneity in the expression of the mRNAs in the primary CD4 T cells from different donors, it also reveals significant differences in the transcriptional pattern between primary CD4 T cells and established cell lines. This is not surprising, as Jurkat and SupT1 are derived from lymphoma, and it has been shown that selenium metabolism could be altered in cancer cells [18]. In addition, the level of several selenoproteins, including TXNRD1 and GPX2 [50,51], have been linked to different cancers. It was, therefore, not surprising to observe a different pattern of selenoprotein mRNA expression in cancer cells compared to primary cells.

2.2. Selenium-Dependent Hierarchy of Selenoprotein Expression Levels in Established and Primary CD4 T-Cells

Our data show, so far, that the level of expression of selenoprotein mRNAs was poorly affected by selenium variations in the culture medium in human T lymphocytes. Thus, the next step was to measure protein expression, as it is expected that selenoprotein production could be more sensitive to selenium addition as previously observed in non-immune cell lines [33,34,42]. Indeed, as described in these previous reports, most of this regulation occurs at the level of translation. This was confirmed by ribosome profiling in animal models fed diets containing different amounts of selenium, where it was observed that the UGA-selenocysteine recoding event was affected by the level of selenium [44,45]. Well-described selenium responders were GPX members and several ER-located selenoproteins [33,34,42]. These proteins are considered 'stress-response' selenoproteins, in contrast to the ones that are less sensitive to selenium variations, which are referred to as 'housekeeping' members. These less sensitive selenoproteins often include the family of TXNRDs. As was done previously, the different established human cell lines (Jurkat and SupT1) and primary CD4 T cells purified from four different donors were grown in culture medium supplemented, or not, with sodium selenite. For the Jurkat and SupT1 cell lines, we used eight different growth media with increasing selenium concentration up to 300 nM. Due to the finite number of cell divisions of donor-derived cells, and the amount of protein extracts necessary for biochemical studies, only two conditions could be imposed for primary cells, Ctrl and 100 nM, respectively. For all these conditions, cell extracts were harvested after three days of growth in these respective media. Cell extracts were analyzed by Western blots (Figures 4 and 5), but also by enzymatic assays that specifically measure glutathione peroxidase and thioredoxin reductase activities (Figure 6).

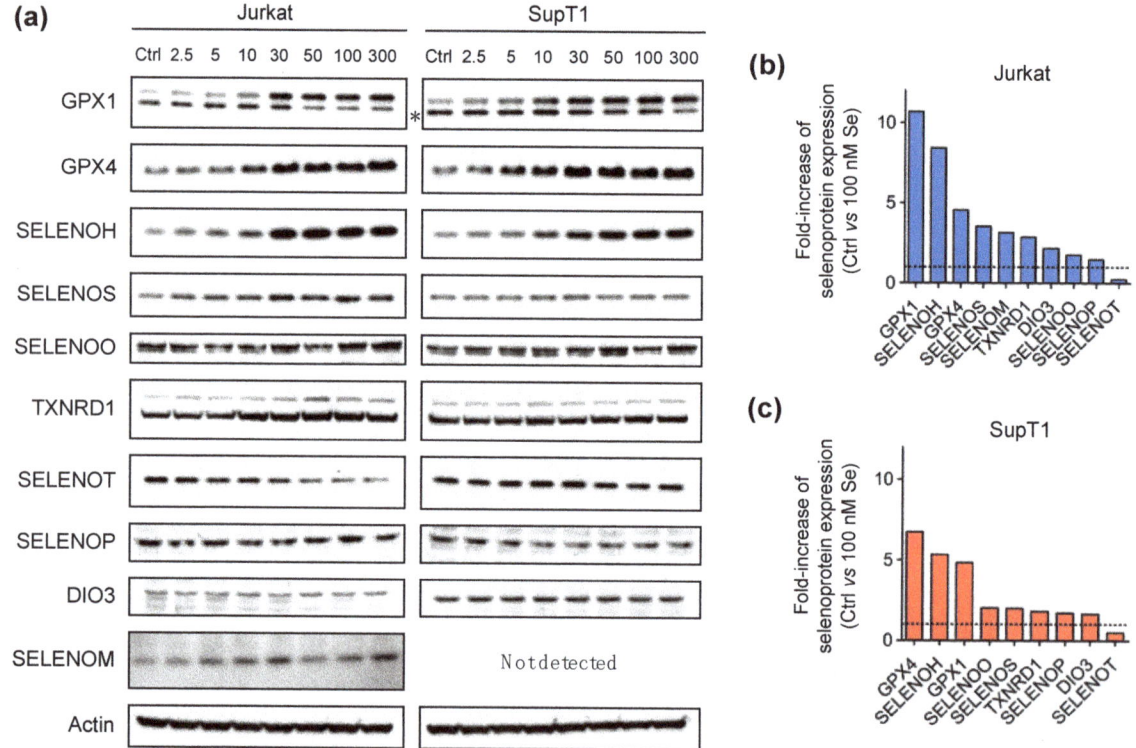

Figure 4. Analysis of selenoprotein expression in Jurkat and SupT1 cells in response to different selenium supplementations of the culture media (in nM) after three days. The results from a representative experiment are shown in (**a**). The values are given in Table S4. The fold-increase in expression of each selenoprotein in response to selenium supplementation was calculated between the value at 100 nM over that at Ctrl. These ratios were plotted from the highest to the lowest for Jurkat (**b**) and SupT1 (**c**) cells. The dotted line represents a fold-increase value of one.

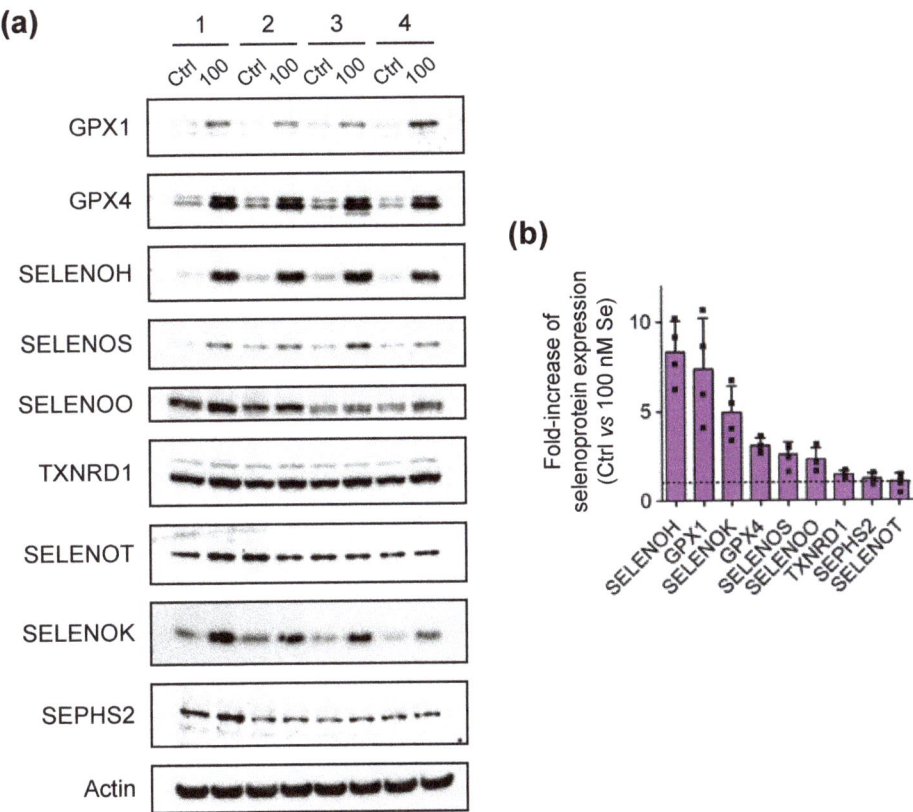

Figure 5. Analysis of selenoprotein expression in CD4 T cells isolated from four different donors in control (Ctrl) and selenium-supplemented (100 nM) conditions over three days. The results from a representative experiment are shown in (**a**). The values are given in Table S4. (**b**) The fold-increase in expression of each selenoprotein in response to selenium supplementation was calculated between the value at 100 nM over that at Ctrl. The average value (±standard deviation) of the four donors was represented by a bar, but the individual value of each donor was indicated by a square. The dotted line represents a fold-increase value of one.

Immunodetection of proteins by Western blot was not as easy as the detection of transcripts by RT-qPCR, as it was highly dependent on the availability and quality of antibodies raised against human proteins. Most selenoproteins are rather small in size, which reduces the probability of having immunogenic regions. We tried to detect 19 selenoproteins for which commercially antibodies were available (see Table S3), with several of them previously validated in other cell lines [33,34,37,38,52]. Among them, 10 selenoproteins were unambiguously detected in, at least, one of the T-cell derived extracts. As illustrated in Figures 4 and 5, seven selenoproteins, that included TXNRD1, GPX4, GPX1, SELENOH, SELENOO, SELENOS and SELENOT, were detected in both lymphoma-derived and primary CD4 T cells. The expression of others was more cell-line specific. SELENOM was only detected in Jurkat (Figure 4a), while SELENOK and SEPHS2 were only detected in primary cells (Figure 5a). Several of them were highly sensitive to selenium supplementation of the culture medium. As illustrated in Figure 4b,c and Figure 5b, selenium was able to stimulate the expression of most of them with the notable exception of SELENOT, whose expression decreased with selenium supplementation in both Jurkat and SupT1 cell extracts. However, in extracts derived from primary cells, SELENOT had a variable response to selenium supplementation. In any case, our data further confirm That the control growth conditions that contain 10% FCS and are typically used in most laboratories, is limiting for the expression of several selenoproteins.

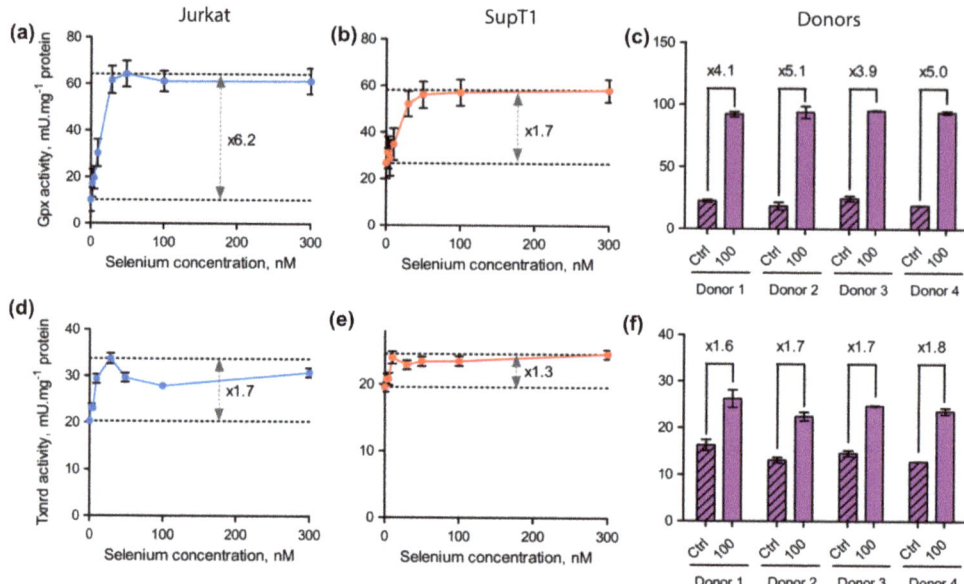

Figure 6. Evaluation of GPX (**a**–**c**) and TXNRD (**d**–**f**) enzymatic activities in protein extracts from various T cell types (Jurkat, SupT1 and the four donors) cultured with different concentrations of supplemented selenium. The GPX activities were represented for Jurkat (**a**) ($n = 1$), SupT1 (**b**) ($n = 1$) and the four donors (**c**) ($n = 2$) for the different growth conditions. The TXNRD activities were represented for Jurkat (**d**) ($n = 1$), SupT1 (**e**) ($n = 1$) and the four donors (**f**) ($n = 2$) for the different growth conditions. The experiments were done in technical triplicates. The differences between the lowest and highest values were indicated by an arrow or a bar (±standard deviation), with the corresponding fold-change factor beside or above.

Immunodetection of proteins by Western blot was not as easy as the detection of transcripts by RT-qPCR, as it was highly dependent on the availability and quality of antibodies raised against human proteins. Most selenoproteins are rather small in size, which reduces the probability of having immunogenic regions. We tried to detect 19 selenoproteins for which commercially antibodies were available (see Table S3), with several of them previously validated in other cell lines [33,34,37,38,52]. Among them, 10 selenoproteins were unambiguously detected in, at least, one of the T-cell derived extracts. As illustrated in Figures 4 and 5, seven selenoproteins, that included TXNRD1, GPX1, GPX4, SELENOH, SELENOO, SELENOS and SELENOT, were detected in both lymphoma-derived and primary CD4 T cells. The expression of others was more cell-line specific. SELENOM was only detected in Jurkat (Figure 4a), while SELENOK and SEPHS2 were only detected in primary cells (Figure 5a). Several of them were highly sensitive to selenium supplementation of the culture medium. As illustrated in Figure 4b,c and Figure 5b, selenium was able to stimulate the expression of most of them with the notable exception of SELENOT, whose expression decreased with selenium supplementation in both Jurkat and SupT1 cell extracts. However, in extracts derived from primary cells, SELENOT had a variable response to selenium supplementation. In any case, our data further confirm That the control growth conditions that contain 10% FCS and are typically used in most laboratories, is limiting for the expression of several selenoproteins.

As observed in many cell lines, the GPX members expressed in T cells, namely GPX1 and GPX4, were among the most responsive to changes in selenium concentration, confirming their place in the family of 'stress-related' selenoproteins. Reactivity to selenium supplementation was further confirmed by the results of the GPX activity assays that are presented in Figure 6. Indeed, we noted a significant increase in GPX activities in response to selenium supplementation in all cell types. In Jurkat and SupT1, the activities reached a plateau from 30 nM of selenium being added, and it remained high up to the maximal concentration used (300 nM); see Figure 6a,b. As such, a 100 nM selenium concentration was expected to reach the highest GPX activity in primary cells. When comparing the GPX activities in control and supplemented primary cell extracts, we observed a 4- to 5-fold increase between Ctrl and 100 nM conditions (Figure 6c). This stimulation of GPX activity was similar between the four different donors and correlates with the level of GPX1 and GPX4 overexpression observed by Western blots shown in Figure 5b. Concerning TXNRD1, which is considered a 'housekeeping' member, it was found to be weakly but significantly stimulated by selenium supplementation in all cell types tested here (Figures 4–6). These data are in good agreement with previously published reports done in several other non-immune cell lines [33,34,42]. Our data in CD4 T cells were further confirmed by enzymatic assays that are shown in Figure 6c,f. In conclusion, the expression of the selenoproteome is tightly regulated by the selenium level of the culture media in immune cells at a translational level.

2.3. Selenium Levels Did Not Affect HIV-1 Replication in Jurkat Cells but Modified the Proportion of Infected Cells

We then investigated whether selenium had an effect on virus replication and whether the viral particles produced were equally infectious. We started with Jurkat cells, which have been widely used to study the mechanisms of HIV replication and which seemed to be slightly more sensitive to selenium than SupT1 cells in terms of selenoprotein expression. Please note that selenium was supplemented three hours post-infection. Jurkat cells were infected with the fully replicative HIV-1-NL4.3 at a low moiety of infection (MOI = 0.01) in order to follow the production of HIV viral particles over a period of 11 days (Figure 7a). The time required for HIV to complete replication and to produce a new generation of virus is 24 h [53]. Thus, if we follow the replication kinetics over several replication cycles (about 10 generations in 11 days), a cumulative effect could emerge. After infection of the Jurkat cells, samples were harvested at different times and analyzed for the presence of virus particles and their infectivity. In parallel, the progression of viral infection was evaluated by monitoring synthesis of the HIV-1 capsid protein, also known as p24, by immunodetection on Western blot. HIV-1 p24 is the result of proteolytic cleavage of the polyprotein precursor p55 Gag (Pr55Gag), which occurs after the release of virus particles from the infected cell during the maturation step. Although the quantification of p24 is a good indicator of the amounts of HIV-1 particles released in the media from infected cells, this does not reflect the infectivity of the viruses produced, which is dependent from many other viral parameters. Thus, it was critical to also measure the infectious nature of the produced particles since it is well established that many defective HIV particles can be produced and released in vitro. The infectivity of the produced viral particles was evaluated by testing their ability to infect TZM-bl reporter cells, a HeLa-derived cell line that expresses the CD4 receptor, CXCR4 co-receptor and a luciferase gene, whose expression is driven by a LTR HIV-1 promotor. Therefore, by measuring p24 amounts and ability to infect TZM-bl reporter cells, we could measure both viral production and infectivity under different cellular growth conditions. In parallel, the proportion of infected cells within the total cell population was evaluated by detecting p24 positive cells by flow cytometry after immunological labeling.

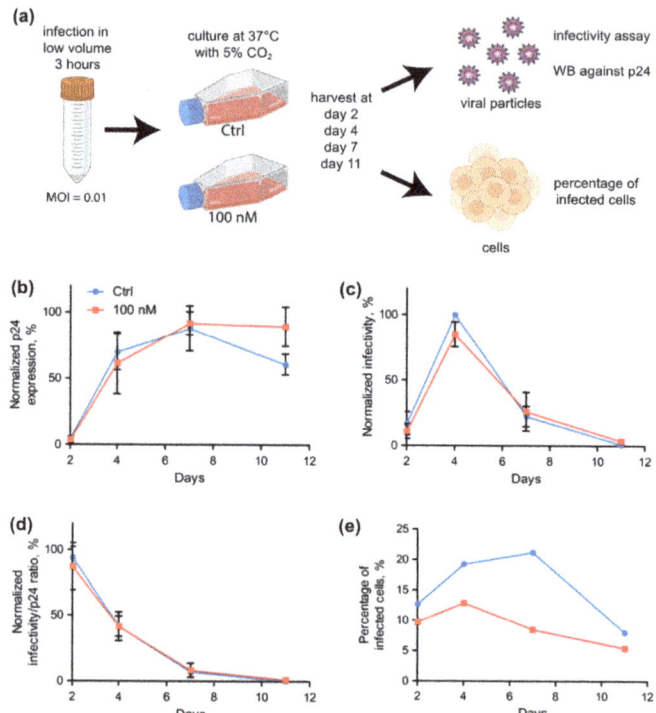

Figure 7. Kinetics of HIV-1 infection of Jurkat cells in control (Ctrl) or 100 nM supplemented conditions. (**a**) Schematic of the experimental procedure used to follow the different parameters of viral production and cell infection during eleven days. The culture media were collected at different time points and analyzed for the levels of HIV-1 p24 capsid protein by Western blot (WB) (**b**) and infectivity of TZM-bl cells (**c**). The anti-p24 Western blots were assessed with biological duplicates, and levels of p24 were expressed relative to the maximum value, arbitrarily set at 100%. The infectivity assays were assessed with biological duplicates in technical triplicates. The maximum value was arbitrarily set as 100%. (**d**) The ratio between infectivity and p24 values were calculated and plotted as a function of time, the maximum value was arbitrarily set as 100%. (**e**) The percentage of infected cell was evaluated in one experiment by immuno-labeling of fixed cells using anti-HIV-1 p24 antibodies coupled with FITC followed by flow cytometry analysis.

As shown in Figure 7b, we found that p24 levels increased during the first few days, post-infection, in a similar manner with or without the addition of selenium to the culture media. The peak of p24 levels was attained at day 7, and this was followed by a significant drop in viral protein amount in control conditions. Interestingly, the infectivity of the particles released in the culture media was virtually identical throughout the kinetics for both cellular growth conditions, with a maximum at day 4 followed by a decrease to reach an almost complete decrease at day 11 (Figure 7c). These data suggest that selenium does not have a significant effect on HIV-1 particle production, neither in terms of quantity nor infectivity as revealed by the ratio representing the infectivity over the level of p24, both detected in the media (Figure 7d). The percentage of infected cells was also monitored at every time point by detecting p24 positive cells by flow cytometry. Interestingly, we found a significant reduction in the percentage of infected cells in selenium supplemented conditions in comparison to cells grown in Ctrl conditions (Figure 7e), at least at days 4 and 7 post-infection. Taken together, our data indicate no difference in HIV-1 replication in

Jurkat by the supplementation of culture media with 100 nM selenium, except for a lower percentage of infected cells.

2.4. Selenium Levels Did Not Affect HIV-1 Replication in Primary Cells but Modified the Proportion of Infected Cells

Since the selenoproteome of primary CD4 T-cells isolated from healthy donors seemed well responsive to selenium level variations and different from Jurkat cells' selenoproteome, we also investigated whether selenium modified HIV-1 infection in this model. As these primary cells were less prone to infection than Jurkat cells, we used a higher MOI (0.1) than with the cell lines. Isolated CD4 T-cells were more susceptible to cell death after HIV-1 infection than Jurkat cells. Therefore, we performed shorter kinetics than with lymphoma-derived T cell infection (Figure 8a). Even then, at day 4 post-infection, for donor 2, cells could not be collected due to the very high rate of mortality. At days 2, 3 and 4 post-infection, aliquots of the culture media were harvested and evaluated for p24 production, and infectivity was assessed by a TZM-bl assay as described in the previous paragraph. The percentage of infected cells was measured when cells were harvested at day 4. As illustrated in Figure 8b, the production of viral particles detected by p24 Western blots were comparable with, or without, addition of selenium in the culture media, and this was observed for the four donors. Then, the infectivity of the collected particles from all donors was assessed on the TZM-bl cell line (Figure 8c). When comparing the particles produced in the presence or absence, of selenium supplementation, no significant difference could be detected for any of the donors. However, we could observe a small, but significant, effect in the proportion of infected cells when they were supplemented with selenium (Figure 8e). Taken together, our data indicate that the amount of virus particles were equally produced and infectious under the different experimental conditions in primary cells, but the percentage of infected cells was reproducibly lower in selenium supplemented conditions four days post infection. These data with primary cells are consistent with what we obtained with Jurkat cells.

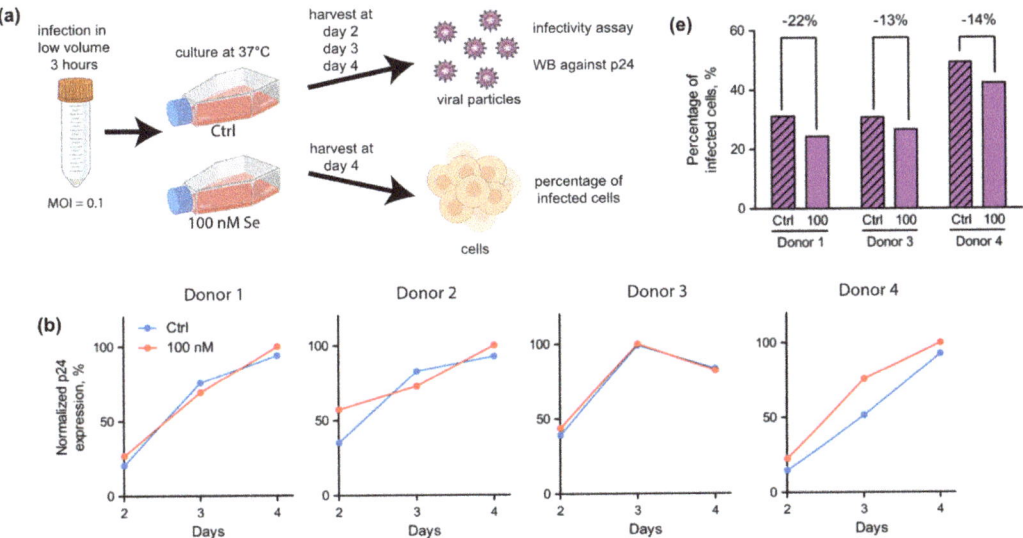

Figure 8. Cont.

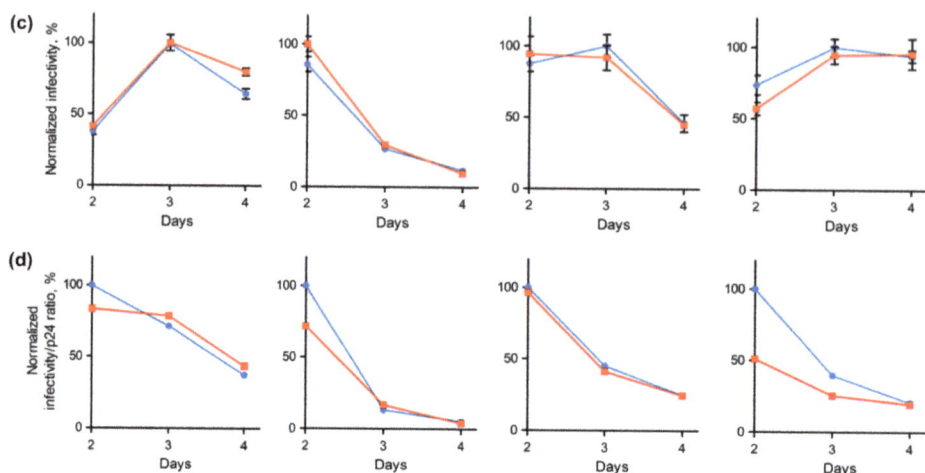

Figure 8. Kinetics of HIV-1 infection of T-cells isolated from donors in Ctrl or 100 nM supplemented conditions (**a–e**). (**a**) Schematic of the experimental procedure used to follow the different parameters of viral production and cell infection over four days. Similar to what was done with Jurkat in Figure 7, the media were collected at different time points and evaluated for the levels of p24 (**b**), infectivity of TZM-bl cells (**c**) and the ratio of infectivity over p24 levels (**d**) ($n = 1$). The infectivity assays were assessed with technical triplicates. For panels (**b–d**), the maximum value was arbitrarily set as 100%.

2.5. The Infection of Primary T Cells with HIV-1 Altered the Levels of Certain Selenoproteins

The fact that the selenium supplementation did not interfere with HIV-1 replication did not prevent the virus from altering the cellular use of selenium in selenoproteins. In this context, we therefore investigated whether HIV-1 infection affected the levels of selected selenoproteins in Ctrl and supplemented conditions (Figure 9). As we were limited in the biological samples available for this experiment, we performed this Western blot analysis with extracts from donors 3 and 4, for which we had a range of p24 positive cells between 30% and 50% (Figure 8e). With this level of infected cells, it would be difficult to visualize subtle effects. For example, in case of an inhibitory effect, the reduction in selenoproteome expression induced by HIV-1 infection would be diluted by the non-infected cells. On the other hand, an increase in expression or the emergence of a band corresponding to a protein isoform would be easier to observe. As illustrated in Figure 9, we monitored the expression of five selenoproteins, namely GPX4, GPX1, SELENOO, TXNRD1 and SELENOS, in response to HIV-1 infection at various time points and in cellular protein extracts. As a general trend, we observed a slight decrease in the expression of GPX1, GPX4, SELENOO and SELENOS at days 2, 3 or 4 post-infection. Interestingly, for SELENOS, an additional shorter form of the protein is visible at every time point post-infection and for the two donors tested here.

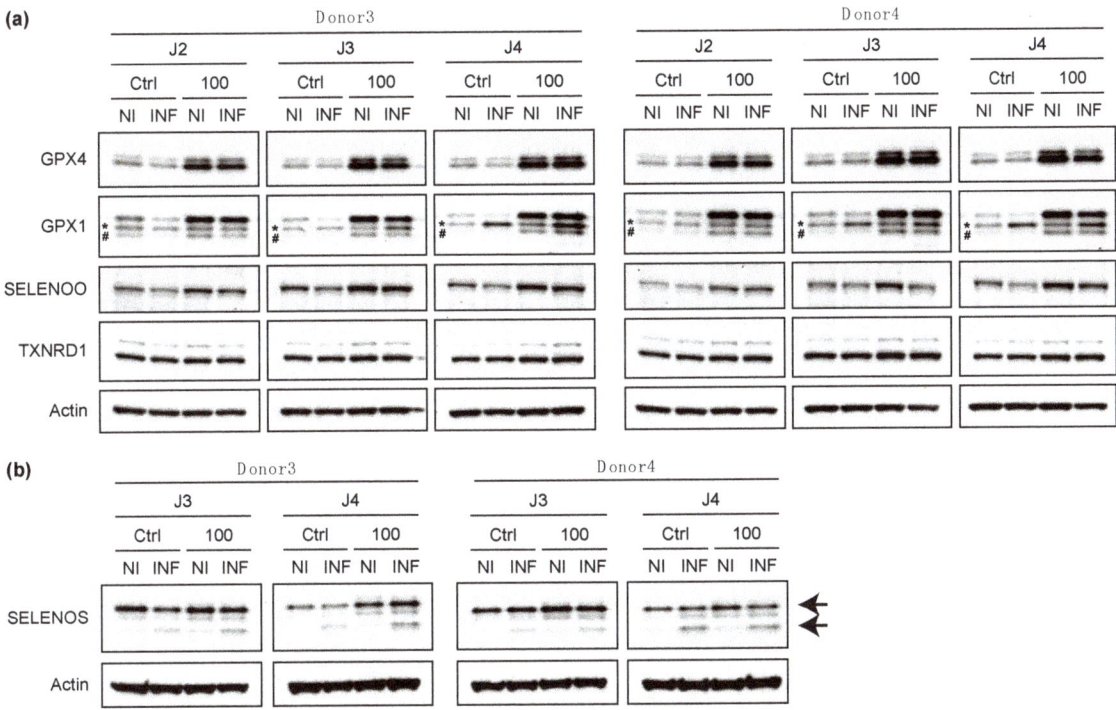

Figure 9. Analysis of selenoprotein expression in response to HIV-1 infection, in control (Ctrl) and 100 nM supplemented conditions. NI, non-infected cell extracts; INF, infected cell extracts. Donor 3 and Donor 4 were analyzed ($n = 1$) for the expression of SELENOO, GPX1, GPX4 and TXNRD1 (**a**) and SELENOS (**b**). Non-specific bands are indicated by an asterisk. The hash indicates an unknown band for GPX1. The arrows indicate the migration of two isoforms for SELENOS. The values are given in Table S5.

3. Discussion

Viruses use their host's cellular machinery and metabolism to replicate and form new infectious particles. In turn, host cells have developed antiviral mechanisms to counteract or restrict viral replication. These anti-viral proteins are often referred to as restriction factors [54]. These proteins, which are considered the first line of defense against viral pathogens, are either upregulated by interferons (IFNs) or constitutively expressed. Interestingly, several of them have additional biological functions outside of immunity and can be envisioned as moonlighting proteins. The identification of cellular proteins able to restrict HIV-1 replication have received enormous research interest over the years, and several antiviral factors have now been clearly defined and include apolipoprotein B mRNA editing enzyme, catalytic polypeptide-like 3G (APOBEC3G), Tetherin, interferon-induced transmembrane proteins (IFITMs), cholesterol-25-hydroxylase (CH25H), KRAB-associated protein 1/ tripartite motif-containing protein 28 (KAP1/TRIM28), 90K, Moloney leukemia virus 10 protein (MOV10), myxovirus resistance gene B (MxB), Schlafen family member 11 (SLFN11), and zinc-finger antiviral protein (ZAP) [55]. However, viruses constantly evolve intricate solutions to evade or directly counteract many restriction factors, leading to continuous virus–host adaptation. The fact that selenium levels were associated with a decrease in CD4 T-cell counts in infected patients prompted us to hypothesize that selenoproteins could, in some way, also be involved in the control of HIV replication. As selenoproteins are principally regulated by selenium levels, we studied the impact of selenium addition on HIV-1 replication in different in vitro models available in the laboratory. Therefore, we started our work with two established cell lines (Jurkat and SupT1). Our data show that infection of a defined number of cells led to the production of similarly infectious viral particles, independent of the selenium concentration added. We then decided to confirm these results in primary CD4 T-cells that were isolated and activated from four healthy donors. Although primary CD4 T cells represent a more relevant model for HIV-1 infection, these cells are very difficult to grow and infect, and they are rapidly killed by the virus, rendering any biochemical analyses technically challenging. Data presented in Figure 8 confirmed that selenium had no significant effect on HIV-1 protein production nor infectivity. However, we observed that the concentration of selenium in the culture medium had a small, but significant, impact on the percentage of the cell population that is infected by the virus. Interestingly, we also confirmed in primary cells that HIV-1 infection altered the expression of cellular selenoproteins, as suggested in a pioneer experiment done in Jurkat with ^{75}Se radioactive labelling with a decreased expression of several selenoproteins, including Gpx [46]. Our data suggest a link between selenium, selenoproteins and viral infection that we will investigate in future experiments.

The rationale of this work was to investigate the T-cell selenoproteome which is poorly described in the literature. Here, we first showed that expression of the selenoprotein mRNAs varied slightly between primary and established T cells, with high expression of GPX4, TXNRD1, SELENOT, GPX1, SELENOO, SELENOK, and SELENOW. As predicted from experiments that we performed in non-immune cellular models, synthesis of some selenoproteins was particularly sensitive to selenium variations in the culture medium at sub-micromolar ranges. This was particularly true for GPX1, GPX4, SELENOH, SELENOK and SELENOS, which are often referred to as stress-response members. On the other side, the selenoproteins that are part of the housekeeping members were less sensitive to changes in selenium concentration, and these include TXNRD1, SELENOP and SEPHS2. These data were then strengthened by measurement of enzymatic activities for GPXs and TXNRDs, which were enhanced by selenium supplementation, confirming that they represent an important line of the antioxidant defense and redox homeostasis in these immune cells. From our experiments, it appeared that SupT1 was less responsive to selenium variation than other cells, and this is why we selected Jurkat and primary cells to investigate the link between selenium and HIV infection in the following experiments. We were surprised to observe no significant effect of selenium on viral replication even when using primary cells, which is the relevant physiological model for HIV infection. It is noteworthy that we applied many experimental designs changing MOI, kinetics, and readouts but the conclusions remained unchanged. Surprisingly, it was previously shown that the overexpression of Gpx1 in SupT1 clones could accelerate HIV-1 replication, using very low MOI and rather long kinetics (14 days) [56]. In patients, infection with HIV-1 is a long and complex process that can be subdivided into three major phases: primary infection, which lasts for the first few weeks and is followed by a long latency phase (10–12 years) before the eventual progression to AIDS. As it was, our current experimental design mimics the first phase of infection (primary infection) with a high concentration of CD4 T cells that are infected in a short period of time and the virus that replicates rapidly. This contrasts with previous epidemiological studies, which investigated the fate of patients and the role of selenium during the latency. Thus, we do not see our results as being contradictory to previously published work. In fact, during HIV-1 latency, selenium most probably plays a role at many levels of the immune system or on many other cell types than CD4 T lymphocytes, such as macrophages, dendritic cells or reservoir cells, which have yet to be unambiguously identified. To our knowledge, this is the first demonstration that selenium concentration does not interfere with HIV-1 replication during the acute phase of replication but does not exclude a role in the later stages of the disease; this deserves further investigation.

Another interesting part of our finding concerns the changes in selenoprotein expression following HIV-1 infection in primary cells. Although we were faced with the technical challenge of working with non-transformed cells for which the culture is delicate and that are very inefficiently infected by the virus, we were still able to detect a slightly lower expression level for SELENOO, SELENOS, GPX4 and GPX1, whereas it was not the case for TXNRD1. In order to confirm these subtle changes at the selenoproteome scale, we compared our results with data obtained from mass spectrometry proteomics studies. Although these proteomics approaches continue to improve, only a few analyses have detected changes of T cells for low or moderately expressed cellular proteins in response to HIV-1 infection using mass spectrometry. Notably, a recent work described an elegant strategy to (i) circumvent confounding effects due to uninfected cells, (ii) limit the viral replication cycle to a single round and (iii) use a low MOI (i.e., ≤ 1) [57]. In this work, the authors infected primary human CD4 T lymphocytes with an HIV reporter virus encoding a cell surface streptavidin-binding affinity tag in place of the envelope (Env) gene, allowing antibody-free magnetic cell sorting of infected cells (AFMACS). This strategy allowed a rapid purification of HIV-infected cells from cultures, avoiding a tedious and potentially stressful cell sorting by flow cytometry. In this work, the authors were able to quantitate approximately 9000 proteins across multiple donors, among which they found 650 HIV-dependent changes. In this high coverage proteomic atlas, available as online supplementary data, we were able to find quantitative data for 15 members of the selenoproteome. In the three different donors, GPX1, GPX4, SELENOF, SELENOH, SELENOK, SELENOM, SELENON, SELENOO, SELENOS, SELENOT, SELENOW, SEPHS2, TXNRD1, TXNRD2, and TXNRD3 were detected, which is remarkable for a mass spectrometry proteomic approach and close to what could be obtained by immunological strategies. Among these selenoproteins, only TXNRD1 changed significantly in infected cell extracts in comparison to control conditions. At 24 and 48 h post-infection, a 15% and 25% decrease in protein expression was observed in infected cells, respectively. This decrease was in agreement with results obtained two decades ago using ^{75}Se radioactive labeling [46]. This change was too subtle to be detected in our experimental design, using Western blots with less than 50% infected cells. However, the advantage of Western blot versus proteomics is the possibility of visualizing the emergence of protein isoforms, such as the one we observed for SELENOS after HIV infection. In their proteomic study, the authors also reported the downregulation of both GPX1 and GPX4 at 24 and 48 h, although this was not considered significant. The authors also studied proteomic changes observed in response to resting cell activation by TCR stimulation [57] and found a dramatic remodeling of the selenoproteome for the different donors with a significant decrease in SELENOW, SELENOH, SELENOM, GPX4, GPX1, SEPHS2, TXNRD2, SELENOF, TXNRD3 and SELENOO, and a significant increase in SELENOS, SELENON and SELENOK in response to T-cell activation. Only SELENOT and TXNRD1 remained unchanged. These data suggest a significant and differential function for several selenoproteins between resting and activated cells. However, it should be noted that detailed information about selenium concentration in the culture media used is missing, which restricts interpretation of the data. It would be interesting to recapitulate these experiments with a tight control of selenium concentration in order to understand its impact both on T-cell activation and selenoproteome remodeling during this process.

Another study reported the effects of HIV-1 infection on transformed T-cell lines (CEM), which were infected at high MOI with an Env-deficient, VSVg-pseudotyped virus, enabling a synchronous single round infection with less than 10% of uninfected cells [58]. In this work, only eight selenoproteins were detected that included GPX1, GPX4, SELENOF, SELENOH, SELENOS, SEPHS2, TXNRD1, and TXNRD2, but only very little changes were found. Only for SELENOH and SELENOS, a significant decrease (between 20% and 30%) was observed in the late phase of infection (48 and 72 h). Only in early time points (6 h) was a very weak decrease (between 12% to 15%) observed for GPX1, GPX4 and SEPHS2. Thus, this report and our data confirm that HIV-1 replication does not disrupt the selenoproteome during the acute phase of infection, but only induces subtle changes. Altogether, our results show that selenium concentration affects the expression of selenoproteins in T cells in a hierarchical manner, as it was described in non-immune cells. Amongst the proteins that are mostly affected, one can find GPX1, GPX4, SELENOH, SELENOK, SELENOS and SELENOT. However, in the context of our experimental setting that mimics only the early phases of HIV-1 infection, only very little effects or interplay with the selenoproteome could be observed.

4. Materials and Methods

This manuscript adopts the systematic nomenclature of selenoprotein names [59].

4.1. Materials

Jurkat, SupT1 and HEK293T cells lines used in this study were obtained from ATCC. TZM-bl cells were obtained from NIH AIDS Reagent program (Manassas, VA, USA). Fresh blood samples were provided by Etablissement francais du sang (EFS) de Lyon. Cell culture media and supplements, NuPAGE 4–12% bis–Tris polyacrylamide gels, MOPS and MES SDS running buffer, Dynabeads Human T-Activator CD3/CD28 were purchased from Life Technologies (ThermoFisher Scientific, Waltham, MA, USA). Fetal calf serum (FCS), sodium selenite, synthetic oligonucleotides, Percoll, Ficoll, t-BHP, NADPH, thioredoxin, L-GSH, glutathione reductase, DTNB, sucrose, DMSO, EDTA, Triton X100, glycerol and DTT were purchased from Merck (Darmstadt, Germany). Interleukin2 was from Eurobio Scientific (Les Ulis, France). The luciferase assay reagent was purchased from Promega (Charbonnières, France). The microplate readers FLUOSTAR OPTIMA and LUMISTAR OPTIMA were from BMG Labtech (Champigny-sur-Marne, France). The list of antibodies used in this study is given in Table S3. The plasmid pNL4.3 was obtained from the NIH AIDS Reagent program.

4.2. Cell Culture

Adherent cells (HEK293T and TZM-bl) were grown and maintained in 75 cm^2 plates in Dulbecco's Modified Eagle Medium (D-MEM). Cells in suspension (Jurkat, SupT1 and primary T cells) were cultured in Roswell Park Memorial Institute medium (RPMI). Media were supplemented with 10% fetal calf serum, 100 µg/mL streptomycin, 100 UI/mL penicillin, and 2 mM L-glutamine. Only for RPMI medium, 1 mM pyruvate and 10 mM HEPES were added. For the culture of primary T-lymphocytes, 30 U/mL Interleukin2 was added to the growth media. Cells were cultivated at 37 °C in humidified atmosphere containing 5% of CO_2. Selenium supplementation was obtained by adding a defined volume of a concentrated solution of sodium selenite (0.1 mM diluted in ultrapure water) in the culture media. In our cell culture experiments, selenium is exclusively provided by the FCS, and can vary widely from providers and even from one lot to another of the same provider [49]. Here, we used a lot of FCS containing 110 nM of selenium, which is in the low range, and kept it for all the experiments of the present study. Therefore, the Ctrl growth conditions that contained 10% FCS (11 nM of selenium) could be considered selenium deficient.

4.3. Isolation and Culture of Primary CD4 T-Cells

We isolated human CD4 T cells from peripheral blood mononuclear cells (PBMC), using a negative selection strategy following the manufacturer's instructions. Our studies were performed with fresh blood samples originating from different healthy donors. Briefly, peripheral blood mononuclear cells (PBMC) were isolated by Ficoll density gradient centrifugation, 25 min at $750\times g$, without the centrifuge brake. The PBMC ring was collected between the plasma and Ficoll and diluted with $1 \times$ PBS. After several washes with $1 \times$ PBS, the cells were counted, placed on a Percoll cushion and centrifuged for 20 min at $930\times g$. The peripheral blood lymphocytes (PBL) fractions were collected in the upper phase while the monocytes formed a ring at the interface. The PBL were counted, aliquoted at approximately 30×10^6 cells/mL and stored at -80 °C in freezing medium composed of 10% DMSO and 90% SVF. The negative isolation of CD4 positive T cells was performed with the EasySep™ Human CD4 T-Cell Isolation Kit (StemCell Technologies, Saint Égrève,

France) according to the manufacturer's protocol. Then, the activation of primary T lymphocytes via the T-cell receptor (TCR) complex was performed with Dynabeads Human T-Activator CD3/CD28 for 72 h according to the manufacturer's protocol. After this time period of activation and expansion, the cells were then readily used for HIV-1 infection and/or selenium supplementation of the culture media.

4.4. HIV-1 Production in HEK293T Cells and Titration

HIV-1 viral particles were produced by transient calcium phosphate transfection of HEK293T cells with pNL4.3 plasmid DNA. The day prior to transfection, HEK293T cells were seeded at 6.5×10^6 cells per 10 cm plates. Ten micrograms of pNL4.3 plasmid were transfected per plate using calcium phosphate precipitation of DNA. The medium was replaced by a fresh one 6 h later. Two days post-transfection, the medium was collected, filtered (0.45 µm cutoff) and concentrated by ultracentrifugation at 4 °C (1 h 30 min, $110,000 \times g$) through a 20% sucrose cushion prepared in TNE buffer (10 mM Tris, pH 8, 100 mM NaCl, 1 mM EDTA). The concentrated viral particles from one 10 cm plate were resuspended in 100 µL RPMI, aliquoted and stored at −80 °C for several months. The viral particles were titrated by serial dilution on Jurkat cells (LTR HIV-1 GFP, from NIH AIDS reagent program).

4.5. Infection of CD4 T-Cells with HIV-1

Jurkat and primary CD4 T cells were counted and resuspended at 4×10^6 cells/mL. A volume of freshly thawed HIV-1 viral particles was added to the cells to obtain a defined MOI (multiplicity of infection), previously calculated in Jurkat cells stably expressing GFP gene in the control of the HIV-1 LTR promoter. Infection occurred in a small volume for 3 h at 37 °C. The cells were then washed with 1 × PBS and resuspended in a culture medium, supplemented or not with 100 nM sodium selenite, and incubated at 37 °C from 2 to 15 days as indicated in every experiment.

4.6. Quantification of Infectious HIV-1 Produced by CD4 T Cells

The viral particles produced by CD4 T cells were analyzed for their capacity to infect TZM-bl cells. Previously designated JC53-bl (clone 13), the TZM-bl are HeLa cells producing large amounts of CD4 and CCR5 and separate integrated copies of the luciferase and beta-galactosidase genes under control of the HIV-1 promoter. We modified this cell line by integrating copies of the mCherry gene, using a lentivirus. The red fluorescence was directly proportional to cell numbers and therefore used to normalize the luciferase relative to cell density.

The day prior to infection, TZM-bl cell were seeded in a 96 well plate at 1×10^4 cells/well. The day of infection, the collected media from CD4 T cells were serial diluted (1/3) in triplicate and added to the TZM-bl cells. After 48 h of growth, the cells were harvested and evaluated for red fluorescence and luciferase activity using a multiplate reader. In each well, the luciferase activity (AU) was expressed relative to the fluorescence intensity (AU). The infectivity of the media was then expressed relative to the maximum value of the experiment set at 100.

4.7. Quantification of p24 Contained in Viral Particles

To evaluate the quantity of viral particles released by CD4 T cells, aliquots of the culture media were collected and concentrated by 30 min ultracentrifugation at g though a 20% sucrose cushion at 4 °C. After removal of the supernatants, the pellets were resuspended in 20 µL of lysis buffer (25 mM Tris-HCl, pH 7.8, 2 mM DTT, 2 mM EDTA, 1% Triton X-100 and 10% glycerol). Equal volumes (10 µL) were separated in 4–12% Bis-Tris NuPAGE Novex Midi Gels and transferred onto nitrocellulose membranes using iBlot DRy blotting System (Themo Fisher Scientific, Waltham, MA, USA). Membranes were probed with indicated p24 antibodies and HRP-conjugated anti-mouse secondary antibodies. The chemiluminescence signal was detected using ECL Clarirty Max detection kit (Biorad, Hercules, CA, USA) in the Chemidoc Imager (Biorad). Data quantifications were performed with ImageLab Software (Biorad, Version 6.0.1, Hercules, CA, USA).

4.8. Total RNA Extraction and Analysis by RT-qPCR

The quantification of selenoprotein mRNAs was performed as described in [31]. Briefly, total RNAs were purified with Tri Reagent® (Molecular Research Center, Cincinnati, OH, USA) according to manufacturer's instructions and resuspended in ultrapure milliQ water. RNAs were reverse transcribed using qScript cDNA Synthesis kit (Quanta Bio, Beverly, MA, USA) according to the manufacturer's instructions. Real-time PCRs were performed in triplicate using FastStart Universal SYBR® Green master (Roche Applied Science, Penzberg, Germany) on a StepOne Real-Time PCR System (Applied Biosystems, Foster City, CA, USA). Primers used are described in [31] and listed in Table S6. Serial dilutions of a cDNA mixture were used to create a standard curve and determine the efficiency of the amplification for each pair of primers. The quantification of selenoprotein mRNAs was performed, using a set of four reference genes.

4.9. Protein Extraction and Analysis by Western Blot

After a wash with 1X PBS, cellular protein extracts were harvested with lysis buffer (described earlier). Then, protein concentrations were measured using the DC™ protein assay kit (Biorad) in microplate assays. Equal protein amounts (30 µg) were separated in 4–12% Bis-Tris NuPAGE Novex Midi Gels and transferred onto nitrocellulose membranes using the iBlot® DRy blotting System. Membranes were probed with indicated primary antibodies and HRP-conjugated anti-rabbit or anti-mouse secondary antibodies. The chemiluminescence signal was detected using an ECL Select detection kit (GEHealthcare, Chicago, IL, USA) in the Chemidoc Imager (Biorad, Hercules, CA, USA). Data quantifications were performed as described earlier.

4.10. GPX and TXNRD Enzymatic Assays

GPX and TXNRD activities were measured in enzymatic coupled assay as described in [37,38,60] with 30 µg of protein extracts. GPX enzymatic activities (U/mg) were measured with t-BHP substrate and expressed as nmol of glutathione/min·mg. TXNRD enzymatic activities (U/mg) were expressed as nmol of NADPH/min·mg.

4.11. Flow Cytometry

The percentage of infected cell was monitored by flow cytometry. Briefly, cells were washed once with PBS and fixed with PBS containing 4% formaldehyde for 20 min at 4 °C. Cells were washed once with PBS and once with staining buffer (PBS containing 2% fetal calf serum and 2 mM EDTA). Then, cells were incubated with anti-HIV-1 p24 antibody coupled with FITC diluted 1/100 in staining buffer, for 1 h at 4 °C. Cells were washed twice with staining buffer and analyzed on a FACS Canto (7 colors) using FACS Diva software (BD Biosciences) and analyzed with FlowJo software (Treestar, Ashland, OR, USA).

4.12. Ethics Statement

Primary blood cells were obtained from the blood of healthy donors (EFS-Lyon) in the form of discarded "leukopacks" obtained anonymously so that gender, race, and age of donors were unknown to the investigator and inclusion of women or minorities cannot be determined. This research is exempt from approval, although written informed consent was obtained from blood donors to allow use of their cells for research purposes.

5. Conclusions

HIV-1 infects CD4 T lymphocytes, monocytes, macrophages and dendritic cells, but only a decline in CD4 T cells reflects the progression of HIV infection. The fact that selenium levels were associated with a decrease in CD4 T-cell counts in infected patients justified the study of the impact of selenium on HIV-1 replication in laboratory models. We primarily characterized the expression pattern of the selenoprotein genes in two established cell lines (Jurkat and SupT1) and primary CD4 T cells isolated from the blood of anonymous healthy donors. We found that the expression of the selenoproteome in human T cells is tightly regulated by the selenium level of the culture media, and this is in agreement with reports from non-immune cells. We showed that selenium had no significant effect on HIV-1 protein production nor infectivity, indicating that selenium may not be involved in the acute phase of replication but does not exclude a role in the later stages of the disease since the selenium levels slightly modified the proportion of infected cells. Finally, the expression of the selenoproteome is slightly modified by HIV-1 infection.

Supplementary Materials: The following are available online at https://www.mdpi.com/article/10.3390/ijms23031394/s1.

Author Contributions: Conceptualization, C.V., O.M.G., T.O. and L.C.; methodology, C.V., O.M.G., T.O. and L.C.; validation, C.V., O.M.G.; formal analysis, C.V., O.M.G.; investigation, C.V., O.M.G.; writing—original draft preparation, C.V., O.M.G., T.O. and L.C.; writing—review and editing, C.V., O.M.G., T.O. and L.C.; visualization, C.V., O.M.G., T.O. and L.C.; supervision, T.O. and L.C.; project administration, T.O. and L.C.; funding acquisition, C.V., O.M.G., T.O. and L.C. All authors have read and agreed to the published version of the manuscript.

Funding: This work was supported by Agence nationale de recherche sur le sida et les hépatites virale (ANRS), INSERM, CNRS and ENS de Lyon 'Emerging Project' (LC in 2016 and TO in 2018), the CNRS MITI programs "Isotop" and "Metallo-Mix". OG was a recipient of a Master fellowship from Labex Ecofect (ANR-11-LABX-0048) of the Université de Lyon, within the program Investissements d'Avenir (ANR-11-IDEX-0007) operated by the French National Research Agency (ANR). OG is a recipient of a PhD fellowship from the Université Claude Bernard Lyon 1. CV is a recipient of a post-doctoral fellowship from ANRS.

Institutional Review Board Statement: Not applicable.

Informed Consent Statement: Informed consent was obtained from all subjects involved in the study.

Data Availability Statement: Requests for further information about resources, reagents, and data availability should be directed to the corresponding author.

Acknowledgments: In several figures and in the graphical abstract, we used illustrations from (https://smart.servier.com/) and (https://biorender.com/) websites. We are grateful to Xuan-Nhi Nguyen, Andrea Cimarelli, Lucie Etienne and Didier Décimo for their technical assistance and scientific advices. We acknowledge the contribution of the Etablissement Français du Sang Auvergne—Rhône-Alpes.

Conflicts of Interest: The authors declare no conflict of interest.

References

1. Ferguson, M.R.; Rojo, D.R.; von Lindern, J.J.; O'Brien, W.A. HIV-1 replication cycle. *Clin. Lab. Med.* **2002**, *22*, 611–635. [CrossRef]
2. Nyamweya, S.; Hegedus, A.; Jaye, A.; Rowland-Jones, S.; Flanagan, K.L.; Macallan, D.C. Comparing HIV-1 and HIV-2 infection: Lessons for viral immunopathogenesis. *Rev. Med. Virol.* **2013**, *23*, 221–240. [CrossRef] [PubMed]
3. Samji, H.; Cescon, A.; Hogg, R.S.; Modur, S.P.; Althoff, K.N.; Buchacz, K.; Burchell, A.N.; Cohen, M.; Gebo, K.A.; Gill, M.J.; et al. Closing the gap: Increases in life expectancy among treated HIV-positive individuals in the United States and Canada. *PLoS ONE* **2013**, *8*, e81355. [CrossRef] [PubMed]
4. Palella, F.J., Jr.; Delaney, K.M.; Moorman, A.C.; Loveless, M.O.; Fuhrer, J.; Satten, G.A.; Aschman, D.J.; Holmberg, S.D. Declining morbidity and mortality among patients with advanced human immunodeficiency virus infection. HIV Outpatient Study Investigators. *N. Engl. J. Med.* **1998**, *338*, 853–860. [CrossRef] [PubMed]
5. Guillin, O.M.; Vindry, C.; Ohlmann, T.; Chavatte, L. Selenium, Selenoproteins and Viral Infection. *Nutrients* **2019**, *11*, 2101. [CrossRef]
6. Pitney, C.L.; Royal, M.; Klebert, M. Selenium supplementation in HIV-infected patients: Is there any potential clinical benefit? *J. Assoc. Nurses AIDS Care* **2009**, *20*, 326–333. [CrossRef]
7. Bogden, J.D.; Oleske, J.M. The essential trace minerals, immunity, and progression of HIV-1 infection. *Nutr. Res.* **2007**, *27*, 69–77. [CrossRef]

8. Kamwesiga, J.; Mutabazi, V.; Kayumba, J.; Tayari, J.-C.K.; Uwimbabazi, J.C.; Batanage, G.; Uwera, G.; Baziruwiha, M.; Ntizimira, C.; Murebwayire, A.; et al. Effect of selenium supplementation on CD4+ T-cell recovery, viral suppression and morbidity of HIV-infected patients in Rwanda: A randomized controlled trial. *AIDS* **2015**, *29*, 1045–1052. [CrossRef]
9. Baum, M.K.; Campa, A.; Lai, S.; Sales Martinez, S.; Tsalaile, L.; Burns, P.; Farahani, M.; Li, Y.; van Widenfelt, E.; Page, J.B.; et al. Effect of micronutrient supplementation on disease progression in asymptomatic, antiretroviral-naive, HIV-infected adults in Botswana: A randomized clinical trial. *JAMA* **2013**, *310*, 2154–2163. [CrossRef]
10. Hurwitz, B.E.; Klaus, J.R.; Llabre, M.M.; Gonzalez, A.; Lawrence, P.J.; Maher, K.J.; Greeson, J.M.; Baum, M.K.; Shor-Posner, G.; Skyler, J.S.; et al. Suppression of human immunodeficiency virus type 1 viral load with selenium supplementation: A randomized controlled trial. *Arch. Intern. Med.* **2007**, *167*, 148–154. [CrossRef]
11. de Menezes Barbosa, E.G.; Junior, F.B.; Machado, A.A.; Navarro, A.M. A longer time of exposure to antiretroviral therapy improves selenium levels. *Clin. Nutr.* **2015**, *34*, 248–251. [CrossRef] [PubMed]
12. Campa, A.; Sales Martinez, S.; Baum, M.K. Selenium in HIV/AIDS. In *Selenium: Its Molecular Biology and Role in Human Health*, 4th ed.; Hatfield, D.L., Schweizer, U., Tsuji, P.A., Gladyshev, V.N., Eds.; Springer Science + Business Media, LLC: New York, NY, USA, 2016; pp. 333–342.
13. Schomburg, L. Selenium Deficiency Due to Diet, Pregnancy, Severe Illness, or COVID-19-A Preventable Trigger for Autoimmune Disease. *Int. J. Mol. Sci.* **2021**, *22*, 8532. [CrossRef] [PubMed]
14. Beck, M.A.; Levander, O.A.; Handy, J. Selenium deficiency and viral infection. *J. Nutr.* **2003**, *133* (Suppl. 1), 1463S–1467S. [CrossRef] [PubMed]
15. Beck, M.A.; Nelson, H.K.; Shi, Q.; Van Dael, P.; Schiffrin, E.J.; Blum, S.; Barclay, D.; Levander, O.A. Selenium deficiency increases the pathology of an influenza virus infection. *FASEB J.* **2001**, *15*, 1481–1483. [CrossRef] [PubMed]
16. Beck, M.A. Rapid genomic evolution of a non-virulent coxsackievirus B3 in selenium-deficient mice. *Biomed. Environ. Sci.* **1997**, *10*, 307–315.
17. Beck, M.A.; Shi, Q.; Morris, V.C.; Levander, O.A. Rapid genomic evolution of a non-virulent coxsackievirus B3 in selenium-deficient mice results in selection of identical virulent isolates. *Nat. Med.* **1995**, *1*, 433–436. [CrossRef] [PubMed]
18. Sonet, J.; Bulteau, A.-L.; Chavatte, L. Selenium and Selenoproteins in Human Health and Diseases. In *Metallomics: Analytical Techniques and Speciation Methods*; Michalke, B., Ed.; 2016 Wiley-VCH Verlag GmbH & Co. KGaA: Weinheim, Germany, 2016; pp. 364–381.
19. Labunskyy, V.M.; Hatfield, D.L.; Gladyshev, V.N. Selenoproteins: Molecular pathways and physiological roles. *Physiol. Rev.* **2014**, *94*, 739–777. [CrossRef]
20. Papp, L.V.; Lu, J.; Holmgren, A.; Khanna, K.K. From selenium to selenoproteins: Synthesis, identity, and their role in human health. *Antioxid. Redox Signal.* **2007**, *9*, 775–806. [CrossRef]
21. Kryukov, G.V.; Castellano, S.; Novoselov, S.V.; Lobanov, A.V.; Zehtab, O.; Guigo, R.; Gladyshev, V.N. Characterization of mammalian selenoproteomes. *Science* **2003**, *300*, 1439–1443. [CrossRef]
22. Driscoll, D.M.; Copeland, P.R. Mechanism and regulation of selenoprotein synthesis. *Annu. Rev. Nutr.* **2003**, *23*, 17–40. [CrossRef]
23. Brigelius-Flohe, R.; Maiorino, M. Glutathione peroxidases. *Biochim. Biophys. Acta* **2012**, *1830*, 3289–3303. [CrossRef] [PubMed]
24. Arner, E.S. Focus on mammalian thioredoxin reductases—Important selenoproteins with versatile functions. *Biochim. Biophys. Acta* **2009**, *1790*, 495–526. [CrossRef] [PubMed]
25. Zhang, X.; Xiong, W.; Chen, L.L.; Huang, J.Q.; Lei, X.G. Selenoprotein V protects against endoplasmic reticulum stress and oxidative injury induced by pro-oxidants. *Free Radic. Biol. Med.* **2020**, *160*, 670–679. [CrossRef] [PubMed]
26. Pothion, H.; Jehan, C.; Tostivint, H.; Cartier, D.; Bucharles, C.; Falluel-Morel, A.; Boukhzar, L.; Anouar, Y.; Lihrmann, I. Selenoprotein T: An Essential Oxidoreductase Serving as a Guardian of Endoplasmic Reticulum Homeostasis. *Antioxid. Redox Signal.* **2020**, *33*, 1257–1275. [CrossRef] [PubMed]
27. Rocca, C.; Pasqua, T.; Boukhzar, L.; Anouar, Y.; Angelone, T. Progress in the emerging role of selenoproteins in cardiovascular disease: Focus on endoplasmic reticulum-resident selenoproteins. *Cell. Mol. Life Sci.* **2019**, *76*, 3969–3985. [CrossRef]
28. Pitts, M.W.; Hoffmann, P.R. Endoplasmic reticulum-resident selenoproteins as regulators of calcium signaling and homeostasis. *Cell Calcium* **2018**, *70*, 76–86. [CrossRef]
29. Addinsall, A.B.; Wright, C.R.; Andrikopoulos, S.; van der Poel, C.; Stupka, N. Emerging roles of endoplasmic reticulum-resident selenoproteins in the regulation of cellular stress responses and the implications for metabolic disease. *Biochem. J.* **2018**, *475*, 1037–1057. [CrossRef]
30. Avery, J.C.; Hoffmann, P.R. Selenium, Selenoproteins, and Immunity. *Nutrients* **2018**, *10*, 1203. [CrossRef]
31. Vindry, C.; Ohlmann, T.; Chavatte, L. Translation regulation of mammalian selenoproteins. *Biochim. Biophys. Acta Gen. Subj.* **2018**, *1862*, 2480–2492. [CrossRef]
32. Bulteau, A.-L.; Chavatte, L. Update on selenoprotein biosynthesis. *Antioxid. Redox Signal.* **2015**, *23*, 775–794. [CrossRef]
33. Sonet, J.; Bulteau, A.L.; Touat-Hamici, Z.; Mosca, M.; Bierla, K.; Mounicou, S.; Lobinski, R.; Chavatte, L. Selenoproteome Expression Studied by Non-Radioactive Isotopic Selenium-Labeling in Human Cell Lines. *Int. J. Mol. Sci.* **2021**, *22*, 7308. [CrossRef] [PubMed]
34. Touat-Hamici, Z.; Bulteau, A.L.; Bianga, J.; Jean-Jacques, H.; Szpunar, J.; Lobinski, R.; Chavatte, L. Selenium-regulated hierarchy of human selenoproteome in cancerous and immortalized cells lines. *Biochim. Biophys. Acta Gen. Subj.* **2018**, *1862*, 2493–2505. [CrossRef] [PubMed]

35. Hammad, G.; Legrain, Y.; Touat-Hamici, Z.; Duhieu, S.; Cornu, D.; Bulteau, A.L.; Chavatte, L. Interplay between Selenium Levels and Replicative Senescence in WI-38 Human Fibroblasts: A Proteomic Approach. *Antioxidants* **2018**, *7*, 19. [CrossRef] [PubMed]
36. Touat-Hamici, Z.; Legrain, Y.; Sonet, J.; Bulteau, A.-L.; Chavatte, L. Alteration of selenoprotein expression during stress and in aging. In *Selenium: Its Molecular Biology and Role in Human Health*, 4th ed.; Hatfield, D.L., Schweizer, U., Tsuji, P.A., Gladyshev, V.N., Eds.; Springer Science + Business Media, LLC: New York, NY, USA, 2016; pp. 539–551.
37. Touat-Hamici, Z.; Legrain, Y.; Bulteau, A.-L.; Chavatte, L. Selective up-regulation of human selenoproteins in response to oxidative stress. *J. Biol. Chem.* **2014**, *289*, 14750–14761. [CrossRef] [PubMed]
38. Legrain, Y.; Touat-Hamici, Z.; Chavatte, L. Interplay between selenium levels, selenoprotein expression, and replicative senescence in WI-38 human fibroblasts. *J. Biol. Chem.* **2014**, *289*, 6299–6310. [CrossRef]
39. Bock, A.; Forchhammer, K.; Heider, J.; Leinfelder, W.; Sawers, G.; Veprek, B.; Zinoni, F. Selenocysteine: The 21st amino acid. *Mol. Microbiol.* **1991**, *5*, 515–520. [CrossRef]
40. Bock, A.; Forchhammer, K.; Heider, J.; Baron, C. Selenoprotein synthesis: An expansion of the genetic code. *Trends Biochem. Sci.* **1991**, *16*, 463–467. [CrossRef]
41. Lejeune, F. Nonsense-mediated mRNA decay at the crossroads of many cellular pathways. *BMB Rep.* **2017**, *50*, 175–185. [CrossRef]
42. Vindry, C.; Guillin, O.; Mangeot, P.E.; Ohlmann, T.; Chavatte, L. A Versatile Strategy to Reduce UGA-Selenocysteine Recoding Efficiency of the Ribosome Using CRISPR-Cas9-Viral-Like-Particles Targeting Selenocysteine-tRNA([Ser]Sec) Gene. *Cells* **2019**, *8*, 574. [CrossRef]
43. Dalley, B.K.; Baird, L.; Howard, M.T. Studying Selenoprotein mRNA Translation Using RNA-Seq and Ribosome Profiling. *Methods Mol. Biol.* **2018**, *1661*, 103–123.
44. Fradejas-Villar, N.; Seeher, S.; Anderson, C.B.; Doengi, M.; Carlson, B.A.; Hatfield, D.L.; Schweizer, U.; Howard, M.T. The RNA-binding protein Secisbp2 differentially modulates UGA codon reassignment and RNA decay. *Nucleic Acids Res.* **2017**, *45*, 4094–4107. [CrossRef] [PubMed]
45. Howard, M.T.; Carlson, B.A.; Anderson, C.B.; Hatfield, D.L. Translational redefinition of UGA codons is regulated by selenium availability. *J. Biol. Chem.* **2013**, *288*, 19401–19413. [CrossRef] [PubMed]
46. Gladyshev, V.N.; Stadtman, T.C.; Hatfield, D.L.; Jeang, K.T. Levels of major selenoproteins in T cells decrease during HIV infection and low molecular mass selenium compounds increase. *Proc. Natl. Acad. Sci. USA* **1999**, *96*, 835–839. [CrossRef] [PubMed]
47. Mariotti, M.; Ridge, P.G.; Zhang, Y.; Lobanov, A.V.; Pringle, T.H.; Guigo, R.; Hatfield, D.L.; Gladyshev, V.N. Composition and evolution of the vertebrate and mammalian selenoproteomes. *PLoS ONE* **2012**, *7*, e33066. [CrossRef] [PubMed]
48. Lobanov, A.V.; Fomenko, D.E.; Zhang, Y.; Sengupta, A.; Hatfield, D.L.; Gladyshev, V.N. Evolutionary dynamics of eukaryotic selenoproteomes: Large selenoproteomes may associate with aquatic life and small with terrestrial life. *Genome. Biol.* **2007**, *8*, R198. [CrossRef] [PubMed]
49. Karlenius, T.C.; Shah, F.; Yu, W.C.; Hawkes, H.J.; Tinggi, U.; Clarke, F.M.; Tonissen, K.F. The selenium content of cell culture serum influences redox-regulated gene expression. *Biotechniques* **2011**, *50*, 295–301. [CrossRef]
50. Brigelius-Flohe, R.; Kipp, A.P. Physiological functions of GPx2 and its role in inflammation-triggered carcinogenesis. *Ann. N. Y. Acad. Sci.* **2012**, *1259*, 19–25. [CrossRef]
51. Stafford, W.C.; Peng, X.; Olofsson, M.H.; Zhang, X.; Luci, D.K.; Lu, L.; Cheng, Q.; Tresaugues, L.; Dexheimer, T.S.; Coussens, N.P.; et al. Irreversible inhibition of cytosolic thioredoxin reductase 1 as a mechanistic basis for anticancer therapy. *Sci. Transl. Med.* **2018**, *10*. [CrossRef]
52. Sonet, J.; Mosca, M.; Bierla, K.; Modzelewska, K.; Flis-Borsuk, A.; Suchocki, P.; Ksiazek, I.; Anuszewska, E.; Bulteau, A.L.; Szpunar, J.; et al. Selenized Plant Oil Is an Efficient Source of Selenium for Selenoprotein Biosynthesis in Human Cell Lines. *Nutrients* **2019**, *11*, 1524. [CrossRef]
53. Mohammadi, P.; Desfarges, S.; Bartha, I.; Joos, B.; Zangger, N.; Munoz, M.; Gunthard, H.F.; Beerenwinkel, N.; Telenti, A.; Ciuffi, A. 24 hours in the life of HIV-1 in a T cell line. *PLoS Pathog.* **2013**, *9*, e1003161. [CrossRef]
54. Prelli Bozzo, C.; Kmiec, D.; Kirchhoff, F. When good turns bad: How viruses exploit innate immunity factors. *Curr. Opin. Virol.* **2021**, *52*, 60–67. [CrossRef] [PubMed]
55. Kluge, S.F.; Sauter, D.; Kirchhoff, F. SnapShot: Antiviral restriction factors. *Cell* **2015**, *163*, 774. [CrossRef] [PubMed]
56. Sandstrom, P.A.; Murray, J.; Folks, T.M.; Diamond, A.M. Antioxidant defenses influence HIV-1 replication and associated cytopathic effects. *Free Radic. Biol. Med.* **1998**, *24*, 1485–1491. [CrossRef]
57. Naamati, A.; Williamson, J.C.; Greenwood, E.J.; Marelli, S.; Lehner, P.J.; Matheson, N.J. Functional proteomic atlas of HIV infection in primary human CD4+ T cells. *eLife* **2019**, *8*, e41431. [CrossRef] [PubMed]
58. Greenwood, E.J.; Matheson, N.J.; Wals, K.; van den Boomen, D.J.; Antrobus, R.; Williamson, J.C.; Lehner, P.J. Temporal proteomic analysis of HIV infection reveals remodelling of the host phosphoproteome by lentiviral Vif variants. *eLife* **2016**, *5*, e18296. [CrossRef] [PubMed]
59. Gladyshev, V.N.; Arner, E.S.; Berry, M.J.; Brigelius-Flohe, R.; Bruford, E.A.; Burk, R.F.; Carlson, B.A.; Castellano, S.; Chavatte, L.; Conrad, M.; et al. Selenoprotein Gene Nomenclature. *J. Biol. Chem.* **2016**, *291*, 24036–24040. [CrossRef]
60. Sonet, J.; Bierla, K.; Bulteau, A.L.; Lobinski, R.; Chavatte, L. Comparison of analytical methods using enzymatic activity, immunoaffinity and selenium-specific mass spectrometric detection for the quantitation of glutathione peroxidase 1. *Anal. Chim. Acta* **2018**, *1011*, 11–19. [CrossRef]

Article

Natural Autoimmunity to Selenoprotein P Impairs Selenium Transport in Hashimoto's Thyroiditis

Qian Sun [1,2], Sebastian Mehl [1,2], Kostja Renko [1,2,3], Petra Seemann [1,2,4], Christian L. Görlich [1,2], Julian Hackler [1,2], Waldemar B. Minich [1,2], George J. Kahaly [5,*] and Lutz Schomburg [1,2,*]

1. Institute for Experimental Endocrinology, Charité-Universitätsmedizin Berlin, 13353 Berlin, Germany; qian.sun@charite.de (Q.S.); sebastian.mehl@mac.com (S.M.); Kostja.Renko@bfr.bund.de (K.R.); seemann@selenomed.com (P.S.); christian.goerlich@charite.de (C.L.G.); Julian.hackler@charite.de (J.H.); Waldemar.minich@charite.de (W.B.M.)
2. Freie Universität Berlin, Humboldt-Universität zu Berlin, Berlin Institute of Health, 13353 Berlin, Germany
3. German Federal Institute for Risk Assessment, Department Experimental Toxicology and ZEBET, 12277 Berlin, Germany
4. selenOmed GmbH, 10965 Berlin, Germany
5. Johannes Gutenberg University Medical Center, Department of Medicine I, 55101 Mainz, Germany
* Correspondence: george.kahaly@unimedizin-mainz.de (G.J.K.); lutz.schomburg@charite.de (L.S.)

Abstract: The essential trace element selenium (Se) is needed for the biosynthesis of selenocysteine-containing selenoproteins, including the secreted enzyme glutathione peroxidase 3 (GPX3) and the Se-transporter selenoprotein P (SELENOP). Both are found in blood and thyroid colloid, where they serve protective functions. Serum SELENOP derives mainly from hepatocytes, whereas the kidney contributes most serum GPX3. Studies using transgenic mice indicated that renal GPX3 biosynthesis depends on Se supply by hepatic SELENOP, which is produced in protein variants with varying Se contents. Low Se status is an established risk factor for autoimmune thyroid disease, and thyroid autoimmunity generates novel autoantigens. We hypothesized that natural autoantibodies to SELENOP are prevalent in thyroid patients, impair Se transport, and negatively affect GPX3 biosynthesis. Using a newly established quantitative immunoassay, SELENOP autoantibodies were particularly prevalent in Hashimoto's thyroiditis as compared with healthy control subjects (6.6% versus 0.3%). Serum samples rich in SELENOP autoantibodies displayed relatively high total Se and SELENOP concentrations in comparison with autoantibody-negative samples ([Se]; 85.3 vs. 77.1 µg/L, $p = 0.0178$, and [SELENOP]; 5.1 vs. 3.5 mg/L, $p = 0.001$), while GPX3 activity was low and correlated inversely to SELENOP autoantibody concentrations. In renal cells in culture, antibodies to SELENOP inhibited Se uptake. Our results indicate an impairment of SELENOP-dependent Se transport by natural SELENOP autoantibodies, suggesting that the characterization of health risk from Se deficiency may need to include autoimmunity to SELENOP as additional biomarker of Se status.

Keywords: antioxidative defense; autoantibody; glutathione peroxidase; Hashimoto's thyroiditis; trace element

1. Introduction

Autoimmune thyroid disease (AITD) is characterized by an inappropriate interaction of immune cells with thyroid proteins. An activated immune system, inflammation, and lymphocytic infiltration into the thyroid gland, accompanied with autoantibodies (aAb) to thyroid antigens, are hallmarks of AITD [1]. Natural aAb levels are commonly determined in AITD as diagnostic markers, which support the diagnosis, correlate to disease severity, and enable the monitoring of treatment success. In Graves' disease, natural aAb to the TSH-receptor (TSHR-aAb) are causative for the clinical phenotype, as they bind as endocrine active agonists to the TSH-receptor and stimulate hyperthyroidism and thyroid eye disease [2]. In comparison, natural aAb to another thyroid autoantigen, namely the thyroperoxidase (TPO-aAb), are detectable in both Graves' disease and Hashimoto's thyroiditis, where they are associated with cell-mediated cytotoxicity [3]. The TPO-aAb are not directly causing the disease but rather reflect disease activity and disease risk in asymptomatic healthy subjects [4]. Accordingly, TPO-aAb-positive healthy women who become pregnant are, for example, at relatively high risk for the development of postpartum thyroiditis, potentially due to the declining selenium (Se) status during pregnancy [5,6]. The presence of natural aAb in a healthy person without obvious clinical symptoms is neither necessarily an indication of disease nor of immediate diagnostic value but may indicate an increased risk of disease.

In general, the pathogenesis of AITD is a multifaceted process, where the essential production of hydrogen peroxide needed for thyroid hormone biosynthesis may contribute to tissue inflammation, gland destruction, and the generation of novel autoantigens [7,8]. Accordingly, anti-inflammatory and antioxidative measures are considered in prevention and treatment of AITD [9,10]. Observational studies have indicated an inverse relationship between the intake of the essential trace element Se with thyroid volume [11], the development of thyroid nodules [12] and thyroid disease [13]. Dietary uptake of Se is needed for the formation of the 21st proteinogenic amino acid selenocysteine [14], and supports the biosynthesis of health-relevant, redox-active selenoproteins, including the secreted plasma proteins extracellular glutathione peroxidase (GPX3) and selenoprotein P (SELENOP) [15–17]. These two selenoproteins are found both in the circulation and in thyroid follicles [18]. While GPX3 is capable of peroxide degradation, SELENOP serves mainly as transporter for the systemic distribution of Se [19,20]. Importantly, renal GPX3 biosynthesis is supported by SELENOP from hepatocytes [21,22].

SELENOP is a most exceptional protein that varies naturally in its primary sequence [23,24]. Translational errors occur during decoding of the ten UGA triplets within the open reading frame [25], and cysteine can replace selenocysteine during biosynthesis [26]. Under antibiotic treatment, Se-free selenoproteins can be synthesized [27]; additionally, tryptophan or arginine may become inserted at UGA codons [28]. Consequently, the average content of Se per SELENOP molecule has been determined at 5.4 ± 0.5 Se/SELENOP in human [29], 7.5 Se/Selenop in rat, and 5 Se/Selenop in mouse [30]. Due to this flexibility in the decoding of UGA, the primary sequence of newly synthesized SELENOP varies. Circulating SELENOP may thus constitute a mixture of slightly different protein variants. For this reason, we hypothesized that autoantibodies to SELENOP (SELENOP-aAb) may develop naturally, particularly in patients with AITD and thyroid inflammation. In order to test this hypothesis, we analyzed control subjects and patients with different thyroid diseases for SELENOP-aAb and correlated the results to biomarkers of Se status.

2. Results

An immunoluminometric assay for the detection and quantification of SELENOP-aAb was established by generating a fusion protein encoding-secreted alkaline phosphatase (SEAP) in frame with full length human SELENOP, where UGA codons had been replaced by cysteine codons. Assay functionality was verified with a SELENOP-specific antibody and showed the expected concentration-dependent signal intensity with dilution (Figure 1A). This result was replicated in dilution experiments with serum samples from patients identified as positive (P01–P05), whereas samples categorized as negative for SELENOP-aAb (N01–N03) showed background signals only (Figure 1B). Unlabeled recombinant SELENOP (1 mg/mL) applied to BSA-free (control 1), or BSA-containing reaction buffer (control 2), was capable of suppressing SELENOP-aAb signals from positive samples, highlighting the specificity of the detection method (Figure 1C).

In order to test for natural SELENOP-aAb in human subjects, serum samples from 2 cohorts of thyroid patients ($n = 423$), along with a collection of healthy controls ($n = 400$), were compared. The results showed a skewed distribution of signals (Figure 1D), and relative binding indices (BI) were calculated by dividing the individual SELENOP-aAb signals by the average signal obtained from the bottom half of all samples. In this first analysis, the thresholds of BI = 5 and BI = 10, respectively, were chosen for the classification of positive vs. negative samples. Irrespective of threshold, SELENOP-aAb were more prevalent in thyroid patients than controls (4.3 vs. 0.3%, or 2.4 vs. 0.3%, respectively) (Figure 1E). Among the patients, elevated SELENOP-aAb was mainly found in patients with Hashimoto's thyroiditis (Figure 1F).

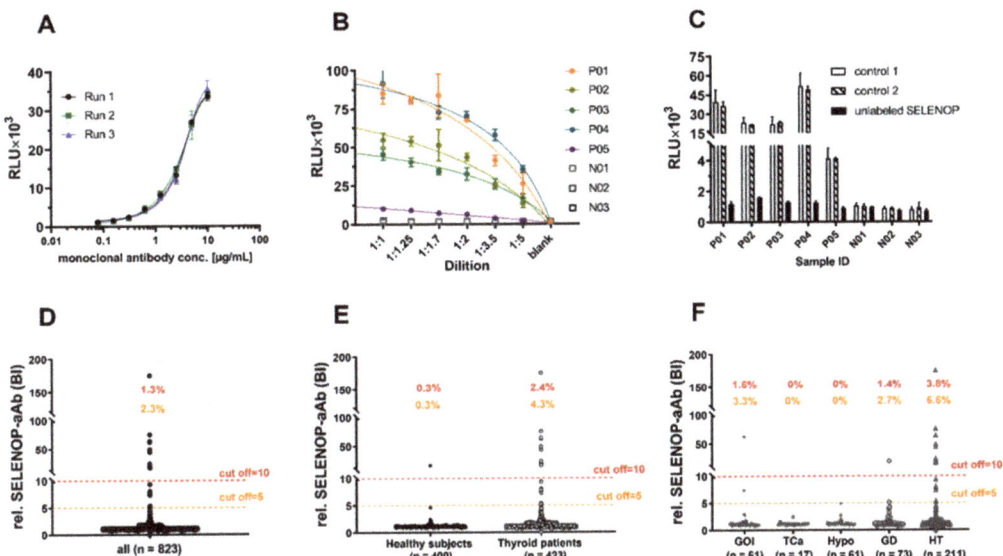

Figure 1. Characterization of the novel autoantibody assay and prevalence of SELENOP-aAb in thyroid patients. (**A**) The SELENOP-aAb assay is characterized by linear response to a SELENOP-specific antibody in gradual dilutions. Signal intensity (RLU—relative light units) correlated to antibody concentration over a range from 0.078–5.0 µg/mL, yielding an R squared by sigmoidal 4 PL curve fitting above 0.99. (**B**) High signal intensities (RLU) decreased gradually with linear dilutions of SELENOP-aAb-positive (P01–P05) samples, but not of negative (N01–N03) samples. (**C**) Suppression of SELENOP-aAb signals by competition with unlabeled SELENOP (1 mg/mL) using equal volumes of sample and unlabeled SELENOP without (control 1) or with BSA (1 mg/mL; control 2). (**D**) Analysis of the full cohort of samples (n = 823) yielded a skewed distribution of SELENOP-aAb signals; the thresholds of BI = 5.0 or BI = 10.0 are indicated by dashed lines. (**E**) SELENOP-aAb-positive samples were more prevalent in thyroid patients than in controls (BI > 5; 4.3 vs. 0.3%, or BI > 10; 2.4 vs. 0.3%). (**F**) Among patients, SELENOP-aAb were most prevalent in Hashimoto's thyroiditis (HT). GOI—goitre; TCa—thyroid carcinoma; Hypo—hypothyroidism; GD—Graves' disease.

Next, the impact of SELENOP-aAb on the Se status was analyzed in the AITD patients. Three complementary biomarkers of Se status, i.e., total serum Se and SELENOP concentrations along with GPX3 activity, were assessed in parallel. Serum SELENOP showed positive and strong linear correlations with total Se and GPX3 activity (Figure 2A,C), while total serum Se correlated only weakly with GPX3 activity (Figure 2B). One particular sample displayed moderate GPX3 activity in combination with exceptionally high Se and SELENOP concentrations (indicated as red dot; Se—495 µg/L; SELENOP—13.2 mg/L; GPX3 activity—179 U/L). This finding may indicate a compromised *GPX3* gene expression, a mutated GPX3 variant with reduced enzymatic activity, accelerated GPX3 degradation, or little renal GPX3 biosynthesis, potentially due to impaired renal Se uptake. This sample displayed highest SELENOP-aAb concentrations (BI = 175), in agreement with the hypothesis of a disturbed Se transport towards the kidney. In order to test whether antibodies to SELENOP are capable of affecting target cell Se uptake, human embryonic kidney cells (HEK293) were transfected with a reporter for selenoprotein biosynthesis, encoding a

Se-dependent luciferase in combination with the selenocysteine insertion sequence (SECIS) element of GPX4. The reporter system showed Se-dependent induction of luciferase activity in response to inorganic selenite (1.0, 5.0 or 10 nM, f.c.), and in response to human serum (0.05% or 0.5%, v/v) as a natural Se source, as expected under regular Se-deficient culture conditions with 10% FBS as sole Se source. A significant decline in the luciferase reporter signal was observed upon adding an antibody to SELENOP (0.8 ng/mL), into the culture medium (Figure 2D). This result is compatible with the notion of SELENOP-aAb disturbing SELENOP uptake into target cells. To further verify the characteristics of natural SELENOP-aAb, immunoglobulins were isolated from positive and negative serum samples by protein A-mediated immunoprecipitation and analyzed for directly associated Se and SELENOP content. The isolates from SELENOP-aAb-positive samples contained measurable amounts of Se (Figure 2E) and/or SELENOP (Figure 2F), whereas isolates from controls were devoid of associated Se or SELENOP.

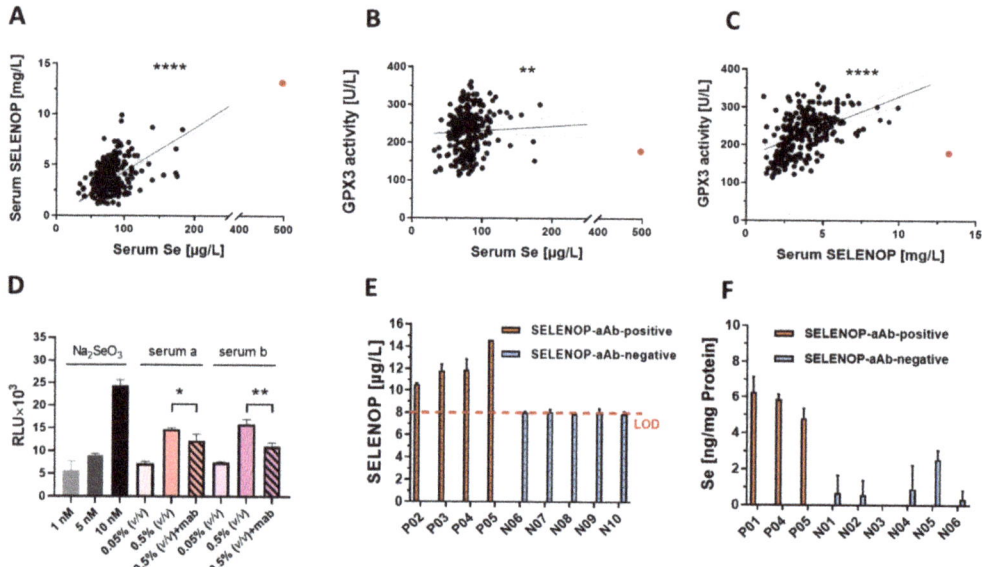

Figure 2. Selenium (Se) status assessment and potential role of natural SELENOP-aAb. Three biomarkers of Se status were determined in serum samples from AITD patients (n = 284). Positive correlations were observed for (**A**) total Se with SELENOP (r = 0.342, p < 0.0001), (**B**) total Se with GPX3 activity (r = 0.171, p = 0.0046), and (**C**) SELENOP with GPX3 activity (r = 0.545, p < 0.0001). The sample with highest SELENOP-aAb displayed exceptionally high Se and SELENOP levels in combination with moderate GPX3 activity (indicated as red dots). (**D**) HEK293 cells expressing a Se-dependent luciferase showed increased reporter activity (RLU) in response to selenite or human serum added to the culture medium. The signal was significantly suppressed by the addition of a SELENOP-specific antibody (n = 3). (**E**) The immunoglobulins from 5 SELENOP-aAb-positive (P01–P05) and 10 SELENOP-aAb-negative samples (N01–N10) were precipitated by protein A. Measurable (**E**) SELENOP or (**F**) Se concentrations were detected in the isolates from SELENOP-aAb-positive samples only, but not in those from SELENOP-aAb-negative samples. Correlations were analyzed by Spearman's correlation test. Two-tailed t-test was used for comparisons between two groups, p-values < 0.05 were considered statistically significant; * indicates p < 0.05; ** indicates p < 0.01; **** indicates p < 0.0001.

When classifying patients as SELENOP-aAb-positive or -negative, by choosing BI > 5.0 as threshold, the positive samples displayed elevated total serum Se (median: 85.3 vs. 77.0 µg/L, p = 0.033), whereas SELENOP concentration or GPX3 activity was not different (Figure 3A–C). A positive correlation was observed between SELENOP-aAb with Se (r = 0.582, p = 0.018) (Figure 3D) and with SELENOP (r = 0.730, p = 0.001) (Figure 3E), but not with GPX3 activity (Figure 3F). When using a higher threshold, i.e., BI > 10 as cut-off, relatively high Se (85.3 vs. 77.1 µg/L, $p = 0.0178$) and SELENOP (5.1 vs. 3.5 mg/L, $p = 0.001$) concentrations were observed in the group of SELENOP-aAb-positive samples (Figure 3A,B), whereas GPX3 activity was not elevated (Figure 3C). Positive correlation was observed between SELENOP-aAb and Se ($r = 0.695$, $p = 0.038$) (Figure 3G) or SELENOP ($r = 0.627$, $p = 0.035$) (Figure 3H), reaching statistical significance when the very positive sample highlighted in Figure 2A–C is included, but not when omitted from the analysis. Notably, the SELENOP-aAb correlated inversely to GPX3 activity ($r = -0.669$, $p = 0.049$) (Figure 3I).

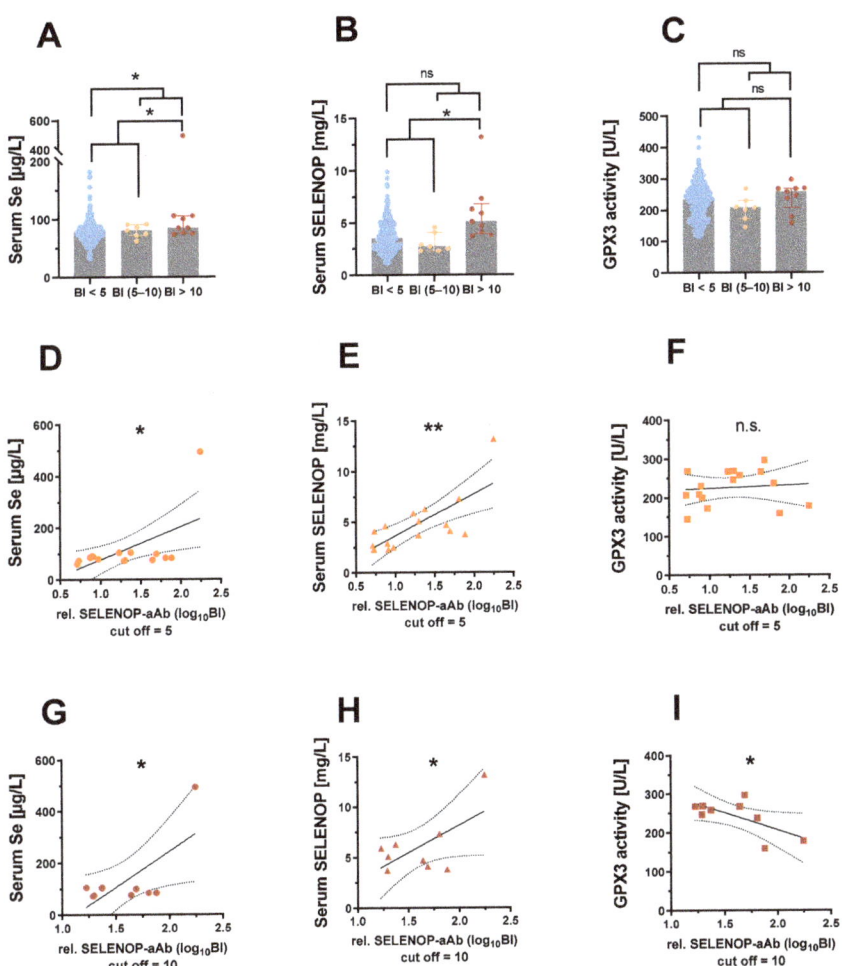

Figure 3. Comparison of three complementary biomarkers of Se-status in relation to SELENOP autoimmunity. The patients with positive SELENOP-aAb (BI > 5) displayed relatively high (**A**) total Se, while (**B**) SELENOP or (**C**) GPX3 activity were not different in comparison to SELENOP-aAb-negative patients. SELENOP-aAb correlated positively to (**D**) serum Se and (**E**) SELENOP, but not to (**F**) GPX3 activity. When applying a more stringent cut-off for positivity (BI > 10), serum samples with positive SELENOP-aAb displayed relatively high (**A**) total Se and (**B**) SELENOP, whereas (**C**) GPX3 activity was not elevated. Positive correlation was observed between SELENOP-aAb and (**G**) serum Se and (**H**) SELENOP. (**I**) GPX3 activity was inversely correlated to SELENOP-aAb. Comparisons between two groups were conducted by Mann–Whitney test. Correlations were tested by Pearson's correlation analysis with two-tailed p-values. p-values < 0.05 were considered statistically significant; n.s. indicates $p \geq 0.05$; * indicates $p < 0.05$, and ** indicates $p < 0.01$.

3. Discussion

This study describes naturally occurring autoantibodies to the Se transport protein SELENOP in human subjects and suggests a potential physiological relevance. Our data indicate that autoimmunity to SELENOP is a rare finding in healthy adult subjects. However, a considerable fraction of thyroid patients express SELENOP-aAb to varying degrees, with some patients being highly positive for SELENOP-aAb. The elevated prevalence in Hashimoto's thyroiditis may either result from SELENOP-aAb predisposing them to the diseases—as low Se has been identified as risk factor for autoimmune thyroid disease [31–33]—or the autoimmunity develops as a consequence of the ongoing inflammation and associated oxidative damage to thyroid proteins, i.e., resulting from the lymphocytic thyroiditis and generation of novel autoantigens as side effect of disease [8,34]. Alternatively, both the predisposition to and consequences of the inflammatory disease may have contributed to the elevated prevalence of SELENOP-aAb observed in the patients [6]. From an etiological perspective, human SELENOP is naturally synthesized in different variants, some of which are potentially prone to becoming easily modified and eliciting an autoimmune response, as the protein carries several highly reactive selenocysteine residues [35,36]. Both notions may offer an explanation for the increased prevalence of SELENOP-aAb in AITD and indicate a diagnostic or even pathophysiological relevance, as just recently documented for the central role of Se status and GPX-dependent protection from neutrophil ferroptosis in systemic autoimmunity [37]. This hypothesis needs to be elucidated in larger studies and ideally with samples from a longitudinal, prospective trial.

Against our expectation, SELENOP-aAb were not associated with SELENOP deficiency, but rather with elevated Se and SELENOP serum concentrations, suggesting some stabilizing effect of the autoantibodies on circulating SELENOP. This interpretation is consistent with the observed inhibition of SELENOP uptake into renal HEK293 target cells by antibodies to SELENOP, nicely echoing published results on SELENOP-neutralizing antibodies antagonizing Se uptake as novel candidates for type 2 diabetes therapy [38,39]. Importantly, this theory is substantiated in human subjects by the relatively low and inadequate GPX3 activity in SELENOP-aAb-positive patients, and the inverse association of GPX3 activity with SELENOP-aAb concentrations, despite an increased Se status as reflected in elevated total Se and SELENOP levels. Transgenic mouse models have shown that hepatic SELENOP biosynthesis is supplying the kidney with Se for GPX3 biosynthesis via specific uptake through the lipoprotein receptor megalin [22,40]. In the case that the antagonistic nature of SELENOP-aAb, impairing Se supply to target cells, becomes substantiated by future research, it would be of far-reaching medical relevance, as also the central nervous system relies on receptor-mediated SELENOP uptake for Se supply and protection of highly active interneurons from death by ferroptosis [41,42]. In the absence of regular and efficient SELENOP supply and uptake, severe neurological symptoms, including epileptic seizures, were observed in transgenic mice [41,43–46], and also recently in a dog model of impaired SELENOP expression, leading to brain atrophy and cerebellar ataxia [47]. It remains to be tested whether SELENOP-aAb are relevant for neurological disease, and whether SELENOP-aAb-positive thyroid patients are at particular risk for neurological sequelae, e.g., seizures, tremors, ataxia, or symptoms of Hashimoto encephalopathy.

Correcting a Se deficit, whatever the cause, needs an adequate supply, as excessive Se intake can be toxic [48]. Some observational studies and systematic analyses have reported an association between elevated Se concentrations and type 2 diabetes [49]. Other systematic reviews, focusing on sufficiently large and well-controlled supplementation studies, show that there is no evidence that Se supplementation increases the risk of type 2 diabetes [50]. The documented association between increased serum Se concentrations and type 2 diabetes could be a consequence of insulin resistance leading to increased hepatic biosynthesis of SELENOP due to abrogation of insulin-mediated SELENOP suppression, i.e., a question of reverse causality, potentially a function of protecting the cardiovascular system [51]. This notion is supported by recent findings of a positive association between increased Se concentrations and protection from all-cause mortality and heart disease mortality in diabetic patients, as determined in NHANES III [52]. The extent to which SELENOP-aAb may alter the protective interaction of Se and SELENOP on the vasculature is unknown, and the question of whether functional Se deficiency by low Se supply, in combination with SELENOP-aAb causes hypoglycemia needs to be investigated [53].

Our findings suggest that Se status assessment by the biomarkers used until now, i.e., total Se and SELENOP concentrations along with GPX3 activity [54–58], may be insufficient when SELENOP-aAb are present. In particular, the poor correlation between Se or SELENOP concentrations and GPX3 activity in a given subject may indicate the presence of SELENOP-aAb impairing SELENOP uptake and GPX3 biosynthesis by kidney, as impressively documented in the positive subject with highest SELENOP-aAb (red dots in Figure 2). This interrelation may also be of relevance for preventive or adjuvant Se therapy [33,59–62], as positive effects may be observed, particularly in the presence of SELENOP-aAb. This hypothesis is based on the findings in transgenic mice, where SELENOP deficiency and resulting symptoms can be successfully compensated for by supplemental selenite or other selenocompounds [45,63–65]. A score combining classical Se status biomarkers with the presence of SELENOP-aAb may be superior in judging the functional Se status and requirements for Se supplementation than a single biomarker alone. Such a scoring system would be capable of identifying Se deficiency also in cohorts of well-supplied subjects with apparently sufficient SELENOP expression, and might indicate subjects with specific requirements to overcome SELENOP-aAb-mediated inhibition of SELENOP-dependent Se supply (Figure 4). Prospective studies of sufficient size and length are needed next to test the clinical relevance of SELENOP-aAb for AITD risk, and for stratifying the results from Se supplementation studies.

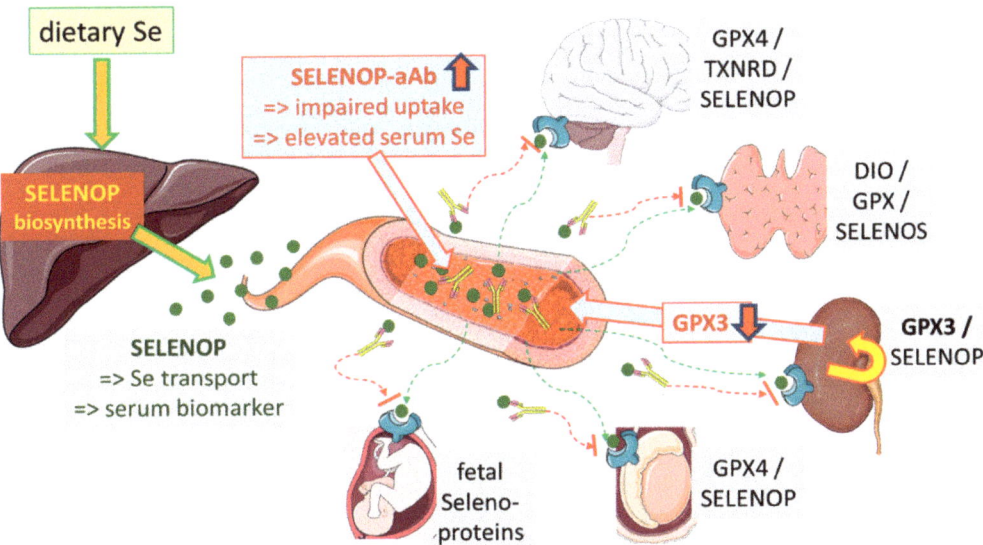

Figure 4. Potential pathophysiological relevance of SELENOP-aAb. Dietary sources of Se are mainly converted in hepatocytes into SELENOP for secretion and systemic supply of target tissues. Thereby, SELENOP constitutes both the major serum Se transporter and a reliable biomarker of Se status. Patients with elevated SELENOP-aAb display increased serum Se and SELENOP concentrations, indicative of SELENOP stabilization or impaired receptor-mediated SELENOP uptake and clearance. Thereby, target tissues may become insufficiently supplied for full biosynthesis of the essential tissue-relevant selenoproteins (examples are indicated). The biosynthesis of circulating GPX3 by kidney cells is known to directly depend on hepatic SELENOP. Accordingly, serum GPX3 activity correlated inversely to SELENOP-aAb in thyroid patients, consistent with target cell Se deficiency. Extrapolating these results, SELENOP-aAb may increase the risk of neurological symptoms, thyroid disease, subfertility, pregnancy problems, and other Se-dependent diseases.

4. Materials and Methods

4.1. Human Samples

Blood samples (n = 423) from patients with different thyroid diseases were collected and serum was prepared, aliquoted, and stored at −80 °C. The study had been approved by the Ethical committee of the Charité-University Medical School, Berlin (#EA2/173/17). Recruitment of the patients proceeded in consecutive manner and diagnostic criteria for thyroid disease were applied as described in [66]. A cohort of serum samples from subjects with a self-reported status as healthy (controls, n = 400) was obtained from a commercial supplier (InVent Diagnostica GmbH, Hennigsdorf, Germany) and served as control (Table 1). All samples included were derived after obtaining written informed consent from the subjects enrolled into the analyses, and the study was conducted in accordance with the declaration of Helsinki on ethical principles for medical research involving human subjects.

Table 1. Characterization of the study cohort.

Healthy Controls	n = 400
sex, female/male [n/n]	200/200
age, median (95% CI) [y]	31 (29–32)
Thyroid Patients	n = 423
sex, female/male [n/n]	362/61
age, median (95% CI) [y]	49 (47–51)
GOI, n (%)	61 (14.4%)
TCa, n (%)	17 (4.0%)
Hypo, n (%)	61 (14.4%)
GD, n (%)	73 (17.3%)
HT, n (%)	211 (49.9%)

GOI—goitre; TCa—thyroid carcinoma; Hypo—hypothyroidism; GD—Graves' disease; HT—Hashimoto's thyroiditis.

4.2. Generation of Recombinant SEAP-SELENOP Reporter Proteins

The cDNA of secreted alkaline phosphatase (SEAP) was amplified by PCR and inserted into plasmid pIRESneo (Clontech, Palo Alto, CA, USA) giving rise to pIRESneo-SEAP. A cDNA of a UGA-free human SELENOP reading frame was synthesized, whereby UGA codons were replaced by cysteine codons by a commercial supplier (Eurofins Genomics GmbH, Ebersberg, Germany). The fragment was amplified by PCR and ligated into pIRESneo-SEAP, amplified in E. coli and sequence verified. Recombinant protein was produced via stable expression of pIRESneo-SEAP-SELENOP in HEK 293 cells. Cell culture supernatants containing the secreted recombinant protein were collected and used as reporter in the SELENOP-aAb detection assay.

4.3. Immunoluminometric Assay for Detection of SELENOP-aAb

The immunoluminometric assay is based on the binding of aAb to recombinant SEAP-SELENOP, followed by precipitation of the antibody–antigen–reporter complex by protein A. To this end, diluted supernatants of HEK293 cells expressing SEAP-SELENOP was incubated with 5 µL serum sample at 4 °C overnight. On the second day, the same volume (40 µL) of a protein A slurry (ASKA Biotech GmbH, Berlin, Germany) was added and incubated for 1 h at RT. Complexes formed were washed 6 times with washing buffer (50 mM Tris HCl, pH 7.4, 100 mM NaCl, 10% glycerol, and 0.5% Triton X-100), removing unbound SEAP-SELENOP and other serum proteins. Finally, reporter activity was detected as luminescence signal by a luminometer (Berthold Technologies GmbH, Bad Wildbad, Germany). Binding index (BI) was calculated for each sample by dividing the relative light units (RLU), obtained with the average RLU from the lowest 50% of signals in the same assay plate, which was set as BI = 1.0. This mathematical method for background signal definition was based on the assumption that SELENOP-aAb were present in less than 50% of the samples tested. During the analyses, the inter- and intra-assay CV was determined to be below 20% based on the analysis of selected positive and negative serum samples included into each assay run.

4.4. Isolation of Immunoglobulins (IgG) from Serum Sample

Total IgGs were isolated from five positive and ten negative serum samples by precipitation with Protein A slurry (ASKA Biotech GmbH, Berlin, Germany). To this end, serum samples were incubated with 2 volumes of protein A in PBS (50%) overnight at 4 °C. The supernatants were discarded, and the pellets were washed 6 times with PBS. Precipitated IgG were eluted with 3-fold volume of citric acid (25 mM, pH 2.0) and neutralized using 1-fold volume of HEPES (1 M, pH 8.0).

4.5. Quantification of GPX3 Activity and Total Se and SELENOP Concentrations

GPX3 activity in serum samples was determined by a coupled enzymatic test monitoring NADPH decline at 340 nm due to glutathione reductase activity catalyzing regeneration of consumed glutathione by GPX during H_2O_2 reduction [67]. Briefly, serum samples of 5 µL were applied to 96-well plates containing 200 µL of a test mixture, including 1 mM NaN_3, 3.4 mM reduced glutathione, 0.3 U/mL glutathione reductase, and 0.27 mg/mL NADPH. The reaction was started by 10 µL of 0.00375% H_2O_2. A constant serum sample was included into each assay run for quality control. The inter- and intra-assay CVs were determined to be below 15% during the analyses.

Se concentrations in serum samples or in the immunoglobulin-isolates were analyzed by total reflection X-ray fluorescence (TXRF) using a benchtop TXRF analyzer (S4 T-STAR, Bruker Nano GmbH, Berlin, Germany), as described previously [68,69]. Briefly, serum samples were diluted with a Ga-standard (1 mg/L, Alfa Aesar GmbH & Co KG, Karlsruhe, Germany), applied to polished quartz glass plates, and dried at 37 °C overnight. Isolated immunoglobulin complexes from SELENOP-aAb-positive and -negative serum samples were precipitated by adding 9 volumes of ice-cold ethanol (100%) overnight at −80 °C, and subsequent centrifugation for 30 min at 4 °C and 14,000× g. The pellets were dissolved at 80 °C for 2 h in 100 µL HNO_3 (61%), the Ga standard was added, and sample was applied to quartz plates and analyzed. A commercial serum standard (Seronorm, Sero AS, Billingstad, Norway) served as control in each analytical run. The determined concentrations of Se were within the specified range of the standard, and the inter-assay coefficient of variation (CV) was below 5% during the analyses.

SELENOP concentrations were determined by a validated commercial SELENOP-specific ELISA (selenOtest ELISA, selenOmed GmbH, Berlin, Germany), essentially as described in [29]. Briefly, 100 µL of each IgG-isolate or 1:33 diluted serum samples were applied to pre-coated 96-well plates. Standards and calibrators were included into each assay run for quality control. The inter-assay CV was determined to be below 10%.

4.6. Luciferase-Based Reporter Gene Assay for Selenoprotein Biosynthesis in Cell Culture

A Se-responsive reporter gene assay was established using stably transfected human embryonic kidney (HEK293) cells, expressing a luciferase reporter containing a fusion protein of Firefly (FLuc) and Renilla (RLuc) luciferase, interrupted by an in-frame UGA codon and a GPX4-derived selenocysteine-insertion sequence (SECIS) element, as described previously [70]. Cells were cultured in Dulbecco's Modified Eagle Medium (DMEM/F12; Pan-Biotech GmbH, Aidenbach, Germany), containing 10% (v/v) FCS. For reporter gene assay, 20,000 cells per well were cultivated in 96-well plates in DMEM/F12 containing 10% (v/v) FCS for 24 h and stimulated by different sources of Se (sodium selenite or human serum) in the absence or presence of an anti-SELENOP antibody (0.8 ng/mL f.c., #SM-MAB-7356, selenOmed GmbH, Berlin, Germany).

The two human sera used in this experiment were tested free of SELENOP-aAb. The Se concentrations of these serum samples used in the in vitro studies were (a) 100.6 µg/L and (b) 109.7 µg/L, corresponding to the final Se concentrations in the cell culture mediums (0.5% serum, v/v), of 6.4 nM and 6.9 nM, respectively. After 48 h of incubation, medium was removed and cells were lysed in 40 µL passive Lysis puffer (PromoCell, Heidelberg, Germany) for 10 min at RT and stored at $-30\,°C$ to support cell lysis until measurement. For the detection, 25 µL of cell lysates were transferred into white 96-well plates, and Renilla luciferase activity was measured 30 s after adding coelenterazine (100 µL/well, 2.5 µg/mL in PBS, Synchem UG, Altenburg, Germany) using a microplate reader (PerkinElmer, Waltham, MA, USA).

4.7. Statistical Analyses

Statistical analyses were performed using SPSS (version 25, SAS Institute, Cary, NC, USA) or GraphPad Prism v.9.1.2 (GraphPad Software Inc., San Diego, CA, USA). The results are represented as mean with SD, as median with interquartile range, or by displaying the individual values. Normal distribution of values was tested by the Shapiro–Wilk test. Comparisons between two groups were conducted by unpaired t-test, and non-normally distributed variables were compared with the Mann–Whitney test. Correlations were tested by Pearson's correlation analysis and for non-normally distributed variables by Spearman's correlation test. p-values < 0.05 were considered significant; * $p < 0.05$, ** $p < 0.01$, *** $p < 0.001$, **** $p < 0.0001$.

5. Conclusions

The results indicated an impairment of SELENOP-dependent Se transport by natural SELENOP autoantibodies, and suggest that the characterization of health risk from Se deficiency may need to include autoimmunity to SELENOP as additional biomarker of Se status. Thereby, patients with "functional Se deficits" may be identified as those who may display a particular requirement for a sufficiently high Se intake and are likely to respond positively to adjuvant Se in clinical supplementation trials.

Author Contributions: Q.S. performed the measurements; S.M., C.L.G. and G.J.K. conducted the clinical sample collection; K.R. designed the reporter experiment; W.B.M. prepared the autoantigen and developed the aAb assay; J.H. and P.S. contributed to the experiments; L.S. designed the study; L.S. supervised and coordinated its execution; G.J.K. and L.S. wrote the manuscript. All authors have read and agreed to the published version of the manuscript.

Funding: The research was supported by Deutsche Forschungsgemeinschaft (DFG), research unit FOR-2558 "TraceAge" (Scho 849/6-2), and CRC/TR 296 "Local control of TH action" (LocoTact, P17).

Institutional Review Board Statement: The study was conducted according to the guidelines of the Declaration of Helsinki, and approved by the Ethics Committee of Charité-University Medical School, Berlin (#EA2/173/17).

Informed Consent Statement: Informed consent was obtained from all subjects involved in the study by the study authors, or by the commercial supplier providing the control samples.

Data Availability Statement: The data presented in this study are available upon reasonable request from the corresponding author.

Acknowledgments: Essential intellectual support was provided by the inspiring colleagues from the International Society for Selenium Research (ISSR). Figure 4 and the graphical abstract were drawn with graphic elements from Servier Medical Art (www.servier.com), provided under a Creative Commons 3.0 license.

Conflicts of Interest: L.S. and P.S. hold shares, and P.S. serves as CEO of selenOmed GmbH, a company involved in Se status assessment. L.S. is listed as inventor on a related patent application.

References

1. Weetman, A.P. An update on the pathogenesis of Hashimoto's thyroiditis. *J. Endocrinol. Investig.* **2021**, *44*, 883–890. [CrossRef] [PubMed]
2. George, A.; Diana, T.; Langericht, J.; Kahaly, G.J. Stimulatory Thyrotropin Receptor Antibodies Are a Biomarker for Graves' Orbitopathy. *Front. Endocrinol.* **2020**, *11*, 629925. [CrossRef]
3. Sinclair, D. Analytical aspects of thyroid antibodies estimation. *Autoimmunity* **2008**, *41*, 46–54. [CrossRef] [PubMed]

4. Prummel, M.F.; Wiersinga, W.M. Thyroid peroxidase autoantibodies in euthyroid subjects. *Best Pract. Res. Clin. Endocrinol. Metab.* **2005**, *19*, 1–15. [CrossRef] [PubMed]
5. Negro, R.; Greco, G.; Mangieri, T.; Pezzarossa, A.; Dazzi, D.; Hassan, H. The influence of selenium supplementation on postpartum thyroid status in pregnant women with thyroid peroxidase autoantibodies. *J. Clin. Endocrinol. Metab.* **2007**, *92*, 1263–1268. [CrossRef] [PubMed]
6. Schomburg, L. Selenium Deficiency Due to Diet, Pregnancy, Severe Illness, or COVID-19—A Preventable Trigger for Autoimmune Disease. *Int. J. Mol. Sci.* **2021**, *22*, 8532. [CrossRef] [PubMed]
7. Di Dalmazi, G.; Hirshberg, J.; Lyle, D.; Freij, J.B.; Caturegli, P. Reactive oxygen species in organ-specific autoimmunity. *Autoimmun. Highlights* **2016**, *7*, 11. [CrossRef] [PubMed]
8. Ralli, M.; Angeletti, D.; Fiore, M.; D'Aguanno, V.; Lambiase, A.; Artico, M.; de Vincentiis, M.; Greco, A. Hashimoto's thyroiditis: An update on pathogenic mechanisms, diagnostic protocols, therapeutic strategies, and potential malignant transformation. *Autoimmun. Rev.* **2020**, *19*, 102649. [CrossRef]
9. Wichman, J.; Winther, K.H.; Bonnema, S.J.; Hegedus, L. Selenium Supplementation Significantly Reduces Thyroid Autoantibody Levels in Patients with Chronic Autoimmune Thyroiditis: A Systematic Review and Meta-Analysis. *Thyroid* **2016**, *26*, 1681–1692. [CrossRef] [PubMed]
10. Filipowicz, D.; Majewska, K.; Kalantarova, A.; Szczepanek-Parulska, E.; Ruchala, M. The rationale for selenium supplementation in patients with autoimmune thyroiditis, according to the current state of knowledge. *Endokrynol. Pol.* **2021**, *72*, 153–162. [CrossRef]
11. Derumeaux, H.; Valeix, P.; Castetbon, K.; Bensimon, M.; Boutron-Ruault, M.C.; Arnaud, J.; Hercberg, S. Association of selenium with thyroid volume and echostructure in 35- to 60-year-old French adults. *Eur. J. Endocrinol.* **2003**, *148*, 309–315. [CrossRef]
12. Rasmussen, L.B.; Schomburg, L.; Kohrle, J.; Pedersen, I.B.; Hollenbach, B.; Hog, A.; Ovesen, L.; Perrild, H.; Laurberg, P. Selenium status, thyroid volume, and multiple nodule formation in an area with mild iodine deficiency. *Eur. J. Endocrinol.* **2011**, *164*, 585–590. [CrossRef]
13. Wu, Q.; Rayman, M.P.; Lv, H.; Schomburg, L.; Cui, B.; Gao, C.; Chen, P.; Zhuang, G.; Zhang, Z.; Peng, X.; et al. Low Population Selenium Status Is Associated with Increased Prevalence of Thyroid Disease. *J. Clin. Endocrinol. Metab.* **2015**, *100*, 4037–4047. [CrossRef]
14. Carlson, B.A.; Lee, B.J.; Tsuji, P.A.; Copeland, P.R.; Schweizer, U.; Gladyshev, V.N.; Hatfield, D.L. Selenocysteine tRNA([Ser]Sec), the Central Component of Selenoprotein Biosynthesis: Isolation, Identification, Modification, and Sequencing. *Methods Mol. Biol.* **2018**, *1661*, 43–60. [CrossRef] [PubMed]
15. Hatfield, D.L.; Tsuji, P.A.; Carlson, B.A.; Gladyshev, V.N. Selenium and selenocysteine: Roles in cancer, health, and development. *Trends Biochem. Sci.* **2014**, *39*, 112–120. [CrossRef]
16. Gladyshev, V.N.; Arner, E.S.; Berry, M.J.; Brigelius-Flohe, R.; Bruford, E.A.; Burk, R.F.; Carlson, B.A.; Castellano, S.; Chavatte, L.; Conrad, M.; et al. Selenoprotein Gene Nomenclature. *J. Biol. Chem.* **2016**, *291*, 24036–24040. [CrossRef] [PubMed]
17. Steinbrenner, H.; Speckmann, B.; Klotz, L.O. Selenoproteins: Antioxidant selenoenzymes and beyond. *Arch. Biochem. Biophys.* **2016**, *595*, 113–119. [CrossRef] [PubMed]
18. Schmutzler, C.; Mentrup, B.; Schomburg, L.; Hoang-Vu, C.; Herzog, V.; Kohrle, J. Selenoproteins of the thyroid gland: Expression, localization and possible function of glutathione peroxidase 3. *Biol. Chem.* **2007**, *388*, 1053–1059. [CrossRef] [PubMed]
19. Saito, Y.; Sato, N.; Hirashima, M.; Takebe, G.; Nagasawa, S.; Takahashi, K. Domain structure of bi-functional selenoprotein P. *Biochem. J.* **2004**, *381*, 841–846. [CrossRef] [PubMed]
20. Burk, R.F.; Hill, K.E. Regulation of Selenium Metabolism and Transport. *Annu. Rev. Nutr.* **2015**, *35*, 109–134. [CrossRef] [PubMed]
21. Schweizer, U.; Streckfuss, F.; Pelt, P.; Carlson, B.A.; Hatfield, D.L.; Kohrle, J.; Schomburg, L. Hepatically derived selenoprotein P is a key factor for kidney but not for brain selenium supply. *Biochem. J.* **2005**, *386*, 221–226. [CrossRef]
22. Renko, K.; Werner, M.; Renner-Muller, I.; Cooper, T.G.; Yeung, C.H.; Hollenbach, B.; Scharpf, M.; Kohrle, J.; Schomburg, L.; Schweizer, U. Hepatic selenoprotein P (SePP) expression restores selenium transport and prevents infertility and motor-incoordination in Sepp-knockout mice. *Biochem. J.* **2008**, *409*, 741–749. [CrossRef] [PubMed]
23. Tanaka, M.; Saito, Y.; Misu, H.; Kato, S.; Kita, Y.; Takeshita, Y.; Kanamori, T.; Nagano, T.; Nakagen, M.; Urabe, T.; et al. Development of a Sol Particle Homogeneous Immunoassay for Measuring Full-Length Selenoprotein P in Human Serum. *J. Clin. Lab. Anal.* **2016**, *30*, 114–122. [CrossRef] [PubMed]
24. Mariotti, M.; Shetty, S.; Baird, L.; Wu, S.; Loughran, G.; Copeland, P.R.; Atkins, J.F.; Howard, M.T. Multiple RNA structures affect translation initiation and UGA redefinition efficiency during synthesis of selenoprotein P. *Nucleic Acids Res.* **2017**, *45*, 13004–13015. [CrossRef] [PubMed]
25. Tobe, R.; Naranjo-Suarez, S.; Everley, R.A.; Carlson, B.A.; Turanov, A.A.; Tsuji, P.A.; Yoo, M.H.; Gygi, S.P.; Gladyshev, V.N.; Hatfield, D.L. High Error Rates in Selenocysteine Insertion in Mammalian Cells Treated with the Antibiotic Doxycycline, Chloramphenicol, or Geneticin. *J. Biol. Chem.* **2013**, *288*, 14709–14715. [CrossRef] [PubMed]
26. Turanov, A.A.; Everley, R.A.; Hybsier, S.; Renko, K.; Schomburg, L.; Gygi, S.P.; Hatfield, D.L.; Gladyshev, V.N. Regulation of Selenocysteine Content of Human Selenoprotein P by Dietary Selenium and Insertion of Cysteine in Place of Selenocysteine. *PLoS ONE* **2015**, *10*, e0140353. [CrossRef] [PubMed]
27. Handy, D.E.; Hang, G.Z.; Scolaro, J.; Metes, N.; Razaq, N.; Yang, Y.; Loscalzo, J. Aminoglycosides decrease glutathione peroxidase-1 activity by interfering with selenocysteine incorporation. *J. Biol. Chem.* **2006**, *281*, 3382–3388. [CrossRef]

28. Renko, K.; Martitz, J.; Hybsier, S.; Heynisch, B.; Voss, L.; Everley, R.A.; Gygi, S.P.; Stoedter, M.; Wisniewska, M.; Kohrle, J.; et al. Aminoglycoside-driven biosynthesis of selenium-deficient Selenoprotein P. *Sci. Rep.* **2017**, *7*, 4391. [CrossRef] [PubMed]
29. Hybsier, S.; Schulz, T.; Wu, Z.; Demuth, I.; Minich, W.B.; Renko, K.; Rijntjes, E.; Kohrle, J.; Strasburger, C.J.; Steinhagen-Thiessen, E.; et al. Sex-specific and inter-individual differences in biomarkers of selenium status identified by a calibrated ELISA for selenoprotein P. *Redox Biol.* **2017**, *11*, 403–414. [CrossRef]
30. Hill, K.E.; Zhou, J.D.; Austin, L.M.; Motley, A.K.; Ham, A.J.L.; Olson, G.E.; Atkins, J.F.; Gesteland, R.F.; Burk, R.F. The selenium-rich C-terminal domain of mouse selenoprotein P is necessary for the supply of selenium to brain and testis but not for the maintenance of whole body selenium. *J. Biol. Chem.* **2007**, *282*, 10972–10980. [CrossRef]
31. Duntas, L.H. The role of selenium in thyroid autoimmunity and cancer. *Thyroid* **2006**, *16*, 455–460. [CrossRef]
32. Schomburg, L. Selenium, selenoproteins and the thyroid gland: Interactions in health and disease. *Nat. Rev. Endocrinol.* **2011**, *8*, 160–171. [CrossRef] [PubMed]
33. Winther, K.H.; Rayman, M.P.; Bonnema, S.J.; Hegedus, L. Selenium in thyroid disorders—Essential knowledge for clinicians. *Nat. Rev. Endocrinol.* **2020**, *16*, 165–176. [CrossRef] [PubMed]
34. Burek, C.L.; Rose, N.R. Autoimmune thyroiditis and ROS. *Autoimmun. Rev.* **2008**, *7*, 530–537. [CrossRef] [PubMed]
35. Arner, E.S.J. Selenoproteins—What unique properties can arise with selenocysteine in place of cysteine? *Exp. Cell Res.* **2010**, *316*, 1296–1303. [CrossRef] [PubMed]
36. Ste Marie, E.J.; Wehrle, R.J.; Haupt, D.J.; Wood, N.B.; van der Vliet, A.; Previs, M.J.; Masterson, D.S.; Hondal, R.J. Can Selenoenzymes Resist Electrophilic Modification? Evidence from Thioredoxin Reductase and a Mutant Containing alpha-Methylselenocysteine. *Biochemistry* **2020**, *59*, 3300–3315. [CrossRef]
37. Li, P.; Jiang, M.; Li, K.; Li, H.; Zhou, Y.; Xiao, X.; Xu, Y.; Krishfield, S.; Lipsky, P.E.; Tsokos, G.C.; et al. Glutathione peroxidase 4-regulated neutrophil ferroptosis induces systemic autoimmunity. *Nat. Immunol.* **2021**, *22*, 1107–1117. [CrossRef] [PubMed]
38. Mita, Y.; Nakayama, K.; Inari, S.; Nishito, Y.; Yoshioka, Y.; Sakai, N.; Sotani, K.; Nagamura, T.; Kuzuhara, Y.; Inagaki, K.; et al. Selenoprotein P-neutralizing antibodies improve insulin secretion and glucose sensitivity in type 2 diabetes mouse models. *Nat. Commun.* **2017**, *8*, 1658. [CrossRef]
39. Takamura, T. Hepatokine Selenoprotein P-Mediated Reductive Stress Causes Resistance to Intracellular Signal Transduction. *Antioxid. Redox Signal.* **2020**, *33*, 517–524. [CrossRef] [PubMed]
40. Olson, G.E.; Winfrey, V.P.; Hill, K.E.; Burk, R.F. Megalin mediates selenoprotein P uptake by kidney proximal tubule epithelial cells. *J. Biol. Chem.* **2008**, *283*, 6854–6860. [CrossRef] [PubMed]
41. Wirth, E.K.; Conrad, M.; Winterer, J.; Wozny, C.; Carlson, B.A.; Roth, S.; Schmitz, D.; Bornkamm, G.W.; Coppola, V.; Tessarollo, L.; et al. Neuronal selenoprotein expression is required for interneuron development and prevents seizures and neurodegeneration. *FASEB J.* **2010**, *24*, 844–852. [CrossRef]
42. Ingold, I.; Berndt, C.; Schmitt, S.; Doll, S.; Poschmann, G.; Buday, K.; Roveri, A.; Peng, X.X.; Freitas, F.P.; Seibt, T.; et al. Selenium Utilization by GPX4 Is Required to Prevent Hydroperoxide-Induced Ferroptosis. *Cell* **2018**, *172*, 409. [CrossRef] [PubMed]
43. Burk, R.F.; Hill, K.E.; Olson, G.E.; Weeber, E.J.; Motley, A.K.; Winfrey, V.P.; Austin, L.M. Deletion of apolipoprotein E receptor-2 in mice lowers brain selenium and causes severe neurological dysfunction and death when a low-selenium diet is fed. *J. Neurosci.* **2007**, *27*, 6207–6211. [CrossRef] [PubMed]
44. Seeher, S.; Carlson, B.A.; Miniard, A.C.; Wirth, E.K.; Mahdi, Y.; Hatfield, D.L.; Driscoll, D.M.; Schweizer, U. Impaired selenoprotein expression in brain triggers striatal neuronal loss leading to co-ordination defects in mice. *Biochem. J.* **2014**, *462*, 67–75. [CrossRef] [PubMed]
45. Byrns, C.N.; Pitts, M.W.; Gilman, C.A.; Hashimoto, A.C.; Berry, M.J. Mice Lacking Selenoprotein P and Selenocysteine Lyase Exhibit Severe Neurological Dysfunction, Neurodegeneration, and Audiogenic Seizures. *J. Biol. Chem.* **2014**, *289*, 9662–9674. [CrossRef]
46. Burk, R.F.; Hill, K.E.; Motley, A.K.; Winfrey, V.P.; Kurokawa, S.; Mitchell, S.L.; Zhang, W. Selenoprotein P and apolipoprotein E receptor-2 interact at the blood-brain barrier and also within the brain to maintain an essential selenium pool that protects against neurodegeneration. *FASEB J.* **2014**, *28*, 3579–3588. [CrossRef] [PubMed]
47. Christen, M.; Hogler, S.; Kleiter, M.; Leschnik, M.; Weber, C.; Thaller, D.; Jagannathan, V.; Leeb, T. Deletion of the SELENOP gene leads to CNS atrophy with cerebellar ataxia in dogs. *PLoS Genet.* **2021**, *17*, e1009716. [CrossRef] [PubMed]
48. Raisbeck, M.F. Selenosis in Ruminants. *Vet. Clin. N. Am. Food Anim. Pract.* **2020**, *36*, 775–789. [CrossRef] [PubMed]
49. Vinceti, M.; Filippini, T.; Wise, L.A.; Rothman, K.J. A systematic review and dose-response meta-analysis of exposure to environmental selenium and the risk of type 2 diabetes in nonexperimental studies. *Environ. Res.* **2021**, *197*, 111210. [CrossRef] [PubMed]
50. Kohler, L.N.; Foote, J.; Kelley, C.P.; Florea, A.; Shelly, C.; Chow, H.S.; Hsu, P.; Batai, K.; Ellis, N.; Saboda, K.; et al. Selenium and Type 2 Diabetes: Systematic Review. *Nutrients* **2018**, *10*, 1924. [CrossRef]
51. Schomburg, L. The other view: The trace element selenium as a micronutrient in thyroid disease, diabetes, and beyond. *Hormones* **2020**, *19*, 15–24. [CrossRef] [PubMed]
52. Qiu, Z.; Geng, T.; Wan, Z.; Lu, Q.; Guo, J.; Liu, L.; Pan, A.; Liu, G. Serum selenium concentrations and risk of all-cause and heart disease mortality among individuals with type 2 diabetes. *Am. J. Clin. Nutr.* **2021**. [CrossRef] [PubMed]
53. Wang, Y.; Rijntjes, E.; Wu, Q.; Lv, H.; Gao, C.; Shi, B.; Schomburg, L. Selenium deficiency is linearly associated with hypoglycemia in healthy adults. *Redox Biol.* **2020**, *37*, 101709. [CrossRef]

54. Hurst, R.; Collings, R.; Harvey, L.J.; King, M.; Hooper, L.; Bouwman, J.; Gurinovic, M.; Fairweather-Tait, S.J. EURRECA-Estimating selenium requirements for deriving dietary reference values. *Crit. Rev. Food Sci. Nutr.* **2013**, *53*, 1077–1096. [CrossRef] [PubMed]
55. Ashton, K.; Hooper, L.; Harvey, L.J.; Hurst, R.; Casgrain, A.; Fairweather-Tait, S.J. Methods of assessment of selenium status in humans: A systematic review. *Am. J. Clin. Nutr.* **2009**, *89*, 2025s–2039s. [CrossRef] [PubMed]
56. Combs, G.F., Jr. Biomarkers of selenium status. *Nutrients* **2015**, *7*, 2209–2236. [CrossRef]
57. Brodin, O.; Hackler, J.; Misra, S.; Wendt, S.; Sun, Q.; Laaf, E.; Stoppe, C.; Bjornstedt, M.; Schomburg, L. Selenoprotein P as Biomarker of Selenium Status in Clinical Trials with Therapeutic Dosages of Selenite. *Nutrients* **2020**, *2*, 1067. [CrossRef]
58. Demircan, K.; Bengtsson, Y.; Sun, Q.; Brange, A.; Vallon-Christersson, J.; Rijntjes, E.; Malmberg, M.; Saal, L.H.; Ryden, L.; Borg, A.; et al. Serum selenium, selenoprotein P and glutathione peroxidase 3 as predictors of mortality and recurrence following breast cancer diagnosis: A multicentre cohort study. *Redox Biol.* **2021**, *47*, 102145. [CrossRef]
59. Muecke, R.; Schomburg, L.; Buentzel, J.; Kisters, K.; Micke, O.; German Working Group Trace Elements and Electrolytes in Oncology-AKTE. Selenium or no selenium—That is the question in tumor patients: A new controversy. *Integr. Cancer Ther.* **2010**, *9*, 136–141. [CrossRef] [PubMed]
60. Aaseth, J.; Alexander, J.; Bjorklund, G.; Hestad, K.; Dusek, P.; Roos, P.M.; Alehagen, U. Treatment strategies in Alzheimer's disease: A review with focus on selenium supplementation. *Biometals* **2016**, *29*, 827–839. [CrossRef] [PubMed]
61. Steinbrenner, H.; Al-Quraishy, S.; Dkhil, M.A.; Wunderlich, F.; Sies, H. Dietary selenium in adjuvant therapy of viral and bacterial infections. *Adv. Nutr.* **2015**, *6*, 73–82. [CrossRef]
62. Tan, H.W.; Mo, H.Y.; Lau, A.T.Y.; Xu, Y.M. Selenium Species: Current Status and Potentials in Cancer Prevention and Therapy. *Int. J. Mol. Sci.* **2018**, *20*, 75. [CrossRef]
63. Schweizer, U.; Michaelis, M.; Kohrle, J.; Schomburg, L. Efficient selenium transfer from mother to offspring in selenoprotein-P-deficient mice enables dose-dependent rescue of phenotypes associated with selenium deficiency. *Biochem. J.* **2004**, *378*, 21–26. [CrossRef]
64. Valentine, W.M.; Abel, T.W.; Hill, K.E.; Austin, L.M.; Burk, R.F. Neurodegeneration in mice resulting from loss of functional selenoprotein P or its receptor apolipoprotein E receptor 2. *J. Neuropathol. Exp. Neurol.* **2008**, *67*, 68–77. [CrossRef]
65. Kremer, P.M.; Torres, D.J.; Hashimoto, A.C.; Berry, M.J. Sex-Specific Metabolic Impairments in a Mouse Model of Disrupted Selenium Utilization. *Front. Nutr.* **2021**, *8*, 682700. [CrossRef]
66. Mehl, S.; Sun, Q.; Gorlich, C.L.; Hackler, J.; Kopp, J.F.; Renko, K.; Mittag, J.; Schwerdtle, T.; Schomburg, L. Cross-sectional analysis of trace element status in thyroid disease. *J. Trace Elem. Med. Biol.* **2020**, *58*, 126430. [CrossRef]
67. Flohé, L.; Gunzler, W.A. Assays of glutathione peroxidase. *Methods Enzymol.* **1984**, *105*, 114–121. [CrossRef] [PubMed]
68. Hoeflich, J.; Hollenbach, B.; Behrends, T.; Hoeg, A.; Stosnach, H.; Schomburg, L. The choice of biomarkers determines the selenium status in young German vegans and vegetarians. *Br. J. Nutr.* **2010**, *104*, 1601–1604. [CrossRef] [PubMed]
69. Jeffery, J.; Frank, A.R.; Hockridge, S.; Stosnach, H.; Costelloe, S.J. Method for measurement of serum copper, zinc and selenium using total reflection X-ray fluorescence spectroscopy on the PICOFOX analyser: Validation and comparison with atomic absorption spectroscopy and inductively coupled plasma mass spectrometry. *Ann. Clin. Biochem.* **2019**, *56*, 170–178. [CrossRef]
70. Martitz, J.; Hofmann, P.J.; Johannes, J.; Kohrle, J.; Schomburg, L.; Renko, K. Factors impacting the aminoglycoside-induced UGA stop codon readthrough in selenoprotein translation. *J. Trace Elem. Med. Biol.* **2016**, *37*, 104–110. [CrossRef] [PubMed]

Article

Loss of SELENOF Induces the Transformed Phenotype in Human Immortalized Prostate Epithelial Cells

Lenny K. Hong [1], Shrinidhi Kadkol [1], Maria Sverdlov [1,2], Irida Kastrati [3], Mostafa Elhodaky [1], Ryan Deaton [1,2], Karen S. Sfanos [4], Heidi Wang [5], Li Liu [5] and Alan M. Diamond [1,*]

[1] Department of Pathology, College of Medicine, University of Illinois at Chicago, Chicago, IL 60612, USA; lennyh@uic.edu (L.K.H.); skadko2@uic.edu (S.K.); mariasve@uic.edu (M.S.); mehod@uic.edu (M.E.); rdeaton@uic.edu (R.D.)
[2] Research Histology Core, Research Resources Center, University of Illinois at Chicago, Chicago, IL 60612, USA
[3] Department of Cancer Biology, Loyola University of Chicago, Cardinal Bernardin Cancer Center, Maywood, IL 60153, USA; ikastrati@luc.edu
[4] Department of Pathology, Johns Hopkins University School of Medicine, Baltimore, MD 21287, USA; ksandel1@jhmi.edu
[5] Department of Epidemiology and Biostatistics, School of Public Health, University of Illinois at Chicago, Chicago, IL 60612, USA; xwang298@uic.edu (H.W.); liliu@uic.edu (L.L.)
* Correspondence: adiamond@uic.edu; Tel.: +1-312-413-8747

Abstract: SELENOF is a member of the class of selenoproteins in which the amino acid selenocysteine is co-translationally inserted into the elongating peptide in response to an in-frame UGA codon located in the 3′-untranslated (3′-UTR) region of the SELENOF mRNA. Polymorphisms in the 3′-UTR are associated with an increased risk of dying from prostate cancer and these variations are functional and 10 times more frequent in the genomes of African American men. SELENOF is dramatically reduced in prostate cancer compared to benign adjacent regions. Using a prostate cancer tissue microarray, it was previously established that the reduction of SELENOF in the cancers from African American men was significantly greater than in cancers from Caucasian men. When SELENOF levels in human prostate immortalized epithelial cells were reduced with an shRNA construct, those cells acquired the ability to grow in soft agar, increased the ability to migrate in a scratch assay and acquired features of energy metabolism associated with prostate cancer. These results support a role of SELENOF loss in prostate cancer progression and further indicate that SELENOF loss and genotype may contribute to the disparity in prostate cancer mortality experienced by African American men.

Keywords: prostate; cancer; selenium; selenoprotein; tumor suppressor

1. Introduction

Prostate cancer remains a significant clinical problem in the US, with an estimated 248,530 men diagnosed with the disease and 34,130 men dying from prostate cancer in 2021, making death from prostate cancer the second leading cause of death among American men [1]. Moreover, genetic factors contribute to prostate cancer risk; individuals with first-degree relatives diagnosed with prostate cancer have a higher risk of disease (as confirmed by meta-analysis [2]). However, most men with prostate cancer do not have a known family history, which strongly suggests that high-risk genetic factors are likely to be of relatively low penetrance and may be influenced by environmental variables.

Epidemiological studies report an inverse association between selenium (Se) levels and risk of several cancer types [3]. There are 25 human selenium-containing proteins or selenoproteins containing selenocysteine [4]. One selenoprotein implicated in prostate cancer etiology is SELENOF [5] which was originally identified as a human T cell 15 kDa protein that labels with ^{75}Se and is expressed at high levels in the prostate [6,7]. SELENOF belongs to the family of selenium-containing proteins in which selenium is inserted co-translationally in response to a UGA codon in the corresponding mRNA [8]. Notably, the function of SELENOF remains unknown. In most cell types, SELENOF resides in the endoplasmic reticulum (ER) in a complex with UDP-glucose: glycoprotein glucosyltransferase (UGGT), which helps to maintain correct protein folding [9]. However, SELENOF is uniquely associated with the plasma membrane in normal prostate epithelial cells [10].

The *SELENOF* gene is polymorphic in the 3′-untranslated region (UTR) that recognizes in-frame UGA codons as the amino acid selenocysteine [6]. Polymorphisms at positions 811 and 1125 form a haplotype where a C at 811 always corresponds to a G at 1125 and a T at 811 always corresponds to an A at 1125. Using two different specialized reporter constructs, these genetic variations were shown to be functional and likely determine the amount of SELENOF made as a function of selenium availability [7,11]. A potential role for SELENOF in carcinogenesis in multiple organ types is indicated by several lines of evidence, including loss of heterozygosity in breast cancers [11,12] and the association of specific polymorphisms of *SELENOF* and the risk of colorectal cancer [13,14]. Genetic data implicate SELENOF in prostate cancer etiology, with a statistically significant association between *SELENOF* polymorphisms, plasma selenium levels, and prostate cancer mortality [15]. Indeed, 811/1125 polymorphisms exhibit a trend toward association with prostate cancer-specific mortality.

Epidemiological data reported by Penney et al. [15] and our *in vitro* data [7,11] indicate an interaction between selenium and SELENOF, and its association with prostate cancer mortality. Selenium is distributed to the body from the liver as a component of selenoprotein P (SELENOP), previously referred to as SEPP1 [5], which contains 10 or more UGA-encoded selenocysteine residues. There are two functional polymorphisms in the gene for SELENOP, one a coding single-nucleotide polymorphism (SNP) (rs3877899, $Ala234^{Thr}$) that influences plasma selenium levels [16] and another in the 3′ UTR that effects SELENOP synthesis (rs7579). These polymorphisms interact with *SELENOF* alleles to impact prostate cancer risk in a cohort of European men [17].

Studies using human tissues have supported a role for the loss of SELENOF in prostate cancer etiology. SELENOF localizes to the plasma membrane in normal human prostate tissues, but not in other tissues, where its location is predominately in the ER and levels are dramatically reduced in prostate cancers compared to adjacent benign tissue [10]. The Gleason Grade Group, an indicator of prostate cancer aggressiveness, negatively associates with *SELENOF* genotype [10]. While these data provide an indication of a contributing role for SELENOF loss in prostate cancer incidence or severity, the studies presented here were designed to obtain evidence that the loss of SELENOF may directly contribute to disease progression.

2. Results

2.1. Lower Levels of SELENOF in Human Prostate Tissues Obtained from African American and Caucasian Men

Previously, we reported that lower levels of SELENOF were found in prostate cancer of African Americans compared to that from Caucasian men [10]. Here, a disparity TMA set was obtained from the Prostate Cancer Biorepository Network (PCBN) containing primarily high-grade tumor and matched benign prostate tissue cores (4 tumor and 4 benign tissue cores) from 60 Caucasian and 60 African American men matched on age +/− 3 years, Gleason grade, and stage. The TMA slides were stained using a validated SELENOF specific antibody to determine SELENOF levels, an E-cadherin antibody to identify regions of epithelial cells and to designate membrane localization, and DAPI to stain the nucleus of each cell. Only areas of the obtained images with both E-cadherin and SELENOF were used in the analysis to exclude stromal regions and non-specific background signal. As we published previously [10], SELENOF was expressed mainly in the epithelial cells and not in the stromal cells and was located near the lateral and basal membrane of the epithelial cells with undetectable levels on the apical side facing the lumen. Prostate cancer had dramatically lower levels of SELENOF compared to adjacent benign tissue with diffuse staining throughout the cytoplasm (See Figure 1 for representative images).

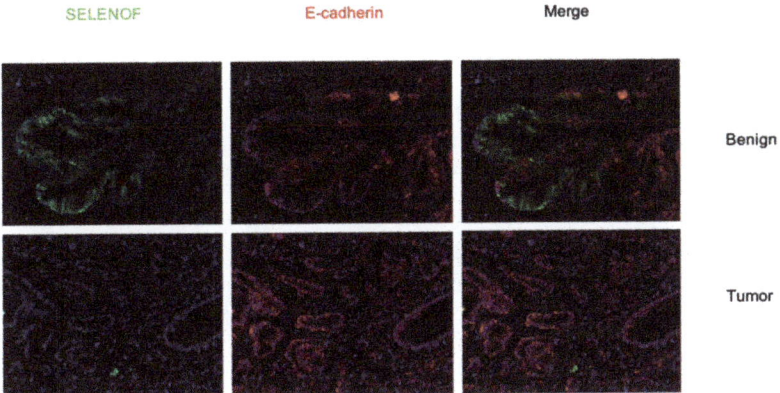

Figure 1. SELENOF levels are lower in prostate cancer compared to benign prostatic tissue. A representative immunofluorescent image of SELENOF (green) and E-cadherin (red). The top images show a region of benign prostatic tissue and the bottom images represent a region of prostate cancer. (Magnification = 20×). Staining with anti-E-cadherin antibodies was included both as a marker for the outer membrane and to restrict quantification of images to epithelial cells.

The comparisons of SELENOF intensities between benign and tumor-appearing regions, and between races are listed in Supplemental Table S1. We analyzed averaged SELENOF levels, and the relative amount or percentages of membrane associated SELENOF, calculated as the amount of SELENOF that co-localized with E-cadherin, compared to that amount determined for the entire cell. Paired t-tests revealed that SELENOF levels and distribution were dramatically lower in cancer cells when compared to benign tissues, consistent with our previous study [10]. Differences in SELENOF levels that co-localized with E-cadherin, presumably at the outer cellular membrane, were analyzed by paired t-test for significance. SELENOF levels that did not co-localize with E-cadherin were considered cytoplasmic. Regardless as to whether signals obtained were from the membrane or cytoplasmic regions, epithelial cells from prostate cancer cores had significantly lower levels and percentages of SELENOF in the membrane than benign prostate cores (Supplemental Table S1, top).

To investigate if there was a difference in SELENOF levels between African Americans and Caucasians, race comparisons on log transformed SELENOF levels, the relative cellular distribution were performed for each tissue and for the benign versus cancer differences (Supplemental Table S1, bottom). There were no significant differences in SELENOF levels or membrane percentages in benign prostate tissues between Caucasians and African Americans, regardless of whether the entire cell, cytoplasmic, or membrane regions were considered. Contrary to the previous study indicating that prostate tissue from African Americans had lower levels of SELENOF compared to that of Caucasians [10], tissue from African Americans had significantly higher levels of SELENOF in prostate cancer cores compared to Caucasians. Possible reasons for the differences in these results include the different designs of the TMAs, the much smaller sample size for the TMA presented here and much fewer low-grade cancers in this analysis compared to the previous one. The higher levels of SELENOF in African American cancer cells were statistically significant whether the entire cell ($p = 0.036$), cytoplasmic ($p = 0.036$), or membrane ($p = 0.024$) regions were compared. Similar race differences were also found in the cytoplasmic ($p = 0.047$), membrane ($p = 0.039$) regions, and the entire cell ($p = 0.036$). The difference in SELENOF levels and distribution between benign and cancer cells were slightly higher in African Americans than Caucasian samples, although these differences were not statistically significant.

To assess if membrane-located SELENOF levels were associated with prostate cancer stage, and if race modified the association, multivariate ordinal logistic regression models for Gleason score were employed. In these models, the effects of SELENOF that co-localized with the outer membrane-located protein E-cadherin in terms of percentage, race, and their interactions were tested. Gleason grading is a scoring system used by pathologists based on histology [18]. Each prostate core is assigned a score ranging from 1 to 5, with 1 representing benign, well-differentiated glands, 3 representing moderately differentiated glands, and 5 representing poorly differentiated glands [18]. Two scores are reported with the primary score being the most prevalent Gleason grade in the biopsied core and a secondary score representing the second most prevalent Gleason grade [19]. The primary and secondary scores are added together to calculate the Gleason score [19]. Out of 60 African American and 59 Caucasian patients, 20 (16.8%) patients had Gleason grade of 3 + 3, 26 (21.9%) had 3 + 4 or 3 + 5, 19 (16.0%) had 4 + 3, and 54 (45.4%) had 4 + 4, 4 + 5, or 5 + 4. In the multivariate ordinal logistic regression models, Gleason grades 4 or higher were treated as an ordered categorical outcome SELENOF distribution (percentages) and interactions with race were examined separately as the SELENOF percent distribution in cancer and benign cells (Supplemental Table S2a), and benign minus cancer percentage differences (Supplemental Table S2b).

The model selection process of SELENOF distributions in benign and cancer cells, and interactions between race and cancer cell membrane distribution indicated significant associations with Gleason grades. Higher cytoplasmic levels decreased the risk of higher Gleason grades (OR = 0.92, p-value = 0.0003). Among Caucasian patients, higher membrane levels strongly reduced the risk of higher Gleason grades (estimate = -0.34, p-value = 0.0006; OR of Caucasian membrane association is 0.71). Such protective effects of membrane localization were significantly reduced in African American patients (estimate of interaction = 0.26, p-value = 0.0132). The effect of SELENOF in the membrane of cancer cells in African American patients had an OR closer to 1 compared to that in Caucasians (OR = 0.92, p-value = 0.015).

The model selection process for the ordinal logistic regression model using benign-cancer SELENOF percent differences, race, and their interactions as predictors for Gleason grades revealed race-differential effects of the benign-cancer difference in SELENOF percentages in the entire cell and in the cytoplasm (Supplemental Table S2b). Benign-cancer cell percent differences did not influence Gleason grade in Caucasians (OR = 0.98, p-value = 0.80). However, an increase in benign-cancer cell SELENOF percent difference in the entire cell significantly increased the risk of having higher Gleason grade for African Americans (OR = 1.57, 95% CI = 1.2, 2.04). These results indicated that African Americans were almost 1.6 times more likely to have a higher Gleason score with each percent increase in the difference in SELENOF levels between benign and prostate cancer. Similarly, the benign-cancer cytoplasmic percent difference did not influence Gleason grades among Caucasians (OR = 0.999, p-value = 0.99). However, an increase in the benign-cancer cell SELENOF cytoplasmic percentage difference significantly reduced the risk of having a higher Gleason grade for African Americans (OR = 0.59, 95% CI = 0.43, 0.80). In other words, among African Americans, each percent increase in the benign-cancer cytoplasmic difference resulted in and odds of 1/0.59 = 1.70 times in having a lower Gleason grade.

2.2. Reducing SELENOF Levels Alters Phenotypes Associated with Transformation

To investigate whether the loss of SELENOF in prostate cancer indicates a tumor suppressor function, immortalized and non-transformed RWPE-1 human prostate cells were used because of their high expression of SELENOF and similar localization of SELENOF near the plasma membrane as seen in human benign tissues [10]. Four shRNA constructs and a scramble construct were obtained, transfected into RWPE-1 cells and stably transfected RWPE-1 cells were selected using puromycin. Pools and individual clones of SELENOF shRNA and SELENOF scramble transfected cells were isolated and expanded. As shown in Figure 2A, parental RWPE-1 cells and the pool of RWPE-1 cells transfected with the scramble shRNA construct exhibit high levels of SELENOF. Two of the SELENOF shRNA transfectant pools, designated shRNA B and shRNA D, displayed reduced levels of SELENOF by approximately 60% and 20%, respectively (Figure 2A). Fluorescent signals from the Western blots were quantified and presented as a fold change compared to the parental RWPE-1 cells below the corresponding Western blot. There was no significant difference in SELENOF levels between the parental RWPE-1 cells and the RWPE-1 scramble pool cells. SELENOF levels were significantly lower in shRNA B RWPE-1 cells and shRNA D RWPE-1 cells compared to the parental RWPE-1 cells ($p < 0.001$ and $p < 0.05$, respectively).

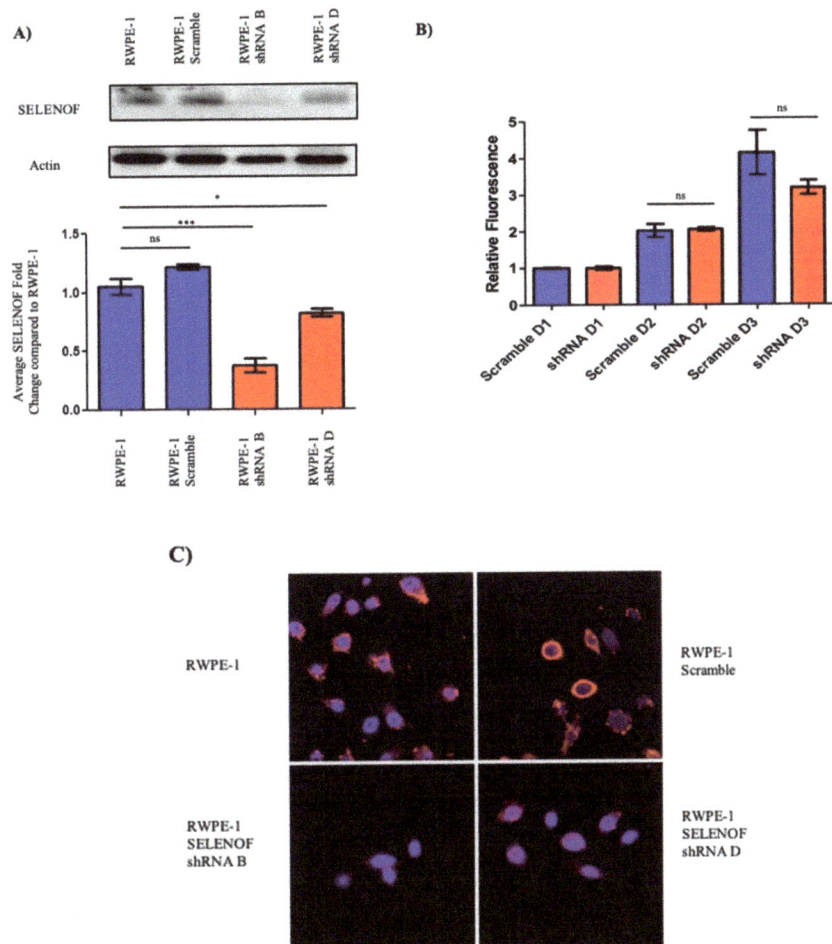

Figure 2. Reduction of SELENOF levels in RWPE-1 cells. (**A**) SELENOF levels were successfully reduced with an shRNA vector. Western blots using anti-SELENOF antibodies were quantified and a two-tailed *t*-test was performed for significant differences parental RWPE-1 cells and transfected RWPE-1 cells. $n = 3$, * $p < 0.05$, *** $p < 0.001$. (**B**) Fluorometric dsDNA quantification was performed after 3 days. Data are represented as the mean ± SEM, ns, not significant, $n = 3$. (**C**) Parental RWPE-1 and RWPE-1 scramble cells exhibit SELENOF membrane-associated localization shown in red, similar to what is seen in human benign tissues. Nuclei are stained blue with DAPI. Both shRNA RWPE-1 transfected clones have diffuse SELENOF staining in the cytoplasm similar to what is seen in prostate cancer tissue and prostate cancer cell lines (magnification is 63×).

The reduction of SELENOF levels did not alter the proliferation of shRNA RWPE-1 cells relative to RWPE-1 scramble cells after 24, 48, and 72 h (Figure 2B). Given the localization of SELENOF in RWPE-1 cells near the plasma membrane [10], SELENOF was visualized by immunofluorescence to determine if any changes in localization occurred with the reduced SELENOF levels (Figure 2C). Both pools of RWPE-1 shRNA cells exhibited lower levels of SELENOF. SELENOF staining was in a diffuse pattern throughout the cytoplasm that was consistent with what was seen previously in prostate cancer tissues and prostate cancer cell lines [10]. With the greatest reduction in SELENOF achieved in the SELENOF shRNA B pool cells compared to either native RWPE-1 cells or an expanded clone transfected with a control, scrambled shRNA construct, further experiments utilized two individual clones derived from the shRNA B pool, each with a similar reduction in SELENOF as seen in the pool samples.

To investigate if the reduction of SELENOF could alter the phenotype of parental RWPE-1 cells, RWPE-1 scramble, and two individual clones of shRNA RWPE-1 cells, were subjected to commonly used assays of cellular transformation. The ability of cells to grow in semi-solid media was assessed as anchorage-independent growth, a common phenotype of transformation [20–22]. Five thousand cells were seeded in semi-solid media, allowed to grow into colonies and imaged. As shown in Figure 3A, parental RWPE-1 and RWPE-1 scramble cells did not form any colonies in the semi-solid media. In contrast, clone 1 of the shRNA B RWPE-1 cells formed approximately 500 colonies per seeded 5000 cells and clone 2 of the shRNA B RWPE-1 cells formed approximately 1000 colonies per seeded 5000. These results indicated that reducing SELENOF levels in non-transformed RWPE-1 cells promoted anchorage-independent growth, a phenotype of transformation.

A scratch or wound healing assay was next used to determine the ability of cells to migrate on a tissue culture dish, a surrogate for aggressive or advanced cancer cells [23]. Parental RWPE-1, RWPE-1 scramble, and two clones of shRNA RWPE-1 cells were plated to form a fully confluent cell monolayer. A scratch was created using a pipette tip and the width of the scratch was measured. Although no differences in proliferation were observed when SELENOF levels were reduced, aphidicolin was used as an anti-proliferative agent to ensure differences in proliferation would not affect the results. Forty-eight hours after the initial scratch, both shRNA RWPE-1 clones nearly closed the wound completely while the parental RWPE-1 and RWPE-1 scramble-transfected cells only decreased the width of the wound slightly (Figure 4A). The widths of the scratch at 24 and 48 h were compared to the initial scratch width made at 0 h as a fold change (Figure 4). No significant difference was observed when the RWPE-1 scramble pool cells when compared to the parental RWPE-1 cells. Both shRNA RWPE-1 clones significantly migrated faster when compared to the RWPE-1 scramble cells at 48 h ($p < 0.0001$, Figure 4B).

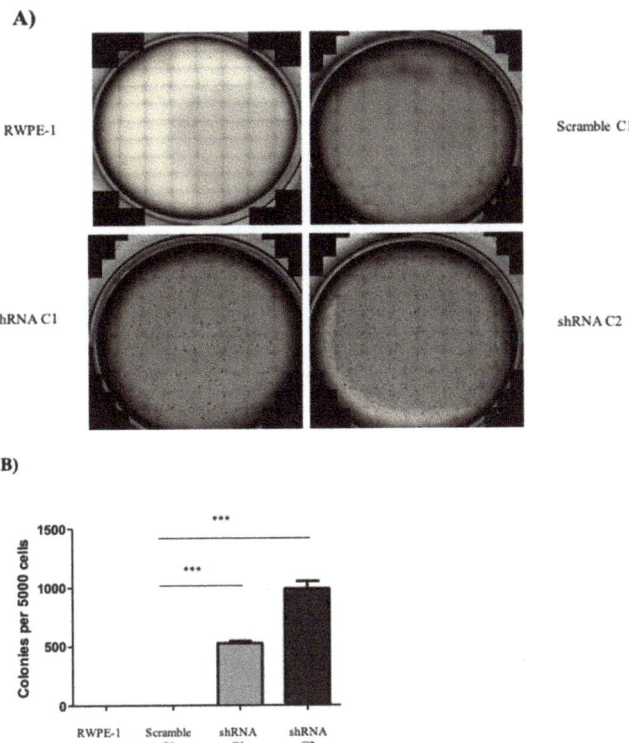

Figure 3. Reduction of SELENOF in RWPE-1 cells results in anchorage-independent growth. (**A**) Representative images of cells cultured in soft agar used to quantify the number of colonies formed. RWPE-1 and RWPE-1 scramble C1 did not form any colonies. (**B**) Quantification of the average number of colonies formed per 5000 cells plated. Data are represented in mean ± SEM, $n = 3$, and a two-tailed paired t-test was used to determine statistical differences *** $p < 0.001$.

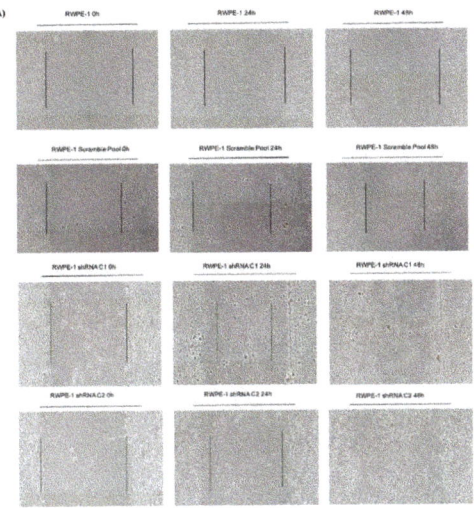

Figure 4. *Cont.*

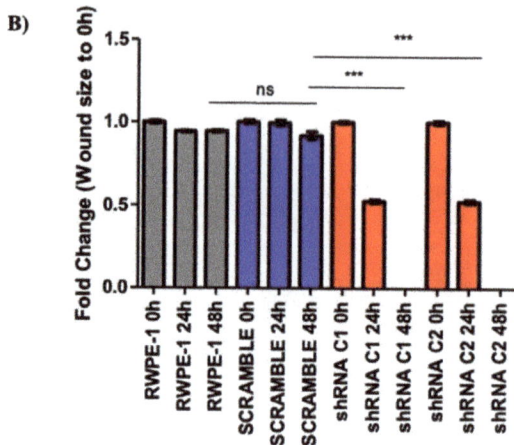

Figure 4. Reduction of SELENOF increases the migration of RWPE-1 cells in a scratch assay. (**A**) Representative images were captured at 0 h, 24 h, and 48 h for RWPE-1, RWPE-1 scramble, and 2 clones of RWPE-1 shRNA SELENOF cells. (**B**) Quantification of the scratch from three independent experiments are shown. The images were captured using the EVOS FL Auto Imaging system (ThermoFisher) and the width was measured using ImageJ. Data are represented as the mean ± SEM, ns, not significant, $n = 3$ and a two-tailed paired t-test was used to determine statistical significance, *** $p < 0.001$.

2.3. Effects of SELENOF on Metabolism

The metabolism of the benign prostate includes a truncated TCA cycle with glycolysis predominating to accumulate high levels of citrate required for sperm health [24]. Prostate carcinogenesis often involves a shift in energy metabolism from glycolysis to oxidative phosphorylation [25,26]. To investigate the effects of SELENOF on mitochondrial respiration in prostate cells, the oxygen consumption rate (OCR) of RWPE-1 cells with reduced SELENOF and control cells were measured in real time using the Seahorse XFe24 platform. The OCR is presented as the fold change of shRNA RWPE-1 cells compared RWPE-1 scramble cells in Figure 5. Reducing the levels of SELENOF significantly increased basal oxidative phosphorylation by approximately 3-fold (Figure 5A) compared to control RWPE-1 scramble cells. Maximal respiration is calculated by measuring the OCR after ATP synthase is inhibited by oligomycin and an uncoupler, FCCP, is injected into the assay. The spare respiratory capacity is the difference between the peak of OCR after FCCP injection and the initial basal OCR. Reducing the levels of SELENOF in RWPE-1 cells resulted in increased maximal respiration and spare respiratory capacity by 3.5-fold (Figure 5B) and 4-fold (Figure 5C), respectively. ATP production (Figure 5D), which is determined as the difference in OCR before and after oligomycin injection that inhibits ATP synthase also increased by 3-fold. Together, these results indicate that the reducing SELENOF levels in RWPE-1 cells increase mitochondrial respiration and presumably ATP synthesis.

A recent study using hepatic cells isolated from SELENOF knockout mice indicated differences in the levels of proteins related to energy metabolism, specifically in pathways involving glycolysis/gluconeogenesis and the TCA cycle compared to control mice [27]. In addition, SELENOF knockout mice experienced a greater frequency of glucose and lipid metabolism disorders [28]. Given these observations, changes in AMP-activated protein kinase (AMPK) in SELENOF knock down cells were explored due to its central role in cellular metabolism. AMPK is activated by phosphorylation at Thr172 when ATP levels decline resulting in the stimulation of glucose utilization for ATP production. Reducing SELENOF in RWPE-1 cells resulted in an approximately 1.7–2-fold increase in pAMPK levels without changes in total AMPK (Figure 6A). Both a representative Western blot and the quantification of the signals are shown in the figure.

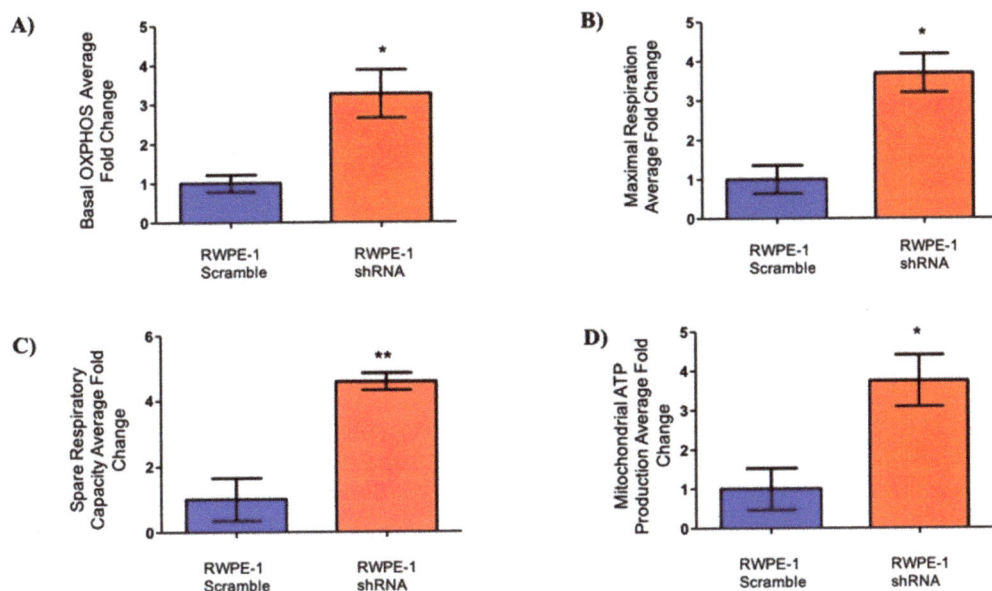

Figure 5. Reduction of SELENOF increases oxygen consumption in RWPE-1 cells. Average fold changes are represented for (**A**) basal, (**B**) maximal respiration after FCCP injection, (**C**) spare respiratory capacity (difference in peak OCR and basal measurements), and (**D**) mitochondrial ATP production (difference before and after oligomycin injection). Data are presented as the mean ± SEM, $n = 3$, two-tailed paired t-test, * $p < 0.05$, ** $p < 0.01$, $n = 3$.

Another key regulator of metabolism and a target of pAMPK is acetyl-CoA carboxylase (ACC), which functions in the production of the malonyl-CoA substrate for the biosynthesis of fatty acids as an alternative energy source [29]. Inhibitory phosphorylation at Ser79, was determined by Western blotting with pACC specific antibodies (Figure 6B). Reducing SELENOF levels resulted in increased phosphorylation of ACC without changes to total ACC levels. Together, the increased phosphorylation of AMPK and ACC when SELENOF levels are reduced indicate a potential function of SELENOF in glycolysis and lipid metabolism.

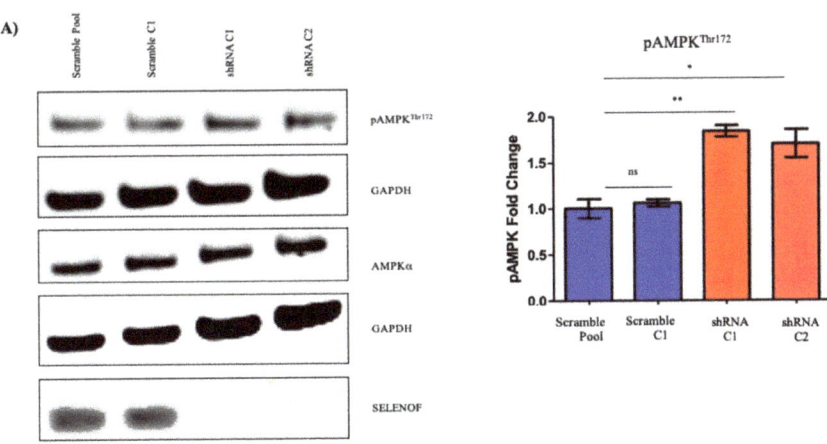

Figure 6. *Cont.*

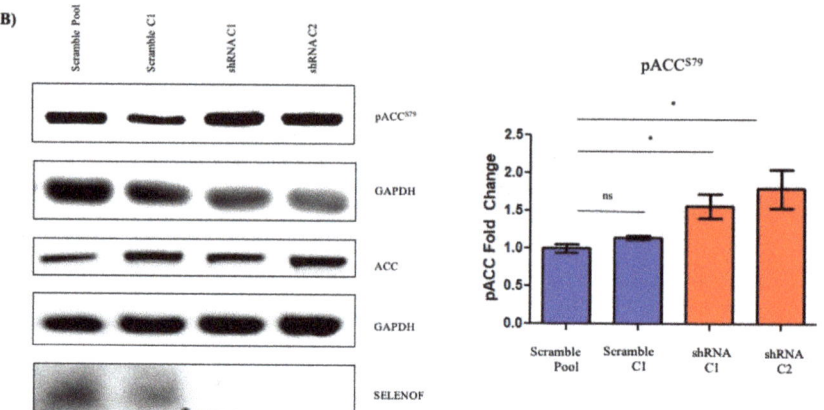

Figure 6. Reducing SELNENOF levels resulted in enhanced phosphorylation of AMPK and ACC in RWPE-1 cells. (**A**) Representative Western blots using anti-SELEONOF antibodies demonstrating the increase in the phosphorylation of AMPK in shRNA RWPE-1 cells without changes in total AMPK. (**B**) Representative Western blot and quantification of the changes in the phosphorylation of ACC and total levels of ACC. Florescent intensities were quantified and are shown next to the corresponding Western blots of three independent biological replicates. The data are presented as the mean ± SEM, $n = 3$, ns, not significant, * $p < 0.05$, ** $p < 0.01$.

3. Discussion

Loss of SELENOF was considered to be a significant occurrence in prostate cancer progression due to the lower levels in prostate cancers and the association of polymorphisms in the SELENOF gene with prostate cancer mortality [10,15]. In order to extend these observational studies to a model that can begin to assess functional consequences to its loss, SELENOF levels were reduced in the RWPE-1 non-tumorigenic prostate cell line. As a result, these cells acquired the characteristics of the transformed phenotype, including growth in soft agar and increased migration. In addition, the shift to a more aerobic respiration from a glycolytic means of energy production mirrors the changes that happen during the progression of normal prostate epithelium to malignancy. Collectively, these cell culture studies indicate that loss of SELENOF is very likely contributing to disease progression and that loss is not a bystander effect of the carcinogenic process.

Neither the function of SELENOF nor the molecular consequences of its loss are understood. In most cell types, SELENOF is located in the ER, and its location in the ER and tight binding to UGGT, a protein that functions in protein disulfide bond formation and quality control in that organelle, indicates that SELENOF may contribute to that process [9]. In contrast, SELENOF appears to reside in the cell membrane of benign prostatic epithelia based on its co-localization with E-cadherin, being excluded from the apical portion facing the lumen [10]. These observations indicate that SELENOF may have a function in regulating the secretion of bioactive proteins or compounds. Studies with SELENOF knockout mice have indicated the enhanced secretion of non-functional IgM, leading the authors to suggest that SELENOF was a "gatekeeper of secreted disulfide-rich glycoproteins" [30]. Determining whether the loss of SELENOF contributes to prostate cancer progression via a secretory mechanism involving the release of proteins that promote cancer initiation or progression will require additional studies.

The data presented herein along with previous results indicate that the loss of SELENOF during prostate cancer progression may contribute to the disparity in prostate cancer among African American men. The cell culture work provided evidence that lower levels of SELENOF have a physiological impact on prostate epithelial cells. Additionally, prostatic cancers seen in African American men have a lower level of SELENOF [10] with a greater difference between the tumor and benign tissue than Caucasian men. Moreover, the haplotype much more frequent among the genomes of African Americans is predicted to result in less SELENOF based on cell culture studies with reporter constructs that specifically determine the impact of these variations on the recognition of in-frame UGA codons as selenocysteine [8,11]. In addition, SELENOF levels are affected by selenium availability, and this is likely to occur to a greater extent among those with the at-risk haplotype, and several reports have indicated that African Americans exhibit lower levels of selenium [10,31]. The loss of SELENOF in prostate cancers may therefore be a risk factor present disproportionally among the genomes of African Americans that can help guide clinicians in their decisions regarding treatment options. Restoring SELENOF levels may also be developed as a new therapy to treat men with the detrimental genetic variation or greatest loss of SELENOF during cancer progression. Rather than an approach to increasing SELENOF levels using selenium supplementation, we suggest that future studies address the reasons for SELENOF loss in prostate cancers to develop a targeted approach, given concerns raised about selenium supplementation and the potential for enhancing the risk of cancer and other diseases [32–34].

4. Materials and Methods

Tissue Microarray (TMA). The "120 Case High Grade Race Disparity TMA" was obtained from the Prostate Cancer Biorepository Network (PCBN). This TMA contains tumor and matched benign prostate tissue from 60 self-identified Caucasian and 60 self-identified African American patients matched by age $+/-$ 3 years, Gleason grade, and stage and is enriched for cases with a Gleason score ≥ 8.

Cell Culture and Plasmid Construction. Immortalized non-transformed RWPE-1 prostate epithelial cells were obtained from ATCC and maintained in keratinocyte serum free media (Gibco. Langley, OK, USA). The media was supplemented with recombinant epidermal growth fact (rEGF) and bovine pituitary extract (BPE). The PC-3 human prostate carcinoma cell line was maintained in RPMI-1640 media (Gibco, Cat#: 11,875) supplemented with 10% fetal bovine serum (Gemini Bio, West Sacramento, CA, USA), 100 U/mL penicillin and 100 µg/mL streptomycin (Gibco). All cells were maintained at 37 °C with 5% CO_2. Identity verification of the cell lines was performed by Genetica Cell Line Testing (Burlington, NJ, USA) and the Clinical Molecular Pathology Laboratory at UIC by short tandem repeat (STR) analyses. SELENOF shRNA and control scramble RNA were purchased from Origene Technologies (Cat#: TG316,856) and transfected into RWPE-1 cells using ContinuumTM transfection reagent (Gemini Bio, Cat#: 400–700). Transfected cells were selected in 1 µg/mL puromycin (Sigma-Aldrich, St. Louis, MO, USA), individual clones were isolated and expanded for RWPE-1 scramble shRNA transfected cells and RWPE-1 shRNA cells.

Quantification of SELENOF mRNA. Cells were allowed to grow to 90% confluency and total RNA was extracted using the RNeasy Plus Mini Kit (Qiagen, Hilden, Germany, Cat#: 74134) according to the manufacturer's protocol. The RNA was reverse-transcribed using the high-capacity cDNA reverse transcription kit (Thermofisher, Waltham, MA, USA, Cat#: 4368814) without changes to the protocol. Utilizing the Fast SYBR Green Master Mix (Thermofisher, Cat#: 4,385,612) and QuantStudio 6 Flex Real-time PCR System (Thermofisher), fold changes were calculated using the delta–delta CT method using GAPDH as the internal amplification control. SELENOF was amplified using 6 pmol of forward (5′–TGCGGAAAATGGTAGCGAT–3′) and reverse (5′–CATGCCTCCGATGAAAACTCT–3′) primers per reaction. GAPDH was amplified as an internal control using 6 pmol of forward (5′–ACCCCTTCATTGACCTCAACTA–3′) and reverse (5′–ATCGCCCCACTTGATTTTG–3′) primers.

Western Blotting. Cells were grown to approximately 90% confluency, harvested, and lysed using 1x cell lysis buffer (Cell Signaling Technology, Danvers, MA, USA) containing both protease and phosphatase inhibitors (Millipore). Protein concentrations were measured using the Bradford assay (Bio-Rad, Hercules, CA, USA) and measured using a spectrophotometer (Bio-Rad). Lysates were combined with NuPAGE LDS loading buffer (Life Technologies, Carlsbad, CA, USA) and 1x reducing agent (Life Technologies) and boiled at 95 °C for 10 min. The prepared lysates were electrophoresed on a gradient 4–12% Bis-Tris denaturing polyacrylamide gel (Life Technologies). After electrophoresis, proteins were transferred to a small pore nitrocellulose membrane (Thermofisher). Membranes were blocked using Licor blocking buffer (LI-COR, Lincoln, NE, USA) for 1 h and incubated overnight with antibodies at 4 °C. SELENOF antibody (Abcam, Waltham, MA, USA) was used at 1:2000, GAPDH (Cell Signaling Technologies, Cat#: 2118) at 1:10,000, β-actin (Abcam) at 1:10,000, pAMPKThr172 at 1:1000 (Cell Signaling Technologies, Cat#: 2535), AMPKα at 1:1000 (Cell Signaling Technologies, Cat#: 5832), pACCS79 at 1:1000 (Cell Signaling Technologies, Cat#: 11,818), ACC at 1:1000 (Cell Signaling Technologies, Cat#: 3676) concentrations. Secondary antibodies, either anti-rabbit or anti-mouse (LI-COR Biosciences), were used at 1:5000 concentrations. The membrane was visualized and analyzed using the Odyssey® CLx imaging system (LI-COR Biosciences).

Cell Proliferation and Growth in Semi-Solid Media. Proliferation was measured using the FluoReporter blue fluorometric dsDNA quantitation kit (Thermofisher, Cat#: F-2962). An equal number of cells (5000 cells/well) were plated on a 96-well plate (Corning Inc., Glendale, AZ, USA) in triplicate and were incubated at 37 °C for 3 days according to the manufacturer's protocol. Growth in semi-sold agar was assayed using a 6-well plate with an equal number of cells in each well (5000 cells/well) containing 0.6% agar (Gibco) and culture media. Cells were incubated at 37 °C for 6 weeks, imaged using an EVOS FL Imaging System (Invitrogen, Waltham, MA, USA and counted using Celeste software (Invitrogen).

Wound Healing Assay. Cells plated for the wound healing or scratch assays were incubated on a 6-well plate until 100% confluency. The wound or scratch was generated by dragging a pipette tip through the middle of the well across the monolayer of cells. The media was replaced after the scratch was created. Aphidicolin (Cayman Chemical, Cat#: 14,007, Ann Arbor, MI, USA), a cell proliferation inhibitor, was added at 1 µ/mL to each of the wells. Images of the wound were captured using an EVOS FL imaging system (Invitrogen, Waltham, MA) at 0 h, 24 h, and 48 h after the initial scratch. Three areas of the wound closure were measured using ImageJ and averaged for each of the conditions.

Oxygen Consumption. Oxygen consumption rate (OCR) was used as a surrogate for oxidative phosphorylation using a Seahorse XF analyzer and Seahorse XF cell mito stress test kit (Agilent Technologies, Inc., Santa Clara, CA, USA) according to the manufacturer's protocol. In short, oxygen consumption was measured by a fluorophore that is quenched by the presence of oxygen. As mitochondrial respiration consumes oxygen, the signal is measured by the instrument. The cell mito stress test kit utilizes specific electron transport chain inhibitors, oligomycin, carbonyl cyanide 4-(trifluoromethoxy)phenylhydrazone (FCCP), and a combination of rotenone and antimycin A to measure various parameters of mitochondrial function with OCR. Basal measurements are first obtained, followed by sequential injections of these inhibitors. Oligomycin is an ATP synthase or complex V inhibitor that decreases the electron flow through the electron transport chain and builds a gradient for the next injection. This measurement after the oligomycin injection is indicative of mitochondrial ATP production. Next, FCCP, an uncoupler of oxidative phosphorylation is injected and disrupts the proton gradient across the mitochondrial membrane. This enables electrons to flow freely, and this parameter is indicative of the maximal respiration of the mitochondria. The difference between maximal respiration and the initial basal measurement is the spare respiratory capacity or the ability of the mitochondria to respond to various increased energy needs of the cell. Lastly, the combination of rotenone and antimycin A, inhibitors of complex I and III respectively, shuts down mitochondrial respiration completely permitting the measurement of non-mitochondrial respiration in the cells.

Immunohistochemistry. Immunofluorescent staining of the TMA was performed by the UIC Research Histology Core using the BondTM Research Detection System (Leica Biosystems, Buffalo Grove, IL, USA, DS9455). A Leica BondTM (Leica Biosystems, Buffalo Grove, IL, USA) was used for deparaffinization, antigen retrieval, and staining.

The tissues were washed with bond dewax solution (Leica Biosystems, AR9222) at 72 °C and then washed with 100% ethanol. Following the ethanol wash and washes with bond wash solution (Leica Biosystems, AR9590), target antigens were unmasked by incubation in bond ER 1 solution at pH 6.0 for 40 min at 98 °C (Leica Biosystems, AR9640). After the incubation with a background sniper protein block (Biocare Medical, #BS966) for 30 min at RT, sequential immunostaining was performed with antibodies directed against SELENOF (Abcam, Cat#: ab124,840, rabbit monoclonal NCIR128A, 1:100 dilution) and E-cadherin (Cell Signaling, Cat#: 14,472, mouse monoclonal 4A2, 1:100). Sections were incubated with the primary antibodies at room temperature for 1 h followed by incubation with Alexa-488 and Alexa-555-conjugated secTondary goat anti-mouse and goat-anti rabbit IgG (Thermofisher Scientific, # A32,734 and #A32,732) at room temperature for 1 h. Lastly, the slides were counterstained with DAPI and mounted with Pro-GoldTM diamond antifade mountant (Thermofisher Scientific, #P36961). Stained tissue slides were scanned at 20x on Vectra 3 automated quantitative imaging system (Akoya Biosciences, Marlborough, MA, USA) and analyzed as previously described [10] by the UIC Research Tissue Imaging Core. Briefly, images were spectrally unmixed and adjusted for tissue autofluorescence. E-cadherin staining was used for tissue segmentation, and after cell segmentation, the levels of SELENOF in epithelial cells were analyzed for each core. Various artifacts were manually excluded.

Statistical Analysis. Statistical analyses were performed in SAS 9.4 by our statistical team. SELENOF levels and percentages in cancer and benign tissues from African American and Caucasian men were compared using paired t-tests for within-individual type of tissue comparison or independent group t-tests for race comparison. Due to skewness in the observed individual SELENOF levels, log-transformed SELENOF measures were used in all t-tests. Multivariate ordinal logistic regression models for Gleason sum categories were employed to assess the association between SELENOF levels, race (African Americans vs. Caucasian), and the potential interactions between them. Two sets of SELENOF percent measures were considered separately in the regression models: 1. SELENOF percentages in cancer and benign cells; the benign minus cancer differences in SELENOF percentages. Interactions between race and SELENOF percent or benign-cancer differences were first tested. Then backward selections were performed for each main SELENOF effects keeping the significant interaction terms in the model. Proportional odds assumptions were tested for each ordinal logistic regression model. In the results, estimate of effects, standard errors, odds ratios with 95% confidence intervals were reported. All statistical tests are two-sided tests that control for a Type I error probability of 0.05. All other experimental data were collected from at least three biologically independent experiments. Results are reported as mean ± SEM and $p < 0.05$ were considered statistically significant.

5. Conclusions

In conclusion, the data provided here indicates that SELENOF is a prostate cancer tumor suppressor able to alter the transformed phenotype of human prostate epithelial cells and may contribute to the disparity in prostate cancer mortality that exists among men of African descent.

Supplementary Materials: The following are available online at https://www.mdpi.com/article/10.3390/ijms222112040/s1.

Author Contributions: Conceptualization, L.K.H. and A.M.D.; methodology, L.K.H. and M.E.; formal analysis, H.W., A.M.D., L.L. and R.D.; investigation, L.K.H., S.K., M.S. and M.E; writing—L.K.H. and A.M.D.; writing—L.K.H., M.S., I.K. and L.L.; funding acquisition, L.K.H., M.E. and A.M.D.; resources, K.S.S. All authors have read and agreed to the published version of the manuscript.

Funding: This work was supported by a Department of Defense Prostate Cancer Research Program Health Disparity Research Award PC170236 to AMD and by the Department of Defense Prostate Cancer Research Program Award No W81XWH-14-2-0182, W81XWH-14-2-0183, W81XWH-14-2-0185, W81XWH-14-2-0186, and W81XWH-15-2-0062 Prostate Cancer Biorepository Network (PCBN).

Data Availability Statement: All data generated from this study are included in the manuscript and supplemental materials.

Acknowledgments: Histology/services were provided by the Research Resources Center—Research Histology and Tissue Imaging Core at the University of Illinois at Chicago established with the support of the Vice Chancellor of Research.

Conflicts of Interest: The authors declare no conflict of interest.

References

1. Siegel, R.L.; Miller, K.D.; Fuchs, H.E.; Jemal, A. Cancer Statistics. *CA Cancer J. Clin.* **2021**, *71*, 7–33. [CrossRef] [PubMed]
2. Zeegers, M.P.; Jellema, A.; Ostrer, H. Empiric risk of prostate carcinoma for relatives of patients with prostate carcinoma: A meta-analysis. *Cancer* **2003**, *97*, 1894–1903. [CrossRef]
3. Vinceti, M.; Filippini, T.; Giovane, D.C.; Dennert, G.; Zwahlen, M.; Brinkman, M.; Zeegers, P.M.; Horneber, M.; D'Amico, R.; Crespi, M.C. Selenium for preventing cancer. *Cochrane Database Syst. Rev.* **2018**, *1*, CD005195. [CrossRef] [PubMed]
4. Kryukov, G.V.; Castellano, S.; Novoselov, S.V.; Lobanov, A.V.; Zehtab, O.; Guigó, R.; Gladyshev, V.N. Characterization of mammalian selenoproteomes. *Science* **2003**, *300*, 1439–1443. [CrossRef]
5. Gladyshev, V.N.; Arnér, E.S.; Berry, M.J.; Brigelius-Flohé, R.; Bruford, E.A.; Burk, R.F.; Carlson, B.A.; Castellano, S.; Chavatte, L.; Conrad, M.; et al. Selenoprotein Gene Nomenclature. *J. Biol. Chem.* **2016**, *291*, 24036–24040. [CrossRef]
6. Gladyshev, V.N.; Jeang, K.-T.; Wootton, J.C.; Hatfield, D.L. A new human selenium-containing protein: Purification, characterization and cDNA sequence. *J. Biol. Chem.* **1998**, *273*, 8910–8915. [CrossRef]
7. Kumaraswamy, E.; Malykh, A.; Korotkov, K.V.; Hatfield, D.L.; Diamond, A.M.; Gladyshev, V.N. Structure-expression relationships of the 15-kDa selenoprotein gene. Possible role of the protein in cancer etiology. *J. Biol. Chem.* **2000**, *275*, 35540–35547. [CrossRef] [PubMed]
8. Kumaraswamy, E.; Korotkov, K.V.; Diamond, A.M.; Gladyshev, V.N.; Hatfield, D.L. Genetic and functional analysis of mammalian Sep15 selenoprotein. *Methods Enzym.* **2002**, *347*, 187–197.
9. Labunskyy, V.M.; Hatfield, D.L.; Gladyshev, V.N. The Sep15 protein family: Roles in disulfide bond formation and quality control in the endoplasmic reticulum. *IUBMB Life* **2007**, *59*, 1–5. [CrossRef]
10. Ekoue, D.N.; Ansong, E.; Liu, L.; Macias, V.; Deaton, R.; Lacher, C.; Picklo, M.; Nonn, L.; Gann, P.H.; Kajdacsy-Balla, A.; et al. Correlations of SELENOF and SELENOP genotypes with serum selenium levels and prostate cancer. *Prostate* **2018**, *78*, 279–288. [CrossRef] [PubMed]
11. Hu, Y.J.; Korotkov, K.V.; Mehta, R.; Hatfield, D.L.; Rotimi, C.N.; Luke, A.; Prewitt, T.E.; Cooper, R.S.; Stock, W.; Vokes, E.E.; et al. Distribution and functional consequences of nucleotide polymorphisms in the 3'-untranslated region of the human Sep15 gene. *Cancer Res.* **2001**, *61*, 2307–2310. [PubMed]
12. Nasr, M.; Hu, Y.J.; Diamond, A.M. Allelic loss of Sep15 in breast cancer. *Cancer Ther.* **2003**, *1*, 307–312.
13. Meplan, C.; Hughes, D.J.; Pardini, B.; Naccarati, A.; Soucek, P.; Vodickova, L.; Hlavatá, I.; Vrána, D.; Vodicka, P.; Hesketh, J.E. Genetic variants in selenoprotein genes increase risk of colorectal cancer. *Carcin* **2010**, *31*, 1074–1079. [CrossRef] [PubMed]
14. Sutherland, A.; Kim, D.H.; Relton, C.; Ahn, Y.O.; Hesketh, J. Polymorphisms in the selenoprotein S and 15-kDa selenoprotein genes are associated with altered susceptibility to colorectal cancer. *Genes Nutr.* **2010**, *5*, 215–223. [CrossRef]
15. Penney, K.L.; Schumacher, F.R.; Li, H.; Kraft, P.; Morris, J.S.; Kurth, T.; Mucci, L.A.; Hunter, D.J.; Kantoff, P.W.; Meir, J.; et al. A large prospective study of SEP15 genetic variation, interaction with plasma selenium levels, and prostate cancer risk and survival. *Cancer Prev. Res.* **2010**, *3*, 604–610. [CrossRef] [PubMed]
16. Karunasinghe, N.; Han, D.Y.; Zhu, S.; Yu, J.; Lange, K.; Duan, H.; Medhora, R.; Singh, N.; Kan, J.; Alzaher, W.; et al. Serum selenium and single-nucleotide polymorphisms in genes for selenoproteins: Relationship to markers of oxidative stress in men from Auckland, New Zealand. *Genes Nutr.* **2012**, *7*, 179–190. [CrossRef] [PubMed]
17. Steinbrecher, A.; Méplan, C.; Hesketh, J.; Schomburg, L.; Endermann, T.; Jansen, E.; Åkesson, B.; Rohrmann, S.; Linseisen, J. Effects of selenium status and polymorphisms in selenoprotein genes on prostate cancer risk in a prospective study of European men. *Cancer Epidem. Bio. Prev.* **2010**, *19*, 2958–2968. [CrossRef]
18. Gleason, D.F.; Mellinger, G.T. Prediction of prognosis for prostatic adenocarcinoma by combined histological grading and clinical staging. *J. Urol.* **1974**, *111*, 58–64. [CrossRef]
19. Epstein, J.I.; Allsbrook, W.C., Jr.; Amin, M.B.; Egevad, L.L.; Committee, I.G. The 2005 International Society of Urological Pathology (ISUP) Consensus Conference on Gleason Grading of Prostatic Carcinoma. *Am. J. Surg. Pathol.* **2005**, *29*, 1228–1242. [CrossRef]
20. Borowicz, S.; Scoyk, M.V.; Avasarala, S.; Rathinam, M.K.K.; Tauler, J.; Bikkavilli, R.K.; Winn, R.A. The soft agar colony formation assay. *J. Vis. Exp.* **2014**, *92*, e51998. [CrossRef]
21. Puck, T.T.; Marcus, P.I.; Cieciura, S.J. Clonal growth of mammalian cells in vitro; growth characteristics of colonies from single HeLa cells with and without a feeder layer. *J. Exp. Med.* **1956**, *103*, 273–283. [CrossRef] [PubMed]
22. Todaro, G.J.; Lazar, G.K.; Green, H. The initiation of cell division in a contact-inhibited mammalian cell line. *J. Cell. Comp. Physiol.* **1965**, *66*, 325–333. [CrossRef] [PubMed]
23. Cory, G. Scratch-wound Assay. *Meth. Mol. Biol.* **2011**, *769*, 25–30.
24. Costello, L.C.; Franklin, R.B.; Feng, P.; Tan, M.; Bagasra, O. Zinc and prostate cancer: A critical scientific, medical, and public interest issue (United States). *Cancer Causes Control.* **2005**, *16*, 901–915. [CrossRef] [PubMed]

25. Cutruzzola, F.; Giardina, G.; Marani, M.; Macone, A.; Paiardini, A.; Rinaldo, S.; Paone, A. Glucose Metabolism in the Progression of Prostate Cancer. *Front. Physiol.* **2017**, *8*, 97. [CrossRef] [PubMed]
26. Costello, L.C.; Franklin, R.B. The intermediary metabolism of the prostate: A key to understanding the pathogenesis and progression of prostate malignancy. *Oncology* **2000**, *59*, 269–282. [CrossRef]
27. Zheng, X.; Ren, B.; Wang, H.; Huang, R.; Zhou, J.; Liu, H.; Tian, J.; Huang, K. Hepatic proteomic analysis of selenoprotein F knockout mice by iTRAQ: An implication for the roles of selenoprotein F in metabolism and diseases. *J. Proteom.* **2020**, *215*, 103653. [CrossRef]
28. Zheng, X.; Ren, B.; Wang, H.; Huang, R.; Zhou, J.; Liu, H.; Tian, J.; Huang, K. Selenoprotein F knockout leads to glucose and lipid metabolism disorders in mice. *J. Biol. Inorg. Chem.* **2020**, *25*, 1009–1022. [CrossRef] [PubMed]
29. Wang, C.; Ma, J.; Zhang, N.; Yang, Q.; Jin, Y.; Wang, Y. The acetyl-CoA carboxylase enzyme: A target for cancer therapy? *Expert Rev. Anticancer Ther.* **2015**, *15*, 667–676. [CrossRef]
30. Yim, S.H.; Everley, R.A.; Schildberg, F.A.; Lee, S.-G.; Orsi, A.; Barbati, Z.R.; Fomenko, D.E.; Tsuji, P.A.; Luo, H.R.; Gygi, S.P.; et al. Role of Selenof as a Gatekeeper of Secreted Disulfide-Rich Glycoproteins. *Cell Rep.* **2018**, *23*, 1387–1398. [CrossRef]
31. Xun, P.; Bujnowski, D.; Liu, K.; Morris, J.S.; Guo, Z.; He, K. Distribution of toenail selenium levels in young adult Caucasians and African Americans in the United States: The CARDIA Trace Element Study. *Environ. Res.* **2011**, *111*, 514–519. [CrossRef] [PubMed]
32. Duffield-Lillico, A.J.; Slate, E.H.; Reid, M.E.; Turnbull, B.W.; Wilkins, P.A.; Combs, G.F., Jr.; Park, H.K.; Gross, E.G.; Graham, G.F.; Stratton, M.S.; et al. Selenium supplementation and secondary prevention of nonmelanoma skin cancer in a randomized trial. *J. Natl. Cancer Inst.* **2003**, *95*, 1477–1481. [CrossRef] [PubMed]
33. Kenfield, S.A.; Blarigan, E.L.V.; DuPre, N.; Stampfer, M.J.; Giovannucci, E.L.; Chan, J.M. Selenium supplementation and prostate cancer risk. *J. Natl. Cancer Inst.* **2015**, *107*, 360–368. [CrossRef] [PubMed]
34. Vinceti, M.; Filipini, T.; Rothman, K.J. Selenium exposure and the risk of type 2 diabetes: A systematic review and meta-analysis. *Eur. J. Epidemiol.* **2018**, *33*, 789–810. [CrossRef] [PubMed]

Article

Identification of Signaling Pathways for Early Embryonic Lethality and Developmental Retardation in *Sephs1*$^{-/-}$ Mice

Jeyoung Bang [1], Minguk Han [1], Tack-Jin Yoo [2], Lu Qiao [2], Jisu Jung [2], Jiwoon Na [2], Bradley A. Carlson [3], Vadim N. Gladyshev [4], Dolph L. Hatfield [3], Jin-Hong Kim [2], Lark Kyun Kim [5,*] and Byeong Jae Lee [1,2,*]

1. Interdisciplinary Program in Bioinformatics, College of Natural Sciences, Seoul National University, Seoul 08826, Korea; 880419@snu.ac.kr (J.B.); han3270@snu.ac.kr (M.H.)
2. School of Biological Sciences, College of Natural Sciences, Seoul National University, Seoul 08826, Korea; yootackjin@snu.ac.kr (T.-J.Y.); qiaolu@snu.ac.kr (L.Q.); tabris0520@snu.ac.kr (J.J.); naji0708@snu.ac.kr (J.N.); jinhkim@snu.ac.kr (J.-H.K.)
3. Mouse Cancer Genetics Program, Center for Cancer Research, National Cancer Institute, National Institutes of Health, Bethesda, MD 20892, USA; carlsonb@dc37a.nci.nih.gov (B.A.C.); hatfielddolph@gmail.com (D.L.H.)
4. Department of Medicine, Brigham and Women's Hospital, Harvard Medical School, Boston, MA 02115, USA; vgladyshev@rics.bwh.harvard.edu
5. Severance Biomedical Science Institute, Graduate School of Medical Science, Brain Korea 21 Project, Gangnam Severance Hospital, Yonsei University College of Medicine, Seoul 06230, Korea
* Correspondence: lkkim@yuhs.ac (L.K.K.); imbglmg@snu.ac.kr (B.J.L.); Tel.: +82-2-880-6775 (B.J.L.)

Abstract: Selenophosphate synthetase 1 (SEPHS1) plays an essential role in cell growth and survival. However, the underlying molecular mechanisms remain unclear. In the present study, the pathways regulated by SEPHS1 during gastrulation were determined by bioinformatical analyses and experimental verification using systemic knockout mice targeting *Sephs1*. We found that the coagulation system and retinoic acid signaling were most highly affected by SEPHS1 deficiency throughout gastrulation. Gene expression patterns of altered embryo morphogenesis and inhibition of Wnt signaling were predicted with high probability at E6.5. These predictions were verified by structural abnormalities in the dermal layer of *Sephs1*$^{-/-}$ embryos. At E7.5, organogenesis and activation of prolactin signaling were predicted to be affected by *Sephs1* knockout. Delay of head fold formation was observed in the *Sephs1*$^{-/-}$ embryos. At E8.5, gene expression associated with organ development and insulin-like growth hormone signaling that regulates organ growth during development was altered. Consistent with these observations, various morphological abnormalities of organs and axial rotation failure were observed. We also found that the gene sets related to redox homeostasis and apoptosis were gradually enriched in a time-dependent manner until E8.5. However, DNA damage and apoptosis markers were detected only when the *Sephs1*$^{-/-}$ embryos aged to E9.5. Our results suggest that SEPHS1 deficiency causes a gradual increase of oxidative stress which changes signaling pathways during gastrulation, and afterwards leads to apoptosis.

Keywords: selenium; selenoprotein; SEPHS1; early embryogenesis; embryonic lethality; reactive oxygen species

1. Introduction

Selenium (Se) is an essential trace element required in the diet of humans and other forms of life. An adequate amount of selenium is essential to good health. For example, Se is implicated in cancer prevention, antiviral response, boosting the immune system, male reproduction, and embryo development [1–3]. In addition to those beneficial effects of Se, it is notable that Se also affects the progression of pregnancy. Specifically, Se levels and the activity of blood glutathione peroxidase were lower than average in women who had experienced miscarriage, premature birth, or preeclampsia [4,5]. Moreover, pregnant women who had low serum Se levels showed a high incidence of spontaneous miscarriage [6].

Se is the only trace element to be specified in the genetic code, and selenocysteine (Sec) is the 21st amino acid, which is incorporated into protein in response to the UGA codon during translation [7]. Selenoproteins contain single or multiple Sec residues in their active sites and are known to carry out important roles, often with the beneficial effects of Se described above [1,2]. One of most common functions of selenoproteins is to protect cells and tissues from oxidative stress by serving as reactive oxygen species (ROS) scavengers [8,9]. During Sec synthesis, selenophosphate serves as the selenium donor. Selenophosphate synthetase (SEPHS) catalyzes the reaction of selenophosphate synthesis, in which inorganic selenium and ATP are used as substrates [10]. There are two types of SEPHSs in higher eukaryotes, SEPHS1 and 2 [11]. Both isotypes have ATP binding and catalytic domains, and there is a high amino acid sequence homology between them. One of the biggest differences between SEPHS1 and 2 is that SEPHS2 contains Sec in the catalytic domain, while SEPHS1 does not. Instead, SEPHS1 contains threonine at the position corresponding to Sec in SEPHS2 [12]. Another feature is that only SEPHS2 has the ability to synthesize selenophosphate [13]. According to data from the International Mouse Phenotyping Consortium, whole-gene deletion of *Sephs2* did not show embryonic lethality but showed abnormalities in heart morphology of the knockout fetuses [14]. SEPHS1, however, is required for cell survival and proliferation [15].

The most prominent role of SEPHS1 is that it participates in the regulation of cellular redox homeostasis [3]. In *Drosophila*, a P-element insertion mutation in *Sephs1* (*SelD*) led to embryonic lethality following the loss of imaginal disc formation [16]. Subsequently, it was demonstrated that the embryonic lethality in *Drosophila* is mediated by ROS-induced apoptosis [17]. An in vitro study using SL2 cells, an embryonic cell line of *Drosophila*, showed that SEPHS1 deficiency induced ROS accumulation which in turn led to the inhibition of cell proliferation and glutamine-dependent megamitochondria formation [18]. In addition to studies in *Drosophila*, systemic knockout mice targeting *Sephs1* showed embryonic lethality and complete resorption by E14.5 [19]. Knockdown of *Sephs1* in a mouse embryonic cancer cell line (F9) showed the inhibition of cell proliferation by increased levels of ROS, specifically hydrogen peroxide [19]. Deficiency of SEPHS1 in F9 cells also reversed cancer malignancy characteristics such as cell invasion and anchorage independence. The expression levels of glutaredoxin 1 (*Glrx1*) and glutathione-s-transferase O1 (*GstO1*), which are involved in redox homeostasis, were significantly decreased by SEPHS1-deficiency in F9 cells.

Oxidative stress causes numerous types of damage during embryonic development by altering cellular macromolecules such as lipids, proteins, and nucleic acids [20]. Consequently, an affected embryo undergoes growth inhibition, development retardation, metabolic dysfunction, and apoptosis [21]. In the case of embryos cultured in vitro, high ROS levels have detrimental effects on embryo growth, but the addition of free radical scavengers recovers cells from the detrimental effects of ROS [22]. The importance of ROS during development was also demonstrated by regulating the expression of antioxidant genes. For example, disruption of thioredoxin 1 (*Txn1*) resulted in embryo hatching failure and lethality shortly after implantation [23]. Thioredoxin 2 (*Txn2*) mutation inhibited neural tube formation and induced massive apoptosis at E10.5 [24].

Although it is known that SEPHS1 plays an essential role for cell survival and proliferation, the underlying molecular and biochemical mechanisms have not been fully clarified. To elucidate the role of SEPHS1 during early embryogenesis, we generated systemic knockout mice targeting *Sephs1*. Pathways and genes regulated by SEPHS1 were predicted using various bioinformatical tools described herein, and the predicted pathways were verified by analyzing the anatomical structure of the developing embryo.

2. Results

2.1. RNA-seq Data Analysis

We previously reported that the systemic knockout targeting *Sephs1* in a mouse model resulted in the embryo beginning to show a difference in size at E7.5 and lethality at E11.5 [19].

In this study, the effect of SEPHS1 deficiency on early embryogenesis was analyzed in more detail during this period. Figure 1A shows whole-embryo images obtained by optical microscopy. There are no differences between wild-type ($Sephs1^{+/+}$ and $Sephs1^{+/-}$) and the $Sephs1^{-/-}$ embryo at E6.5. The $Sephs1^{-/-}$ embryo began to show size differences from wild-type at E7.5, and the difference extended at E8.5, wherein head folds were observed to be less developed than in wild-type. In addition, the $Sephs1^{-/-}$ embryo does not turn in the final fetal position at E9.5 and is smaller than wild-type. Unlike in the wild-type embryo, optic and otic vesicles were not observed, although the allantois still remained, and three primary brain vesicles (prosencephalon, mesencephalon, and rhombencephalon) were observed as being immature. At E9.5, no vitelline vessel was found in the yolk sac (arrow in Figure S1A). At E10.5, the size of the $Sephs1^{-/-}$ embryo was dramatically reduced from that at E9.5 (data not shown) and the embryo appeared to be fully absorbed at E11.5 (arrowhead in panel E11.5 of Figure 1A). We used 32, 24, 34, 20, 17, and 10 embryos at E6.5, E7.5, E8.5, E9.5, E10.5, and E11.5, respectively (See Materials and Methods for more detailed information). All the embryos that had the same genotype and were prepared at the same embryonic day showed similar phenotypes both in size and in morphology.

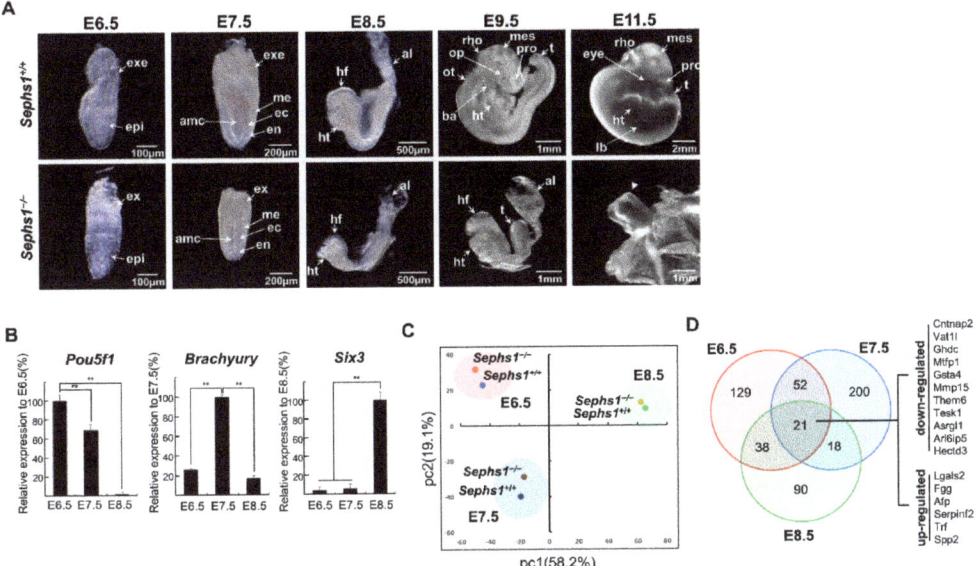

Figure 1. Transcriptome analysis of $Sephs1^{-/-}$ embryos. (**A**) Morphology of wild-type and $Sephs1^{-/-}$ embryos after dissection. The yolk sac was removed after E8.5. Arrowhead indicates absorbed $Sephs1^{-/-}$ embryo at E11.5 that attached to the yolk sac. (**B**) Real-time quantitative PCR of embryonic stage marker genes. Expression levels of *Pou5f1* (*Oct4*), *Brachyury* (*T*), and *Six3* were measured in wild-type embryos as described in Materials and Methods. ** indicates $p < 0.01$. (**C**) Principal component analysis of RNA-seq data. The percentages represent the variance captured by each principal components 1 and 2 in analysis. (**D**) Venn diagram of the number of DEGs overlapping between embryonic day E6.5-E8.5. DEG cut-off: max(FPKM) > 1, |Log2(Fold Change)| > 1 and $p < 0.01$. al, allantois; amc, amniotic cavity; ba, first branchial arch; ec, ectoderm; en, endoderm; epi, epiblast; exe, extra embryonic region; hf, head fold; ht, heart; lb, limb bud; me, mesoderm; mes, mesencephalon; op, optic vesicle; ot, otic vesicle; rho, rhombencephalon; t, tail; tel, telencephalon.

In order to assess the effect of SEPHS1 deficiency on embryonic development and the related signaling pathways responsible for phenotypic changes, RNA-seq was performed using purified RNA from wild-type and $Sephs1^{-/-}$ embryos at the E6.5, E7.5, and E8.5 stages.

As shown in Figure 1B, *Pou5f1* (*Oct4*) and *Brachyury* (*T*) were expressed most abundantly at E6.5 and at E7.5, respectively, and *Six3* was expressed only at E8.5 in wild-type embryo. These results indicate that the pooled RNAs isolated from embryos at the same embryonic day were highly homogeneous. In addition, principal component analysis (PCA) was performed to examine the relationship between the read sets. PCA revealed that the differences in developmental stages are much more distinct than genotypic differences between wild-type and *Sephs1* knockout (Figure 1C). Differentially expressed genes (DEGs) were obtained at E6.5, E7.5 and E8.5 with the $\log_2$(fold change) cut off of ± 1 at $p < 0.01$ (Table S1). There are 21 genes commonly affected by SEPHS1 deficiency at all three stages (Figure 1D). The functions of the down-regulated genes include those relating to organogenesis (*Cntnap2, Gata4, Mmp15, Asrgl1,* and *Arl6ip5*) and cell survival (*Hectd3*) [25–27]. Interestingly, the up-regulated genes showed region-specificity; *Fgg, Afp, Trf,* and *Serpinf2* in the extraembryonic region and *Spp2* in the placenta. Notably, *Galectin 2* (*Lgals2*), the most up-regulated DEG, is reported to be an oxidative stress-responsive gene shown to be up-regulated under H_2O_2 treatment [28]. These results suggest that SEPHS1-deficiency causes developmental retardation and the induction of oxidative stress.

2.2. Pathway Analysis of Differentially Expressed Genes

In order to identify the pathways that are regulated by SEPHS1, pathway enrichment analysis using Metascape was performed with DEGs (Figure 2A), with the $\log_2$(fold change) cut off of ± 0.5 at $p < 0.01$. At E6.5, embryo and tissue morphogenesis were significantly affected by *Sephs1* knockout. At E7.5 and E8.5, development of differentiated tissues such as 'vascular morphogenesis', 'epithelial cell differentiation', 'mesenchymal development', 'head' development', and 'heart development' were greatly affected. Some of the predicted results in Figure 2A were consistent with the results of Ingenuity Pathway Analysis (IPA) (Table S2). For example, the genes included in 'LXR/RXR activation' predicted by IPA were also found in the 'Regulation of body fluid level' and 'Plasma lipoprotein assembly, remodeling, and clearance' predicted by the Metascape analysis. These genes participate in retinoic acid (RA) signaling. In addition, most of the genes in the 'Coagulation system' of IPA were included in the Metascape category of 'Hemostasis'.

In order to identify the transcription factors targeting the DEGs during gastrulation, transcription factors that are known to be activated during gastrulation were selected from Transcriptional Regulatory Relationships Unraveled by Sentence-based Text mining (TRRUST) database. Among the gastrulation-specific transcription factors, the transcription factors that contain target genes (of which more than 50% are DEGs) were further selected to identify the transcription factors governing the expression of target DEGs (Figure 2B). Among the selected transcription factors, 12 transcription factors were DEG themselves (asterisks in Figure 2B). We defined these transcription factor genes as differentially expressed transcription factor genes (DTFGs). Interestingly, DTFGs were apt to be clustered together. The expression of DEGs by non-DTFGs, the transcription factors whose expression was not changed by *Sephs1* knockout, may be regulated indirectly by changes in protein stability, phosphorylation, and interaction with other co-regulators.

To analyze the expression pattern of DEGs, hierarchical clustering was performed (Figure 2C). Expression patterns of DTFGs (asterisks in Figure 2B) were indicated on the right of each cluster to which they belong. The DTFGs in the same cluster in Figure 2B were in the same or neighboring cluster in Figure 2C. For example, *Eomes, Tal1, Hnf4α, Gata4, Stat3,* and *Pitx2* were included in C3 and C4. This suggests that DTFGs are expressed in the same pattern with their target genes.

To identify biological processes that were most highly affected by SEPHS1 deficiency, gene ontology (GO) analysis was performed with DEGs belonging to each cluster (Figure 2D). As a result, genes in each cluster were found to regulate distinct pathways. For example, the genes in cluster 1 were predicted to regulate pathways participating in axis formation and the genes in cluster 2 in neuron development.

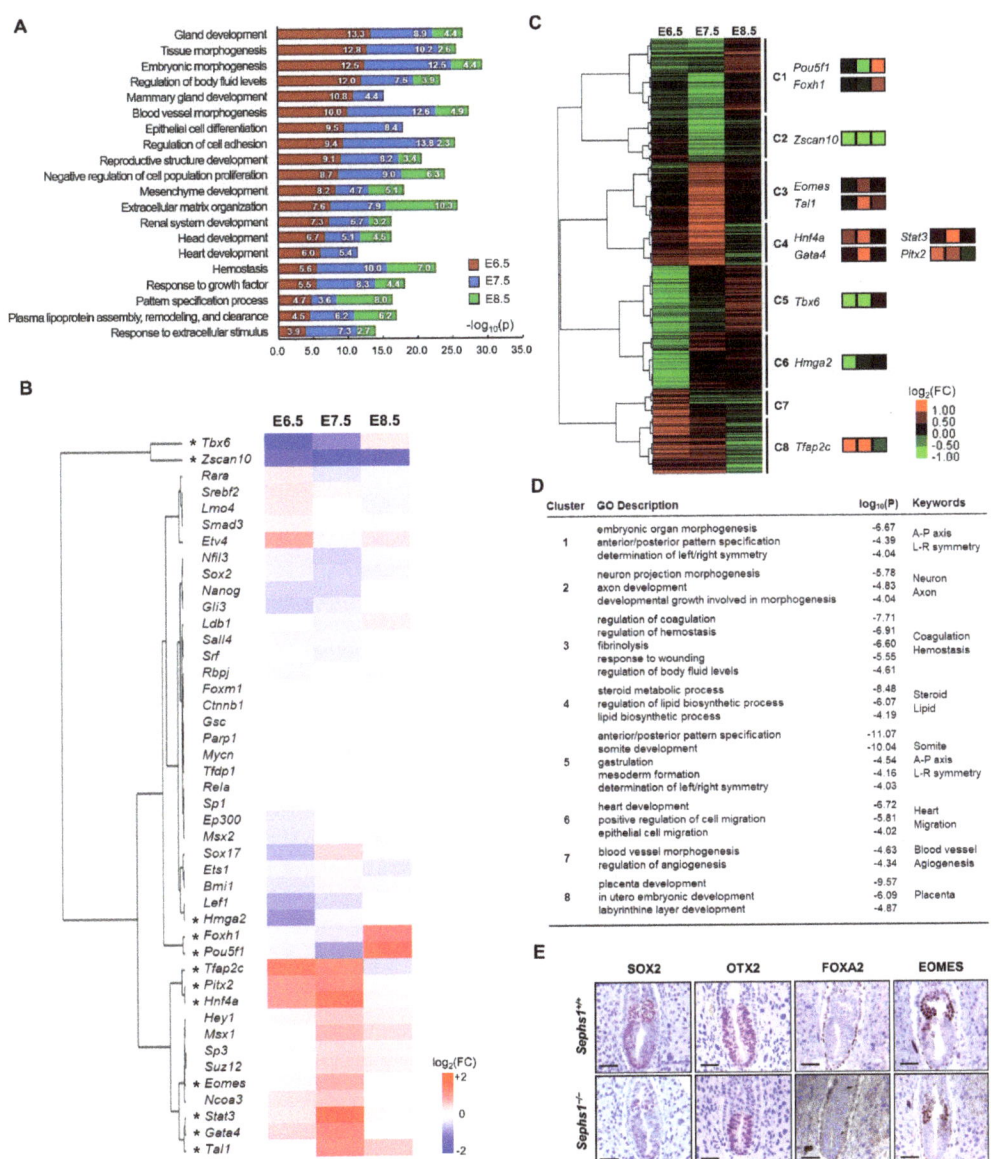

Figure 2. Pathway analysis of *Sephs1*$^{-/-}$ embryo and transcription factor prediction. (**A**) Metascape pathway enrichment analysis. Each gene ontology group was integrated into summary terms according to the significance. $p < 0.01$. (**B**) Transcription factors that regulate gastrulation-related DEGs in the *Sephs1*$^{-/-}$ embryo. Fold-change of transcription factor between wild-type and *Sephs1*$^{-/-}$ embryos was illustrated by heat map. * indicates DEGs. (**C**) Hierarchical clustering of DEGs according to their fold-change. The distance between DEGs was calculated via Pearson correlation, and clustered by the pairwise complete linkage method. Clusters were defined at the third branch from the base trunk. (**D**) GO analysis using clustered DEG sets. Cut off at $p < 0.001$. (**E**) Expression patterns of dermal layer marker genes at E6.5. Immunohistochemistry was performed with SOX2, OTX2, FOXA2, and EOMES as described in Materials and Methods. Scale bars represent 100 μm.

Since dermal layers of the embryos at the early gastrula stage determine the cell lineage leading to tissue differentiation, the gene expression pattern and/or morphological feature of each dermal layer are important to organ development at the following stages. We examined the expression levels and locations of dermal layer markers (*Sox2*, *Otx2*, *Foxa2*, and *Eomes*) selected from the transcription factors shown in Figure 2B.

Sex determining region Y-box 2 (*Sox2*) encodes a transcription factor that is essential for maintaining pluripotency of undifferentiated embryonic stem cells, but is also known to be expressed specifically in the ectoderm of both embryonic and extraembryonic regions during the gastrulation stage [29]. As shown in Figure 2E, the levels of *Sox2* expression were significantly decreased in the $Sephs1^{-/-}$ embryo compared to wild-type ($\log_2$(Fold Change) at E6.5 = −0.42). Orthodenticle homeobox 2 (*Otx2*), which encodes a member of the bicoid subfamily of homeodomain-containing transcription factors, is an embryonic mesoderm-specific gene that plays a key role in nervous system development [30]. The area in which *Otx2* was expressed was reduced to the bottom half of the $Sephs1^{-/-}$ embryo, while the size of the embryo was similar with that of wild-type. The fact that the structure of mesodermal layer was changed by SEPHS1-deficiency suggests that the development of connective tissues such as blood, blood vessels, muscles, and heart will proceed abnormally in the $Sephs1^{-/-}$ embryos, because these tissues are differentiated from the mesodermal lineage cells. Notably, the expression level of *Otx2* was not changed by *Sephs1* knockout ($\log_2$(Fold Change) at E6.5 = −0.22). Forkhead box protein A2 (*Foxa2*) encodes a protein belonging to a subfamily of the Forkhead box transcription factors and is an endoderm-specific gene [31]. There was no difference in expression levels of *Foxa2* and location of expression between wild-type and the $Sephs1^{-/-}$ embryo. We observed that only the rate of differentiation of the endoderm-originated organs, such as the digestive and respiratory systems, was retarded and these organs appeared to be at the E9.5 stage in the $Sephs1^{-/-}$ embryo. It appears that SEPHS1-deficiency does not affect the morphology of endodermal lineage tissues during development. Eomesodermin (*Eomes*), also referred to as T-box brain protein 2 (*Tbr2*), is a member of the T-box family of transcription factors initially expressed in the extraembryonic ectoderm and is known to play an important role in anterior visceral endoderm formation and epithelial-mesenchymal transition (EMT) [32,33]. In the wild-type embryo, *Eomes* was expressed throughout the extraembryonic ectoderm. However, in the $Sephs1^{-/-}$ embryo, the expression area was limited to the bottom of the extraembryonic ectoderm region, wherein the expression levels were slightly decreased ($\log_2$(Fold Change) at E6.5 = −0.15), suggesting that extraembryonic lineage tissues, such as the yolk sac and placenta, may differentiate abnormally after E6.5. In conclusion, expression levels and locations of the dermal layer marker genes provide evidence that knockout of *Sephs1* causes morphological changes as well as the retardation of development during dermal layer formation.

2.3. Morphological Changes in the Embryonic Region by Sephs1 Knockout

We confirmed the developmental abnormalities predicted in Figure 2 by examining morphological changes in the $Sephs1^{-/-}$ embryos using X-ray microscopy (XRM) technology (Figure 3). The central nervous system of the $Sephs1^{-/-}$ embryo manifested retarded development and abnormal shape compared to those of the wild-type embryo. At E7.75, head fold (hf) was found in the wild-type embryo (panel (a) of Figure 3A) but not in the $Sephs1^{-/-}$ embryo (panel (c) in Figure 3A). At E8.5 and E9.5, although three primary brain vesicles (prosencephalon (pro) and mesencephalon (ms) and rhombencephalon (rho)) were formed, the size of brain tissue in the $Sephs1^{-/-}$ embryo was much smaller than that in wild-type, and the neural groove between prosencephalon (pro) and mesencephalon (ms) was not closed (arrowheads in Figure 3B). In the transverse section of the cervical region in E8.5, the closures of the neural grooves in wild-type embryos were progressing, whereas in the $Sephs1^{-/-}$ embryo, it was open to the outside (Figure 3C). The neural plate (blue) was closed around somites (yellow) in the wild-type embryo (Video S1), but remained open in the $Sephs1^{-/-}$ embryo (Video S2). In addition, the development of somites was delayed. Segmentation of somites was clearly observed in the wild-type embryo at E8.5 and E9.5, while somite segmentation did not occur in the $Sephs1^{-/-}$ embryo until E9.5 (see yellow-colored regions in Figure 3B). Segmented somites (yellow) were straight and parallel with the neural plate (blue), and clearly distinguished from the neural plate (blue) in wild type at E9.5 (Video S3). On the other hand, somites (yellow) in the *Sephs1*$^{-/-}$ embryo were severely twisted and had ambiguous boundaries (Video S4).

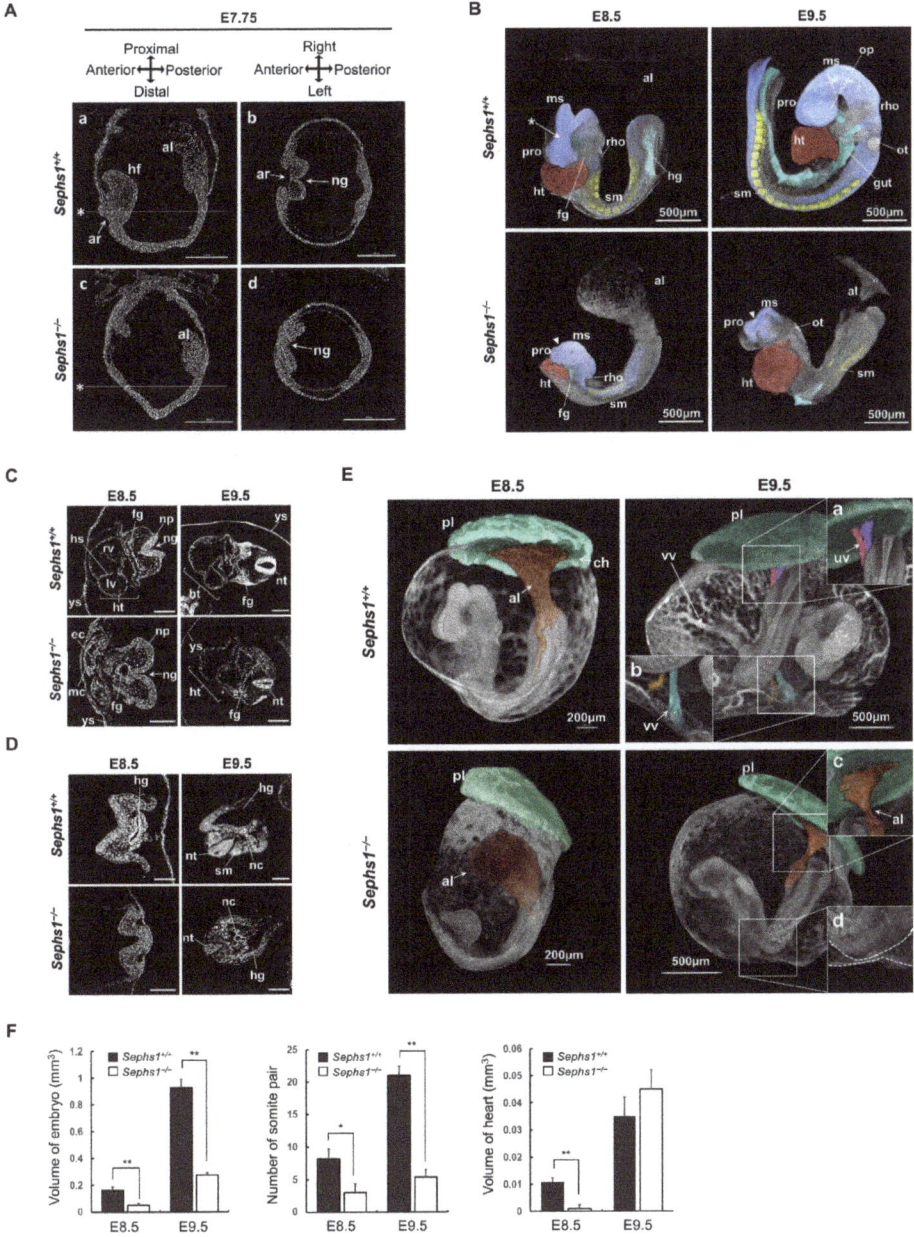

Figure 3. Embryonic defects in *Sephs1*$^{-/-}$ embryos. (**A**) Virtual view of embryos cut in sagittal (**a,c**) and transverse (**b,d**) directions at E7.75. Transverse views shown in b and d were generated at positions marked with an asterisk in a and c, respectively. Scale bars represent 200 μm. (**B**) 3-dimensional reconstruction of XRM data. Each organ was indicated by a different color using the Region of Interest tool in Dragonfly software (red, heart; yellow, somite; blue, neural plate; light

blue, gut). (**C,D**) Transverse section of XRM data. The upper embryonic body (**C**) and the lower half of the embryonic body (**D**) were virtually sectioned. Scale bars represent 200 μm. (**E**) Comparison of detailed structure of allantois. The junction sites of the umbilical vein and vitelline vein are more magnified in the box. (**F**) Statistical analysis of 3-dimensional reconstruction. Volume was calculated by Dragonfly, ORS. * indicates $p < 0.05$. ** indicates $p < 0.01$. orange, allantois; green, chorion; pink and purple, umbilical vessel; blue and yellow, vitelline vessel. al, allantois; ar, archenteron; ch, chorion; ec, endocardium; fg, foregut; hg, hindgut; hs, heart sectum; ht, heart; mc, myocardium; lv, left ventricle; ms, mesencephalon; ng, neural groove; np, neural plate; nt, neural tube; ot, otic vesicle; pl, placenta; pro, prosencephalon; rho, rhombencephalon; rv, right ventricle; sm, somite; uv, umbilical vessel; vv, vitelline vessel; and ys, yolk sac.

Heart development was also inhibited by SEPHS1-deficiency. As shown in Figure 3B, the differentiation of heart (ht) in the $Sephs1^{-/-}$ embryo at E8.5 proceeded slowly compared to wild-type. Heart structure of $Sephs1^{-/-}$ embryo remained in the cardiac crescent stage, while that in wild-type embryo developed into the heart tube stage showing ventricles and heart septum at E8.5 (Figure 3C). Notably, the size of the heart in both wild-type- and $Sephs1^{-/-}$ embryos was similar at E9.5, but the cell density was significantly reduced and the morphology of the heart was irregular in $Sephs1^{-/-}$ embryos (compare $Sephs1^{+/+}$ with $Sephs1^{-/-}$ of E9.5 in Figure 3C).

The archenteron (ar), which differentiates into foregut (fg) was not observed in the $Sepsh1^{-/-}$ embryo at E7.75 (Figure 3A), but observed at E8.5 and E9.5 (Figure 3C). Hindgut (hg) development seemed to proceed more slowly than foregut formation in the $Sephs1^{-/-}$ embryo. Hindgut formation was observed only at E9.5 in the $Sephs1^{-/-}$ embryo (Figure 3D). These data suggest that gut formation was not significantly affected by $Sephs1$ knockout and this phenomenon is consistent with the result of FOXA2 expression patterns in Figure 2E.

The process that connects allantois (al) with the chorion (ch) was completed in wild-type at E8.5, but al and ch were not connected in the $Sephs1^{-/-}$ embryo at E8.5 (Figure 3E). At E9.5, umbilical (uv) and vitelline vessels (vv, insets (a) and (b) in Figure 3E) were formed in wild-type, but not in the $Sephs1^{-/-}$ embryo (insets (c) and (d) in Figure 3E). The fact that the formation of vv and uv is inhibited by SEPHS1-deficiency suggests that SEPHS1 plays an essential role in the transport of substances between mother and embryo.

In addition, the total volume of the embryo was measured from the 3-dimensional structure obtained by reconstructing XRM images using a volume calculation software (Dragonfly). The volume ratio between the $Sephs1^{-/-}$ embryo and the wild-type decreased by approximately 3.2 times and 3.4 times at E8.5 and E9.5, respectively (Figure 3F). At E8.5, the number of somites (sm) in $Sephs1^{-/-}$ and wild-type embryos was 3 and 8.2, respectively. The gap in the somite number between $Sephs1^{-/-}$ embryo and wild-type embryos was increased at E9.5 (5.3 somites in the $Sephs1^{-/-}$ embryo and 21 somites in the wild-type embryo). The heart volume of the wild-type embryo at E8.5 was 10.1 times greater than that of the $Sephs1^{-/-}$ embryo. However, at E9.5, the heart volume in the $Sephs1^{-/-}$ embryo became abnormally enlarged (Figure 3C). Unlike the wild-type embryo, the $Sephs1^{-/-}$ embryo did not exhibit axial rotation at E9.5 and the embryonic axis was oriented in the same direction as at E8.5 (compare the E9.5 wild-type with the E9.5 $Sephs1^{-/-}$ embryos in Figure 3B). No further axial rotation was observed at E10.5 (data not shown) suggesting that axial rotation stopped at E8.5.

2.4. Pattern Analysis of DEGs Expressed in Extraembryonic Region

Since biological processes (for example, 'placenta development' in Figure 2D) occurring in the extraembryonic region were also predicted, we hypothesized that development of the extraembryonic region also would proceed abnormally by SEPHS1 deficiency. To test if there is any abnormality of development in the extraembryonic region in the $Sephs1^{-/-}$ embryo, pathways involved in the development of extraembryonic region were identified and the pathway-related morphological changes were examined. We first selected DEGs among targets of extraembryonic region-specific DTFGs listed in Figure 2C. Four DTFGs expressed in the extraembryonic region and 70 of their target DEGs were identified. GO analysis was performed for the identified genes (Figure 4A). As a result, four pathways (Retinoid metabolism and transport (RMT), Vitamin transport (VT), Embryonic morphogenesis (EM), and Epithelial cell differentiation (ECD)) were predicted with high probability ($\log_{10}(p) < -4.9$).

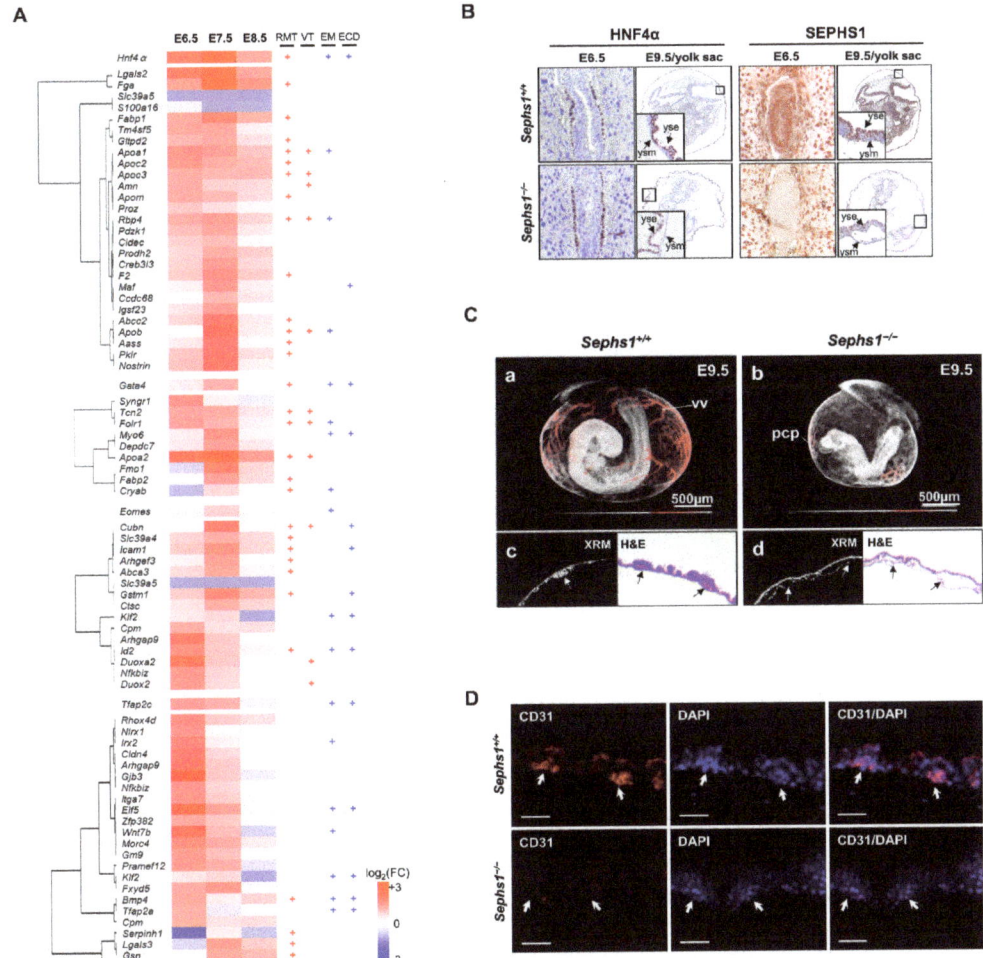

Figure 4. Expression patterns of transcription factors and its target DEGs in extraembryonic region and extraembryonic defects in *Sephs1*$^{-/-}$ embryo. (**A**) Expression pattern of *Hnf4α*, *Gata4*, *Eomes*, *Tfap2c*, and their target genes. Target genes obtained from the Chip-Atlas were filtered by expression specificity to the extraembryonic region. Among these genes, only DEGs were selected and subjected to hierarchical clustering. Target gene sets were subjected to GO analysis. On the right of the heat map, the GO term(s) assigned by each gene was (were) marked with '+'. RMT, Retinoid metabolism and transport; VIT, Vitamin transport; EM, Embryo morphogenesis; ECD, Epithelial cell differentiation. (**B**) Expression patterns of HNF4α and SEPHS1. Immunohistochemistry was performed against HNF4α and SEPHS1 as described in Materials and Methods. Small box was enlarged to inbox showing the yolk sac layer. (**C**) 3D structure of E9.5 embryos reconstructed from XRM images as described in Materials and Methods (**a**,**b**). The virtual section (XRM) and H&E staining image of paraffin section at the yolk sac (**c**,**d**). Arrow indicates hemato-endothelial progenitors. Scale bars represent 500 μm. (**D**) CD31 expression on yolk sac at E9.5. The images were acquired with a confocal microscope. Scale bars represent 50 μm. pcp, primary capillary plexus; vv, vitelline vessel; yse, and yolk sac ensoderm; ysm, yolk sac mesoderm.

Hnf4α and target genes are predicted to participate mainly in 'Retinoid metabolism and transport' and 'Vitamin transport' (RMT and VT in Figure 4A). Most of the genes in 'Vitamin transport' are included in 'Retinoid metabolism and transport' except *Duox2*, *Duoxa2*, *Cubn*, and *Amn*. *Duox2* is target of retinoic acid signaling and *Duoxa2* is a maturation factor of *Duox2* [34]. *Cubn*, which is the cobalamin (vitamin B12) receptor gene, has an inhibitory function against retinoic acid signaling and *Amn* encodes the protein that facilitates uptake of vitamin B12. Therefore, genes in 'Vitamin transport' are related with retinoic acid signaling and can be categorized into the same group with 'Retinoid metabolism and transport'. It should be noted that GO analysis using total DEGs by IPA also predicted the retinoic acid signaling with the highest probability (Table S2). Retinoic acid is a morphogen participating in axis formation, embryo growth, and cell fate determination during early embryogenesis [35]. In addition, *Hnf4α* and its target genes are expressed in extraembryonic endoderm which develops into the yolk sac at a later stage suggesting that HNF4α plays an essential role in yolk sac development and its function via retinoic acid signaling.

Gata4 and *Eomes*, and their target genes did not show enrichment in any specific pathway. Instead, they commonly participate in the pathways regulated by both *Hnf4α* and *Tfap2c*. It should be noted that *Gata4* is an activator of *Hnf4α* and is expressed more widely than *Hnf4α* [36].

Tfap2c and target genes are predicted to participate mainly in embryo morphogenesis and epithelial cell differentiation (EM and ECD in Figure 4A). During organogenesis, embryonic morphogenesis and epithelial cell differentiation should occur together. Therefore, embryo morphogenesis and epithelial cell differentiation can be categorized into organogenesis. In addition, both embryo morphogenesis and epithelial cell differentiation share BMP4 as a key protein suggesting BMP4 is used as a common morphogen for embryo morphogenesis and epithelial cell differentiation. Recently, it was reported that *Tfap2c*, which is expressed in extraembryonic ectoderm, plays an essential role in the development of trophoblast and placenta [37]. Chorioallantoic fusion is a process in trophoblast differentiation. As described in the previous section, we found chorioallantoic fusion did not occur in the $Sephs1^{-/-}$ embryo (Figure 3E) suggesting that *Tfap2c* and its target genes inhibit chorioallantoic fusion in the SEPHS1-deficient embryo.

Since *Hnf4α* is known to be expressed in extraembryonic endoderm, we examined the region where this gene was expressed during gastrulation. At E6.5, HNF4α was distributed more widely in the extraembryonic region of the $Sephs1^{-/-}$ embryo than that of wild-type, and at E9.5, the cell density of HNF4α-expressing yolk sac endoderm was lower in the $Sephs1^{-/-}$ embryo than that of the wild-type (Figure 4B). The yolk sac endoderm morphology was concavo-convex in the wild-type but became flat in the $Sephs1^{-/-}$ embryo. Another significant difference in the structure of the yolk sac between wild-type and $Sephs1^{-/-}$ embryo was that the yolk sac endoderm and mesoderm were separated in the $Sephs1^{-/-}$ embryo at E9.5. XRM image showed that vitelline vessels (vv), which are normally generated in the yolk sac at E9.5 stage, were not found in the $Sephs1^{-/-}$ embryo (compare panels (a) and (b) of Figure 4C). Separation of the yolk sac endoderm and mesoderm was also observed in the $Sephs1^{-/-}$ embryo (panel (d) of Figure 4C). At the same time point, expression of CD31, which is known as a hemato-endothelial progenitor marker [38], was not detected in the $Sephs1^{-/-}$ embryo, while it was detected in the wild-type embryo suggesting that both vitelline vessel and blood progenitors were not formed due to SEPHS1 deficiency (Figure 4D).

It should be noted that SEPHS1, which is known to be expressed in all tissues, was expressed only in yolk sac endoderm, not yolk sac mesoderm in the wild-type embryo (Figure 4B) suggesting that SEPHS1 is expressed cell-type specifically in the same tissue and its deficiency affects mainly the function of yolk sac endoderm at late gastrulation.

2.5. Pathway Prediction through Protein–Protein Interaction and Gene Set Enrichment Analysis

In order to identify the signaling pathway and molecular mechanism that participate in gastrulation, additional analyses were performed. By analyzing protein–protein interaction

(PPI) of DEGs at E6.5, we could obtain three modules with high probability ($\log_{10}(p) < -9.5$); Wnt signaling pathway, Glutathione metabolism, and the post-translational protein phosphorylation pathway (Figure 5A).

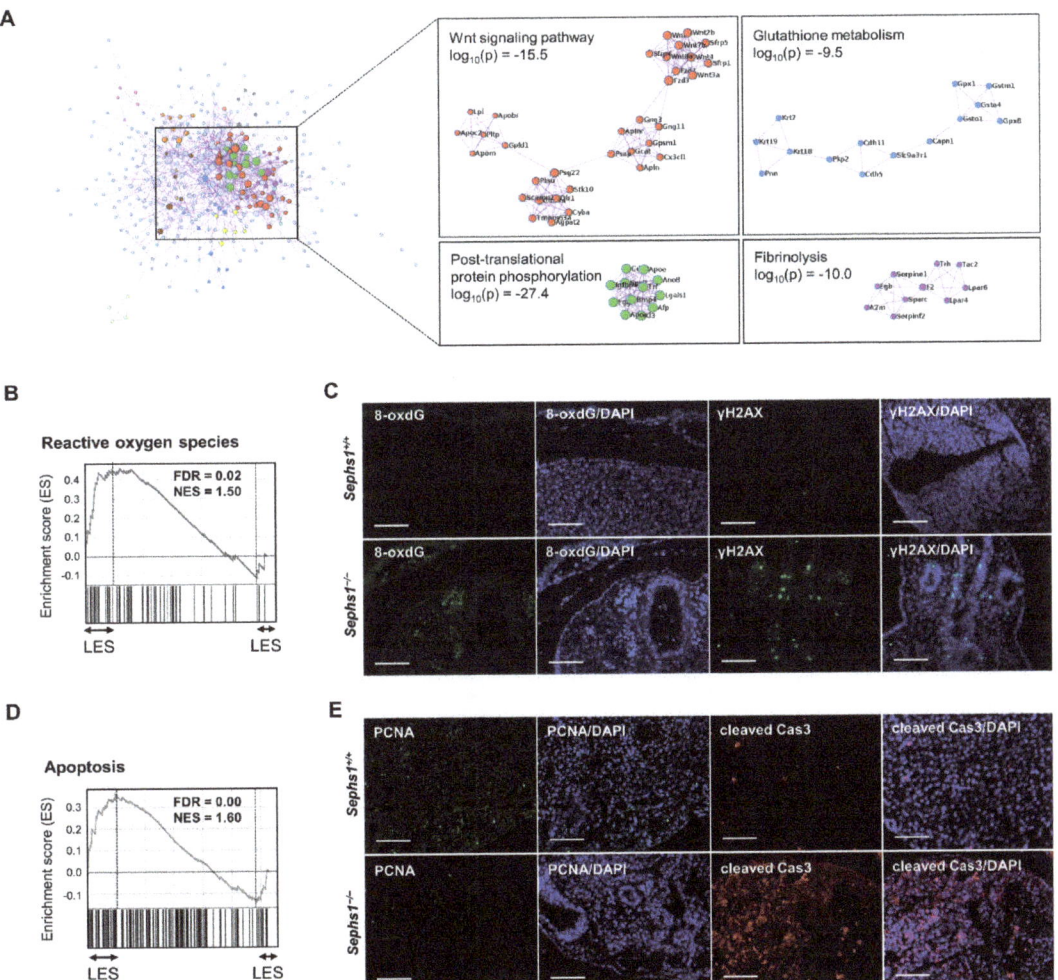

Figure 5. *Sephs1*$^{-/-}$ embryo shows growth retardation, apoptosis and ROS stress. (**A**) Protein–protein interaction map at E6.5 using PPI analysis tool plugged-in Metascape. (**B**) Enrichment score plot of Reactive oxygen species obtained by gene set enrichment analysis (GSEA). The ranked list was created by sorting the gene list according to fold-change. LES, Leading edge subset. (**C**) Immunohistochemistry of 8-oxo-guanine and γH2AX. (**D**) Enrichment score plot of Apoptosis obtained by GSEA. The ranked list was created as described above. (**E**) Protein level of PCNA and cleaved caspase 3. Immunohistochemistry was performed as described in Materials and Methods. The images were acquired with an Axiovert 200M inverted microscope. Scale bars represent 100 μm.

Wnt signaling regulates mainly morphogenesis, such as growth in the early embryo, axis formation, and pattern formation [39]. Wnt signaling was inhibited in *Sephs1*$^{-/-}$ embryos at E6.5, suggesting that pathways determining embryonic morphology were affected by *Sephs1* knockout before the organogenesis stage. Glutathione is one of the molecules that plays an essential role in maintaining redox homeostasis. Imbalance in glutathione metabolism by *Sephs1* knockout leads to the accumulation of ROS that causes oxidative stress in the cell [19]. Prediction of post-translational protein phosphorylation with high probability suggests that signaling pathways are actively regulated, since almost all signaling pathways are regulated through phosphorylation of proteins participating in each cognate pathway. These results of PPI analysis suggest that SEPHS1 deficiency causes oxidative stress by disrupting redox homeostasis, and that the oxidative stress will primarily affect Wnt signaling.

Besides GO analysis, another useful method for pathway prediction is Gene Set Enrichment Analysis (GSEA). Because GO analysis uses only a gene list to predict pathways or processes without giving weight according to fold change value of each gene, we cannot determine to what extent a specific gene contributes to those pathways or processes. The advantage of GO analysis is that one can identify all possible pathways or processes enriched by DEGs. On the other hand, with GSEA methods, we can determine how a specific gene contributes to a specific pathway or process, since DEGs are ranked according to their fold change value and the ranked DEGs are used to calculate enrichment score for a pathway or process of interest [40].

GSEA with total genes using the Hallmark Gene Sets from the Molecular Signatures Database revealed that 'Reactive Oxygen Species' was predicted with high significance where normalized enrichment score (NES) and false discovery rate (FDR) were 1.50 and 0.02, respectively (Figure 5B). There are 28 genes within the leading-edge subset (LES) of 'Reactive Oxygen Species' (Table S3). The genes include *Glrx1*, *Prdx2*, *Prdx6*, *Txnrd1*, *Txnrd2*, and *Nqo1*, and these genes participate in redox homeostasis or ROS generation. We then examined the gene number within LES and NES of 'Reactive Oxygen Species' in more detail at each embryonic day. The gene numbers within LES were 25, 27, and 31 at E6.5, E7.5, and E8.5, respectively (Table S3). In addition, the NES of 'Reactive Oxygen Species' at E6.5, E7.5, and E8.5 were 0.56, 1.0, and 1.42, respectively (Figure S2B). These results suggest that ROS levels in the embryo are gradually increased. DNA damage, such as formation of γH2AX and 8-oxoguanine, is commonly used as an oxidative stress marker.

In order to determine whether DNA damage due to oxidative stress occurred in the *Sephs1* knockout, immunohistochemistry (IHC) using an antibody against 8-oxoguanine and γH2AX (phosphorylated H2AX) was performed. Neither formation of γH2AX nor of 8-oxoguanine were detected from E6.5 to E8.5 (Figure S2C) in $Sephs1^{+/-}$ or $Sephs1^{-/-}$ embryos. However, both 8-oxoguanine and γH2AX were generated only in the $Sephs1^{-/-}$ embryo at E9.5 (Figure 5C). These results indicate that oxidative stress was too mild to cause DNA damage in $Sephs1^{-/-}$ embryos by E8.5, but was strong enough to cause DNA damage at E9.5.

In addition to 'Reactive Oxygen Species', 'Apoptosis' was also predicted with high significance where NES and FDR were 1.6 and 0.00, respectively (Figure 5D). The leading-edge subset of 'Apoptosis' consists of 63 genes (Table S4) including *Sqstm1*, *Pdcd4*, *Smad7*, *Hspb1*, *Faslg*, *Bax*, *Madd*, *Bmf*, and *Hgf*, which are known to participate in apoptotic signaling. The gene numbers within LES and the NESs were gradually increased as embryogenesis proceeded (Figure S2B and Table S4). Unexpectedly, we could not detect the activated form of caspase-3 in SEPHS1-deficient embryos until E8 (Figure S2D). However, the activated form of caspase-3 was detected in the $Sephs1^{-/-}$ embryo at E9.5, suggesting that SEPHS1 deficiency induced cell death through apoptosis in the embryo (Figure 5D). Notably, genes such as *Gadd45b*, *cdkn1a*, and *Btg2*, which are related to cell proliferation, were also included in the apoptosis pathway. Therefore, we assumed that cell proliferation was also inhibited by *Sephs1* knockout. PCNA is used as the most reliable marker for evaluating cell proliferation. As shown in Figure 5E, the level of PCNA was significantly decreased in the $Sephs1^{-/-}$ embryo. These results strongly suggest that SEPHS1-deficiency causes inhibition of cell proliferation followed by apoptosis.

3. Discussion

SEPHS1 regulates various cellular functions such as redox homeostasis, and thus is known to be essential for embryo survival and growth. However, the detailed mechanisms of which genes or pathways are controlled by *Sephs1* during development and how they result in embryonic lethality have not been fully elucidated. Through transcriptome analysis, the pathways that are responsible for morphological defects in *Sephs1*$^{-/-}$ embryo were predicted. To predict pathways affected by *Sephs1* knockout, a combination of various bioinformatic tools (AmiGO and Metascape analysis, IPA analysis and GSEA) with various databases for protein–protein interaction (String and BioGrid), transcription factor (TRRUST and ChIP-Atlas) and cell type (Mouse Gastrulation Atlas and MGI) were applied.

Bioinformatic analyses suggested several interesting features of the effect of SEPHS1-deficiency on the development at early embryonal stages. Throughout the gastrulation stage, coagulation was predicted as the most highly affected pathway ($\log_{10}(p) = -11.9$). Genes included in this GO term are implicated in cell migration and adhesion which are the most prominent features of cells involved in the dermal layer and in axis formation [41]. Among signaling pathways participating in gastrulation, retinoic acid signaling was predicted to be activated throughout gastrulation with high significance ($\log_{10}(p) = -8.9$). Retinoic acid is a morphogen, and its signaling pathway is known to regulate axis formation and cell fate determination [35]. The upregulation of *Rbp4*, which mediates the transport of retinoic acid into the cell suggests that the levels of intracellular retinoic acids were increased by *Sephs1* knockout. In addition, we found that the targets of retinoic acid pathway were also increased in their expression levels, suggesting that retinoic acid signaling was hyperactivated. Hyperactivation of retinoic acid signaling, due to imbalance of retinoic acid concentration in the cells, may adversely affect embryonal axis formation and cell fate determination. We observed various abnormalities during gastrulation and early organogenesis of SEPHS1-deficient embryo, including the absence of axial rotation.

PPI analysis produced other interesting findings. Wnt, prolactin and insulin-like growth factor (IGF) signaling pathways were predicted to be affected by *Sephs1* knockout at E6.5, E7.5, and E8.5, respectively (Figure 5A and Figure S2A). Wnt signaling is known to be involved in regulation of growth in early embryo development, axis formation, and pattern formation [39,42]. Since the expression of *Wnt3a*, *Wnt2b*, *Wnt8a*, and *Fzd3* were significantly decreased in *Sephs1*$^{-/-}$ embryo at E6.5, Wnt signaling seemed to be inhibited. Among the direct targets of Wnt signaling listed in the database 'the Wnt' [43], the expression of 85% of the targets was down-regulated in the *Sephs1*$^{-/-}$ embryos. At E6.5, *Sephs1*$^{-/-}$ embryos showed changes in the expression level and location of dermal layer markers, such as *Sox*, *Otx2*, and *Eomes*, which may lead to the retardation of embryonic growth and axis formation at E7.5 (see Figures 1A and 2E). Prolactin is known to play an important role in the trophoblast growth and the development and differentiation of neural crest cells in neurulation stage, but its role in the gastrulation stage has not been clearly identified [44]. Since the expression of *prl2c2*, *prl2c3*, *prl4a1*, and *prl5a1* was increased in the *Sephs1*$^{-/-}$ embryos, we hypothesized that prolactin signaling was hyperactivated. Among the target genes of the prolactin pathway in KEGG DB (map04917; Prolactin signaling pathway), the expression of *Socs3*, *Elf5*, *Prlr*, and *Slc2a2* were increased by more than 1.5-fold in the *Sephs1*$^{-/-}$ embryos at E7.5. Interestingly, although not at E7.5, abnormal development of the central nervous system was observed at E8.5 (Figure 3B,C). One of the important roles of prolactin is to control the development of the nervous system [44]. IGF signaling is known to be involved in promoting growth of embryos and organ development [45]. The expression of IGFBP3 and other proteins including proteases that binds to IGFBP3 were up-regulated in the *Sephs1*$^{-/-}$ embryos (Figure S2A). Binding of IGF to IGFBP3 and binding of IGFBP3 to proteases inhibit IGF signaling [46]. Therefore, SEPHS1-deficiency seems to inhibit IGF signaling during late gastrulation and early neurulation. The *Sephs1*$^{-/-}$ embryos showed growth retardation, structural brain abnormalities and disrupted cardiac development at E8.5 and these phenotypes are consistent with those described in another study [45]. These results suggest that SEPHS1-deficiency affects embryo morphogenesis by regulating Wnt and retinoic acid signaling and then organogenesis by regulating retinoic acid, prolactin, and IGF signal pathways in combination.

It should be noted that most of the genes responsible for reception of retinoic acid signaling (apolipoprotein and retinoic acid binding protein genes) are expressed in the extraembryonic region specifically, but its target genes are expressed both in the embryonic and extraembryonic regions. This finding is consistent with the experimental results. For example, both vitelline vessel formation and chorioallantoic fusion did not occur in the $Sephs1^{-/-}$ embryo (Figures 3 and 4). Abnormalities in the yolk sac and placental development will inhibit nutrient uptake and waste disposal from the embryo.

Taking the results from both the bioinformatic analyses and experimental evidence into consideration, we propose the role of SEPHS1 during early mouse embryogenesis as follows (Figure 6). SEPHS1 regulates intracellular ROS levels through regulating the redox homeostasis system. The deficiency of SEPHS1 causes disruption of the redox homeostasis system and leads to a gradual increase in ROS levels. Low levels of ROS will cause mild oxidative stress, leading to abnormalities in signaling pathways. During early gastrulation, abnormalities in embryonic morphogenesis, such as dermal layer and axis formation, occur presumably through Wnt and retinoic acid signaling. Abnormalities in organogenesis occur presumably through prolactin and retinoic acid signaling, followed by insulin-like growth hormone and retinoic acid signaling during mid- and late-gastrulation, respectively. ROS accumulate sufficiently to cause cell death through DNA damage at E9.5, followed by embryonic death. Dead embryos then undergo resorption by E14.5 [19]. Since the signaling pathways were predicted using bioinformatic tools, future molecular studies will be needed to further validate our results.

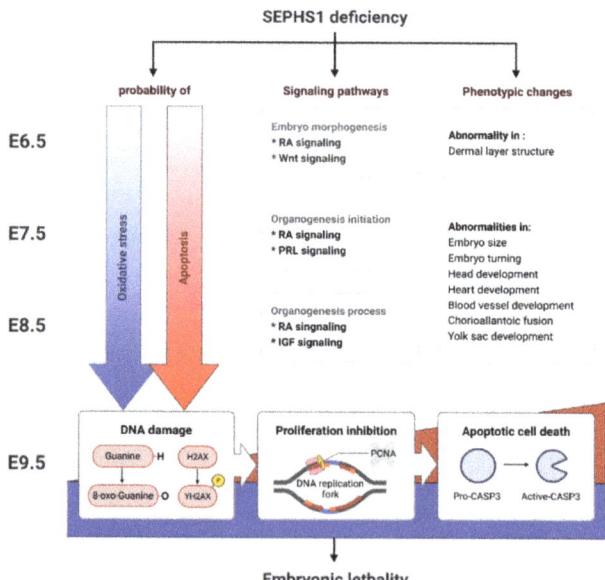

Figure 6. Schematic diagram of effect of SEPHS1 deficiency on early embryogenesis.

In this study, we elucidated how SEPHS1-deficiency induces developmental abnormality and embryonic lethality. Our results provide evidence that SEPHS1-deficiency is one of the contributors of natural miscarriages, and that the levels of SEPHS1 can be used as a marker for diagnosis or prognosis of natural miscarriages.

4. Materials and Methods

4.1. Materials

Antibodies against SEPHS1, γH2AX, 8-oxo-guanine, POU5F1, CFL488-conjugated mouse IgG, and CFL488-conjugated rabbit IgG were purchased from Santa Cruz Biotech

(Dallas, TX, USA). Antibodies against CD31, EOMES, FOXA2, HNF4α, cleaved caspase3, PCNA and cy3 conjugated rabbit IgG, AEC substrate, and Mayer's modified hematoxylin were purchased from Abcam (Cambridge, UK). Antibody against SOX2 and OTX2 were purchased from R&D Systems (Minneapolis, MN, USA). Biotin-conjugated antibody against mouse IgG and rabbit IgG were purchased from Jackson Immunoresearch (West Grove, PA, USA). MG^{TM} Tissue SV kit was purchased from MG Med (Seoul, Korea). Trizol^{TM} reagent, DNase I, PowerUp^{TM} SYBR^{TM} Green Master Mix were purchased from Thermo Fisher (Waltham, MA, USA). MMLV-RT was purchased from Enzynomics (Daejeon, Korea). TruSeq RNA Sample Prep Kit v2 was purchased from Illumina (San Diego, CA, USA). Lugol's solution, streptavidin-HRP, DAPI and paraformaldehyde were purchased from Sigma-Aldrich (Burlington, MA, USA). Slide glasses coated with 3-aminopropyl triethoxysilane was purchased from Matsunami Glass (Osaka, Japan).

Programs and their websites are as follows:

AmiGO [47]: http://amigo.geneontology.org/amigo

Ballgown [48], DESeq2 [49] and edgeR [50]: plugged in R: https://www.r-project.org/

BioRender (Toronto, Ontario, Canada): Figure 6 was created with https://biorender.com/

ChIP-Atlas [51]: https://chip-atlas.org/

Dragonfly (ORS, Montreal, Quebec, Canada): https://www.theobjects.com/index.html

FastQC (Babraham Bioinformatics, Babraham Institute, Cambridge, UK): https://www.bioinformatics.babraham.ac.uk/projects/fastqc

GenePattern 2.0. [52]: https://www.genepattern.org/

Gene Set Enrichment Analysis [40]: https://www.gsea-msigdb.org/gsea/index.jsp

HISAT2 [53]: http://daehwankimlab.github.io/hisat2/

Ingenuity Pathway Analysis (QIAGEN, Germantown, MD, USA): http://www.ingenuity.com

Metascape [54]: https://metascape.org/gp/index.html

Mouse Gastrulation Atlas [55]: https://marionilab.cruk.cam.ac.uk/MouseGastrulation2018/

Mouse Genome Informatics (MGI) [56]: http://www.informatics.jax.org/index.shtml

MsigDB [57]: https://www.gsea-msigdb.org/gsea/msigdb/

PANTHER [58]: http://www.pantherdb.org/

PRISM (GraphPad Software, San Diego, CA, USA): https://www.graphpad.com/scientific-software/prism/

StringTie [59]: https://ccb.jhu.edu/software/stringtie/

TreeView [60]: http://jtreeview.sourceforge.net/

Trim Galore (Babraham Bioinformatics, Babraham Institute, Cambridge, UK): https://www.bioinformatics.babraham.ac.uk/projects/trim_galore/

TRRUST v2 [61]: https://www.grnpedia.org/trrust/

4.2. Generation of SEPHS1 Total Knockout (Sephs1$^{-/-}$) Mice

All mice used in this study were on a C57BL/6J background. The generation of SEPHS1 total knockout (Sephs1$^{-/-}$) mice have been described previously [19]. Female Sephs1$^{+/-}$ mice were crossed with male Sephs1$^{+/-}$ mice, and then dissected to obtain Sephs1$^{-/-}$ embryos. Embryonic day (days post-coitus) E0.5 was defined as noon on the day that a mating plug was detected. We used 32 embryos (25 Sephs1$^{+/-}$ and 7 Sephs1$^{-/-}$), 24 embryos (17 Sephs1$^{+/-}$ and 7 Sephs1$^{-/-}$), 34 embryos (26 Sephs1$^{+/-}$ and 8 Sephs1$^{-/-}$), 20 embryos (16 Sephs1$^{+/-}$ and 4 Sephs1$^{-/-}$), 17 embryos (14 Sephs1$^{+/-}$ and 3 Sephs1$^{-/-}$), and 10 embryos (8 Sephs1$^{+/-}$ and 2 Sephs1$^{-/-}$) for morphological observation and histological analysis at E6.5, E7.5, E8.5, E9.5, E10.5 and E11.5, respectively. All procedures performed involving the mice were conducted in accordance with the Institutional Guidelines of the Institute of Laboratory Animal Resources (Seoul National University, Seoul, Korea). All mouse experiments were approved by the Institutional Animal Care and Use Committee at Seoul National University. All mice used in this study were bred and cared for under sterile conditions with constant temperature and humidity in the specific pathogen free animal facility at Seoul National University.

4.3. Genotyping

For genotyping using DNA, genomic DNA was extracted using the MG Genomic DNA Purification kit following the manufacturer's instruction and then subjected to PCR. The wild-type (WT) allele was amplified by gWT genotyping primer set and the knockout (KO) allele was amplified by gKO genotyping primer set. For genotyping using RNA, total RNA was extracted with Trizol^{TM} reagent following the manufacturer's instruction. Total RNA was reverse-transcribed by MMLV-RT and then cDNA was subjected to PCR. The wild-type (WT) allele was amplified by cWT genotyping primer set and the KO allele was amplified by cKO genotyping primer set. Primer sequences are as follows:

gWT_genotype, forward: 5′-GAGATGCGTTTGTGTCCTCC-3′
gWT_genotype, reverse: 5′-AGTGAGTGCCCGCCTTTA-3′
gKO_genotyping, forward: 5′-GTGTCCTCCATAACTTCGTATAGC-3′
gKO_genotyping, reverse: 5′-GAGAGCAGCAGTGTAGAGGTC-3′
cWT_genotype, forward: 5′-GAGAGTCCTTTAACCCGGAG-3′
cWT_genotype, reverse: 5′-AGGAAAGACCACCATGCCTC-3′
cKO_genotyping, forward: 5′-ATTCAGGAGACGCTTAAGG-3′
cKO_genotyping, reverse: 5′-AGGAAAGACCACCATGCCTC-3′

For genotyping using paraffin-embedded sections, immunohistochemistry against anti-SEPHS1 antibody was performed.

4.4. Embryo Preparation and RNA Extraction for RNA-seq

Embryos were dissected from the uterus at E6.5, E7.5 and E8.5 with the aid of super-fine forceps and a 29-gauge needle. RNA was extracted from whole embryos using Trizol^{TM} reagent with glycogen carrier. After genotyping, $Sephs1^{+/+}$, $Sephs1^{+/-}$, or $Sephs1^{-/-}$ embryos were pooled separately according to their genotype. To remove genomic DNA (gDNA), RNA samples were treated with DNase I following the manufacturer's instruction. mRNA quality was assessed using the RNA 6000 Nano-Assay on a BioAnalyser 2100 (Agilent Technologies, Waltham, MA, USA). A total of 118 embryos at E6.5 (30 $Sephs1^{+/+}$, 63 $Sephs1^{+/-}$, and 25 $Sephs1^{-/-}$ embryos), 86 embryos at E7.5 (21 $Sephs1^{+/+}$, 39 $Sephs1^{+/-}$, and 26 $Sephs1^{-/-}$ embryos), and 25 embryos at E8.5 (6 $Sephs1^{+/+}$, 15 $Sephs1^{+/-}$, and 4 $Sephs1^{-/-}$ embryos) were used.

4.5. Realtime PCR

Realtime PCR was performed as described previously [19] with minor modifications. Total RNA isolated using Trizol^{TM} reagent was reverse-transcribed by M-MLV reverse transcriptase, and then subjected to real-time PCR using PowerUp^{TM} SYBR^{TM} Green Master with the StepOne^{TM} system (Applied Biosystems, Waltham, MA, USA) according to the manufacturer's instructions. Primer sequences are listed below:

Pou5f1 (Oct4), forward: 5′-AAGTTGGCGTGGAGACTTTG-3′
Pou5f1 (Oct4), reverse: 5′-CCGCAGCTTACACATGTTCT-3′
Nanog, forward: 5′-CGGCTCACTTCCTTCTGACT-3′
Nanog, reverse: 5′-GCGTTCCCAGAATTCGATGC-3′
Brachyury (T), forward: 5′-CTGTGAGTCATAACGCCAGC-3′
Brachyury (T), reverse: 5′-AGATCCAGTTGACACCGGTT-3′
Six3, forward: 5′-TTGCTCTCTCTAACTCGCTGG-3′
Six3, reverse: 5′-CCCGACCCTTGTTCATCTGG-3′

4.6. Transcriptome Analysis

Transcriptome analyses were performed using the TruSeq RNA Sample Prep Kit v2 and paired-end sequencing (Illumina HiSeq 4000). We acquired 101-bp reads in 6.3–7.4 Gb per sample. Data yield was approximately 7.0×10^7 raw reads, and approximately 98.5% reads were mapped. Sequencing quality was assessed by FastQC and low-quality bases were trimmed using Trim Galore. Alignment of reads to the mouse reference genome (mm10) was carried out by HISAT2 and STAR aligner. Gene quantification was performed by StringTie, gene counts by HTseq-count and differential expression analysis by Ballgown and DESeq2 packages in R. Differential gene expression was evaluated using max (FPKM) > 1, $p < 0.01$ and $|\log_2(\text{Fold Change})| > 1$. To reduce the sample-to-sample systematic bias that may affect the interpretation, the data were calibrated by Trimmed Mean of M-values (TMM) normalization and estimating the size factor using count data 'edgeR' in R.

4.7. Pathway Analyses and Transcription Factor Prediction

Ingenuity Pathway Analysis (IPA) was carried out using the canonical pathway module in IPA and Gene Set Enrichment Analysis (GSEA) using pre-ranked modules. Functional enrichment analysis was carried out using the specified gene lists by PANTHER or Metascape. Transcription factor analysis was performed using TRRUST and PaGenBase modules in Metascape. Protein–protein interaction analysis was carried out using BioGrid and String modules in Metascape.

Manual transcription factor analysis was performed using TRRUST database. All transcription factors and their target lists were recruited from the TRRUST database, and then filtered into the 'gastrulation' GO category. The ratio of DEGs among these target genes was calculated. Transcription factors were filtered with max (FPKM) > 10, the number of target gene (s) in 'gastrulation' GO category > 1 and DEGs in target gene (%) > 50%.

Hierarchical clustering was performed using GenePattern. Distance was calculated by Euclidean distance or Pearson correlation, and clustered by pairwise complete-linkage method. Heatmap was illustrated by Treeview.

4.8. X-ray Microscopy Imaging

4.8.1. Sample Preparation

Embryos were prepared, stained, and imaged as described previously with slight modification [62]. Embryos were dissected within decidua to fully preserve embryo to extra-embryonic association without disturbing the orientation or morphology. Embryos were fixed in 4% paraformaldehyde overnight at 4 °C and were washed with 1X PBS three times for 1 h each. Fixed samples were immersed in 0.1 N (v/v) Lugol's solution at room temperature for varying time according to the size of the embryo, 3–5 days. After staining, samples were mounted individually within a microcentrifuge tube filled with 0.5% w/v agarose and imaged immediately.

4.8.2. Image Acquisition

The raw data for 3D imaging of the samples were acquired via Xradia 620 versa (Carl Zeiss, Oberkochen, Baden-Württemberg, Germany). Each data set was acquired with the X-ray source at 60 kV and 142 µA with a 0.5 mm aluminum attenuation filter. The acquisition time/isotropic voxel is 2 h/1.3 voxel to 4 h/0.8 voxel. Each sample was rotated 360° along the anterior–posterior axis, and a projection image at 2016 × 1344 pixels was generated every 0.3° at an average of 4 images. Acquired projection images were reconstructed and analyzed via Dragonfly (Version: 2021.1; ORS, Montreal, Quebec, Canada) software.

4.9. Histological Analysis

Embryos were dissected and fixed with 4% paraformaldehyde overnight at 4 °C. Fixed embryos were washed in running tap water for at least 6 hr and then paraffin embedded. The paraffin embedding process was performed automatically by Tissue processor (Leica, Wetzlar, Germany) following the routine overnight protocol. Paraffin blocks were sliced to 5 µm thickness, and then the slices attached to slide glasses coated with 3-aminopropyl triethoxysilane. After overnight drying at 37 °C on a slide warmer, the sample slides were used for further analysis. Hematoxylin and eosin staining was processed automatically by Autostainer (Leica).

4.10. Immunohistochemistry

Immunohistochemistry was performed following standard protocols using primary antibodies: anti-SEPHS1 (1:100), anti-EOMES (1:200), anti-HNF4α (1:100), anti-FOXA2 (1:200), anti-SOX2 (1:200), anti-OTX2 (1:100), anti-8-oxo-guanine (1:100), anti-γH2AX (1:100), anti-cleaved caspase 3 (1:100), anti-PCNA (1:200), and anti-CD31 (1:100). Briefly, sample slides were incubated with xylene for paraffin removal and ethanol for rehydration. Antigen retrieval was performed with sodium citrate (pH 5.0) or Tris-EDTA (pH 9.0) buffer, according to the manufacturer's instruction for the primary antibody. Slides were blocked with 5% normal goat serum in Tris-buffered saline/Tween for 1 h at room temperature. After the incubation with primary antibody overnight at 4 °C, CFL488-conjugated mouse IgG, Cy3-conjugated rabbit IgG, biotin-conjugated mouse IgG or biotin-conjugated rabbit IgG was applied for 1 h at room temperature, and biotin-conjugated IgGs were incubated with streptavidin-HRP for 1 h at room temperature. Slides were counter-stained with DAPI and observed using a fluorescence microscope (Axiovert 200M, Carl Zeiss) or confocal microscope (LSM710, Carl Zeiss). In the case of chromogen dye, color development was performed using an AEC substrate and samples were counter-stained with Mayer's modified hematoxylin. Slides were observed using a stereo microscope (SMZ18, Nikon, Tokyo, Japan).

4.11. Statistical Analysis

Statistical analyses were performed using an unpaired Student's *t*-test or one-way ANOVA with GraphPad PRISM 7.0. Statistical analysis for DEGs were performed by exact test using edgeR. A value of $p < 0.01$ was considered significant.

Supplementary Materials: The following are available online at https://www.mdpi.com/article/10.3390/ijms222111647/s1.

Author Contributions: Conceptualization, J.B., L.K.K. and B.J.L.; investigation, J.B., M.H., T.-J.Y., L.Q., J.J. and J.N.; data curation, J.B., J.-H.K., L.K.K. and B.J.L.; resources, J.-H.K., V.N.G., B.A.C. and D.L.H.; writing—original draft preparation, J.B., L.K.K. and B.J.L.; writing—review and editing, L.K.K., B.A.C., D.L.H. and B.J.L.; project administration, B.J.L.; funding acquisition, B.J.L. All authors have read and agreed to the published version of the manuscript.

Funding: This research was supported by the National Research Foundation of Korea (NRF) funded by the Basic Science Research Program (NRF-2018R1D1A1A09083692 to B.J.L.), NIH grants (to V.N.G.) and by the Korea Foundation for International Cooperation of Science & Technology (NRF-2016K1A3A1A20006069 to B.J.L.).

Institutional Review Board Statement: The study was conducted according to the guidelines of the Institutional Guidelines of the Institute of Laboratory Animal Resources (Seoul National University, Seoul, Korea) and approved by the Institutional Animal Care and Use Committee at Seoul National University (IACUC; IACUC Nos. SNU-150313-1-3 and SNU-200923-7) from 3 March 2019 to 23 March 2022.

Data Availability Statement: All of the raw datasets can be found in the Short Read Archive (SRA) database of the National Center for Biotechnology Information (NCBI) under accession number PRJNA764535.

Acknowledgments: We thank Petra Tsuji for her critical review of the manuscript.

Conflicts of Interest: The authors declare no conflict of interest.

References

1. Brigelius-Flohé, R.; Sies, H. *Diversity of Selenium Functions in Health and Disease*; CRC Press: Boca Raton, FL, USA, 2016; pp. 109–170.
2. Gladyshev, V.N.; Hatfield, D.L.; Schweizer, U.; Tsuji, P.A. *Selenium: Its Molecular Biology and Role in Human Health*, 4th ed.; Hatfield, D.L., Ed.; Springer: Cham, Germany, 2016; pp. 399–531.
3. Na, J.; Jung, J.; Bang, J.; Lu, Q.; Carlson, B.A.; Guo, X.; Gladyshev, V.N.; Kim, J.-H.; Hatfield, D.L.; Lee, B.J. Selenophosphate synthetase 1 and its role in redox homeostasis, defense and proliferation. *Free. Radic. Biol. Med.* **2018**, *127*, 190–197. [CrossRef]

4. Ghaemi, S.Z.; Forouhari, S.; Dabbaghmanesh, M.H.; Sayadi, M.; Bakhshayeshkaram, M.; Vaziri, F.; Tavana, Z. A Prospective Study of Selenium Concentration and Risk of Preeclampsia in Pregnant Iranian Women: A Nested Case–Control Study. *Biol. Trace Elem. Res.* **2013**, *152*, 174–179. [CrossRef] [PubMed]
5. Mihailović, M.; Cvetkovic, M.; Ljubić, A.; Kosanović, M.; Nedeljković, S.; Jovanovic, I.; Pesut, O.; Cvetkovč, M. Selenium and Malondialdehyde Content and Glutathione Peroxidase Activity in Maternal and Umbilical Cord Blood and Amniotic Fluid. *Biol. Trace Elem. Res.* **2000**, *73*, 47–54. [CrossRef]
6. Abdulah, R.; Noerjasin, H.; Septiani, L.; Mutakin; Defi, I.R.; Suradji, E.W.; Puspitasari, I.M.; Barliana, M.I.; Yamazaki, C.; Nakazawa, M.; et al. Reduced serum selenium concentration in miscarriage incidence of Indonesian subjects. *Biol. Trace Elem. Res.* **2013**, *154*, 1–6. [CrossRef]
7. Lee, B.J.; Worland, P.J.; Davis, J.N.; Stadtman, T.C.; Hatfield, D.L. Identification of a selenocysteyl-tRNA(Ser) in mammalian cells that recognizes the nonsense codon, UGA. *J. Biol. Chem.* **1989**, *264*, 9724–9727. [CrossRef]
8. Rayman, M.P. The importance of selenium to human health. *Lancet* **2000**, *356*, 233–241. [CrossRef]
9. Lu, J.; Holmgren, A. Selenoproteins. *J. Biol. Chem.* **2009**, *284*, 723–727. [CrossRef]
10. Glass, R.S.; Singh, W.P.; Jung, W.; Veres, Z.; Scholz, T.; Stadtman, T. Monoselenophosphate: Synthesis, characterization, and identity with the prokaryotic biological selenium donor, compound SePX. *Biochemistry* **1993**, *32*, 12555–12559. [CrossRef]
11. Guimarães, M.J.; Peterson, D.; Vicari, A.; Cocks, B.G.; Copeland, N.G.; Gilbert, D.J.; Jenkins, N.A.; Ferrick, D.A.; Kastelein, R.A.; Bazan, J.F.; et al. Identification of a novel selD homolog from Eukaryotes, Bacteria, and Archaea: Is there an autoregulatory mechanism in selenocysteine metabolism? *Proc. Natl. Acad. Sci. USA* **1996**, *93*, 15086–15091. [CrossRef] [PubMed]
12. Low, S.C.; Harney, J.W.; Berry, M.J. Cloning and Functional Characterization of Human Selenophosphate Synthetase, an Essential Component of Selenoprotein Synthesis. *J. Biol. Chem.* **1995**, *270*, 21659–21664. [CrossRef]
13. Xu, X.M.; Carlson, B.A.; Irons, R.; Mix, H.; Zhong, N.; Gladyshev, V.N.; Hatfield, D.L. Selenophosphate synthetase 2 is essential for selenoprotein biosynthesis. *Biochem. J.* **2007**, *404*, 115–120. [CrossRef]
14. Dickinson, M.E.; Flenniken, A.M.; Ji, X.; Teboul, L.; Wong, M.D.; White, J.K.; Meehan, T.F.; Weninger, W.J.; Westerberg, H.; Adissu, H.; et al. High-throughput discovery of novel developmental phenotypes. *Nature* **2016**, *537*, 508–514. [CrossRef]
15. Alsina, B.; Serras, F.; Baguñà, J.; Corominas, M. patufet, the gene encoding the Drosophila melanogaster homologue of selenophosphate synthetase, is involved in imaginal disc morphogenesis. *Mol. Genet. Genom.* **1998**, *257*, 113–123. [CrossRef]
16. Serras, F.; Morey, M.; Alsina, B.; Baguna, J.; Corominas, M. The Drosophila selenophosphate synthetase (selD) gene is required for development and cell proliferation. *Biofactors* **2001**, *14*, 143–149. [CrossRef] [PubMed]
17. Morey, M.; Corominas, M.; Serras, F. DIAP1 suppresses ROS-induced apoptosis caused by impairment of the selD/sps1 homolog in Drosophila. *J. Cell Sci.* **2003**, *116*, 4597–4604. [CrossRef] [PubMed]
18. Shim, M.S.; Kim, J.Y.; Jung, H.K.; Lee, K.H.; Xu, X.-M.; Carlson, B.A.; Kim, K.W.; Kim, I.Y.; Hatfield, D.L.; Lee, B.J. Elevation of Glutamine Level by Selenophosphate Synthetase 1 Knockdown Induces Megamitochondrial Formation in Drosophila Cells. *J. Biol. Chem.* **2009**, *284*, 32881–32894. [CrossRef] [PubMed]
19. Tobe, R.; Carlson, B.A.; Huh, J.H.; Castro, N.P.; Xu, X.-M.; Tsuji, P.A.; Lee, S.-G.; Bang, J.; Na, J.-W.; Kong, Y.-Y.; et al. Selenophosphate synthetase 1 is an essential protein with roles in regulation of redox homoeostasis in mammals. *Biochem. J.* **2016**, *473*, 2141–2154. [CrossRef] [PubMed]
20. Noda, Y.; Matsumoto, H.; Umaoka, Y.; Tatsumi, K.; Kishi, J.; Mori, T. Involvement of superoxide radicals in the mouse two-cell block. *Mol. Reprod. Dev.* **1991**, *28*, 356–360. [CrossRef]
21. Guérin, P.; El Mouatassim, S.; Ménézo, Y. Oxidative stress and protection against reactive oxygen species in the pre-implantation embryo and its surroundings. *Hum. Reprod. Update* **2001**, *7*, 175–189. [CrossRef]
22. Rodríguez-González, E.; López-Bejar, M.; Mertens, M.-J.; Paramio, M.-T. Effects on in vitro embryo development and intracellular glutathione content of the presence of thiol compounds during maturation of prepubertal goat oocytes. *Mol. Reprod. Dev.* **2003**, *65*, 446–453. [CrossRef]
23. Matsui, M.; Oshima, M.; Oshima, H.; Takaku, K.; Maruyama, T.; Yodoi, J.; Taketo, M.M. Early Embryonic Lethality Caused by Targeted Disruption of the Mouse Thioredoxin Gene. *Dev. Biol.* **1996**, *178*, 179–185. [CrossRef]
24. Nonn, L.; Williams, R.R.; Erickson, R.P.; Powis, G. The Absence of Mitochondrial Thioredoxin 2 Causes Massive Apoptosis, Exencephaly, and Early Embryonic Lethality in Homozygous Mice. *Mol. Cell. Biol.* **2003**, *23*, 916–922. [CrossRef]
25. Xin, M.; Davis, C.A.; Molkentin, J.; Lien, C.-L.; Duncan, S.; Richardson, J.A.; Olson, E.N. A threshold of GATA4 and GATA6 expression is required for cardiovascular development. *Proc. Natl. Acad. Sci. USA* **2006**, *103*, 11189–11194. [CrossRef]
26. Huang, S.; Shen, Q.; Mao, W.-G.; Li, A.-P.; Ye, J.; Liu, Q.-Z.; Zou, C.-P.; Zhou, J.-W. JWA, a novel signaling molecule, involved in the induction of differentiation of human myeloid leukemia cells. *Biochem. Biophys. Res. Commun.* **2006**, *341*, 440–450. [CrossRef]
27. Tao, G.; Levay, A.; Gridley, T.; Lincoln, J. Mmp15 is a direct target of Snai1 during endothelial to mesenchymal transformation and endocardial cushion development. *Dev. Biol.* **2011**, *359*, 209–221. [CrossRef] [PubMed]
28. Li, H.; Zhao, L.; Lau, Y.S.; Zhang, C.; Han, R. Genome-wide CRISPR screen identifies LGALS2 as an oxidative stress-responsive gene with an inhibitory function on colon tumor growth. *Oncogene* **2021**, *40*, 177–188. [CrossRef]
29. Wood, H.; Episkopou, V. Comparative expression of the mouse Sox1, Sox2 and Sox3 genes from pre-gastrulation to early somite stages. *Mech. Dev.* **1999**, *86*, 197–201. [CrossRef]
30. Ip, C.K.; Fossat, N.; Jones, V.; Lamonerie, T.; Tam, P.P.L. Head formation: OTX2 regulates Dkk1 and Lhx1 activity in the anterior mesendoderm. *Development* **2014**, *141*, 3859–3867. [CrossRef] [PubMed]

31. Burtscher, I.; Lickert, H. Foxa2 regulates polarity and epithelialization in the endoderm germ layer of the mouse embryo. *Development* **2009**, *136*, 1029–1038. [CrossRef] [PubMed]
32. Arnold, S.J.; Hofmann, U.K.; Bikoff, E.K.; Robertson, E.J. Pivotal roles for eomesodermin during axis formation, epithelium-to-mesenchyme transition and endoderm specification in the mouse. *Development* **2008**, *135*, 501–511. [CrossRef]
33. Nowotschin, S.; Costello, I.; Piliszek, A.; Kwon, G.S.; Mao, C.-A.; Klein, W.H.; Robertson, E.J.; Hadjantonakis, A.-K. The T-box transcription factor Eomesodermin is essential for AVE induction in the mouse embryo. *Genes Dev.* **2013**, *27*, 997–1002. [CrossRef] [PubMed]
34. Linderholm, A.L.; Onitsuka, J.; Xu, C.; Chiu, M.; Lee, W.-M.; Harper, R.W. All-trans retinoic acid mediates DUOX2 expression and function in respiratory tract epithelium. *Am. J. Physiol. Cell. Mol. Physiol.* **2010**, *299*, L215–L221. [CrossRef]
35. Kam, R.K.T.; Deng, Y.; Chen, Y.; Zhao, H. Retinoic acid synthesis and functions in early embryonic development. *Cell Biosci.* **2012**, *2*, 11. [CrossRef] [PubMed]
36. Simeonov, K.P.; Uppal, H. Direct Reprogramming of Human Fibroblasts to Hepatocyte-Like Cells by Synthetic Modified mRNAs. *PLoS ONE* **2014**, *9*, e100134. [CrossRef]
37. Kuckenberg, P.; Kubaczka, C.; Schorle, H. The role of transcription factor Tcfap2c/TFAP2C in trophectoderm development. *Reprod. Biomed. Online* **2012**, *25*, 12–20. [CrossRef]
38. Teichweyde, N.; Kasperidus, L.; Carotta, S.; Kouskoff, V.; Lacaud, G.; Horn, P.A.; Heinrichs, S.; Klump, H. HOXB4 Promotes Hemogenic Endothelium Formation without Perturbing Endothelial Cell Development. *Stem Cell Rep.* **2018**, *10*, 875–889. [CrossRef]
39. Steinhart, Z.; Angers, S. Wnt signaling in development and tissue homeostasis. *Development* **2018**, *145*, dev146589. [CrossRef] [PubMed]
40. Subramanian, A.; Tamayo, P.; Mootha, V.K.; Mukherjee, S.; Ebert, B.L.; Gillette, M.A.; Paulovich, A.; Pomeroy, S.L.; Golub, T.R.; Lander, E.S.; et al. Gene set enrichment analysis: A knowledge-based approach for interpreting genome-wide expression profiles. *Proc. Natl. Acad. Sci. USA* **2005**, *102*, 15545–15550. [CrossRef]
41. Aman, A.; Piotrowski, T. Cell migration during morphogenesis. *Dev. Biol.* **2010**, *341*, 20–33. [CrossRef]
42. Berge, D.T.; Koole, W.; Fuerer, C.; Fish, M.; Eroglu, E.; Nusse, R. Wnt Signaling Mediates Self-Organization and Axis Formation in Embryoid Bodies. *Cell Stem Cell* **2008**, *3*, 508–518. [CrossRef]
43. Nusse, R.; Lim, X. The Wnt. Available online: http://web.stanford.edu/group/nusselab/cgi-bin/wnt/ (accessed on 20 October 2021).
44. Martinez-Alarcon, O.; García-Lopez, G.; Mora, J.R.G.; Molina-Hernandez, A.; Diaz-Martinez, N.E.; Portillo, W.; Diaz, N.F. Prolactin from Pluripotency to Central Nervous System Development. *Neuroendocrinology* **2021**. [CrossRef]
45. Hartnett, L.; Glynn, C.; Nolan, C.M.; Grealy, M.; Byrnes, L. Insulin-like growth factor-2 regulates early neural and cardiovascular system development in zebrafish embryos. *Int. J. Dev. Biol.* **2010**, *54*, 573–583. [CrossRef] [PubMed]
46. Shrivastav, S.V.; Bhardwaj, A.; Pathak, K.A.; Shrivastav, A. Insulin-Like Growth Factor Binding Protein-3 (IGFBP-3): Unraveling the Role in Mediating IGF-Independent Effects within the Cell. *Front. Cell Dev. Biol.* **2020**, *8*, 286. [CrossRef]
47. Carbon, S.; Ireland, A.; Mungall, C.J.; Shu, S.; Marshall, B.; Lewis, S.; Ami, G.O.H.; Web Presence Working Group. AmiGO: Online access to ontology and annotation data. *Bioinformatics* **2009**, *25*, 288–289. [CrossRef]
48. Frazee, A.C.; Pertea, G.; Jaffe, A.E.; Langmead, B.; Salzberg, S.L.; Leek, J.T. Ballgown bridges the gap between transcriptome assembly and expression analysis. *Nat. Biotechnol.* **2015**, *33*, 243–246. [CrossRef] [PubMed]
49. Love, M.I.; Huber, W.; Anders, S. Moderated estimation of fold change and dispersion for RNA-seq data with DESeq2. *Genome Biol.* **2014**, *15*, 550. [CrossRef]
50. Robinson, M.D.; McCarthy, D.J.; Smyth, G.K. edgeR: A Bioconductor package for differential expression analysis of digital gene expression data. *Bioinformatics* **2010**, *26*, 139–140. [CrossRef]
51. Oki, S.; Ohta, T.; Shioi, G.; Hatanaka, H.; Ogasawara, O.; Okuda, Y.; Kawaji, H.; Nakaki, R.; Sese, J.; Meno, C. ChIP-Atlas: A data-mining suite powered by full integration of public ChIP-seq data. *EMBO Rep.* **2018**, *19*, e46255. [CrossRef] [PubMed]
52. Reich, M.; Liefeld, T.; Gould, J.; Lerner, J.; Tamayo, P.; Mesirov, J.P. GenePattern 2.0. *Nat. Genet.* **2006**, *38*, 500–501. [CrossRef]
53. Kim, D.; Paggi, J.M.; Park, C.; Bennett, C.; Salzberg, S.L. Graph-based genome alignment and genotyping with HISAT2 and HISAT-genotype. *Nat. Biotechnol.* **2019**, *37*, 907–915. [CrossRef]
54. Zhou, Y.; Zhou, B.; Pache, L.; Chang, M.; Khodabakhshi, A.H.; Tanaseichuk, O.; Benner, C.; Chanda, S.K. Metascape provides a biologist-oriented resource for the analysis of systems-level datasets. *Nat. Commun.* **2019**, *10*, 1523. [CrossRef]
55. Pijuan-Sala, B.; Griffiths, J.A.; Guibentif, C.; Hiscock, T.W.; Jawaid, W.; Calero-Nieto, F.J.; Mulas, C.; Ibarra-Soria, X.; Tyser, R.C.V.; Ho, D.L.L.; et al. A single-cell molecular map of mouse gastrulation and early organogenesis. *Nature* **2019**, *566*, 490–495. [CrossRef]
56. Bult, C.J.; Blake, J.A.; Smith, C.L.; Kadin, J.A.; Richardson, J.E.; Mouse Genome Database, G. Mouse Genome Database (MGD) 2019. *Nucleic Acids Res.* **2019**, *47*, D801–D806. [CrossRef] [PubMed]
57. Liberzon, A.; Birger, C.; Thorvaldsdottir, H.; Ghandi, M.; Mesirov, J.P.; Tamayo, P. The Molecular Signatures Database (MSigDB) hallmark gene set collection. *Cell Syst.* **2015**, *1*, 417–425. [CrossRef]
58. Mi, H.; Ebert, D.; Muruganujan, A.; Mills, C.; Albou, L.P.; Mushayamaha, T.; Thomas, P.D. PANTHER version 16: A revised family classification, tree-based classification tool, enhancer regions and extensive API. *Nucleic Acids Res.* **2021**, *49*, D394–D403. [CrossRef] [PubMed]

59. Pertea, M.; Pertea, G.M.; Antonescu, C.M.; Chang, T.C.; Mendell, J.T.; Salzberg, S.L. StringTie enables improved reconstruction of a transcriptome from RNA-seq reads. *Nat. Biotechnol.* **2015**, *33*, 290–295. [CrossRef]
60. Page, R.D. TreeView: An application to display phylogenetic trees on personal computers. *Comput. Appl. Biosci.* **1996**, *12*, 357–358. [CrossRef]
61. Han, H.; Cho, J.W.; Lee, S.; Yun, A.; Kim, H.; Bae, D.; Yang, S.; Kim, C.Y.; Lee, M.; Kim, E.; et al. TRRUST v2: An expanded reference database of human and mouse transcriptional regulatory interactions. *Nucleic Acids Res.* **2018**, *46*, D380–D386. [CrossRef]
62. Hsu, C.-W.; Wong, L.; Rasmussen, T.L.; Kalaga, S.; McElwee, M.L.; Keith, L.C.; Bohat, R.; Seavitt, J.R.; Beaudet, A.L.; Dickinson, M.E. Three-dimensional microCT imaging of mouse development from early post-implantation to early postnatal stages. *Dev. Biol.* **2016**, *419*, 229–236. [CrossRef] [PubMed]

International Journal of Molecular Sciences

Article

Constitutive Oxidative Stress by SEPHS1 Deficiency Induces Endothelial Cell Dysfunction

Jisu Jung [1,†], Yoomin Kim [1,†], Jiwoon Na [1,†], Lu Qiao [1], Jeyoung Bang [2], Dongin Kwon [1], Tack-Jin Yoo [1], Donghyun Kang [1], Lark Kyun Kim [3], Bradley A. Carlson [4], Dolph L. Hatfield [4], Jin-Hong Kim [1,*] and Byeong Jae Lee [1,2,*]

1. School of Biological Sciences, College of Natural Sciences, Seoul National University, Seoul 08826, Korea; tabris0520@snu.ac.kr (J.J.); foxmin12@snu.ac.kr (Y.K.); naji0708@snu.ac.kr (J.N.); qiaolu@snu.ac.kr (L.Q.); okkdi@snu.ac.kr (D.K.); yootackjin@snu.ac.kr (T.-J.Y.); kangd@snu.ac.kr (D.K.)
2. Interdisciplinary Program in Bioinformatics, College of Natural Sciences, Seoul National University, Seoul 08826, Korea; 880419@snu.ac.kr
3. Severance Biomedical Science Institute, Graduate School of Medical Science, Brain Korea 21 Project, Gangnam Severance Hospital, Yonsei University College of Medicine, Seoul 06230, Korea; LKKIM@yuhs.ac
4. Mouse Cancer Genetics Program, Center for Cancer Research, National Cancer Institute, National Institutes of Health, Bethesda, MD 20892, USA; carlsonb@dc37a.nci.nih.gov (B.A.C.); hatfielddolph@gmail.com (D.L.H.)
* Correspondence: jinhkim@snu.ac.kr (J.-H.K.); imbglmg@snu.ac.kr (B.J.L.); Tel.: +82-2-880-6775 (B.J.L.)
† Contributed equally.

Abstract: The primary function of selenophosphate synthetase (SEPHS) is to catalyze the synthesis of selenophosphate that serves as a selenium donor during selenocysteine synthesis. In eukaryotes, there are two isoforms of SEPHS (SEPHS1 and SEPHS2). Between these two isoforms, only SEPHS2 is known to contain selenophosphate synthesis activity. To examine the function of SEPHS1 in endothelial cells, we introduced targeted null mutations to the gene for SEPHS1, *Sephs1*, in cultured mouse 2H11 endothelial cells. SEPHS1 deficiency in 2H11 cells resulted in the accumulation of superoxide and lipid peroxide, and reduction in nitric oxide. Superoxide accumulation in *Sephs1*-knockout 2H11 cells is due to the induction of xanthine oxidase and NADPH oxidase activity, and due to the decrease in superoxide dismutase 1 (SOD1) and 3 (SOD3). Superoxide accumulation in 2H11 cells also led to the inhibition of cell proliferation and angiogenic tube formation. *Sephs1*-knockout cells were arrested at G2/M phase and showed increased gamma H2AX foci. Angiogenic dysfunction in *Sephs1*-knockout cells is mediated by a reduction in nitric oxide and an increase in ROS. This study shows for the first time that superoxide was accumulated by SEPHS1 deficiency, leading to cell dysfunction through DNA damage and inhibition of cell proliferation.

Keywords: selenium; selenoprotein; selenophosphate synthetase; endothelial cell; reactive oxygen species; cell growth; angiogenesis

1. Introduction

Selenium is an essential trace element that provides many health benefits. For example, selenium has been shown to prevent heart disease, have antiviral effects, and to boost the immune system, when it is consumed in adequate amounts, as discussed in [1] and references therein. This element also plays important roles in animal development and in the male reproductive system. Most of the beneficial effects of selenium are likely mediated by selenoproteins, which contain selenocysteine (Sec) at the active site [2]. Selenocysteine, the 21st amino acid in the genetic code, can be incorporated into a growing peptide in response to UGA codon translation [3,4]. Sec is produced by replacing the hydroxyl group of serine that is aminoacylated on tRNA[Ser]Sec with inorganic selenium [2]. Selenophosphate serves as a selenium donor during Sec synthesis. Selenophosphate synthetase (SEPHS) catalyzes the reaction of selenophosphate synthesis from selenide at an ATP [5]. There are two isoforms of SEPHSs, SEPHS1 and 2, in eukaryotes, while only one form of SEPHS (SelD) exists in prokaryotes and Archaea. Mouse and human SEPHS1 are composed of 392 amino acids, and only two amino acid residues are different between human and mouse SEPHS1 at position 11 (serine in humans and threonine in mice) and 121 (methionine in humans and isoleucine in mice) according to the NCBI database (https://www.ncbi.nlm.nih.gov/protein/term=sephs1, accessed on 20 October 2021). These mammalian SEPHS1s have a high sequence homology with their SEPHS2s counterparts. SEPHS in *E. coli* (SelD) is composed of 347 amino acids (~37 kDa). The functions of prokaryotic SelD and eukaryotic SEPHS2 have been well established. SelD and SEPHS2 synthesize selenophosphate using inorganic selenium and ATP as substrates. The gamma phosphate of ATP is cleaved and attached to selenium to form selenophosphate. Interestingly, the beta phosphate on the remaining ADP is further cleaved, leaving free inorganic phosphate and AMP as final products [6]. Although SEPHS1s have high sequence homology with their SEPHS2 paralogues, they do not generate selenophosphate. It is of interest that SEPHS1 still retains the ability to cleave the gamma phosphate from ATP [7]. Furthermore, SEPHS1 plays an essential role in cell proliferation and survival. Knockout of the SEPHS1 gene, *Sephs1*, by P-element insertion in *Drosophila* resulted in embryonic lethality at the third instar larval/pupal stage [8]. When this gene was disrupted at the 5′-untranslated region, the imaginal disc was subject to aberrant formation. The cell number in mutant imaginal discs and in the brain was reduced, and apoptotic cells were observed in the abnormal disc. In another P-element mutant, the larval brain size was reduced, and DNA synthesis decreased significantly [9]. In *Drosophila* embryo-derived SL2 cells, the deficiency of SEPHS1 (SelD) led to a significant reduction in cell proliferation, and interestingly, also to the formation of megamitochondria [10]. These phenotypic changes occurred through the inhibition of pyridoxal phosphate synthesis. In mammals, SEPHS1 also plays key roles in cell proliferation and survival. In mice, systemic *Sephs1* knockout led to embryonic lethality. The knockout embryos were clearly underdeveloped by day E8.5 and virtually resorbed by day E14.5 [11]. Knockdown of *Sephs1* mRNA in both mouse and human cells also suppressed cell proliferation. Malignant properties, including cell invasion and foci formation, were inhibited by SEPHS1 deficiency in F9 cells [11], which are a mouse embryonic cancer cell line. Notably, accumulation of reactive oxygen species (ROS), especially hydrogen peroxide (H_2O_2), was observed in *Sephs1*-knockout F9 cells.

ROS include peroxides, superoxide, hydroxyl radicals, singlet oxygen, and alpha-oxygen species. Among these ROS, superoxide is frequently used as the precursor of most other ROS. Dismutation of superoxide produces H_2O_2. Partial reduction of H_2O_2 forms a hydroxide ion and a hydroxyl radical and full reduction of H_2O_2 produces water. Although ROS can be provided exogenously [12,13], it can also be produced endogenously. The cellular sources of ROS include electron transport chain complexes in mitochondria, as well as NADPH oxidase, the cytochrome P450 system and xanthine oxidoreductase (XOR) [14]. Among these, the electron chain complexes are the main source of ROS generation. Complex I and III produce superoxide, while complex IV produces H_2O_2. NADPH oxidases (NOXs) are a complex of enzymes that produce superoxide by oxidizing NADPH to NADP+. NOX was originally found in phagocytic cells and is highly expressed in immune cells. However, NOXs are also expressed in many other different tissues [15]. Four different NOX isoforms and two dual oxidases (DUOX) were discovered. These NOX family genes are expressed in a tissue-specific manner and localized in specific subcellular organelles. For example, endothelial cells express NOX1 and NOX4. NOX4 is localized in the mitochondria and produces H_2O_2 [16,17]. DUOXs are mainly expressed in thyroid cells, and also oxidize NADPH to produce NADP and protons, as well as producing H_2O_2. XOR has dual enzyme activity that produces xanthine from hypoxanthine and uric acid from xanthine. XOR can be reversibly converted into two different forms, xanthine dehydrogenase (XDH) and xanthine oxidase (XO) in mammals. Of these, XO uses oxygen to produce H_2O_2 and superoxide by transferring monovalent and divalent electrons to O_2, respectively [18]. It is known that the ratio of XDH/XO is affected by cellular conditions.

For example, in healthy tissues, the XDH form is most abundant, and this form uses NAD^+ as a cofactor. However, in diseased cells, calcium-activated proteinases cleave XDH to XO, and XO uses oxygen as a cofactor [19,20]. It was also shown that the conversion of XDH to XO is accelerated under ischemic conditions [21].

The levels of ROS produced endogenously are tightly controlled by diverse ROS scavengers within the cell under normal conditions. Natural intracellular ROS scavengers include enzymes such as superoxide dismutases (SODs), catalases, glutathione peroxidases (GPXs), thioredoxin reductases (TXNRDs), and glutaredoxins (GLRXs). In normal cells, ROS can be used as a signaling molecule to activate cell proliferation and defense. However, various ROS species can yield cytotoxic effects, such as DNA damage and cell death, when the levels are high enough to cause oxidative stress [22].

Although SEPHS1 has been implicated in cell proliferation and oxidation/reduction homeostasis, the understanding of the detailed mechanism of how SEPHS1 functions is still limited. In our previous study, SEPHS1 deficiency was shown to inhibit cell growth and malignancy in F9 cancer cells [11]. It is well known that endothelial cells are involved in the formation of new blood vessels, i.e., angiogenesis. Therefore, we hypothesized that SEPHS1 deficiency in endothelial cells would reduce the ability of angiogenesis in tumors. A cultured endothelial cell line model is well suited to elucidate a more in depth understanding of the underlying mechanism(s) of SEPHS1 effects on angiogenesis. Since 2H11 is a commonly used cancer cell line derived from mouse endothelial cells, we chose 2H11 to investigate the function of SEPHS1 in endothelial cells. In this study, we examined the effect of SEPHS1 loss in endothelial cells by the targeted removal of *Sephs1*, and found that superoxide, not H_2O_2, was accumulated. We also observed that superoxide led to morphological changes of the cell through reducing focal adhesion, growth inhibition by oxidative-stress-mediated DNA damage, and loss of angiogenic ability by downregulating nitric oxide which was induced by oxidative stress.

2. Results
2.1. Sephs1 Knockout Leads to Morphological Changes

To study SEPHS1 function in endothelial cells, we constructed a *Sephs1*-knockout cell line targeting exon 8, using an endothelial cancer cell line 2H11, combined with CRISPR/Cas9 technology (see Figure S1 for the position of the target site). 2H11 is derived from an immortalized cell line established by transformation of mouse lymphoid endothelial cells with SV40 large T antigen. Among 14 independent clones that were puromycin-resistant, four possible mutant candidates were selected and subjected to sequencing to identify mutations. Interestingly, although positions and sequences of mutations were different among the clones, all knockout clones showed four different mutations each. For example, the knockout cell line 8-22 used in this study showed four different mutations, including a base insertion and deletions leading to a frameshift mutation in exon 8 (Figure S1). In addition to exon 8, we constructed knockout cell lines targeting exons 3 and 7. All the knockout clones also contained four different mutations in each single clone (data not shown). These results suggest that there are four copies of *Sephs1*. As shown in Figure 1, no SEPHS1 protein was detected by immunocytochemistry or Western blot analysis in the knockout cells (Figure 1A–C). To exclude off-target effects by the knockout, a rescue construct was produced by introducing silent mutations into the guide RNA target site. The levels of SEPHS1 expressed in this rescue cell line increased slightly compared with those of wild-type 2H11 cells, suggesting the rescue *Sephs1* construct recovered from the knockout effect. Unlike in other cell lines, such as the embryonic cancer F9 cell line in which SEPHS1 expression was ablated, knockout of *Sephs1* in 2H11 cells led to a morphological change from a fibroblast-like shape to a spindle shape with long, thin cytoplasm (Figure 1D). Additionally, the number of focal adhesions at lamellipodia was significantly reduced (lower panel of Figure 1D). For these reasons, the knockout cells appeared to have weak attachment abilities and exhibited a thinner and longer morphology.

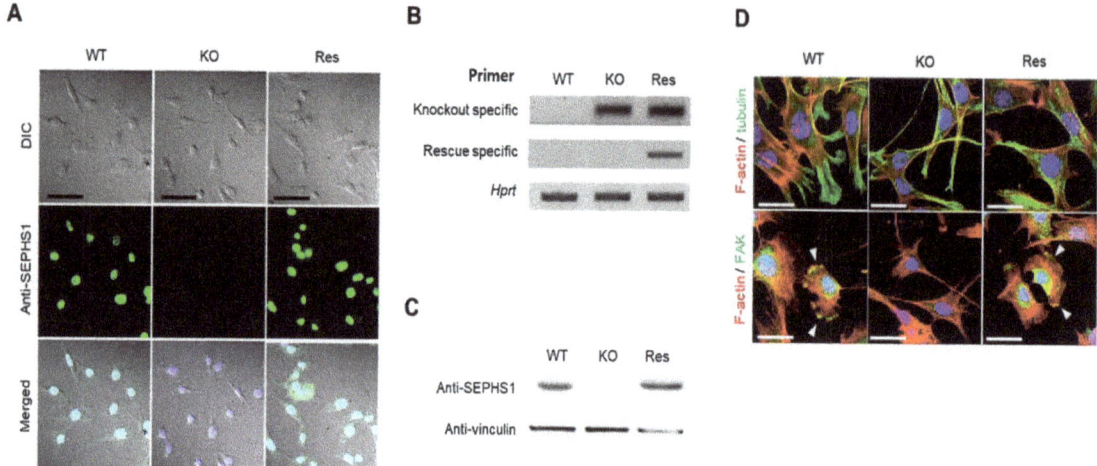

Figure 1. Construction of *Sephs1* knockout and rescue 2H11 endothelial cancer cells. Confirmation of *Sephs1* knockout by (**A**) immunocytochemistry (Scale bars represent 100 μm) and by (**B**) RT-PCR (the knockout-specific and rescue-specific primer sets are described in Materials and Methods) or (**C**) Western blot analysis. *Hprt* and anti-vinculin were used as internal controls. (**D**) Immunostaining of cytoskeletons and focal adhesion. Red and green color designate F-actin and α-tubulin, respectively (upper panel). In the lower panel, F-actin is in red and FAK in green. Arrowheads in the lower panel designate focal adhesion. Since most FAK is overlapped with F-actin, it appears as a yellow color. Scale bars represent 50 μm. DIC, digital image correlation; WT, wild type; Res, rescue; KO, knockout; FAK, focal adhesion kinase.

2.2. Superoxide Is Accumulated by SEPHS1 Deficiency in Endothelial Cells

In this study, the type of ROS accumulated in *Sephs1*-knockout cells was determined by staining with fluorescent dye or a GFP probe (Figure 2). Staining with CM-DCFDA showed that total ROS levels were increased by approximately 2.4-fold in *Sephs1*-knockout cells compared to wild-type controls (Figure 2A and Figure S2A). Furthermore, DHE staining demonstrated that superoxide levels were significantly increased (approximately 2.2-fold) in *Sephs1-knockout cells* (Figure 2A and Figure S2B). To detect the levels of cytosolic H_2O_2, cells were transfected with a roGFP-Orp1 probe [23,24], and both oxidized and reduced forms of Orp1 were measured. The ratio of the oxidized/reduced form was 0.96, 0.95, and 0.96 in wild-type control, knockout, and rescue cells, respectively (Figure S2C). This result clearly indicates that the levels of H_2O_2 were not changed by SEPHS1 deficiency in 2H11 cells. With these cytological data, we can conclude that superoxide, not H_2O_2, was accumulated by SEPHS1 deficiency in endothelial cells.

ROS can be accumulated in cells by overproduction of ROS and/or by reducing their scavengers. As shown in Figure 2B, supplementation of the cell culture medium with either SOD or N-acetyl cysteine (NAC) reduced intracellular ROS levels, while addition of catalase did not (Figure S2D). These data suggest that accumulation of superoxide in *Sephs1*-knockout cells was due to the lack of SOD and possibly other superoxide scavengers. This idea was confirmed by examining the RNA levels of ROS scavengers. As shown in Figure 2C, the RNA expressions of *Sod1* and *Sod3* decreased significantly in *Sephs1*-knockout cells. SOD1 destroys superoxides that are normally produced within the cells, while SOD3 catalyzes the dismutation of superoxide in the extracellular space secretion. However, the expression of enzymes such as catalase and glutathione peroxidase 1 (GPX1) that catalyze the reduction of H_2O_2, was not changed. These results indicate that superoxide accumulation in *Sephs1*-knockout endothelial cells is mediated by downregulating the expression of superoxide scavengers. Notably, *Sod2* levels were not decreased by *Sephs1* knockout. In addition, mitochondrial superoxide levels were not changed between wild-type and knockout cells when the cells were stained with MitoSOX™ Red (data not shown).

These data suggest that superoxides generated in mitochondria were converted to H_2O_2 immediately after being produced.

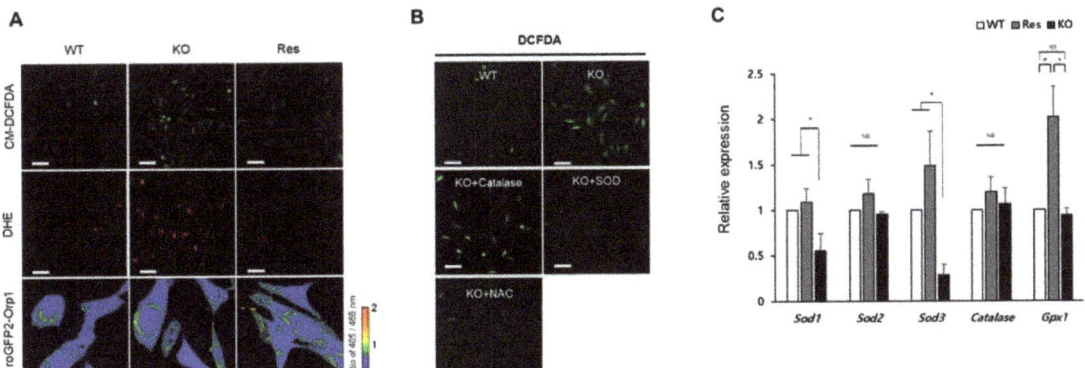

Figure 2. Identification of ROS type accumulated in *Sephs1*-knockout 2H11 cells. (**A**) Cells were stained with CM-DCFDA for general ROS and DHE for superoxide, and hydrogen peroxide levels were measured using roGFP2-Orp1 probe. The ratio between oxidized Orp (405 nm) and reduced Orp (488 nm) was calculated as described in Materials and Methods. Scale bars represent 100 µm for CM-DCFDA and DHE, and 50 µm for roGFP2-Orp1, respectively. (**B**) Confirmation of superoxide accumulation by treatment of scavengers. Scale bars represent 100 µm. (**C**) Measuring expression levels of ROS-scavenging enzymes by real-time PCR. The primers used in this experiment are shown in Table S1. NS, and * indicate not significant, and *p*-value < 0.05, respectively. WT, wild type; Res, rescue; KO, knockout.

To identify which synthesis pathway contributes to the accumulation of superoxide in the *Sephs1*-knockout 2H11 cells, cells were treated with chemicals that inhibit individual pathways, and superoxide levels were observed using DHE staining. As shown in Figure 3A, the addition of allopurinol (XO inhibitor) and GKT137831 (NOX1 and 4 inhibitor) decreased superoxide levels in *Sephs1*-knockout cells near to those levels observed in wild-type control cells (see also Figure S3). However, the treatment with VAS2780 (NOX2 inhibitor), ML171 (NOX1 inhibitor) or Mito-TEMPO (mitochondrial superoxide scavenger) did not reduce superoxide levels. These results suggest that XO and NOX4 are the major sources of superoxide production in *Sephs1*-knockout endothelial cells. The increase in superoxide levels caused by XO can be achieved by increasing the expression of XO. Generation of XO, which is produced by cleaving xanthine oxidoreductase (XOR), was dramatically induced by *Sephs1* knockout (Figure 3B). In *Sephs1*-knockout cells, the ratio between XOR and XO was approximately 1, while the XO form was not detectable in wild-type control and rescue cells. Although GKT137831 is an inhibitor of both NOX1 and NOX4, only *Nox4* mRNA levels were increased significantly by *Sephs1* knockout (Figure 3C). These results indicate that superoxide accumulation in *Sephs1*-knockout cells occurs both by a reduction in superoxide scavengers and by an increase in superoxide-producing enzymes such as XO and NOX4.

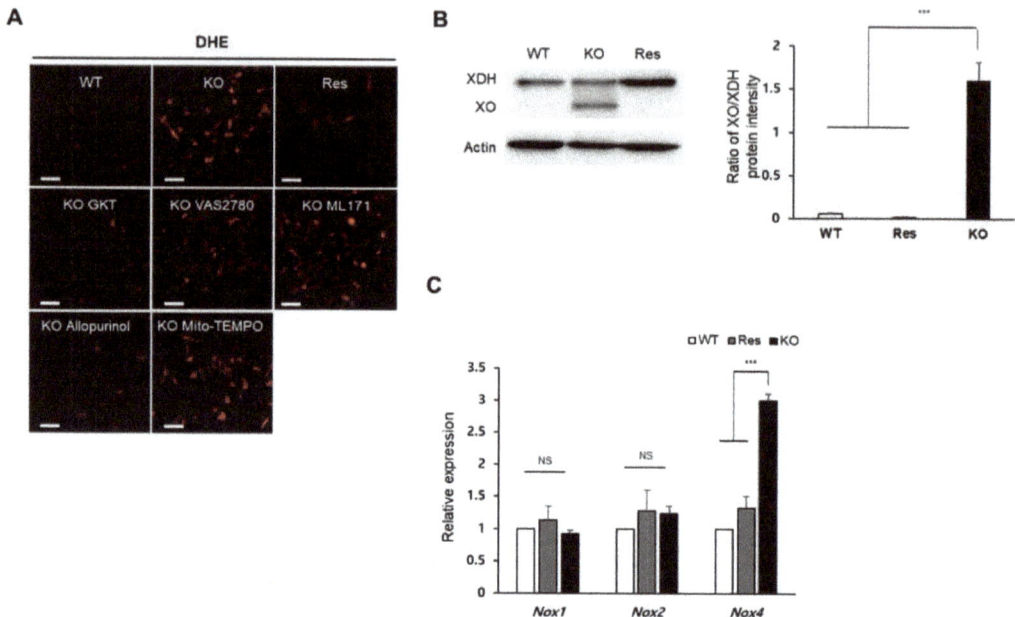

Figure 3. Identification of superoxide-generating sources in *Sephs1*-knockout 2H11 endothelial cells. (**A**) Cells were treated with selective inhibitors of superoxide production (allopurinol, GKT, VAS2780 and ML171) or superoxide scavenger (Mito-TEMPO). Scale bars represent 100 μm. (**B**) Western blot analysis using anti-XOR antibody. Actin was used as an internal control. XOR, xanthine oxidoreductase; XO, xanthine oxidase. (**C**) Measuring relative expression of NADPH oxidases by real-time PCR. Relative expression represents the ratio of ΔCt between wild type (WT) and knockout (KO) or rescue (Res) cells. NS and *** indicate not significant and *p*-value < 0.001, respectively.

2.3. SEPHS1 Regulates the Levels of Reactive Nitrogen Species and Lipid Peroxidation

In addition to the production of H_2O_2, superoxide can also be used as a substrate for the production of other free radicals and reactive species (FRRS) such as reactive nitrogen species (RNS) and peroxidated lipids. We examined how the accumulated superoxide affects the formation of these FRRS. As shown in Figure 4A, nitric oxide (NO) levels (DAF-FM) and peroxynitrite levels (DHR123) were significantly reduced in *Sephs1*-knockout cells (Figure S4A,B). In addition, the mRNA levels of nitric oxide synthase 2 and 3 (*Nos2* and *Nos3*) were also significantly reduced (Figure 4B,C). Since NO is a substrate of peroxynitrite production, the decrease in peroxynitrite levels in *Sephs1*-knockout cells is likely due to the shortage of a substrate. Conversely, the lipid peroxidation was dramatically increased (see 4HNE staining in Figure 4A and Figure S4C) suggesting that the lipids used as substrates for lipid peroxidation were sufficiently present to react with superoxide. In addition, we found that mRNA levels of scavengers of lipid peroxidation products such as glutaredoxin 1 (*Glrx1*), peroxiredoxin 1 (*Prdx1*), glutathione S-transferase a4 (*Gsta4*) and glutathione peroxidase 4 (*Gpx4*) were downregulated in the *Sephs1*-knockout cells (Figure 4B). Therefore, it seems that the lipid peroxidation in the *Sephs1*-knockout cells was induced by both sufficient substrate and a reduction in scavengers.

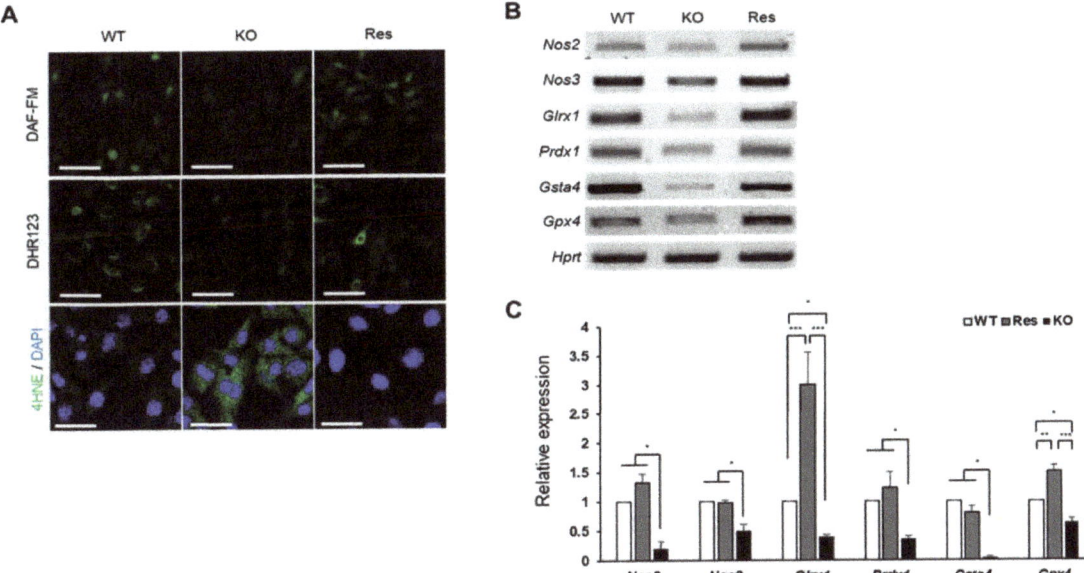

Figure 4. Effect of SEPHS1 deficiency on reactive nitrogen species generation and lipid peroxidation in 2H11 cells. (**A**) Staining of cells with fluorescent dyes to detect nitric oxide (DAF-FM), peroxynitrite (DHR123), and lipid peroxidation (4HNE). Scale bars in DAF-FM and DHR123 represent 100 μm, and 50 μm in in 4HNE. (**B**) Measuring expression levels of redox regulators participating in RNS generation and lipid oxidation by RT-PCR and (**C**) real-time PCR. *, ** and *** indicate p-value < 0.05, 0.01 and 0.001, respectively. The primers used in this experiment are shown in Table S1. WT, wild type; Res, rescue; KO, knockout.

2.4. Superoxide Inhibits Cell Proliferation at G2/M Phase

In addition to ROS regulation, another common feature of SEPHS1 is that it is required for cell proliferation and viability [1]. BrdU incorporation assays to assess DNA synthesis revealed that cell proliferation was reduced by approximately 5-fold in *Sephs1*-knockout cells compared to that in wile-type control and rescue cells (Figure 5A and Figure S5A). BrdU incorporation was restored to normal levels by the addition of NAC or SOD, suggesting that accumulated superoxide was the main cause of the inhibition of cell proliferation.

Flow cytometric analysis showed that the number of cells in the G2/M phase was significantly increased (approximately 2-fold) by SEPHS1 deficiency (Figure 5B and Figure S5B). However, the number of ROS scavenger-treated knockout cells arrested in G2/M was decreased similar to that of wile-type control and rescue cells (Figure 5B and Figure S5B). Generally, ROS affect cell cycle progression by causing DNA damage. For example, ROS induce the formation of gamma H2AX (a phosphorylated form of H2AX) foci that represent a double-strand DNA break marker, and subsequently leads to G2/M phase arrest [25,26]. As shown in Figure 5C, the number of gamma H2AX foci was increased approximately 3.5-fold in *Sephs1*-knockout cells compared to that of the wile-type control (Figure S5C). The phosphorylation levels of gamma H2AX were decreased in rescue cells and ROS scavenger-treated knockout cells. In addition, we further examined the levels of G2/M checkpoint markers such as cyclin A2 and B1, and growth-arrest and DNA-damage protein beta (GADD45β). As expected, the protein levels of cyclin A2 were increased, but cyclin B1 decreased in *Sephs1*-knockout cells and ROS scavenger-treated cells (Figure 5D), suggesting that *Sephs1*-knockout cells were arrested at the G2/M phase. The mRNA expression of GADD45β showed a similar pattern to cyclin A2, suggesting G2/M phase arrest occurred through DNA damage (Figure 5E). These results strongly suggest that the superoxide accumulated by *Sephs1* knockout resulted in gamma H2AX formation through a double-strand break, and that this DNA damage subsequently led to G2/M phase arrest.

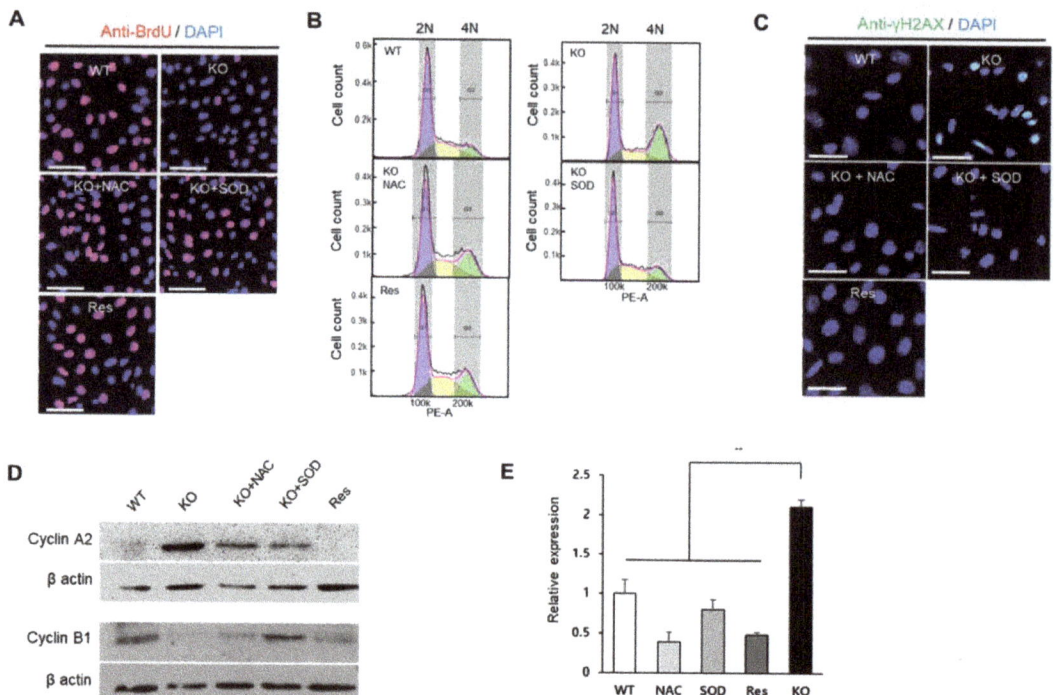

Figure 5. SEPHS1 deficiency affects cell proliferation, cell cycle and DNA damage in 2H11 cells. (**A**) Cell proliferation assay by BrdU incorporation. BrdU incorporation signal is in red, and the nucleus was counterstained with DAPI (blue). Scale bars represent 100 μm. (**B**) The effect of SEPHS1 deficiency on cell cycle progression. Cells were stained with PI and subjected to FACS analysis. DNA content of each cell was visualized as a histogram. 2N and 4N ploidy were grouped by interval gate. (**C**) Detection of DNA damage response by immunostaining with anti-gamma H2AX antibody. Nucleus was counterstained with DAPI. Scale bars represent 50 μm. (**D**) Expression pattern of G2/M arrest markers by Western blotting. (**E**) Measuring expression levels of GADD45β by real-time PCR. ** designates p-value < 0.01. WT, wild type; Res, rescue; KO, knockout.

2.5. SEPHS1 Deficiency Inhibits Angiogenic Activity of Endothelial Cells

Because endothelial cells are the main cell type of blood vessels, dysfunction of endothelial cells will cause the loss of angiogenic ability of an organism. Excessive ROS have been known to inactivate NO by oxidoreduction, and the reduction of NO causes endothelial dysfunction [27]. As described in the previous sections, the targeted removal of *Sephs1* in endothelial cells led to growth inhibition, reduction in focal adhesion, ROS accumulation and reduction in RNS levels. We, therefore, examined whether SEPHS1 deficiency affects the ability of endothelial cells to carry out angiogenesis. Cell migration is one of the important characteristics of angiogenesis. Wound healing is often used to measure migration ability of cells. As shown in Figure 6A, *Sephs1* knockout caused significant reduction in the endothelial cells wound-healing ability in scratch-wound assays. The wound-healing ability was reduced by approximately 2-fold in *Sephs1*-knockout cells compared to that in wild-type cells after 12 h incubation (Figure S6A). Both *Sephs1*-rescue- and ROS-scavenger (NAC and/or SOD)-treated *Sephs1*-knockout cells recovered wound-healing abilities with similar levels to those observed in wild-type 2H11 cells. This suggests that inhibition of wound-healing ability in *Sephs1*-knockout cells is mediated by superoxide accumulation. Future experiments to determine whether this inhibition of wound healing was due to the loss of migration ability or the inhibition of proliferation will require additional control experiments using a proliferation inhibitor such as aphidicolin.

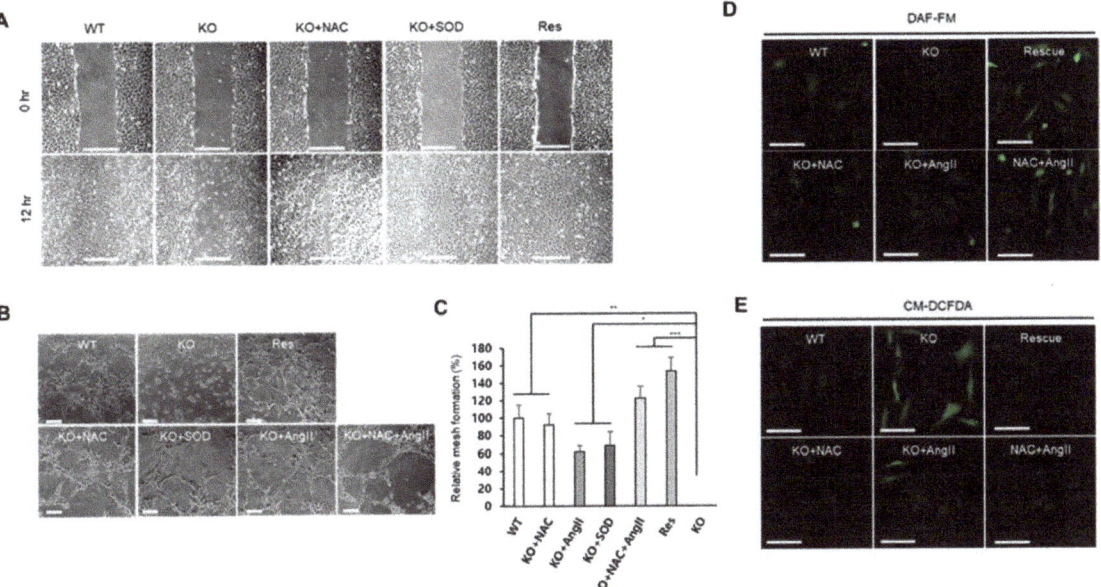

Figure 6. SEPHS1 deficiency impairs wound-healing ability and angiogenesis in 2H11 cells. (**A**) Wound-healing ability measured by scratch-wound assay. Scale bars represent 500 µm. (**B**) Tube formation assay to measure angiogenic ability. The images obtained by optical microscopy were analyzed using Angiogenesis Analyzer in ImageJ. Scale bars represent 100 µm. (**C**) Measurement of relative mesh formation. Mesh counts and relative mesh formation were obtained as described in Materials and Methods. *, ** and *** indicate p-value < 0.05, 0.01 and 0.001, respectively. (**D**) Detection of NO by staining with DAF-FM. Scale bars represent 100 µm. (**E**) ROS staining with CM-DCFDA. Scale bars represent 100 µm. WT, wild type; Res, rescue; KO, knockout.

One of the commonly used methods for detecting angiogenic ability of endothelial cells is the tube formation assay [28]. As expected, targeted removal of *Sephs1* deprived the knockout cells of tube forming ability (Figure 6B). However, tube forming ability was recovered by the addition of NAC, SOD, or angiotensin II. Mesh formation is the last stage of tube formation. The number of meshes observed after NAC, SOD, angiotensin II, or NAC plus angiotensin II treatment of knockout cells was increased by 80%, 65%, 60% and 120%, respectively (Figure 6C). These data suggest that treatment of ROS scavengers or angiotensin II alone is not sufficient to recover angiogenic ability, but mixed treatment of a ROS scavenger and angiotensin II is enough for full recovery. Interestingly, tube formation in rescue cells was similarly increased compared with wile-type control cells (130%), suggesting that the ability of endothelial cells to carry out angiogenesis is correlated with the intracellular levels of SEPHS1 (Figure S6A–C). Since angiotensin II is an inducer of NO synthesis, NO levels were examined after treatment of angiotensin II in the knockout cells. As shown in Figure 6D, angiotensin II treatment increased NO levels in *Sephs1*-knockout cells similarly to those of wild type cells. Unexpectedly, the levels were also increased in the cells by NAC treatment. Treatment of NAC combined with angiotensin II increased NO level more than wild-type, and at similar levels to that of rescue cells. ROS levels were also examined after treatment of NAC and/or angiotensin II. The levels of ROS were negatively correlated to NO levels, suggesting ROS affects NO synthesis (Figure 6E). Notably, angiotensin II treatment reduced the ROS levels in the *Sephs1*-knockout cells, although not as much as with NAC treatment. Since regulation of intracellular NO and ROS levels determines the angiogenic ability in endothelial cells, these results suggest that SEPHS1 plays an essential role in angiogenesis by regulating the NO and ROS levels in endothelial cells.

3. Discussion

Although previous studies showed that the deficiency of SEPHS1 led to the accumulation of ROS, the types of ROS were not determined [10]. Recently it was reported that H_2O_2 was accumulated in *Sephs1*-knockout, F9, embryonic carcinoma cells [11]. In the *Sephs1*-knockout F9 cells, the expression of redox-homeostasis-related genes encoding such proteins as GLRX1 and various GSTs was dysregulated [11]. However, the levels of SODs and catalases were not changed by SEPHS1 deficiency in F9 cells, suggesting that H_2O_2 was accumulated mainly by dysregulation of genes involved in redox homeostasis. In this study, we found that superoxide, rather than H_2O_2, was accumulated in *Sephs1*-knockout 2H11 endothelial cancer cells, indicating a cell-type specificity for ROS types controlled by SEPHS1. As shown in the NIH database (https://www.ncbi.nlm.nih.gov/gene/22929, accessed on 20 October 2021), there is low cell-type specificity in *Sephs1* expression. The accumulation of different kinds of ROS in different cell types is, therefore, not likely dependent on the expression levels of *Sephs1*, but is likely dependent on the interaction of SEPHS1 with a different set of cellular components. It appears that SEPHS1 interacts with different ROS scavengers and/or producers depending on the cell type, since the expression levels of genes participating in oxidation/reduction homeostasis appear to differ. The mechanism of how SEPHS1 interacts with and regulates those proteins needs to be defined. It is interesting that among six pathways that produce superoxide, only XO and NOXs are the main sources of superoxide production in *Sephs1*-knockout endothelial cells. In the case of XO, deficiency of SEPHS1 induced the processing of XOR to produce XO, and enzymatic activity of XO was increased accordingly. Another important feature of *Sephs1*-knockout endothelial cells is the downregulation of *Sod1* and *Sod3*, which are localized in the cytoplasm. SOD2 expression was not decreased by SEPHS1 deficiency, suggesting that the mitochondrial electron transfer chain did not release superoxide into the cytoplasm. In conclusion, superoxide accumulation in *Sephs1*-knockout endothelial cells occurs both by increasing the superoxide-producing system and by reducing SOD expression.

Although the accumulated ROS types are different in *Sephs1*-knockout cells depending on cell type, cell proliferation is commonly inhibited. *Sephs1*-knockout endothelial cells were arrested at the G2/M phase, and phosphorylation of H2AX was increased by ROS accumulation. This modification of H2AX is a known marker of double-strand DNA breaks [29]. Therefore, it is highly likely that the superoxide accumulated in the cell induces DNA damage and arrests cells at the G2/M checkpoint (Figure 7).

Superoxide can induce the production of reactive nitrogen species such as peroxynitrite and lipid peroxy radicals. In *Sephs1*-knockout endothelial cells, only lipid peroxidation was induced and, unexpectedly, peroxynitrite production was inhibited through the inhibition of NO production. These reduced NO levels were due to the downregulation of NOSs. Superoxide accumulation in endothelial cells seems to lead to the inhibition of NO levels, because NO levels could be increased by ROS scavenger treatment [29]. The mechanism of how SEPHS1 deficiency inhibits the expression of NOSs is unclear.

Endothelial cells form the lining of the interior surface of blood and lymphatic vessels. Therefore, endothelial cells play key barrier roles between vessels and neighboring tissues and in controlling the flow of blood and lymph. In this study, we revealed that SEPHS1 plays an important role in maintaining NO levels in endothelial cells, possibly by regulating ROS homeostasis. Furthermore, we showed that NO is not the only factor for angiogenic functions of endothelial cells, but that other factor(s) is (are) required for SEPHS1-mediated angiogenesis. Since angiogenesis is the most prominent feature of tumor growth, and since we have previously shown that SEPHS1 deficiency inhibits tumor-cell malignancy [11], targeted removal of SEPHS1 in endothelial cells may provide a potential new measure for antitumor therapy.

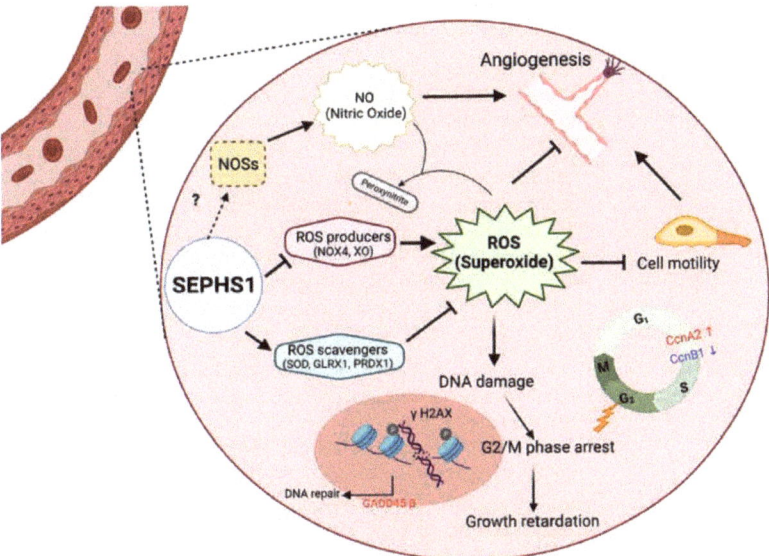

Figure 7. A schematic model for SEPHS1 function in mammalian endothelial cells. The cell represents an endothelial cell in the blood vessel. In normal conditions, SEPHS1 expresses ROS scavengers at appropriate levels, but inhibits superoxide generators such as NOX4 and XO. Superoxide is accumulated when SEPHS1 is deficient, and the superoxide causes endothelial cell dysfunctions such as growth retardation by G2/M phase arrest and loss of angiogenic ability by a decrease in NO levels. NO levels are decreased by downregulation of NOSs. Oxidative stress by superoxide accumulation leads to DNA damage and then G2/M phase arrest. The mechanism of how SEPHS1 inhibits the expression of NOSs is unclear. Arrows (→) designate activation and barred-lines (⊥) designate inhibition.

4. Materials and Methods

4.1. Materials

Dulbecco's modified Eagle's medium (DMEM) was purchased from Hyclone (Logan, UT, USA). Fetal bovine serum (FBS) was purchased from Serana (Bunbury, Australia). Antibiotic–antimycotic, Dulbecco's phosphate-buffered saline (PBS), trypan blue solution, rhodamine phalloidin, Lipofectamine® Reagent, blasticidin S HCl and puromycin were purchased from Life Technologies (Waltham, MA, USA). Neomycin was purchased from AG Scientific (San Diego, CA, USA). LentiCRISPR v2 and psPAX2 were purchased from Addgene (Watertown, MA, USA). Superoxide dismutase, catalase, N-acetyl cysteine (NAC), apocynin, allopurinol, Mito-TEMPO, angiotensin II, dihydroethidium (DHE), dihydrorhodamine 123 (DHR123), propidium iodide (PI), PMSF cocktail (protease inhibitor) and DAPI were purchased from Sigma (St. Louis, MI, USA). GKT137831 was purchased from Cayman (Ann Arbor, MI, USA). 5-(and-6)-chloromethyl-2′,7′-dichlorodihydrofluorescein diacetate, acetyl ester (CM-DCFDA) and diaminofluorescein-FM diacetate (DAF-FM) were purchased from Molecular Probes (Eugene, OR, USA). 5-Bromo-2′-deoxy-uridine (BrdU) Labeling and Detection Kit was purchased from Roche (Basel, Switzerland). ECL reagent was purchased from Amersham (Buckinghamshire, UK). Antibodies against SEPHS1(sc-365945), BrdU(sc-32323), gamma H2AX(sc-517348), Cyclin B1(sc-245), Xanthine oxidase(sc-398548), CFL488 conjugated mouse IgG(sc-533653) and CFL488 conjugated rabbit IgG(sc-516248) were purchased from Santa Cruz Biotech (Dallas, TX, USA). Anti-alpha tubulin(ab15246), anti-4HNE(ab46545), anti-cyclin A2(ab137769), anti-beta actin (ab8227), antimouse and rabbit IgG cy3(ab97035 and ab97075) antibodies were purchased from Abcam (Cambridge, UK). HRP conjugated anti-rabbit IgG and HRP conjugated anti-mouse IgG(GTX213111-01) were purchased from Genetex (Irvine, CA, USA). Matrigel used for tube formation assay was purchased from Corning (Corning, NY, USA). MG^{TM} Tissue SV kit was purchased from MG Med, (Seoul, Korea). PVDF membranes was purchased from GE Healthcare (Chicago, IL, USA) FlowJo™ Software Version 10.8.1 from BD (Ashland, OR, USA).

4.2. Cell Culture

The 2H11, HEK293T and GP2-293 cells were cultured as described previously with minor modifications [11]. Cells were incubated in DMEM with 10% FBS and 1% antibiotic–antimycotic in a humidified atmosphere containing 5% CO_2 at 37 °C.

4.3. CRISPR-Based Knockout Cell Line Construction

sgRNA (single guide RNA sequences) targeting a region in exon 8 of *Mus musculus Sephs1* was designed using the CRISPR online design tool (http://crispr.mit.edu, accessed on 26 September 2021). The sequences of sgRNA_E8 were: 5′-CACCGTAGGCCGAACATGT TTCCGC-3′/5′-AAACGCGGAAACATGTTCGGCCTAC-3′. These complementary oligonucleotides of sgRNA were annealed and cloned into LentiCRISPR v2 vector as described previously [30].

For lentivirus production, HEK293T cells were transfected with both the constructed sgRNA-containing LentiCRISPR v2 vector and virus packaging plasmid psPAX2 and pMD2.G. After a 48hr incubation, lentiviruses were harvested by filtration through 0.45 μm filter. 2H11 cells were infected with the harvested lentiviruses by incubating for 48hr, and then the infected cells were selected with 2 mg/mL of puromycin for a further 6 days. Single clones were obtained using a 96-well plate. From each clone, genomic DNA was extracted and subjected to PCR amplification using primers (forward: 5′-ACAAAGT GGGTGTTGGGTGT-3′; reverse: 5′-AGCCTTGTAACCATCCTGCC-3′). The amplified DNA fragments were cloned into TA-cloning vector, and then each clone was sequenced. The knockout cell line was further confirmed by PCR using the primer set (forward: 5′-AAGCATGTGGCAATATGTTTGGAT-3′, reverse: 5′-GTGGCACCAGGTGTGGG-3′).

4.4. CRISPR-Based Rescue Cell Line Construction

To exclude any off-target effect of the *Sephs1* gene knockout, a rescue cell line was constructed as follows. First, silent mutations in sgRNA regions (see Figure S1) were introduced to wild-type *Sephs1* which were resistant to Cas9 cleavage and expressed wild-type SEPHS1 proteins. Site-directed mutagenesis was carried out by two-step PCR methods with primer sets (primer set 1 forward: 5′-TACCGAGCTCGGATCCGAAC-3′, reverse: 5′-CAATCCAAACATATTGCCACATGCTTTGCTCACAGCGGCCAT-3′; primer set 2 forward: 5′-CATGTGGCAATATGTTTGGATTGATGCATGGGACCTGCCAGA-3′, reverse: 5′-GGTTTAAACGGGCCCTCTAG-3′. The PCR products were cloned into a retroviral vector at the BamHI/EcoRI site (pRv.neo; [31]), and the plasmid was delivered to retroviral packaging cells (GP2-293). After incubating for 48 h, viral particles were harvested and used to infect 2H11 cells where *Sephs1* was knocked out. After infection, a rescue cell line was selected with G418 (400 μg/mL) and confirmed by PCR with primer set (forward: 5′-GCATTCCCCACAAAGGCAA-3′, reverse: 5′-AGCAAAGCCTGACACCCA T-3′), Western blot analysis and immunocytochemistry.

4.5. Real-Time PCR

Real-time PCR was performed as described previously [11] with minor modifications. Total RNA was isolated using TRIZOL reagent. Total RNA (2000 ng) was reverse-transcribed by Mo-MuLV reverse transcriptase, and real-time PCR was performed in triplicate using PowerUp™ SYBR™ Green Master Mix and Prism7300 (Applied Biosystems) according to the manufacturer's instructions. Sequences of primers used in this study are provided in Table S1. The annealing temperature was set as 3 °C below the T_m of the primer set. The *Hprt* gene was used as an internal control.

4.6. Western Blot Analysis

Western blot analysis was carried out as described previously [11,32] with slight modifications. Briefly, cells were washed twice with PBS and harvested in ice-cold lysis buffer (PBS with 0.5% Triton X-100 and 0.1% PMSF cocktail). The protein concentrations of the resulting cell extracts were measured by Bradford dye-binding method and 20 µg of total protein from each sample were subjected to 10% SDS-polyacrylamide gel electrophoresis, then transferred to PVDF membranes. The membranes were incubated overnight at 4°C with primary antibodies against SEPHS1 (1:1000), vinculin, xanthine oxidase (1:1000 each), actin (1:5000), cyclin A2 (1:1000) or cyclin B1 (1:2000). Membranes were washed with Tris-buffered saline (TBS) containing 0.1% Tween 20 and incubated with secondary antibodies for 30 min at room temperature. Immunolabeling was detected using ECL reagent, and luminescence signal was detected using Chemi-Doc (Luminograph II, ATTO). The band intensities on each blot were quantified using ImageJ software (NIH).

4.7. Immunocytochemistry

Immunocytochemistry was carried out as described previously [10], with modifications. Briefly, 1.5×10^4 cells were seeded on a 9 mm coverslip and fixed with 4% paraformaldehyde in PBS, washed with PBS twice, and then permeabilized with 0.1% Triton X-100 in PBS for 10 min. The permeabilized cells were blocked with 5% FBS in PBS for 1 h at room temperature and then incubated with primary antibodies with appropriate dilution folds; anti-SEPHS1 (1:100), anti-alpha-tubulin (1:200), anti-4-hydroxynonenal (4-HNE) (1:50), anti-BrdU (1:100), anti-gamma H2AX (1:100), and anti-CD31 (1:50), respectively. Primary antibody binding was performed at 4 °C overnight, and then secondary antibody conjugated with fluorescent dye; anti-rabbit IgG conjugated with Cy3 and anti-mouse IgG conjugated with CFL488 was incubated (1:100) for 30 min at room temperature. Cells were observed by Diaphot 300 fluorescence microscope (Nikon FL, Tokyo, Japan) or LSM700 confocal microscope (Carl Zeiss, Jena, Germany).

4.8. Determining ROS Types

The detection of intracellular ROS was carried out with CM-DCFDA as described previously [10] with minor modifications. 2H11 cells were seeded at a density of 5×10^4 cells/well in a 12-well plate 1 day before staining. The cells were incubated with 5 µM CM-DCFDA in DMEM containing 1% antibiotic–antimycotic without FBS for 30 min at 37 °C in 5% CO_2, washed twice with PBS, and then observed under fluorescence microscope (Nikon FL) at an excitation wavelength of 470 nm.

Superoxide was stained with DHE as described previously [11], with slight modifications. Then, 5×10^4 cells were prepared as above and stained by incubating cells in 10 µM DHE in DMEM containing 1% antibiotic–antimycotic and 10% FBS for 15 min at 37 °C. After washing with PBS, the fluorescence signals were observed under fluorescence microscope (Nikon FL) at an excitation wavelength of 531 nm.

Detection of hydrogen peroxide was carried out as described previously [11], with minor modifications. The cytosolic roGFP2-Orp1 vector [23,24] was transfected into each cell. After incubation for 24 h at 37 °C in 5% CO_2, 1.5×10^4 cells were seeded on a 9 mm coverslip and incubated for 12 h, washed with PBS, fixed with 4% paraformaldehyde in PBS and then observed under an LSM 700 confocal microscope (Carl Zeiss). The ratio between the oxidized (405 nm) and reduced (488 nm) forms of the probes was calculated for each cell according to Morgan et al. [23]. The intensities of the 405 nm and 488 nm image from the same original field (100× magnification) were obtained separately as described [11]. The intensity of the 405 nm images was divided by the intensity of the 488 nm images to calculate the ratio. This procedure was repeated in six different fields for each cell line, and the ratio images were created by dividing the 405 nm image by the 488 nm image pixel by pixel. The ImageJ 'Blue Green Red' Look Up Table (LUT) was used for creating false-color ratio pictures.

4.9. Measurement of ROS Levels with Fluorescence Activated Cell Sorting

Cells were seeded at 2×10^5 cells per well in a 6-well plate. On the following day, cells were stained with a ROS probe. After staining, cells were harvested and intracellular fluorescence intensity of the probe was quantified using a fluorescence-activated cell sorter (FACS, Canto II, BD Biosciences, Franklin Lakes, NJ, USA). A minimum of 2×10^4 cells were counted from each sample and fluorescence distribution of cells was analyzed and displayed as a histogram.

4.10. Scratch-Wound Assay

Scratch-wound assay was carried out as described previously [31], with minor modifications. Cells were seeded at a density of 3×10^5 cells per well in a 6-well plate. On the following day, cells were scratched with a yellow pipette tip, washed with PBS twice, and then further incubated in DMEM with 10% FBS at 37 °C for 12 h. The movement of cells into the wound area was measured after photographing the cells. The covered area was calculated by subtracting initial wound area from the remaining wound area at the end of the assay. To avoid scratch-width variation, the relative covered area (RCA) was calculated by RCA[%] = covered area $\times$ 100 [%]/initial area. Experiments were performed in triplicate.

4.11. Tube Formation Assay

Tube formation assay was carried out as described in Cao et al. [33] with modifications. Cells were seeded at a density of 1.5×10^5 cells per well in a 6-well plate. The cells were incubated with trypsin EDTA, neutralized with DMEM with 10% FBS and 1% antibiotic–antimycotic, and harvested by centrifugation. Cells were washed and resuspended with serum-free DMEM. Cells were then seeded at a density of 3×10^4 cells per well in a Matrigel-precoated 96-well plate. A Matrigel-precoated 96-well plate was prepared a day before use by incubating the plate with 100 µL of Matrigel per well for 1 h at 37 °C. After incubating cells for 6 h at 37 °C, tube formations were observed with an optical microscope. The extent of tube formation was analyzed using "Angiogenesis Analyzer" software plugin for ImageJ as described previously [28].

4.12. Administration of ROS Scavengers and Inhibitor Treatments

ROS scavengers were treated as described by Kate et al. [34] with minor modifications. In this study, NAC, catalase, SOD and Mito-TEMPO were used in a concentration of 1 mM, 300 units/mL, 300 units/mL, and 50 µM, respectively. Selective inhibitors for xanthine oxidase and NADPH oxidases were administrated according to Augsburger et al. [35]. In this study, allopurinol, GKT136901, VAS2780, and ML171 were used at a concentration of 50 µM, 50 µM, 5 µM, and 5 µM, respectively. ROS scavengers and selective inhibitors were administrated overnight, and on the next day, cells were stained with an appropriate ROS probe observed with fluorescent microscope or confocal microscope.

4.13. Detection of Reactive Nitrogen Species

Reactive nitrogen species (RNS) analysis was carried out as described by Handa et al. [36] with modifications. For the detection of nitric oxide, cells were stained by incubating with 5 µM at 37 °C for 30 min. Peroxynitrite was detected by staining cells with DHR123, 10 µM at 37 °C for 30 min. Stained cells were observed with a fluorescence microscope (Nikon FL). The intensity of the RNS signal was quantified using FACS (Canto II, BD Biosciences) and FlowJo software.

4.14. Cell Cycle Analysis

Cell cycle analysis was performed by measuring DNA content after staining cells with propionic iodide (PI) [37]. Cells were seeded onto a 12-well plate at a density of 3×10^4 cells/well. After a 12 h incubation, ROS scavengers such as SOD or NAC were administered into *Sephs1*-knockout cell overnight. On the following day, cells were harvested, washed once with PBS, and fixed in 70% ethanol at 4 °C overnight. The fixed cells were washed with PBS twice, resuspended with PBS containing RNase A (200 µg/mL) and incubated for 10 min at room temperature. Cells were then further incubated with 100 µg/mL of PI in PBS for 30 min at room temperature. Stained cells were analyzed using FACS (Canto II, BD Biosciences) and FlowJo software.

4.15. BrdU Incorporation Assay

BrdU incorporation assay was carried out using BrdU Labeling and Detection Kit (Roche) according to the manufacturer's instructions. Briefly, cells were seeded onto a 24-well plate at a density of 1.5×10^4 cells per well overnight. On the following day, BrdU was added to the growth media at the concentration of 10 µg/mL for 2 h. Incorporated BrdU was detected by immunocytochemistry using anti-BrdU antibody, and the fluorescent signals conjugated to the secondary antibody were observed using a confocal microscope (Carl Zeiss).

4.16. Statistics

Each experiment was performed in biological triplicate for statistical analysis. Statistical significance was tested by one-way ANOVA followed by Tukey's multiple comparison test.

Supplementary Materials: The following are available online at https://www.mdpi.com/article/10.3390/ijms222111646/s1, Figure S1: Sequences of sgRNA, mutations in a knockout cell line (8–22) and rescue construct, Figure S2: Quantification of intensity of ROS probes. Figure S3: Quantification of superoxide accumulation in scavenger and selective inhibitor treated cells. Figure S4: Measurement of signal intensities of RNS probe and 4-HNE. Figure S5. Analysis of cell proliferation. Figure S6: Measurement of cell motility and angiogenesis. Table S1: Primer sequences used in this study.

Author Contributions: Conceptualization J.-H.K. and B.J.L.; Data curation, J.J., Y.K., J.N. and B.J.L.; Investigation, L.Q., J.B., D.K. (Dongin Kwon), T.-J.Y., D.K. (Donghyun Kang); Project administration, B.J.L.; Funding acquisition, B.J.L.; writing—review and editing, L.K.K., B.A.C., D.L.H. and B.J.L.; writing—original draft, J.J., J.N., J.-H.K. and B.J.L. All authors have read and agreed to the published version of the manuscript.

Funding: This work has supported by the National Research Foundation of Korea (NRF), funded by the Basic Science Research Program (NRF-2018R1D1A1A09083692 to B.J.L.) and by the Korea Foundation for International Cooperation of Science & Technology (NRF-2016K1A3A1A20006069 to B.J.L.).

Data Availability Statement: All data are contained within the article or Supplementary Material are available for anyone to utilize without violating participant confidentiality. Additional information can be requested from authors upon reasonable request.

Acknowledgments: We thank Joon Kim for providing 2H11 and Petra Tsuji for her in depth review.

Conflicts of Interest: The authors declare no conflict of interest.

References

1. Na, J.; Jung, J.; Bang, J.; Qiao, L.; Carlson, B.A.; Xiong, G.; Gladyshev, V.N.; Kim, J.; Hatfield, D.L.; Lee, B.J. Selenophosphate synthetase 1 and its role in redox homeostasis, defense and proliferation. *Free Radic. Biol. Med.* **2018**, *127*, 190–197. [CrossRef] [PubMed]
2. Hatfield, D.L.; Gladyshev, V.N. How selenium has altered our understanding of the genetic code. *Mol. Cell. Biol.* **2002**, *22*, 3565–3576. [CrossRef] [PubMed]
3. Lee, B.J.; Worland, P.J.; Davis, J.N.; Stadtman, T.C.; Hatfield, D.L. Identification of a selenocysteyl-tRNA (Ser) in mammalian cells that recognizes the nonsense codon, UGA. *J. Biol. Chem.* **1989**, *264*, 9724–9727. [CrossRef]
4. Longtin, R. A forgotten debate: Is selenocysteine the 21st amino acid? *J. Natl. Cancer Inst.* **2004**, *96*, 504–505. [CrossRef] [PubMed]

5. Glass, R.S.; Singh, W.P.; Jung, W.; Veres, Z.; Scholz, T.D.; Stadtman, T.C. Monoselenophosphate: Synthesis, Characterization, and Identity with the Prokaryotic Biological Selenium Donor, Compound SePX. *Biochemistry* **1993**, *32*, 12555–12559. [CrossRef]
6. Veres, Z.; Kim, I.Y.; Scholz, T.D.; Stadtman, T.C. Selenophosphate synthetase. Enzyme properties and catalytic reaction. *J. Biol. Chem.* **1994**, *269*, 10597–10603. [CrossRef]
7. Xu, X.-M.; Carlson, B.A.; Irons, R.; Mix, H.; Zhong, N.; Gladyshev, V.N.; Hatfield, D.L. Selenophosphate synthetase 2 is essential for selenoprotein biosynthesis. *Biochem. J.* **2007**, *404*, 115–120. [CrossRef] [PubMed]
8. Alsina, B.; Serras, F.; Baguñà, J.; Corominas, M. patufet, the gene encoding the Drosophila melanogaster homologue of selenophosphate synthetase, is involved in imaginal disc morphogenesis. *Mol. Gen. Genet.* **1998**, *257*, 113–123. [CrossRef]
9. Serras, F.; Morey, M.; Alsina, B.; Baguñà, J.; Corominas, M. The *Drosophila selenophosphate synthetase* (*selD*) gene is required for development and cell proliferation. *Biofactors* **2001**, *14*, 143–149. [CrossRef]
10. Shim, M.S.; Kim, J.Y.; Jung, H.K.; Lee, K.H.; Xu, X.-M.; Carlson, B.A.; Kim, K.W.; Kim, I.Y.; Hatfield, D.L.; Lee, B.J. Elevation of glutamine level by selenophosphate synthetase 1 knockdown induces megamitochondrial formation in *Drosophila* cells. *J. Biol. Chem.* **2009**, *284*, 32881–32894. [CrossRef]
11. Tobe, R.; Carlson, B.A.; Huh, J.H.; Castro, N.P.; Xu, X.-M.; Tsuji, P.A.; Lee, S.-G.; Bang, J.; Na, J.-W.; Kong, Y.-Y.; et al. Selenophosphate synthetase 1 is an essential protein with roles in regulation of redox homoeostasis in mammals. *Biochem. J.* **2016**, *473*, 2141–2154. [CrossRef] [PubMed]
12. Riley, P.A. Free radicals in biology: Oxidative stress and the effects of ionizing radiation. *Int. J. Radiat. Biol.* **1994**, *65*, 27–33. [CrossRef] [PubMed]
13. Yildirim, A.; Mavi, A.; Oktay, M.; Kara, A.A.; Algur, O.F.; Bilaloglu, V. Comparison of antioxidant and antimicrobial activities of tilia (Tilia argentea Desf ex DC), sage (Salvia triloba l.), and black tea (Camellia sinensis) extracts. *J. Agric. Food Chem.* **2000**, *48*, 5030–5034. [CrossRef]
14. Hopkins, R.Z.; Li, Y.R. Sources of free radicals and related reactive species. In *Essentials of Free Radical Biology and Medicine*, 1st ed.; Cell Med Press AIMSCI, Inc.: Raleigh, NC, USA, 2017; pp. 49–58.
15. Cifuentes, M.E.; Pagano, P.J. Targeting reactive oxygen species in hypertension. *Curr. Opin. Nephrol. Hypertens.* **2006**, *15*, 179–186. [CrossRef] [PubMed]
16. Panday, A.; Sahoo, M.K.; Osorio, D.; Batra, S. NADPH oxidases: An overview from structure to innate immunity-associated pathologies. *Cell. Mol. Immunol.* **2015**, *12*, 5–23. [CrossRef]
17. Shanmugasundaram, K.; Nayak, B.K.; Friedrichs, W.E.; Kaushik, D.; Rodriguez, R.; Block, K. NOX4 functions as a mitochondrial energetic sensor coupling cancer metabolic reprogramming to drug resistance. *Nat. Commun.* **2017**, *8*, 997. [CrossRef]
18. Nishino, T.; Okamoto, K. Mechanistic insights into xanthine oxidoreductase from development studies of candidate drugs to treat hyperuricemia and gout. *J. Biol. Inorg. Chem.* **2015**, *20*, 195–207. [CrossRef]
19. Schaffer, S.W.; Roy, R.S.; McMcord, J.M. Possible role for calmodulin in calcium paradox-induced heart failure. *Eur. Heart J.* **1983**, *4*, 81–87. [CrossRef]
20. Corte, E.D.; Stirpe, F. The regulation of rat liver xanthine oxidase. Involvement of thiol groups in the conversion of the enzyme activity from dehydrogenase (type D) into oxidase (type O) and purification of the enzyme. *Biochem. J.* **1972**, *126*, 739–745. [CrossRef]
21. Lee, M.C.I.; Velayutham, M.; Komatsu, T.; Hille, R.; Zweier, J.L. Measurement and Characterization of Superoxide Generation from Xanthine Dehydrogenase: A Redox-Regulated Pathway of Radical Generation in Ischemic Tissues. *Biochemistry* **2014**, *53*, 6615–6623. [CrossRef]
22. Schieber, M.; Chandel, N.S. ROS Function in Redox Signaling and Review Oxidative Stress. *Curr. Biol.* **2014**, *24*, R453–R462. [CrossRef]
23. Morgan, B.; Sobotta, M.C.; Dick, T.P. Measuring E(GSH) and H_2O_2 with roGFP2-based redox probes. *Free Radic. Biol. Med.* **2011**, *51*, 1943–1951. [CrossRef] [PubMed]
24. Gutscher, M.; Pauleau, A.L.; Marty, L.; Brach, T.; Wabnitz, G.H.; Samstag, Y.; Meyer, A.J.; Dick, T.P. Real-time imaging of the intracellular glutathione redox potential. *Nat. Methods* **2008**, *5*, 553–559. [CrossRef]
25. Xu, L.; Wu, T.; Lu, S.; Hao, X.; Qin, J.; Wang, J.; Zhang, X.; Liu, Q.; Kong, B.; Gong, Y.; et al. Mitochondrial superoxide contributes to oxidative stress exacerbated by DNA damage response in RAD51-depleted ovarian cancer cells. *Redox Biol.* **2020**, *36*, 101604. [CrossRef] [PubMed]
26. Fernandez-Capetillo, O.; Chen, H.T.; Celeste, A.; Ward, I.; Romanienko, P.J.; Morales, J.C.; Naka, K.; Xia, Z.; Camerini-Otero, R.D.; Motoyama, N.; et al. DNA damage-induced G2-M checkpoint activation by histone H2AX and 53BP1. *Nat. Cell. Biol.* **2002**, *4*, 993–997. [CrossRef] [PubMed]
27. Lubos, E.; Handy, D.E.; Loscalzo, J. Role of oxidative stress and nitric oxide in atherothrombosis. *Biosciences* **2009**, *13*, 5323–5344. [CrossRef] [PubMed]
28. Carpentier, G.; Berndt, S.; Ferratge, S.; Rasband, W.; Cuendet, M.; Uzan, G.; Albanese, P. Angiogenesis Analyzer for ImageJ—A comparative morphometric analysis of "Endothelial Tube Formation Assay" and "Fibrin Bead Assay". *Sci. Rep.* **2020**, *10*, 11568. [CrossRef] [PubMed]
29. Wang, Q.; Mazur, A.; François Guerrero, X.; Lambrechts, K.; Buzzacott, P.; Belhomme, M.; Theron, M. Antioxidants, endothelial dysfunction, and DCS: In vitro and in vivo study. *J. Appl. Physiol.* **2015**, *119*, 1355–1362. [CrossRef] [PubMed]

30. Ran, F.A.; Hsu, P.D.; Wright, J.; Agarwala, V.; Scott, D.A.; Zhang, F. Genome engineering using the CRISPR-Cas9 system. *Nat. Protoc.* **2013**, *8*, 2281–2308. [CrossRef]
31. Bang, J.; Huh, J.H.; Na, J.W.; Lu, Q.; Carlson, B.A.; Tobe, R.; Tsuji, P.A.; Gladyshev, V.N.; Hatfield, D.L.; Lee, B.J. Cell proliferation and motility are inhibited by G1phase arrest in 15-kDa selenoprotein-deficient Chang liver cells. *Mol. Cells.* **2015**, *38*, 457–465. [CrossRef]
32. Kim, M.; Chen, Z.; Shim, M.S.; Lee, M.S.; Kim, J.E.; Kwon, Y.E.; Yoo, T.J.; Kim, J.Y.; Bang, J.Y.; Carlson, B.A.; et al. SUMO Modification of NZFP Mediates Transcriptional Repression through TBP Binding. *Mol. Cells* **2013**, *35*, 70–78. [CrossRef] [PubMed]
33. Cao, J.; Liu, X.; Yang, Y.; Wei, B.; Li, Q.; Mao, G.; He, Y.; Li, Y.; Zheng, L.; Zhang, Q.; et al. Decylubiquinone suppresses breast cancer growth and metastasis by inhibiting angiogenesis via the ROS/p53/BAI1 signaling pathway. *Angiogenesis* **2020**, *23*, 325–338. [CrossRef] [PubMed]
34. Kate, M.; van der Wal, J.B.C.; Sluiter, W.; Hofland, L.J.; Jeekel, J.; Sonneveld, P.; van Eijck, C.H.J. The role of superoxide anions in the development of distant tumour recurrence. *Br. J. Cancer* **2006**, *95*, 1497–1503. [CrossRef] [PubMed]
35. Augsburger, F.; Filippova, A.; Rasti, D.; Seredenina, T.; Lam, M.; Maghzal, G.; Mahiout, Z.; Jansen-Dürr, P.; Knaus, U.G.; Doroshow, J.; et al. Pharmachological characterization of the seven human NOX isoforms and their inhibitors. *Redox. Biol.* **2019**, *26*, 2213–2317. [CrossRef] [PubMed]
36. Handa, O.; Stephen, J.; Cepinskas, G. Role of endothelial nitric oxide synthase-derived nitric oxide in activation and dysfunction of cerebrovascular endothelial cells during early onsets of sepsis. *Am. J. Physiol. Heart Circ. Physiol.* **2008**, *295*, H1712–H1719. [CrossRef]
37. Kim, J.Y.; Lee, K.H.; Shim, M.S.; Shin, H.I.; Xu, X.M.; Carlson, B.A.; Hatfield, D.L.; Lee, B.J. Human selenophosphate synthetase 1 has five splice variants with unique interactions, subcellular localizations and expression patterns. *Biochem. Biophys. Res. Commun.* **2010**, *397*, 53–58. [CrossRef]

Article

The Effect of tRNA$^{[Ser]Sec}$ Isopentenylation on Selenoprotein Expression

Noelia Fradejas-Villar [1,†], Simon Bohleber [1,†], Wenchao Zhao [1], Uschi Reuter [1], Annika Kotter [2], Mark Helm [2], Rainer Knoll [1], Robert McFarland [3], Robert W. Taylor [3], Yufeng Mo [4], Kenjyo Miyauchi [4], Yuriko Sakaguchi [4], Tsutomu Suzuki [4] and Ulrich Schweizer [1,*]

1. Institut für Biochemie und Molekularbiologie, Rheinische Friedrich-Wilhelms-Universität Bonn, D-53115 Bonn, Germany; fradejan@uni-bonn.de (N.F.-V.); sbohl@uni-bonn.de (S.B.); wenchao.zhao@ki.se (W.Z.); ureuter@uni-bonn.de (U.R.); rainer.knoll@uni-bonn.de (R.K.)
2. Institute of Pharmacy and Biochemistry, Johannes Gutenberg University of Mainz, Staudingerweg 5, D-55128 Mainz, Germany; akotter@uni-mainz.de (A.K.); mhelm@uni-mainz.de (M.H.)
3. Wellcome Centre for Mitochondrial Research, Clinical and Translational Research Institute, Faculty of Medical Sciences, Newcastle University, Newcastle upon Tyne NE2 4HH, UK; robert.mcfarland@ncl.ac.uk (R.M.); robert.taylor@ncl.ac.uk (R.W.T.)
4. Department of Chemistry and Biotechnology, Graduate School of Engineering, University of Tokyo, Tokyo 113-8656, Japan; moyufeng@g.ecc.u-tokyo.ac.jp (Y.M.); mkenjyo@chembio.t.u-tokyo.ac.jp (K.M.); yurikos@chembio.t.u-tokyo.ac.jp (Y.S.); ts@chembio.t.u-tokyo.ac.jp (T.S.)
* Correspondence: uschweiz@uni-bonn.de
† These authors contributed equally to this work.

Abstract: Transfer RNA$^{[Ser]Sec}$ carries multiple post-transcriptional modifications. The A37G mutation in tRNA$^{[Ser]Sec}$ abrogates isopentenylation of base 37 and has a profound effect on selenoprotein expression in mice. Patients with a homozygous pathogenic p.R323Q variant in tRNA-isopentenyl-transferase (*TRIT1*) show a severe neurological disorder, and hence we wondered whether selenoprotein expression was impaired. Patient fibroblasts with the homozygous p.R323Q variant did not show a general decrease in selenoprotein expression. However, recombinant human TRIT1^{R323Q} had significantly diminished activities towards several tRNA substrates in vitro. We thus engineered mice conditionally deficient in *Trit1* in hepatocytes and neurons. Mass-spectrometry revealed that hypermodification of U$_{34}$ to mcm^{5}Um occurs independently of isopentenylation of A$_{37}$ in tRNA$^{[Ser]Sec}$. Western blotting and ^{75}Se metabolic labeling showed only moderate effects on selenoprotein levels and ^{75}Se incorporation. A detailed analysis of *Trit1*-deficient liver using ribosomal profiling demonstrated that UGA/Sec re-coding was moderately affected in *Selenop*, *Txnrd1*, and *Sephs2*, but not in *Gpx1*. 2′O-methylation of U$_{34}$ in tRNA$^{[Ser]Sec}$ depends on FTSJ1, but does not affect UGA/Sec re-coding in selenoprotein translation. Taken together, our results show that a lack of isopentenylation of tRNA$^{[Ser]Sec}$ affects UGA/Sec read-through but differs from a A37G mutation.

Keywords: *Trit1*; isopentenylation; tRNA$^{[Ser]Sec}$; selenoproteins

1. Introduction

Selenoproteins are proteins containing the rare and essential amino acid selenocysteine (Sec), which is co-translationally inserted into proteins. Hierarchical expression of selenoproteins depends on the availability of selenium (Se) both among organs and among individual selenoproteins [1]. Moreover, at lower Se availability, selenoprotein expression is more robust in female than in male mammals [2]. The hierarchy among organs is established by provision of selenoprotein P (SELENOP) by the liver and its receptor-mediated uptake through endocytic receptors [3,4]. Several mechanisms cooperate to establish a second hierarchy among selenoproteins in one cell. For example, glutathione peroxidase 1 (GPX1) and SELENOW levels closely reflect bioavailability of Se, while GPX4 and thioredoxin reductases (TXNRD) remain stably expressed at lower Se levels. This hierarchy has

been correlated with the affinities of selenocysteine insertion sequences (SECIS) present in the 3′-untranslated regions of selenoprotein mRNAs to the SECIS-binding protein 2 (SECISBP2) [5–7]. These correlations, however, are not perfect, and binding and competition of other mRNA binding proteins such as RPL30, NUCLEOLIN, and eIF4A3 have been invoked to explain aspects of the hierarchy [8–10]. Moreover, selenoprotein mRNAs may be subject to mRNA surveillance pathways if Se levels are limiting, in particular, *GPX1* and *SELENOW*.

Sec is encoded by the UGA codon, and thus translation involves a competition between elongation and termination. Central to this process is tRNA[Ser]Sec. This tRNA was discovered as a rare seryl-tRNA that recognizes a UGA codon [11,12]. Accordingly, it is amino-acylated by SerRS. The 3′-Ser is subsequently phosphorylated by PSTK and further converted to Sec-tRNA[Ser]Sec by selenocysteine synthase [13–16]. Unlike other tRNAs, there is only one gene encoding tRNA[Ser]Sec in mammals, *Trsp* (in mice), and *TRU-TCA1-1* (in humans). Transfer RNA[Ser]Sec carries several modifications (Figure 1). In both bacteria and in vertebrates, the anticodon loop carries a hypermodified 5-methylcarboxymethyl (mcm^5)U$_{34}$, which may be further methylated on the 2′O-position of the ribose (mcm^{5}Um$_{34}$), and a N^6-isopentenyl(i^6)A$_{37}$ [17–21].

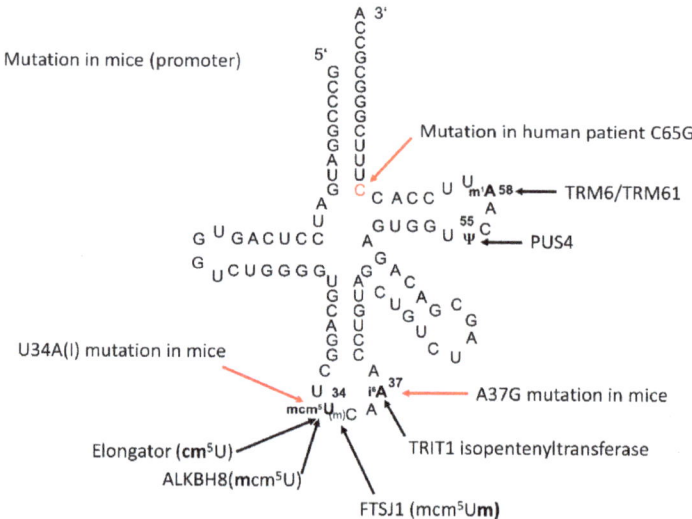

Figure 1. Modifications and mutations in tRNA[Ser]Sec. Sequence of murine tRNA[Ser]Sec with post-transcriptional modifications indicated [20]. Where proposed, we mentioned the respective enzymes responsible for the modifications (black). Labelled in red are mutations in the primary sequence of tRNA[Ser]Sec in transgenic mouse models or observed in a human patient. The U34A mutant tRNA[Ser]Sec is further deaminated in vivo to inosine (I).

Mutations in tRNA[Ser]Sec affect selenoprotein expression [22]. In transgenic mice, expression of a hypomorphic tRNA[Ser]Sec with a promotor mutation leads to a neurological phenotype [23], and gene targeting of *Trsp* leads to complete abrogation of selenoprotein expression [24–26]. A pathogenic homozygous c.C65G variant in *TRU-TCA1-1* causes a phenotype resembling the phenotype of patients with pathogenic *SECISBP2* variants [27–29]. Thus, the level and integrity of tRNA[Ser]Sec modulates selenoprotein expression. Interestingly, mutations in A$_{37}$ and U$_{34}$ in tRNA[Ser]Sec affect not only the levels but also the hierarchy of selenoprotein expression. Both mutations reduce expression of GPX1 effectively, while GPX4 and TNXRD1 remain more stably expressed [30,31]. Another publication presented data that support the notion that the A37G mutation acts, in part, as a dominant negative [32]. In a pioneering study interrogating selenoprotein translation using ribosomal profiling, it was shown that Se availability modulates the efficiency of UGA/Sec recoding [33]. This study further showed that the A37G mutant tRNA[Ser]Sec was not efficiently supporting selenoprotein translation, even in the presence of supra-nutritional selenium. Therefore, it is evident that modification of tRNA[Ser]Sec has a major impact on the process of UGA/Sec re-coding. In fact, the crystal structure of hypomodified tRNA[Ser]Sec showed a disordered anticodon stem loop [34]. In the cryo-EM structure of a bacterial ribosome in complex with mRNA and elongation factor SELB, the modified tRNA[Ser]Sec shows stacking of i^6A_{37} on the anticodon:codon minihelix; however, modification of U_{34} was not resolved [35,36].

It has been observed that 2′O-methylation of mcm^5U_{34} (mcm^5Um) in tRNA[Ser]Sec correlates with Se bioavailability [19,37]. Due to the correlation with hierarchical selenoprotein expression, a role for tRNA modification was proposed, and the effect of A37G and U34A(I) mutations were explained with the lack of 2′O-methylation of nucleoside 34 in both mutant tRNAs [30]. Interference with 5-methylcarboxymethylation of tRNA[Ser]Sec by mutation of the enzyme ALKBH8 reduced selenoprotein expression, supporting a role for U_{34} modification in UGA/Sec recoding [38,39].

The observation that treatment with lovastatin affected selenoprotein expression in cultured cells suggested that isopentenylation of tRNA[Ser]Sec was important for its function [40,41]. Later, it was shown that tRNA-isopentenyltransferase (TRIT1) was the enzyme modifying tRNA[Ser]Sec, and knock-down of *Trit1* reduced GPX1 expression in NIH 3T3 cells under the condition of low Se availability [42]. Patients carrying pathogenic variants in *TRIT1* show microcephaly with epilepsy that was primarily explained by a mitochondrial disease associated with deficient isopentenylation of mitochondrial tRNAs [21,43]. Since neurological disorders including seizures are also phenotypes observed in several mouse models carrying mutations in tRNA[Ser]Sec [23,44], we wondered whether selenoprotein expression was also affected in patients harboring pathogenic *TRIT1* variants.

We therefore studied selenoprotein biosynthesis in *TRIT1*-deficient human fibroblasts, recombinant human TRIT1, and in mice with inactivation of *Trit1*.

2. Results

2.1. TRIT1-Mutant Human Fibroblasts

Studies of Kim and colleagues suggested that the acquisition of post-transcriptional modifications in tRNA[Ser]Sec was sequential and interdependent in *Xenopus* oocytes [20]. Likewise, profound changes in selenoprotein expression were described in mouse models, wherein A_{37} in tRNA[Ser]Sec was not isopentenylated due to a A37G mutation [30]. Hence, we wondered whether fibroblasts derived from a patient carrying a homozygous pathogenic variant in *TRIT1* represented an excellent model to study the role of i^6A in tRNA[Ser]Sec for selenoprotein expression [43]. To our surprise, Western blot against selenoproteins did not reveal any reduction in the patient fibroblasts (Figure 2A), despite the fact that the TRIT1 protein appeared greatly reduced. We subsequently metabolically labelled the fibroblasts with ^{75}Se-selenite, finding no reduction in ^{75}Se incorporation into selenoproteins (Figure 2B). We then asked whether another unidentified A_{37}-tRNA-isopentenyltransferase activity was expressed in these cells. We therefore determined the modification indices of several tRNAs that are normally isopentenylated, using an established RT-qPCR technique [45]. This assay exploits the sensitivity of a reverse transcriptase reaction to the presence of 2-methylthio-i^6A (ms^2i^6A) in the tRNA substrate. The resulting cDNA is then quantified by qPCR (Figure 2C). This technique independently confirmed the results obtained before with a positive hybridization assay [43] and showed that those tRNAs that are normally containing $ms^2i^6A_{37}$ are hypomodified in *TRIT1*-mutant cells (Figure 2D,E). In order to obtain an overview of the gene regulation of selenoproteins and NRF2-depedent anti-oxidative genes, we performed RNA sequencing in *TRIT1*-mutant fibroblasts (Figure 2F). Some NRF2 target genes were up-regulated (e.g., *MT2*), but others were down-regulated (e.g., *GSTM4*, *GSTM5*, MGST1). Induction of mitochondrial transcripts is in line with the mitochondriopathy of the patient. Among selenoproteins, only

TXNRD1 and *SELENOW* were decreased at the mRNA level, but this was not reflected at the protein level (Figure 2A), suggesting that reduced mRNAs are a result of gene-specific regulation rather than an effect on selenoprotein translation. Taken together, the p.R322Q variant in *TRIT1* did not show a general deficiency in the function of tRNA[Ser]Sec in selenoprotein translation.

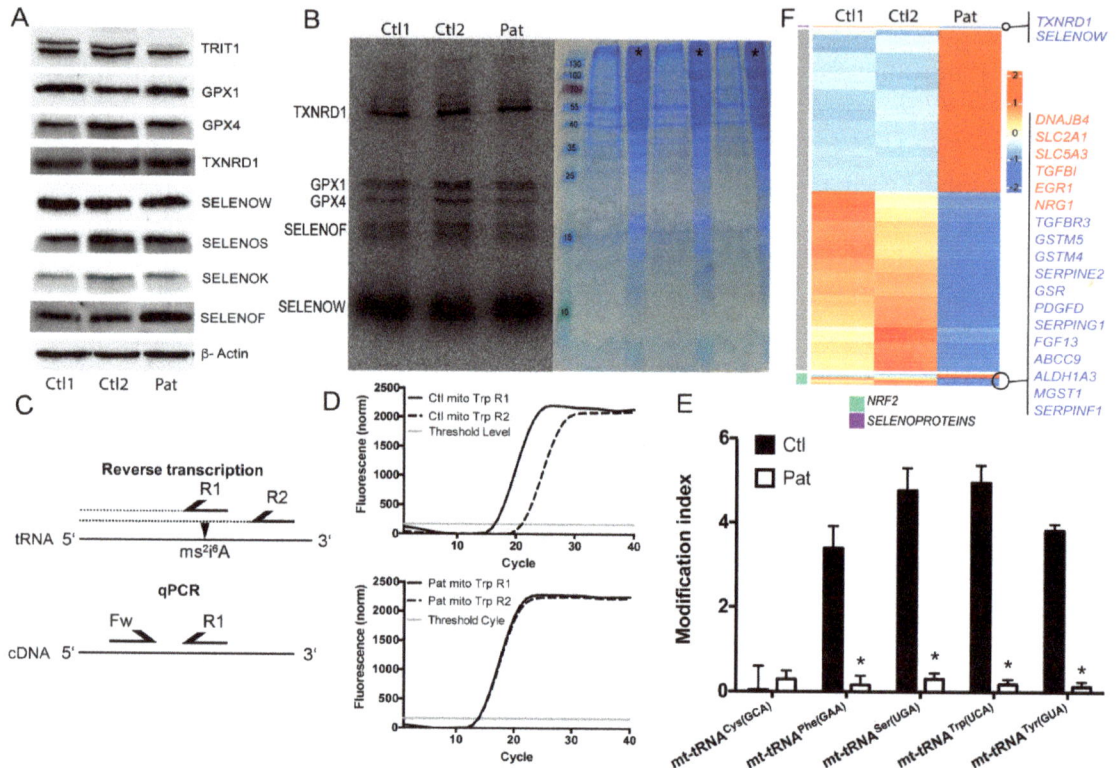

Figure 2. Selenoprotein expression in patient fibroblasts carrying a pathogenic homozygous TRIT1[R323Q] variant. (**A**) Western blot comparing selenoprotein expression in *TRIT1* patient fibroblasts with two control fibroblast lines. The signal corresponding to TRIT1 protein is reduced almost to the detection limit in the *TRIT1* patient cells, while the unspecific (lower) band suggests equal loading. β-Actin served as control. (**B**) Metabolic ^{75}Se-labeling of cultured fibroblasts reveals normal ^{75}Se incorporation in selenoproteins. Coomassie brilliant blue stained gel shows equal protein loading. Asterisks represent wells loaded with un-labelled protein to avoid diffusion (**C**) RT-PCR to determined ms^{2}i^{6}A in tRNAs. The two steps, reverse transcription of tRNA and qPCR of cDNA, are depicted. Primers are represented as half arrows (R1 and R2 are reverse primers and Fw is the forward primer) Arrowhead shows the position of the ms^{2}i^{6}A. (**D**) Determination of tRNA modification index based on RT-PCR. Traces from mt-tRNA[Trp] analysis. (**E**) Modification index of several mt-tRNAs normally containing ms^{2}i^{6}A$_{37}$ modifications depends on functional TRIT1. (**F**) Heatmap of significantly regulated genes from human fibroblasts focused on selenoprotein and NRF2 target genes. Up-regulated and down-regulated genes in the patient fibroblasts are depicted in red and blue, respectively.

2.2. In Vitro Activity of TRIT1 and TRIT1R323Q

The exact function of Arg323 in human TRIT1 is not known, but a crystal structure of the yeast tRNA-isopentenyltransferase MOD5 suggested that the amino acid is involved in substrate binding [36,46]. Thus, we wondered whether Arg323 might interact only with some, but not all substrates, and a substitution to Gln might specifically not affect isopentenylation of tRNA[Ser]Sec. We therefore recombinantly expressed human TRIT1 protein along with a p.R323Q variant and subjected the purified proteins to biochemical activity assays. Recombinant TRIT1 with and without the p.R323Q variant transferred ^{14}C-labelled dimethylallylpyrophosphate (DMAPP) to in vitro transcribed (IVT) tRNA$^{[Ser]Sec}$, while a tRNA$^{[Ser]Sec}$ mutant with A37 replaced by G was not isopentenylated, as expected (Figure 3A). In order to test a battery of cytosolic and mitochondrial tRNAs in the following isopentenylation assays, we used synthetic anticodon-containing fragment (ACF) oligonucleotides as substrates in a filter-binding assay. As a first step, we determined for each substrate the K_M values along with the respective V_{max} towards recombinant human TRIT1 (Figure 3B; Table 1). In order to assess the effect of the p.R323Q variant, we then determined in a separate experiment the specific activities of recombinant TRIT1 and the p.R323Q variant protein against eight ACF substrates (Figure 3C). The variant protein was significantly less active towards each four mitochondrial and four cytosolic tRNA ACF substrates, including tRNA$^{[Ser]Sec}$. This finding suggests that the p.R323Q variant affects activity towards all tRNA substrates.

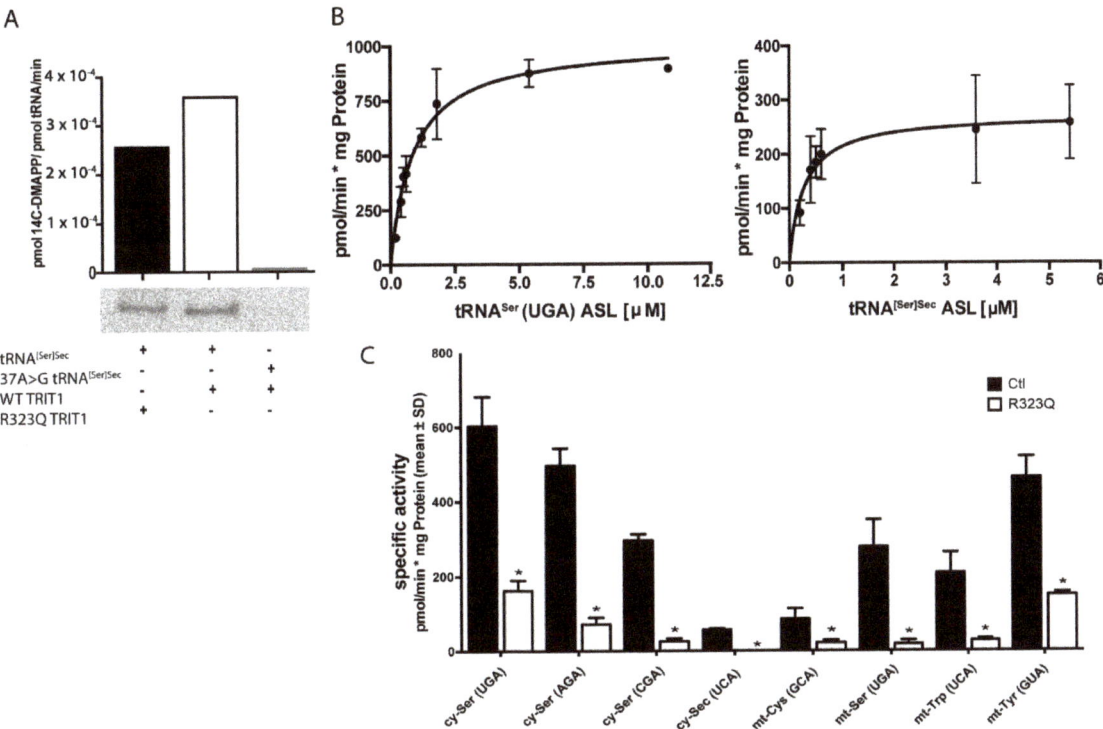

Figure 3. Activity assays using recombinant TRIT1. (**A**) In vitro assay using wild type and p.R323Q variant TRIT1 recombinant proteins and in vitro transcribed tRNA$^{[Ser]Sec}$. Isopentenylated tRNA was also detected in a urea-acrylamide gel. (**B**) Representative results of kinetic analyses of TRIT1 with ACF substrates corresponding to mt-tRNA^{Ser}(UGA) and tRNA$^{[Ser]Sec}$. (**C**) Specific activities determined for eight substrates using TRIT1 (Ctl) and p.R323Q. N = 3. * $p < 0.05$, Student's t-test. The ACF oligonucleotide concentration in the endpoint assay corresponded to the K_M of the oligonucleotide with the wild-type enzyme (Table 1).

Table 1. Determination of kinetic parameters of recombinant human TRIT1 with anticodon-containing fragment substrates.

	tRNA	K_M [µM]	V_{max} [pmol/min * mg Protein]	Sequence
cytosolic	Ser AGA	0.7980 ± 0.1091	989.0 ± 45.27	GA-UGG-ACU-AGA-AAU-CCA-UU
	Ser CGA	0.4384 ± 0.0849	465.5 ± 27.17	GU-UGG-ACU-CGA-AAU-CCA-AU
	Ser UGA	0.8690 ± 0.1210	1016 ± 54.85	GA-UGG-ACU-UGA-AAU-CCA-UU
	Sec UCA	0.3848 ± 0.0934	321.5 ± 25.72	UG-CAG-GCU-UCA-AAC-CUG-UA
mitochondrial	Cys GCA	5.293 ± 4.294	73.94 ± 28.37	AU-UGA-AUU-GCA-AAU-UCG-AA
	Ser UGA	1.673 ± 0.2683	523.6 ± 32.23	GG-UUG-GCU-UGA-AAC-CAG-CU
	Trp UCA	3.710 ± 0.5627	442.2 ± 31.37	AA-GAG-CCU-UCA-AAG-CCC-UC
	Tyr GUA	0.7735 ± 0.1127	549.7 ± 26.37	AU-UGG-ACU-GUA-AAU-CUA-AA

2.3. Inactivation of Trit1 in the Mouse

The severity of missense mutations in the selenoprotein biosynthesis pathway may depend on the cell type [47]. Hence, we created conditional *Trit1*-knockout mice and crossed them with an *Alb-Cre* transgene abrogating *Trit1* expression in hepatocytes, an established model for selenoprotein expression analyses. Western blot against TRIT1 shows a greatly diminished signal in livers from *Alb-Cre; Trit1*$^{fl/fl}$ (KO) mice (Figure 4A). Accordingly, the abundance of i^{6}A in the tRNA fraction isolated from *Trit1* KO liver was less than 10% of the controls (Ctl) compatible with preserved TRIT1 expression in endothelial cells and liver macrophages (Figure 4B). Northern blot against tRNA[Ser]Sec demonstrated unchanged levels in the *Trit1* KO (Figure 4C). We then specifically isolated tRNA[Ser]Sec from *Trit1* KO and Ctl livers by reciprocal circulating chromatography [48], followed by RNase T1 digestion, and subjected the fragments to capillary LC/nanoESI mass spectrometry to analyze its tRNA modifications [49,50]. In Ctl liver, we detected several species of the anticodon-containing fragments with different modification status (Figure 4D). The fully modified fragment with mcm^{5}Um at position 34 and i^{6}A at position 37 is a major fragment (53.3%), and the same fragment with mcm^{5}U at position 34 and i^{6}A at position 37 is the second major fragment (40.0%). In *Trit1* KO, both fragments decreased significantly, and instead, the hypomodified fragment with mcm^{5}Um$_{34}$ and A$_{37}$ increased drastically (74.2%). The result demonstrated that TRIT1 is responsible for i^{6}A$_{37}$ formation in tRNA[Ser]Sec. In addition, 5-methylcarboxymethylation of U$_{34}$ does not require prior i^{6}A modification. Curiously, the hypomodified fragment with mcm^{5}U$_{34}$ and A$_{37}$ was not accumulated in *Trit1* KO (Figure 4D), indicating that 2′O-methylation of mcm^{5}Um$_{34}$ is promoted in the absence of i^{6}A$_{37}$. In other words, i^{6}A$_{37}$ might have an inhibitory effect on FTSJ1-mediated 2′O-methylation. Importantly, although previous studies using A37G mutant tRNA[Ser]Sec suggested that Um$_{34}$ formation depended on prior i^{6}A$_{37}$ formation [20,30], our data clearly showed that mcm^{5}Um$_{34}$ formation was promoted in the absence of i^{6}A$_{37}$ (Figure 4D). Thus, it appears as if the mutant G37 nucleotide in the transgenic mouse model prevented Um$_{34}$ formation and not the lack of isopentenylation of A$_{37}$.

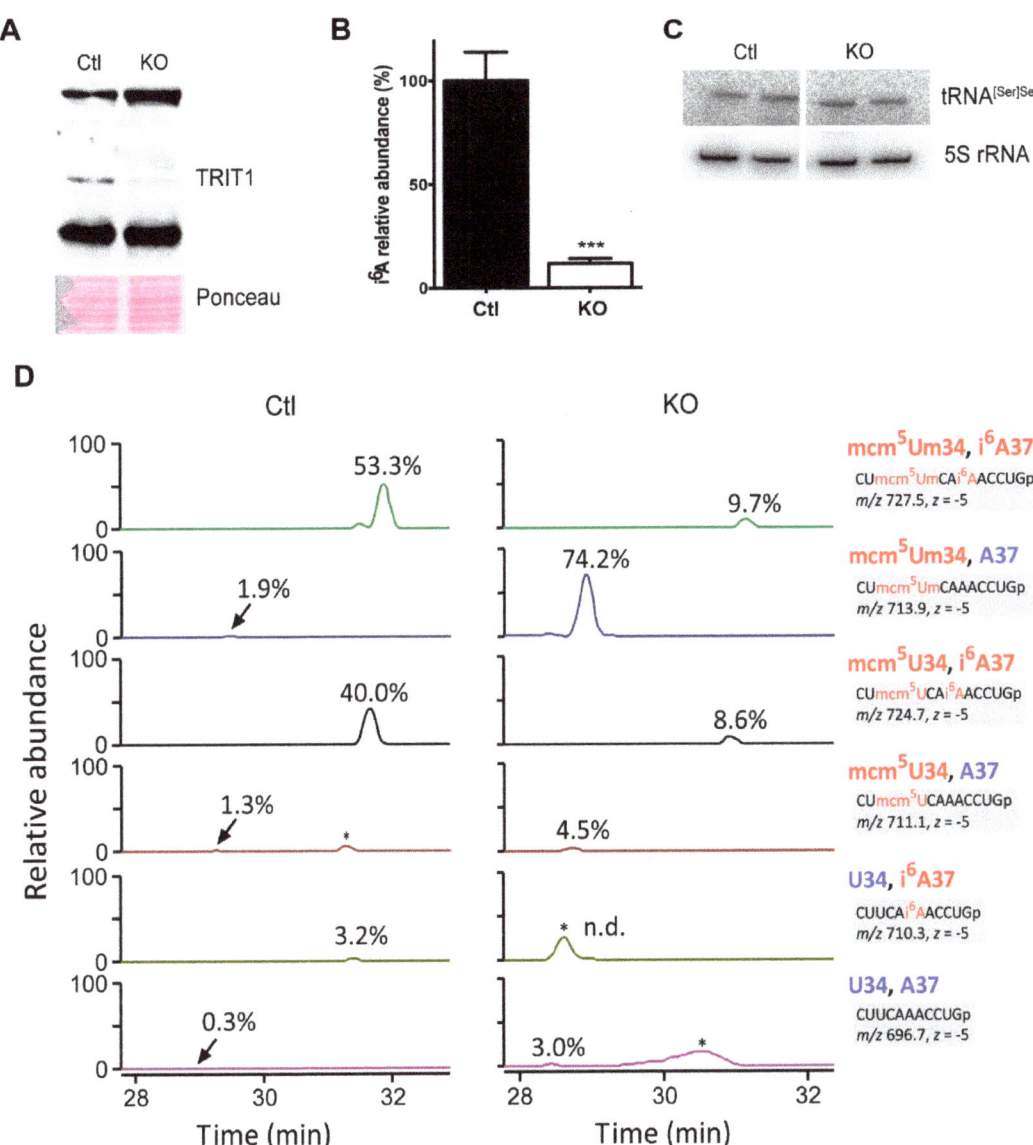

Figure 4. Knockout of *Trit1* in liver abrogates formation of i⁶A in tRNA and tRNA[Ser]Sec isopentenylation. (**A**) Western blot on liver extract using an antibody against TRIT1. (**B**) Levels of i⁶A in the tRNA fraction isolated from liver are significantly reduced in *Trit1* KO. N = 3. *** $p < 0.001$, Student's *t*-test. (**C**) Northern blot against tRNA[Ser]Sec and 5S rRNA as control. (**D**) Mass spectrometric analysis of the tRNA[Ser]Sec isolated from Ctl (left panels) and *Trit1* KO (right panels) livers. Each panel from top to bottom shows an extracted-ion chromatogram for the RNase T1-digested anticodon-containing fragments with different modification status at positions 34 and 37. Modification status, sequence of the fragment, *m/z* value, and charge state (z) are shown on the right for each panel. Relative abundance of each fragment is denoted in each panel. Non-specific peaks are marked with asterisks.

2.4. Selenoprotein Expression in Trit1-KO Mice

Being confident that TRIT1 is the only available tRNA-isopentenyltransferase in mouse hepatocytes and having ascertained that *Trit1* was quantitatively inactivated in our mouse model, we returned to the question whether the hierarchy of selenoprotein expression in hepatocytes depends on tRNA$^{[Ser]Sec}$ A$_{37}$ isopentenylation. To assess the expression of selenoproteins by Western blotting, we focused on those selenoproteins that are easily detected with a panel of antibodies that work well in our hands. There was no general reduction in selenoprotein expression in *Trit1* KO mouse liver, as studied by Western blot against eight selenoproteins (Figure 5A). In particular, GPX1 and SELENOW, which are known to respond sensitively to changes in Se availability, were not changed. In contrast, GPX4 was increased, and SEPHS2 was reduced, as confirmed by densitometric analysis of Western- blots (Figure 5B). Metabolic labeling of primary hepatocytes from wild-type and *Trit1* KO mice did not show diminished ^{75}Se incorporation into selenoproteins (Figure 5C). Since selenoprotein expression is organ-dependent, we also tested selenoprotein expression in the brain by Western blot. In neuron-specific *Trit1* KO brains, we detected a reduction in SELENOW, but not of any other selenoproteins (Figure 5D). This suggested that the regulation of SELENOW was gene-specific and not a general effect on selenoproteins. In order to directly assess the neuronal Sec-incorporation machinery, we isolated primary cortical neurons from newborn mice and metabolically labeled them in vitro with ^{75}Se. Again, there was no change of ^{75}Se incorporation into proteins (Figure 5E). We thus have to conclude that there is no general defect in selenoprotein expression, if tRNA$^{[Ser]Sec}$ lacks i^{6}A.

2.5. Ribosomal Profiling for Selenoproteins in Trit1-KO

We reasoned, that moderate effects on UGA/Sec re-coding may be better revealed by ribosomal profiling in *Trit1* KO mouse liver. Thus, we isolated polysomes from *Trit1* KO and Ctl livers and performed ribosomal profiling. When we plotted all ribosome-protected fragments (RPF) associated with all selenoprotein transcripts around the UGA/Sec codon, we noticed in the *Trit1* KO liver a small reduction in ribosomes sitting with the A-site on the UGA/Sec codon (Figure 6A). Based on a footprint size of 28 nucleotides, these ribosomes mostly represented ribosomes with a tRNA in the A-site. We then calculated the differential UGA re-coding efficiency (ΔURE) for individual selenoproteins, a measure that represents how a condition affects UGA/Sec re-coding in a given selenoprotein [47,51]. According to ΔUGA, effects on selenoprotein translation seemed rather mild; just for *Selenop*, there was a significant change (Figure 6B). SELENOP is unique among mammalian selenoproteins for containing more than one Sec codon per polypeptide. Inspection of the ribosomal coverage along the mRNA revealed a reduced density 3' of the first UGA/Sec codon (Figure 6C). Similarly, a cumulative sum plot supported this finding in the *Trit1*-KO liver. In contrast to our expectations, no such effect was seen for *Gpx1* whatsoever (Figure 6D). Because the UGA/Sec codon resides in the penultimate position of the *Txnrd1* mRNA, ΔURE cannot be calculated for this selenoprotein. In the ribosomal coverage and cumulative sum plots, however, an impairment of UGA/Sec recoding was apparent (Figure 6E). In Figure 5A,B, SEPHS2 was clearly reduced in *Trit1*-KO liver. Similarly, UGA/Sec re-coding in *Sephs2* was reduced according to the ribosomal coverage and cumulative sum plots (Figure 6F). In agreement with higher protein amounts, we observed a slightly higher coverage on *Gpx4* after the UGA/Sec in the *Trit1* KO compared to Ctl (Figure 6G). Thus, under conditions of adequate dietary Se supply, only moderate effects were found on selenoprotein expression, when tRNA$^{[Ser]Sec}$ was lacking the i^{6}A$_{37}$ modification.

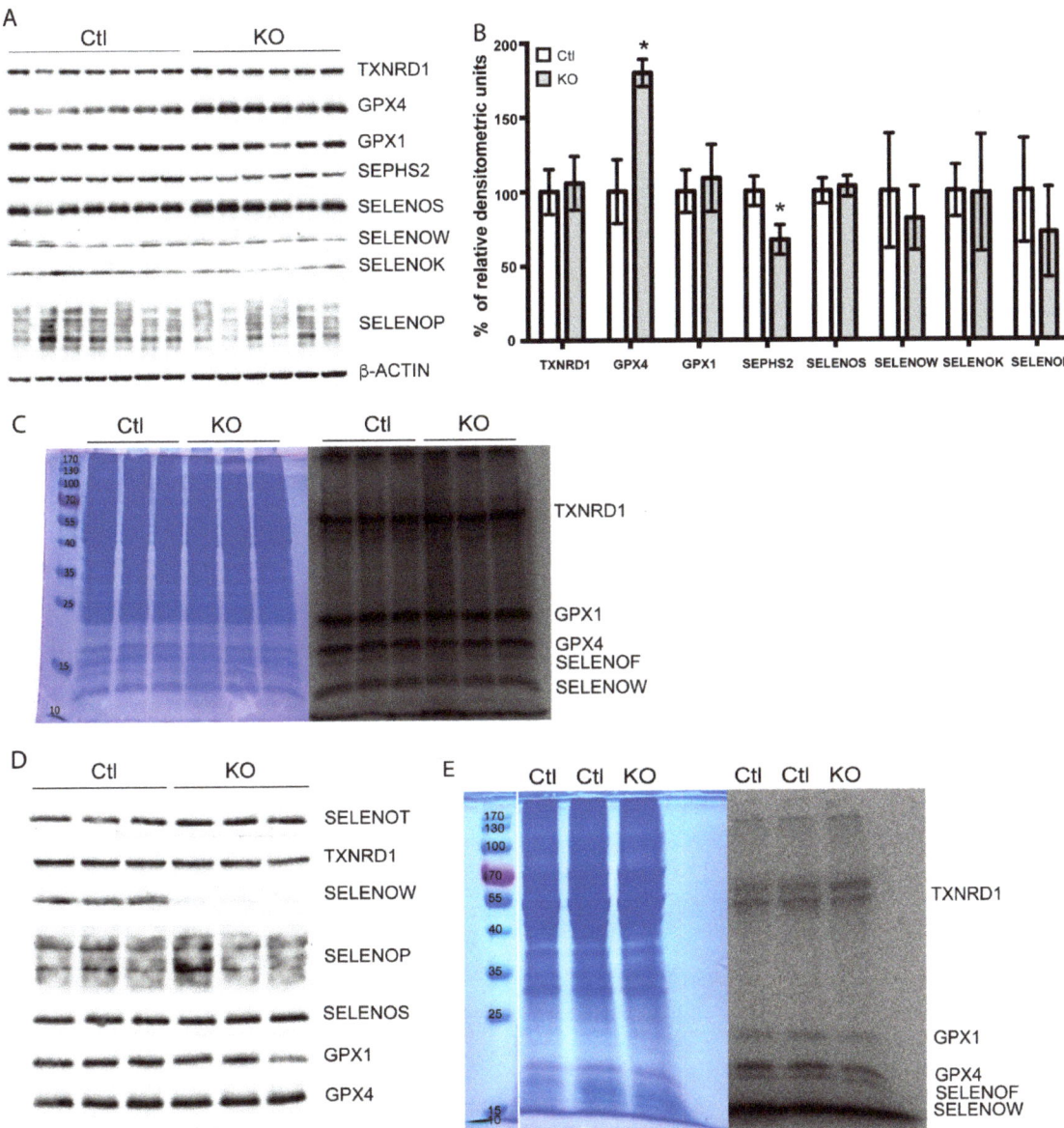

Figure 5. Selenoprotein expression in *Trit1*-knockout (KO) mice. (**A**) Western blot against a panel of 8 selenoproteins in mouse liver. N = 6–7 individual mice. Liver protein, 50 µg, separated on SDS-PAGE. (**B**) Densitometric analysis of the western blot in (**A**). Ponceau was used for normalization. Results are expressed as mean ± SD of the percentage relative to the control (Ctl). GPX4 and SEPHS2 showed significant differences according two-tailed *t*-test. * $p < 0.05$. *p*-values of GPX4 and SEPHS2 were 4×10^{-6} and 2.25×10^{-4}, respectively. (**C**) Metabolic labeling with ^{75}Se-selenite of isolated primary hepatocytes. Coomassie brilliant blue-stained gel for loading control (left) and autoradiogram (right). N = 3 individual cultures. (**D**) Selenoprotein western blot from cortices of neuron-specific *Trit1* KO mice. (**E**) ^{75}Se-labeling of *Trit1* KO and Ctl neuron cultures (a representative experiment). Coomassie showed equal loading.

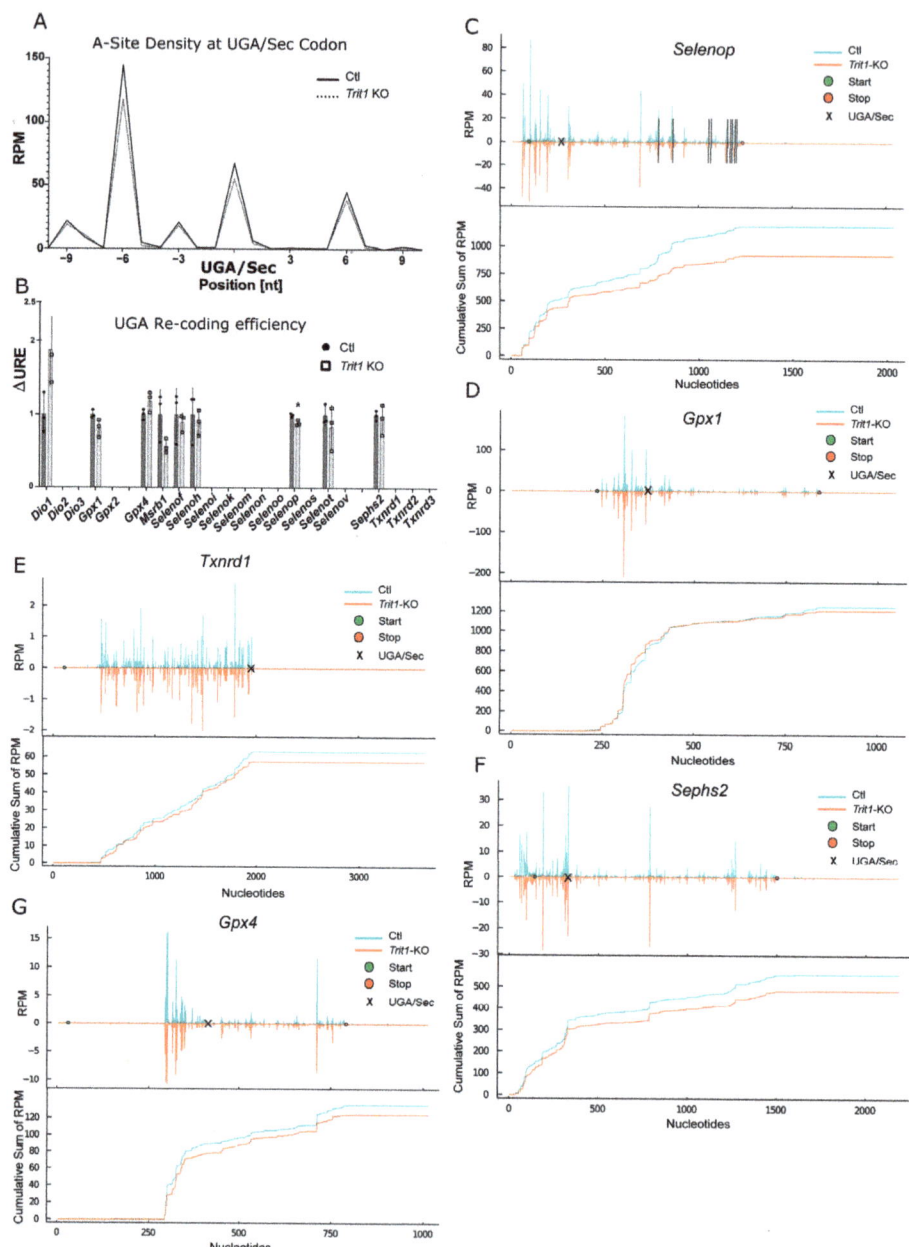

Figure 6. Selenoprotein RiboSeq analysis of *Trit1* knockout (KO) liver. (**A**) RPFs with the UGA/Sec in the A-site expressed as reads per million mapped reads (RPM) over all selenoproteins. (**B**) UGA recoding efficiency (URE, 3′RPF/5′RPF) calculated for selenoproteins with UGA/Sec far from the termination codon. ΔURE is calculated as URE(KO)/URE(Ctl). (**C–G**) RPF coverage of selected selenoprotein mRNAs in *Trit1* KO mouse liver. The mean values of the groups were plotted. Start and stop positions are marked as green and red circles. Reads are plotted in blue for control (Ctl) and in orange for *Trit1* KO livers. The position of the UGA/Sec codon is indicated by a black "x" mark. In the case of *Selenop*, following UGA codons after the first are displayed as black vertical lines. Cumulative sums of RPF are shown below the corresponding profiles. RPM: reads per million mapped reads.

2.6. Effect of 2′O-Methylation of U_{34} in tRNA[Ser]Sec

It has been proposed that 2′O-methylation is a Se-dependent process, and methylated tRNA[Ser]Sec is superior to less modified tRNA[Ser]Sec in supporting Sec incorporation into GPX1 and other stress-related selenoproteins. Because selenoprotein expression is generally more stable in females than in males, we speculated that FTSJ1, which is associated with X-linked mental disability in humans [52], might represent the elusive 2′-O-methyltransferase. We recently inactivated the *Ftsj1* gene in mice and demonstrated by mass-spectrometry that 2′O-methylation of U_{34} in tRNA[Ser]Sec is entirely undetectable in tRNA[Ser]Sec isolated from *Ftsj1* mutant mice [53]. This work has included ribosomal profiling of *Ftsj1*-deficient brain, but expression of selenoproteins was not specifically investigated. Here, we subjected the dataset from the earlier study to our analysis pipeline regarding selenoprotein expression (Figure 7). Plotting the density of RPFs around the UGA/Sec codon of all selenoproteins showed absolutely no difference between controls and *Ftsj1*-KO mice (Figure 7A). Calculation of ΔURE likewise showed no differences, in particular, for *Gpx1*, the selenoprotein best known for its response to Um_{34} modification in tRNA[Ser]Sec (Figure 7B). The ribosomal coverage and cumulative sum plots of *Selenop* did not show any impact of the *Ftsj1* inactivation despite 10 UGA/Sec codons in the open reading frame (Figure 7C). Finally, ribosomal coverage of *Gpx1* was not reduced either (Figure 7D).

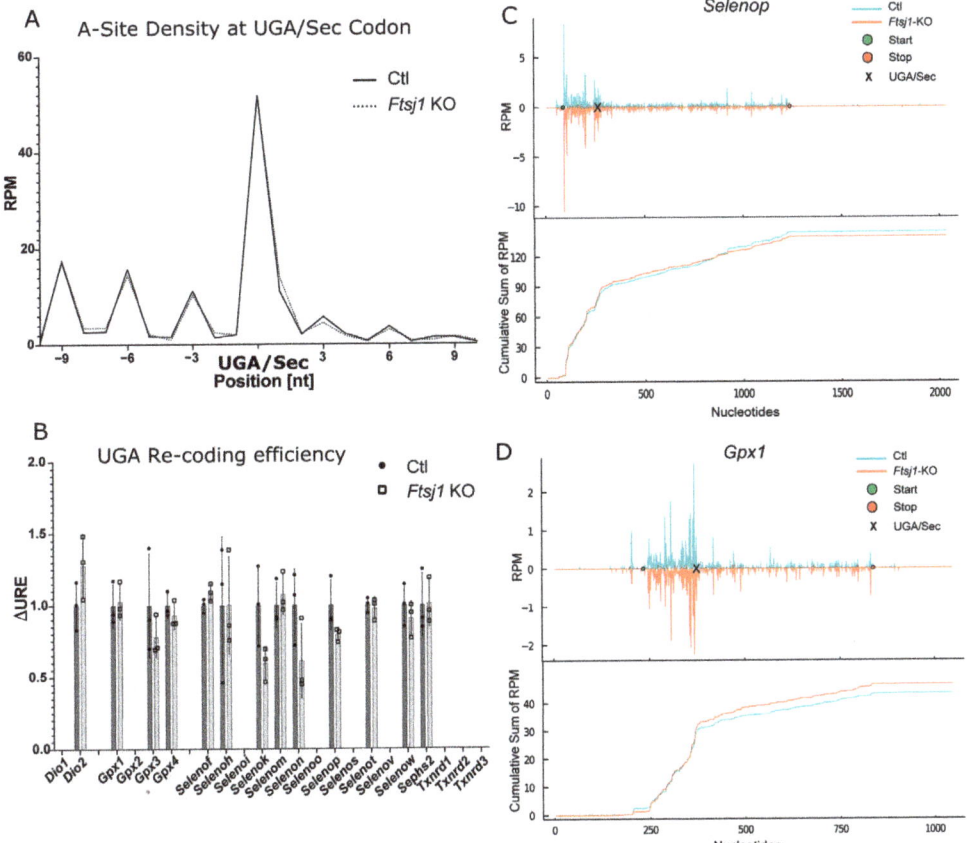

Figure 7. Re-analysis of *Fstj1* knockout (KO) brain RiboSeq data focussed on selenoproteins. (**A**) RPFs with the UGA/Sec in the A-site expressed as reads per million mapped reads (RPM) over all selenoproteins. (**B**) UGA recoding efficiency

(URE, 3′RPF/5′RPF) calculated for selenoproteins with UGA/Sec far from the termination codon. ΔURE is calculated as URE(KO)/URE(Ctl). (**C**,**D**) RPF coverage of selected selenoprotein mRNAs in *Ftsj1* KO mouse brain. The mean values of the groups were plotted. Start and stop positions are marked as green and red circles. Reads are plotted in blue for control (Ctl) and in orange for *Ftsj1* KO brains. The position of the UGA/Sec codon is indicated by a black "x" mark. Cumulative sums of RPF are shown below the corresponding profiles. RPM: reads per million mapped reads.

3. Discussion

Expression of selenoproteins is governed by the availability of Se. Dietary Se restriction, interference with Se transport within the body or pathogenic variations in genes encoding certain biosynthesis factors have a major impact on selenoprotein biosynthesis [4]. The above effects all converge on the availability of amino-acylated tRNA[Ser]Sec (Sec-tRNA[Ser]Sec). This notion is supported by a hypomorphic mouse model with a promoter mutation in the gene encoding tRNA[Ser]Sec [23].

A large body of evidence suggests that hierarchical expression of selenoproteins is modulated, perhaps governed, by modification of tRNA[Ser]Sec. However, most of the studies delineating the function of tRNA[Ser]Sec in selenoprotein expression were based on (over-)expression of mutant tRNA[Ser]Sec in the presence or not of endogenous, functional tRNA[Ser]Sec. In particular, the mouse model expressing A37G mutant tRNA[Ser]Sec has been the subject of many studies [30,33]. However, multiple copies of the mutant transgene have integrated into the mouse genome, and a direct effect of the base exchange on tRNA structure may also affect tRNA charging, binding to the elongation factor, or decoding in the ribosome. Hence, we wanted to address the question of tRNA[Ser]Sec modification from the side of the modifying enzyme and studied cell and animal models deficient in the tRNA-isopentenyltransferase TRIT1.

Besides tRNA[Ser]Sec, this enzyme modifies several substrates, among them, cytosolic tRNA^{Ser}(UCN) and several mitochondrial tRNAs [21,36,42,54]. In fact, patients carrying pathogenic *TRIT1* variants show a mitochondrial phenotype [43,55]. Yet, although the p.R323Q variant greatly diminished TRIT1 activity towards tRNA[Ser]Sec in vitro, we found no evidence that selenoprotein expression was generally reduced in patient fibroblasts. In fact, a deficiency of selenoproteins is usually reflected by an induction of NRF2-target genes [56,57]. In these cells, however, many genes known to be induced by NRF2 in selenoprotein deficiency are not up- but down-regulated.

The most direct way to assess the effect of tRNA isopentenylation in translation of selenoproteins is gene targeting of the responsible enzyme, TRIT1. We have generated conditional *Trit1*-knockout mice and analyzed selenoprotein expression in mouse liver and cultured hepatocytes. In liver, Western blotting showed only SELENOP and SEPHS2 levels moderately reduced, while GPX4 was even increased. We thus used ribosomal profiling to assess translation through UGA/Sec in mouse liver. When we summed up all ribosome protected fragments of selenoproteins with the UGA/Sec codon in the A-site, we found a small decrease in *Trit1*-KO liver. Individual analyses of all selenoproteins expressed in mouse liver supported reduced translation of *Sephs2*, *Selenop*, and *Txnrd1* after the UGA/Sec codons. GPX1 protein level and *Gpx1* translation were not altered in the *Trit1* mutant. The decrease in GPX1 and the preservation of TXNRD1, however, were among the key observations in the A37G mutant tRNA mouse model. Thus, we conclude that the effect of the A37G mutation does not result from the lack of isopentenylation of base 37, but from the base exchange.

It has been proposed that the A37G mutation in tRNA[Ser]Sec leads to hypomodification of U_{34}. Base 34 carries two modifications: mcm^{5}U and 2′-O methylation (mcm^{5}Um$_{34}$). Since the U34A(I) mutated tRNA[Ser]Sec also shows massively reduced selenoprotein expression, it was concluded that lack of ribose 2′-O methylation was the reason for reduced UGA/Sec translation in both mouse models [31]. In fact, exposure of endothelial cells to an inhibitor of S-adenosylhomocysteine (S-Ado-Hcy) hydrolase reduced both GPX1 and TXNRD1 expression [58]. The authors showed that increased levels of S-Ado-Hcy increased the level of mcm^{5}U tRNA[Ser]Sec at the expense of the mcm^{5}Um isoform, possibly through inhibition of the elusive 2′O-methyltransferase [58]. Gene targeting in mice showed that FTSJ1 is the 2′O-methylase of U_{34} in tRNA$^{[Ser]Sec}$ [53]. However, our data do not support a role of FTSJ1 in UGA/Sec re-coding during selenoprotein translation in mice fed adequate Se levels in their diet. It is still possible that analysis of mice fed a Se-deficient diet may reveal an effect of Um_{34} in tRNA$^{[Ser]Sec}$ on selenoprotein translation. Apart from this possibility, is there any other way to reconcile these seemingly conflicting observations?

It is interesting that UGA/Sec re-coding in *Gpx1* is not (always) sensitive to full mcm^{5}U modification [59]. So, we wonder whether one could look at the available data in another way: formation of mcm^{5}Um$_{34}$ in tRNAs is not unique for tRNA$^{[Ser]Sec}$, but may represent a general mechanism to cope with oxidative stress [60]. Impaired expression of selenoproteins, in turn, leads to oxidative stress [61]. Thus, mcm^{5}Um methylation and GPX1 activity may correlate, but not necessarily through tRNA modification. This idea would explain why co-administration of the antioxidant N-acetylcysteine with the S-Ado-Hcy hydrolase inhibitor rescued GPX1 expression [58]. An antioxidant should not be able to replace a specific methylase activity.

The modification of U_{34} is, in fact, important, as shown by two groups that independently targeted the *Alkbh8* gene in mice [38,39]. ALKBH8 is the methylase forming the methyl-ester in mcm^{5}U. Lack of this modification clearly impairs translation of GPX1 in liver and fibroblasts [38,39], while in lungs, TXRND1 is more affected [59]. We hypothesized that inactivation of the elongator complex, which initiates the mcm^{5}U modification, should impair selenoprotein translation. A paper targeting *Elp3* in mouse developing cortex has provided ribosomal profiling data [62]. We analyzed this dataset using the methodology presented here and found that *Gpx1* indeed shows decreased UGA/Sec read-through (Figure 8). Due to the low sequencing depth of this experiment, it is difficult to make statements on less abundantly expressed selenoproteins, and a future experiment should analyze a hepatocyte-specific *Elp3* knockout model.

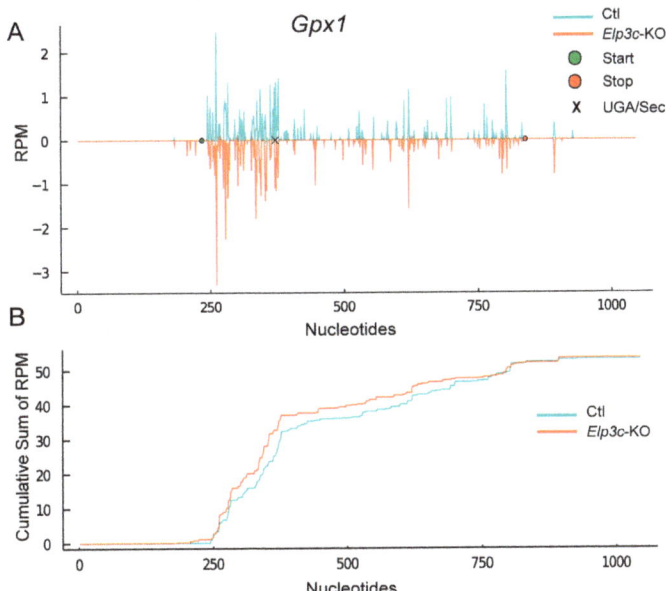

Figure 8. Ribosomal profiling of *Gpx1* in *Elp3* knockout (KO) developing brain. (**A**) Ribosomal coverage plot. The position of the UGA/Sec codon is indicated by a black "x" mark. Note the decreased ribosomal coverage in *Elp3* KO 3′ from the UGA/Sec codon. (**B**) Cumulative sum plot. The net translation of *Gpx1* in *Elp3* KO is apparently adjusted by increased translation/initiation 5′ of the UGA/Sec. Data from [62] were re-analyzed with the methods presented here.

It is an intriguing observation that interference with different positions in tRNA[Ser]Sec and different modifications leads to very specific effects on the expression of only a subset of selenoproteins: lack of i^6A_{37} affects TXNRD1 and GPX4, but not GPX1. Inactivation of *Alkbh8* affects GPX1 in liver and fibroblasts, but TXNRD1 in lung. Mutation of A_{37} to G reduces GPX1 expression but not TXNRD1. Research into this mechanism will profit from using the same type of cell or organ and the same methodology. Finally, Se availability may mitigate or potentiate effects of tRNA modification [33]. What remains beyond is the question of how tRNA modification can differentially affect the UGA/Sec re-coding event in different selenoproteins. Here, we are lacking data on the mammalian ribosome in complex with SECISBP2, mRNA, and tRNA[Ser]Sec. It is conceivable that codon context, i.e., bases 5′ and 3′ from the UGA will modulate how the codon and the anticodon accommodate in the ribosomal decoding center. Transfer RNA[Ser]Sec modifications may enhance decoding or not, e.g., in the bacterial cryo-EM structure, a hydrogen bond is observed between 2′O-U_{34} and 5′O-C_{35} [35]. Hence, a 2′O methylated U_{34} may be able to modulate codon:anticodon interactions. In the bacterial situation, the two bases following the UGA codon engage in stacking interactions with bases from the 16S ribosomal RNA in which the hypermodified U_{34} is involved. Thus, a sequence-specific communication between codon context and tRNA modification is conceivable and awaits experimental verification.

4. Materials and Methods

4.1. Mouse Model

The generation and further characterization of the conditional *Trit1* mouse model will be described elsewhere in the context of the impact of TRIT1 on translational fidelity (Bohleber, Fradejas, Suzuki, Schweizer et al., in preparation). Animal experiments were performed according to approval by the LANUV Recklinghausen (AZ 84-02.04.2014.A436 and 81-02.04.2020.A042).

4.2. Human Fibroblast Culture

Cells were cultured following the same procedures described [43].

4.3. Hepatocyte Culture

Procedure was described in [51]. After perfusion of mice, livers were mechanically disaggregated into DMEM high Glucose, 10% FBS, 1% Glutamine and 1% Penicillin/Streptomycin. Cell suspension was passed through a cell strainer before going through serial centrifugation/resuspension steps. Finally, cells were counted and seeded in collagen-coating plates. Experiments were performed the following day.

4.4. Neuron Culture

The procedure was followed as indicated in Beaudoin et al. [63]. The cortices of each pup (P1) were dissected and individually kept in separate Eppendorf tubes with 1 mL dissection medium. After trypsinization for 20 min at 37 °C, trypsin was removed and cortices were washed with plating medium and triturated with a polished Pasteur pipette against a Petri dish. Cells were passed through a cell strainer, counted, and plated with 1 million cells per 100 mm plate coated with poly-L-lysine. The following day medium was replaced with maintenance medium. After two days in culture, Ara-C was added to a final concentration of 5 µM and kept for one day until half of the medium was replaced by fresh maintenance medium. Cultures were used for experiments around 10 days post-plating.

4.5. Western Blot

Mouse tissues (liver and cortex) and confluent human fibroblasts were homogenized in RIPA lysis buffer containing protease inhibitors (Roche, Basel, Switzerland). Protein extracts were resolved by 12% SDS-PAGE, transferred to nitrocellulose membranes (GE Healthcare, Chicago, IL, USA) and immunoblotted using the antibodies against listed in Table S1. Detection was performed with horseradish peroxidase-conjugated anti-mouse or anti-rabbit

IgG (Jackson ImmunoResearch, West Grove, PA, USA) and an enhanced chemiluminescence detection system (Supersignal West Dura, Thermo Scientific, Waltham, MA, USA) using Fusion Solo detector (Vilber Lourmat Deutschland GmbH, Eberhardzell, Germany).

4.6. ^{75}Se Labeling

Confluent human fibroblast, neuron, and hepatocyte cultures grown in 100 mm plates were labelled overnight with radioactive sodium selenite (Na$_2$[^{75}Se]O$_3$) (10 µCi/plate). Cells were washed with 1× PBS and lysed in RIPA buffer. Then, 50 µg of lysate were separated by SDS-PAGE (12% gel). Coomassie blue staining was performed before gel drying (Gel dryer Bio-Rad, Bio-Rad, Hercules, CA, USA). Autoradiography was obtained using a BAS-1800 II (Fujifilm, Tokyo, Japan) Phosphoimager.

4.7. Transfer RNA Modification Index by RT-PCR

The qualitative determination of the ms^{2}i^{6}A modification in human mitochondrial tRNAs was adapted from [45]. Total RNA was extracted with Trizol (Invitrogen, Waltham, MA, USA) following the manufacturer's instructions. DNase treatment and cDNA synthesis were performed according to Xie et al. [45], using RQ1 RNase-free DNase (Promega, Madison, WI, USA) and the Transcriptor First Strand cDNA Synthesis Kit (Roche, Basel, Switzerland). Primers used for cDNA synthesis and qPCR were previously published in [64]. qPCR was performed using Absolute qPCR SYBER Green according to the manufacturer's instructions in a Mastercycler epgradient S realplex (Eppendorf, Hamburg, Germany). Specific primer annealing temperatures were determined by gradient PCR (see Table S2). Amplified products were verified by melting curve analysis and gel electrophoresis. Modification indexes were calculated as in [45].

4.8. TRIT1 In Vitro Assay

Human TRIT1 gene was cloned, the p.R323Q variant introduced, and recombinant protein (wild type or variant TRIT1) purified using the same methods as for the mouse TRIT1 in [42]. Primers used for cloning and site direct mutagenesis are shown in Table S3. Reactions were performed using the same conditions as in [42], but with slight variations. Different anticodon stem loop RNA primers were used as substrate (Table S4). Then, 2.5 U of pyrophosphatase (Genecraft, Cologne, Germany) were added to the mixture. The reaction was stopped after 10 min by adding 100 µL of ice-cold 10% TCA. Precipitation was done via a modified TCA precipitation protocol [65]. The whole volume of reaction tube was transferred to a Whatman filter paper and air dried for 15 min. Afterwards, it was washed in TCA (10%), in EtOH (95%), and in diethyl ether. The filter paper was air dried for 30 min between the washing steps and for 60 min after the last one. Scintillation liquid was added to the filter papers and a measurement was measured in a LS 6500 scintillation counter (Beckman, Pasadena, CA, USA).

4.9. tRNA$^{[Ser]Sec}$ Northern Blot

The procedure and probes used were previously described [51].

4.10. Quantification of i^{6}A by LC-MS

First, 500 ng tRNA were digested into nucleotides using 0.3 U nuclease P1 from *P. citrinum* (Sigma-Aldrich, St. Louis, MI, USA), 0.1 U snake venom phosphodiesterase from *C. adamanteus* (Worthington, Columbus, OH, USA), 200 ng Pentostatin (Sigma-Aldrich, St. Louis, MI, USA), and 500 ng Tetrahydrouridine (Merck-Millipore, Burlington, MA, USA) in 5 mM ammonium acetate (pH 5.3; Sigma-Aldrich, St. Louis, MI, USA) for two hours at 37 °C. The remaining phosphates were removed by 1 U FastAP (Thermo Scientific, Waltham, MA, USA) in 10 mM ammonium acetate (pH 8) for one hour at 37 °C. The nucleosides were then spiked with internal standard (^{13}C stable isotope-labeled nucleosides from *E. coli*, SIL-IS) and subjected to analysis. Technical triplicates with 26.6 ng digested RNA and 20 ng internal standard were analyzed via LC–MS (Agilent 1260 series and Agilent 6460

Triple Quadrupole mass spectrometer equipped with an electrospray ion source (ESI)). The solvents consisted of 5 mM ammonium acetate buffer (pH 5.3; solvent A) and LC–MS grade acetonitrile (solvent B; Honeywell, Charlotte, NC, USA). The elution started with 100% solvent A with a flow rate of 0.35 mL/min, followed by a linear gradient to 10% solvent B at 20 min, 25% solvent B at 30 min and 80% solvent B after 40 min. Initial conditions were regenerated with 100% solvent A for 14 min. The column used was a Synergi Fusion (4 µM particle size, 80 Å pore size, 250 × 2.0 mm; Phenomenex, Torrance CA, USA). The UV signal at 254 nm was recorded via a diode array detector (DAD) to monitor the main nucleosides. ESI parameters were as follows: gas temperature 350 °C, gas flow 8 L/min, nebulizer pressure 50 psi, sheath gas temperature 300 °C, sheath gas flow 12 L/min, and capillary voltage 3500 V. The MS was operated in the positive ion mode using Agilent MassHunter software in the dynamic MRM (multiple reaction monitoring) mode. For relative quantification, the signals of i^6A were normalized to the ^{13}C-labeled signal and then normalized to the UV signal of guanosine.

4.11. Isolation and LC/MS Analysis of tRNA$^{[Ser]Sec}$

Mouse liver total RNA was separated by anion exchange chromatography with DEAE Sepharose Fast Flow (GE Healthcare, Chicago, IL, USA) to obtain crude tRNAs with removal of polysaccharides and rRNA [66]. Cytoplasmic tRNA$^{[Ser]Sec}$ was isolated from the crude tRNAs by reciprocal circulating chromatography, as described in [48]. The 5'-EC amino-modified DNA probe (Sigma-Aldrich, St. Louis, MI, USA), TGGGCCCGAAAGGTG-GAATTGAACCACTCTGTCGCTAGAC was covalently immobilized on NHS-activated Sepharose 4 Fast Flow (GE Healthcare, Chicago, IL, USA). About 6 µg of tRNA$^{[Ser]Sec}$ were obtained from 1.4 mg crude tRNAs. Mouse tRNA$^{[Ser]Sec}$ was digested by RNase T1 (Thermo Fisher Scientific, Waltham, MA, USA), and subjected to capillary liquid chromatography (LC) coupled to nano electrospray (ESI)/mass spectrometry (MS) on a linear ion trap-Orbitrap hybrid mass spectrometer (LTQ Orbitrap XL; Thermo Fisher Scientific, Waltham, MA, USA), as described in [49,50]. The RNA fragments were scanned in a negative polarity mode over a range of m/z 600–2000.

4.12. 3'-. RNA Sequencing

RNA was extracted from human fibroblasts with TRIzol Reagent (Invitrogen, Waltham, MA, USA) according to the manufacturer protocol. Approximately, 500 ng of RNA were used for library preparation with QuantSeq 3'-mRNA Library Prep (Lexogen, Vienna, Austria). Sequencing was performed by the Illumina HiSeq 2500 instrument on 50-cycle single-end mode.

4.13. RiboSeq

Treatment of the samples was performed as described previously [47,51], with some changes for generating the RPF. Cycloheximide was omitted from the lysis buffer. All steps were carried out on ice. Then, 50 mg of frozen mouse liver was crushed in 1000 µL ice-cold lysis buffer using a pellet pestle. Lysate was pipetted 3 times up and down with a 1000 µL pipette before it was passed through a 26-gauge needle for 10 times. After 10 min incubation on ice, the lysate was centrifuged at $20.000 \times g$ for 10 min at 4 °C. The supernatant was transferred to a new 1.5 mL reaction tube. Then, 200 µL of the lysate were incubated for 60 min with 1000 U of RNase I at 25 °C and at 1300 rpm in a thermoblock. Pre-processing and alignments of the reads were performed as described before [47]. For analysis, 28 nt and 29 nt read sizes and an offset of 12 nt to the P-site were used. Three individual mouse livers were used per genotype.

Supplementary Materials: All data are available online at https://www.mdpi.com/article/10.3390/ijms222111454/s1.

Author Contributions: Conceptualization, N.F.-V., S.B. and U.S.; Data curation, S.B.; Formal analysis, N.F.-V. and S.B.; Funding acquisition, R.M., R.W.T., T.S. and U.S.; Investigation, N.F.-V., S.B., W.Z., U.R., A.K., R.K., Y.M., K.M. and Y.S.; Methodology, N.F.-V. and S.B.; Resources, R.M. and R.W.T.; Software, S.B.; Supervision, M.H. and U.S.; Visualization, N.F.-V. and S.B.; Writing—original draft, U.S.; Writing—review and editing, N.F.-V., S.B., R.M., R.W.T., T.S. and U.S. All authors have read and agreed to the published version of the manuscript.

Funding: This research was funded by Deutsche Forschungsgemeinschaft (DFG), SCHW914/2-2, SCHW914/5-1, and Uniklinikum Bonn. R.M. and R.W.T. are supported by the Wellcome Centre for Mitochondrial Research (203105/Z/16/Z), the Medical Research Council (MRC) International Centre for Genomic Medicine in Neuromuscular Disease (MR/S005021/1), the Mitochondrial Disease Patient Cohort (UK) (G0800674), the UK NIHR Biomedical Research Centre for Ageing and Age-related disease award to the Newcastle upon Tyne Foundation Hospitals NHS Trust, the Lily Foundation, and the UK NHS Specialist Commissioners which funds the "Rare Mitochondrial Disorders of Adults and Children" Diagnostic Service in Newcastle upon Tyne. R.W.T. also receives financial support from the Pathology Society. This work was supported by Grants-in-Aid for Scientific Research on Priority Areas from the Ministry of Education, Culture, Sports, Science, and Technology of Japan (MEXT), Japan Society for the Promotion of Science (JSPS) [20292782 and 18H05272 to T.S.], and Exploratory Research for Advanced Technology (ERATO, JPMJER2002 to T.S.) from Japan Science and Technology Agency (JST).

Institutional Review Board Statement: The study was conducted according to German law and was approved by the local authorities (LANUV Recklinghausen AZ 84-02.04.2014.A436 on 7.10.2013, and 81-02.04.2020.A042 on 16.04.2020).

Informed Consent Statement: Informed consent was obtained from all subjects involved in the study.

Data Availability Statement: The ribosomal profiling data on *Trit1*-KO liver was submitted to GEO (GSE183923).

Acknowledgments: We would like to dedicate this article to Dolph Hatfield, Bethesda, Maryland, who identified tRNA[Ser]Sec and several biosynthetic co-factors of selenocysteine formation. N.F.-V. and U.S. thank him for his support during their scientific careers.

Conflicts of Interest: The authors declare no conflict of interest.

References

1. Behne, D.; Hilmert, H.; Scheid, S.; Gessner, H.; Elger, W. Evidence for specific selenium target tissues and new biologically important selenoproteins. *Biochim. Biophys. Acta* **1988**, *966*, 12–21. [CrossRef]
2. Schomburg, L.; Schweizer, U. Hierarchical regulation of selenoprotein expression and sex-specific effects of selenium. *Biochim. Biophys. Acta* **2009**, *1790*, 1453–1462. [CrossRef]
3. Burk, R.F.; Hill, K.E. Regulation of selenium metabolism and transport. *Annu. Rev. Nutr.* **2015**, *35*, 109–134. [CrossRef] [PubMed]
4. Schweizer, U.; Bohleber, S.; Zhao, W.; Fradejas-Villar, N. The neurobiology of selenium: Looking back and to the future. *Front. Neurosci* **2021**, *15*, 652099. [CrossRef]
5. Wingler, K.; Bocher, M.; Flohe, L.; Kollmus, H.; Brigelius-Flohe, R. Mrna stability and selenocysteine insertion sequence efficiency rank gastrointestinal glutathione peroxidase high in the hierarchy of selenoproteins. *Eur. J. Biochem.* **1999**, *259*, 149–157. [CrossRef] [PubMed]
6. Müller, C.; Wingler, K.; Brigelius-Flohé, R. 3′utrs of glutathione peroxidases differentially affect selenium-dependent mrna stability and selenocysteine incorporation efficiency. *Biol. Chem.* **2003**, *384*, 11–18. [CrossRef]
7. Squires, J.E.; Stoytchev, I.; Forry, E.P.; Berry, M.J. Sbp2 binding affinity is a major determinant in differential selenoprotein mrna translation and sensitivity to nonsense-mediated decay. *Mol. Cell. Biol.* **2007**, *27*, 7848–7855. [CrossRef]
8. Chavatte, L.; Brown, B.A.; Driscoll, D.M. Ribosomal protein l30 is a component of the uga-selenocysteine recoding machinery in eukaryotes. *Nat. Struct. Mol. Biol.* **2005**, *12*, 408–416. [CrossRef]
9. Budiman, M.E.; Bubenik, J.L.; Miniard, A.C.; Middleton, L.M.; Gerber, C.A.; Cash, A.; Driscoll, D.M. Eukaryotic initiation factor 4a3 is a selenium-regulated rna-binding protein that selectively inhibits selenocysteine incorporation. *Mol. Cell* **2009**, *35*, 479–489. [CrossRef]
10. Wu, R.; Shen, Q.; Newburger, P.E. Recognition and binding of the human selenocysteine insertion sequence by nucleolin. *J. Cell. Biochem.* **2000**, *77*, 507–516. [CrossRef]
11. Hatfield, D.; Portugal, F.H. Seryl-trna in mammalian tissues: Chromatographic differences in brain and liver and a specific response to the codon, uga. *Proc. Natl. Acad. Sci. USA* **1970**, *67*, 1200–1206. [CrossRef]

12. Lee, B.J.; Worland, P.J.; Davis, J.N.; Stadtman, T.C.; Hatfield, D.L. Identification of a selenocysteyl-trna(ser) in mammalian cells that recognizes the nonsense codon, uga. *J. Biol. Chem.* **1989**, *264*, 9724–9727. [CrossRef]
13. Carlson, B.A.; Xu, X.M.; Kryukov, G.V.; Rao, M.; Berry, M.J.; Gladyshev, V.N.; Hatfield, D.L. Identification and characterization of phosphoseryl-trna[Ser]sec kinase. *Proc. Natl. Acad. Sci. USA* **2004**, *101*, 12848–12853. [CrossRef] [PubMed]
14. Xu, X.M.; Mix, H.; Carlson, B.A.; Grabowski, P.J.; Gladyshev, V.N.; Berry, M.J.; Hatfield, D.L. Evidence for direct roles of two additional factors, secp43 and soluble liver antigen, in the selenoprotein synthesis machinery. *J. Biol. Chem.* **2005**, *280*, 41568–41575. [CrossRef] [PubMed]
15. Xu, X.M.; Carlson, B.A.; Mix, H.; Zhang, Y.; Saira, K.; Glass, R.S.; Berry, M.J.; Gladyshev, V.N.; Hatfield, D.L. Biosynthesis of selenocysteine on its trna in eukaryotes. *PLoS Biol.* **2007**, *5*, e4. [CrossRef]
16. Yuan, J.; Palioura, S.; Salazar, J.C.; Su, D.; O'Donoghue, P.; Hohn, M.J.; Cardoso, A.M.; Whitman, W.B.; Söll, D. Rna-dependent conversion of phosphoserine forms selenocysteine in eukaryotes and archaea. *Proc. Natl. Acad. Sci. USA* **2006**, *103*, 18923–18927. [CrossRef]
17. Sturchler, C.; Lescure, A.; Keith, G.; Carbon, P.; Krol, A. Base modification pattern at the wobble position of xenopus selenocysteine trna(sec). *Nucleic Acids Res.* **1994**, *22*, 1354–1358. [CrossRef] [PubMed]
18. Schön, A.; Böck, A.; Ott, G.; Sprinzl, M.; Söll, D. The selenocysteine-inserting opal suppressor serine trna from e. Coli is highly unusual in structure and modification. *Nucleic Acids Res.* **1989**, *17*, 7159–7165. [CrossRef]
19. Diamond, A.M.; Choi, I.S.; Crain, P.F.; Hashizume, T.; Pomerantz, S.C.; Cruz, R.; Steer, C.J.; Hill, K.E.; Burk, R.F.; McCloskey, J.A. Dietary selenium affects methylation of the wobble nucleoside in the anticodon of selenocysteine trna([Ser]sec). *J. Biol. Chem.* **1993**, *268*, 14215–14223. [CrossRef]
20. Kim, L.K.; Matsufuji, T.; Matsufuji, S.; Carlson, B.A.; Kim, S.S.; Hatfield, D.L.; Lee, B.J. Methylation of the ribosyl moiety at position 34 of selenocysteine trna[Ser]sec is governed by both primary and tertiary structure. *RNA* **2000**, *6*, 1306–1315. [CrossRef] [PubMed]
21. Suzuki, T. The expanding world of trna modifications and their disease relevance. *Nat. Rev. Mol. Cell. Biol.* **2021**, *22*, 375–392. [CrossRef] [PubMed]
22. Carlson, B.A.; Lee, B.J.; Tsuji, P.A.; Copeland, P.R.; Schweizer, U.; Gladyshev, V.N.; Hatfield, D.L. Selenocysteine trna([Ser]sec), the central component of selenoprotein biosynthesis: Isolation, identification, modification, and sequencing. *Methods Mol. Biol.* **2018**, *1661*, 43–60.
23. Carlson, B.A.; Schweizer, U.; Perella, C.; Shrimali, R.K.; Feigenbaum, L.; Shen, L.; Speransky, S.; Floss, T.; Jeong, S.J.; Watts, J.; et al. The selenocysteine trna staf-binding region is essential for adequate selenocysteine trna status, selenoprotein expression and early age survival of mice. *Biochem. J.* **2009**, *418*, 61–71. [CrossRef] [PubMed]
24. Bösl, M.R.; Takaku, K.; Oshima, M.; Nishimura, S.; Taketo, M.M. Early embryonic lethality caused by targeted disruption of the mouse selenocysteine trna gene (trsp). *Proc. Natl. Acad. Sci. USA* **1997**, *94*, 5531–5534. [CrossRef]
25. Kumaraswamy, E.; Carlson, B.A.; Morgan, F.; Miyoshi, K.; Robinson, G.W.; Su, D.; Wang, S.; Southon, E.; Tessarollo, L.; Lee, B.J.; et al. Selective removal of the selenocysteine trna [Ser]sec gene (trsp) in mouse mammary epithelium. *Mol. Cell. Biol.* **2003**, *23*, 1477–1488. [CrossRef] [PubMed]
26. Schweizer, U.; Streckfuss, F.; Pelt, P.; Carlson, B.A.; Hatfield, D.L.; Köhrle, J.; Schomburg, L. Hepatically derived selenoprotein p is a key factor for kidney but not for brain selenium supply. *Biochem. J.* **2005**, *386*, 221–226. [CrossRef] [PubMed]
27. Dumitrescu, A.M.; Liao, X.H.; Abdullah, M.S.; Lado-Abeal, J.; Majed, F.A.; Moeller, L.C.; Boran, G.; Schomburg, L.; Weiss, R.E.; Refetoff, S. Mutations in secisbp2 result in abnormal thyroid hormone metabolism. *Nat. Genet.* **2005**, *37*, 1247–1252. [CrossRef] [PubMed]
28. Schoenmakers, E.; Agostini, M.; Mitchell, C.; Schoenmakers, N.; Papp, L.; Rajanayagam, O.; Padidela, R.; Ceron-Gutierrez, L.; Doffinger, R.; Prevosto, C.; et al. Mutations in the selenocysteine insertion sequence-binding protein 2 gene lead to a multisystem selenoprotein deficiency disorder in humans. *J. Clin. Investig.* **2010**, *120*, 4220–4235. [CrossRef]
29. Schoenmakers, E.; Carlson, B.; Agostini, M.; Moran, C.; Rajanayagam, O.; Bochukova, E.; Tobe, R.; Peat, R.; Gevers, E.; Muntoni, F.; et al. Mutation in human selenocysteine transfer rna selectively disrupts selenoprotein synthesis. *J. Clin. Investig.* **2016**, *126*, 992–996. [CrossRef] [PubMed]
30. Carlson, B.A.; Xu, X.M.; Gladyshev, V.N.; Hatfield, D.L. Selective rescue of selenoprotein expression in mice lacking a highly specialized methyl group in selenocysteine trna. *J. Biol. Chem.* **2005**, *280*, 5542–5548. [CrossRef]
31. Carlson, B.A.; Moustafa, M.E.; Sengupta, A.; Schweizer, U.; Shrimali, R.; Rao, M.; Zhong, N.; Wang, S.; Feigenbaum, L.; Lee, B.J.; et al. Selective restoration of the selenoprotein population in a mouse hepatocyte selenoproteinless background with different mutant selenocysteine trnas lacking um34. *J. Biol. Chem.* **2007**, *282*, 32591–32602. [CrossRef] [PubMed]
32. Kasaikina, M.V.; Turanov, A.A.; Avanesov, A.; Schweizer, U.; Seeher, S.; Bronson, R.T.; Novoselov, S.N.; Carlson, B.A.; Hatfield, D.L.; Gladyshev, V.N. Contrasting roles of dietary selenium and selenoproteins in chemically induced hepatocarcinogenesis. *Carcinogenesis* **2013**, *34*, 1089–1095. [CrossRef] [PubMed]
33. Howard, M.T.; Carlson, B.A.; Anderson, C.B.; Hatfield, D.L. Translational redefinition of uga codons is regulated by selenium availability. *J. Biol. Chem.* **2013**, *288*, 19401–19413. [CrossRef]
34. Ganichkin, O.M.; Anedchenko, E.A.; Wahl, M.C. Crystal structure analysis reveals functional flexibility in the selenocysteine-specific trna from mouse. *PLoS ONE* **2011**, *6*, e20032. [CrossRef]

35. Fischer, N.; Neumann, P.; Bock, L.V.; Maracci, C.; Wang, Z.; Paleskava, A.; Konevega, A.L.; Schroder, G.F.; Grubmuller, H.; Ficner, R.; et al. The pathway to gtpase activation of elongation factor selb on the ribosome. *Nature* **2016**, *540*, 80–85. [CrossRef] [PubMed]
36. Schweizer, U.; Bohleber, S.; Fradejas-Villar, N. The modified base isopentenyladenosine and its derivatives in trna. *RNA Biol.* **2017**, 1–12. [CrossRef] [PubMed]
37. Hatfield, D.; Lee, B.J.; Hampton, L.; Diamond, A.M. Selenium induces changes in the selenocysteine trna[Ser]sec population in mammalian cells. *Nucleic Acids Res.* **1991**, *19*, 939–943. [CrossRef]
38. Songe-Moller, L.; van den Born, E.; Leihne, V.; Vagbo, C.B.; Kristoffersen, T.; Krokan, H.E.; Kirpekar, F.; Falnes, P.O.; Klungland, A. Mammalian alkbh8 possesses trna methyltransferase activity required for the biogenesis of multiple wobble uridine modifications implicated in translational decoding. *Mol. Cell. Biol.* **2010**, *30*, 1814–1827. [CrossRef] [PubMed]
39. Endres, L.; Begley, U.; Clark, R.; Gu, C.; Dziergowska, A.; Malkiewicz, A.; Melendez, J.A.; Dedon, P.C.; Begley, T.J. Alkbh8 regulates selenocysteine-protein expression to protect against reactive oxygen species damage. *PLoS ONE* **2015**, *10*, e0131335. [CrossRef] [PubMed]
40. Diamond, A.M.; Jaffe, D.; Murray, J.L.; Safa, A.R.; Samuels, B.L.; Hatfield, D.L. Lovastatin effects on human breast carcinoma cells. Differential toxicity of an adriamycin-resistant derivative and influence on selenocysteine trnas. *Biochem. Mol. Biol. Int.* **1996**, *38*, 345–355.
41. Warner, G.J.; Berry, M.J.; Moustafa, M.E.; Carlson, B.A.; Hatfield, D.L.; Faust, J.R. Inhibition of selenoprotein synthesis by selenocysteine trna[Ser]sec lacking isopentenyladenosine. *J. Biol. Chem.* **2000**, *275*, 28110–28119. [CrossRef]
42. Fradejas, N.; Carlson, B.A.; Rijntjes, E.; Becker, N.P.; Tobe, R.; Schweizer, U. Mammalian trit1 is a trna([Ser]sec)-isopentenyl transferase required for full selenoprotein expression. *Biochem. J.* **2013**, *450*, 427–432. [CrossRef] [PubMed]
43. Yarham, J.W.; Lamichhane, T.N.; Pyle, A.; Mattijssen, S.; Baruffini, E.; Bruni, F.; Donnini, C.; Vassilev, A.; He, L.; Blakely, E.L.; et al. Defective i6a37 modification of mitochondrial and cytosolic trnas results from pathogenic mutations in trit1 and its substrate trna. *PLoS Genet* **2014**, *10*, e1004424. [CrossRef] [PubMed]
44. Wirth, E.K.; Conrad, M.; Winterer, J.; Wozny, C.; Carlson, B.A.; Roth, S.; Schmitz, D.; Bornkamm, G.W.; Coppola, V.; Tessarollo, L.; et al. Neuronal selenoprotein expression is required for interneuron development and prevents seizures and neurodegeneration. *FASEB J.* **2010**, *24*, 844–852. [CrossRef]
45. Xie, P.; Wei, F.Y.; Hirata, S.; Kaitsuka, T.; Suzuki, T.; Suzuki, T.; Tomizawa, K. Quantitative pcr measurement of trna 2-methylthio modification for assessing type 2 diabetes risk. *Clinical. Chem.* **2013**, *59*, 1604–1612. [CrossRef] [PubMed]
46. Zhou, C.; Huang, R.H. Crystallographic snapshots of eukaryotic dimethylallyltransferase acting on trna: Insight into trna recognition and reaction mechanism. *Proc. Natl. Acad. Sci. USA* **2008**, *105*, 16142–16147. [CrossRef]
47. Zhao, W.; Bohleber, S.; Schmidt, H.; Seeher, S.; Howard, M.T.; Braun, D.; Arndt, S.; Reuter, U.; Wende, H.; Birchmeier, C.; et al. Ribosome profiling of selenoproteins in vivo reveals consequences of pathogenic secisbp2 missense mutations. *J. Biol. Chem.* **2019**, *294*, 14185–14200. [CrossRef]
48. Miyauchi, K.; Ohara, T.; Suzuki, T. Automated parallel isolation of multiple species of non-coding rnas by the reciprocal circulating chromatography method. *Nucleic Acids Res.* **2007**, *35*, e24. [CrossRef]
49. Suzuki, T.; Ikeuchi, Y.; Noma, A.; Suzuki, T.; Sakaguchi, Y. Mass spectrometric identification and characterization of rna-modifying enzymes. *Methods Enzym.* **2007**, *425*, 211–229.
50. Suzuki, T.; Yashiro, Y.; Kikuchi, I.; Ishigami, Y.; Saito, H.; Matsuzawa, I.; Okada, S.; Mito, M.; Iwasaki, S.; Ma, D.; et al. Complete chemical structures of human mitochondrial trnas. *Nat. Commun.* **2020**, *11*, 4269. [CrossRef] [PubMed]
51. Fradejas-Villar, N.; Seeher, S.; Anderson, C.B.; Doengi, M.; Carlson, B.A.; Hatfield, D.L.; Schweizer, U.; Howard, M.T. The rna-binding protein secisbp2 differentially modulates uga codon reassignment and rna decay. *Nucleic Acids Res.* **2017**, *45*, 4094–4107. [CrossRef] [PubMed]
52. Freude, K.; Hoffmann, K.; Jensen, L.R.; Delatycki, M.B.; des Portes, V.; Moser, B.; Hamel, B.; van Bokhoven, H.; Moraine, C.; Fryns, J.P.; et al. Mutations in the ftsj1 gene coding for a novel s-adenosylmethionine-binding protein cause nonsyndromic x-linked mental retardation. *Am. J. Hum. Genet.* **2004**, *75*, 305–309. [CrossRef] [PubMed]
53. Nagayoshi, Y.; Chujo, T.; Hirata, S.; Nakatsuka, H.; Chen, C.W.; Takakura, M.; Miyauchi, K.; Ikeuchi, Y.; Carlyle, B.C.; Kitchen, R.R.; et al. Loss of ftsj1 perturbs codon-specific translation efficiency in the brain and is associated with x-linked intellectual disability. *Sci. Adv.* **2021**, *7*, eabf3072. [CrossRef]
54. Suzuki, T.; Suzuki, T. A complete landscape of post-transcriptional modifications in mammalian mitochondrial trnas. *Nucleic Acids Res.* **2014**, *42*, 7346–7357. [CrossRef] [PubMed]
55. Kernohan, K.D.; Dyment, D.A.; Pupavac, M.; Cramer, Z.; McBride, A.; Bernard, G.; Straub, I.; Tetreault, M.; Hartley, T.; Huang, L.; et al. Matchmaking facilitates the diagnosis of an autosomal-recessive mitochondrial disease caused by biallelic mutation of the trna isopentenyltransferase (trit1) gene. *Hum. Mutat.* **2017**, *38*, 511–516. [CrossRef] [PubMed]
56. Sengupta, A.; Carlson, B.A.; Weaver, J.A.; Novoselov, S.V.; Fomenko, D.E.; Gladyshev, V.N.; Hatfield, D.L. A functional link between housekeeping selenoproteins and phase ii enzymes. *Biochem. J.* **2008**, *413*, 151–161. [CrossRef]
57. Seeher, S.; Atassi, T.; Mahdi, Y.; Carlson, B.A.; Braun, D.; Wirth, E.K.; Klein, M.O.; Reix, N.; Miniard, A.C.; Schomburg, L.; et al. Secisbp2 is essential for embryonic development and enhances selenoprotein expression. *Antioxid. Redox. Signal* **2014**, *21*, 835–849. [CrossRef]

58. Barroso, M.; Florindo, C.; Kalwa, H.; Silva, Z.; Turanov, A.A.; Carlson, B.A.; Tavares de Almeida, I.; Blom, H.J.; Gladyshev, V.N.; Hatfield, D.L.; et al. Inhibition of cellular methyltransferases promotes endothelial cell activation by suppressing glutathione peroxidase-1 expression. *J. Biol. Chem.* **2014**. [CrossRef] [PubMed]
59. Leonardi, A.; Kovalchuk, N.; Yin, L.; Endres, L.; Evke, S.; Nevins, S.; Martin, S.; Dedon, P.C.; Melendez, J.A.; Van Winkle, L.; et al. The epitranscriptomic writer alkbh8 drives tolerance and protects mouse lungs from the environmental pollutant naphthalene. *Epigenetics* **2020**, *15*, 1121–1138. [CrossRef]
60. Endres, L.; Dedon, P.C.; Begley, T.J. Codon-biased translation can be regulated by wobble-base trna modification systems during cellular stress responses. *RNA Biol.* **2015**, *12*, 603–614. [CrossRef]
61. Lee, M.Y.; Leonardi, A.; Begley, T.J.; Melendez, J.A. Loss of epitranscriptomic control of selenocysteine utilization engages senescence and mitochondrial reprogramming. *Redox Biol.* **2020**, *28*, 101375. [CrossRef]
62. Laguesse, S.; Creppe, C.; Nedialkova, D.D.; Prevot, P.P.; Borgs, L.; Huysseune, S.; Franco, B.; Duysens, G.; Krusy, N.; Lee, G.; et al. A dynamic unfolded protein response contributes to the control of cortical neurogenesis. *Dev. Cell.* **2015**, *35*, 553–567. [CrossRef]
63. Beaudoin, G.M., 3rd; Lee, S.H.; Singh, D.; Yuan, Y.; Ng, Y.G.; Reichardt, L.F.; Arikkath, J. Culturing pyramidal neurons from the early postnatal mouse hippocampus and cortex. *Nat. Protoc.* **2012**, *7*, 1741–1754. [CrossRef] [PubMed]
64. Wei, F.; Tomizawa, K. Measurement of 2-methylthio modifications in mitochondrial transfer rnas by reverse-transcription quantitative pcr. *Bio-Protoc.* **2016**, *6*, e1695. [CrossRef]
65. Igloi, G.L.; von der Haar, F.; Cramer, F. Experimental proof for the misactivation of amino acids by aminoacyl-trna synthetases. *Methods Enzym.* **1979**, *59*, 282–291.
66. Yoshida, M.; Kataoka, N.; Miyauchi, K.; Ohe, K.; Iida, K.; Yoshida, S.; Nojima, T.; Okuno, Y.; Onogi, H.; Usui, T.; et al. Rectifier of aberrant mrna splicing recovers trna modification in familial dysautonomia. *Proc. Natl. Acad. Sci. USA* **2015**, *112*, 2764–2769. [CrossRef] [PubMed]

International Journal of
Molecular Sciences

Article

The Impact of ZIP8 Disease-Associated Variants G38R, C113S, G204C, and S335T on Selenium and Cadmium Accumulations: The First Characterization

Zhan-Ling Liang [†], Heng Wee Tan [†], Jia-Yi Wu, Xu-Li Chen, Xiu-Yun Wang, Yan-Ming Xu * and Andy T. Y. Lau *

Laboratory of Cancer Biology and Epigenetics, Department of Cell Biology and Genetics, Shantou University Medical College, Shantou 515041, China; 16zlliang@stu.edu.cn (Z.-L.L.); hwtan@stu.edu.cn (H.W.T.); jiayiwu1@163.com (J.-Y.W.); 19xlchen@stu.edu.cn (X.-L.C.); 19xywang@stu.edu.cn (X.-Y.W.)
* Correspondence: amyymxu@stu.edu.cn (Y.-M.X.); andytylau@stu.edu.cn (A.T.Y.L.);
  Tel.: +86-754-8890-0437 (Y.-M.X.); +86-754-8853-0052 (A.T.Y.L.)
† These authors contributed equally to this work.

Abstract: The metal cation symporter ZIP8 (SLC39A8) is a transmembrane protein that imports the essential micronutrients iron, manganese, and zinc, as well as heavy toxic metal cadmium (Cd). It has been recently suggested that selenium (Se), another essential micronutrient that has long been known for its role in human health and cancer risk, may also be transported by the ZIP8 protein. Several mutations in the ZIP8 gene are associated with the aberrant ion homeostasis of cells and can lead to human diseases. However, the intricate relationships between ZIP8 mutations, cellular Se homeostasis, and human diseases (including cancers and illnesses associated with Cd exposure) have not been explored. To further verify if ZIP8 is involved in cellular Se transportation, we first knockout (KO) the endogenous expression of ZIP8 in the HeLa cells using the CRISPR/Cas9 system. The elimination of ZIP8 expression was examined by PCR, DNA sequencing, immunoblot, and immunofluorescence analyses. Inductively coupled plasma mass spectrometry indicated that reduced uptake of Se, along with other micronutrients and Cd, was observed in the ZIP8-KO cells. In contrast, when ZIP8 was overexpressed, increased Se uptake could be detected in the ZIP8-overexpressing cells. Additionally, we found that ZIP8 with disease-associated single-point mutations G38R, G204C, and S335T, but not C113S, showed reduced Se transport ability. We then evaluated the potential of Se on Cd cytotoxicity prevention and therapy of cancers. Results indicated that Se could suppress Cd-induced cytotoxicity via decreasing the intracellular Cd transported by ZIP8, and Se exhibited excellent anticancer activity against not all but only selected cancer cell lines, under restricted experimental conditions. Moreover, clinical-based bioinformatic analyses revealed that up-regulated ZIP8 gene expression was common across multiple cancer types, and selenoproteins that were significantly co-expressed with ZIP8 in these cancers had been identified. Taken together, this study concludes that ZIP8 is an important protein in modulating cellular Se levels and provides insights into the roles of ZIP8 and Se in disease prevention and therapy.

Keywords: cadmium cytotoxicity; cancer therapy; cisplatin; ICP-MS; nonsynonymous mutation; selenium homeostasis; selenoproteins; ZIP8

Citation: Liang, Z.-L.; Tan, H.W.; Wu, J.-Y.; Chen, X.-L.; Wang, X.-Y.; Xu, Y.-M.; Lau, A.T.Y. The Impact of ZIP8 Disease-Associated Variants G38R, C113S, G204C, and S335T on Selenium and Cadmium Accumulations: The First Characterization. *Int. J. Mol. Sci.* **2021**, *22*, 11399. https://doi.org/10.3390/ijms222111399

Academic Editors: Petra A. Tsuji and Dolph L. Hatfield

Received: 7 September 2021
Accepted: 14 October 2021
Published: 22 October 2021

Publisher's Note: MDPI stays neutral with regard to jurisdictional claims in published maps and institutional affiliations.

Copyright: © 2021 by the authors. Licensee MDPI, Basel, Switzerland. This article is an open access article distributed under the terms and conditions of the Creative Commons Attribution (CC BY) license (https://creativecommons.org/licenses/by/4.0/).

1. Introduction

Selenium (Se) is an essential micronutrient critical for maintaining normal cellular function in human and animal cells [1]. It is an integral component of selenoproteins which are involved in a wide range of cellular physiological processes, including but not limited to antioxidant defense, inflammatory response, immune regulation, and maintenance of cardiovascular and reproductive system [2–4]. A balanced Se level is vital to human health as it has been broadly recognized that Se deficiency is one of the causative factors for many human diseases (e.g., heart failure, male infertility, neurodegenerative disease, Keshan disease, and Kashin-Beck disease) [5–7] while excessive Se intake is associated with acute or chronic Se poisoning, the selenosis [8].

Se also plays an important role in cancer prevention and therapy [9]. Although some conflicting results have been obtained over the years, epidemiologic studies generally agree that Se deficiency is significantly associated with greater cancer risk, especially in gastrointestinal and prostate cancer [10]. Se and Se compounds have displayed promising anticancer activity against some cancer cell types [11]. However, none of these compounds have yet been clinically recognized as anticancer agents mostly due to inconsistent outcomes within and between clinical trials and laboratory studies [12–14]. In addition, Se compounds may be useful in the field of chemoprevention as it has been reported that Se could potentiate the efficacy of some chemotherapeutic drugs, for example, the first-line chemotherapeutic drug cisplatin [15].

Although the effects of Se on human health have received attraction over the past few decades, how Se is being transported into the human cells remains largely unknown. Previous studies have revealed that Se in the body is usually delivered from the liver to other organs or tissues by plasma transporter selenoprotein P (SELENOP/SelP) via interaction with transmembrane receptors such as lipoproteins (e.g., LRP1, LRP2, and LRP8) [16]. Recently, McDermott et al. suggested that another transmembrane protein, metal cation symporter ZIP8, was able to transport Se (in the form of selenite). Furthermore, it was shown that the transport of Se via ZIP8 required zinc (Zn) ion and bicarbonate as co-substrates [17].

ZIP8 is a member of the solute carrier gene family (encoded by the gene SLC39A8) that facilitates the cellular uptake of several essential divalent metals such as iron (Fe), manganese (Mn), and Zn [18–20], and it is also responsible for the uptake of toxic heavy metal cadmium (Cd) [19]. Researchers have identified several mutations in the ZIP8 gene that can cause aberrant ion homeostasis of cells and lead to human diseases. For example, mutations G38R (c. 112G>C), G204C (c. 610G>T), and S335T (c. 1004G>C) in the ZIP8 gene are known to be associated with type II congenital disorder of glycosylation, which is characterized by intellectual disability, profound psychomotor retardation, hypotonia, strabismus, loss of hearing, and short stature [21,22]. Leigh syndrome, a rare inherited neurodegenerative disease, is caused by C113S (c. 338G>C) mutation in the ZIP8 [23]. However, the intricate relationships between ZIP8 mutations, cellular Se homeostasis, and human diseases (including cancers and diseases associated with Cd exposure) have not been previously explored.

In the current study, we successfully established a ZIP8-knockout (KO) human cell model using the CRISPR/Cas9 system and confirmed that human ZIP8 is involved in the intracellular transportation of Se. ZIP8-KO cells also showed a decreased ability to transport Mn, Zn, and Cd. We then examined the effects of four selected disease-associated ZIP8 single-point mutations (G38R, C113S, G204C, and S335T) on intracellular Se uptake and assessed the relationship between Se and Cd cytotoxicity in these ZIP8-variant and ZIP8-KO cells. Furthermore, the potential anticancer and synergistic effect of Se combined with cisplatin was determined. Lastly, clinical datasets from TCGA database [24] were analyzed for gene expressions of ZIP8 and 25 genes coding for selenocysteine-containing proteins in multiple cancer types. Overall, findings from this study suggest that ZIP8 is an important protein in modulating cellular Se levels that may have implications for disease prevention and therapy.

2. Results

2.1. Generation and Verification of a ZIP8 Gene Knockout Human Cell Model

To investigate the relationship between ZIP8 and Se transport, we used the CRISPR/Cas9 system to KO the ZIP8 gene in human cervical cancer HeLa cells. Specifically, we designed two sgRNAs targeting the genomic region of ZIP8 among exon 1 and intron 1, and gene sequence analysis indicated that 845 bp of genomic DNA was successfully deleted in that region (Figure 1A). PCR and immunoblot analysis were used to detect the gene and protein status of ZIP8 in the ZIP8-KO HeLa cell model. Gel electrophoresis assay showed that the HeLa parental cells with wildtype (WT) ZIP8 contained a full-length ZIP8 PCR product (1207 bp) whereas the ZIP8-KO cells generated a ZIP8 PCR product with only 362 bp (Figure 1B, Supplementary Table S1). On the other hand, immunoblot analysis showed that the ZIP8 protein expression was dramatically decreased in the ZIP8-KO cells when compared with the HeLa parental cells (Figure 1C). Furthermore, we performed an immunofluorescence assay to visualize the expression and localization of ZIP8 in the ZIP8-WT and ZIP8-KO cells. Microscopic results indicated that ZIP8 in the ZIP8-KO cells had substantially weaker overall expression and did not co-localize with the membrane protein marker DMT1 (Figure 1D). The above results showed that we have successfully generated a ZIP8-KO human cell model.

2.2. Involvement of Human ZIP8 in Intracellular Mn, Zn, Cd, and Se Uptakes

ZIP8 is a well-known transporter for certain divalent metal ions, including essential micronutrients Zn and Mn and toxic heavy metal Cd. Thus, we hypothesized that the transport capability of these metals should, to some extent, be affected in the ZIP8-KO cells. To test this hypothesis, we treated the cells with Mn, Zn, or Cd and used inductively coupled plasma mass spectrometry (ICP-MS) to quantify the intracellular uptake of these metals (Figure 2A–C). We found that when compared with HeLa parental cells, the levels of Mn uptake were significantly decreased in the ZIP8-KO cells following the treatment with 100 μM $MnCl_2$ at two different time points (Figure 2A). Although the levels of Zn detected in ZIP8-KO cells appeared to be slightly lower than the HeLa parental cells when treated with or without $ZnCl_2$, the differences were not statistically significant (Figure 2B). Also, the level of Cd was reduced in the ZIP8-KO cells treated with 1 μM $CdCl_2$ for 24 h (Figure 2C). We then used naphthol blue-black (NBB) staining assay to check whether the ZIP8-KO cells, which showed reduced Cd uptake, could survive better against Cd cytotoxicity. As expected, when ZIP8-KO and HeLa parental cells were treated with various concentrations of $CdCl_2$ for 12, 24, or 36 h (Figure 2D), greater cell viability was generally observed in ZIP8-KO cells in comparison to HeLa parental cells. These findings strongly suggest that our ZIP8-KO cell line is a reliable cell model to study the biological functions of human ZIP8 further. Also, it is worth noting that a gene sequencing assay was performed to ensure that the ZIP8 gene in HeLa parental cells used in this study has no mutation (data not shown).

The relationship between ZIP8 and Se transport in human cells remains largely unknown. Here, ZIP8-KO cells were used to investigate the effects of ZIP8 on regulating selenite homeostasis. Briefly, HeLa parental and ZIP8-KO cells were exposed to either 200 μM Na_2SeO_3 for 10 min or 4 μM Na_2SeO_3 for 12 h to mimic acute Se exposure or moderate Se exposure, respectively. The 4 μM Na_2SeO_3 treatment was selected to represent "moderate Se exposure" because no cytotoxicity was observed at this concentration, at least for 12 h (Supplementary Figure S1). Results from the ICP-MS indicated that although Se level appeared to be reduced in the ZIP8-KO cells after supplement with 200 μM Na_2SeO_3 for 10 min or 4 μM Na_2SeO_3 for 12 h, but only the Se treatment from the latter showed statistical significance (Figure 2E,F). Conversely, the intracellular Se content was enhanced in the ZIP8-overexpressed ZIP8-KO cells (transient-transfected with pcDNA3.1-ZIP8-WT) when compared to the ZIP8-KO control cells expressing pcDNA3.1 empty vector upon Se treatments (Figure 2G). Overall, these results indicated that human ZIP8 is involved in the intracellular uptake of Se.

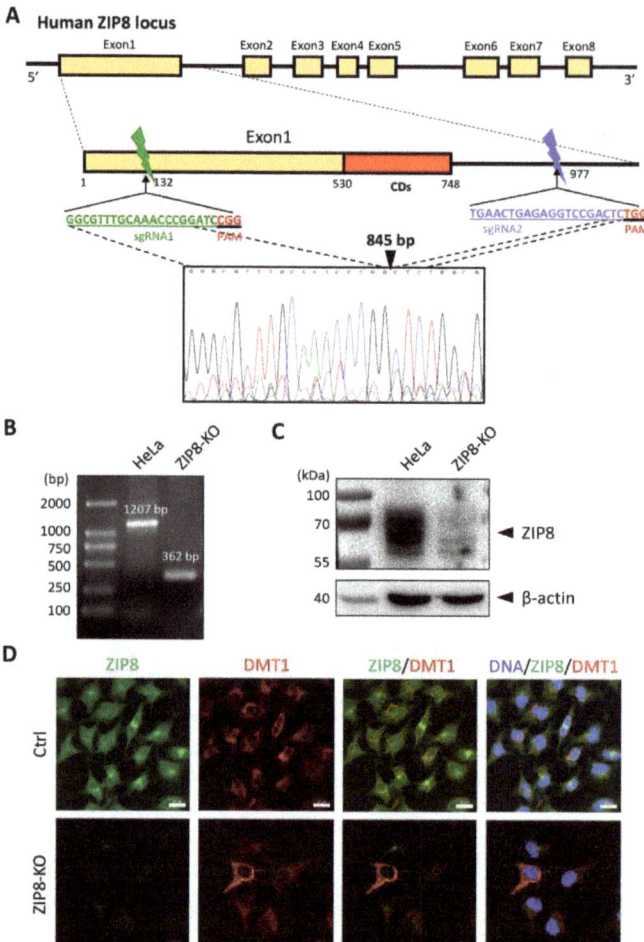

Figure 1. Generation of a ZIP8-KO HeLa cell line using the CRISPR/Cas9 genome editing technology. (**A**) Schematic of the CRISPR/Cas9 system to KO ZIP8. Two sgRNAs (green and blue bars) were designed to pair with the targeted DNA in order to delete a desire region in the human ZIP8 locus (part of exon1 and intron1). Each of the sgRNAs contains 20 bp single guide sequence followed by an essential PAM sequence (5′-NGG adjacent motif), and the site of double-stranded breaks induced by Cas9 is located about 3 nt upstream of the PAM sequence. DNA sequencing chromatogram shows that 845 bp of the ZIP8 gene has been deleted and the DNA junctions are connected via non-homologous end-joining (filled triangle). (**B**) Genomic DNA of HeLa and ZIP8-KO cells were extracted and amplified by PCR. (**C**) ZIP8 protein levels in HeLa and ZIP8-KO cells were analyzed by immunoblot analysis. Total cell lysates were resolved by SDS-PAGE, transferred onto polyvinylidene difluoride membrane, and incubated with antibody against ZIP8. β-actin was used as the indicated loading control. (**D**) The localization of ZIP8 was verified by immunofluorescence analysis in HeLa parental and HeLa ZIP8-KO cells using ZIP8 and DMT1 antibodies. DMT1 is a marker located in the membrane. The images were captured by confocal fluorescence microscopy. The merged images of ZIP8 (green) and DMT1 (red) reflect the colocalization area (yellow), and the blue color represents the nuclei stained with Hoechst 33258. Scale bar = 20 μm.

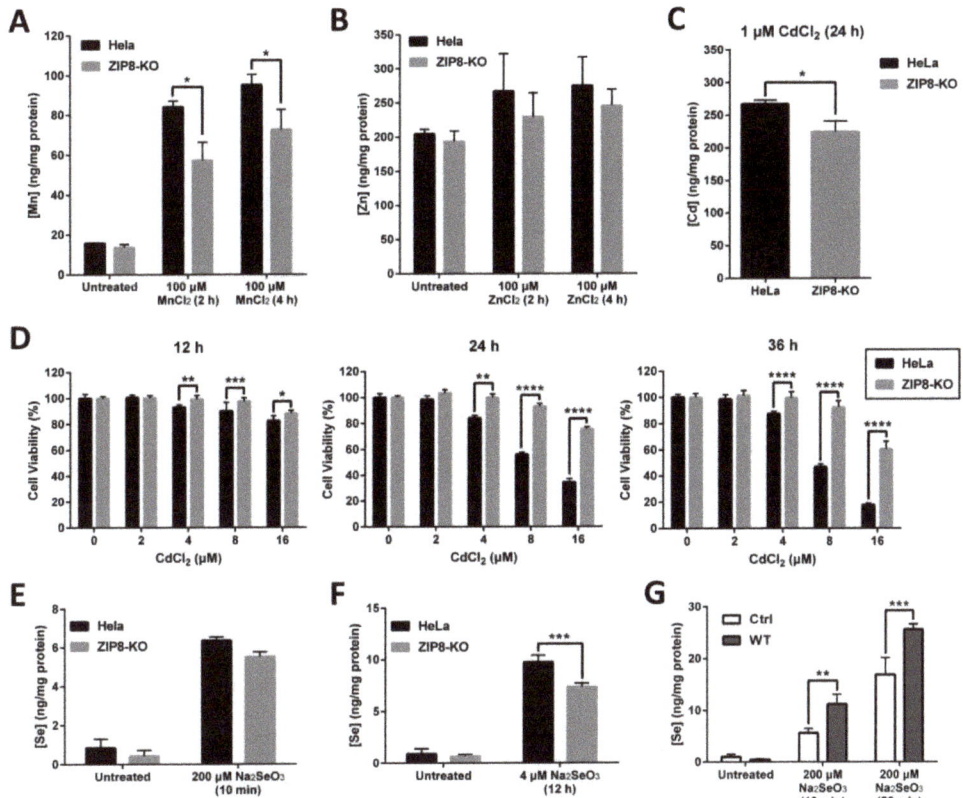

Figure 2. ZIP8-KO cells show decreased ability in intracellular Mn, Zn, Cd, and Se uptakes. (**A–C**) Intracellular uptake of Mn (**A**), Zn (**B**), or Cd (**C**) in HeLa parental and ZIP8-KO cells. After treating the cells with 100 μM MnCl$_2$ or ZnCl$_2$ for 2 or 4 h or 1 μM Cd for 24 h, the intracellular content of Mn, Zn, or Cd was assessed by ICP-MS. (**D**) The HeLa parental and ZIP8-KO cells were exposed to increasing amounts of CdCl$_2$ (0, 2, 4, 8, or 16 μM) for 12, 24, and 36 h. NBB staining assay was performed to determine the cell viability. Data are representative of three experiments, and error bars represent standard deviation of six replicates. (**E,F**) HeLa parental and ZIP8-KO cells were treated with 200 μM Na$_2$SeO$_3$ for 10 min (**E**) or 4 μM Na$_2$SeO$_3$ for 12 h (**F**), followed by measurement of Se cellular contents using ICP-MS. (**G**) HeLa ZIP8-KO cells were transfected with pcDNA3.1 empty vector (as a control) or pcDNA3.1-ZIP8-WT. Cells were exposed to 200 μM Na$_2$SeO$_3$ for 10 or 20 min, and then the intracellular Se concentration was measured by ICP-MS. Data are representative of two experiments, and error bars represent standard deviation of three replicates. p-value less than 0.05 was considered to be statistically significant. * $p < 0.05$, ** $p < 0.01$, *** $p < 0.001$, **** $p < 0.0001$.

2.3. The Effects of Disease-Associated ZIP8 Single-Point Mutations on Cellular Se Uptake Ability

Studies have shown that single-point mutations, including SNPs, on a single gene can cause abnormal functions of the protein and lead to human diseases [25–27]. Several mutations in the ZIP8 have been implicated in the occurrence of diseases related to the dysregulation of ion homeostasis [23]. Here, we selected four disease-associated ZIP8 mutations (G38R, C113S, G204C, and S335T; Table 1) and tested their Se uptake abilities. The selected ZIP8 single-point mutations are illustrated in Figure 3A. ICP-MS was performed to detect the intracellular level of Se in ZIP8-KO cells transiently transfected with pcDNA3.1 empty vector (as ZIP8-negative control) or pcDNA3.1 vector encoding ZIP8-WT or ZIP8-mutant upon acute (Figure 3B) or moderate (Figure 3C) Se exposure. It was demonstrated that with the exception of C113S, the Se transport abilities in all the cells with ZIP8 single-point mutations (G38R, G204C, and S335T) were remarkably suppressed when compared with the ZIP8-WT cells (Figure 3B,C). Particularly, the ZIP8 proteins with G38R, G204C, and S335T mutations appeared to have lost their ability to transport Se during acute Se exposure because the Se concentrations in these mutants were similar to the ZIP8-negative control cells (Figure 3B). However, when these cells were exposed to a moderate concentration of Se (e.g., 4 µM for 12 h), ZIP8-G38R and ZIP8-G204C mutants showed a significant increase in Se uptake compared with the ZIP8-negative control cells despite the overall Se concentration still lower than the ZIP8-WT and ZIP8-C113S (Figure 3C). Nevertheless, the Se uptake ability of ZIP8 in the S335T mutant remains negligible in the moderate Se exposure setting (Figure 3C).

Table 1. Information of selected disease-associated ZIP8 single-point mutations examined in this study.

Site(aa)	Type	Variant	Disease	References
38	Homozygous variant	c.112G>C (p. Gly38Arg)	Type II congenital disorder of glycosylation	[21,23,28]
113	Homozygous variant	c.338G>C (p. Cys113Ser)	Leigh syndrome	[23]
204	Heterozygous variant	c.610G>T (p. Gly204Cys)	Type II congenital disorder of glycosylation	[22,23]
335	Heterozygous variant	c.1004G>C (p. Ser335Thr)	Type II congenital disorder of glycosylation	[22,23]

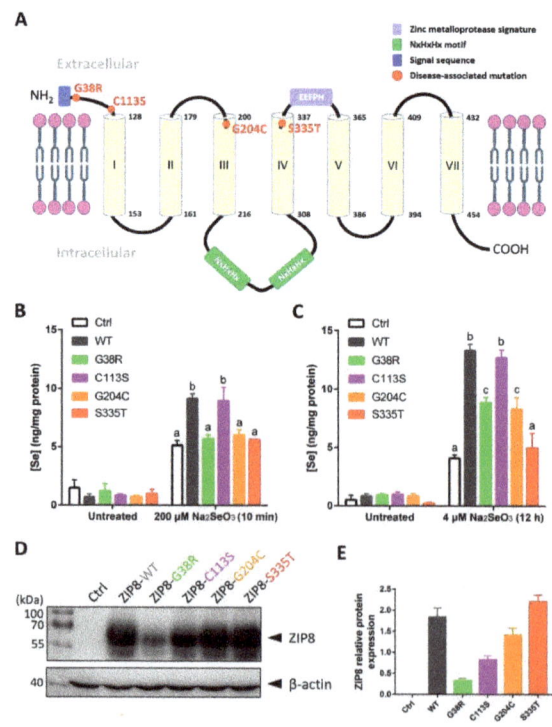

Figure 3. Se transport ability and protein expression level of ZIP8 variants. (**A**) Structural diagram of ZIP8 protein based on in silico predictions. It is predicted that the ZIP8 protein has seven transmembrane domains (TMD, yellow cylinder). The sites of four mutations (G38R, C113S, G204C, and S335T) in ZIP8 were labeled in red color. The N-terminus of ZIP8 has a signal sequence containing

22 amino acids (blue frame), and the protein possesses a long intracellular loop between TMD III and TMD IV containing two copied of NxHxHx domains (green frame) responsible for interacting with transition metals such as copper, nickel, and zinc [29,30]. (**B,C**) The HeLa ZIP8-KO cells were transient-transfected with pcDNA3.1-ZIP8-WT or ZIP8 mutants (G38R, C113S, G204C, and S335T) for 36 h, and then cells were treated with 200 µM Na$_2$SeO$_3$ for 10 min (**B**) or 4 µM Na$_2$SeO$_3$ for 12 h (**C**), followed by detection of intracellular Se concentration using ICP-MS. A significant difference of $p < 0.05$ between any two data columns marked with different alphabets. (**D**) The HeLa ZIP8-KO cells were transient-transfected with pcDNA3.1 (Ctrl) or pcDNA3.1-ZIP8-WT or mutants (G38R, C113S, G204C, and S335T) for 24 h, and proteins from these cells were harvested for immunoblot analysis to detect the protein expression level of ZIP8. β-actin was used as the loading control. (**E**) The quantifications of ZIP8-WT and other mutants (G38R, C113S, G204C, and S335T) in immunoblot result.

To examine whether the abilities of ZIP8 mutants to transport Se are associated with their ZIP8 protein expression, we performed immunoblot analysis to detect ZIP8 protein expression in these mutants. Results indicated that when compared with the ZIP8-WT, most of the mutants (ZIP8-C113S, ZIP8-G204C, and ZIP8-S335T) showed equal or stronger ZIP8 expressions where only ZIP8-G38R showed substantially weaker expression of ZIP8 (Figure 3D,E). Therefore, it appears that there is no direct connection between the ZIP8 protein expression level and Se transport ability.

2.4. The Role of ZIP8 and Se in Cd Cytotoxicity

It has been previously recognized by researchers that Se can effectively counteract the cytotoxic effect of Cd [31]. Since we have identified ZIP8 as a transporter of both Se and Cd, we next sought to investigate whether ZIP8 and/or Se play a role in counteracting Cd cytotoxicity. As expected, when exposed to CdCl$_2$ (2 µM for 12 h), HeLa ZIP8-KO cells transfected with pcDNA3.1-ZIP8-WT showed a remarkable increment (875.4-fold) of intracellular Cd as compared with the cells transfected with pcDNA3.1 empty vector (only up by 205-fold) (Figure 4A). Addition of 4 µM Na$_2$SeO$_3$ differentially reduced the Cd concentrations in both ZIP8-WT (by 73.6%) and empty vector control cells (by 44.5%) (Figure 4A). Consistent with these findings, cell viability results in Figure 4B indicated that the cell toxicity mediated by Cd was suppressed in the presence of Se, in particular, the suppressive effect of Cd-induced cell cytotoxicity can be found more obviously in the cells expressing ZIP8 protein (Supplementary Figure S2). To conclude, these results demonstrate that ZIP8 has an essential regulatory effect on cellular Cd and Se uptake that is associated with Cd-induced cell toxicity.

We then used ICP-MS to further investigate the Cd transport ability of the four selected ZIP8 mutations (G38R, C113S, G204C, and S335T), and at the same time, we also assessed whether Se could suppress Cd cytotoxicity in the cells with these ZIP8 mutations. Overall, the results indicated that when treated with 2 µM CdCl$_2$ for 12 h, the Cd concentrations were increased by 757 to 1053-fold for ZIP8-WT, -G38R, -C113S, and -G204C cells and only increased by 402-fold for ZIP8-S335T cells (Figure 4C). When treated the cells together with Cd and Se, a significant reduction of Cd levels was observed in the ZIP8-WT, -G38R, -C113S, and -G204C cells but not in the ZIP8-S335T cells (Figure 4D). Surprisingly, we discovered that higher Se levels were detected in all the cells treated with Se and Cd as compared with treated with Se alone (Figure 4E), and among them, the ZIP8-S335T cells which previously showed poor to no Se and Cd transport ability has the highest intracellular Se concentration (Figure 4E,F).

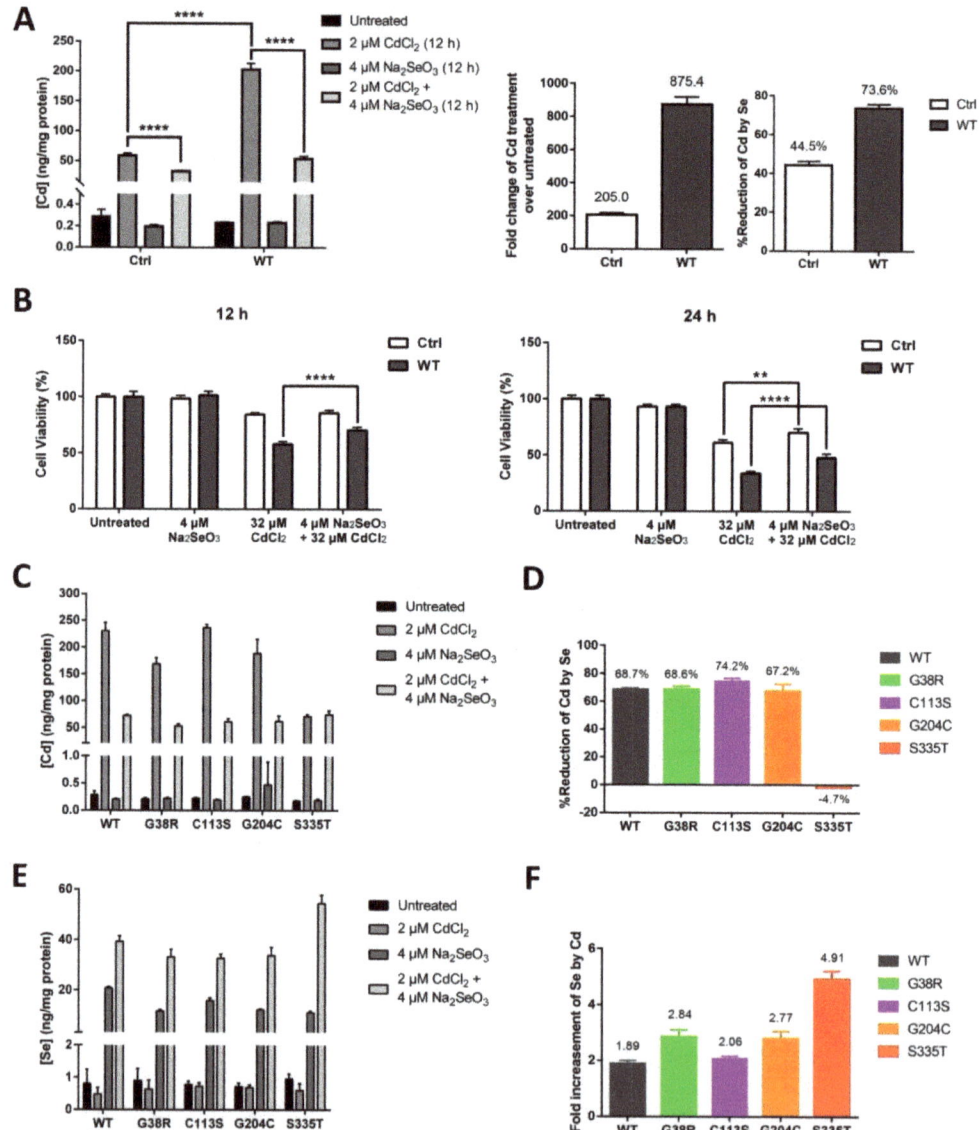

Figure 4. Se reduces Cd-induced cytotoxicity via decreasing the level of intracellular Cd transported by ZIP8. (**A**,**B**) The HeLa ZIP8-KO cells were transfected with a control plasmid (pcDNA3.1 empty vector) or pcDNA3.1-ZIP8-WT overexpression plasmid for 36 h, and cells were treated with various concentrations of $CdCl_2$ and/or Na_2SeO_3 for 12 h. Intracellular Cd content was measured by the ICP-MS (**A**), and cell viability was determined by the NBB staining assay (**B**). (**C**,**E**) Intracellular Cd (**C**) or Se (**E**) concentration was determined by ICP-MS after exposed to 2 μM $CdCl_2$ and/or 4 μM Na_2SeO_3 for 12 h in HeLa ZIP8-KO cells overexpressing ZIP8-WT or ZIP8-mutant. (**D**) Percentage reduction of intracellular Cd level between the Cd and Cd + Se treatment groups from (**C**). (**F**) Fold change of intracellular Se level between the Se and Cd + Se treatment groups from (**E**). Data in (**B**) are representative of three experiments, and the error bars represent standard deviation of six replicates. Two-way ANOVA (A, B) and one-way ANOVA (**A**) were performed, and a p-value less than 0.05 was considered to be statistically significant. ** $p < 0.01$, **** $p < 0.0001$.

2.5. Investigating the Potential Anticancer Effects of Se and Synergistic Anticancer Effect of Se and Cisplatin in Cancer Therapy

The potential of Se in cancer prevention and therapy has been extensively discussed. Although it remains controversial, some studies have reported that Se is an effective anticancer agent [3,32]. In our study, we found that even though 4 µM of Na_2SeO_3 showed no cytotoxic effect on the HeLa cells when treated for 12 h (Figure 4B), the same concentration of Se would kill about 75.72% of the cells at 48 h (Figure 5A). The dramatic drop of cell viability at 3 to 4 µM of Se treatment indicated that the anticancer property of Se is not only highly dependent on treatment duration but the concentration of Se is also very critical.

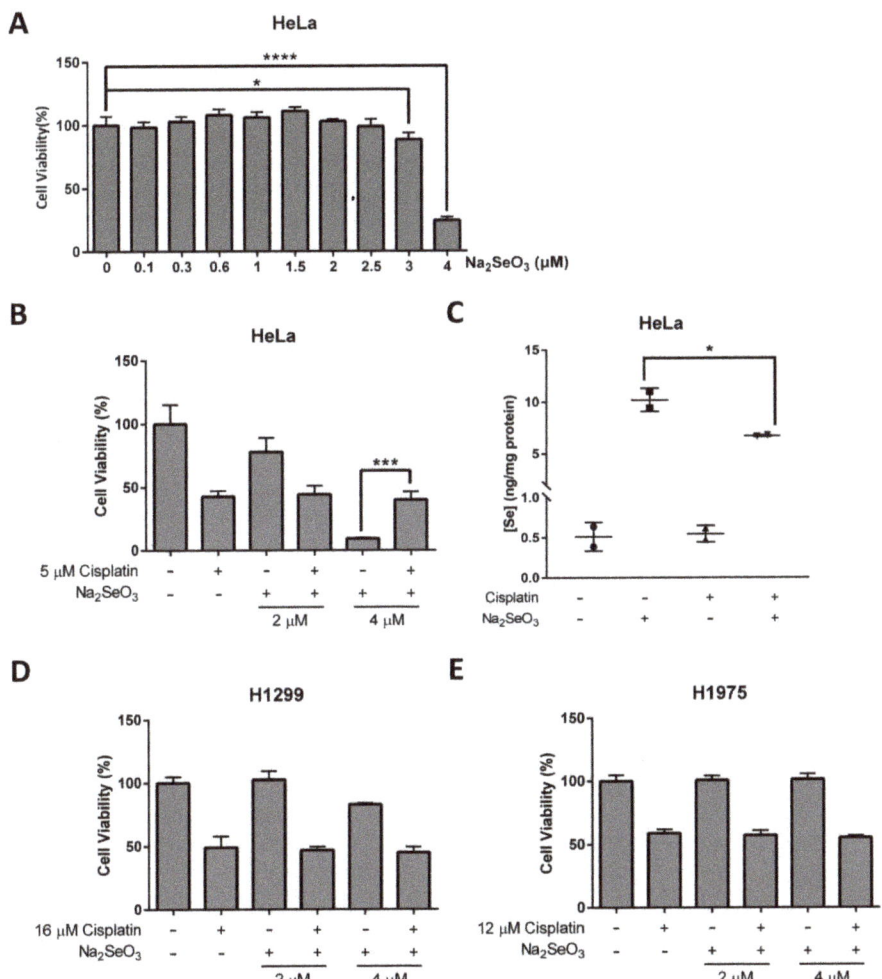

Figure 5. Anticancer effects of Se and cisplatin on HeLa, H1299, and H1975 cells. (**A**) Cell viability of HeLa cells treated with increasing doses of Na_2SeO_3 for 48 h. (**B**) Cell viability of HeLa cells incubated with 5 µM cisplatin and/or different doses of Na_2SeO_3 (2–4 µM) for 48 h. (**C**) Se concentration in HeLa cells treated with 5 µM cisplatin and/or 4 µM Na_2SeO_3. (**D,E**) Cell viability of H1299 cells (**D**) and H1975 cells (**E**) incubated with 16 µM (D) or 12 µM (E) cisplatin and/or different doses of Na_2SeO_3 (2–4 µM) for 48 h. Cell viability was measured by NBB staining assay (**A,B,D,E**), and Se concentration was measured by ICP-MS (**C**). One-way ANOVA (**A–C**) was performed, and a *p*-value less than 0.05 was considered to be statistically significant. * $p < 0.05$, *** $p < 0.001$, **** $p < 0.0001$.

Research has shown that Se can potentiate the anticancer effects of some chemotherapeutic agents, including the first-line chemotherapeutic drug cisplatin [15]. Herein, we further explore the potential of Se on cancer therapy, and evaluate whether Se in combination with cisplatin shows a synergistic effect on HeLa cells. Explicitly, cell viability results obtained from NBB staining assay demonstrated that 4 µM Na_2SeO_3 was more effective in killing HeLa cells than 5 µM cisplatin (concentration was selected based on IC50 determined previously) and no synergistic effect was observed between Se and cisplatin (Figure 5B and Supplementary Figure S3). On the contrary, the HeLa cells survived better in the combined treatment of Na_2SeO_3 (4 µM) and cisplatin (5 µM) than the Na_2SeO_3 (4 µM) alone treatment (Figure 5B). This unexpected finding implies that cisplatin may alleviate the Se-induced cell toxicity in the HeLa cells. ICP-MS was then performed to see if cisplatin affects Se transport ability in the HeLa cells, and it was revealed that cisplatin treatment resulted in a reduced intracellular Se uptake (Figure 5C).

Another two human lung cancer cell lines (H1299 and H1975) were also tested to verify whether the anticancer effect of Se and cisplatin are cell type-dependent. Collectively, H1299 and H1975 cells did not respond to Se and cisplatin the same way as the HeLa cells: treatment of 4 µM Na_2SeO_3 for 48 h was toxic for the HeLa cells but not so much for H1299 and H1975 cells (Figure 5D,E). Again, no synergistic effect between Se and cisplatin was observed in H1299 and H1975 cells (Figure 5D,E).

2.6. Clinical Database Analysis: Gene Expressions of ZIP8 and Selenoproteins in Cancers

We next evaluate how relevant is ZIP8 in cancers by utilizing TCGA database [24]. We analyzed the mRNA expression of ZIP8 in 40 types of cancers, and found that for those that showed significant differences in their ZIP8 expressions between tumor and healthy tissues, almost all were up-regulated and only five cancer types (acute myeloid leukemia, kidney renal clear cell carcinoma, liver hepatocellular carcinoma, thymoma, and uveal melanoma) contained a small percentage of tumor samples with down-regulated ZIP8 expression (Figure 6). To further verify that up-regulated ZIP8 expression is a common feature of cancer cells, we detected the protein expression of ZIP8 in normal lung epithelial BEAS-2B cells along with several other cancer cell lines (A549, H1299, H1975, H358, H460, and HeLa), and found the ZIP8 expressions in all these cancer cell lines were indeed higher than the BEAS-2B cells (Supplementary Figure S4). Given the fact that ZIP8 is involved in cellular Se homeostasis and that Se exhibits antitumor properties, the above findings suggest that the expression of ZIP8 in cancer tissues could be an important predictive factor for Se-based anticancer therapy.

The biological function of Se, as an essential micronutrient, is principally regulated by the synthesis of selenoproteins [3,33]. So far, 25 genes coding for selenocysteine-containing proteins have been identified in humans, and some of these proteins are involved in tumorigenesis [34,35]. To evaluate the potential connections between ZIP8 expression and the selenocysteine-containing proteins in cancers, we obtained and summarized the co-expression data of ZIP8 and selenoprotein genes in a range of cancer types from TCGA database (Figure 7). Overall, it is demonstrated that two selenoproteins, SELENOF (SelF) and SELENOP (SelP), are only positively correlated with ZIP8 across multiple cancer types (23/40 and 19/40, respectively). On the other hand, SELENOV (SelV) is identified as the selenoprotein that only negatively correlated with ZIP8 in 10 out of 40 cancer types. Interesting results can also be found if we focus on individual cancer types. For example, the majority of selenoproteins (18/25) in the upper tract urothelial carcinoma show significant correlations with ZIP8 and these correlations are all positive. Furthermore, little or no correlation between ZIP8 and selenoprotein is observed in certain cancers such as kidney renal papillary cell carcinoma and adrenocortical carcinoma. Taken together, the above data provide new insights into molecular and clinical aspects revolving around ZIP8, Se, selenoproteins, and cancer. Nevertheless, substantial studies are required in the future to assess if Se-based therapy could be more effective in certain cancer types with altered ZIP8 expressions.

Cancer type	Sample (n)	Alteration (n)	Total (%) % Down	Total (%) % Up	Cancer type	Sample (n)	Alteration (n)	Total (%) % Down	Total (%) % Up
Acral Melanoma	38	2		5.26 100.00	Metastatic Prostate Adenocarcinoma	208	9		4.33 100.00
Acute Myeloid Leukemia	173	18	22.22	10.40 77.78	Ovarian Serous Cystadenocarcinoma	300	19		6.33 100.00
Adrenocortical Carcinoma	78	9		11.54 100.00	Pancreatic Adenocarcinoma	177	11		6.21 100.00
Bladder Urothelial Carcinoma	407	23		5.65 100.00	Papillary Thyroid Carcinoma	498	16		3.21 100.00
Brain Lower Grade Glioma	514	38		7.39 100.00	Pediatric Acute Lymphoid Leukemia (Phase II)	203	13		6.4 100.00
Breast Invasive Carcinoma	1082	31		2.87 100.00	Pediatric Neuroblastoma	143	12		8.39 100.00
Cervical Squamous Cell Carcinoma	294	9		3.06 100.00	Pediatric Rhabdoid Tumor	43	2		4.65 100.00
Cholangiocarcinoma	36	1		2.78 100.00	Pediatric Wilms' Tumor	130	9		6.92 100.00
Colorectal Adenocarcinoma	592	35		5.91 100.00	Miscellaneous Neuroepithelial Tumor	184	8		4.35 100.00
Diffuse Large B-Cell Lymphoma	48	3		6.25 100.00	Prostate Adenocarcinoma	493	40		8.11 100.00
Esophageal Adenocarcinoma	181	9		4.97 100.00	Skin Cutaneous Melanoma	443	16		3.61 100.00
Glioblastoma Multiforme	160	12		7.50 100.00	Small Cell Lung Cancer	81	6		7.41 100.00
Head and Neck Squamous Cell Carcinoma	515	22		4.27 100.00	Soft Tissue Pheochromocytoma and Paraganglioma	178	8		4.49 100.00
Kidney Renal Clear Cell Carcinoma	510	38		7.45 100.00	Stomach Adenocarcinoma	412	20		4.85 100.00
Kidney Renal Papillary Cell Carcinoma	283	22	11.76	7.77 89.47	Testicular Germ Cell Tumors	149	10		6.71 100.00
Liver Hepatocellular Carcinoma	366	17	5.88	4.64 94.12	Thymoma	119	7	28.57	5.88 71.43
Lung Adenocarcinoma	510	26		5.10 100.00	Upper Tract Urothelial Carcinoma	32	6		19 100.00
Lung Squamous Cell Carcinoma	484	18		3.72 100.00	Uterine Carcinosarcoma	57	4		7.02 100.00
Melanoma	21	2		9.52 100.00	Uterine Corpus Endometrial Carcinoma	527	23		4.36 100.00
Mesothelioma	87	6		6.90 100.00	Uveal Melanoma	80	11	18.18	13.75 81.82

Figure 6. ZIP8 is up-regulated in cancer tissues and cell lines. ZIP8 mRNA expressions in 40 types of cancer were obtained from TCGA database. z-score threshold was set at ±2.0. The number of % alteration in each of the analyzed datasets was listed and further divided into the ratio of down-regulated and or up-regulated expression.

Cancers \ Selenoproteins	GPx1	GPx2	GPx3	GPx4	GPx6	DIO1	DIO2	DIO3	TXNRD1	TXNRD2	TXNRD3	SEPHS2	MsrB1	SelF	SelH	SelI	SelK	SelM	SelN	SelO	SelP	SelS	SelT	SelV	SelW	+ve	-ve	ns
Adrenocortical Carcinoma	ns	ns	ns	ns	ns	0.259	ns	ns	ns	ns	ns	ns	ns	ns	ns	ns	ns	ns	ns	ns	ns	ns	ns	ns	ns	1	0	24
Acral Melanoma	ns	-0.378	ns	-0.561	ns	ns	ns	ns	0.513	-0.503	ns	-0.417	-0.425	N/A	-0.447	0.501	N/A	N/A	ns	N/A	ns	ns	N/A	N/A	-0.636	2	7	10
Acute Myeloid Leukemia	-0.395	ns	ns	ns	N/A	ns	0.168	N/A	0.352	-0.167	ns	0.190	-0.156	0.258	ns	0.384	ns	ns	ns	-0.273	ns	ns	0.173	N/A	ns	6	4	12
Bladder Urothelial Carcinoma	ns	-0.545	0.179	ns	ns	-0.174	0.299	ns	0.325	ns	ns	ns	ns	ns	ns	ns	-0.186	0.282	ns	-0.393	ns	ns	ns	ns	-0.111	4	5	16
Brain Lower Grade Glioma	ns	ns	-0.163	0.118	ns	ns	0.199	ns	0.175	-0.191	ns	0.298	0.151	0.207	-0.090	ns	ns	0.278	ns	-0.257	0.148	0.209	0.311	ns	0.122	11	4	10
Cervical Squamous Cell Carcinoma	0.124	-0.309	0.154	ns	ns	0.169	0.171	ns	0.233	0.167	-0.153	ns	ns	0.275	ns	ns	ns	ns	ns	-0.222	ns	ns	-0.125	ns	ns	7	4	14
Cholangiocarcinoma	ns	-0.390	ns	ns	N/A	ns	0.331	ns	ns	ns	ns	-0.336	ns	ns	ns	ns	0.349	ns	0.430	ns	ns	ns	0.372	ns	ns	4	1	18
Colorectal Adenocarcinoma	ns	ns	-0.146	-0.140	ns	ns	ns	-0.099	0.120	ns	0.084	ns	ns	0.351	ns	0.161	0.118	ns	-0.101	-0.154	ns	ns	0.336	N/A	0.096	7	5	12
Diffuse Large B-Cell Lymphoma	ns	ns	0.415	ns	ns	0.481	ns	-0.335	0.290	ns	ns	ns	0.303	ns	ns	ns	ns	ns	0.409	-0.359	ns	0.450	ns	ns	ns	6	2	17
Esophagogastric Adenocarcinoma	ns	ns	ns	ns	ns	0.197	0.152	ns	ns	ns	-0.186	0.302	-0.204	0.255	ns	ns	0.178	N/A	0.156	-0.269	0.192	0.271	-0.206	-0.172	ns	8	5	11
Glioblastoma Multiforme	0.284	ns	ns	ns	ns	0.303	0.335	ns	0.338	-0.234	0.306	0.322	ns	0.255	ns	-0.171	ns	0.317	0.195	-0.195	0.216	ns	ns	-0.366	ns	10	4	11
Head and Neck Squamous Cell Carcinoma	0.099	-0.253	ns	ns	ns	ns	0.193	0.239	ns	ns	0.090	ns	ns	ns	-0.161	ns	ns	0.143	0.289	-0.190	0.247	ns	ns	-0.100	-0.140	7	5	13
Invasive Breast Carcinoma	ns	ns	0.113	-0.177	ns	-0.212	ns	ns	0.153	-0.194	0.091	-0.156	-0.117	0.187	-0.133	0.070	ns	-0.067	ns	-0.129	0.064	ns	ns	ns	-0.151	6	9	10
Kidney Renal Papillary Cell Carcinoma	ns	ns	ns	ns	ns	ns	ns	ns	ns	ns	ns	ns	ns	ns	ns	ns	ns	ns	ns	ns	ns	ns	ns	ns	ns	0	0	25
Liver Hepatocellular Carcinoma	ns	ns	0.110	ns	ns	0.312	0.110	-0.288	0.187	ns	0.164	0.183	0.162	0.125	-0.213	0.317	ns	-0.367	-0.236	ns	ns	-0.208	0.205	ns	-0.431	10	6	9
Lung Adenocarcinoma	0.225	-0.117	0.345	ns	ns	0.200	-0.205	0.108	-0.088	ns	ns	-0.187	ns	0.104	ns	-0.351	ns	ns	0.111	-0.094	0.338	-0.158	ns	-0.265	0.194	8	8	9
Lung Squamous Cell Carcinoma	0.325	-0.343	0.511	ns	-0.108	0.329	ns	0.199	-0.112	ns	ns	ns	ns	0.319	ns	-0.267	0.139	0.231	0.136	-0.287	0.488	0.091	ns	-0.168	0.218	11	6	8
Melanoma	ns	0.597	0.513	ns	ns	ns	0.595	ns	-0.471	ns	ns	ns	ns	ns	ns	ns	ns	ns	ns	ns	ns	ns	0.704	ns	ns	4	1	20
Mesothelioma	ns	ns	ns	ns	ns	ns	ns	ns	0.240	-0.233	ns	0.262	ns	0.458	ns	ns	ns	-0.234	-0.250	ns	ns	0.378	0.507	-0.416	-0.256	5	5	15
Metastatic Prostate Adenocarcinoma	0.233	ns	ns	ns	ns	ns	ns	ns	0.163	ns	ns	ns	0.239	0.335	0.296	ns	0.216	0.295	-0.166	ns	ns	0.160	0.220	-0.269	0.189	10	2	13
Miscellaneous Neuroepithelial Tumor	0.269	ns	ns	ns	ns	ns	0.168	0.145	ns	ns	ns	ns	ns	ns	ns	ns	ns	ns	ns	ns	0.259	ns	ns	ns	ns	4	0	21
Ovarian Serous Cystadenocarcinoma	0.161	ns	0.206	ns	-0.135	ns	0.281	0.127	ns	ns	ns	ns	ns	0.126	ns	ns	ns	0.145	0.133	ns	0.149	ns	ns	ns	ns	8	1	16
Pancreatic Adenocarcinoma	ns	-0.405	0.166	-0.202	ns	ns	0.202	ns	ns	ns	0.350	-0.211	ns	0.298	-0.150	ns	0.154	ns	0.282	-0.203	0.151	ns	0.199	-0.176	-0.275	8	7	10
Papillary Thyroid Carcinoma	ns	-0.241	0.196	-0.138	-0.139	ns	ns	ns	0.293	-0.158	0.175	0.234	-0.162	ns	-0.181	0.255	-0.235	-0.127	-0.096	-0.310	0.172	ns	0.196	ns	-0.348	7	11	7
Pediatric Acute Lymphoid Leukemia (Phase II)	-0.213	0.262	ns	-0.307	ns	0.173	0.179	ns	ns	-0.270	-0.202	ns	-0.343	ns	-0.270	0.309	ns	N/A	-0.379	0.368	ns	ns	ns	ns	-0.483	5	8	11
Pediatric Neuroblastoma	0.409	ns	0.249	0.190	ns	ns	0.230	0.239	ns	ns	0.169	ns	0.200	N/A	ns	-0.223	N/A	N/A	0.235	N/A	0.387	ns	N/A	N/A	0.194	10	1	8
Pediatric Rhabdoid Tumor	0.310	ns	0.357	0.475	ns	ns	ns	ns	ns	ns	ns	ns	ns	ns	ns	ns	ns	N/A	ns	ns	0.359	ns	ns	ns	ns	4	0	20
Pediatric Wilms' Tumor	ns	ns	ns	ns	ns	ns	0.283	0.241	ns	ns	ns	ns	ns	N/A	ns	ns	N/A	N/A	0.251	N/A	ns	ns	ns	N/A	N/A	3	0	16
Prostate Adenocarcinoma	ns	ns	ns	ns	ns	0.263	ns	ns	ns	ns	ns	ns	0.167	ns	-0.093	ns	ns	ns	-0.175	ns	ns	ns	ns	ns	ns	2	2	21
Renal Cell Carcinoma	-0.275	-0.321	ns	-0.260	ns	ns	-0.116	ns	0.119	-0.244	-0.090	0.131	ns	ns	-0.226	0.266	-0.138	-0.240	-0.142	-0.392	0.100	ns	ns	-0.100	-0.162	4	13	8
Skin Cutaneous Melanoma	ns	ns	ns	ns	0.110	0.202	0.105	0.105	0.102	-0.237	ns	ns	ns	0.354	ns	0.116	ns	ns	ns	0.116	ns	0.095	0.322	ns	-0.180	10	2	13
Small Cell Lung Cancer	0.376	-0.269	0.457	ns	ns	ns	0.275	ns	-0.265	ns	ns	ns	ns	N/A	ns	ns	ns	0.484	ns	ns	0.531	0.307	ns	ns	ns	6	2	16
Soft Tissue Pheochromocytoma and Paraganglioma	0.275	ns	ns	ns	ns	ns	0.192	0.165	ns	ns	ns	ns	-0.149	0.181	ns	ns	ns	ns	0.230	ns	0.265	ns	ns	ns	ns	6	1	18
Stomach Adenocarcinoma	ns	ns	ns	ns	ns	-0.130	ns	-0.108	ns	ns	ns	-0.116	ns	0.182	ns	0.100	ns	ns	ns	-0.126	ns	0.192	0.109	ns	-0.124	4	5	16
Testicular Germ Cell Tumors	ns	-0.288	0.382	0.523	-0.247	0.388	ns	0.504	-0.377	ns	ns	0.272	0.552	0.507	ns	ns	0.221	ns	-0.402	-0.248	ns	-0.248	0.575	ns	0.225	10	6	9
Thymoma	-0.390	ns	ns	-0.366	ns	ns	ns	ns	0.274	-0.404	ns	ns	-0.285	0.264	-0.338	0.196	ns	-0.214	ns	-0.359	ns	ns	0.365	ns	-0.350	4	8	13
Upper Tract Urothelial Carcinoma	0.359	ns	ns	ns	ns	0.408	0.776	0.354	0.872	ns	0.601	0.797	0.441	0.715	0.541	0.688	0.621	0.530	0.363	ns	0.665	0.714	0.706	ns	0.487	18	0	7
Uterine Carcinosarcoma	ns	ns	ns	ns	ns	0.452	0.333	ns	ns	ns	ns	ns	ns	0.396	ns	ns	ns	ns	ns	ns	0.331	ns	ns	ns	-0.395	4	1	20
Uterine Corpus Endometrial Carcinoma	-0.090	ns	-0.113	-0.177	ns	ns	0.183	ns	0.111	-0.097	ns	ns	-0.092	0.244	ns	0.125	ns	-0.088	ns	-0.183	0.116	ns	0.129	ns	ns	6	7	12
Uveal Melanoma	ns	0.278	-0.282	ns	N/A	0.242	ns	ns	0.359	ns	-0.290	0.437	ns	0.542	ns	0.398	ns	0.288	-0.223	ns	ns	0.493	0.391	-0.285	ns	9	4	11
+ve correlation	13	3	15	4	1	15	22	11	19	1	9	11	8	23	2	14	8	10	13	2	19	11	17	0	8			
-ve correlation	5	12	4	9	4	3	2	4	5	12	5	6	9	0	11	4	3	7	10	19	0	3	2	10	14			
ns	22	25	21	27	32	22	16	24	16	27	26	23	23	13	27	22	26	17	17	16	21	26	19	25	17	-1	0	+1

Figure 7. Co-expressions of ZIP8 and selenoprotein genes in cancers analyzed by spearman's correlation. Co-expression data of ZIP8 and 25 selenocysteine-containing protein genes in 40 types of cancers were obtained from TCGA database. A spearman's correlation value greater than 0 indicates a positive correlation (red) and value less than 0 indicates a negative correlation (green). Only values with *p*-value less than 0.05 were shown. ns: non-significant; N/A: not available.

3. Discussion

ZIP8 was recently reported as a Se transport protein. McDermott et al. [17] used multiple model systems to study the Se transport ability of ZIP8, and they showed that Se uptake was closely associated with ZIP8 expression in all the models tested, including cell cultures and transgenic mice. However, unlike McDermott et al. [17] where the ZIP8 expression of arsenic-transformed human bronchial epithelial BEAS-2B cells was knockdown by shRNA, in the current study, we used CRISPR/Cas9-based KO system to eliminate the endogenous expression of ZIP8 in the HeLa cells. This ZIP8-KO cell model allows us to more explicitly study the functions of ZIP8 and ZIP8 mutations because no interference will occur between the endogenous ZIP8 and ectopic ZIP8-WT or ZIP-mutation. Using the established ZIP8-KO and ZIP8-overexpressed human cell models, we confirmed and verified that ZIP8 is involved in the intracellular transportation of Se.

It is not uncommon that a single-point mutation on a single gene can cause the protein to function abnormally, leading to human diseases [27]. Over the years, a number of human disease-related ZIP8 mutations have been identified. These mutations include V33M (c. 97G>A), G38R (c. 112G>C), C113S (c. 338G>C), G204C (c. 610G>T), S335T (c. 1004G>C), I340N (c. 1019T>A), and A391T (c. 1171G>A) [21,23,36], and cells with ZIP8 containing these mutations appear to have dysregulated ion homeostasis (especially Mn) which can lead to diseases such as type II congenital disorder of glycosylation [37], Leigh syndrome [37], cardiovascular diseases [38], severe idiopathic scoliosis [39], and schizophrenia [40]. Here, for the first time, we discovered that single-point mutations in the ZIP8 gene could differentially affect the uptake of Se: ZIP8 variants G38R, G204C, and S335T showed reduced Se uptake ability whereas variant C113S had a Se transport ability comparable to the ZIP8-WT. These results indicate that ZIP8 is participating in Se homeostasis and that certain mutations in the ZIP8 could disrupt intracellular Se levels, and consequently, may lead to diseases related to Se deficiency or selenosis (Se overdose).

The molecular and cellular mechanisms underlying how mutations in ZIP8 affect the transportation of Se and other essential and nonessential metals remain largely unknown. Studies have shown that some of the nonsynonymous mutations will cause the protein structure of ZIP8 to change and alter the metal binding affinity. For example, it was shown that mutation A391T induced a structural change of the ZIP8 protein and resulted in a less intracellular Cd uptake [38]. Moreover, a report has indicated that disease-associated human ZIP8 mutants V33M, G38R, C113S, G204C, S335T, and I340N are unable to localize to the plasma membrane of the cells. Therefore, the mislocalization of ZIP8 with nonsynonymous mutations may be one of the reasons that some of the ZIP8-mutant cells in this study showed dysregulated Se homeostasis, but further ZIP8 localization test has to be carried out to check this possibility.

Results obtained from the ICP-MS (Figure 2A–C and Figure 3B,C) and immunoblot analysis (Figure 3D,E) indicated the metal transport abilities of the ZIP8-WT and ZIP8-mutant cells are not entirely correlated with the ZIP8 protein expressions. This is partly because the ZIP8 protein is not the sole transporter for the metals tested—the nutrient and metal transportation networks in the cells are extremely complicated and virtually each of the elements can be transported by more than one transporter [41]. Also, we previously chose the first 100 amino acids of ZIP8 N-terminal region as the immunogen for antibody production, and therefore, we cannot exclude the possibility that some mutations (e.g., the G38R mutant) may affect the affinity of ZIP8 antibody. Nevertheless, the critical role of human ZIP8 in transporting Se has been clearly demonstrated in the current study.

We next explored the potential of Se in disease prevention (Cd cytotoxicity) and therapy (cancer), from a ZIP8 point of view. Here, we found that Se could suppress Cd-induced cytotoxicity by reducing the entry of Cd via ZIP8 (Figure 4A–C). These results generally coincide with recent studies which reported that Se mitigated Cd-induced toxicity by regulating selenoprotein synthesis through promoting selenoprotein transcription [31,42]. In addition, an interesting discovery has been observed: Cd could promote intracellular Se uptake (Figure 4F,G). The above findings warrant further epidemiological and clinical studies to assess whether Se-deficient people are more prone to Cd-induced diseases and if Cd exposure is linked to selenosis. Also, since Cd is a widespread environmental pollutant and carcinogen that poses a serious threat to human health, especially in smokers and people working or living in Cd-polluted environments [43], research on the mechanistic association between ZIP8, Se, and Cd is urgently needed.

Se and Se-based compounds are considered promising candidates in chemoprevention and cancer therapy. Some of the Se compounds exhibit not only excellent anticancer properties but also remarkable tumor specificity [44]. The anticancer abilities of Se are mainly achieved by the direct or indirect antioxidant properties of Se [45]. Studies have indicated that Se may lower the adverse effects and increase treatment efficacy when it is used in combination with other anticancer agents [3]. Specifically, it has been reported that Se could potentiate the efficacy of cisplatin, a first-line chemotherapeutic drug for many cancer types [8,32]. However, despite the benefits of Se in chemoprevention and cancer therapy mentioned above, the utility and safety of Se and Se compounds for chemoprevention and cancer treatment remains uncertain due to the two-sided effect on cancer cells: on the one hand, Se shows cancer cell inhibitory properties, but on the other hand, Se may also promote cancer cell growth [9,46]. These contradictory results may be due in part to different chemical forms and bioavailability of Se, cancer cell types, and experimental conditions. Here, we showed that the anticancer effects of Se could be higher than the chemotherapeutic drug cisplatin in selected cancer cell lines and that the efficacy of Se is strictly concentration-dependent (Figure 5). Also, we did not observe a synergistic effect of Se and cisplatin when both were used together on the HeLa, H1299, and H1975 cell lines. However, this could be due to the limited range of Se concentrations tested in our study. Therefore, for Se-based compounds to be clinically used in cancer therapy, systematic and comprehensive screenings are required to obtain the optimal treatment conditions of Se against cancer cells with various phenotypic and genotypic backgrounds. Additionally, since ZIP8 is involved in the uptake of Se, further studies should be also considered for the potential involvement of ZIP8 in Se-based anticancer therapy.

Our study also provides clinical insights on how ZIP8 may be involved in Se-based cancer therapy. In Figures 6 and 7, we analyzed the gene expression of ZIP8 and 25 selenocysteine-containing proteins in clinical samples across multiple cancer types. We found that cancer tissues in almost all cancer types are inclined to have up-regulated ZIP8 expressions, and this may have implications for precision medicine since Se may more precisely target the tumor cells instead of the healthy cells. Selenium nanoparticles (SeNPs) possess a strong pro-oxidant property, and hyper-accumulation of SeNPs can generate potent therapeutic effects for cancer [47]. In addition, we identified selenoproteins that are positively or negatively co-expressed with ZIP8 in multiple cancer types. The roles of selenoproteins in cancer have been reported [36,48]. Overall, data summarized in Figure 7 provide guidance for future research by showing which selenoproteins in a specific cancer type are worthy of further investigation.

4. Materials and Methods

4.1. Cell Lines and Culture Conditions

Human cervical cancer (HeLa) and lung cancer (H1299 and H1975) cell lines were purchased from the American Type Culture Collection (ATCC) (Rockville, MD, USA). All cells were routinely grown in MEM or RPMI medium containing 10% fetal bovine serum (FBS) and 1% penicillin/streptomycin at 37 °C in a 5% CO_2 incubator as recommended by ATCC. For experiments containing additional Se, 2.5 mM $NaHCO_3$ and 150 µM $ZnCl_2$ were added into the culture medium.

4.2. Generation of ZIP8-KO HeLa Cell Model Using CRISPR/Cas9 Genome Editing Technology

HeLa cells were transfected using the pSpCas9(BB)-2A-Puro (PX459) plasmid plus the sgRNA sequence. Primers of sgRNA are listed in Supplementary Table S2. The cells were digested and selected by puromycin (1 µg/mL, Sigma-Aldrich, Taufkirchen, Germany) after transfection for 48 h, and refreshed with MEM medium containing FBS and puromycin every day. After several passages, clones were picked, seeded into a 12-well plate (one clone each), and cultured. The efficiency of KO was detected using immunoblot analysis and PCR. Genomic DNA was extracted with Thermo Genomic DNA Kit (Thermo Fisher Scientific, Waltham, MA, USA) according to the manufacturer's instructions, and gene KO condition was verified by touchdown PCR using Premix Taq Hot Start on a PCR. The PCR conditions were 94 °C for 3 min, a touchdown step of 15 cycles at 94 °C for 30 s, 65 °C for 30 s, and 72 °C for 1 min with a 1 °C decrease every cycle in annealing temperature, followed by 25 cycles at 94 °C for 30 s, 50 °C for 30 s, and 72 °C for 1 min, and a final extension at 72 °C for 10 min. The PCR products were examined by 1% agarose gel electrophoresis with gel-red staining, and visualized by a UV illuminator; subsequently, the PCR products were sent for sequencing to further confirm the homozygous KO of the ZIP8 gene.

4.3. Plasmids and Cell Transfection

pSpCas9(BB)-2A-Puro (PX459) plasmid, obtained from Addgene, was used to KO ZIP8. Single guide RNAs (sgRNAs), listed in Supplementary Table S2, were designed using an online web-based tool (http://crispor.tefor.net/). To overexpress the ZIP8, the coding sequence of human ZIP8 was amplified by PCR using cDNA of normal human bronchial epithelial BEAS-2B cells as a template, and inserted into pcDNA3.1 vector through BamHI and EcoRI. Its mutations were constructed following the molecular cloning guidelines and using the QuickChange site-directed mutagenesis kit (Agilent#210519, Palo Alto, Santa Clara, CA, USA) according to manufacturer's instructions. The primers used are shown in Supplementary Table S3. The verification of all plasmids was subjected to sequencing. Plasmid transfection was carried out in HeLa ZIP8-KO cells following the manufacturer's instructions.

4.4. Cytotoxicity Assay

Cell viability was performed by the NBB staining assay. Briefly, 100 µL cell suspension containing 1×10^6 cells/mL was added in each well of a 96-well plate, and cells were cultured in an incubator for 24 h. Cells were then exposed to drugs of interest for a duration as indicated prior to NBB staining assay. During NBB assay, cells were fixed with 4% formaldehyde solution (10% formalin) for 8 min, followed by staining with 0.05% NBB solution for 25 min at room temperature. Then, cells were washed with distilled water thrice, and finally, 50 µL of 50 mM NaOH was added. The absorbance of the cell suspension was measured at 595 nm using a 96-well multiscanner (Thermo Scientific Multiskan FC, Thermo Scientific, Waltham, MA, USA).

4.5. Immunoblot Analysis

Cells were lysed with NP-40 lysis buffer containing a protease inhibitor cocktail and sonicated for 6 s each for three times. The cell lysates were then centrifuged at $16900 \times g$ for 10 min at 4 °C. Then, 30 µg protein were quantified and used for immunoblot analysis as described previously [49]. The following antibodies were used: anti-ZIP8 (1:500; produced by our lab [50]), anti-β-actin (1:10,000; #A5441, Sigma-Aldrich, Taufkirchen, Germany), anti-rabbit IgG-HRP (1:10,000; #sc-2004, Santa Cruz Biotechnology, Santa Cruz, CA, USA), and anti-mouse IgG-HRP (1:10,000; #sc-2005, Santa Cruz Biotechnology, Santa Cruz, CA, USA).

4.6. Inductively Coupled Plasma Mass Spectrometry (ICP-MS)

The quantifications of intracellular Se, Mn, Zn, and Cd levels were performed using Agilent 7900 ICP-MS (Agilent Technologies Inc., Santa Clara, CA, USA) using argon as a plasma gas. Briefly, the cells to be measured were digested by trypsin, washed with $1 \times$ PBS thrice and lysed in Milli-Q H_2O, followed by quantification of cell lysate. 700 µg of cell lysates were taken out and subsequently added into a polytetrafluoroethylene tube for air drying at 60 °C in an oven overnight. In order to decompose the organic matrix, samples were processed in 100 µL 65% HNO$_3$ and 100 µL H$_2$O$_2$ at 80 °C for nitrification, then cooled to room temperature. Then, all samples were diluted in 2% HNO$_3$ to a final volume of 5 mL and eventually quantification of intracellular elements by ICP-MS. To validate the method, we evaluated the linearity, limit of detection (LOD), and the limit of quantification (LOQ) [51]. When cell samples were assessed by ICP-MS, a blank and a trace elements serum L-2 RUO were measured for each time as quality control.

At least seven calibration points were used for external calibration of each of the elements, all elemental concentrations of samples have been considered. The multi-element calibration standard solution (10 µg/mL) was diluted by 2% HNO$_3$ solution to prepare the working standards at multiple concentrations of 1, 5, 10, 50, 100, and 500 µg/L, which would be re-prepared for each time before measuring the samples. The internal standards were applied to correct signal drift. The multi-element solution (100 µg/mL) was diluted with 2% HNO$_3$ solution to prepare the internal standard solutions, germanium, indium, and scandium were used as the internal isotopes for quantification. In addition, the LOQ and the LOD of each element were determined by measuring elemental concentrations until detecting digestion blanks for each time (Supplementary Figure S5). Correlation coefficients of all elemental calibration curves were above 0.999.

4.7. Immunofluorescence Microscopy

Cells were seeded on coverslips in a 6-well plate, fixed with 4% paraformaldehyde for 15 min at room temperature, and rinsed with phosphate-buffered saline (PBS) containing Ca^{2+} and Mg^{2+} thrice. Then, cells were permeabilized with pre-cooled 0.02% Triton X-100 for 10 min on ice and again rinsed with PBS containing Ca^{2+} and Mg^{2+} thrice. Next, cells were blocked with 5% bovine serum albumin (BSA) in PBS at room temperature for 2 h. After blockage with 5% BSA, cells were incubated in the dark with ZIP8 (1:200) and DMT1 (1:150, #166884, Santa Cruz Biotechnology, Santa Cruz, CA, USA) primary antibodies for more than 8 h at 4 °C. Subsequently, cells were incubated with secondary antibodies (Alexa Fluor secondary 488 and 595; Invitrogen) for 2 h at 1:800 dilution in the dark; The nuclei were stained with Hoechst 33258 (#B1155, Sigma–Aldrich, Taufkirchen, Germany) for 15 min. Immunofluorescence images were taken by a LSM800 confocal laser scanning microscope (Zeiss, Oberkochen, Germany) at 400× magnification.

4.8. Clinical-Based Bioinformatics Analysis

mRNA expression data of ZIP8 and 25 genes coding for selenocysteine-containing proteins in patients from 40 cancer types were obtained from TCGA database. The mRNA expression z-score threshold was set at ±2.0. For co-expression analysis, only data with $p \leq 0.05$ and spearman's correlation > 0 (positive correlation between ZIP8 and selenoprotein) or <0 (negative correlation between ZIP8 and selenoprotein) were considered significant.

4.9. Statistical Analysis

Statistical analysis was performed using the GraphPad Prism® 7 software (v6.02, GraphPad Software Inc., San Diego, CA, USA). All quantitative data were expressed as means ± standard deviation of at least three independent experiments unless mentioned otherwise. Statistical significance between groups was analyzed using a two-tailed Student's *t-test*, one-way ANOVA, or two-way ANOVA. A $p \leq 0.05$ was used as the criterion for statistical significance.

Supplementary Materials: The following are available online at https://www.mdpi.com/article/10.3390/ijms222111399/s1.

Author Contributions: Conceptualization, Y.-M.X. and A.T.Y.L.; Data curation, Z.-L.L., H.W.T., J.-Y.W., X.-L.C. and X.-Y.W.; Formal analysis, Z.-L.L., H.W.T., Y.-M.X. and A.T.Y.L.; Funding acquisition, Y.-M.X. and A.T.Y.L.; Investigation, Z.-L.L., H.W.T., Y.-M.X. and A.T.Y.L.; Methodology, Z.-L.L., H.W.T., Y.-M.X. and A.T.Y.L.; Project administration, Y.-M.X. and A.T.Y.L.; Resources, Y.-M.X. and A.T.Y.L.; Supervision, Y.-M.X. and A.T.Y.L.; Validation, Z.-L.L.; Visualization, Z.-L.L. and H.W.T.; Writing—original draft, Z.-L.L., H.W.T., Y.-M.X. and A.T.Y.L.; Writing—review & editing, Z.-L.L., H.W.T., Y.-M.X. and A.T.Y.L. All authors have read and agreed to the published version of the manuscript.

Funding: This work was supported by the grants from the National Natural Science Foundation of China (Nos. 31771582 and 31271445), the Guangdong Natural Science Foundation of China (No. 2017A030313131), the "Thousand, Hundred, and Ten" project of the Department of Education of Guangdong Province of China, the Basic and Applied Research Major Projects of Guangdong Province of China (2017KZDXM035 and 2018KZDXM036), the "Yang Fan" Project of Guangdong Province of China (Andy T. Y. Lau-2016; Yan-Ming Xu-2015), and the Shantou Medical Health Science and Technology Plan (200624165260857).

Institutional Review Board Statement: Not applicable.

Informed Consent Statement: Not applicable.

Data Availability Statement: All data generated or analyzed during this study are included in the main manuscript and Supplementary Materials.

Acknowledgments: We would like to thank members of the Lau and Xu laboratory for critical reading of this manuscript.

Conflicts of Interest: The authors declare no conflict of interest.

References

1. Lau, A.T.Y.; Tan, H.W.; Xu, Y.M. Epigenetic effects of dietary trace elements. *Curr. Pharmacol. Rep.* **2017**, *3*, 232–241. [CrossRef]
2. Forootanfar, H.; Adeli-Sardou, M.; Nikkhoo, M.; Mehrabani, M.; Amir-Heidari, B.; Shahverdi, A.R.; Shakibaie, M. Antioxidant and cytotoxic effect of biologically synthesized selenium nanoparticles in comparison to selenium dioxide. *J. Trace Elem. Med. Biol.* **2014**, *28*, 75–79. [CrossRef]
3. Razaghi, A.; Poorebrahim, M.; Sarhan, D.; Bjornstedt, M. Selenium stimulates the antitumour immunity: Insights to future research. *Eur. J. Cancer* **2021**, *155*, 256–267. [CrossRef]
4. Alehagen, U.; Aaseth, J.; Lindahl, T.L.; Larsson, A.; Alexander, J. Dietary supplementation with selenium and coenzyme Q10 prevents increase in plasma d-dimer while lowering cardiovascular mortality in an elderly swedish population. *Nutrients* **2021**, *13*, 1344. [CrossRef]
5. Ellwanger, J.H.; Franke, S.I.; Bordin, D.L.; Pra, D.; Henriques, J.A. Biological functions of selenium and its potential influence on Parkinson's disease. *An. Acad. Bras. Ciências* **2016**, *88*, 1655–1674. [CrossRef]
6. Fairweather-Tait, S.J.; Bao, Y.; Broadley, M.R.; Collings, R.; Ford, D.; Hesketh, J.E.; Hurst, R. Selenium in human health and disease. *Antioxid. Redox Signal.* **2011**, *14*, 1337–1383. [CrossRef]
7. Mistry, H.D.; Pipkin, F.B.; Redman, C.W.; Poston, L. Selenium in reproductive health. *Am. J. Obstet. Gynecol.* **2012**, *206*, 21–30. [CrossRef]
8. Raisbeck, M.F. Selenosis. *Vet. Clin. N. Am. Food Anim. Pract.* **2000**, *16*, 465–480. [CrossRef]
9. Tan, H.W.; Mo, H.Y.; Lau, A.T.Y.; Xu, Y.M. Selenium species: Current status and potentials in cancer prevention and therapy. *Int. J. Mol. Sci.* **2018**, *20*, 75. [CrossRef] [PubMed]
10. Cai, X.; Wang, C.; Yu, W.; Fan, W.; Wang, S.; Shen, N.; Wu, P.; Li, X.; Wang, F. Selenium exposure and cancer risk: An updated meta-analysis and meta-regression. *Sci. Rep.* **2016**, *6*, 19213. [CrossRef] [PubMed]
11. Kieliszek, M.; Lipinski, B.; Blazejak, S. Application of sodium selenite in the prevention and treatment of cancers. *Cells* **2017**, *6*, 39. [CrossRef] [PubMed]
12. Ramoutar, R.R.; Brumaghim, J.L. Antioxidant and anticancer properties and mechanisms of inorganic selenium, oxo-sulfur, and oxo-selenium compounds. *Cell Biochem. Biophys.* **2010**, *58*, 1–23. [CrossRef]
13. Nicastro, H.L.; Dunn, B.K. Selenium and prostate cancer prevention: Insights from the selenium and vitamin E cancer prevention trial (SELECT). *Nutrients* **2013**, *5*, 1122–1148. [CrossRef]
14. Lippman, S.M.; Klein, E.A.; Goodman, P.J.; Lucia, M.S.; Thompson, I.M.; Ford, L.G.; Parnes, H.L.; Minasian, L.M.; Gaziano, J.M.; Hartline, J.A.; et al. Effect of selenium and vitamin E on risk of prostate cancer and other cancers: The Selenium and Vitamin E Cancer Prevention Trial (SELECT). *JAMA* **2009**, *301*, 39–51. [CrossRef]

15. Sakalli Cetin, E.; Naziroglu, M.; Cig, B.; Ovey, I.S.; Aslan Kosar, P. Selenium potentiates the anticancer effect of cisplatin against oxidative stress and calcium ion signaling-induced intracellular toxicity in MCF-7 breast cancer cells: Involvement of the TRPV1 channel. *J. Recept. Sig. Transd.* **2017**, *37*, 84–93. [CrossRef] [PubMed]
16. Saito, Y. Selenium transport mechanism via selenoprotein P—Its physiological role and related diseases. *Front. Nutr.* **2021**, *8*, 685517. [CrossRef]
17. McDermott, J.R.; Geng, X.; Jiang, L.; Gálvez-Peralta, M.; Chen, F.; Nebert, D.W.; Liu, Z. Zinc- and bicarbonate-dependent ZIP8 transporter mediates selenite uptake. *Oncotarget* **2016**, *7*, 35327–35340. [CrossRef]
18. Bonaventura, E.; Barone, R.; Sturiale, L.; Pasquariello, R.; Alessandri, M.G.; Pinto, A.M.; Renieri, A.; Panteghini, C.; Garavaglia, B.; Cioni, G.; et al. Clinical, molecular and glycophenotype insights in SLC39A8-CDG. *Orphanet. J. Rare. Dis.* **2021**, *16*, 307. [CrossRef]
19. Xu, Y.M.; Gao, Y.M.; Wu, D.D.; Yu, F.Y.; Zang, Z.S.; Yang, L.; Yao, Y.; Cai, N.L.; Zhou, Y.; Chiu, J.F.; et al. Aberrant cytokine secretion and zinc uptake in chronic cadmium-exposed lung epithelial cells. *Proteomics Clin. Appl.* **2017**, *11*, 1600059. [CrossRef]
20. Ji, C.; Kosman, D.J. Molecular mechanisms of non-transferrin-bound and transferring-bound iron uptake in primary hippocampal neurons. *J. Neurochem.* **2015**, *133*, 668–683. [CrossRef] [PubMed]
21. Boycott, K.M.; Beaulieu, C.L.; Kernohan, K.D.; Gebril, O.H.; Mhanni, A.; Chudley, A.E.; Redl, D.; Qin, W.; Hampson, S.; Küry, S.; et al. Autosomal-recessive intellectual disability with cerebellar atrophy syndrome caused by mutation of the manganese and zinc transporter gene SLC39A8. *Am. J. Hum. Genet.* **2015**, *97*, 886–893. [CrossRef]
22. Park, J.H.; Hogrebe, M.; Gruneberg, M.; DuChesne, I.; von der Heiden, A.L.; Reunert, J.; Schlingmann, K.P.; Boycott, K.M.; Beaulieu, C.L.; Mhanni, A.A.; et al. SLC39A8 deficiency: A disorder of manganese transport and glycosylation. *Am. J. Hum. Genet.* **2015**, *97*, 894–903. [CrossRef]
23. Choi, E.-K.; Nguyen, T.-T.; Gupta, N.; Iwase, S.; Seo, Y.A. Functional analysis of SLC39A8 mutations and their implications for manganese deficiency and mitochondrial disorders. *Sci. Rep.* **2018**, *8*, 3163. [CrossRef] [PubMed]
24. Gao, J.; Aksoy, B.A.; Dogrusoz, U.; Dresdner, G.; Gross, B.; Sumer, S.O.; Sun, Y.; Jacobsen, A.; Sinha, R.; Larsson, E.; et al. Integrative analysis of complex cancer genomics and clinical profiles using the cBioPortal. *Sci. Signal.* **2013**, *6*, pl1. [CrossRef] [PubMed]
25. Yao, Y.; Tan, H.W.; Liang, Z.L.; Wu, G.Q.; Xu, Y.M.; Lau, A.T.Y. The impact of coilin nonsynonymous SNP Variants E121K and V145I on cell growth and cajal body Formation: The first characterization. *Genes* **2020**, *11*, 895. [CrossRef] [PubMed]
26. Meng, L.; Song, Z.; Liu, A.; Dahmen, U.; Yang, X.; Fang, H. Effects of lipopolysaccharide-binding protein (LBP) single nucleotide polymorphism (SNP) in infections, inflammatory diseases, metabolic disorders and cancers. *Front. Immunol.* **2021**, *12*, 681810. [CrossRef]
27. Peng, Y.; Liang, C.; Xi, H.; Yang, S.; Hu, J.; Pang, J.; Liu, J.; Luo, Y.; Tang, C.; Xie, W.; et al. Case report: Novel NIPBL variants cause cornelia de lange syndrome in Chinese patients. *Front. Genet.* **2021**, *12*, 699894. [CrossRef] [PubMed]
28. Park, J.H.; Hogrebe, M.; Fobker, M.; Brackmann, R.; Fiedler, B.; Reunert, J.; Rust, S.; Tsiakas, K.; Santer, R.; Grüneberg, M.; et al. SLC39A8 deficiency: Biochemical correction and major clinical improvement by manganese therapy. *Genet. Med.* **2018**, *20*, 259–268. [CrossRef]
29. Kerkeb, L.; Mukherjee, I.; Chatterjee, I.; Lahner, B.; Salt, D.E.; Connolly, E.L. Iron-induced turnover of the Arabidopsis IRON-REGULATED TRANSPORTER1 metal transporter requires lysine residues. *Plant. Physiol.* **2008**, *146*, 1964–1973. [CrossRef]
30. Zang, Z.S.; Xu, Y.M.; Lau, A.T.Y. Molecular and pathophysiological aspects of metal ion uptake by the zinc transporter ZIP8 (SLC39A8). *Toxicol. Res.* **2016**, *5*, 987–1002. [CrossRef]
31. Zhang, C.; Ge, J.; Lv, M.; Zhang, Q.; Talukder, M.; Li, J.L. Selenium prevent cadmium-induced hepatotoxicity through modulation of endoplasmic reticulum-resident selenoproteins and attenuation of endoplasmic reticulum stress. *Environ. Pollut.* **2020**, *260*, 113873. [CrossRef]
32. Zhao, P.; Li, M.; Chen, Y.; He, C.; Zhang, X.; Fan, T.; Yang, T.; Lu, Y.; Lee, R.J.; Ma, X.; et al. Selenium-doped calcium carbonate nanoparticles loaded with cisplatin enhance efficiency and reduce side effects. *Int. J. Pharm.* **2019**, *570*, 118638. [CrossRef]
33. Reeves, M.A.; Hoffmann, P.R. The human selenoproteome: Recent insights into functions and regulation. *Cell. Mol. Life Sci.* **2009**, *66*, 2457–2478. [CrossRef]
34. Hughes, D.J.; Kunicka, T.; Schomburg, L.; Liska, V.; Swan, N.; Soucek, P. Expression of selenoprotein genes and association with selenium status in colorectal adenoma and colorectal cancer. *Nutrients* **2018**, *10*, 1812. [CrossRef] [PubMed]
35. Wu, B.K.; Chen, Q.H.; Pan, D.; Chang, B.; Sang, L.X. A novel therapeutic strategy for hepatocellular carcinoma: Immunomodulatory mechanisms of selenium and/or selenoproteins on a shift towards anti-cancer. *Int. Immunopharmacol.* **2021**, *96*, 107790. [CrossRef]
36. Riley, L.G.; Cowley, M.J.; Gayevskiy, V.; Roscioli, T.; Thorburn, D.R.; Prelog, K.; Bahlo, M.; Sue, C.M.; Balasubramaniam, S.; Christodoulou, J. A SLC39A8 variant causes manganese deficiency, and glycosylation and mitochondrial disorders. *J. Inherit. Metab. Dis.* **2017**, *40*, 261–269. [CrossRef]
37. Li, D.; Achkar, J.P.; Haritunians, T.; Jacobs, J.P.; Hui, K.Y.; D'Amato, M.; Brand, S.; Radford-Smith, G.; Halfvarson, J.; Niess, J.H.; et al. A pleiotropic missense variant in SLC39A8 is associated with crohn's disease and human gut microbiome composition. *Gastroenterology* **2016**, *151*, 724–732. [CrossRef] [PubMed]
38. Zhang, R.; Witkowska, K.; Afonso Guerra-Assunção, J.; Ren, M.; Ng, F.L.; Mauro, C.; Tucker, A.T.; Caulfield, M.J.; Ye, S. A blood pressure-associated variant of the SLC39A8 gene influences cellular cadmium accumulation and toxicity. *Hum. Mol. Genet.* **2016**, *25*, 4117–4126. [CrossRef] [PubMed]

39. Haller, G.; McCall, K.; Jenkitkasemwong, S.; Sadler, B.; Antunes, L.; Nikolov, M.; Whittle, J.; Upshaw, Z.; Shin, J.; Baschal, E.; et al. A missense variant in SLC39A8 is associated with severe idiopathic scoliosis. *Nat. Commun.* **2018**, *9*, 4171. [CrossRef] [PubMed]
40. McCoy, T.H., Jr.; Pellegrini, A.M.; Perlis, R.H. Using phenome-wide association to investigate the function of a schizophrenia risk locus at SLC39A8. *Transl. Psychiat.* **2019**, *9*, 45. [CrossRef] [PubMed]
41. Zaugg, J.; Melhem, H.; Huang, X.; Wegner, M.; Baumann, M.; Surbek, D.; Körner, M.; Albrecht, C. Gestational diabetes mellitus affects placental iron homeostasis: Mechanism and clinical implications. *FASEB J.* **2020**, *34*, 7311–7329. [CrossRef]
42. Tang, L.; Hamid, Y.; Liu, D.; Shohag, M.J.I.; Zehra, A.; He, Z.; Feng, Y.; Yang, X. Foliar application of zinc and selenium alleviates cadmium and lead toxicity of water spinach—Bioavailability/cytotoxicity study with human cell lines. *Environ. Int.* **2020**, *145*, 106122. [CrossRef] [PubMed]
43. Tan, H.W.; Liang, Z.L.; Yao, Y.; Wu, D.D.; Mo, H.Y.; Gu, J.; Chiu, J.F.; Xu, Y.M.; Lau, A.T.Y. Lasting DNA damage and aberrant DNA repair gene expression profile are associated with post-chronic cadmium exposure in human bronchial epithelial cells. *Cells* **2019**, *8*, 842. [CrossRef] [PubMed]
44. Misra, S.; Boylan, M.; Selvam, A.; Spallholz, J.E.; Bjornstedt, M. Redox-active selenium compounds—From toxicity and cell death to cancer treatment. *Nutrients* **2015**, *7*, 3536–3556. [CrossRef] [PubMed]
45. Rahmanto, A.S.; Davies, M.J. Selenium-containing amino acids as direct and indirect antioxidants. *IUBMB Life* **2012**, *64*, 863–871. [CrossRef]
46. Lee, K.H.; Jeong, D. Bimodal actions of selenium essential for antioxidant and toxic pro-oxidant activities: The selenium paradox (Review). *Mol. Med. Rep.* **2012**, *5*, 299–304. [PubMed]
47. Zhao, G.; Wu, X.; Chen, P.; Zhang, L.; Yang, C.S.; Zhang, J. Selenium nanoparticles are more efficient than sodium selenite in producing reactive oxygen species and hyper-accumulation of selenium nanoparticles in cancer cells generates potent therapeutic effects. *Free Radic Biol. Med.* **2018**, *126*, 55–66. [CrossRef]
48. Bevinakoppamath, S.; Saleh Ahmed, A.M.; Ramachandra, S.C.; Vishwanath, P.; Prashant, A. Chemopreventive and anticancer property of selenoproteins in obese breast cancer. *Front. Pharmacol.* **2021**, *12*, 618172. [CrossRef]
49. Liang, Z.L.; Wu, D.D.; Yao, Y.; Yu, F.Y.; Yang, L.; Tan, H.W.; Hylkema, M.N.; Rots, M.G.; Xu, Y.M.; Lau, A.T.Y. Epiproteome profiling of cadmium-transformed human bronchial epithelial cells by quantitative histone post-translational modification-enzyme-linked immunosorbent assay. *J. Appl. Toxicol.* **2018**, *38*, 888–895. [CrossRef]
50. Zang, Z.; Xu, Y.; Lau, A.T.Y. Preparation of highly specific polyclonal antibody for human zinc transporter ZIP8. *Acta Biochim. Biophys. Sin.* **2015**, *47*, 946–949. [CrossRef] [PubMed]
51. Riisom, M.; Gammelgaard, B.; Lambert, I.H.; Sturup, S. Development and validation of an ICP-MS method for quantification of total carbon and platinum in cell samples and comparison of open-vessel and microwave-assisted acid digestion methods. *J. Pharm. Biomed. Anal.* **2018**, *158*, 144–150. [CrossRef] [PubMed]

Article

The Trace Element Selenium Is Important for Redox Signaling in Phorbol Ester-Differentiated THP-1 Macrophages

Theresa Wolfram [1], Leonie M. Weidenbach [1], Johanna Adolf [1], Maria Schwarz [1], Patrick Schädel [2], André Gollowitzer [3], Oliver Werz [2], Andreas Koeberle [3], Anna P. Kipp [1,*] and Solveigh C. Koeberle [1,4,*]

1. Department of Nutritional Physiology, Institute of Nutritional Sciences, Friedrich Schiller University Jena, 07743 Jena, Germany; theresa.wolfram@uni-jena.de (T.W.); leonie.weidenbach@uni-jena.de (L.M.W.); johanna232@gmx.net (J.A.); schwarz.maria@uni-jena.de (M.S.)
2. Department of Pharmaceutical/Medicinal Chemistry, Institute of Pharmacy, University of Jena, 07743 Jena, Germany; patrick.schaedel@uni-jena.de (P.S.); oliver.werz@uni-jena.de (O.W.)
3. Michael Popp Institute and Center for Molecular Biosciences Innsbruck (CMBI), University of Innsbruck, 6020 Innsbruck, Austria; andre.gollowitzer@uibk.ac.at (A.G.); andreas.koeberle@uibk.ac.at (A.K.)
4. Institute of Pharmacy/Pharmacognosy and Center for Molecular Biosciences Innsbruck (CMBI), University of Innsbruck, Innrain 80-82, 6020 Innsbruck, Austria
* Correspondence: anna.kipp@uni-jena.de (A.P.K.); solveigh.koeberle@uibk.ac.at (S.C.K.); Tel.: +49-3641-9-49609 (A.P.K.); +43-512-507-58704 (S.C.K.)

Abstract: Physiological selenium (Se) levels counteract excessive inflammation, with selenoproteins shaping the immunoregulatory cytokine and lipid mediator profile. How exactly differentiation of monocytes into macrophages influences the expression of the selenoproteome in concert with the Se supply remains obscure. THP-1 monocytes were differentiated with phorbol 12-myristate 13-acetate (PMA) into macrophages and (i) the expression of selenoproteins, (ii) differentiation markers, (iii) the activity of NF-κB and NRF2, as well as (iv) lipid mediator profiles were analyzed. Se and differentiation affected the expression of selenoproteins in a heterogeneous manner. GPX4 expression was substantially decreased during differentiation, whereas GPX1 was not affected. Moreover, Se increased the expression of selenoproteins H and F, which was further enhanced by differentiation for selenoprotein F and diminished for selenoprotein H. Notably, LPS-induced expression of NF-κB target genes was facilitated by Se, as was the release of COX- and LOX-derived lipid mediators and substrates required for lipid mediator biosynthesis. This included TXB_2, TXB_3, 15-HETE, and 12-HEPE, as well as arachidonic acid (AA), eicosapentaenoic acid (EPA), and docosahexaenoic acid (DHA). Our results indicate that Se enables macrophages to accurately adjust redox-dependent signaling and thereby modulate downstream lipid mediator profiles.

Keywords: selenium; selenoprotein; macrophage; differentiation; inflammation; redox signaling; NRF2; NF-κB; lipid mediators

1. Introduction

Selenium (Se) is an essential trace element that is important for human health. It is required for various physiological processes including immune function and mammalian development, as well as thyroid hormone metabolism, and Se deficiency has been implicated in various pathologies including inflammation, heart disease, and cancer [1,2]. The biological effects of Se are mainly mediated by selenoproteins, which contain the amino acid selenocysteine, the Se-containing analogue of cysteine, in their polypeptide chain [2]. Most commonly, selenocysteine is part of the catalytic center and provides the selenoproteins with unique chemical activities due to the superior reactivity of selenocysteine. Accordingly, many selenoproteins regulate redox-dependent processes [3] and adequate levels of Se and selenoproteins have been reported to be critical for proliferation, differentiation, and inflammatory processes [4]. The impact of individual selenoproteins is thereby dependent on the cell type and differentiation state. Hence, expression levels of selenoproteins largely differ between undifferentiated and differentiated cells, which suggests that distinct selenoproteins inherit specific functions depending on the differentiation state. Knock-down of selenoprotein (SELENO)H, for example, triggers proliferation of undifferentiated human HT-29 colorectal carcinoma cells and decreases their differentiation, while SELENOH is significantly downregulated in differentiated HT-29 cells [5]. In contrast, thioredoxin reductases (TXNRDs) are upregulated in differentiated adiopocytes and inhibit the differentiation process [6,7]. Moreover, various other selenoproteins, such as glutathione peroxidases (GPXs), SELENOO, and SELENOK, have been shown to be implicated in proliferation and differentiation of distinct cell types [4,8,9].

Macrophages, which are derived from circulating monocytes, are important cells of the innate immune system [10]. Depending on the microenvironmental signals, macrophages give rise to a variety of different tissue-resident-activated macrophage sub-populations, including pro-inflammatory M1- or anti-inflammatory M2-type macrophages [11,12]. Two redox-regulated transcription factors, nuclear factor κ-light-chain-enhancer (NF-κB) and nuclear factor erythroid 2-related factor 2 (NRF2), are key regulators of inflammation and oxidative stress-induced cellular responses in macrophages. These play pivotal roles in host defense and inflammatory processes but are also important for tissue homeostasis and repair [13,14]. While NF-κB mainly triggers gene expression of inflammatory proteins like cyclooxygenase (COX)2 and tumor necrosis factor (TNF)α, thereby enhancing oxidative stress, NRF2 activates the antioxidant response, which includes the upregulation of TXNRD1, GPX4, superoxide dismutase (SOD)1, heme oxygenase (HMOX)1, and catalase (CAT), as well as proteins involved in glutathione (GSH) homeostasis, including solute carrier family 7 member 11 (SLC7A11), glutamate-cysteine ligase catalytic subunit (GCLC), and glutamate-cysteine ligase modifier subunit (GCLM). The two pathways are crosslinked by a complex network, thereby allowing fine-tuned inflammatory responses and preventing exacerbated oxidative stress and inflammation [15]. Various studies suggest that adequate Se levels limit excessive inflammation [2], and genetic inhibition studies identified specific functions of individual selenoproteins in the macrophage maturation process (GPX1 and SELENOP), as well as in the limitation of the oxidative burst (SELENOK) [16].

In mammals, the hierarchy of selenoproteins ensures that indispensable selenoproteins are expressed at the expense of less important ones when the Se supply is limited [17]. A second hierarchical principle ensures the transport of Se to privileged tissues, e.g., the brain. The expression pattern of selenoproteins is highly dynamic, tightly regulated, and fine-tuned depending on the developmental stage, physiological conditions, and the availability of Se [1,18,19].

Given that adequate expression of selenoproteins (i) depends on Se supply, (ii) varies widely between different tissue types, (iii) can be altered by the differentiation process, and (iv) is required for an appropriate inflammatory response, this raises the questions of how the Se status affects monocyte/macrophage function and how selenoprotein expression levels are regulated in macrophages during the differentiation process. Human THP-1 monocytes have been suggested to be a valuable model for macrophage differentiation because phorbol 12-myristate 13-acetate (PMA)-differentiated THP-1 macrophages resemble the native monocyte-derived macrophages with regard to (i) morphology, (ii) expression of membrane antigens and receptors, (iii) the release of secretory factors, and (iv) transient induction of proto-oncogenes [20]. We established a THP-1 model to investigate the effect of Se on (i) cell proliferation and differentiation of monocytes and (ii) the expression of selenoproteins in undifferentiated monocytes as compared to differentiated macrophages, as well as (iii) the inflammatory response upon lipopolysaccharide (LPS) stimulation. Interestingly, Se treatment neither significantly affected the proliferation of human THP-1 monocytes nor the differentiation process itself. However, expression of almost all selenoproteins was modulated by the differentiation state of THP-1 cells. Adequate levels of Se were moreover important for the LPS-induced inflammatory response, resulting in an upregulation of NF-κB target gene expression as well as increased biosynthesis of lipid mediators (LMs). In conclusion, the tight regulation of the selenoproteome by both Se and differentiation processes together with the dependency of the inflammatory response on Se levels strongly suggest that an adequate Se status is indispensable for proper functioning of macrophages.

2. Results

2.1. Se Does Not Affect Differentiation of THP-1 Cells into Macrophages

To investigate the effect of Se on the differentiation process of human THP-1 leukemic monocytes into macrophages, we initially determined the Se concentration and incubation time required for adequate expression of Se-sensitive selenoproteins. Concentration-dependent studies with 0–500 nM sodium selenite for 72 h confirmed a significant upregulation of the protein expression levels of the Se-sensitive selenoproteins GPX1 and SELENOH (Figure S1a,b), while expression of TXNRD1 was not affected (Figure S1c). Moreover, protein expression of GPX1, SELENOH, and TXNRD1 was time-dependently upregulated, reaching a maximum after 72 h (Figure S1d–f). Cell number was not significantly affected up to 72 h by 50 nM Se treatment (Figure S1g). Pre-treatment with 50 nM Se for 72 h was therefore used for subsequent experiments.

THP-1 differentiation into macrophages can be induced by PMA and is associated with increased cell size and adherence [21,22]. Concentration–response studies (0–100 nM PMA) showed that treatment with low (5 nM) PMA concentrations for 48 h was not sufficient to induce complete differentiation, as indicated by loosely attached cells with monocytic morphology (Figure S1h). Incubation with 25 nM for 48 h induced a pronounced adherence of the cells, with significantly increased cell sizes (Figure S1h). Higher concentrations of PMA (100 nM) did not further increase adherence or cell size. Recent studies indicate that high PMA concentrations induce a rather pro-inflammatory type of macrophage which is less responsive towards stimuli [22,23]. Hence, 25 nM of PMA was used for differentiation. In addition, the cell cycle inhibitor p21WAF1/Cip1 was induced by PMA reaching a maximum at 25 nM PMA (Figure 1a). This effect was independent of Se treatment. Time-dependent studies (2–72 h) with 25 nM PMA confirmed a progressive increase in relative cell adherence, reaching a maximum after 48 h (Figure 1b). To further explore the influence of Se on the PMA-induced THP-1 differentiation process, cellular autofluorescence and granularity, as well as extracellular CD68 levels, were analyzed. All three differentiation markers increased with PMA treatment, whereas Se did not significantly affect differentiation (Figure 1c–e). In summary, differentiation of monocytes into macrophages was not significantly affected by the Se status of the cells.

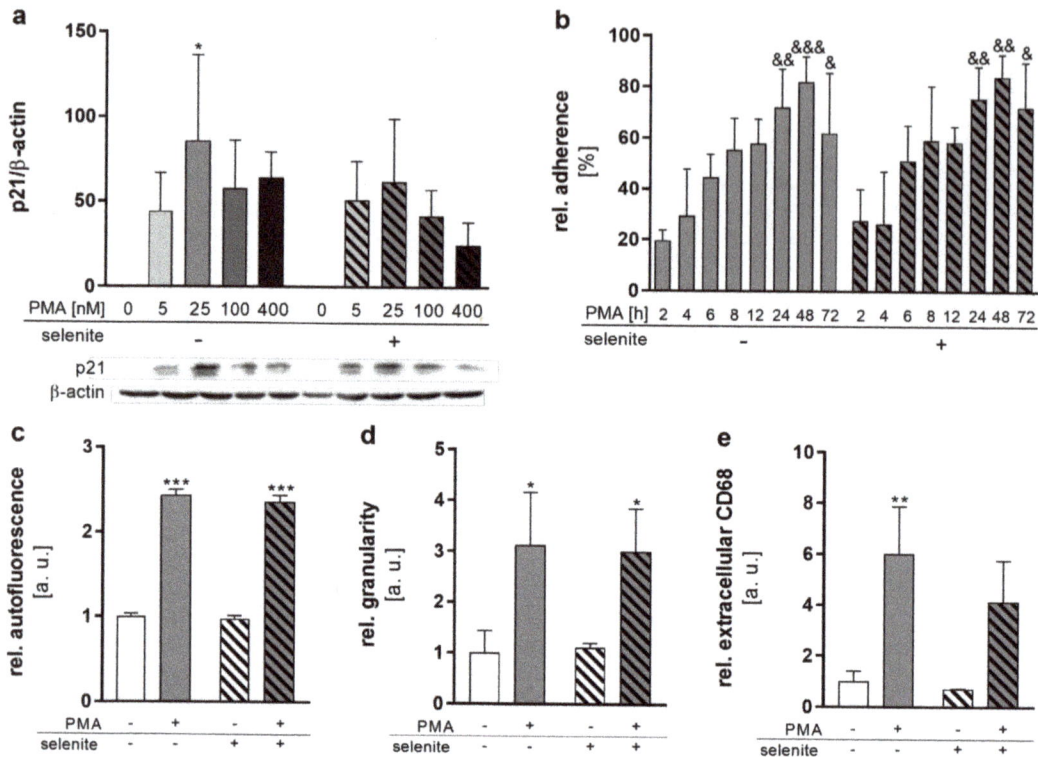

Figure 1. Selenium does not affect differentiation. THP-1 monocytes were pre-treated for 72 h with or without 50 nM sodium selenite (selenite) and differentiated into macrophages by 25 nM PMA or the indicated concentration with or without 50 nM selenite for the indicated time points (**b**) or for 48 h (**a,c–e**). (**a**) Protein expression of p21 was normalized to β-actin. -PMA/-selenite samples were set to 1. A representative blot is shown. (**b**) The relative adherence was calculated as (number of adherent cells/total number of cells) × 100. (**c–e**) The autofluorescence (**c**), intracellular granularity (**d**), and extracellular CD68 (**e**) were determined using flow cytometry. Data are given as means + SD (n = 3–4). Two-way ANOVA with Bonferroni's post-test. Significant outliers were determined by Grubbs' test (α = 0.05). * $p < 0.05$, ** $p < 0.01$, *** $p < 0.001$ vs. cells without PMA (PMA effect); $^\&$ $p < 0.05$, $^{\&\&}$ $p < 0.01$, $^{\&\&\&}$ $p < 0.001$ vs. 2 h PMA.

2.2. PMA-Induced Differentiation of Monocytes to Macrophages Alters the Selenoprotein Expression Profile

To investigate whether selenoprotein expression is affected by differentiation in combination with varying Se supply, we analyzed the protein levels of selenoproteins under Se-deficient, as well as Se-adequate, conditions at increasing PMA concentrations. As expected, selenoprotein expression was upregulated in Se-treated cells, as shown for GPX4, SELENOH, SELENOS, and SELENOF (Figure 2a–d). Most pronounced Se-dependent effects were obtained for SELENOF in undifferentiated THP-1 monocytes with a 12-fold upregulation, followed by SELENOS and SELENOH. GPX4 expression was only moderately affected, being upregulated by 1.8-fold in Se-treated cells. In contrast, PMA substantially reduced expression levels of the selenoproteins SELENOH, SELENOS, and GPX4 in both Se-treated as well as untreated cells (Figure 2a–c). However, the protein expression of SELENOF increased upon PMA-induced differentiation (Figure 2d), while other selenoproteins, such as GPX1, SELENOO, TXNRD1, and TXNRD2, were almost unaffected by PMA-induced differentiation (Figure S2a–d). Accordingly, total activities of the TXNRD and GPX selenoprotein families were also not modulated by PMA (Figure S2e,f).

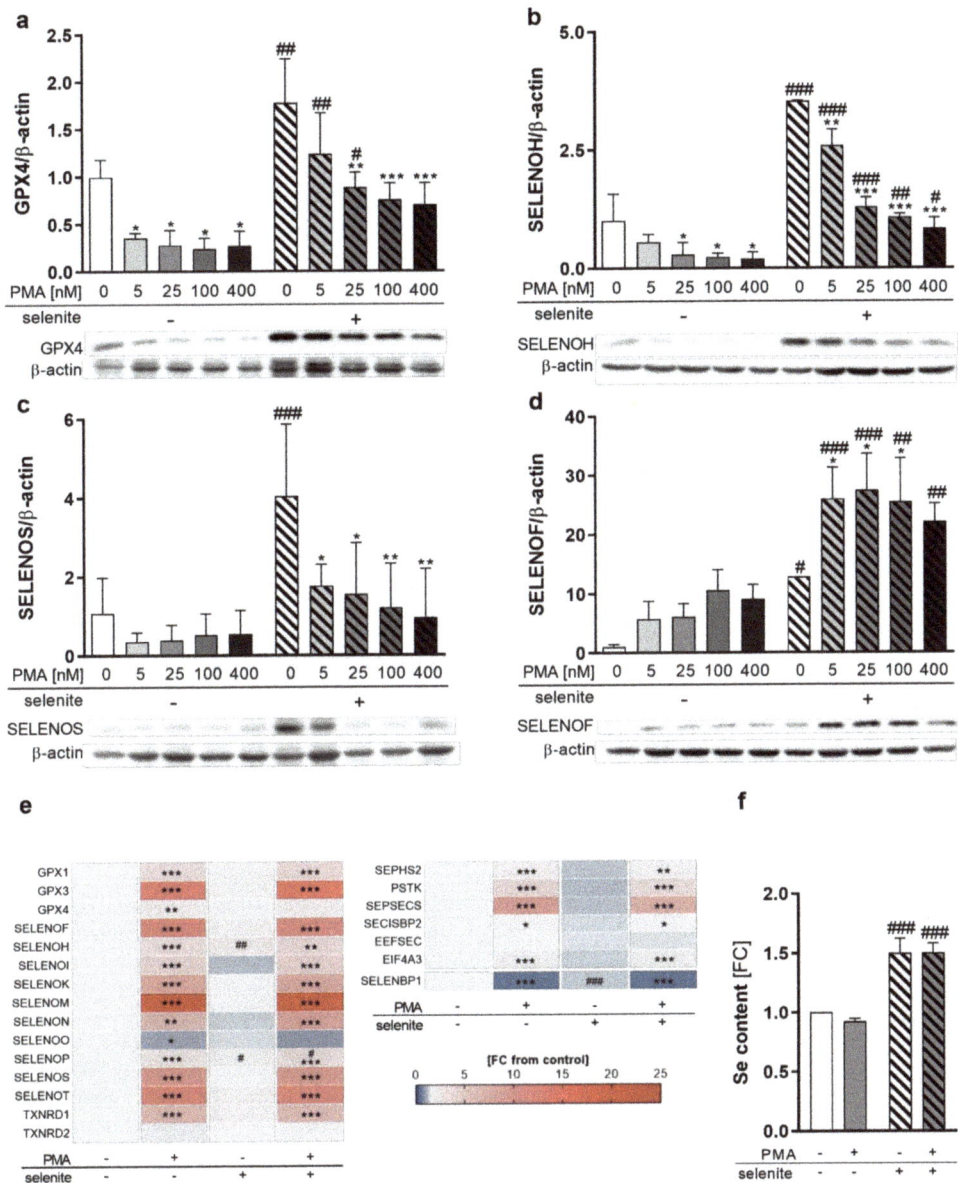

Figure 2. Expression of selenoproteins as well as proteins involved in selenoprotein biosynthesis during differentiation. THP-1 monocytes were pre-treated for 72 h with or without 50 nM sodium selenite (selenite) and differentiated into macrophages using 5–400 nM PMA (a–d) or 25 nM PMA (e,f) with or without 50 nM sodium selenite treatment for 48 h. (a–d) Protein expressions of GPX4 (a), SELENOH (b), SELENOS (c), and SELENOF (d) were normalized to β-actin. -PMA/-selenite samples were set to 1. Representative blots are shown. (e) Heatmap of mRNA expression of different selenoproteins and proteins involved in the biosynthesis of selenoproteins analyzed by qRT-PCR. The color scale indicates the mean difference as fold change (FC) of target genes vs. -PMA/-selenite (FC = 1). (f) Se content of the cells. -PMA/-selenite samples were set to 1. Data are given as means + SD (n = 3–4). Two-way ANOVA with Bonferroni's post-test. Significant outliers were determined by Grubbs' test ($\alpha = 0.05$). * $p < 0.05$, ** $p < 0.01$, *** $p < 0.001$ vs. cells without PMA (PMA effect); # $p < 0.05$, ## $p < 0.01$, ### $p < 0.001$ vs. cells without selenite (Se effect).

As changes in selenoprotein expression upon PMA treatment can be regulated at either the transcriptional or translational levels, we investigated mRNA levels of the selenoproteins in untreated and PMA-differentiated THP-1 cells. Interestingly, mRNA levels were not related to protein expression. The only selenoprotein which was downregulated by PMA on mRNA level was SELENOO (Figure 2e). Significant upregulation of mRNA levels was observed for all other selenoproteins investigated, with SELENOM showing the strongest effect with a 20-fold increase in macrophages as compared to monocytes (Figure 2e). Moreover, mRNA levels of Se-binding protein 1 (SELENBP1), a non-selenocysteine-containing protein that covalently binds Se [24,25], were downregulated by PMA (Figure 2e), which was also observed upon differentiation of other cell types [26,27].

Based on this discrepancy between mRNA and protein expression of selenoproteins, the mRNA levels of proteins involved in selenoprotein synthesis were analyzed. While expression of genes responsible for the biosynthesis of selenocysteine at the t-RNA$^{[Ser]Sec}$ (e.g., selenophosphate synthetase 2 (SEPHS2), phosphoseryl-tRNA kinase (PSTK), sep (o-phosphoserine) tRNA:sec (selenocysteine) tRNA synthase (SEPSECS)) were significantly upregulated, mRNA expression of proteins involved in selenoprotein translation (SECIS binding protein 2 (SECISBP2) and eukaryotic elongation factor, selenocysteine-tRNA specific (EEFSEC)) were less affected by PMA treatment (Figure 2e). Interestingly, eukaryotic translation initiation factor 4A3 (EIF4A3), which inhibits the biosynthesis of Se-sensitive selenoproteins, was significantly upregulated by PMA (Figure 2e). Moreover, cellular Se levels were independent from the differentiation state and upregulated by Se treatment by 1.5-fold in undifferentiated and differentiated cells, which excludes the possibility that Se availability limited protein expression of selenoproteins in PMA-treated cells (Figure 2f). In conclusion, PMA-induced differentiation downregulated selenoprotein expression of GPX4, SELENOH, and SELENOS but upregulated SELENOF, indicating differentiation-induced shifts of the selenoproteome.

2.3. PMA and Se Treatment Modulate the Cellular Redox Status and the Activity of the Redox-Sensitive Transcription Factors NRF2 and NF-κB

Based on the observed changes in selenoprotein expression induced by PMA differentiation together with the well-known impact of many selenoproteins on the regulation of the cellular redox homeostasis [18,19], we wondered whether Se treatment affects the redox status in PMA-differentiated THP-1 macrophages. Surprisingly, total GSH levels decreased from PMA-induced differentiation independent of the Se status (Figure 3a). NRF2 and NF-κB are two transcription factors which are well-known to be regulated in a redox-dependent manner and to activate gene expression of antioxidant and inflammatory proteins, respectively [28]. The activity of NAD(P)H quinone oxidoreductase 1 (NQO1), a NRF2 target gene, was significantly upregulated in differentiated compared to undifferentiated cells (Figure 3b), which hints towards a putative activation of the transcription factor NRF2 in differentiated cells. To induce inflammatory conditions and activate the NF-κB signaling pathway, we included LPS stimulation for further experiments. NF-κB activation was studied by nuclear translocation of the subunit p65. In macrophages, NF-κB was activated by LPS stimulation in Se-treated but not in Se-deficient cells (Figure 3c). The mRNA expression of the NF-κB target genes COX2 and TNFα was increased by PMA-induced differentiation and further enhanced in LPS-stimulated THP-1 macrophages, with more pronounced effects under Se treatment (Figure 3d). Se effects were present in neither the release of TNFα nor COX2 protein levels (Figures 3e and S3a). Moreover, proteins that are involved in the biosynthesis of prostaglandins (e.g., prostaglandin E synthase (PGES) and thromboxane (TX)A synthase 1 (TXAS1)), as well as leukotriens and resolvins (arachidonate 5-lipoxygenase (ALOX5) and arachidonate 15-lipoxygenase (ALOX15)), were upregulated by PMA but not affected by Se (Figure 3d). Interestingly, ALOX5 mRNA expression was significantly upregulated in undifferentiated THP-1 cells and downregulated in the differentiated cells by LPS. Otherwise, LPS did not significantly affect the protein expression of the two selenoproteins GPX4 and SELENOF (Figure S3c,d).

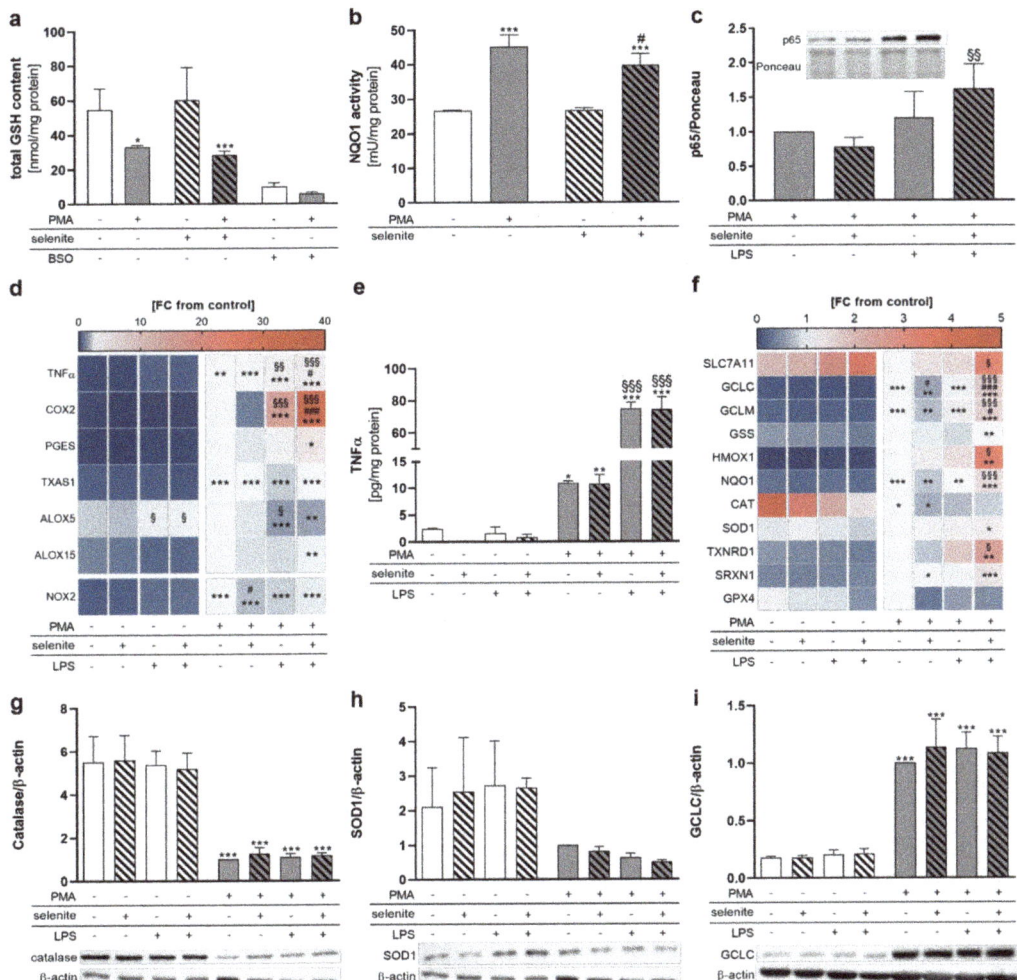

Figure 3. Redox-dependent regulation of NF-κB and NRF2 target gene transcription in THP-1 monocytes and macrophages. THP-1 monocytes were pre-treated with or without 50 nM sodium selenite (selenite) for 72 h and differentiated into macrophages with 25 nM PMA with or without 50 nM selenite treatment for 48 h. Cells were stimulated with 1 µg/mL lipopolysaccharide (LPS) (c–i) for 1 h (c), 6 h (d,f), or 24 h (e,g–i). (a) Total GSH content. The GSH synthesis inhibitor buthionine-sulfoximine (0.25 mM, BSO) was used as positive control. (b) NQO1 activity. (c) The protein expression of nuclear p65 was normalized to Ponceau staining. The -selenite/-LPS sample was set to 1. (d) Heatmap of mRNA expression of NF-κB target genes and enzymes involved in the LM biosynthesis investigated by qRT-PCR. mRNA expression of genes is given as fold change (FC) vs. +PMA/-selenite/-LPS (FC = 1). The color scale indicates the mean difference as the FC of target genes vs. +PMA/-selenite/-LPS (FC = 1). (e) TNFα release was analyzed by ELISA. (f) Heatmap of mRNA expression of NRF2 target genes investigated by qRT-PCR. mRNA expression of genes is given as fold change (FC) vs. +PMA/-selenite/-LPS (FC = 1). The color scale indicates the mean difference as the FC of target genes vs. +PMA/-selenite/-LPS (FC = 1). (g–i) The protein expressions of CAT (g), SOD1 (h), and GCLC (i) were normalized to β-actin. The +PMA/-selenite/-LPS samples were set to 1. The data of (a–c,g–i) are given as means + SD (n = 3–4). Two-way ANOVA (a–c) or three-way ANOVA (d–i) with Bonferroni´s post-test. Significant outliers were determined by Grubbs´ test (α = 0.05). * $p < 0.05$, ** $p < 0.01$, *** $p < 0.001$ vs. cells without PMA (PMA effect); # $p < 0.05$, ### $p < 0.001$ vs. cells without selenite (Se effect); § $p < 0.05$, §§ $p < 0.01$, §§§ $p < 0.001$ vs. cells without LPS (LPS effect).

For mRNA levels, PMA treatment upregulated most NRF2 target genes (GCLC, GCLM, glutathione synthetase (GSS), HMOX1, NQO1, SOD1, TXNRD1, SRXN1) with the exception of SLC7A11, CAT, and GPX4 (Figure 3f). The strong PMA-dependent downregulation of SLC7A11, a subunit of the cystine/glutamate antiporter system X_c^- that is essential for the intracellular supply with cysteine and subsequent GSH biosynthesis, might explain the decreased levels of GSH in the PMA-differentiated cells, although mRNA levels of the proteins of GSH biosynthesis (GCLC, GCLM, GSS) were upregulated. Notably, Se treatment makes THP-1 macrophages more responsive to LPS-induced NRF2 gene expression, with up to threefold higher mRNA levels of NRF2 target gene expression (e.g., HMOX1) under Se-treated conditions as compared to Se-deficient conditions. Protein expressions of NRF2 target proteins CAT, SOD1, and GCLC were not affected by LPS stimulation and Se treatment (Figure 3g–i). However, CAT expression was significantly downregulated and GCLC was upregulated by PMA, whereas no pronounced PMA effects were observed on SOD1 expression (Figure 3g–i).

2.4. Effects of LPS Treatment on LM profiles in THP-1 Monocytes and Macrophages

Given that Se-treated PMA-differentiated macrophages were more responsive to LPS-induced NF-κB activation, we speculated that an adequate Se status might also affect LM biosynthesis. A metabolipidomics approach was employed to investigate the impact of Se on LPS-induced changes on the LM profile. Similar to what we observed for mRNA expression of COX2 and TNFα (Figure 3d), nuclear translocation of the p65 subunit of NF-κB (Figure 3d), and the expression of NRF2 target genes (Figure 3f), Se supported the production of COX- and LOX-derived LMs as well as the release of the fatty acid substrates required for LM biosynthesis, reaching significance for the COX-derived LM TXB_2 as well as docosahexaenoic acid (DHA) in differentiated macrophages. Interestingly, in undifferentiated Se-deficient cells, almost all lipid mediators were downregulated by LPS by trend but upregulated by Se with significant effects on COX-derived TXB_2 and TXB_3, and the LOX-derived products 15-hydroxyeicosatetraenoic acid (15-HETE) and 12-hydroxypentaenoic acid (12-HEPE), as well as arachidonic acid (AA), eicosapentaenoic acid (EPA), and docosahexaenoic acid (DHA) (Figure 4). Together, our data clearly indicate that Se facilitates LM biosynthesis by modulating the enzymes for LM biosynthesis.

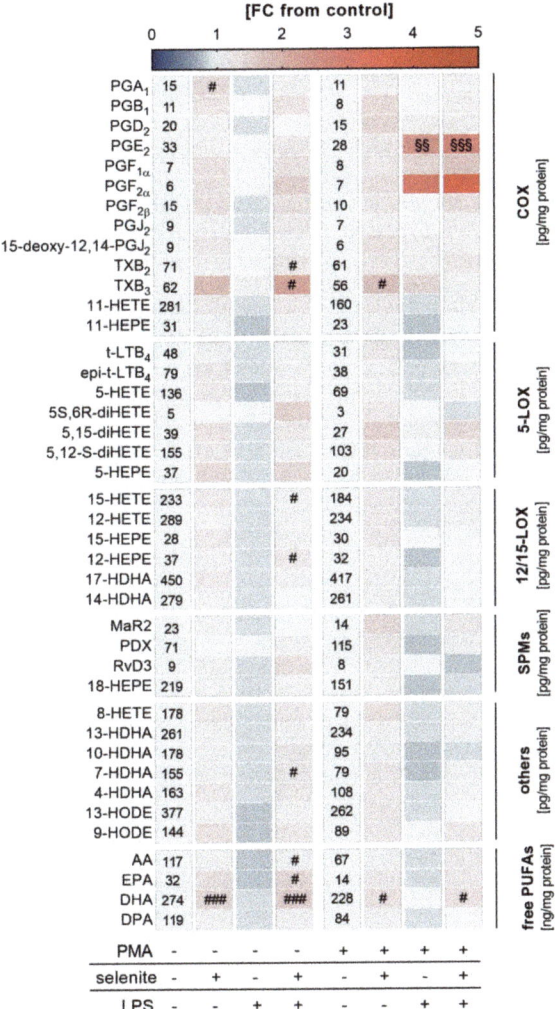

Figure 4. Effect of Se, PMA-induced differentiation, and LPS on lipid mediator (LM) profiles. THP-1 cells were pre-treated with or without 50 nM sodium selenite (selenite) for 72 h and differentiated into macrophages by treatment with 25 nM PMA in co-treatment with or without 50 nM sodium selenite (selenite) for 48 h and with or without 1 μg/mL lipopolysaccharide (LPS) for an additional 24 h. LM profiles of the supernatant were analyzed by UPLC-MS/MS. The heatmap organizes LM according to key biosynthetic enzymes (n = 4). The values in the columns give the concentrations of LMs in pg/mg protein (COX, 5-LOX, 12/15-LOX, SPMs, others) or ng/mg protein (free PUFAs). The color scale indicates the mean difference as fold change (FC) of control cells vs. LPS and/or selenite-treated cells (FC = 1). Two-way ANOVA with Bonferroni's post-test. Significant outliers were determined by Grubbs' test (α = 0.05). # $p < 0.05$, ### $p < 0.001$ vs. cells without selenite (selenite effect); §§ $p < 0.01$, §§§ $p < 0.001$ vs. cells without LPS (LPS effect). AA, arachidonic acid; COX, cyclooxygenase; DHA, docosahexaenoic acid; DPA, docosapentaenoic acid; EPA, eicosapentaenoic acid; HDHA, hydroxy-docosahexaenoic acid; HEPE, hydroxypentaenoic acid; HETE, hydroxyeicosatetraenoic acid; HODE, hydroxyoctadecadienoic acid; LOX, lipoxygenase; LT, leukotriene; MaR, maresin; PD, protectin D; PG, prostaglandin; PUFA, polyunsaturated fatty acid; Rv, resolvin; SPM, specialized pro-resolving mediator; TX, thromboxane.

3. Discussion

Since Se was recognized as an essential trace element, the role played by its adequate supply has been intensively studied. Various studies have shown that adequate levels of Se are important for the immune function and are able to prevent excessive inflammation [3]. Hence, Se was found to limit the expression of pro-inflammatory cytokines under inflammatory conditions in vitro at relatively high concentrations of 2 µM [29–31]. In vivo effects of Se are more distinct [32,33]. While long-term treatment of mice with 0.6 mg/kg Se (which was given as sodium selenite and corresponded to approximately four times the adequate intake) had no effects on dextran sulfate sodium (DSS)-induced colitis, short-term treatment with 0.6 mg/kg Se actually enhanced inflammation scores, as well as TNFα and COX2 mRNA expression. Notably, 0.6 mg/kg Se given in the form of selenomethionine had no effect [32]. Otherwise, rats that received 0.2–20 µg/kg organic selenocompounds during pregnancy developed less severe *Staphylococcus aureus*-induced mastitis with lower NF-κB activation and TNFα mRNA expression [33]. Within the innate immune system, macrophages inherit key functions in neutralizing pathogens and regulating inflammatory processes and individual selenoproteins have been shown to affect the macrophage maturation process (GPX1 and SELENOP) and limit oxidative burst (SELENOK) [34–38]. However, little is known about how Se deficiency impacts redox-dependent proliferation, differentiation and signaling cascades in monocytes and differentiated macrophages, nor about how selenoprotein expression affects these processes.

Human THP-1 monocytes were used as a model system under Se-deficient as well as Se-adequate conditions and were differentiated into macrophages by PMA. Interestingly, the hierarchy of selenoprotein expression seems to be different in monocytes/macrophages as compared to other cell types, such as epithelial cells. While the Se-sensitive selenoprotein SELENOH was significantly upregulated in a time- and concentration-dependent manner, with fourfold higher expression levels at 500 nM Se, the expression of another Se-sensitive selenoprotein, GPX1, was only slightly affected by Se, being upregulated by 1.5-fold (Figure S1a,b). Notably, the protein expression of GPX1 and of the protein family member GPX4, which ranks higher in hierarchy, were comparably upregulated by 1.5–1.8-fold by Se, which indicates that GPX1 is more indispensable in THP-1 monocytes compared to other cell types (Figures 2a and S2a). This might be due to the absence of another GPX isoenzyme in monocytes, GPX2 [35], which has been previously shown to have overlapping functions with GPX1 in limiting the redox-dependent activation of the NF-κB pathway and LM biosynthesis in human epithelial-derived cancer cells [29]. It is therefore tempting to speculate that the absence of the epithelial GPX2 makes GPX1 more indispensable in THP-1 monocytes, thereby advancing GPX1 in hierarchy.

Interestingly, the PMA-induced differentiation process was also not significantly affected by Se status, as suggested from cell adherence, granularity, autofluorescence, and surface marker expression (Figure 1b–e). However, the differentiation process had a strong impact on the protein expression levels of various selenoproteins (Figures 2 and S2). Expression of the two GPX isoenzymes GPX1 and GPX4 was differently regulated by PMA. Whereas GPX1 protein expression was not affected by PMA treatment (Figure S2a), GPX4 protein expression was downregulated under both Se-deficient as well as Se-adequate conditions by PMA-induced differentiation (Figure 2a). Notably, mRNA expression levels of selenoproteins did not correlate with protein expression levels and were upregulated by PMA treatment for all selenoproteins except for SELENOO (Figure 2e). We could exclude (i) Se levels, (ii) impaired selenocysteine biosynthesis, and (iii) selenoprotein translation as limiting variables for the PMA-induced reduction of several selenoproteins since the cellular Se content as well as mRNA expression of proteins responsible for selenocysteine biosynthesis and selenoprotein translation were not affected or even upregulated by PMA (Figure 2e,f). We therefore speculate that the upregulation of selenoprotein mRNAs is due to the special M0 state of macrophages, which has to be considered as a transitional rather than a terminal stage. Depending on the milieu in the tissue, M0 macrophages polarize into various subpopulations with either pro- or anti-inflammatory properties. Hence, mRNA levels of the selenoproteins might be kept at relatively high levels to rapidly adapt the protein expression of selenoproteins to the requirement of the respective macrophage subtype after maturation.

GPX4 has been shown to inhibit NF-κB activation [39], and the downregulation of GPX4 in differentiated THP-1 macrophages might be essential to enable inflammatory responses in these cells. Accordingly, NF-κB target genes TNFα and COX2 were upregulated by PMA (Figure 3d). Moreover, TNFα and COX2 mRNA expression as well as the release of COX-derived PGE$_2$ and PGF$_{2\alpha}$ were increased by LPS stimulation in macrophages, with more distinct effects under Se treatment (Figures 3 and 4). However, the Se effect on TNFα and COX2 expression was not present for protein levels (Figures 3e and S3a). Possibly, Se effects on protein expression are only visible within another time frame which was not captured by our study (24 h). Essentially, previous studies indicate that the Se effects on biosynthesis of pro-inflammatory cytokines differ at different time points [30,40]. Importantly, a rapid and adequate secretion of inflammatory cytokines upon bacterial or viral infection is essential in pathogen removal and to prevent sustained inflammation. Optimal Se supply might be important to maintain homeostasis between the formation of inflammation-dependent production of reactive oxygen species and the antioxidant NRF2 system, thereby enabling cells' fast upregulation of inflammatory response upon stimulation, and to prevent chronic inflammation [41]. Inversely, physiological changes in Se levels as studied herein only moderately impacted on NF-κB signaling and LM biosynthesis, which indicates that THP-1 cells are able to compensate for the reduced expression of selenoproteins. Otherwise, expression of selenoproteins that are important for the regulation of inflammatory signaling might be less dependent on Se in macrophages, as demonstrated for GPX1 and TXNRD1 (Figure S2a,c).

Interestingly, the release of other COX-derived products was enhanced neither by PMA nor by LPS (Figure 4), which indicates that LM biosynthesis is efficiently channeled into PGE$_2$ and PGF$_{2\alpha}$ production. Notably, previous studies reported that Se treatment decreased the expression of pro-inflammatory cytokines such as TNFα and the release of LMs including PGE$_2$ in macrophages [29–31,33,40,42,43]. Differences in the study outcomes might have been related to different experimental settings. While our study used human THP-1 macrophages treated with physiological concentrations of Se (50 nM), previous studies employed murine RAW264.7 and bone marrow-derived macrophages (BMDM) treated with 2 µM Se (given as sodium selenite). Hence, even striking differences in the Se-dependent regulation of COX2 expression are observed in RAW264.7 and BMDM cells under similar experimental conditions (2 µM Se), with strong effects of Se (2 µM) on COX2 expression in RAW264.7 and noticeably less pronounced effects in the BMDM cells [30]. Moreover, another study showed that TNFα mRNA expression in *Staphylococcus aureus*-stimulated RAW264.7 macrophages was significantly reduced by 2 µM Se (given as sodium selenite) and unaffected by 1–1.5 µM [40], which underlines the concentration dependency of the Se effects. Notably, while an adequate supply with Se is supposed to have antioxidant properties via the upregulation of selenoproteins, Se given at concentrations that exceed the dosage required to maximize selenoprotein expression most likely does not exert its effects via selenoproteins and can even have pro-oxidant effects, which underscores the necessity of a well-balanced Se status [44–46]. Another critical aspect of an appropriate immune reaction might be that related to an accurate time-dependent regulation of the inflammatory response. While an upregulation of COX2 expression after infection is required to remove pathogens, an excessively prolonged activation might result in chronic inflammation. Overall, it appears that moderate changes in the Se supply do not overtly change the expression of NF-κB-responsive proteins and thus the immune response of THP-1 monocytes and M0 macrophages.

Analog with NF-κB, mRNA expression levels of NRF2 target genes were increased by LPS stimulation and were dependent on adequate Se levels in macrophages (Figure 3f). The increase of NRF2 target gene transcription by LPS is in line with a previous report which also observed an upregulation of NRF2 upon activation of the toll-like receptor 1/2 with Pam3CSK4 in THP-1 cells [47]. Given that the NF-κB and NRF2 signaling pathways are interconnected via a complex signaling network, the coordinated increase in NF-κB and NRF2 target gene expression upon LPS stimulation by Se hints towards an important function in maintaining a balance of NF-κB-dependent reactive oxygen species/reactive nitrogen species formation and NRF2-dependent hydrogen peroxide detoxification in THP-1-derived macrophages. Otherwise, Se downregulated most of the NRF2 target genes under unstimulated conditions in macrophages, reaching significance for GCLC (Figure 3f). This is in line with a previous in vivo study which found that mRNA expression of the NRF2 target genes GCLC, NQO1, SOD1, and SRXN1 was significantly downregulated in mice that received a Se-adequate diet (0.15 mg/kg) compared to mice with a Se-poor diet (0.086 mg/kg) [48]. Notably, while most of the NRF2 target genes were upregulated by PMA-induced differentiation, the two target genes CAT and SLC7A11 were downregulated (Figure 3f). NRF2-dependent target gene expression is regulated by complex mechanisms, which include: (i) different cysteine sensors within the NRF2 regulatory protein KEAP1 (Cys151, Cys257, Cys273, Cys288, and Cys297) that can be targeted by distinct endogenous and exogenous ligands, (ii) phosphorylation of NRF2, and (iii) transactivation via dimerization with different transcription factors (e.g., activating transcription factor 4 (ATF4), ATF3 or CREB binding protein (CBP)) and nuclear receptors (e.g., peroxisome proliferator-activated receptor gamma (PPARγ), estrogen receptor α, and retinoic X receptor α (RXRα)) [15]. How exactly PMA stimulates NRF2 activation is not known. However, a similar controversial regulation of NRF2 target gene subsets has also been observed with other small molecule inducers of NRF2, such as Bay 11-7085 [49] and sulforaphane [50], which stimulate ferroptosis and induce a strong upregulation of HMOX1 with concomitant downregulation of other NRF2 target genes like GPX4 and SLC7A11, respectively. Depending on the point of attack, small molecule inducers of NRF2 might therefore reveal a specific profile of NRF2 target gene regulation.

In conclusion, we anticipate that adequate Se levels as well as differentiation impact on selenoprotein expression, whereas the Se status only moderately affected NF-κB signaling and biosynthesis of inflammatory LMs. Of particular interest are the discrepancies with previous findings that indicated an inhibitory effect of Se on NF-κB-dependent inflammatory signaling; our results indicate that Se facilitates LPS-induced NF-κB activation. We speculate that differences arise from (i) different Se concentrations and (ii) timing and maybe even the species (e.g., human vs. mouse). Notably, PMA, which was used for differentiation, had significant effects on NF-κB as well as NRF2 that might have impacted the overall effects of Se in our THP-1 model system. However, the well-known effects of physiological Se levels for a fine-tuned adaptive immune response upon bacterial and viral infections [51,52], as well as during inflammation resolution [53–55], might be mediated by other mechanisms as well. Further concentration- and time-dependent studies on inflammatory responses upon inflammatory stimulation are required, as are investigations in primary cells which more closely reflect in vivo conditions and research on the function of specific selenoproteins in order to address these issues.

4. Materials and Methods

4.1. Materials

Materials are listed in Table S1.

4.2. Cell Culture

Human THP-1 acute monocytic leukemia cells (monocytes, ATCC® TIB-202™) were grown in RPMI completed with 10% (v/v) FCS, 1% (v/v) penicillin-streptomycin and 1% GlutaMAX at 37 °C in 5% CO_2 and sub-cultured every three to five days. Cell treatment: THP-1 monocytes were incubated with different concentrations of Se (0–500 nM) for 72 h or for varying incubation times (0–96 h) with 50 nM Se. THP-1 monocytes were differentiated with different PMA (0–400 nM) concentrations as indicated or with 25 nM PMA for 48 h or as indicated. THP-1 monocytes were preincubated for 72 h with or without 50 nM

Se, seeded at densities of 0.5×10^5 cells/mL THP-1 monocytes (DMSO control) and 2.5×10^5/mL for PMA-induced differentiation with or without Se (50 nM) treatment, and harvested after 48 h. LPS (1 µg/mL) stimulation was performed with undifferentiated and differentiated THP-1 cells with or without Se treatment for 1 h (nuclear extracts for Western blot analysis), 6 h (quantitative real-time PCR analysis) or 24 h (Western blot and LM analysis). The protein content of the cell lysates was measured using Bradford analysis (Bio-Rad Laboratories).

4.3. Determination of Adherence, Morphology and Cell Number

Adherence of PMA-treated THP-1 cells was determined at the indicated time points. Cells were harvested by trypsinization and resuspended in RPMI. The cell number was determined using 0.4% trypan blue staining with a Vi-Cell XR Cell Viability Analyzer (Beckman Coulter, Brea, CA, USA). To determine the relative adherence, adherent cells were collected and percentage adherence was calculated as the number of adherent cells divided by the starting cell number and multiplied by 100.

Images of cells were collected with an Eclipse Ti-S microscope (Nikon GmbH, Tokyo, Japan) equipped with a Nikon Digital Sight DS-Fi1c color camera (Nikon GmbH) and a $20\times$ objective (NA = 0.45, S Plan Fluor, Nikon GmbH) or a $10\times$ objective (NA = 0.3, Plan Fluor, Nikon GmbH).

4.4. Flow Cytometry

For flow cytometry experiments, cells were treated as described above and harvested. Granularity was measured using forward and side scatter light and autofluorescence analyzed in unstained cells in the FL-1 channel (blue laser, excitation 488, emission 525/40). For detection of the cell surface marker CD68, cells were resuspended in ice cold primary antibody solution (containing phosphate-buffered saline (PBS, 140 mM NaCl, 10 mM Na_2HPO_4, and 2.99 mM KH_2PO_4, pH 7.4) with 10% FCS and 1% NaN_3), fixed with PBS and 0.01% formaldehyde (diluted in PBS) for 15 min, and washed with primary antibody solution. Cells were resuspended in primary antibody solution with 0.5% (v/v) Tween® 20 and incubated with primary antibody mouse anti-CD68 (1:200) for 30 min. Samples were centrifuged for 3 min at $500\times g$ and washed twice in primary antibody solution and once in secondary antibody solution (containing PBS with 3% BSA and 1% NaN_3) with the same centrifugation steps in between. Samples were incubated with the secondary antibody (anti-mouse IgG Alexa Fluor® 594 Conjugate (1:1000)) for 30 min protected from light. Analysis was performed on a CytoFLEX flow cytometer (Beckman Coulter, excitation: 488 nm; emission: 525/40 nm) and data were analyzed using CytExpert software version 2.2 and version 2.3 (Beckman Coulter).

4.5. Nuclear Isolation

The nuclear lysates of 3.0×10^5/mL PMA-differentiated macrophages were harvested after 1 h of LPS stimulation. Lysis buffer I (10 mM HEPES, 1.5 mM $MgCl_2$, 10 mM KCl, 0.5 mM DTT, 0.5 mM PMSF, and 0.1% NP-40 Alternative, pH 7.9) was used. After incubation for 7 min at 4 °C under shaking, cells were centrifuged for 1 min at 4 °C and $6800\times g$. Thereafter, NaCl (5 M) was added to the lysis buffer II (40 mM HEPES, 400 mM KCl, 10% glycerol, 1 mM DTT, and 0.1 mM PMSF, pH 7.9) and the cell pellet was lysed by ultrasonic treatment (80% amplitude, 0.5 s cycle), and centrifuged for 30 min at 4 °C and $20,000\times g$ to obtain nuclear lysates. The protein content was measured and the nuclear lysate was used for Western blot analysis.

4.6. Western Blot

Harvested THP-1 monocytes and macrophages were lysed with RIPA buffer (50 mM Tris, 150 mM NaCl, 2 mM EDTA, 0.5% sodium deoxycholate, 0.1% SDS, and 1% NP-40 Alternative, pH 7.7 with 0.1% (v/v) protease inhibitor) for 15 min at 4 °C and 1200 rpm using the ThermoMixer® (Eppendorf AG, Hamburg, Germany). The samples were centrifuged for 15 min at 4 °C and 14,000× g, and protein concentration was determined in the supernatant. Samples were mixed with loading buffer (1×; 41.7 mM Tris pH 6.8, 10% glycerin, 2% SDS, 0.125% bromophenol blue, and 2.5% β-mercaptoethanol) and heated for 5 min at 95 °C. Proteins were separated on SDS polyacrylamide gels (10–15%) and immunoblotted on Amersham™ Protran® nitrocellulose membrane. Membranes were incubated for 2 min in Ponceau S solution (0.2% (w/v) with 3% (w/v) trichloroacetic acid) and bands were recorded by a ChemiDoc™ MP Imaging System (Bio-Rad Laboratories). Subsequently, membranes were blocked in 5% (w/v) non-fat dry milk in Tris-buffered saline (5 mM Tris, 15 mM NaCl) with 0.1% (v/v) Tween® 20 (T-TBS) for 1 h at room temperature and incubated with primary antibodies overnight at 4 °C in T-TBS: Mouse anti-γ-GCSc (H-5) (1:10,000), mouse anti-mPGES1 (1:300); rabbit anti-beta actin (1:10,000); rabbit anti-oxidative stress defense (catalase, SOD1) (1:500); rabbit anti-COX2 (D5H5) XP® (1:1000); rabbit anti-GPX1 (1:5000); rabbit anti-GPX4 (1:5000); rabbit anti-NF-κB p65 (D14E12) XP® (1:4000); rabbit anti-p21WAF1/Cip1 (12D1) (1:1000); rabbit anti-SEP15 (1:3000); rabbit anti-SELH (1:1000); rabbit anti-SELO (1:2500); rabbit anti-VIMP (D1D1M)) (1:1000); rabbit anti-TXNRD1 (1:10,000); and rabbit anti-TXNRD2 (1:1000). HRP-coupled goat anti-rabbit IgG (1:50,000) or HRP-coupled horse anti-mouse IgG (1:3000) were used as secondary antibody. Protein bands were detected using SuperSignal™ West Dura and band intensities were densitometrically quantified using a ChemiDoc™ MP Imaging System (Bio-Rad Laboratories) and data analyzed using Image Lab software version 5.0 (Bio-Rad Laboratories). Protein expression was normalized to ß-actin or Ponceau S staining as indicated.

4.7. Quantitative Real-Time PCR

Total mRNA of THP-1 cells was isolated using the Dynabeads™ mRNA DIRECT™ Purification Kit according to the manufacturer's protocol. The SensiFAST™ cDNA Synthesis Kit was used for cDNA transcription. For real-time PCR, cDNA were combined with Master mix (PerfeCTa SYBR Green Supermix, forward and reverse primer (each 250 nM final concentration)) in a 96-well plate as previously described [56]. Primer sequences are given in Table 1. PCR was performed on a real-time PCR system (MX3005P, Agilent, Santa Clara, CA, USA) with heat cycle to 95 °C for 3 min, 40 cycles of denaturation at 95 °C for 15 sec, annealing at 60 °C for 20 s, and elongation at 72 °C for 30 sec. Standard curves from diluted PCR products were used for quantification and all samples and standards were measured in triplicate. For normalization, sample values were normalized to a composite factor based on the reference genes RPL13a and RPS9. MIQE guidelines served as reference for the quantification procedure.

Table 1. Human-specific primers for quantitative real-time PCR.

Gene	RefSeq-ID	Sequence (5′→3′)
ALOX15 (arachidonate 15-lipoxygenase)	NM_001140.3	TGGAGCCTTCCTAACCTACAG TCCACATACCGATAGATGATTTCC
ALOX5 (arachidonate 5-lipoxygenase)	NM_000698.3	GCTGCAACCCTGTGTTGATCC AAATGTTCCCTTGCTGGACCTC
CAT (catalase)	NM_001752.4	CCTATCCTGACACTCACCGCCA GAGCACCACCCTGATTGTCCTG
COX2 (cyclooxygenase 2; prostaglandin-endoperoxide synthase 2 (PTGS2))	NM_000963.2	CCCAGCACTTCACGCATCAG CTGTCTAGCCAGAGTTTCACCGT
EEFSEC (eukaryotic elongation factor, selenocysteine-tRNA-specific)	NM_021937.3	CCCTAGAGAACACCAAGTTCCGAG TCAATGAGCTCTGGAATGCCCT
EIF4A3 (eukaryotic translation initiation factor 4A3)	NM_014740.3	AAAGAAAGGTGGACTGGCTGACGG ACTCCTTCATGATGGACTCCCGCT
GCLC (glutamate-cysteine ligase catalytic subunit)	NM_001498.3	TGCTGTCTCCAGGTGACATTCCA GGAGATGCAGCACTCAAAGCCA
GCLM (glutamate-cysteine ligase modifier subunit)	NM_002061.3	GTTGACATGGCCTGTTCAGTCCT CCCAGTAAGGCTGTAAATGCTCCA
GPX1 (glutathione peroxidase 1)	NM_000581.2	TACTTATCGAGAATGTGGCGTCCC TTGGCGTTCTCCTGATGCCC

Table 1. *Cont.*

Gene	RefSeq-ID	Sequence (5′→3′)
GPX3 (glutathione peroxidase 3)	NM_002084.5	GTCGAAGATGGACTGCCATGGT AGCTGGCCACGTTGACAAAGAG
GPX4 (glutathione peroxidase 4)	NM_002085.3	AGGCAAGACCGAAGTAAACTACAC TCTCTTCGTTACTCCCTGGCT
GSS (glutathione synthesis)	NM_000178.2	CCAAGTGCCCAGACATTGCCA CCTTCTTCACCCACATCCAGTGAG
HMOX1 (heme oxygenase 1)	NM_002133.2	CAACAAAGTGCAAGATTCTGCCC CTACAGCAACTGTCGCCACC
NOX2 (NADPH oxidase 2)	NM_000397.3	TCACCAAGGTGGTCACTCACCC TGCCACTCCAGCTTGGACAC
NQO1 (NAD(P)H quinone oxidoreductase 1)	NM_001025434.1	CATCACAGGTAAACTGAAGGACCC CTCTGGAATATCACAAGGTCTGCG
PGES (prostaglandin E synthase)	NM_004878.3	ACGCTGCTGGTCATCAAGATG TGGCAAAGGCCTTCTTCCGC
PSTK (phosphoseryl-tRNA kinase)	NM_153336	TTTGAGGCCCAGTCTTGCTACC GCCCAACGAATATTTCCGAGCC
RPL13a (ribosomal protein L13a)	NM_012423.4	GAGGTTGGCTGGAAGTACCAGG TGTTTCCGTAGCCTCATGAGCTG
RPS9 (ribosomal protein S9)	NM_001013.4	CCATATCAGGGTCCGCAAGCA GGCCCTTCTTGGCATTCTTCCT
SECISBP2 (SECIS-binding protein 2)	NM_024077.4	TGAAGAGCCACCAGGCACAG GCATCTGGCTGCAGTAATCCCT
SELENBP1 (selenium-binding protein 1)	NM_009150.3	TTCCCTTGGAGATCCGCTTCCT ACTGACCATGTACCTCCCTCGT
SELENOF (selenoprotein F)	NM_203341.1	TGATCTTCTCGGACAGTTCAACCT CACGGACATACTTGGACTTGAGGG
SELENOH (C11orf31, chromosome 11 open reading frame 31; selenoprotein H)	NM_170746.2	GCTTCCAGTAAAGGTGAACCCGA TCAGGGAATTTGAGTTTGCGTGG
SELENOI (selenoprotein I)	NM_033505.4	ATGCCTCAGCACCAGGTCAC GTTCTGCGAGCTTGCTTTCCGT
SELENOK (selenoprotein K)	NM_021237.3	GATGATGGAAGAGGGCCACCAG CGCATGTCCGGTTGTCTGCT
SELENOM (selenoprotein M)	NM_080430.2	TGAAGGCTTTCGTCACGCAG AGTGGGATGCGCTCTAGTTCCT
SELENON (selenoprotein N)	NM_206926.2	TGTGATGTTCCGGATCCATGCC TGTGGTTGGGCACGAAGAGC
SELENOO (selenoprotein O)	NM_031454.1	TGACGCCGAGTTCCAAAGGCA TTGTGAAGTCGGCACCGGTCAG
SELENOP (selenoprotein P)	NM_005410	GAAACTCCATCGCCTCATTACCAT CTGCCTATGCTGACCCTTGTG
SELENOS (selenoprotein S, VIMP)	NM_203472.1	GCTGCATCCTTCTCTACGTGGTC CAACAACATCAGGTTCCACAGCA
SELENOT (selenoprotein T)	NM_016275.3	CGATCATAGCACCACCTATCAGCA GAGCCTGCCAAGAAAGCATCTG
SEPHS2 (selenophosphate synthetase 2)	NM_012248.4	GACGGTTTGGGCTTCTTCAAGG TCCACAATGCCAACGATCCAC
SEPSECS (sep (o-phosphoserine) tRNA:sec (selenocysteine) tRNA synthase)	NM_016955.3	CTAGTGCTCCCGCTTATTCGCC CTGGACACTTGCCCTTCTCCAG
SLC7A11 (solute carrier family 7 member 11)	NM_014331.3	TGCTCTTCTCTGGAGACCTCGAC ACAGTGGCACCTTGAAAGGACG
SOD1 (superoxide dismutase 1)	NM_000454.4	TACAGCAGGCTGTACCAGTGCA TCGGCCACACCATCTTTGTCAG
SRXN1 (sulfiredoxin 1)	NM_080725.1	CTCAGTGCTCGTTACTTCATGGTC GTTTGGCCCTTCCTCTTCCTCC
TNFA (tumor necrosis factor α)	NM_000594.2	AGCCCATGTTGTAGCAAACCCT GGAGTAGATGAGGTACAGGCCC

Table 1. *Cont.*

Gene	RefSeq-ID	Sequence (5′→3′)
TXAS1 (thromboxane A synthase 1)	NM_001061.4	CCTGAAAGGTTCACGGCTGAG CAACTTGACCTCAAGCAGCCCT
TXNRD1 (thioredoxin reductase 1)	NM_182742.1	GTGTTGTGGGCTTTCACGTACTG TGTTGTGAATACCTCTGCACAGAC
TXNRD2 (thioredoxin reductase 2)	NM_006440.5	GTTCCCACGACCGTCTTCAC GTGATAGACCTCAACATGCTCCTG

4.8. ELISA

Incubation medium from seeded cells was collected and the content of TNFα was determined according to the manufacturer's protocol using a microplate reader (Synergy H1, BioTek, Bad Friedrichshall, Germany). A Human TNFα Pre-Coated ELISA Kit was used. The incubation medium of THP-1 monocytes (DMSO control) was diluted 1:2 and for PMA-induced differentiated macrophages 1:50. A standard curve was used to generate a four-parameter logistic curve-fit and to calculate the content of TNFα which was normalized to the protein content of the cells.

4.9. Trace Element Analysis by Total Reflection X-ray Fluorescence (TXRF) Spectroscopy

TXRF was used to determine the intracellular levels of Se. Harvested THP-1 monocytes and macrophages were collected and lysed as described in Section 4.6. Cellular debris was removed by centrifugation for 10 min at 4 °C and 15,000× g. Trace elements analysis was performed using the supernatant (S2 Picofox™, Bruker Nano GmbH, Berlin, Germany), and 1 mg/L yttrium was used as internal standard. Ten microliters of each sample was placed on a siliconized quartz glass carrier and dried for 30 min at 40 °C. Samples were measured randomized in triplicate for 1000 s respectively. The Se content was normalized to the protein content of the samples.

4.10. GSH Assay

For the measurement of the total GSH content, cells were harvested, resuspended in 150 µL ice-cold 10 mM HCl, and lysed by ultrasonic treatment (80% amplitude, 0.5 s). Cellular debris was removed by centrifugation for 30 s at 8000× g at room temperature. Then, 30 µL (w/v) 5% SSA was added to the supernatant, incubated for 10 min, and centrifuged for 15 min at 4 °C and 8000× g at 4 °C, and supernatant was used for GSH measurement. After addition of 10 mM DTNB, the formed 2-nitro-5-thiobenzoic acid (TNB) was measured on a microplate reader (Synergy H1) at 412 nm for 5 min with path-length correction, as previously described [57]. BSO (0.25 mM) was used as control for GSH depletion and added 24 h before the cell harvest. A standard curve was used to calculate the total GSH content, which was normalized to the protein content of the samples.

4.11. Enzyme Activities

Cells were homogenized with a TissueLyser II (Qiagen, Hilden, Germany) in Tris buffer (100 mM Tris, 300 mM KCl, pH 7.6 with 0.1% (v/v) Triton X-100, and 0.1% (v/v) protease inhibitor) twice for 60 s at 30 Hz and centrifuged for 10 min at 4 °C and 14,000× g. The enzymatic activity was normalized to the protein content of the samples.

NQO1 activity was measured via menadione-dependent reduction of MTT as previously described [58]. Briefly, cell lysates were mixed with reaction buffer. Incubation mixtures contained 25 mM Tris, pH 7.4, 0.665 mg/l BSA, 0.01% Tween® 20, 5 µM FAD, 1 mM D-glucose-6-phosphate, 35 µM NADP, 0.72 mM MTT, 0.3 U/mL glucose-6-phosphate dehydrogenase, and 50 µM menadione. Absorbance was measured at 590 nm for 5.5 min using a microplate reader (Synergy H1) with path-length correction. Differences between the MTT reduction rates with and without dicumarol (final concentration 60 µM in phosphate buffer (1 mM KH_2PO_4, pH 7.4 and 0.1% DMSO) were used for calculation of NQO1 activity ($\varepsilon_{590\,nm}$ = 11.961 $mM^{-1} \times cm^{-1}$). Data are given as mUnit(mU)/mg protein.

GPX activity was measured by a GR-coupled test with NADPH consumption using H_2O_2 as substrate. Briefly, cell lysate was mixed with reaction buffer. Incubation mixtures contained Tris buffer (96 mM Tris, 4.8 mM EDTA, 960 µM NaN_3, pH 7.6) with 0.11% Triton X-100, 224 µM NADPH, 3.37 mM GSH, and 78 mU/mL GR and were incubated for 10 min at 37 °C. Ten microliters H_2O_2 (0.00375%) was added and the reaction was monitored for 2 min at 340 nm and 37 °C using a microplate reader (Synergy H1) with path-length correction. Data are given as mU/mg protein ($\varepsilon_{340\,nm}$ = 6.22 $mM^{-1} \times cm^{-1}$).

TXNRD activity was assessed using DTNB. Cell lysates were mixed with reaction buffer. Incubation mixtures contained 185 µL phosphate buffer (92.5 mM KH_2PO_4, 1.85 mM EDTA, pH 7.4) and 15 µL DTNB (3.75 mM in DMSO)). Twenty-five microliters NADPH (final concentration of 200 µM in phosphate buffer) was used to start the reaction, and TNB formation was monitored at 412 nm for 5 min using a microplate reader (Synergy H1) with path-length correction. TXNRD-independent formation of TNB was determined in the absence of NADPH and was subtracted. Data are given as mU/mg protein ($\varepsilon_{412\,nm}$ = 13.6 $mM^{-1} \times cm^{-1}$).

4.12. Solid Phase Extraction of LMs

LMs were extracted from cell culture medium as previously described [29]. Briefly, medium was mixed with a twofold amount of ice-cold methanol containing deuterium-labeled internal standards (200 nM d8-5S-HETE, d4-LTB4, d5-LXA4, d5-RvD2, d4-PGE2, and 10 µM d8-AA). Proteins were precipitated at −20 °C overnight and clear supernatant was collected after centrifugation (10 min, 1200× g, 4 °C). LMs were extracted from supernatant as previously described [59]. Briefly, the supernatant was acidified with water, pH 3.5, and LMs were purified by solid phase extraction (Sep-Pak® Vac 6 cc 500 mg/6 mL C18; Waters, Milford, CT, USA) using MilliQ water and n-hexane for washing steps. LMs were eluted with methyl formate, eluates were evaporated to dryness using nitrogen (TurboVap LV, Biotage, Uppsala, Sweden), and LMs were resolved in 100 µL methanol/water (50/50) for UPLC-MS/MS analysis [29].

4.13. LM Analysis by UPLC-MS/MS

Separation of LMs was performed at 50 °C on an Acquity UPLC® BEH C18 column (1.7 µm, 2.1 × 100 mm; Waters, Eschborn, Germany) with an Acquity™ UPLC system (Waters) as described previously [29]. In brief, methanol-water-acidic acid was used as mobile phase starting at a ratio of 42:58:0.01, which was gradually changed to 86:14:0.01 over 12.5 min and then to 98:2:0.01 over 3 min using a flow rate of 0.3 mL/min. LMs were analyzed on a QTRAP 5500 ESI tandem mass spectrometer (AB Sciex, Darmstadt, Germany) which was coupled to the LC system. After electrospray ionization operated in negative mode, multiple reaction monitoring was performed in the negative ionization mode with parameters adjusted as previously described [59]. External standards were used to confirm the retention time, and calibration curves for each analyte were prepared. For quantification of LMs, analyte levels were normalized by calculating the ratio of internal standard and directly measured deuterium-labeled standards to account for loss of analytes during purification and measurement, as reported previously [60], and they were normalized to the protein content of the cells.

4.14. Statistics

Data are given as means + SD of independent experiments. The statistical analysis was performed with GraphPad Prism 8.4.3 Software (San Diego, CA, USA) using one-way, two-way or three-way analysis of variance (ANOVA) with Bonferroni´s post-test. Outliers were determined by Grubbs' test, with significance level α = 0.05. p-values < 0.05 (*), (#), (§), (&), ($), (+); p < 0.01 (**), (##), (§§), (&&), ($$), and for p < 0.001 (***), (###), (§§§), (&&&),

($$$), (+++) were considered as statistically significant. Figures were created with GraphPad Prism 8.4.3 Software.

Supplementary Materials: The following are available online at https://www.mdpi.com/article/10.3390/ijms222011060/s1.

Author Contributions: T.W., data curation, original draft preparation, formal analysis, visualization; L.M.W., data curation, formal analysis, visualization; J.A., data curation, formal analysis, visualization; M.S., methodology, formal analysis; P.S., methodology, formal analysis, visualization; A.G., methodology, formal analysis; O.W., conceptualization, methodology, writing—review and editing; A.K., methodology, formal analysis, writing—review and editing; A.P.K., project administration, data curation, writing—review and editing, supervision, funding acquisition; S.C.K., conceptualization, data curation, investigation, project administration, writing—original draft preparation, writing—review and editing, supervision, funding acquisition. All authors have read and agreed to the published version of the manuscript.

Funding: This research was funded by the Austrian Science Fund (grant number I4968), the German Research Council (grant numbers KO4589/7-1, FOR 2558 (KI 1590/3-2)), SFB 1127 ChemBioSys 239748522), the Carl Zeiss Foundation (IMPULS), and the University of Jena (grant number AZ2.113-A1_2018-02).

Data Availability Statement: The data presented in this study are available in Supplementary Materials.

Acknowledgments: We would like to thank Thomas Schneider and Michael Glei for their support concerning FACS analysis and Kristina Lossow for her support regarding the ELISA measurement. The technical assistance of Doreen Ziegenhardt and Alrun Schumann is strongly acknowledged.

Conflicts of Interest: The authors declare no conflict of interest.

References

1. Labunskyy, V.; Hatfield, D.L.; Gladyshev, V.N. Selenoproteins: Molecular Pathways and Physiological Roles. *Physiol. Rev.* **2014**, *94*, 739–777. [CrossRef] [PubMed]
2. Huang, Z.; Rose, A.H.; Hoffmann, P.R. The Role of Selenium in Inflammation and Immunity: From Molecular Mechanisms to Therapeutic Opportunities. *Antioxid. Redox Signal.* **2012**, *16*, 705–743. [CrossRef] [PubMed]
3. Avery, J.C.; Hoffmann, P.R. Selenium, Selenoproteins, and Immunity. *Nutrients* **2018**, *10*, 1203. [CrossRef] [PubMed]
4. Steinbrenner, H.; Speckmann, B.; Klotz, L.-O. Selenoproteins: Antioxidant selenoenzymes and beyond. *Arch. Biochem. Biophys.* **2016**, *595*, 113–119. [CrossRef]
5. Bertz, M.; Kühn, K.; Koeberle, S.C.; Müller, M.F.; Hoelzer, D.; Thies, K.; Deubel, S.; Thierbach, R.; Kipp, A.P. Selenoprotein H controls cell cycle progression and proliferation of human colorectal cancer cells. *Free. Radic. Biol. Med.* **2018**, *127*, 98–107. [CrossRef]
6. Peng, X.; Giménez-Cassina, A.; Petrus, P.; Conrad, M.; Rydén, M.; Arnér, E.S.J. Thioredoxin reductase 1 suppresses adipocyte differentiation and insulin responsiveness. *Sci. Rep.* **2016**, *6*, 28080. [CrossRef]
7. Rajalin, A.-M.; Micoogullari, M.; Sies, H.; Steinbrenner, H. Upregulation of the thioredoxin-dependent redox system during differentiation of 3T3-L1 cells to adipocytes. *Biol. Chem.* **2014**, *395*, 667–677. [CrossRef]
8. Wang, C.; Li, R.; Huang, Y.; Wang, M.; Yang, F.; Huang, D.; Wu, C.; Li, Y.; Tang, Y.; Zhang, R.; et al. Selenoprotein K modulate intracellular free Ca^{2+} by regulating expression of calcium homoeostasis endoplasmic reticulum protein. *Biochem. Biophys. Res. Commun.* **2017**, *484*, 734–739. [CrossRef]
9. Yan, J.; Fei, Y.; Han, Y.; Lu, S. Selenoprotein O deficiencies suppress chondrogenic differentiation of ATDC5 cells. *Cell Biol. Int.* **2016**, *40*, 1033–1040. [CrossRef]
10. Geissmann, F.; Manz, M.G.; Jung, S.; Sieweke, M.H.; Merad, M.; Ley, K. Development of Monocytes, Macrophages, and Dendritic Cells. *Science* **2010**, *327*, 656–661. [CrossRef]
11. Dall'Asta, M.; Derlindati, E.; Ardigò, D.; Zavaroni, I.; Brighenti, F.; Del Rio, D. Macrophage polarization: The answer to the diet/inflammation conundrum? *Nutr. Metab. Cardiovasc. Dis.* **2012**, *22*, 387–392. [CrossRef]
12. Gordon, S.; Taylor, P. Monocyte and macrophage heterogeneity. *Nat. Rev. Immunol.* **2005**, *5*, 953–964. [CrossRef]
13. Dorrington, M.G.; Fraser, I.D.C.; Dorrington, M.G.; Fraser, I.D.C. NF-κB Signaling in Macrophages: Dynamics, Crosstalk, and Signal Integration. *Front. Immunol.* **2019**, *10*, 705. [CrossRef]
14. Saha, S.; Buttari, B.; Panieri, E.; Profumo, E.; Saso, L. An Overview of Nrf2 Signaling Pathway and Its Role in Inflammation. *Molecules* **2020**, *25*, 5474. [CrossRef]
15. He, F.; Ru, X.; Wen, T. NRF2, a Transcription Factor for Stress Response and Beyond. *Int. J. Mol. Sci.* **2020**, *21*, 4777. [CrossRef]
16. Michalke, B. *Selenium*; Springer: Berlin/Heidelberg, Germany, 2018. [CrossRef]

17. Schomburg, L.; Schweizer, U. Hierarchical regulation of selenoprotein expression and sex-specific effects of selenium. *Biochim. Biophys. Acta* **2009**, *1790*, 1453–1462. [CrossRef]
18. Papp, L.V.; Lu, J.; Holmgren, A.; Khanna, K.K. From Selenium to Selenoproteins: Synthesis, Identity, and Their Role in Human Health. *Antioxid. Redox Signal.* **2007**, *9*, 775–806. [CrossRef]
19. Reeves, M.A.; Hoffmann, P.R. The human selenoproteome: Recent insights into functions and regulation. *Cell. Mol. Life Sci.* **2009**, *66*, 2457–2478. [CrossRef]
20. Auwerx, J. The human leukemia cell line, THP-1: A multifacetted model for the study of monocyte-macrophage differentiation. *Cell. Mol. Life Sci.* **1991**, *47*, 22–31. [CrossRef]
21. Daigneault, M.; Preston, J.; Marriott, H.; Whyte, M.K.B.; Dockrell, D.H. The Identification of Markers of Macrophage Differentiation in PMA-Stimulated THP-1 Cells and Monocyte-Derived Macrophages. *PLoS ONE* **2010**, *5*, e8668. [CrossRef]
22. Lund, M.E.; To, J.; O'Brien, B.A.; Donnelly, S. The choice of phorbol 12-myristate 13-acetate differentiation protocol influences the response of THP-1 macrophages to a pro-inflammatory stimulus. *J. Immunol. Methods* **2016**, *430*, 64–70. [CrossRef]
23. Maeß, M.B.; Wittig, B.; Cignarella, A.; Lorkowski, S. Reduced PMA enhances the responsiveness of transfected THP-1 macrophages to polarizing stimuli. *J. Immunol. Methods* **2014**, *402*, 76–81. [CrossRef]
24. Li, T.; Yang, W.; Li, M.; Byun, D.-S.; Tong, C.; Nasser, S.; Zhuang, M.; Arango, D.; Mariadason, J.M.; Augenlicht, L.H. Expression of selenium-binding protein 1 characterizes intestinal cell maturation and predicts survival for patients with colorectal cancer. *Mol. Nutr. Food Res.* **2008**, *52*, 1289–1299. [CrossRef]
25. Steinbrenner, H.; Micoogullari, M.; Hoang, N.A.; Bergheim, I.; Klotz, L.-O.; Sies, H. Selenium-binding protein 1 (SELENBP1) is a marker of mature adipocytes. *Redox Biol.* **2018**, *20*, 489–495. [CrossRef]
26. Chen, G.; Wang, H.; Miller, C.T.; Thomas, D.G.; Gharib, T.G.; E Misek, D.; Giordano, T.J.; Orringer, M.B.; Hanash, S.M.; Beer, D.G. Reduced selenium-binding protein 1 expression is associated with poor outcome in lung adenocarcinomas. *J. Pathol.* **2004**, *202*, 321–329. [CrossRef]
27. Wang, N.; Chen, Y.; Yang, X.; Jiang, Y. Selenium-binding protein 1 is associated with the degree of colorectal cancer differentiation and is regulated by histone modification. *Oncol. Rep.* **2014**, *31*, 2506–2514. [CrossRef]
28. Sivandzade, F.; Prasad, S.; Bhalerao, A.; Cucullo, L. NRF2 and NF-κB interplay in cerebrovascular and neurodegenerative disorders: Molecular mechanisms and possible therapeutic approaches. *Redox Biol.* **2018**, *21*, 101059. [CrossRef]
29. Koeberle, S.C.; Gollowitzer, A.; Laoukili, J.; Kranenburg, O.; Werz, O.; Koeberle, A.; Kipp, A.P. Distinct and overlapping functions of glutathione peroxidases 1 and 2 in limiting NF-κB-driven inflammation through redox-active mechanisms. *Redox Biol.* **2019**, *28*, 101388. [CrossRef]
30. Vunta, H.; Davis, F.; Palempalli, U.D.; Bhat, D.; Arner, R.J.; Thompson, J.T.; Peterson, D.G.; Reddy, C.C.; Prabhu, K.S. The Anti-inflammatory Effects of Selenium Are Mediated through 15-Deoxy-Δ12,14-prostaglandin J2 in Macrophages. *J. Biol. Chem.* **2007**, *282*, 17964–17973. [CrossRef]
31. Wang, H.; Bi, C.; Wang, Y.; Sun, J.; Meng, X.; Li, J. Selenium ameliorates Staphylococcus aureus-induced inflammation in bovine mammary epithelial cells by inhibiting activation of TLR2, NF-κB and MAPK signaling pathways. *BMC Vet. Res.* **2018**, *14*, 197. [CrossRef]
32. Hiller, F.; Oldorff, L.; Besselt, K.; Kipp, A.P. Differential Acute Effects of Selenomethionine and Sodium Selenite on the Severity of Colitis. *Nutrients* **2015**, *7*, 2687–2706. [CrossRef] [PubMed]
33. Liu, K.; Ding, T.; Fang, L.; Cui, L.; Li, J.; Meng, X.; Zhu, G.; Qian, C.; Wang, H.; Li, J. Organic Selenium Ameliorates Staphylococcus aureus-Induced Mastitis in Rats by Inhibiting the Activation of NF-κB and MAPK Signaling Pathways. *Front. Vet. Sci.* **2020**, *7*, 443. [CrossRef] [PubMed]
34. Barrett, C.W.; Reddy, V.K.; Short, S.; Motley, A.K.; Lintel, M.K.; Bradley, A.M.; Freeman, T.; Vallance, J.; Ning, W.; Parang, B.; et al. Selenoprotein P influences colitis-induced tumorigenesis by mediating stemness and oxidative damage. *J. Clin. Investig.* **2015**, *125*, 2646–2660. [CrossRef] [PubMed]
35. A Carlson, B.; Yoo, M.-H.; Sano, Y.; Sengupta, A.; Kim, J.Y.; Irons, R.; Gladyshev, V.N.; Hatfield, D.L.; Park, J.M. Selenoproteins regulate macrophage invasiveness and extracellular matrix-related gene expression. *BMC Immunol.* **2009**, *10*, 57. [CrossRef]
36. Nelson, S.M.; Lei, X.; Prabhu, K.S. Selenium Levels Affect the IL-4–Induced Expression of Alternative Activation Markers in Murine Macrophages. *J. Nutr.* **2011**, *141*, 1754–1761. [CrossRef]
37. Norton, R.L.; Fredericks, G.J.; Huang, Z.; Fay, J.D.; Hoffmann, F.W.; Hoffmann, P.R. Selenoprotein K regulation of palmitoylation and calpain cleavage of ASAP2 is required for efficient FcγR-mediated phagocytosis. *J. Leukoc. Biol.* **2016**, *101*, 439–448. [CrossRef]
38. Verma, S.; Hoffmann, F.W.; Kumar, M.; Huang, Z.; Roe, K.; Nguyen-Wu, E.; Hashimoto, A.S.; Hoffmann, P.R. Selenoprotein K Knockout Mice Exhibit Deficient Calcium Flux in Immune Cells and Impaired Immune Responses. *J. Immunol.* **2011**, *186*, 2127–2137. [CrossRef]
39. Brigelius-Flohé, R.; Friedrichs, B.; Maurer, S.; Schultz, M.; Streicher, R. Interleukin-1-induced nuclear factor κB activation is inhibited by overexpression of phospholipid hydroperoxide glutathione peroxidase in a human endothelial cell line. *Biochem. J.* **1997**, *328*, 199–203. [CrossRef]
40. Bi, C.-L.; Wang, H.; Wang, Y.-J.; Sun, J.; Dong, J.-S.; Meng, X.; Li, J.-J. Selenium inhibits Staphylococcus aureus-induced inflammation by suppressing the activation of the NF-κB and MAPK signalling pathways in RAW264.7 macrophages. *Eur. J. Pharmacol.* **2016**, *780*, 159–165. [CrossRef]

41. Mao, H.; Zhao, Y.; Li, H.; Lei, L. Ferroptosis as an emerging target in inflammatory diseases. *Prog. Biophys. Mol. Biol.* **2020**, *155*, 20–28. [CrossRef]
42. Mattmiller, S.A.; Carlson, B.A.; Gandy, J.C.; Sordillo, L.M. Reduced macrophage selenoprotein expression alters oxidized lipid metabolite biosynthesis from arachidonic and linoleic acid. *J. Nutr. Biochem.* **2014**, *25*, 647–654. [CrossRef]
43. Zamamiri-Davis, F.; Lu, Y.; Thompson, J.T.; Prabhu, K.; Reddy, P.V.; Sordillo, L.M.; Reddy, C. Nuclear factor-κB mediates over-expression of cyclooxygenase-2 during activation of RAW 264.7 macrophages in selenium deficiency. *Free. Radic. Biol. Med.* **2002**, *32*, 890–897. [CrossRef]
44. Wan, J.M.-F.; Lee, C.-Y.J. Immunoregulatory and Antioxidant Performance of α-Tocopherol and Selenium on Human Lymphocytes. *Biol. Trace Elem. Res.* **2002**, *86*, 123–136. [CrossRef]
45. Spallholz, J.E. Free radical generation by selenium compounds and their prooxidant toxicity. *Biomed. Environ. Sci.* **1997**, *10*, 260–270.
46. Stewart, M.S.; E Spallholz, J.; Neldner, K.H.; Pence, B.C. Selenium compounds have disparate abilities to impose oxidative stress and induce apoptosis. *Free. Radic. Biol. Med.* **1998**, *26*, 42–48. [CrossRef]
47. Karwaciak, I.; Gorzkiewicz, M.; Bartosz, G.; Pulaski, L. TLR2 activation induces antioxidant defence in human monocyte-macrophage cell line models. *Oncotarget* **2017**, *8*, 54243–54264. [CrossRef]
48. Müller, M.; Banning, A.; Brigelius-Flohé, R.; Kipp, A. Nrf2 target genes are induced under marginal selenium-deficiency. *Genes Nutr.* **2010**, *5*, 297–307. [CrossRef]
49. Chang, L.-C.; Chiang, S.-K.; Chen, S.-E.; Yu, Y.-L.; Chou, R.-H.; Chang, W.-C. Heme oxygenase-1 mediates BAY 11–7085 induced ferroptosis. *Cancer Lett.* **2018**, *416*, 124–137. [CrossRef]
50. Iida, Y.; Okamoto-Katsuyama, M.; Maruoka, S.; Mizumura, K.; Shimizu, T.; Shikano, S.; Hikichi, M.; Takahashi, M.; Tsuya, K.; Okamoto, S.; et al. Effective ferroptotic small-cell lung cancer cell death from SLC7A11 inhibition by sulforaphane. *Oncol. Lett.* **2020**, *21*, 71. [CrossRef]
51. Steinbrenner, H.; Al-Quraishy, S.; Dkhil, M.; Wunderlich, F.; Sies, H. Dietary Selenium in Adjuvant Therapy of Viral and Bacterial Infections. *Adv. Nutr.* **2015**, *6*, 73–82. [CrossRef]
52. Zhang, J.; Saad, R.; Taylor, E.W.; Rayman, M.P. Selenium and selenoproteins in viral infection with potential relevance to COVID-19. *Redox Biol.* **2020**, *37*, 101715. [CrossRef]
53. Diwakar, B.T.; Korwar, A.M.; Paulson, R.F.; Prabhu, K.S. The Regulation of Pathways of Inflammation and Resolution in Immune Cells and Cancer Stem Cells by Selenium. *Adv. Cancer Res.* **2017**, *136*, 153–172. [CrossRef]
54. Kaur, S.; Harjai, K.; Chhibber, S. In Vivo Assessment of Phage and Linezolid Based Implant Coatings for Treatment of Methicillin Resistant S. aureus (MRSA) Mediated Orthopaedic Device Related Infections. *PLoS ONE* **2016**, *11*, e0157626. [CrossRef]
55. Nettleford, S.K.; Zhao, L.; Qian, F.; Herold, M.; Arner, B.; Desai, D.; Amin, S.; Xiong, N.; Singh, V.; Carlson, B.A.; et al. The Essential Role of Selenoproteins in the Resolution of Citrobacter rodentium-Induced Intestinal Inflammation. *Front. Nutr.* **2020**, *7*, 96. [CrossRef]
56. Schwarz, M.; Lossow, K.; Schirl, K.; Hackler, J.; Renko, K.; Kopp, J.F.; Schwerdtle, T.; Schomburg, L.; Kipp, A.P. Copper interferes with selenoprotein synthesis and activity. *Redox Biol.* **2020**, *37*, 101746. [CrossRef]
57. Winther, J.R.; Thorpe, C. Quantification of thiols and disulfides. *Biochim. Biophys. Acta (BBA)—Gen. Subj.* **2013**, *1840*, 838–846. [CrossRef]
58. Prochaska, H.J.; Santamaria, A.B. Direct measurement of NAD(P)H:quinone reductase from cells cultured in microtiter wells: A screening assay for anticarcinogenic enzyme inducers. *Anal. Biochem.* **1988**, *169*, 328–336. [CrossRef]
59. Werz, O.; Gerstmeier, J.; Libreros, S.; De La Rosa, X.; Werner, M.; Norris, P.; Chiang, N.; Serhan, C.N. Human macrophages differentially produce specific resolvin or leukotriene signals that depend on bacterial pathogenicity. *Nat. Commun.* **2018**, *9*, 59. [CrossRef]
60. Schädel, P.; Troisi, F.; Czapka, A.; Gebert, N.; Pace, S.; Ori, A.; Werz, O. Aging drives organ-specific alterations of the inflammatory microenvironment guided by immunomodulatory mediators in mice. *FASEB J.* **2021**, *35*, e21558. [CrossRef]

Article

Female Mice with Selenocysteine tRNA Deletion in Agrp Neurons Maintain Leptin Sensitivity and Resist Weight Gain While on a High-Fat Diet

Daniel J. Torres [1,2,*], Matthew W. Pitts [2], Lucia A. Seale [1], Ann C. Hashimoto [2], Katlyn J. An [2], Ashley N. Hanato [2], Katherine W. Hui [2], Stella Maris A. Remigio [2], Bradley A. Carlson [3], Dolph L. Hatfield [3] and Marla J. Berry [1]

[1] Pacific Biosciences Research Center, School of Ocean and Earth Science and Technology, University of Hawaii at Manoa, Honolulu, HI 96822, USA; lseale@hawaii.edu (L.A.S.); mberry@hawaii.edu (M.J.B.)
[2] Department of Cell and Molecular Biology, John A. Burns School of Medicine, University of Hawaii at Manoa, Honolulu, HI 96813, USA; mwpitts@hawaii.edu (M.W.P.); ahashimo@hawaii.edu (A.C.H.); ankatlyn@hawaii.edu (K.J.A.); ahanato@hawaii.edu (A.N.H.); katherinewkhui@gmail.com (K.W.H.); stellamaris.remigio@gmail.com (S.M.A.R.)
[3] Molecular Biology of Selenium Section, Mouse Cancer Genetics Program, National Cancer Institute, National Institutes of Health, Bethesda, MD 20892, USA; carlsonb@dc37a.nci.nih.gov (B.A.C.); hatfielddolph@gmail.com (D.L.H.)
* Correspondence: djtorr@hawaii.edu

Abstract: The role of the essential trace element selenium in hypothalamic physiology has begun to come to light over recent years. Selenium is used to synthesize a family of proteins participating in redox reactions called selenoproteins, which contain a selenocysteine residue in place of a cysteine. Past studies have shown that disrupted selenoprotein expression in the hypothalamus can adversely impact energy homeostasis. There is also evidence that selenium supports leptin signaling in the hypothalamus by maintaining proper redox balance. In this study, we generated mice with conditional knockout of the selenocysteine tRNA$^{[Ser]Sec}$ gene (*Trsp*) in an orexigenic cell population called agouti-related peptide (Agrp)-positive neurons. We found that female *Trsp*^{Agrp} KO mice gain less weight while on a high-fat diet, which occurs due to changes in adipose tissue activity. Female *Trsp*^{Agrp} KO mice also retained hypothalamic sensitivity to leptin administration. Male mice were unaffected, however, highlighting the sexually dimorphic influence of selenium on neurobiology and energy homeostasis. These findings provide novel insight into the role of selenoproteins within a small yet heavily influential population of hypothalamic neurons.

Keywords: selenium; selenoprotein; *Trsp*; hypothalamus; Agrp neuron; sex differences; diet-induced obesity; leptin resistance

1. Introduction

As the global obesity pandemic worsens [1], gaining a better understanding of the molecular mechanisms involved is necessary for developing effective treatments. The hypothalamus of the brain, which regulates energy homeostasis throughout the body, becomes impaired in the obese state and is a potential therapeutic target [2]. In recent years, the role of the essential trace element selenium in supporting hypothalamic function, particularly in response to a high-fat diet (HFD), has begun to come to light [3]. Selenium is used to synthesize selenoproteins, a family of proteins that promote redox balance and support physiological processes such as thyroid hormone metabolism [4] and the inflammatory response [5]. Within the hypothalamus, multiple selenoproteins exhibit high levels of expression that appear to be influenced by changes in nutritional status [6,7]. Mouse models with genetic manipulation of selenoprotein expression targeting the hypothalamus have exhibited significant metabolic repercussions [8–10].

Previously, Yagishita et al. [8] demonstrated that broad hypothalamic deletion of the selenocysteine tRNA[Ser]Sec gene (*Trsp*), which is required for selenoprotein biosynthesis, increases the susceptibility to diet-induced obesity (DIO) in mice. Knockout of *Trsp* in rat-insulin-promoter (RIP)-positive cells induced oxidative stress and resistance to insulin and leptin, highlighting the critical nature of hypothalamus-resident selenoproteins in energy homeostasis. Conditional *Trsp* knockout was noted to affect a wide range of neuronal cell types, including anorexigenic pro-opiomelanocortin (Pomc) neurons and astrocytes [8]. Among the cell types not affected were the appetite-stimulating agouti-related peptide (Agrp)-positive neurons, a group of 'first order' neurons contained within the arcuate nucleus (Arc) of the hypothalamus that is capable of detecting circulating nutrients and hormones such as leptin. This small neuronal subpopulation releases the inhibitory neurotransmitter γ-Aminobutyric acid (GABA) as well as the neuropeptides Agrp and Neuropeptide y (Npy) to suppress the release of thyrotropin-releasing hormone (TRH) and corticotropin-releasing hormone (CRH), which signal to the pituitary gland to promote energy expenditure from the paraventricular nucleus of the hypothalamus. Due to their unique location at the interface between the brain and the bloodstream and their pronounced influence on energy homeostasis, Agrp neurons have increasingly garnered attention as a potential therapeutic target in treating metabolic disease. To assess the role of selenoproteins in Agrp neurons, we created mice with *Agrp-Cre*-driven knockout of *Trsp*. We report that the loss of *Trsp* from Agrp neurons conferred protection from DIO in a sex-dependent manner.

2. Results

2.1. In Vivo Metabolic Assessment of Mice with Agrp Neuron-specific Ablation of Trsp

To generate mice with Agrp neurons lacking selenoprotein biosynthesis, we crossed *Trsp*-floxed mice with *Agrp-Cre* mice, resulting in mice with a deletion of the *Trsp* gene in Agrp neurons, referred to as $Trsp^{Agrp}KO$ mice. Control mice consisted of age-matched *Agrp-Cre* mice. Both groups were fed an HFD beginning at 4 weeks of age, after which we monitored body weight and performed metabolic phenotyping. Female $Trsp^{Agrp}KO$ mice displayed resistance to DIO, gaining roughly 20% less weight than controls by 14 weeks of age (Figure 1a). Weight gained by male $Trsp^{Agrp}KO$ mice, on the other hand, did not differ from controls (Figure 1b). Female $Trsp^{Agrp}KO$ mice also displayed reduced adiposity (Figure 1c) and improved glucose tolerance (Figure 1d,e). Consistent with overall body weight trends, male $Trsp^{Agrp}KO$ mice exhibited similar levels of adiposity (Figure 1c) and glucose sensitivity (Figure 1f) in comparison to control mice.

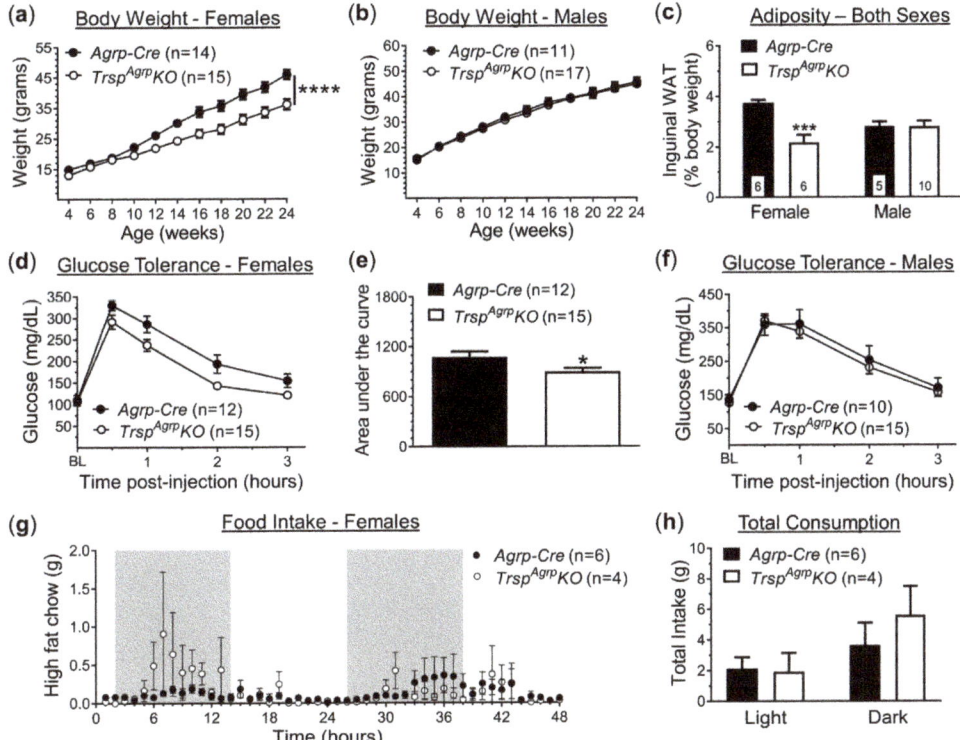

Figure 1. Metabolic profile of *TrspAgrpKO* mice of both sexes raised on a high-fat diet. Female *TrspAgrpKO* mice (**a**) gained less weight while on a high-fat diet compared to *Agrp-Cre* controls; however, there was no effect of *TrspAgrpKO* in the body weight of male mice (**b**) (Sidak's multiple comparisons test following repeated measures: **** $p < 0.0001$). Inguinal white adipose tissue (WAT) deposits weighed less in female *TrspAgrpKO* mice (**c**) and were unchanged in male *TrspAgrpKO* mice (Tukey's multiple comparisons test following two-way analysis of variance: *** $p = 0.0008$). Glucose tolerance was elevated in female *TrspAgrpKO* mice (**d,e**) compared to controls (unpaired *t*-test comparing area under the curve: * $p = 0.03$) and remained unaffected in males (**f**). Feeding behavior (**g,h**) was not significantly changed in *TrspAgrpKO* mice of either sex during 48 h recordings. Samples sizes are displayed in the graphs. All values shown are mean ± standard error of the mean.

The primary mechanism through which Agrp neurons control energy homeostasis is the promotion of feeding behavior [11]. Thus, we used specialized metabolic chambers (OxyletPro Physiocage System, Harvard Apparatus) that allow for constant monitoring of single-housed subjects to measure food intake, physical activity, and respiratory metabolism. Although female $Trsp^{Agrp}KO$ mice were underweight, no significant difference in total food intake was detected compared to controls in either the light or dark phase (Figure 1g,h). Meal analysis revealed a significantly larger inter-meal interval in female $Trsp^{Agrp}KO$ mice during the light phase (Supplementary Figure S1a), but no changes were observed in any of the other metrics analyzed, which included meal size, meal duration, meal count, and eating rate (data not shown). There were also no detectable changes in physical activity in female $Trsp^{Agrp}KO$ mice in terms of either total locomotion (Supplementary Figure S1b,c) or rearing events (Supplementary Figure S1d).

Since neither feeding behavior nor physical activity seemed to be able to account for resistance to DIO observed in female $Trsp^{Agrp}KO$ mice, we also probed for changes in respiratory metabolism using the metabolic chamber system. Female $Trsp^{Agrp}KO$ mice displayed increased rates of oxygen consumption (Figure 2a,b) and carbon dioxide production (Figure 2c,d) compared to control *Agrp-Cre* mice. These differences persisted throughout the duration of both the light and dark cycles. Energy expenditure was estimated using indirect calorimetry and was found to be significantly elevated in female $Trsp^{Agrp}KO$ mice (Figure 2e,f). Male $Trsp^{Agrp}KO$ mice did not display any significant changes in respiratory metabolism (Supplementary Figure S1e,f).

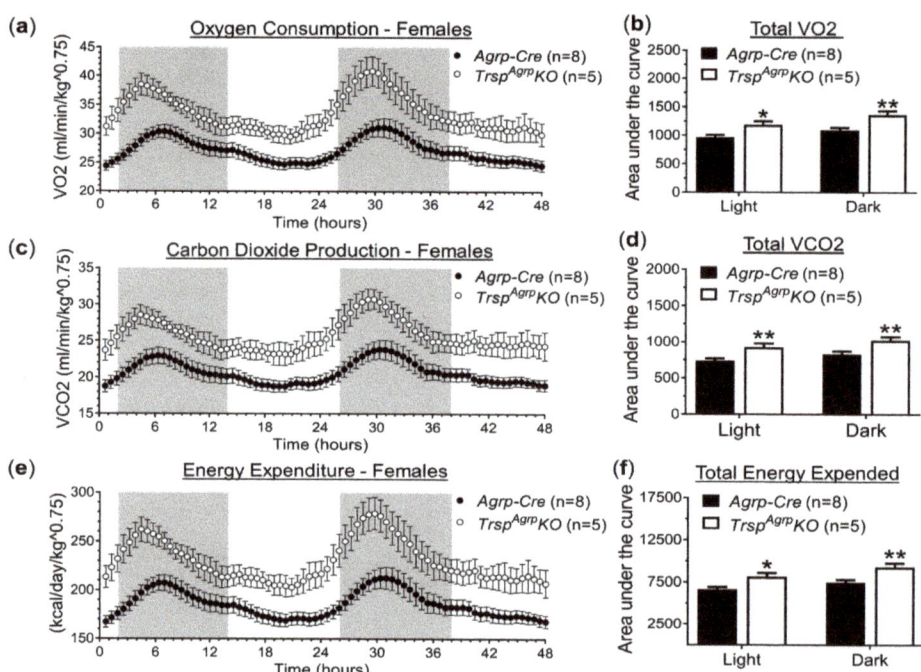

Figure 2. Respiratory metabolism of female $Trsp^{Agrp}KO$ mice raised on a high-fat diet. Female $Trsp^{Agrp}KO$ mice displayed elevated oxygen consumption (VO2, volume of oxygen) (**a,b**), carbon dioxide production (VCO2, volume of carbon dioxide) (**c,d**), and energy expenditure (**e,f**) compared to *Agrp-Cre* control mice. Two-way repeated-measures analysis of variance was used to compare the area under the curve during both light and dark cycles. Sidak's multiple comparisons test: * $p < 0.05$, ** $p < 0.01$. Samples sizes are displayed in the graphs. All values shown are mean ± standard error of the mean.

2.2. Measurement of Hormones in Serum from $Trsp^{Agrp}KO$ Mice

Despite having reduced adiposity, female $Trsp^{Agrp}KO$ mice did not present altered levels of circulating leptin, while insulin levels trended downward (Table 1). Agrp neurons affect the peripheral energy metabolism processes by indirectly suppressing the activity of the pituitary gland. There were no significant differences in serum levels of pituitary hormones between $Trsp^{Agrp}KO$ and control females. There was a downward trend in follicle-stimulating hormone (FSH) that nearly reached statistical significance, however. Male $Trsp^{Agrp}KO$ mice had reduced circulating levels of growth hormone (GH), while all other measured hormones were not affected (Table 1).

Table 1. Serum expression levels of hormones $Trsp^{Agrp}KO$ and *Agrp-Cre* mice.

Hormone Type and Name	*Agrp-Cre*	$Trsp^{Agrp}KO$	*p* Value
Females:			
Metabolic hormones			
Leptin (ng/mL)	32.78 ± 1.58 ($n = 7$)	31.54 ± 2.18 ($n = 9$)	0.67
Insulin (ng/mL)	0.54 ± 0.11 ($n = 7$)	0.32 ± 0.04 ($n = 8$)	0.06

Table 1. Cont.

Hormone Type and Name	Agrp-Cre	Trsp^{Agrp}KO	p Value
Pituitary hormones			
ACTH (pg/mL)	20.06 ± 4.69 (n = 7)	21.75 ± 4.11 (n = 8)	0.79
FSH (ng/mL)	1.41 ± 0.54 (n = 6)	0.39 ± 0.11 (n = 8)	*0.05*
GH (ng/mL)	1.09 ± 0.43 (n = 6)	1.97 ± 0.74 (n = 9)	0.39
LH (pg/mL)	125.65 ± 43.95 (n = 7)	248.90 ± 76.29 (n = 8)	0.20
Prolactin (ng/mL)	9.72 ± 2.10 (n = 7)	8.36 ± 1.49 (n = 8)	0.60
TSH (pg/mL)	49.40 ± 5.27 (n = 6)	61.57 ± 25.53 (n = 8)	0.67
Males:			
Metabolic hormones			
Leptin (ng/mL)	32.58 ± 2.26 (n = 8)	28.38 ± 2.15 (n = 9)	0.20
Insulin (ng/mL)	0.93 ± 0.26 (n = 8)	1.30 ± 0.37 (n = 9)	0.44
Pituitary hormones			
ACTH (pg/mL)	17.94 ± 3.46 (n = 8)	22.05 ± 5.53 (n = 9)	0.55
FSH (ng/mL)	3.19 ± 0.67 (n = 8)	2.77 ± 0.41 (n = 9)	0.59
GH (ng/mL)	0.54 ± 0.09 (n = 6)	2.14 ± 0.54 (n = 9)	**0.03**
LH (pg/mL)	392.17 ± 166.81 (n = 8)	653.28 ± 200.41 (n = 9)	0.34
Prolactin (ng/mL)	1.15 ± 0.33 (n = 8)	2.07 ± 0.85 (n = 9)	0.35
TSH (pg/mL)	173.10 ± 43.22 (n = 8)	258.30 ± 53.23 (n = 9)	0.24

Boldface indicates significance and italicized text indicates a statistically non-significant trend. Comparisons were made between genotypes of the same sex using unpaired *t*-tests; n = 6–9 per group. All values shown are mean ± S.E.M. Abbreviations: ACTH, adrenocorticotropic hormone; FH, follicle-stimulating hormone; GH, growth hormone; LH, luteinizing hormone; TSH, thyroid-stimulating hormone.

2.3. Histological Analysis of Trsp^{Agrp}KO Mouse Brown Adipose Tissue

In addition to regulating appetite, Agrp neurons are able to influence other metabolic processes, including the thermogenic activity of interscapular brown adipose tissue (BAT) [11]. To evaluate the potential influence of Agrp neuron-specific *Trsp* ablation on BAT thermogenesis, we performed histological analysis of BAT morphology. Lipid droplet size was significantly decreased in BAT sections from female Trsp^{Agrp}KO mouse BAT, while male Trsp^{Agrp}KO mice were unaffected (Figure 3a–c). Lipid deposits also occupied a smaller fraction of BAT section surface area on average in female Trsp^{Agrp}KO mice (Figure 3d). Expression of uncoupling protein-1 (UCP1), a marker of thermogenesis, was not significantly changed in Trsp^{Agrp}KO mouse BAT sections (Supplementary Figure S2a,b).

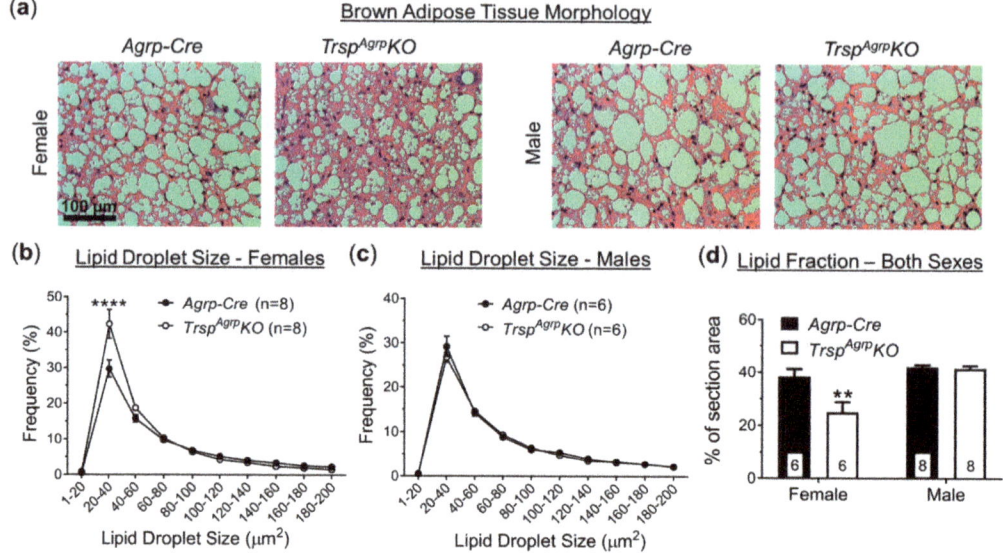

Figure 3. Brown adipose tissue morphology of $Trsp^{Agrp}KO$ mice of both sexes raised on a high-fat diet. (**a**) Sample images of hematoxylin and eosin-stained brown adipose tissue sections from female and male $Trsp^{Agrp}KO$ and *Agrp-Cre* control mice. (**b**) The size of lipid droplets measured sections from female $Trsp^{Agrp}KO$ mice was more frequently present in smaller sizes compared to female *Agrp-Cre* control mice (two-way repeated-measures analysis of variance followed by Sidak's multiple comparisons: **** $p < 0.0001$). No changes were seen in male mice. (**c**) Average lipid fraction, which is the percentage of the section occupied by lipid droplet, was significantly lower in female $Trsp^{Agrp}KO$ mice compared to *Agrp-Cre* control mice (**d**) (two-way analysis of variance, followed by Tukey's multiple comparisons test: ** $p = 0.007$). Samples sizes are displayed in the graphs. All values shown are mean ± standard error of the mean.

2.4. Assessment of Leptin Sensitivity in the Hypothalamus of $Trsp^{Agrp}KO$ Mice

To investigate the neural mechanism underlying the DIO resistance phenotype of female $Trsp^{Agrp}KO$ mice, we performed post-mortem tissue analysis. In response to HFD, the rodent hypothalamus typically develops leptin resistance due to inflammation and oxidative stress [12]. Accumulating evidence suggests that selenoproteins play a major role in regulating hypothalamic leptin signaling [3,13,14]. Therefore, we challenged the mice with intraperitoneal leptin injection prior to sacrifice. Western blot analysis of the whole hypothalamus revealed that female $Trsp^{Agrp}KO$ mice maintained leptin sensitivity, demonstrated by a significant increase in expression of the leptin signaling protein signal transducer and activator of transcription 3 (STAT3) in response to leptin administration (Figure 4a,b), while control mice developed leptin resistance. As expected, male control and $Trsp^{Agrp}KO$ mice developed leptin resistance in response to HFD (Figure 4c,d).

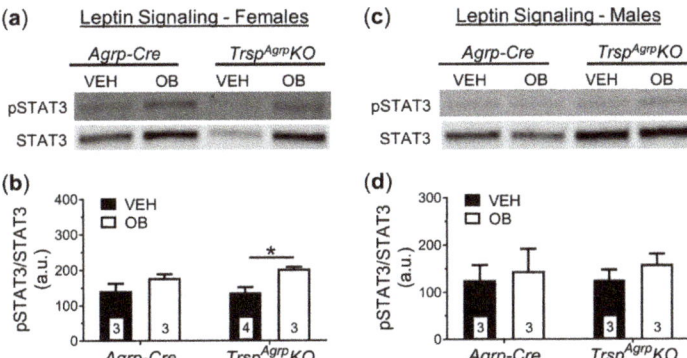

Figure 4. Western blot analysis of broad hypothalamic leptin signaling in $Trsp^{Agrp}KO$ mice of both sexes raised on a high-fat diet. Leptin (OB) injection did not elicit a significant increase in phosphorylated STAT3 (pSTAT3) protein levels in female *Agrp-Cre* control mice compared to vehicle (VEH)-injected *Agrp-Cre* mice (**a,b**). Female $Trsp^{Agrp}KO$ mice, however, displayed an increase in pSTAT3 levels in response to leptin injection (two-way analysis of variance, followed by Tukey's multiple comparisons test: * $p = 0.04$). Leptin injection failed to significantly elevate pSTAT3 protein levels in the hypothalamus of either *Agrp-Cre* or $Trsp^{Agrp}KO$ male mice (**c,d**). Samples sizes are displayed in the graphs. All values shown are mean ± standard error of the mean.

Immunohistochemical measurement of STAT3 expression in hypothalamic sections confirmed that the Arc of the hypothalamus, where Agrp neurons reside, maintains leptin sensitivity in female $Trsp^{Agrp}KO$ mice (Figure 5a,b). The number of STAT3-positive cells did not increase in response to leptin, however, suggesting a more robust response within a comparable population of leptin-receptor-expressing neurons within the Arc (Figure 5c). Hypothalamic sections from male $Trsp^{Agrp}KO$ mice showed no significant response to leptin (Figure 5d,e). Importantly, the number of STAT3-positive cells in $Trsp^{Agrp}KO$ males was comparable to the number present in control mouse sections (Figure 5f).

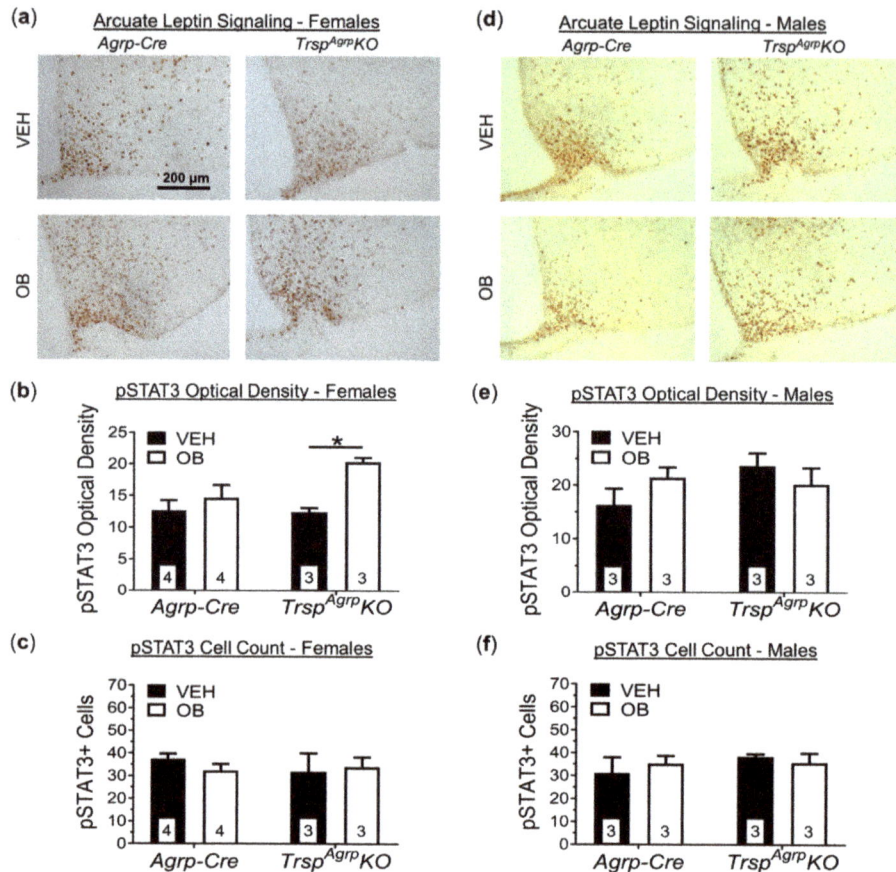

Figure 5. Leptin signaling in brain sections of $Trsp^{Agrp}KO$ mice of both sexes raised on a high-fat diet. Sample images of hypothalamic sections stained for phosphorylated STAT3 (pSTAT3) following in vivo challenge with either vehicle (VEH, saline) or leptin (OB) (**a**). Images are of the medio-basal hypothalamic area containing the arcuate nucleus and were captured at 20× magnification. Optical density of pSTAT3 immunoreactivity, visualized via 3,3′-diaminobenzidine staining, in the arcuate nucleus of female $Trsp^{Agrp}KO$ mice was significantly increased by leptin challenge, but not in *Agrp-Cre* control mice (**b**) (two-way analysis of variance, followed by Tukey's multiple comparisons test: * $p = 0.03$). No change in the number of pSTAT3-positive cells in female $Trsp^{Agrp}KO$ mice was detected (**c**), however. (**d**) Neither genotype nor leptin administration affected pSTAT3 optical density (**e**) or cell count (**f**) in male mice. Samples sizes are displayed in the graphs. All values shown are mean ± standard error of the mean.

2.5. Analysis of Neuropeptide Expression in the $Trsp^{Agrp}KO$ Mouse Hypothalamus

No significant changes in the protein expression of Agrp were observed as a result of *Trsp* deletion in Agrp neurons of female mice (Figure 6a–c). Interestingly, Pomc was upregulated in the hypothalamus of female $Trsp^{Agrp}KO$ mice in response to leptin (Figure 6c). Male $Trsp^{Agrp}KO$ mice did not display any significant changes in the levels of hypothalamic neuropeptides (Figure 6d–f).

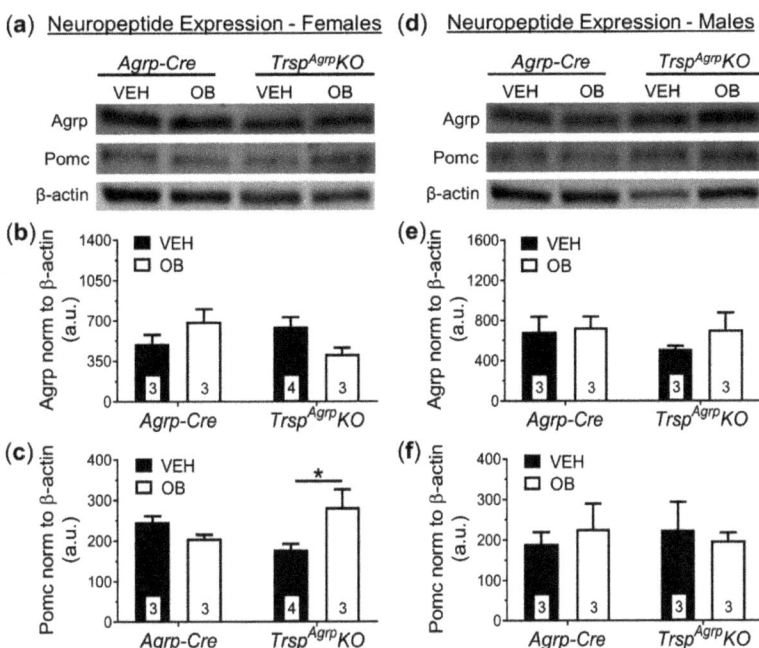

Figure 6. Western blot analysis of broad hypothalamic neuropeptide expression in $Trsp^{Agrp}KO$ mice of both sexes raised on a high-fat diet. Female $Trsp^{Agrp}KO$ mice displayed no significant changes in expression of agouti-related peptide (Agrp), but pro-opiomelanocortin (Pomc) was significantly elevated in female $Trsp^{Agrp}KO$ mice in response to leptin (OB) compared to vehicle (VEH)-injected $Trsp^{Agrp}KO$ mice (**a–c**) (Two-way analysis of variance, followed by Tukey's multiple comparisons test: * $p = 0.04$). Male $Trsp^{Agrp}KO$ mice showed no significant changes in Agrp or Pomc protein levels as a result of either genotype or leptin injection (**d–f**). Samples sizes are displayed in the graphs. All values shown are mean ± standard error of the mean.

3. Discussion

We report here that the ablation of selenoprotein synthesis via *Trsp* deletion in the Agrp neurons of mice results in a phenotype that is protected against DIO and leptin resistance. This protection was only observed in female mice, however, as male $Trsp^{Agrp}KO$ mice gained as much weight and adiposity as controls while on an HFD and developed leptin resistance. Female $Trsp^{Agrp}KO$ mice gained approximately 20% less weight on average than control *Agrp-Cre* mice while on an HFD. Inguinal white adipose tissue (WAT) deposits were about 40% smaller than that found in controls, indicating a leaner body composition, which was accompanied by slightly better glucose sensitivity. Despite the fact that Agrp neurons heavily influence food-seeking behavior, we found that female $Trsp^{Agrp}KO$ mice consumed similar amounts of food as their control counterparts. The amount of time that elapsed between meals was significantly larger in female $Trsp^{Agrp}KO$ mice during the light phase, but no other metrics of meal consumption, such as meal size or meal count, were changed. Thus, although there appears to be some modification of feeding behavior, it is unclear whether an increased inter-meal interval in the light phase could account for the 20% decrease in weight gain observed in female $Trsp^{Agrp}KO$ mice.

It is worth noting that female $Trsp^{Agrp}KO$ mice were undersized compared to their *Agrp-Cre* counterparts at as early as 4 weeks of age, just prior to HFD exposure (Supplementary Figure S2c). Although these data suggest a developmental growth deficit in female mice due to *Trsp* deletion in Agrp neurons, it is unlikely that it can explain the difference in body weight in adulthood following HFD administration, considering the striking differences in adiposity. Moreover, body lengths measured at 24 weeks of age were not significantly different between groups (Supplementary Figure S2d). A major limitation of the current study is that it did not include an experiment with mice on a control diet. Further investigation that includes a control diet will provide additional insight on this and other aspects of the lean phenotype of female $Trsp^{Agrp}$ KO mice.

Energy expenditure was found to be significantly elevated in female $Trsp^{Agrp}$ KO mice. This finding could not be accounted for by a change in physical activity, which remained unchanged, however, suggesting the upregulation of an energy-consuming process within female $Trsp^{Agrp}$ KO mice. Increased BAT thermogenesis is a potential candidate as lipid depositions were reduced in female $Trsp^{Agrp}$ KO mice BAT and the influence of Agrp neural activity on thermogenesis is well established [11]. This is consistent with our previous findings in mice with Agrp neurons lacking the selenium recycling enzyme selenocysteine lyase (Scly), $Scly^{Agrp}$ KO mice, which displayed smaller BAT lipid droplets and increased UCP1 expression in the BAT. We did not observe a change in UCP1 expression in the BAT of $Trsp^{Agrp}$ KO mice, however, which may suggest a UCP1-independent thermogenic process such as BAT Ca^{2+} cycling thermogenesis or WAT lipolysis [15]. Interestingly, recent studies have depicted the ability of the hypothalamus to increase energy expenditure via sympathetic tone to WAT without affecting BAT UCP1 expression [16].

We did not detect any significant changes in pituitary gland hormones in female $Trsp^{Agrp}$ KO mice that can explain their resistance to DIO. Circulating levels of FSH, which plays a role in limiting thermogenesis and WAT browning in mice [17], trended downwards in female $Trsp^{Agrp}$ KO mice, although the change was not quite statistically significant (p = 0.05). Interestingly, blocking FSH activates BAT and reduces adiposity in female mice [17]. Since the thyroid-stimulating hormone (TSH) was unchanged, Agrp neurons lacking $Trsp$ may be promoting a lean phenotype through a pathway that involves sympathetic innervation rather than endocrine means. In addition to synapsing on the neurosecretory neurons of the paraventricular nucleus, Agrp neurons project to multiple other regions within the hypothalamus. One of these regions, the dorsomedial hypothalamus (DMH), is known to regulate thermogenesis and lipolysis directly via the brain stem [18–20]. This DMH-mediated sympathetic pathway is regulated by Npy release from Agrp neurons into the DMH and is thus potentially affected by the genetic deletion of $Trsp$. Therefore, the sympathetic pathway may be involved in the process through which Agrp neurons lacking $Trsp$ promote energy expenditure.

Past evidence suggests that selenoproteins play an important role in supporting leptin signaling in the hypothalamus [3]. For example, the endoplasmic reticulum (ER)-resident selenoprotein M (SELENOM) was found to promote intracellular leptin receptor signaling by upregulating the thioredoxin system and protecting against ER stress [13], a known causative factor in diet-induced leptin resistance [21]. Leptin resistance primarily affects the Arc. While the pathology of leptin resistance is not completely known, there is evidence that Agrp neurons develop leptin resistance prior to other cell types, which may serve as an 'initiating' event [22]. We therefore expected that $Trsp$ ablation would make Agrp neurons more vulnerable to developing leptin resistance. We observed the opposite in female $Trsp^{Agrp}$ KO mice fed an HFD, however, as the hypothalamus maintained leptin responsivity as a whole and the leptin response in the Arc remained robust. At the surface, it would appear that the loss of the $Trsp$ gene prevents Agrp neurons from developing HFD-induced leptin resistance.

The main driver of diet-induced leptin resistance is thought to be hyperleptinemia secondary to excess weight gain [23]. Therefore, since female $Trsp^{Agrp}$ KO mice remained underweight compared to controls throughout HFD administration, they may not have experienced hyperleptinemia comparable to controls throughout the duration of the HFD regimen. Serum leptin levels of female $Trsp^{Agrp}$ KO mice were similar to those of female controls, however, despite their lean body composition and heightened leptin sensitivity. One possible explanation for this paradoxical finding is that Agrp neurons in female $Trsp^{Agrp}$ KO mice develop some level of leptin resistance initially but are replaced by leptin-responsive adult-born neurons [24]. Whereas ablation of Agrp neurons in adult mice causes severe anorexia and rapid starvation [25], genetically induced progressive degeneration of Agrp neurons results in mice that are viable and slightly underweight despite losing ~85% of Agrp neurons by adulthood [26]. Interestingly, according to this report by Xu et al., progressive Agrp neuron ablation caused a decrease in adiposity in female mice, but not males, similar to the results of our study on $Trsp^{Agrp}KO$ mice. It was eventually uncovered that progressive degeneration of Agrp neurons induces a compensatory mechanism of neurogenesis that includes adult-born Agrp neurons that are responsive to leptin [24]. Past research has shown that global $Trsp$ deletion is embryonic lethal [27] and whole-brain neuron-specific knockout of $Trsp$ leads to neurodegeneration and seizures [28]. Therefore, it is plausible that the oxidative stress caused by ablating selenoproteins in Agrp neurons may cause a progressive degeneration that induces compensatory neurogenesis similar to that observed by Xu and colleagues [26]. It is important to note, however, that these studies were performed on mice fed a standard diet rather than an HFD, as used in our current study. Interestingly, Pomc protein levels were significantly increased in female $Trsp^{Agrp}KO$ mice in response to leptin, which may indicate a shift towards greater reliance on Pomc neurons. Whether that is the case and whether those Pomc neurons are new adult-born cells produced in response to Agrp neuron degeneration remains to be investigated.

From the current data, it is not possible to deduce whether the ability of Agrp neurons to avoid leptin resistance drives the lean phenotype of female $Trsp^{Agrp}KO$ mice or if the adiposity of female $Trsp^{Agrp}KO$ mice simply does not reach a level necessary to cause leptin resistance in the first place in the time frame used in our study. In the case of the latter scenario, the implication would be that some other factor is driving the lean phenotype of female $Trsp^{Agrp}KO$ mice. As mentioned above, this could occur in the form of upregulated BAT thermogenesis. Either scenario implies that $Trsp$-lacking Agrp neurons are generally less active than Agrp neurons in control mice as Agrp neurons exert an inhibitory influence on both processes. Indeed, the firing rate of Agrp neurons decreases in response to exogenous reactive oxygen species (ROS) application [29]. Conversely, the anorexigenic Pomc neurons can be activated by ROS [30], suggesting that an oxidative Arc environment promotes a shift towards a negative energy balance. The occurrence of an oxidative suppression of Agrp neuron activity resulting from $Trsp$ deletion would reconcile our results with previous reports.

At first glance, our findings may appear to conflict with previous studies that have reported obesogenic phenotypes resulting from models of hypothalamic selenoprotein disruption. For example, broad hypothalamic silencing of selenoprotein synthesis via RIP-Cre-mediated $Trsp$ deletion was shown by Yagishita et al. [8] to induce oxidative stress in the hypothalamus and result in an increased susceptibility to DIO accompanied by leptin resistance, insulin resistance, and other metabolic disturbances. It is important to note, however, that the RIP-Cre-driven mouse model used in this study did not appear to affect Agrp neurons specifically, which was the target cell population of the current study. Moreover, the finding that Agrp-specific $Trsp$ deletion confers protection against DIO is not surprising, as we have previously reported that $Scly^{Agrp}KO$ mice are protected against DIO and leptin resistance [10]. The loss of Scly impairs the process of synthesizing selenocysteine residues for de facto selenoprotein synthesis, and global $Scly$ knockout reduces selenoprotein expression in multiple tissues, including the hypothalamus [31], in addition to producing an obesogenic phenotype [32,33]. We found that Agrp neuron-specific knockout of $Scly$ resulted in a lean phenotype similar to, but milder than, what we have observed with the $Trsp^{Agrp}KO$ mice. While the loss of Scly might result in a reduction in selenoprotein expression, $Trsp$ deletion prevents the synthesis of selenoproteins and, presumably, causes even greater vulnerability to oxidative stress. Thus, comparing these two mouse models reveals a potential correlation between the severity of antioxidant incapacitation in Agrp neurons and the extent to which the animal is protected from DIO. It is important to note that there is evidence that alternative amino acids, such as cysteine, can become incorporated rather than selenocysteine [34–37]. Such a phenomenon might result in non-selenium-containing selenoproteins with diminished functional efficacy. The main difference between the metabolic outcome of these genetic models is that, while both sexes were equally affected by *Scly* deletion in Agrp neurons, only female *Trsp^{Agrp}KO* mice were protected from DIO and leptin resistance, whereas any differences between male *Trsp^{Agrp}KO* and control mice were barely detectable.

Studies in mice involving selenium and selenoproteins have oftentimes revealed sex differences [33,38–40]. To our knowledge, our study is the first to report sex differences in selenoprotein action in the hypothalamus. Although sexual dimorphism within the hypothalamic–pituitary axis has been reported [41], the only potential changes in pituitary-secreted hormones in female *Trsp^{Agrp}KO* mice we observed was a decrease in FSH. There are a few studies on hypothalamic physiology involving neurogenesis that might provide clues about the sex differences observed in *Trsp^{Agrp}KO* mice. A study by Lee et al. found that an HFD activates neurogenesis in the median eminence (ME) of the hypothalamus of female but not male mice [42]. The researchers found that irradiating the ME, thereby inhibiting neurogenesis, reduced DIO in female mice. This suggests that neurogenic processes mediate DIO in female but not male mice. Therefore, if adult-born Agrp neurons comprise an essential component of this mechanism in female mice, then a lack of selenoprotein action may impair their integration into the local homeostatic circuitry. Interestingly, separate work by Bless and colleagues demonstrated that HFD-induced neurogenesis of leptin-responsive neurons is negatively regulated by estradiol [43] and that an HFD leads to estrogen receptor α (ERα)-positive adult-born neurons [44]. It has also been reported that while Agrp neurons are devoid of ERα expression [45], Pomc neurons express ERα [46]. Therefore, if ERα-positive Pomc neurons are able to exert more homeostatic influence as a result of Agrp neuron degeneration, this may serve to limit neurogenesis-mediated DIO in female mice.

Another possible explanation for the observed sex differences is that the localization of Agrp neurons may contribute to the sex differences in *Trsp^{Agrp}KO* mice. One unique quality of Agrp neurons is that a significant portion of them seem to be positioned outside the blood–brain barrier [22]. Moreover, this sub-population seems particularly susceptible to insult and may have a high turnover rate fueled by adult neurogenesis [47]. As mentioned previously, Agrp neurons may develop leptin resistance before other neuronal cell types, and it has been suggested that those lying outside the blood–brain barrier are the first to become desensitized to leptin [22]. If such an event is necessary for the development of leptin resistance, then the ability of female *Trsp^{Agrp}KO* mice fed an HFD to maintain leptin sensitivity could be due to these particular Agrp neurons dying or failing to be replaced via neurogenesis before leptin resistance fully develops. It is not clear whether there are any sex differences in Agrp neurons that reside outside the blood–brain barrier. Thus, further investigation of this unique sub-population may provide insight into the origins of the sex differences seen in *Trsp^{Agrp}KO* mice. Finally, it is also possible that sexual dimorphism beyond the Arc itself within neural circuitry downstream of Agrp neuron activity may play a role in the sex-specific results obtained from *Trsp^{Agrp}KO* mice. Indeed, male and female mice have been found to demonstrate divergence in the sympathetic innervation of adipose tissue [48]. Future studies should unveil the molecular factors and brain circuitry involved.

The results described herein demonstrate the essential nature of selenoproteins in hypothalamic function and energy homeostasis. Remarkably, conditional ablation of selenoprotein synthesis through *Trsp* gene knockout within a small neural population resulted in substantial changes in the ability of mice to gain excess weight while fed an HFD. The sexual dimorphism displayed by *Trsp^{Agrp}KO* mice may have implications for the apparent sex differences in lipid metabolism that have been observed in humans, particularly with regard to sympathetic activation of adipose tissue and thermogenesis [49,50]. We believe our findings provide novel insight into the interaction between selenium biology and energy homeostasis. Further investigation of hypothalamic selenoproteins may help identify molecular targets that can be leveraged to develop effective therapeutic treatments for obesity and other metabolic diseases.

4. Materials and Methods

4.1. Animals

$Trsp^{Agrp}$ KO mice were generated by cross-breeding Agrp$^{tm1(Cre)Lowl}$/J mice [51] (purchased from The Jackson Laboratory, Bar Harbor, ME, USA) with C57/BL6J mice containing loxP sites flanking the gene for selenocysteine tRNA$^{[Ser]Sec}$, designated Trsp [52]. Age-matched Agrp$^{tm1(Cre)Lowl}$/J mice, referred to as Agrp-Cre mice, were used as controls. Mice were allowed ad libitum food and water access and maintained on a 12 h light/dark cycle. All experiments and procedures were conducted with approval from the University of Hawaii's Institutional Animal Care and Use Committee (IACUC), protocol: APN 16-2375, last approved 22 September 2021.

4.2. Experimental Design

Mice were fed a high-fat diet containing 45% kcal fat and 4.7 kcal/g (Research Diets, New Brunswick, NJ, USA; D12451) beginning at 4 weeks of age. This diet contains ~0.2 ppm of sodium selenite (Mineral Mix S10026). Past studies using this diet resulted in serum total selenium content levels around 0.4 µg/gm measured via inductively coupled plasma-mass spectrometry [32]. Body weight was measured every 2 weeks and measured at 10:00 a.m. Mice were placed in metabolic cages for feeding behavior and respiratory metabolic assessment at 16 weeks of age. At 18 weeks of age, a glucose tolerance test was performed on each mouse, and body weight monitoring continued until mice were sacrificed at age 24 weeks following leptin challenge. Mice were anesthetized either by CO_2 asphyxiation for fresh tissue harvest or via transcardial perfusion with 4% paraformaldehyde following intraperitoneal injection of tribromoethanol for fixed tissue collection.

4.3. Metabolic Chambers

Feeding and drinking behavior, physical activity, and respiratory metabolism were monitored using the PanLab OxyletPro^{TM} System (Harvard Apparatus, Barcelona, Spain) as previously described [10]. Mice were acclimated to the metabolic chambers for 24 h prior to 48 h of data collection. Oxygen and carbon dioxide concentrations were measured for 7 min periods every 35 min and used to calculate oxygen consumption (VO_2) and carbon dioxide production (VCO_2). Data were collected and analyzed using the Panlab METABOLISM software (Vídeňská, Prague, Czech Republic). Energy expenditure (EE) was calculated via indirect calorimetry using the following equation:

$$EE = (3.815 + (1.232 \times RQ)) \times VO_2 \times 1.44 \quad (1)$$

where RQ (Respiratory quotient) is VCO_2/VO_2.

4.4. Glucose Tolerance Test

Mice were fasted overnight for 16 h. At 10:00 a.m. the next day, blood was accessed via tail vein puncture to measure baseline glycemia. Mice were administered glucose (Sigma; 1 g/kg body weight in sterile phosphate-buffered saline) via intraperitoneal injection. Glycemia was then measured at timepoints of 30 min, 1 h, 2 h, and 3 h post-injection using strips inserted into a glucometer (OneTouch Ultra, LifeScan, Milpitas, CA, USA).

4.5. Leptin Challenge and Tissue Collection

On the day of sacrifice, mice were subjected to a leptin challenge. Mice were fasted overnight for 16 h the night before and injected intraperitoneally with leptin (1 mg/kg body weight; R & D Systems, Minneapolis, MN, USA; 498-OB) or sterilized phosphate-buffered saline as a vehicle control. Mice were sacrificed exactly 1 h post-injection. Tissue was either fixed via transcardial perfusion with 4% paraformaldehyde for immunohistochemical analysis or fresh-frozen in liquid nitrogen to be used for Western blots. Prior to perfusion, mice were euthanized with tribromoethanol (1%, 0.1 mL/g body weight), after which blood was collected via cardiac puncture before perfusion with phosphate buffer, followed by 4%

paraformaldehyde in phosphate buffer. Brains were collected and stored overnight in 4% paraformaldehyde for overnight fixation, after which they were dehydrated with daily increased sucrose concentrations. Brains were cut into 40 μm coronal sections using a cryostat and stored in floating fashion in a cryoprotective solution (50% 0.1 M phosphate buffer, 25% glycerol, 25% ethylene glycol). BAT was collected and fixed in 4% paraformaldehyde for 1 week, then paraffin-embedded and cut into 5 μm sections. For collection of fresh tissue, mice were euthanized via CO_2 asphyxiation. Blood was collected via cardiac puncture, and inguinal WAT was removed and weighed on an analytical scale. Brains were placed in 30% sucrose on ice for 1 min, after which the hypothalamus was dissected and immediately frozen in liquid nitrogen.

4.6. Gel Electrophoresis and Western Blotting

Frozen tissue was homogenized with a CryoGrinder kit (OPS Diagnostics, Lebanon, NJ, USA; CG 08-01) as previously described [10]. Protein lysate samples containing 40 μg of protein were separated via electrophoresis using a 4–20% gradient polyacrylamide TGX gel (BIO-RAD, Hercules, CA, USA; 5671094) and then transferred onto a 0.45 μm pore size Immobilon-FL polyvinylidene difluoride membrane (Millipore, Burlington, MA, USA; IPFL00010). Membranes were incubated in PBS-based blocking buffer (LI-COR Biosciences, Lincoln, NE, USA; P/N 927) for 1 h, followed by overnight incubation in primary antibody at 4 °C with slow shaking. Blots were washed using PBS containing 0.01% Tween 20 (PBS-T) and incubated with infrared fluorophore-bound secondary antibodies in the dark, washed again with PBS-T, and analyzed using the infrared scanner Odyssey CLx Imaging System (LI-COR Biosciences).

4.7. Immunohistochemistry and Histology

For the visualization of target proteins, a 3,3′-diaminobenzidine (DAB) kit was used (DAB Substrate Kit; Vector Labs, Burlingame, CA, USA; H-2200) in conjunction with an avidin-biotin-peroxidase complex (Elite ABC Kit; Vector Labs; PK-6100). Endogenous peroxidases were first sequestered with 1% H_2O_2 in methanol, after which sections were blocked in normal goat serum and incubated overnight at 4 °C in primary antibody. Sections were then probed with the appropriate biotinylated secondary antibodies prior to visualization via DAB-mediated reaction. Sections were finally rinsed with phosphate-buffered saline, mounted on glass slides, and dehydrated with an ethanol gradient followed by xylene and cover-slipped.

To visualize BAT morphology, 5 μm sections were stained with hematoxylin and eosin. For measurement of UCP1 immunoreactivity levels, sections were baked in a 60 °C oven for 30 min, deparaffinized using xylene and ethanol, and then incubated in a 1% H_2O_2/methanol solution for 30 min. Antigen retrieval was performed using 0.01 M citric acid (pH 6), and sections were blocked with an avidin/biotin blocking kit (Vector Labs; SP-2001), followed by incubation with primary antibody. Sections were then incubated in a biotinylated secondary antibody, followed by DAB-mediated staining and mounting as described above.

4.8. Stereology and Data Quantification

Sections were analyzed at bregma −1.46 mm for analysis of the arcuate nucleus. Analysis was performed using the Stereo Investigator Software (MBF Bioscience, Williston, VT, USA) and an upright microscope (Axioskop2; Zeiss, Oberkochen, Germany). For measurement of phospho-STAT3 optical density, a magnification of 20× was used to capture brightfield images which were then imported into ImageJ software for analysis. For quantification, images were first converted into black-and-white, inverted, and then the mean value per pixel was measured within the arcuate nucleus. The Cell Counter ImageJ plugin was used to count the number of phospho-STAT3-positive cells.

For analysis of BAT sections, the simple random sampling function was utilized in Stereo Investigator to capture 20× images. The FIJI (ImageJ v2) plugin Adiposoft was used to count and measure lipid droplet size. To measure UCP1 optical density, images were converted to black-and-white, inverted, and the mean pixel value for the entire image was measured. Assessment of UCP1 optical density was performed as described above for phospho-STAT3.

4.9. Measurement of Serum Hormone Levels

Circulating levels of leptin and insulin were assessed by analyzing serum samples using ELISA kits: Mouse Leptin ELISA Kit (Crystal Chem, Elk Grove Village, IL, USA; 90030) and STELLUX Chemiluminescent Rodent Insulin ELISA Kit (Alpco, Salem, NH, USA; 80-INSMR-CH01). Pituitary hormone levels were measured in serum samples using the Mouse Pituitary Magnetic Bead Panel Milliplex Assay (EMD Millipore, Burlington, MA, USA; MPTMAG-49K), performed with the Luminex 200 Instrument System.

4.10. Antibodies

The primary antibodies used were: a rabbit anti-phospho-STAT3 (Tyr705) (D3A7) (1:1000; Cell Signaling, Danvers, MA, USA; 9145), a rabbit anti-STAT3 (D1A5) (1:1000; Cell Signaling, Danvers, MA, USA; 8768), a mouse anti-Agrp (1:2000; Alpha Diagnostics, San Antonio, TX, USA; AGRP-11S), a rabbit anti-Pomc (27–52, porcine) (Phoenix Pharmaceuticals, Burlingame, CA, USA; H-029-30), a mouse anti-β-actin (8H10D10) (1:5000; Cell Signaling, Danvers, MA, USA; 3700S), and a rabbit anti-UCP1 (1:500; Abcam, Cambridge, MA, USA; ab10983).

4.11. Statistical Analysis

Statistical tests used sample numbers are indicated within each figure legend. Two-way ANOVA, followed by Tukey's multiple comparisons test, was used to make comparisons using sex and genotype as factors. For data collected over time, including body weight and metabolic cage data, a repeated measures two-way ANOVA was used with Sidak's multiple comparisons test. An unpaired t-test was used to compare glucose tolerance test results. The graphical representations of data reflect the exact statistical comparisons made. For data affected by leptin challenge, comparisons were made within each sex and used genotype and leptin treatment as factors. Data were analyzed, and graphs were generated using GraphPad Prism version 7 software. All data are presented as mean ± standard error of the mean. Significance was determined by a p-value of <0.05. Sample sizes represent biological replicates.

Supplementary Materials: The following are available online at https://www.mdpi.com/article/10.3390/ijms222011010/s1.

Author Contributions: Conceptualization, D.J.T., M.W.P. and M.J.B.; Data curation, D.J.T., A.C.H., K.W.H. and S.M.A.R.; Formal analysis, D.J.T., M.W.P., L.A.S., K.J.A. and A.N.H.; Funding acquisition, D.J.T., L.A.S. and M.J.B.; Investigation, D.J.T.; Methodology, D.J.T., M.W.P., L.A.S. and A.C.H.; Project administration, M.J.B.; Resources, B.A.C. and D.L.H.; Supervision, M.W.P. All authors have read and agreed to the published version of the manuscript.

Funding: This research was funded by The National Institutes of Health, grant numbers R01DK047320 (M.J.B.), F32DK124963 (D.J.T.), R01DK128390 (L.A.S.), U54MD007601, and P30GM114737 that supported both core facilities, and the Hawaii Community Foundation, grant number 20ADVC-102166 (L.A.S.).

Institutional Review Board Statement: The study was conducted according to the guidelines of the Declaration of Helsinki and approved by the Institutional Review Board, the University of Hawaii's Institutional Animal Care and Use Committee (Protocol: APN 16-2375, last approved: 20 May 2021).

Informed Consent Statement: Not applicable.

Data Availability Statement: The data presented in this study are contained within the main article and Supplementary Materials. Inquiries about data should be directed to the corresponding author.

Acknowledgments: Histological techniques and metabolic experiments were conducted with the help of the Histopathology Core and the Murine Behavior and Metabolic Phenotyping Research

Support Facility, respectively, at the John A. Burns School of Medicine (JABSOM), which are supported by Ola HAWAII (NIH grant U54MD007601). Milliplex assays were conducted in the Molecular and Cellular Immunology Core at JABSOM, which is supported in part by NIH grant P30GM114737 from the Centers of Biomedical Research Excellence (COBRE) program of the National Institute of General Medical Sciences, a component of the National Institutes of Health. The authors wish to thank Miyoko Bellinger and Kristen Ewell of the Histopathology Core, as well as Alexandra Gurary of the Molecular and Cellular Immunology Core, for their excellent technical assistance.

Conflicts of Interest: The authors declare no conflict of interest.

Abbreviations

Agrp	agouti-related peptide
Arc	arcuate nucleus
BAT	brown adipose tissue
CRH	corticotropin-releasing hormone
DAB	3,3′-diaminobenzidine
ER	endoplasmic reticulum
ERα	estrogen receptor α
FSH	follicle-stimulating hormone
GABA	γ-Aminobutyric acid
GH	growth hormone
DIO	diet-induced obesity
DMH	dorsomedial hypothalamus
HFD	high-fat diet
ME	median eminence
Npy	neuropeptide y
Pomc	pro-opiomelanocortin
RIP	rat-insulin-promoter
ROS	reactive oxygen species
Scly	selenocysteine lyase
SELENOM	selenoprotein M
STAT3	signal transducer and activator of transcription 3
TRH	thyrotropin-releasing hormone
Trsp	selenocysteine tRNA$^{[Ser]Sec}$ gene
UCP1	uncoupling protein-1
WAT	white adipose tissue

References

1. Caballero, B. Humans against Obesity: Who Will Win? *Adv. Nutr.* **2019**, *10*, S4–S9. [CrossRef]
2. Zagmutt, S.; Mera, P.; Soler-Vazquez, M.C.; Herrero, L.; Serra, D. Targeting AgRP neurons to maintain energy balance: Lessons from animal models. *Biochem. Pharmacol.* **2018**, *155*, 224–232. [CrossRef] [PubMed]
3. Gong, T.; Torres, D.J.; Berry, M.J.; Pitts, M.W. Hypothalamic redox balance and leptin signaling—Emerging role of selenoproteins. *Free Radic. Biol. Med.* **2018**, *127*, 172–181. [CrossRef]
4. Schomburg, L. Selenium, selenoproteins and the thyroid gland: Interactions in health and disease. *Nat Rev. Endocrinol* **2011**, *8*, 160–171. [CrossRef]
5. Youn, H.S.; Lim, H.J.; Choi, Y.J.; Lee, J.Y.; Lee, M.Y.; Ryu, J.H. Selenium suppresses the activation of transcription factor NF-kappa B and IRF3 induced by TLR3 or TLR4 agonists. *Int. Immunopharmacol.* **2008**, *8*, 495–501. [CrossRef]
6. Zhang, Y.; Zhou, Y.; Schweizer, U.; Savaskan, N.E.; Hua, D.; Kipnis, J.; Hatfield, D.L.; Gladyshev, V.N. Comparative analysis of selenocysteine machinery and selenoproteome gene expression in mouse brain identifies neurons as key functional sites of selenium in mammals. *J. Biol. Chem.* **2008**, *283*, 2427–2438. [CrossRef]
7. Henry, F.E.; Sugino, K.; Tozer, A.; Branco, T.; Sternson, S.M. Cell type-specific transcriptomics of hypothalamic energy-sensing neuron responses to weight-loss. *eLife* **2015**, *4*, e09800. [CrossRef]
8. Yagishita, Y.; Uruno, A.; Fukutomi, T.; Saito, R.; Saigusa, D.; Pi, J.; Fukamizu, A.; Sugiyama, F.; Takahashi, S.; Yamamoto, M. Nrf2 Improves Leptin and Insulin Resistance Provoked by Hypothalamic Oxidative Stress. *Cell. Rep.* **2017**, *18*, 2030–2044. [CrossRef]
9. Schriever, S.C.; Zimprich, A.; Pfuhlmann, K.; Baumann, P.; Giesert, F.; Klaus, V.; Kabra, D.G.; Hafen, U.; Romanov, A.; Tschop, M.H.; et al. Alterations in neuronal control of body weight and anxiety behavior by glutathione peroxidase 4 deficiency. *Neuroscience* **2017**, *357*, 241–254. [CrossRef] [PubMed]

10. Torres, D.J.; Pitts, M.W.; Hashimoto, A.C.; Berry, M.J. Agrp-Specific Ablation of Scly Protects against Diet-Induced Obesity and Leptin Resistance. *Nutrients* **2019**, *11*, 1693. [CrossRef] [PubMed]
11. Deem, J.D.; Faber, C.L.; Morton, G.J. AgRP neurons: Regulators of feeding, energy expenditure, and behavior. *FEBS J.* **2021**. [CrossRef] [PubMed]
12. Cavaliere, G.; Viggiano, E.; Trinchese, G.; De Filippo, C.; Messina, A.; Monda, V.; Valenzano, A.; Cincione, R.I.; Zammit, C.; Cimmino, F.; et al. Long Feeding High-Fat Diet Induces Hypothalamic Oxidative Stress and Inflammation, and Prolonged Hypothalamic AMPK Activation in Rat Animal Model. *Front. Physiol.* **2018**, *9*, 818. [CrossRef]
13. Gong, T.; Hashimoto, A.C.; Sasuclark, A.R.; Khadka, V.S.; Gurary, A.; Pitts, M.W. Selenoprotein M Promotes Hypothalamic Leptin Signaling and Thioredoxin Antioxidant Activity. *Antioxid. Redox. Signal.* **2019**, *35*, 775–787. [CrossRef]
14. Pitts, M.W.; Reeves, M.A.; Hashimoto, A.C.; Ogawa, A.; Kremer, P.; Seale, L.A.; Berry, M.J. Deletion of selenoprotein M leads to obesity without cognitive deficits. *J. Biol. Chem.* **2013**, *288*, 26121–26134. [CrossRef]
15. Phillips, K.J. Beige Fat, Adaptive Thermogenesis, and Its Regulation by Exercise and Thyroid Hormone. *Biology* **2019**, *8*, 57. [CrossRef]
16. Orthofer, M.; Valsesia, A.; Magi, R.; Wang, Q.P.; Kaczanowska, J.; Kozieradzki, I.; Leopoldi, A.; Cikes, D.; Zopf, L.M.; Tretiakov, E.O.; et al. Identification of ALK in Thinness. *Cell* **2020**, *181*, 1246–1262.e22. [CrossRef] [PubMed]
17. Liu, P.; Ji, Y.; Yuen, T.; Rendina-Ruedy, E.; DeMambro, V.E.; Dhawan, S.; Abu-Amer, W.; Izadmehr, S.; Zhou, B.; Shin, A.C.; et al. Blocking FSH induces thermogenic adipose tissue and reduces body fat. *Nature* **2017**, *546*, 107–112. [CrossRef]
18. Oldfield, B.J.; Giles, M.E.; Watson, A.; Anderson, C.; Colvill, L.M.; McKinley, M.J. The neurochemical characterisation of hypothalamic pathways projecting polysynaptically to brown adipose tissue in the rat. *Neuroscience* **2002**, *110*, 515–526. [CrossRef]
19. Bamshad, M.; Song, C.K.; Bartness, T.J. CNS origins of the sympathetic nervous system outflow to brown adipose tissue. *Am. J. Physiol.* **1999**, *276*, R1569–R1578. [CrossRef]
20. Zaretskaia, M.V.; Zaretsky, D.V.; Shekhar, A.; DiMicco, J.A. Chemical stimulation of the dorsomedial hypothalamus evokes non-shivering thermogenesis in anesthetized rats. *Brain Res.* **2002**, *928*, 113–125. [CrossRef]
21. Ye, Z.; Liu, G.; Guo, J.; Su, Z. Hypothalamic endoplasmic reticulum stress as a key mediator of obesity-induced leptin resistance. *Obes. Rev.* **2018**, *19*, 770–785. [CrossRef]
22. Olofsson, L.E.; Unger, E.K.; Cheung, C.C.; Xu, A.W. Modulation of AgRP-neuronal function by SOCS3 as an initiating event in diet-induced hypothalamic leptin resistance. *Proc. Natl Acad. Sci. USA* **2013**, *110*, E697–E706. [CrossRef]
23. Friedman, J.M. Leptin and the endocrine control of energy balance. *Nat. Metab.* **2019**, *1*, 754–764. [CrossRef] [PubMed]
24. Pierce, A.A.; Xu, A.W. De novo neurogenesis in adult hypothalamus as a compensatory mechanism to regulate energy balance. *J. Neurosci.* **2010**, *30*, 723–730. [CrossRef]
25. Luquet, S.; Perez, F.A.; Hnasko, T.S.; Palmiter, R.D. NPY/AgRP neurons are essential for feeding in adult mice but can be ablated in neonates. *Science* **2005**, *310*, 683–685. [CrossRef] [PubMed]
26. Xu, A.W.; Kaelin, C.B.; Morton, G.J.; Ogimoto, K.; Stanhope, K.; Graham, J.; Baskin, D.G.; Havel, P.; Schwartz, M.W.; Barsh, G.S. Effects of hypothalamic neurodegeneration on energy balance. *PLoS Biol.* **2005**, *3*, 415. [CrossRef] [PubMed]
27. Bosl, M.R.; Takaku, K.; Oshima, M.; Nishimura, S.; Taketo, M.M. Early embryonic lethality caused by targeted disruption of the mouse selenocysteine tRNA gene (Trsp). *Proc. Natl Acad. Sci. USA* **1997**, *94*, 5531–5534. [CrossRef] [PubMed]
28. Wirth, E.K.; Conrad, M.; Winterer, J.; Wozny, C.; Carlson, B.A.; Roth, S.; Schmitz, D.; Bornkamm, G.W.; Coppola, V.; Tessarollo, L.; et al. Neuronal selenoprotein expression is required for interneuron development and prevents seizures and neurodegeneration. *FASEB J.* **2010**, *24*, 844–852. [CrossRef] [PubMed]
29. Diano, S.; Liu, Z.W.; Jeong, J.K.; Dietrich, M.O.; Ruan, H.B.; Kim, E.; Suyama, S.; Kelly, K.; Gyengesi, E.; Arbiser, J.L.; et al. Peroxisome proliferation-associated control of reactive oxygen species sets melanocortin tone and feeding in diet-induced obesity. *Nat. Med.* **2011**, *17*, 1121–1127. [CrossRef]
30. Kuo, D.Y.; Chen, P.N.; Yang, S.F.; Chu, S.C.; Chen, C.H.; Kuo, M.H.; Yu, C.H.; Hsieh, Y.S. Role of reactive oxygen species-related enzymes in neuropeptide y and proopiomelanocortin-mediated appetite control: A study using atypical protein kinase C knockdown. *Antioxid. Redox Signal.* **2011**, *15*, 2147–2159. [CrossRef] [PubMed]
31. Ogawa-Wong, A.N.; Hashimoto, A.C.; Ha, H.; Pitts, M.W.; Seale, L.A.; Berry, M.J. Sexual Dimorphism in the Selenocysteine Lyase Knockout Mouse. *Nutrients* **2018**, *10*, 159. [CrossRef] [PubMed]
32. Seale, L.A.; Gilman, C.L.; Hashimoto, A.C.; Ogawa-Wong, A.N.; Berry, M.J. Diet-induced obesity in the selenocysteine lyase knockout mouse. *Antioxid Redox Signal.* **2015**, *23*, 761–774. [CrossRef] [PubMed]
33. Seale, L.A.; Hashimoto, A.C.; Kurokawa, S.; Gilman, C.L.; Seyedali, A.; Bellinger, F.P.; Raman, A.V.; Berry, M.J. Disruption of the selenocysteine lyase-mediated selenium recycling pathway leads to metabolic syndrome in mice. *Mol. Cell. Biol.* **2012**, *32*, 4141–4154. [CrossRef] [PubMed]
34. Romanov, G.A.; Sukhoverov, V.S. Arginine CGA codons as a source of nonsense mutations: A possible role in multivariant gene expression, control of mRNA quality, and aging. *Mol. Genet. Genom.* **2017**, *292*, 1013–1026. [CrossRef]
35. Lu, J.; Zhong, L.; Lonn, M.E.; Burk, R.F.; Hill, K.E.; Holmgren, A. Penultimate selenocysteine residue replaced by cysteine in thioredoxin reductase from selenium-deficient rat liver. *FASEB J.* **2009**, *23*, 2394–2402. [CrossRef]
36. Tobe, R.; Naranjo-Suarez, S.; Everley, R.A.; Carlson, B.A.; Turanov, A.A.; Tsuji, P.A.; Yoo, M.H.; Gygi, S.P.; Gladyshev, V.N.; Hatfield, D.L. High error rates in selenocysteine insertion in mammalian cells treated with the antibiotic doxycycline, chloramphenicol, or geneticin. *J. Biol. Chem.* **2013**, *288*, 14709–14715. [CrossRef]

37. Xu, X.M.; Turanov, A.A.; Carlson, B.A.; Yoo, M.H.; Everley, R.A.; Nandakumar, R.; Sorokina, I.; Gygi, S.P.; Gladyshev, V.N.; Hatfield, D.L. Targeted insertion of cysteine by decoding UGA codons with mammalian selenocysteine machinery. *Proc. Natl Acad. Sci. USA* **2010**, *107*, 21430–21434. [CrossRef]
38. Pitts, M.W.; Kremer, P.M.; Hashimoto, A.C.; Torres, D.J.; Byrns, C.N.; Williams, C.S.; Berry, M.J. Competition between the Brain and Testes under Selenium-Compromised Conditions: Insight into Sex Differences in Selenium Metabolism and Risk of Neurodevelopmental Disease. *J. Neurosci.* **2015**, *35*, 15326–15338. [CrossRef]
39. Kremer, P.M.; Torres, D.J.; Hashimoto, A.C.; Berry, M.J. Sex-Specific Metabolic Impairments in a Mouse Model of Disrupted Selenium Utilization. *Front. Nutr.* **2021**, *8*, 682700. [CrossRef]
40. Seale, L.A.; Ogawa-Wong, A.N.; Berry, M.J. Sexual Dimorphism in Selenium Metabolism and Selenoproteins. *Free Radic. Biol. Med.* **2018**, *127*, 198–205. [CrossRef]
41. Oyola, M.G.; Handa, R.J. Hypothalamic-pituitary-adrenal and hypothalamic-pituitary-gonadal axes: Sex differences in regulation of stress responsivity. *Stress* **2017**, *20*, 476–494. [CrossRef] [PubMed]
42. Lee, D.A.; Yoo, S.; Pak, T.; Salvatierra, J.; Velarde, E.; Aja, S.; Blackshaw, S. Dietary and sex-specific factors regulate hypothalamic neurogenesis in young adult mice. *Front. Neurosci.* **2014**, *8*, 157. [CrossRef]
43. Bless, E.P.; Reddy, T.; Acharya, K.D.; Beltz, B.S.; Tetel, M.J. Oestradiol and diet modulate energy homeostasis and hypothalamic neurogenesis in the adult female mouse. *J. Neuroendocrinol.* **2014**, *26*, 805–816. [CrossRef]
44. Bless, E.P.; Yang, J.; Acharya, K.D.; Nettles, S.A.; Vassoler, F.M.; Byrnes, E.M.; Tetel, M.J. Adult Neurogenesis in the Female Mouse Hypothalamus: Estradiol and High-Fat Diet Alter the Generation of Newborn Neurons Expressing Estrogen Receptor alpha. *eNeuro* **2016**, *3*. [CrossRef] [PubMed]
45. Olofsson, L.E.; Pierce, A.A.; Xu, A.W. Functional requirement of AgRP and NPY neurons in ovarian cycle-dependent regulation of food intake. *Proc. Natl Acad. Sci. USA* **2009**, *106*, 15932–15937. [CrossRef]
46. De Souza, F.S.; Nasif, S.; Lopez-Leal, R.; Levi, D.H.; Low, M.J.; Rubinsten, M. The estrogen receptor alpha colocalizes with proopiomelanocortin in hypothalamic neurons and binds to a conserved motif present in the neuron-specific enhancer nPE2. *Eur. J. Pharmacol.* **2011**, *660*, 181–187. [CrossRef]
47. Yulyaningsih, E.; Rudenko, I.A.; Valdearcos, M.; Dahlen, E.; Vagena, E.; Chan, A.; Alvarez-Buylla, A.; Vaisse, C.; Koliwad, S.K.; Xu, A.W. Acute Lesioning and Rapid Repair of Hypothalamic Neurons outside the Blood-Brain Barrier. *Cell. Rep.* **2017**, *19*, 2257–2271. [CrossRef]
48. Kim, S.N.; Jung, Y.S.; Kwon, H.J.; Seong, J.K.; Granneman, J.G.; Lee, Y.H. Sex differences in sympathetic innervation and browning of white adipose tissue of mice. *Biol. Sex Differ.* **2016**, *7*, 67. [CrossRef] [PubMed]
49. Toth, M.J.; Gardner, A.W.; Arciero, P.J.; Calles-Escandon, J.; Poehlman, E.T. Gender differences in fat oxidation and sympathetic nervous system activity at rest and during submaximal exercise in older individuals. *Clin. Sci.* **1998**, *95*, 59–66. [CrossRef]
50. Herz, C.T.; Kulterer, O.C.; Prager, M.; Marculescu, R.; Langer, F.B.; Prager, G.; Kautzky-Willer, A.; Haug, A.R.; Kiefer, F.W. Sex differences in brown adipose tissue activity and cold-induced thermogenesis. *Mol. Cell. Endocrinol.* **2021**, *534*, 111365. [CrossRef]
51. Tong, Q.; Ye, C.P.; Jones, J.E.; Elmquist, J.K.; Lowell, B.B. Synaptic release of GABA by AgRP neurons is required for normal regulation of energy balance. *Nat. Neurosci.* **2008**, *11*, 998–1000. [CrossRef] [PubMed]
52. Kumaraswamy, E.; Carlson, B.A.; Morgan, F.; Miyoshi, K.; Robinson, G.W.; Su, D.; Wang, S.; Southon, E.; Tessarollo, L.; Lee, B.J.; et al. Selective removal of the selenocysteine tRNA [Ser]Sec gene (Trsp) in mouse mammary epithelium. *Mol. Cell. Biol.* **2003**, *23*, 1477–1488. [CrossRef] [PubMed]

Article

Initial Step of Selenite Reduction via Thioredoxin for Bacterial Selenoprotein Biosynthesis

Atsuki Shimizu [1,†], Ryuta Tobe [1,†], Riku Aono [1,†], Masao Inoue [1,2], Satoru Hagita [1], Kaito Kiriyama [1], Yosuke Toyotake [3], Takuya Ogawa [3], Tatsuo Kurihara [3], Kei Goto [4], N. Tejo Prakash [5] and Hisaaki Mihara [1,*]

1. College of Life Sciences, Ritsumeikan University, 1-1-1 Nojihigashi, Kusatsu 525-8577, Shiga, Japan; 0229atk33@gmail.com (A.S.); tober0925@gmail.com (R.T.); raono@fc.ritsumei.ac.jp (R.A.); mainoue@fc.ritsumei.ac.jp (M.I.); satoru.hagita@daikin.co.jp (S.H.); kirikaiw@gmail.com (K.K.)
2. R-GIRO, Ritsumeikan University, 1-1-1 Nojihigashi, Kusatsu 525-8577, Shiga, Japan
3. Institute for Chemical Research, Kyoto University, Gokasho, Uji 611-0011, Kyoto, Japan; toyotake@fc.ritsumei.ac.jp (Y.T.); ogawa.tky@mbc.kuicr.kyoto-u.ac.jp (T.O.); kurihara@scl.kyoto-u.ac.jp (T.K.)
4. Department of Chemistry, School of Science, Tokyo Institute of Technology, 2-12-1 Ookayama, Meguro-ku, Tokyo 152-8551, Japan; goto@chem.titech.ac.jp
5. School of Energy and Environment, Thapar Institute of Engineering and Technology, Patiala 147004, Punjab, India; ntejoprakash@thapar.edu
* Correspondence: mihara@fc.ritsumei.ac.jp; Tel.: +81-(0)77-561-2732
† These authors contributed equally to this work.

Abstract: Many organisms reductively assimilate selenite to synthesize selenoprotein. Although the thioredoxin system, consisting of thioredoxin 1 (TrxA) and thioredoxin reductase with NADPH, can reduce selenite and is considered to facilitate selenite assimilation, the detailed mechanism remains obscure. Here, we show that selenite was reduced by the thioredoxin system from *Pseudomonas stutzeri* only in the presence of the TrxA (PsTrxA), and this system was specific to selenite among the oxyanions examined. Mutational analysis revealed that Cys33 and Cys36 residues in PsTrxA are important for selenite reduction. Free thiol-labeling assays suggested that Cys33 is more reactive than Cys36. Mass spectrometry analysis suggested that PsTrxA reduces selenite via PsTrxA-SeO intermediate formation. Furthermore, an in vivo formate dehydrogenase activity assay in *Escherichia coli* with a gene disruption suggested that TrxA is important for selenoprotein biosynthesis. The introduction of PsTrxA complemented the effects of TrxA disruption in *E. coli* cells, only when PsTrxA contained Cys33 and Cys36. Based on these results, we proposed the early steps of the link between selenite and selenoprotein biosynthesis via the formation of TrxA–selenium complexes.

Keywords: bacteria; selenite; selenium delivery system; selenoprotein; thioredoxin

1. Introduction

Selenium is an essential trace element in many organisms [1–3]. Most of its important roles in cells are exerted as the 21st amino acid selenocysteine (Sec) [4], which is translationally incorporated into selenoproteins such as formate dehydrogenase (FDH), glycine reductase, and hydrogenase in bacteria as well as glutathione peroxidases, selenoprotein P, and thioredoxin reductase (TXNRD) in mammals [5,6]. Bacterial selenoproteins mostly function in anaerobic energy metabolism, while those of mammals generally play antioxidant roles [5,6]. Compared with cysteine, that contains a thiol group, Sec with a selenol group is more nucleophilic and, thus, often serves as a catalytic residue in enzymes with redox activity [7].

In bacteria, selenide with ATP and water is converted to selenophosphate together with AMP and phosphate by selenophosphate synthetase (SPS) for Sec synthesis [8]. Seryl-tRNA[Sec], formed by the aminoacylation of tRNA[Sec] with serine by seryl-tRNA synthetase, is nucleophilically attacked by selenophosphate through the catalysis of selenocysteine synthase (SelA), resulting in selenocysteyl-tRNA[Sec] generation [9]. Sec is incorporated

into selenoproteins at UGA codons [10–13]. The specific translation elongation factor, SelB, delivers selenocysteyl-tRNA^{Sec} to the ribosome by recognizing the Sec insertion sequence (SECIS) located immediately downstream of the UGA codon on the mRNA. Mammalian selenoprotein synthetic machinery is slightly different from that of bacteria. First, seryl-tRNA^{Sec} is further converted to O-phosphoseryl-tRNA^{Sec} by O-phosphoseryl-tRNA ^{Sec} kinase, then selenocysteyl-tRNA^{Sec} is produced by Sep-tRNA:Sec-tRNA synthase using selenophosphate [14]. Second, mammalian SECIS is present in the 3'-untranslated region, and the SelB recognizes SECIS via a SECIS binding protein [6].

Although selenium plays essential roles in many organisms, it is toxic when present in excess [1,2,15]. Therefore, selenium delivery systems have been proposed to sequester toxic selenium intermediates to utilize the toxic element [16]. Glutathione (GSH) system and/or thioredoxin system (Trx system) are supposed to be involved in the reduction of selenite to selenide via NADPH [17–19]. GSH is the most abundant thiol in many organisms including *Escherichia coli* [20], and it is considered a major facilitator of selenite assimilation. However, selenium and GSH metabolites such as glutathione selenotrisulfide, are unstable [21]. Moreover, GSH is not found in most Gram-positive bacteria [20], and GSH reductase knockout mutants of *E. coli* produce a selenoprotein, FDH, suggesting that selenite assimilation via the GSH system is not a universal mechanism [22]. In contrast to GSH reductase, gene disruption of thioredoxin reductase (TrxR) decreases the FDH activity of *E. coli* cells [22]. TrxR, with thioredoxin 1 (TrxA) and NADPH, comprises the Trx system, which appears to be ubiquitous in many bacteria and participates in various redox reactions [23,24]. TrxA reduces oxidized substrate using two vicinal Cys residues in the active center, and the oxidized form of TrxA is reduced by TrxR using NADPH [23]. TXNRD can directly reduce selenite to selenide without mammalian thioredoxin, whereas that from *E. coli* cannot act without TrxA [25].

Although the previous study using ^{75}Se-labeled selenite demonstrated that TrxA was labeled with ^{75}Se even in the absence of TrxR [26], the mechanism of selenite reduction by TrxA has not been established. In this study, we focused on the reaction between selenite and TrxA from *Pseudomonas stutzeri* F2a [27] (PsTrxA), which is 70% identical to that of *E. coli* (EcTrxA). Since *P. stutzeri* F2a was isolated from seleniferous soil, the strain may serve as an interesting model for studying bacterial selenium metabolism. We observed the formation of a PsTrxA–SeO complex, implicating the early steps of selenite reduction via TrxA as a selenium delivery system in bacteria.

2. Results
2.1. Reduction Activity of the Trx System from P. stutzeri

The insulin disulfide-reductive cleaving activity of PsTrxA with dithiothreitol (DTT) was evaluated using the method previously described [28]. The turbidity of the reaction mixture increased due to the fact of precipitation of the free insulin B chain (Figure 1A), showing that PsTrxA reduced the disulfide bonds in insulin. The Cys33 and Cys36 residues of PsTrxA are broadly conserved in other TrxAs (Figure S1). To examine the involvement of these Cys residues in the disulfide-reducing activity, we constructed the PsTrxA mutants, C33A, C36A, and C33A/C36A, in which each Cys was substituted by Ala, and measured their activities. Unlike the wild-type PsTrxA, the mutants did not reduced insulin (Figure 1A), suggesting that the Cys residues are important active site residues in PsTrxA.

We next examined the selenite reduction activity of the Trx system using PsTrxA and TrxR from *P. stutzeri* F2a (PsTrxR) (Figure 1B). The Trx system with the wild-type PsTrxA exhibited selenite reduction activity, whereas that with the PsTrxA mutants did not, indicating that the active site Cys residues of PsTrxA play an essential role in the selenite reduction activity. These results are consistent with the previous findings that selenite is not reduced by bacterial TrxR alone, and that TrxA is required for selenite reduction [25]. We also examined whether the Trx system reduced other oxyanions, such as selenate, sulfite, sulfate, thiosulfate, nitrite, and nitrate. However, none of them served as a substrate

(Figure S2). These results indicated that the Trx system is specific to selenite among the oxyanions examined in this study.

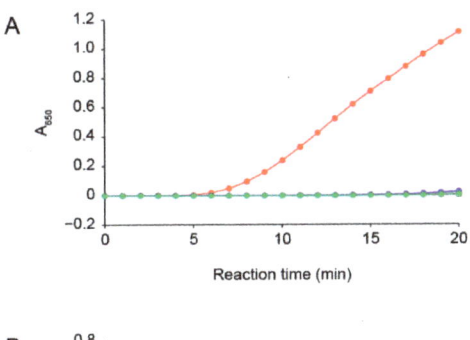

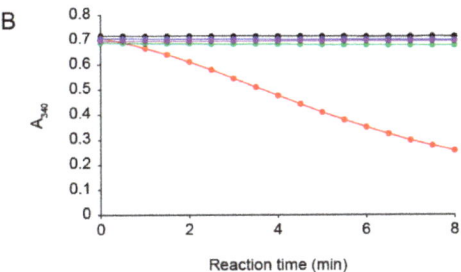

Figure 1. Reduction activity of the Trx system from *P. stutzeri* F2a. (**A**) Insulin disulfide-reducing activity of PsTrxA with DTT as measured by the increase in A_{650} due to the fact of insulin precipitation. (**B**) Selenite reducing activity of PsTrxA with PsTrxR and NADPH as measured by the decrease in A_{340} due to NADPH oxidation. PsTrxA proteins used in the assays were wild type (red), C33A (blue), C36A (green), and C33A/C36A (purple). Assays without PsTrxA are shown in black. Representative data obtained for each experiment are shown.

2.2. Number of Free Thiols in PsTrxA Incubated with Selenite

In the Trx system, selenite may be reduced by the reduced form of TrxA, resulting in the formation of selenide and oxidized TrxA with an intramolecular disulfide bond, which is re-converted to the reduced form by TrxR. Tamura et al. reported that EcTrxA was radiolabeled with ^{75}Se derived from [^{75}Se] selenite [26], implying the formation of a selenium-bound TrxA intermediate. In the GSH system, GSSeSG is formed by binding of selenium to the thiol groups of GSH [17]. We speculated that a somewhat similar selenium-bound intermediate could also occur in the Trx system, and that selenium would bind to TrxA via the thiol groups of the Cys residues. We further examined the involvement of the thiol groups of PsTrxA in selenite reduction by gel retardation assays using maleimide-conjugated polyethylene glycol (PEG-PCMal), which labels cysteine-thiol groups. Reduced PsTrxA was incubated with a five-fold molar excess of selenite, labeled with PEG-PCMal, and resolved by sodium dodecyl sulfate (SDS)-polyacrylamide gel electrophoresis (PAGE) (Figure 2).

The molecular mass of the wild-type and mutant PsTrxA proteins without PEG-PCMal labeling did not significantly differ according to SDS-PAGE (Figure 2). In contrast, various bands shifted in SDS-PAGE when proteins were labeled with PEG-PCMal. Based on the fact that the wild-type PsTrxA has three Cys residues, Cys33, Cys36, and Cys57 (Figure S1), whereas C33A and C36A has two and C33A/C36A has one, the changes in molecular mass apparently corresponded to the number of thiol groups labeled by PEG-PCMal.

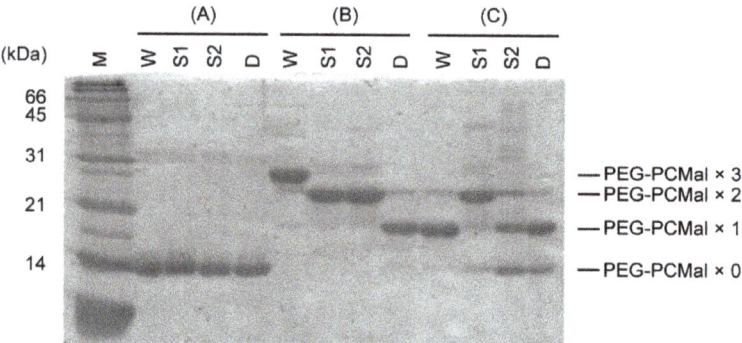

Figure 2. Band shifts of PsTrxAs on SDS-PAGE. PsTrxAs were labeled with PEG-PCMal depending on the numbers of free thiol groups. Recombinant PsTrxA proteins (**A**) were reduced by incubation with DTT, then DTT was removed by size exclusion chromatography. The resulting reduced proteins were incubated without (**B**) or with (**C**) selenite, labeled with PEG-PCMal, then resolved by SDS-PAGE. W, wild type; S1, C33A; S2, C36A; D, C33A/C36A.

In contrast, when the wild-type PsTrxA incubated with selenite was labeled with PEG-PCMal, the band shift diminished, and the molecular mass corresponded to the PsTrxA protein labeled with only one PEG-PCMal, indicating that the two thiol groups were not labeled (Figure 2). This result could be explained by the loss of reactivity between PEG-PCMal and the Cys residues due to the fact of their oxidation or modification with selenite. A weak band corresponding to the protein labeled with two PEG-PCMal was observed. This may be a non-specific product, because the identical band was also seen in the Cys33A/Cys36A, which has only one Cys residue, after incubation with PEG-PCMal in the absence of selenite.

We also incubated PsTrxA mutants with selenite, then performed a PEG-PCMal labeling assay (Figure 2). The molecular masses of most C33A and C33A/C36A were not changed, irrespective of selenite. In contrast, a large portion of C36A incubated with selenite was labeled with only one PEG-PCMal, whereas the mutant without selenite was labeled with two PEG-PCMal. Since Cys32 of EcTrxA corresponding to Cys33 of PsTrxA has a lower pK_a and is more reactive [29], the thiol group of Cys33 might attack selenite nucleophilically to produce a thioselenite moiety (–S-SeO$_2^-$) as previously suggested [26], then the thiol group would not be labeled with PEG-PCMal. The lower bands were also observed in C33A and C33A/C36A incubated with selenite followed by PEG-PCMal labeling. These bands were most likely due to the non-specific binding of reactive selenite to Cys residues, then the thiol group(s) would not be labeled with PEG-PCMal.

2.3. Formation of PsTrxA Complex with Selenium

We assumed that selenite oxidized or modified PsTrxA to prevent labeling with PEG-PCMal. To gain insight into the molecular state of PsTrxA reacted with selenite, the molecular mass of PsTrxA was analyzed using electrospray ionization (ESI)-mass spectrometry (MS) (Figure 3). A predominant protein species with a molecular mass of 13,839 Da corresponded with the recombinant PsTrxA lacking the N-terminal Met, which was calculated to be 13,837 Da from its amino acid sequence (Figure 3A). The intact PsTrxA with the N-terminal Met was also observed as a second major species with 13,970 Da, which is close to the calculated mass of 13,968 Da. When DTT-reduced PsTrxA was incubated with a five-fold molar excess of selenite, a protein species with a mass of 13,936 Da appeared as a new second major peak, while PsTrxA without the N-terminal Met (13,839 Da) remained predominant (Figure 3B). The shift of 97 Da was larger than the mass of selenium (79 Da), but close to that of SeO (95 Da). These data suggested that selenium bound to a significant fraction of PsTrxA in the form of SeO upon reaction with selenite.

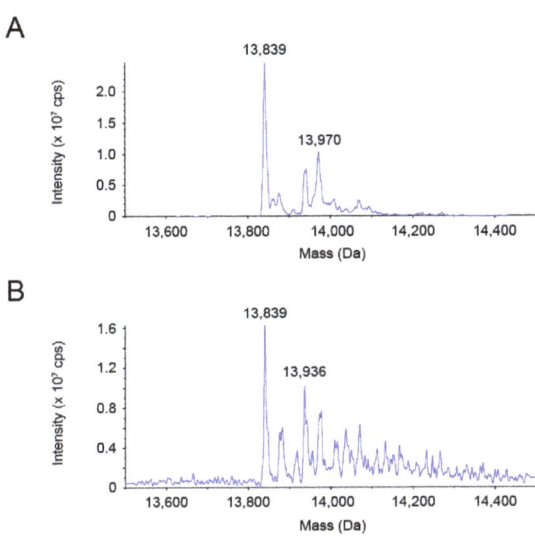

Figure 3. Reconstructed ESI-MS spectra of PsTrxA. PsTrxA protein was incubated without (**A**) or with (**B**) DTT and selenite, followed by ESI-MS analysis.

2.4. Involvement of PsTrxA in Selenoprotein Synthesis

To explore whether TrxA is involved in delivering selenide for selenoprotein biosynthesis in vivo, the activity of the selenoprotein FDH was assayed in whole *E. coli* cells anaerobically cultured on solid medium using benzyl viologen as previously described [22]. The benzyl viologen assay directly reflects FDH activity in cells and indirectly reflects selenoprotein biosynthetic activity. Figure 4 shows that the wild-type *E. coli* cells stained purple, indicating FDH activity, whereas the cells with a disrupted SelA gene (ΔselA) were not stained as they could not express selenoproteins. *E. coli* with a disrupted EcTrxA gene (ΔEctrxA) had low levels of activity, suggesting that EcTrxA is a major facilitator of selenoprotein biosynthesis in this bacterium. Introducing the wild-type PsTrxA gene into the ΔEctrxA strain recovered FDH activity (Figure 4), suggesting that PsTrxA can complement the deletion of EcTrxA. In contrast, introducing the PsTrxA mutants, C33A, C36A, and C33A/C36A, did not complement EcTrxA disruption. These results suggest that Cys33 and Cys36 residues in PsTrxA are important for selenoprotein biosynthesis.

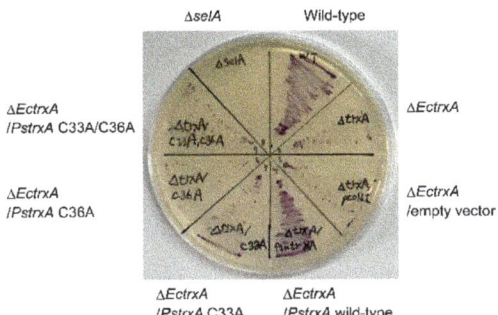

Figure 4. Whole cell FDH assay using *E. coli*. *E. coli* cells of the wild type, a SelA gene disruptant (ΔselA), and EcTrxA disruptants (ΔEctrxA) complemented without (none or empty vector) or with PsTrxA variants (wild type, C33A, C36A, or C33A/C36A) were anaerobically cultured on Luria–Bertani medium containing 0.5% glucose, then FDH activity was assayed using benzyl viologen.

3. Discussion

Selenite can serve as a nutritional source of selenium for bacteria. Selenite is also provided from another inorganic selenium source, selenate, by selenate reductases such as *E. coli* YnfEFGH [30]. Selenite is then reduced to selenide in cells. Selenophosphate, which is essential for selenoprotein biosynthesis, is produced from selenide, ATP, and water by SPS [8]. However, since the K_m values of SPS for selenide reside in the toxic range (20–46 μM) for many organisms [31,32], it has been thought that selenium delivery systems may sequester the toxic element [16]. Some proteins, such as rhodanese, glyceraldehyde-3-phosphate dehydrogenase, and 3-mercaptopyruvate sulfurtransferase, are able to bind selenium and, therefore, debated as possible candidates for selenium delivery proteins [33,34]. However, their physiological relevance to selenium assimilation remains unclear.

The Trx system, which is distributed in many bacterial phyla, appeared as a promising candidate for selenium delivery system in bacteria [16]. The Cys33 and Cys36 residues in PsTrxA are broadly conserved in TrxAs (Figure S1). Disrupting these residues resulted in the loss of not only insulin, but also selenite reduction activity (Figure 1), suggesting that the conserved Cys residues are important for selenite reduction. The results of gel-shift assays suggested that Cys33 and Cys36 were oxidized and/or modified with selenite (Figure 2). A comparison of the two Cys residues suggested that Cys36 was not reactive to selenite without Cys33, due to the fact having less reactivity than Cys33. The ESI-MS analysis suggested that a specific fraction of PsTrxAs proteins formed a PsTrxA–SeO complex, which might be an intermediate in the initial step of selenite reduction by PsTrxA (Figure 3). The results of the benzyl viologen assays indicated that disrupted EcTrxA in *E. coli* led to a decrease in FDH activity (Figure 4). Considering the previous report that the disruption of TrxR in *E. coli* results in a decrease in FDH activity [22], the Trx system functions as the main selenium delivery system from selenite to selenoprotein synthesis in *E. coli*. In addition, since the Cys33 and Cys36 mutants did not complement EcTrxA gene deletion, these residues are important for selenoprotein biosynthesis from selenite in vivo. Taken together, these results indicate that the Trx system actually functions in the selenite assimilation pathway for selenoprotein biosynthesis in bacteria.

A reaction mechanism for alkylthiols with selenite has been proposed [35], in which alkylthioselenic acid (R-S-SeO$_2$H) is generated first, then attacked by another alkylthiol, resulting in the formation of dithioselenite (R-S-Se(O)-S-R), which is further converted to the isomerized form (R-S-Se-O-S-R) [36]. Based on that mechanistic proposal [35,36] and the present results, we propose the early steps of the selenite delivery system with TrxA (Figure 5). First, selenite is attacked by the higher reactive thiol group of Cys33 to form thioselenite (Cys-S-SeO$_2$H) which is suggested by the band shift of C36A caused by the incubation with selenite (Figure 2). Then, further nucleophilic attack by another thiol group of Cys36 results in the formation of dithioselenite (Cys-S-Se(O)-S-Cys), which is supported by our ESI-MS results (Figure 3). This TrxA–SeO complex can also be isomerized (Cys-S-Se-O-S-Cys), and these complexes could be dedicated to further reduction to selenide in later steps of the selenite delivery system. How the TrxA–SeO complex is further reduced to provide a selenide substrate for SPS remains an open question. Other TrxA molecules may be involved in the reduction of the TrxA–SeO complex, and the resulting oxidized TrxA would be reduced by TrxR in an NADPH-dependent manner, or alternatively, it may also be possible that TrxR directly reduces TrxA–SeO to produce selenide. Future studies will focus on a comprehensive understanding of the selenium delivery mechanisms.

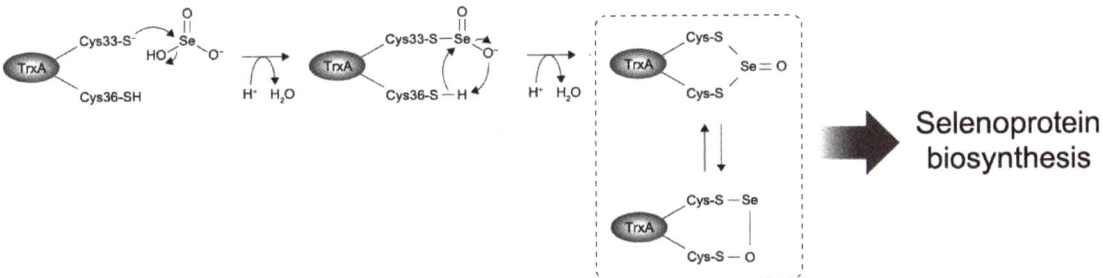

Figure 5. Proposed mechanism for early steps of selenite reduction via TrxA in the Trx system-dependent selenium delivery for selenoprotein biosynthesis.

4. Materials and Methods

4.1. Preparation of the Recombinant Proteins

To prepare His-tagged recombinant proteins, we constructed plasmids carrying the PsTrxA and PsTrxR genes as follows. The coding region of each gene (PszF2a_05700 for PsTrxA and PszF2a_19560 for PsTrxR) [27] was amplified from the genomic DNA of *P. stutzeri* F2a by PCR using the primer sets PsTrxA-f/PsTrxA-r and PsTrxR-f/PsTrxR-r for PsTrxA and PsTrxR, respectively (Table S1). After digestion with NdeI (New England Biolabs, Ipswich, MA, USA) and BamHI (New England Biolabs), the fragments were individually inserted into the same restriction sites of pCold I (Takara Bio Inc., Kusatsu, Japan) to generate pCold-PsTrxA and pCold-PsTrxR. For site-directed mutagenesis, PCR proceeded using pCold-PsTrxA and the primer sets PsTrx_C33A-f/PsTrx_C33A-r and PsTrx_C36A-f/PsTrx_C36A-r for the PsTrxA mutants, C33A and C36A, respectively (Table S1). After digestion of the template plasmid with DpnI (New England Biolabs), the mutated constructs were introduced into *E. coli* DH5α, resulting in pCold-PsTrxA_C33A and pCold-PsTrxA_C36A. The plasmid for gene expression of the PsTrxA C33A/C36A mutant, pCold-PsTrxA_C33A/C36A, was constructed using pCold-PsTrxA_C36A and the primer set PsTrx_C33A/C36A-f/PsTrx_C33A/C36A-r using the same procedure described above (Table S1).

Plasmids for the expression of PsTrxA and its mutants were introduced into *E. coli* DH5α, and pCold-TrxR was introduced into *E. coli* BL21(DE3). The cells were grown at 37 °C in Luria–Bertani (LB) medium containing 100 μg mL^{-1} ampicillin (Nacalai Tesque, Kyoto, Japan) [37] until their optical density at 660 nm reached 0.4. The cells were cooled on ice for 30 min, then gene expression was induced by 0.2 mM isopropyl 1-thio-β-D-galactopyranoside (Protein Ark, Sheffield, UK), and the cells were further incubated at 16 °C for 24 h. The cells were harvested by centrifugation (10,000× g, 5 min, 4 °C), washed with phosphate-buffered saline [37], and collected again by centrifugation (10,000× g, 5 min, 4 °C). The cells were resuspended in 20 mM potassium phosphate (pH 7.4) containing 500 mM NaCl, 20 mM imidazole, and 5 mM 2-mercaptoethanol, sonicated, then centrifuged (15,000× g, 20 min, 4 °C). The crude extract was applied to a Ni-NTA Super Flow column (Thermo Fisher Scientific, Waltham, MA, USA), and the recombinant proteins were eluted by a stepwise increase in the imidazole concentration up to 500 mM. The purified proteins were buffer-exchanged to 40 mM potassium phosphate (pH 7.0) containing 5 mM 2-mercaptoethanol by ultrafiltration using an Amicon Ultra (Merck, Darmstadt, Germany). Protein concentration was determined using Protein Assay CBB Solution (Nacalai Tesque) by a Bradford protein assay [38].

4.2. Reduction Activity of the Recombinant Proteins

The disulfide-reducing activity of PsTrxA was evaluated by insulin reduction assays [28]. The reaction mixture (400 μL) contained 100 mM potassium phosphate (pH 7.0), 2 mM ethylenediaminetetraacetate, 150 μM insulin (Merck), 0.5 mM DTT, and 1 μM PsTrxA or its mutant protein. After preincubation at 37 °C for 3 min followed by addition of DTT and further incubation for 3 min, the reaction was initiated by adding PsTrxA proteins. The increase in absorbance at 650 nm due to the fact of precipitation of insulin by reductive cleavage of the disulfide bond by DTT reduced PsTrxA was measured.

The selenite-reducing activity of the Trx system of *P. stutzeri* F2a was assayed as follows. The reaction mixture (700 µL) contained 50 mM potassium phosphate (pH 7.0), 100 µM selenite, 300 µM NADPH, 2 µM PsTrxR, and 5 µM PsTrxA or its mutant protein. After preincubation at 37 °C for 5 min, the reaction was initiated by adding selenite, and the decrease in absorbance at 340 nm due to the decrease in NADPH was measured. The reducing activity towards other oxyanions was tested with 2 µM PsTrxA and 100 µM selenate, sulfite, sulfate, thiosulfate, nitrite, or nitrate instead of 100 µM selenite.

4.3. Cysteine–Thiol Group Labeling with PEG-PCMal

The numbers of free thiol groups in PsTrxA and its mutants were determined by labeling them with PEG-PCMal (Dojindo, Kumamoto, Japan) followed by SDS-PAGE. The PsTrxA and its mutants were reduced by incubation with 5 mM DTT in 50 mM potassium phosphate (pH 7.0) at 25 °C for 15 min. The mixture was applied to a Micro Bio-Spin 6 size exclusion column (Bio-Rad, Hercules, CA, USA) to remove DTT. The reduced PsTrxAs were incubated in a mixture containing 50 mM potassium phosphate (pH 7.0), 50 µM selenite, and 10 µM PsTrxA or its mutants at 25 °C for 15 min, followed by incubation with 1 mM PEG-PCMal at 37 °C for 20 min. Labeled samples were mixed with 17% (v/v) of loading buffer containing 10% SDS, 50% glycerol, 0.2 M tris(hydroxymethyl)aminomethane (Tris)-HCl (pH 6.8), and 0.05% bromophenol blue and separated on an 18% polyacrylamide gel by SDS-PAGE analysis.

4.4. ESI-MS Analysis

PsTrxA was DTT-reduced and incubated with selenite in the same manner as described in Section 4.3. The buffer of the PsTrxA mixture was replaced with sterile water to remove excess selenite using the Micro Bio-Spin 6. The protein samples were then mixed with the same volume of 98% methanol containing 2% formic acid and analyzed using a triple-quadrupole Sciex API 3000™ mass spectrometer (Applied Biosystems, Foster City, CA, USA) equipped with an electrospray ionization source in positive mode.

4.5. FDH Activity in Whole E. coli Cells

The FDH activity of the whole *E. coli* cells was examined as an indicator of selenoprotein biosynthesis using the strains, BW25113, JW5856-KC, and JW3564-KC as the wild type, Δ*EctrxA*, and Δ*selA* strains, respectively, from the Keio collection [39]. For complementation analysis of the decrease in FDH activity by PsTrxA and its mutants, the Δ*EctrxA* strain was transformed using pColdI, pCold-PsTrxA, pCold-PsTrxA_C33A, pCold-PsTrxA_C36A, or pCold-PsTrxA_C33A/C36A. The activity of FDH was assayed using the benzyl viologen agar overlay method [40]. *E. coli* cells were anaerobically cultivated overnight on LB solid medium containing 0.5% glucose at 37 °C. The medium was then overlayed with 0.75% agar containing 1.0 mg mL^{-1} benzyl viologen, 3.4 mg mL^{-1} KH$_2$PO$_4$, and 17 mg mL^{-1} sodium formate. The agar solidified within a few minutes, and cells with FDH activity were stained purple.

Supplementary Materials: The following are available online at https://www.mdpi.com/article/10.3390/ijms222010965/s1.

Author Contributions: Conceptualization, R.T. and H.M.; methodology, R.T. and H.M.; validation, A.S., R.T., S.H., K.K. and Y.T.; formal analysis, A.S. and R.A.; writing—original draft preparation, R.A.; writing—review and editing, R.A., M.I. and H.M.; visualization, R.A.; supervision, K.G., T.O., T.K. and N.T.P.; project administration, H.M.; funding acquisition, H.M. All authors have read and agreed to the published version of the manuscript.

Funding: This research was funded by the JSPS (KAKENHI 20H02907 to H.M.), Japan-India Cooperative Science Program between JSPS and DST (JPJSBP120217716 to H.M. and N.T.P.), and the Ritsumeikan Global Innovation Research Organization, the Program for the R-GIRO Research (to H.M.).

Institutional Review Board Statement: Not applicable.

Informed Consent Statement: Not applicable.

Data Availability Statement: The data presented in this study are available upon request from the corresponding author.

Conflicts of Interest: The authors declare no conflict of interest.

References

1. Hatfield, D.L.; Tsuji, P.A.; Carlson, B.A.; Gladyshev, V.N. Selenium and selenocysteine: Roles in cancer, health, and development. *Trends Biochem. Sci.* **2014**, *39*, 112–120. [CrossRef]
2. Rayman, M.P. Selenium and human health. *Lancet* **2012**, *379*, 1256–1268. [CrossRef]
3. Schomburg, L. Selenium deficiency due to diet, pregnancy, severe illness, or COVID-1—A preventable trigger for autoimmune disease. *Int. J. Mol. Sci.* **2021**, *22*, 8532. [CrossRef]
4. Böck, A.; Forchhammer, K.; Heider, J.; Leinfelder, W.; Sawers, G.; Veprek, B.; Zinoni, F. Selenocysteine: The 21st amino acid. *Mol. Microbiol.* **1991**, *5*, 515–520. [CrossRef] [PubMed]
5. Zhang, Y.; Gladyshev, V.N. An algorithm for identification of bacterial selenocysteine insertion sequence elements and selenoprotein genes. *Bioinformatics* **2005**, *21*, 2580–2589. [CrossRef] [PubMed]
6. Labunskyy, V.M.; Hatfield, D.L.; Gladyshev, V.N. Selenoproteins: Molecular pathways and physiological roles. *Physiol. Rev.* **2014**, *94*, 739–777. [CrossRef] [PubMed]
7. Reich, H.J.; Hondal, R.J. Why nature chose selenium. *ACS Chem. Biol.* **2016**, *11*, 821–841. [CrossRef]
8. Ehrenreich, A.; Forchhammer, K.; Tormay, P.; Veprek, B.; Böck, A. Selenoprotein synthesis in *E. coli*. Purification and characterisation of the enzyme catalysing selenium activation. *Eur. J. Biochem.* **1992**, *206*, 767–773. [CrossRef]
9. Itoh, Y.; Bröcker, M.J.; Sekine, S.; Hammond, G.; Suetsugu, S.; Söll, D.; Yokoyama, S. Decameric SelA•tRNA^{Sec} ring structure reveals mechanism of bacterial selenocysteine formation. *Science* **2013**, *340*, 75–78. [CrossRef]
10. Hüttenhofer, A.; Heider, J.; Böck, A. Interaction of the *Escherichia coli fdhF* mRNA hairpin promoting selenocysteine incorporation with the ribosome. *Nucleic Acids Res.* **1996**, *24*, 3903–3910. [CrossRef]
11. Paleskava, A.; Konevega, A.L.; Rodnina, M.V. Thermodynamic and kinetic framework of selenocysteyl-tRNA^{Sec} recognition by elongation factor SelB. *J. Biol. Chem.* **2010**, *285*, 3014–3020. [CrossRef] [PubMed]
12. Fischer, N.; Neumann, P.; Bock, L.V.; Maracci, C.; Wang, Z.; Paleskava, A.; Konevega, A.L.; Schröder, G.F.; Grubmüller, H.; Ficner, R.; et al. The pathway to GTPase activation of elongation factor SelB on the ribosome. *Nature* **2016**, *540*, 80–85. [CrossRef] [PubMed]
13. Thanbichler, M.; Böck, A.; Goody, R.S. Kinetics of the interaction of translation factor SelB from *Escherichia coli* with guanosine nucleotides and selenocysteine insertion sequence RNA. *J. Biol. Chem.* **2000**, *275*, 20458–20466. [CrossRef] [PubMed]
14. Yuan, J.; Palioura, S.; Salazar, J.C.; Su, D.; O'Donoghue, P.; Hohn, M.J.; Cardoso, A.M.; Whitman, W.B.; Söll, D. RNA-dependent conversion of phosphoserine forms selenocysteine in eukaryotes and archaea. *Proc. Natl. Acad. Sci. USA* **2006**, *103*, 18923–18927. [CrossRef]
15. Rayman, M.P. The importance of selenium to human health. *Lancet* **2000**, *356*, 233–241. [CrossRef]
16. Tobe, R.; Mihara, H. Delivery of selenium to selenophosphate synthetase for selenoprotein biosynthesis. *Biochim. Biophys. Acta Gen. Subj.* **2018**, *1862*, 2433–2440. [CrossRef] [PubMed]
17. Turner, R.J.; Weiner, J.H.; Taylor, D.E. Selenium metabolism in *Escherichia coli*. *BioMetals* **1998**, *11*, 223–227. [CrossRef]
18. Lu, J.; Berndt, C.; Holmgren, A. Metabolism of selenium compounds catalyzed by the mammalian selenoprotein thioredoxin reductase. *Biochim. Biophys. Acta* **2009**, *1790*, 1513–1519. [CrossRef]
19. Lu, J.; Holmgren, A. The thioredoxin antioxidant system. *Free Radic. Biol. Med.* **2014**, *66*, 75–87. [CrossRef]
20. Fahey, R.C.; Brown, W.C.; Adams, W.B.; Worsham, M.B. Occurrence of glutathione in bacteria. *J. Bacteriol.* **1978**, *133*, 1126–1129. [CrossRef]
21. Sandholm, M.; Sipponen, P. Formation of unstable selenite-glutathione complexes in vitro. *Arch. Biochem. Biophys.* **1973**, *155*, 120–124. [CrossRef]
22. Takahata, M.; Tamura, T.; Abe, K.; Mihara, H.; Kurokawa, S.; Yamamoto, Y.; Nakano, R.; Esaki, N.; Inagaki, K. Selenite assimilation into formate dehydrogenase H depends on thioredoxin reductase in *Escherichia coli*. *J. Biochem.* **2008**, *143*, 467–473. [CrossRef] [PubMed]
23. Holmgren, A. Thioredoxin. *Annu. Rev. Biochem.* **1985**, *54*, 237–271. [CrossRef] [PubMed]
24. Potamitou, A.; Holmgren, A.; Vlamis-Gardikas, A. Protein levels of *Escherichia coli* thioredoxins and glutaredoxins and their relation to null mutants, growth phase, and function. *J. Biol. Chem.* **2002**, *277*, 18561–18567. [CrossRef] [PubMed]

25. Kumar, S.; Björnstedt, M.; Holmgren, A. Selenite is a substrate for calf thymus thioredoxin reductase and thioredoxin and elicits a large non-stoichiometric oxidation of NADPH in the presence of oxygen. *Eur. J. Biochem.* **1992**, *207*, 435–439. [CrossRef]
26. Tamura, T.; Sato, K.; Komori, K.; Imai, T.; Kuwahara, M.; Okugochi, T.; Mihara, H.; Esaki, N.; Inagaki, K. Selenite reduction by the thioredoxin system: Kinetics and identification of protein-bound selenide. *Biosci. Biotechnol. Biochem.* **2011**, *75*, 1184–1187. [CrossRef]
27. Inoue, M.; Hirose, Y.; Tobe, R.; Saito, S.; Aono, R.; Prakash, N.T.; Mihara, H. Complete genome sequence of *Pseudomonas stutzeri* strain F2a, isolated from seleniferous soil. *Microbiol. Resour. Announc.* **2021**, *10*, e0063121. [CrossRef]
28. Holmgren, A. Thioredoxin catalyzes the reduction of insulin disulfides by dithiothreitol and dihydrolipoamide. *J. Biol. Chem.* **1979**, *254*, 9627–9632. [CrossRef]
29. Kallis, G.B.; Holmgren, A. Differential reactivity of the functional sulfhydryl groups of cysteine-32 and cysteine-35 present in the reduced form of thioredoxin from *Escherichia coli*. *J. Biol. Chem.* **1980**, *255*, 10261–10265. [CrossRef]
30. Fujita, D.; Tobe, R.; Tajima, H.; Anma, Y.; Nishida, R.; Mihara, H. Genetic analysis of tellurate reduction reveals the selenate/tellurate reductase genes *ynfEF* and the transcriptional regulation of *moeA* by NsrR in *Escherichia coli*. *J. Biochem.* **2021**, *169*, 477–484. [CrossRef]
31. Veres, Z.; Tsai, L.; Scholz, T.D.; Politino, M.; Balaban, R.S.; Stadtman, T.C. Synthesis of 5-methylaminomethyl-2-selenouridine in tRNAs: ^{31}P NMR studies show the labile selenium donor synthesized by the *selD* gene product contains selenium bonded to phosphorus. *Proc. Natl. Acad. Sci. USA* **1992**, *89*, 2975–2979. [CrossRef]
32. Lacourciere, G.M.; Stadtman, T.C. Catalytic properties of selenophosphate synthetases: Comparison of the selenocysteine-containing enzyme from *Haemophilus influenzae* with the corresponding cysteine-containing enzyme from *Escherichia coli*. *Proc. Natl. Acad. Sci. USA* **1999**, *96*, 44–48. [CrossRef] [PubMed]
33. Ogasawara, Y.; Lacourciere, G.; Stadtman, T.C. Formation of a selenium-substituted rhodanese by reaction with selenite and glutathione: Possible role of a protein perselenide in a selenium delivery system. *Proc. Natl. Acad. Sci. USA* **2001**, *98*, 9494–9498. [CrossRef] [PubMed]
34. Ogasawara, Y.; Lacourciere, G.M.; Ishii, K.; Stadtman, T.C. Characterization of potential selenium-binding proteins in the selenophosphate synthetase system. *Proc. Natl. Acad. Sci. USA* **2005**, *102*, 1012–1016. [CrossRef]
35. Kice, J.L.; Lee, T.W.S.; Pan, S.T. Mechanism of the reaction of thiols with selenite. *J. Am. Chem. Soc.* **1980**, *102*, 4448–4455. [CrossRef]
36. Kice, J.L.; Wilson, D.M.; Espinola, J.M. Oxidation of bis(*tert*-butylthio) selenide at low-temperatures—Search for a bis(alkylthio) selenoxide. *J. Org. Chem.* **1991**, *56*, 3520–3524. [CrossRef]
37. Sambrook, J.; Russel, D. *Molecular Cloning: A Laboratory Manual*, 3rd ed.; Cold Spring Harbor Laboratory Press: Cold Spring Harbor, NY, USA, 2001.
38. Bradford, M.M. A rapid and sensitive method for the quantitation of microgram quantities of protein utilizing the principle of protein-dye binding. *Anal. Biochem.* **1976**, *72*, 248–254. [CrossRef]
39. Baba, T.; Ara, T.; Hasegawa, M.; Takai, Y.; Okumura, Y.; Baba, M.; Datsenko, K.A.; Tomita, M.; Wanner, B.L.; Mori, H. Construction of *Escherichia coli* K-12 in-frame, single-gene knockout mutants: The Keio collection. *Mol. Syst. Biol.* **2006**, *2*, 2006.0008. [CrossRef]
40. Mandrand-Berthelot, M.A.; Wee, M.Y.K.; Haddock, B.A. An improved method for the identification and characterization of mutants of *Escherichia coli* deficient in formate dehydrogenase activity. *FEMS Microbiol. Lett.* **1978**, *4*, 37–40. [CrossRef]

Article

Selenium and the 15 kDa Selenoprotein Impact Colorectal Tumorigenesis by Modulating Intestinal Barrier Integrity

Jessica A. Canter [1], Sarah E. Ernst [1], Kristin M. Peters [1], Bradley A. Carlson [2], Noelle R. J. Thielman [1,3], Lara Grysczyk [1], Precious Udofe [1], Yunkai Yu [4], Liang Cao [4], Cindy D. Davis [5], Vadim N. Gladyshev [6], Dolph L. Hatfield [2] and Petra A. Tsuji [1,*]

[1] Department of Biological Sciences, Towson University, Towson, MD 21252, USA; Jessica.Canter@usda.gov (J.A.C.); sgalinn1@jhmi.edu (S.E.E.); kristin.peters718@gmail.com (K.M.P.); nthielman@lecom.edu (N.R.J.T.); lara@grysczyk.de (L.G.); pudofe1@students.towson.edu (P.U.)
[2] Mouse Cancer Genetics Program, Center for Cancer Research, National Cancer Institute, National Institutes of Health, Bethesda, MD 20892, USA; carlsonb@dc37a.nci.nih.gov (B.A.C.); hatfielddolph@gmail.com (D.L.H.)
[3] Lake Erie College of Osteopathic Medicine, Erie, PA 16509, USA
[4] Genetics Branch, Center for Cancer Research, National Cancer Institute, National Institutes of Health, Bethesda, MD 20892, USA; yuyun@mail.nih.gov (Y.Y.); caoli@mail.nih.gov (L.C.)
[5] Office of Dietary Supplements, National Institutes of Health, Bethesda, MD 20817, USA; Cindy.Davis2@usda.gov
[6] Brigham and Women's Hospital, Harvard Medical School, Boston, MA 02215, USA; vgladyshev@rics.bwh.harvard.edu
* Correspondence: ptsuji@towson.edu; Tel.: +1-410-704-4117

Abstract: Selenoproteins play important roles in many cellular functions and biochemical pathways in mammals. Our previous study showed that the deficiency of the 15 kDa selenoprotein (*Selenof*) significantly reduced the formation of aberrant crypt foci (ACF) in a mouse model of azoxymethane (AOM)-induced colon carcinogenesis. The objective of this study was to examine the effects of *Selenof* on inflammatory tumorigenesis, and whether dietary selenium modified these effects. For 20 weeks post-weaning, Selenof-knockout (KO) mice and littermate controls were fed diets that were either deficient, adequate or high in sodium selenite. Colon tumors were induced with AOM and dextran sulfate sodium. Surprisingly, KO mice had drastically fewer ACF but developed a similar number of tumors as their littermate controls. Expression of genes important in inflammatory colorectal cancer and those relevant to epithelial barrier function was assessed, in addition to structural differences via tissue histology. Our findings point to *Selenof*'s potential role in intestinal barrier integrity and structural changes in glandular and mucin-producing goblet cells in the mucosa and submucosa, which may determine the type of tumor developing.

Keywords: Selenof; selenium; selenoprotein; colon cancer; inflammation; barrier integrity

1. Introduction

Colon cancer remains the second leading cause of cancer-related deaths in the United States with an estimated 104,270 new cases and 52,980 deaths in 2021 [1]. One of the earliest indicators of colorectal cancer development is the formation of aberrant crypt foci (ACF), which are pre-neoplastic lesions in the form of abnormal tube-like glands in the colorectal lining tissue. The number of ACF is thought to have a strong relationship with the number of tumors formed in the colon [2], with about 20–30% of ACF predicted to develop into tumors. Intestinal inflammation is known to promote colorectal cancer through a variety of different mechanisms [3–5]. These include pro- and anti-inflammatory cytokines, oxidative stress, and even the composition of the intestinal microbiota [6]. Many of these mechanisms are thought to be modulated by dietary selenium [7–9].

Selenium is an essential trace mineral found in many foods commonly consumed in the U.S. diet as organic forms, such as selenocysteine and selenomethionine, and inorganic forms, such as sodium selenite [10,11]. Much of selenium's role in health and disease has been attributed to its incorporation into selenoproteins, which are encoded by 25 different genes in humans and 24 genes in mice [12]. Selenoproteins play crucial roles in cellular processes such as DNA synthesis, apoptosis, and protection from oxidative damage [13–15]. Previous studies have shown an inverse relationship between dietary selenium levels and the risk of colon cancer, as well as the functional role of selenoproteins in colorectal cancer (reviewed in [16]).

Among the many selenoproteins, the 15 kDa selenoprotein (*Selenof*, formerly known as *Sep15* or *Sel15*) is expressed in high levels in liver, prostate, kidney, testis, and brain. It is furthermore expressed at very high levels in colon cancer cells [17,18]. SELENOF's molecular function appears to be in quality control of oxidative protein folding in the endoplasmic reticulum and signaling in the cellular misfolded protein response [15,19–22]. Recently, a function as a molecular gate keeper and redox quality control role for immunoglobulins has been described [23]. However, the physiological functions of SELENOF and its role in human health, particularly in inflammation and colorectal cancer, are not well understood. Human and mouse colon cancer cell lines with a targeted downregulation of the *Selenof* gene have been generated previously. Our findings suggested a role for *Selenof* in cell replication, invasion and metastasis, as well as a potential regulation of interferon (IFN)-γ-mediated signaling pathways [17,18,24]. To investigate the role of *Selenof* in health and disease in vivo, a Selenof-knockout (Selenof-KO) model was created using C57BL/6 mice. Systemic SELENOF expression was inhibited in these mice by the targeted insertion of a transcriptional terminator in exon 2 of the *Selenof* gene [20,25]. To create littermate controls for comparison with these KO mice, heterozygous mice were backcrossed to create a pseudo-wild type (WT) mouse group, as well as a Selenof-KO mouse group from the same set of parents. This preserved any genetic background as well as environmental factors that may influence the development of the animals. These Selenof-KO mice have a typical C57BL/6 morphology with no visible phenotypic abnormalities. They do, however, appear to have increased levels of inflammation in the form of elevated serum interferon (IFN)-γ expression [26], and develop cataracts early in life [20]. Despite the apparent increase in basal inflammation, we showed in a previous study that these Selenof-KO mice produce significantly fewer ACF than littermate control mice when exposed to the colon-specific chemical carcinogen azoxymethane (AOM) [26]. These results agreed with the findings in cell culture, where a targeted down-regulation of *Selenof* expression resulted in a reversal of the colon cancer phenotype: reduced cell proliferation, reduced ability to grow anchorage-independently, with a concomitant increase in expression of IFN-γ-regulated guanylate binding protein (GBP)-1 [17,18,26]. In vivo, the effects were modified by dietary selenium, where Selenof-KO mice showed a modest increase in the number of ACF under conditions of selenium-deficiency [26].

In this subsequent study, we were interested to assess whether Selenof-KO mice were also protected against the development of tumors in an inflammatory colon tumorigenesis model, the possible impact dietary selenium had, and whether the colon cancer-specific signaling mechanisms impacted by *Selenof* could be further elucidated. Therefore, Selenof-KO mice and their wildtype (WT) littermates were injected with AOM and exposed to the inflammatory agent, dextran sulfate salt (DSS), and were compared to untreated controls. The addition of DSS allowed us to observe tumors formed, in addition to the ACF expected from AOM-treatment alone. The number of ACF, tumor incidence and mass, gene expression of cell signaling pathways, and production of serum cytokines were analyzed to examine responses in mice from each group. Various factors thought to contribute to the development of inflammatory colon cancer, including the enzymes responsible for bioactivation of the carcinogen, inflammatory cytokines, and measures of the barrier integrity of the intestinal epithelium, were investigated. The results of this study contribute to understanding the role of *Selenof* in the development of inflammatory colon cancer. This knowledge may be useful in further investigation into human health, where functional single nucleotide polymorphisms for *SELENOF* have been reported [27–29]. The allele frequency of such single nucleotide polymorphisms in the *SELENOF* gene appear to differ by ethnicity [27]. Because the identity of nucleotides at the polymorphic sites has been shown to influence selenocysteine insertion during translation in a selenium-dependent manner, differentially expressed *SELENOF* may influence health outcomes or susceptibility to cancer in specific populations.

2. Results

Post-weaning, male Selenof-KO and WT littermate mice were maintained on a Torula yeast-based diet (Teklad Harlan Laboratories, Madison, WI, USA) with deficient (measured amounts were 0.02 µg/g diet), adequate (0.1 µg/g diet) or high (2.0 µg/g diet) levels of sodium selenite for the duration of the study. Animals were injected subcutaneously with either AOM (10 mg/kg) or saline at six weeks of age, and subsequently subjected to two one-week rounds of drinking water with or without 2% DSS, respectively (Figure S1). Mice were sacrificed after 20 weeks, and tissue samples and serum were collected.

2.1. Growth Metrics

All mice were weighed upon entry into the study, twice weekly thereafter, and sacrificed after 20 weeks. Weight gain (Figure S2a,b) was calculated by subtracting the mass determined at entry into the study from the final mass determined at sacrifice, and analyzed with a 2-way ANOVA followed by Tukey's multiple comparisons (N = 10–12/group). Under selenium-deficient conditions, control Selenof-KO mice gained significantly more weight (mean weight gain = 29.17 g) than control WT mice (mean weight gain = 14.76 g; $p < 0.0001$), and also compared to control Selenof-KO mice on selenium-adequate ($p = 0.0009$) or high selenium ($p = 0.0012$) diets. AOM/DSS treatment affected all mice, as generally a lower weight gain was observed (Figure S2b). Surprisingly, only dietary selenium (ANOVA, $p < 0.0001$) but not *Selenof* genotype (ANOVA, $p = 0.1094$) affected weight gain under these conditions, with a higher weight gain observed in WT mice on a high selenium diet compared to WT mice on a selenium-deficient diet (Tukey's, $p < 0.001$).

Absolute colon length from anus to caecum was greatest in Selenof-KO control mice, which correlated with a greater body mass of these animals. Colon length (cm) was normalized against body mass (g), which was determined at sacrifice to compare relative colon length of the animals. Analyses of these data did not indicate any statistically significant differences in weight-normalized colon lengths among control animals (Figure S2c). However, dietary selenium affected AOM/DSS-treated animals (ANOVA, $p = 0.0003$), wherein WT mice on a selenium-deficient diet had a slightly ($p = 0.0329$) greater relative colon length in comparison to WT mice on a high selenium diet. No such increase was observed in Selenof-KO mice. Spleen mass (g) was also determined and expressed relative to total body mass (g). A trend of higher relative spleen mass was observed in Selenof-KO animals exposed to AOM/DSS (ANOVA, $p = 0.0208$ for genotype; Figure S2f), though post hoc analyses failed to reach statistical significance for individual comparisons. Overall, Selenof-KO mice and their WT littermate controls were very similar in terms of growth metrics, whereas dietary selenium levels appeared to exert a modest influence.

2.2. Aberrant Crypt Foci Formation and Tumorigenesis

Although only a small percentage of ACF are thought to become malignant [30], ACF are much more prevalent in colorectal cancer cases and therefore often regarded as biomarkers for colon tumors [31]. None of the untreated (control) Selenof-KO mice spontaneously developed tumors; however, one Selenof-KO mice on the selenium-deficient diet spontaneously developed one ACF, though no ACF were detected among Selenof-KO mice on selenium-adequate or high-selenium diets (Table 1). Among the untreated (control) WT mice, no spontaneous ACF were detected in the eight mice on selenium-deficient diets (Table 1). However, 25% of mice on selenium-adequate diets developed three and four ACF, respectively, and 22% of mice on high-selenium diets developed one ACF each.

Furthermore, one WT mouse on selenium-adequate diet spontaneously developed two small tumors.

Table 1. Incidence of aberrant crypt foci in untreated (control) and AOM/DSS-treated mice.

Selenium (µg/g Diet)	WT Control [1]	WT AOM/DSS [1]	Selenof-KO Control [1]	Selenof-KO AOM/DSS [1]
0.02	0/8 (0%)	13/15 (86.7%)	1/9 (11.1%)	4/14 (28.6%)
0.1	2/8 (25%)	13/18 (72.2%)	0/9 (0%)	3/11 (27.3%)
2.0	2/9 (22%)	10/13 (76.9%)	0/12 (0%)	6/14 (42.9%)

[1] number of mice with ACF/total number of mice in group (percentage).

Among the AOM/DSS-treated mice, as expected based on our previous observations in AOM-treated mice [26], Selenof-KO mice also formed significantly fewer pre-neoplastic lesions than WT mice when exposed to AOM/DSS, regardless of dietary selenium levels (Table 1, Figure 1a). ACF were detected in 86.7% of WT mice on selenium-deficient diet, 72.2% on selenium-adequate diet, and 76.9% on high-selenium diet, respectively. In contrast, only 28.6%, 27.3%, and 42.9% of Selenof-KO mice developed ACF, respectively. As anticipated, we found that 70–80% of the WT mice exposed to AOM/DSS developed colorectal tumors, with a slightly greater number of mice developing tumors under selenium-deficient conditions. Surprisingly, a similar number of Selenof-KO mice developed tumors, regardless of dietary selenium levels (Figure 1b). Similarly, a comparison of the number of tumors per tumor-bearing animal showed no statistically significant differences between Selenof-KO and WT mice, regardless of dietary selenium levels (Figure 1c). Furthermore, no differences were detected in absolute tumor mass among the various groups. However, it is interesting to note that the average tumor mass in all animals on a selenium-deficient diet was 50–100% greater than in animals with adequate or high selenium levels ($p > 0.05$; Figure 1d), providing support that adequate selenium consumption may be helpful in mitigating diseases such as colorectal cancer (reviewed in [16]). Therefore, it appears that while lack of Selenof still resulted in significantly decreased numbers of pre-neoplastic lesions, the number or size of actual tumors formed was not influenced by Selenof expression.

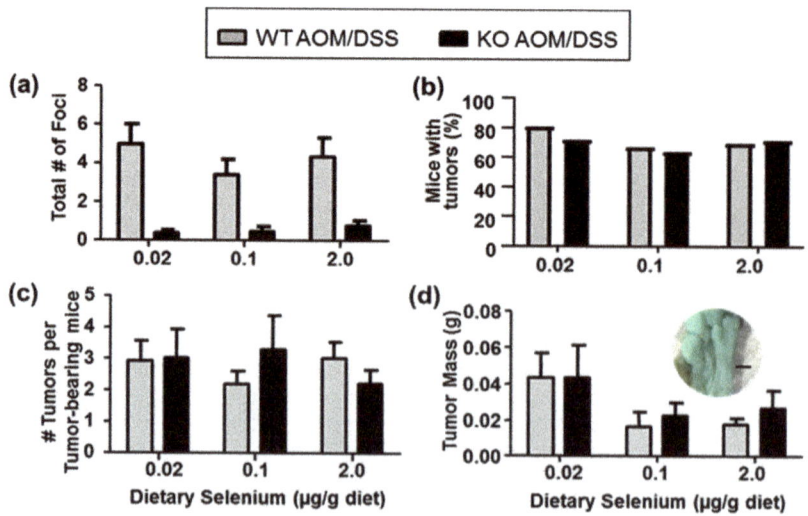

Figure 1. Aberrant crypt foci (ACF) and tumor formation modulated by Selenof genotype and dietary selenium. (a) AOM/DSS-treated Selenof-KO mice formed fewer chemically induced ACF than WT littermates (two-way ANOVA, genotype $p < 0.0001$), independent of dietary selenium levels. (b) Development of macroscopic tumors, and (c) number of tumors per colon in Selenof-KO and WT

mice did not differ. (**d**) Tumor mass appeared slightly modified by dietary selenium (two-way ANOVA, p = 0.117); insert shows stained colon tissue of WT mouse with tumors on selenium-deficient diet (size bar indicates 1 mm).

2.3. Expression and Catalytic Activity of Carcinogen-Activating Enzymes

The formation of chemically induced ACF is well established. Bioactivation of AOM occurs primarily in the liver through hydroxylation via hepatic cytochrome P450 (CYP) 2E1. Subsequently, methylazoxymethanol is formed, which can lead to DNA guanine alkylation and formation of persistent DNA adducts in the colon. Alcohol dehydrogenase (ADH1) and UDP-glucuronosyltransferases (UGT) may additionally modify the activation pathway in liver and colon tissues [32,33]. As a result of AOM metabolism, early neoplastic lesions, ACF, appear in colons.

The catalytic activity of CYP2E1 in liver microsomes (two-way ANOVA, N = 8/group) suggested that dietary selenium affected catalytic activity of CYP2E1 in both untreated (Figure 2a) and AOM/DSS-treated (Figure 2b) animals, with a visible decrease in CYP2E1 activity at adequate and high selenium levels. As part of the bioactivation of AOM via CYP2E1 and ADH1, the oxidized product, methylazoxyformaldehyde, through further modifications yields the methyldiazonium ion. In turn, this ion is thought to methylate DNA bases in AOM- and methylazoxymethanol-target tissues and elicit oxidative stress [33,34]. Dietary selenium has also been shown to affect DNA-methylation in various in vitro [35] and in vivo models [35–37], and the effects of *Selenof* status on DNA methylation was unknown. Therefore, we also assessed the global DNA methylation (Figure 2c,d) in hepatic tissues of Selenof-KO mice and WT littermates. As expected based on other studies [37,38], global DNA methylation in liver tissues positively correlated with increasing dietary selenium in our animals albeit differences not being statistically significant (Figure 2c). This trend was no longer detectable in AOM/DSS-treated animals (Figure 2d). Additionally, statistically significant differences between Selenof-KO and WT mice were not detected. We also assessed mRNA expression of hepatic *Cyp2e1*, *Adh1*, and DNA methyltransferase 1 (*Dnmt1*) and 3a (*Dnmt3a*) (Figure S3). Although *Adh1* expression appeared to increase with dietary selenium (2-way ANOVA, p = 0.0041, Figure S3), mRNA expression of AOM-metabolizing enzymes remained largely unaffected by genotype and dietary selenium in control or AOM/DSS-treated animals. Therefore, it appears that overall, the ability to metabolize AOM via the hepatic CYP2E1 pathway only minimally differs between mice with and without functional SELENOF, with an interesting effect of dietary selenium observed.

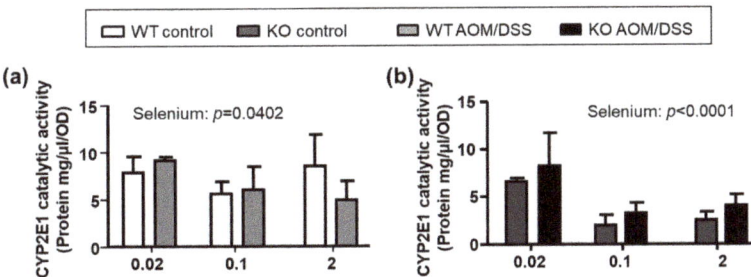

Figure 2. *Cont.*

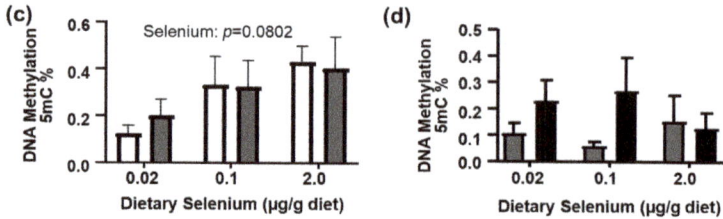

Figure 2. Hepatic AOM-metabolism. Catalytic activity of CYP2E1 in hepatic microsomes was affected by dietary selenium levels, but not by the *Selenof* genotype in (**a**) untreated or (**b**) AOM/DSS treated animals. (**c**) Global 5-mC DNA methylation in liver increased with dietary selenium in control animals. (**d**) AOM/DSS-treated animals displayed greater variability, but no statistically significant differences. Mean (N = 4–8) + SEM, analyzed by 2-way ANOVA, followed by Tukey's post hoc analyses.

Because the metabolism of AOM may continue in colon tissues, where CYP2E1, ADH1 and UGT isoforms process metabolites generated in the liver [32], expression of these genes was assessed in colon scrapes of control animals and in tumors of AOM/DSS-treated animals (Figure S4). The mRNA expression of *Cyp2e1* in colon scrapes of WT and Selenof-KO mice was over 1000-fold less than observed in liver, and were at the limit of detection for AOM/DSS-treated mice on selenium-deficient diets, so we were unable to assess catalytic activity of CYP2E1 in colon tissues. *Cyp2e1* mRNA expression was modestly decreased at high dietary selenium levels in untreated control animals (Figure S4a), and appeared to positively correlate with increasing dietary selenium in colon tumors of AOM/DSS-treated animals (Figure S4b). However, no statistically significant differences were detected for mRNA expression of *Cyp2e1*, *Adh1* (Figure S4c,d) or *Ugt1a* (Figure S4e) in colons among mice with and without *Selenof* expression. *Ugt1a* mRNA levels were below levels of detection in tumors of AOM/DSS-treated mice. Therefore, it appears that the general ability to metabolize AOM does not differ between WT and Selenof-KO mice.

2.4. Serum Inflammatory Markers

Our previous study suggested an increased basal inflammatory state in mice lacking *Selenof* expression [26], especially as it relates to interferon (IFN)-γ and interleukin (IL)-6. Therefore, serum levels of several inflammatory markers were determined using an ELISA-based immunoassay (Figure S5). Overall, modest increases in circulating IFN-γ, IL-10, IL-12p70, IL-1β, C-X-C motif ligand 1 (CXCL1), and tumor necrosis factor (TNF)-α were observed in WT and Selenof-KO mice when treated with AOM/DSS in comparison to their untreated controls, respectively. This suggests, that AOM/DSS treatment resulted in a general increase in production of inflammatory cytokines as would be expected. Systemic *Selenof* expression also appeared to impact production of some circulating serum cytokines. Levels of IL-10 (Figure S5c, $p < 0.05$) and IL-1β (Figure S5g, $p > 0.05$) decreased in control Selenof-KO mice, but only under selenium-deficient conditions. Levels of IL-12p70 (Figure S5f, $p < 0.05$) significantly decreased in AOM/DSS treated mice, but only under selenium-deficient conditions, making interpretations difficult. Therefore, it appears that, as expected, both dietary selenium and AOM/DSS treatment impact serum levels of cytokines relevant to inflammation and cancer. However, mice without *Selenof* expression may be showing some sensitivity to selenium-deficiency, where IL-10 was detected in lower amounts in Selenof-KO control mice compared to their WT littermates. Given that IL-10 plays a dual role in tumor development, these results remain inconclusive. Thus, we continued to focus on tissue-specific differences between WT and Selenof-KO mice that might explain the differences in ACF and tumor burden.

2.5. Colorectal Cancer Cell Signaling Pathways

The primary signaling pathway of interest in colorectal cancer development is the canonical Wnt/β-catenin signaling pathway. We quantitatively assessed mRNA expression of the Wnt/β-catenin complex in colon tumors (Figure 3) to assess whether differences in regulation of cell proliferation, invasion, and metastatic potential in colon tumors excised from both WT and Selenof-KO mice could be detected. This included adenomatous polyposis coli (*Apc*), axin1/2, glycogen synthase kinase 3β (*Gsk3β*), casein kinase 1 (*CK1*), β-transducin repeat containing gene (*bTrCP*1), the dishevelled segment polarity protein 1 (*Dvl1*), and the transcription factor T cell factor 1 (*Tcf1*). Systemic expression of *Selenof* had little to no effect on mRNA expression of genes associated with Wnt-signals or the β-catenin signaling/destruction complex. Dietary selenium exerted a very modest effect ($p > 0.05$), with expression of several genes in the Wnt/β-catenin signaling pathway suggesting a negative correlation with dietary selenium.

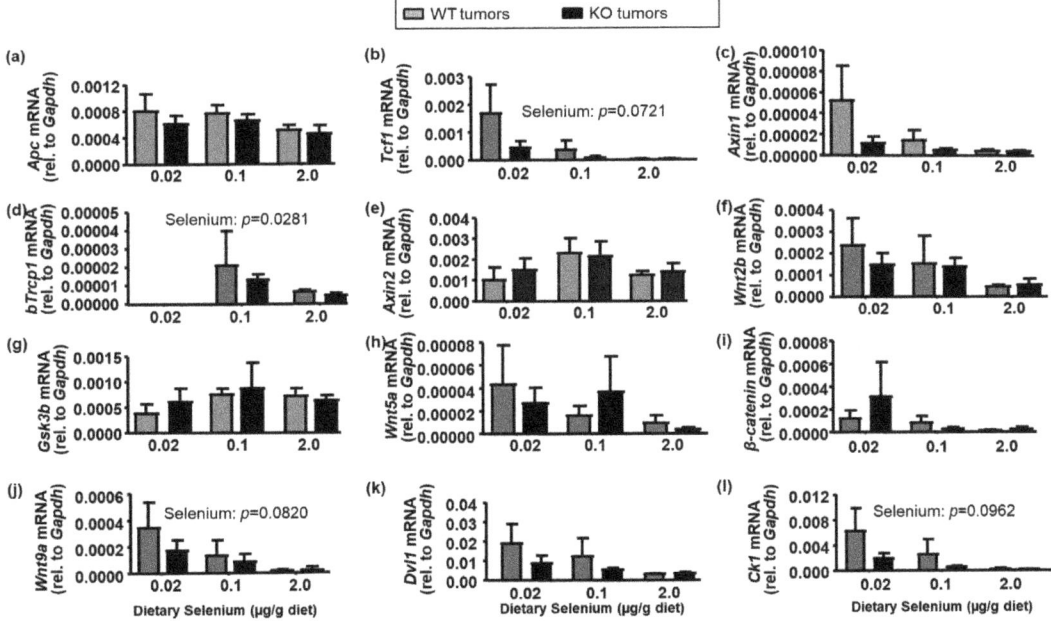

Figure 3. Wnt/β-catenin signaling pathway in colon tumors. mRNA was isolated from colorectal tumors of AOM/DSS-treated mice, reverse-transcribed to cDNA, and quantitatively assessed with gene-specific primers for (**a**) adenomatous polyposis coli (*Apc*), (**b**) T cell factor 1 (*Tcf1*), (**c**) *Axin1*, (**d**) β-transducin repeat containing E3 ubiquitin protein ligase pseudogene 1 (*bTrcp1*), (**e**) Axin 2, (**f**) wingless-type MMTV integration site 2b (*Wnt2b*), (**g**) glycogen synthase kinase 3β (*Gsk3b*), (**h**) *Wnt5a*, (**i**) β-catenin, (**j**) *Wnt9a*, (**k**) human homolog of the Drosophila dishevelled gene (*Dvl1*), and (**l**) casein kinase I (*Ck1*). mRNA levels for *bTrcp1* in selenium-deficient mice were below limits of detection. Means are presented (N = 4 per bar) with SEM, and were analyzed with two-way ANOVA, followed by Tukey's *post hoc* comparisons.

Additionally, we quantitated the expression of downstream targets of the Wnt/β-catenin signaling pathway in tumor tissues. Whereas *Selenof* genotype did not appear to significantly impact mRNA expression of these downstream targets (Figure S6), dietary selenium did in many cases. This was especially evident in the matrix metalloproteinases (*Mmp*7 and *Mmp9*), cyclin D1 (*Ccnd1*), cyclooxygenase 2 (*Cox2*), and the Jun proto-oncogene (*Jun*). Here, insufficient selenium levels resulted in a higher mRNA expression of downstream targets in tumors of AOM/DSS-treated animals.

Several other signaling pathways and molecules are known to directly or indirectly interact with the Wnt signaling pathway. This includes the Nicotinamide adenine dinucleotide phosphate (NADPH) oxidases (*Nox*) and Notch, as well as lysyl oxidase (*Lox*), and Collagen type Iα1 (*Col1a1*). The NOX are transmembrane proteins with diverse physiological functions, including playing roles in cell proliferation [39]. Notch is not only known to activate the WNT/β-catenin signaling pathway, but also interacts with NF-κB, TGFβ and Stat3, which themselves have also been shown to interact with WNT/β-catenin signaling. LOX has been implicated in the inhibition of β-catenin signaling in some cancers [40], and COL1A1 appears upregulated in colorectal cancer tissues and promotes metastasis via Wnt signaling [41]. We therefore assessed mRNA expression of these genes in tumor tissues of AOM/DSS-treated WT and Selenof-KO mice (Figure S7). mRNA expression of *Notch1* modestly correlated negatively with dietary selenium levels ($p = 0.0655$), but no statistically significant differences were observed between tumors of WT or Selenof-KO mice. Similarly, differences between WT or Selenof-KO mice were absent for *Notch2, Nox1, Stat3*, nuclear factor κ-light-chain-enhancer of activated B cells (*NF-κB*), and transforming growth factor β (*Tgfβ,*). *Col1a1* showed a slight increase in Selenof-KO tumors under selenium-deficient conditions (Figure S7), though it failed to reach statistical significance. Overall, we were unable to detect strong differences between Selenof-KO mice and WT controls in canonical signaling pathways relevant to colon carcinogenesis that would possibly have helped explain the dichotomy between ACF and tumor formation in Selenof-KO mice.

2.6. Intestinal Barrier Integrity

Given the very modest changes in expression of the investigated genes and regulatory pathways typically associated with colorectal cancer, we were interested in determining whether Selenof-KO mice exhibited differences in their mucosal morphology and expression of proteins important to barrier integrity instead. Both cross-sectional and longitudinal colon tissue sections of control WT and Selenof-KO animals maintained on adequate selenium diets were prepared with hematoxylin and eosin (H&E, Figure 4a–d) and Masson's Trichrome stains (Figure 4e,f). Although the *muscularis externa* appeared thicker in Selenof-KO mice (Figure 4b,d,f), differences in immune cell infiltration or collagen deposition or fibrosis were not apparent in these samples. However, especially noticeable was the dramatic increase in the size of goblet cells in Selenof-KO mice (Figure 4b,d), suggesting a structural change resulting in ability of increased glycoprotein production for the mucus layer in the intestinal tract.

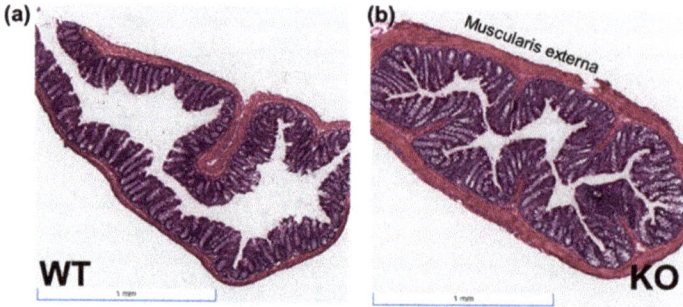

Figure 4. *Cont.*

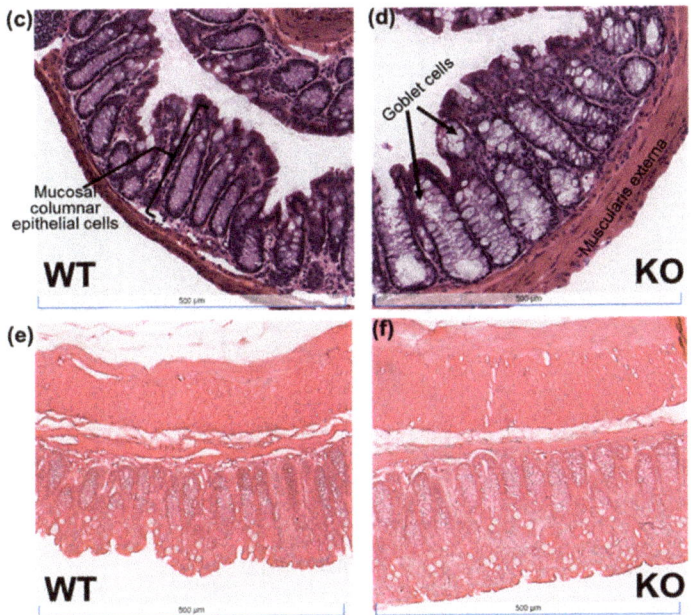

Figure 4. H&E and Masson's Trichrome stains of colon tissues of WT and Selenof-KO animals. Tissue sections of untreated (control) WT and Selenof-KO animals maintained at adequate selenium levels were prepared with (**a–d**) hematoxylin and eosin (H&E) or (**e,f**) Masson's Trichrome stains.

We furthermore investigated the expression of tight junction and other genes known to contribute to intestinal epithelial barrier integrity in colon scrapes of untreated mice, colon tumors of AOM/DSS-treated mice (Figure 5). We did observe a significantly decreased Claudin-1 (*Cldn-1*) mRNA expression in SelenoF-KO mice under high selenium conditions in untreated animals (Figure 5a), a trend that was also seen for Claudin-2 expression (Figure 5d, $p > 0.05$). However, overall, in our in vivo model, the *Selenof* genotype showed little to no effect on mRNA expression of tight junction proteins Claudin-1 (*Cldn-1*), 2 (*Cldn-2*) and 15 (*Cldn-15*). Western blot analyses showed low expression of claudin-2 overall, and no visible differences in protein expression for Claudin-1 or Claudin-3 (Figure 5g) or Claudin-2 (Figure 5h) between WT and KO mice. It should be noted that mRNA expression of these tight junction genes in AOM/DSS-treated animals, interestingly, showed a positive correlation with dietary selenium, with significant impact on expression of *Cldn-2* ($p = 0.0016$) and *Cldn-15* ($p = 0.0008$).

In addition to tight junction genes, we also evaluated the mRNA expression of genes typically associated with adherens junctions and other barrier integrity functions in control animals' colon scrapes and in colon tumor tissues (Figure S8). Dietary selenium levels appeared to affect mRNA expression of the transmembrane glycoprotein epithelial cell adhesion molecule (*EpCAM*), Nectin cell adhesion molecule (*Nectin*)-2, membrane-associated carbonic anhydrase 4 (*Car4*), and the secreted glycoprotein mucin 2 (*Muc2*) in either WT or KO mice, or both. Interestingly, *Selenof*-genotype did not seem to significantly affect mRNA expression of the investigated genes in colons of mice, except for *Epcam*, which was significantly lower in tumors of Selenof-KO mice compared to WT mice, but only at high selenium levels. However, though gene expression of tight junction and adherens junction genes were not significantly altered between Selenof-KO mice and their WT littermates, the dramatically increased size of goblet cells in KO mice suggest structural changes relevant to colon tumorigenesis.

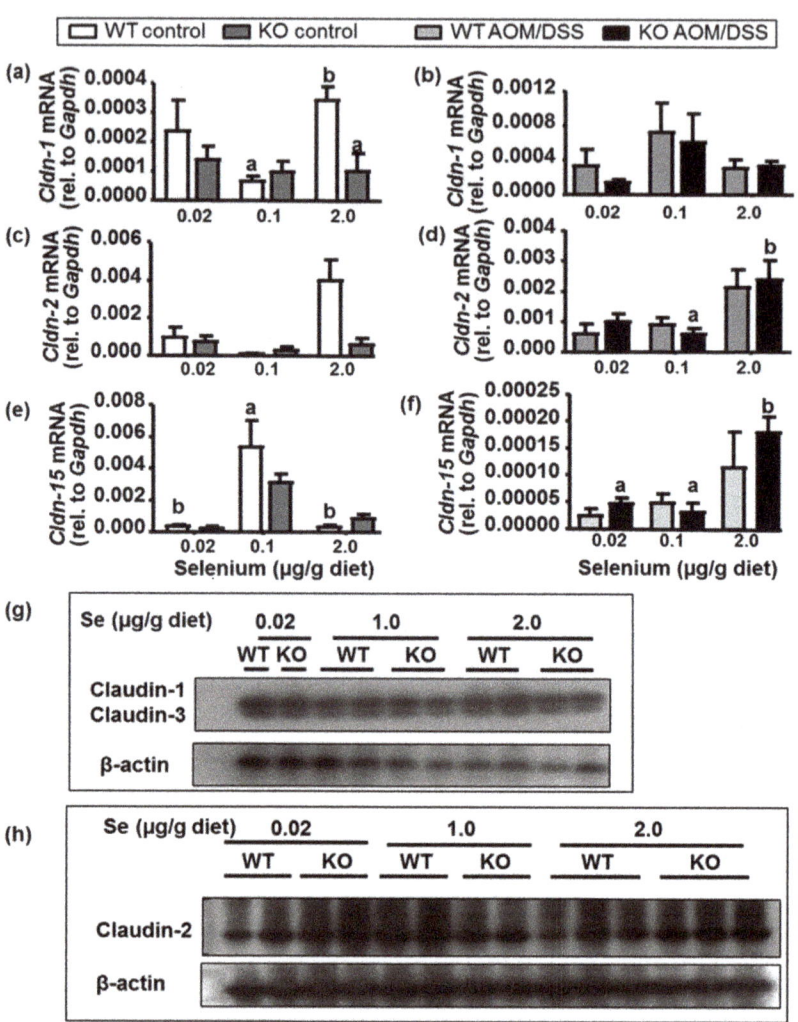

Figure 5. Expression of tight junction Claudin genes. mRNA expression was measured with qPCR in (**a,c,e**) colon scrapes of control mice and (**b,d,f**) colon tumors of AOM/DSS-treated mice. Mean (N = 4) + SEM, 2-way ANOVA, followed by Tukey's post hoc analyses to compare KO vs. WT by diet; letters indicate statistically significant differences. Protein expression of (**g**) Claudin-1 and Claudin-3, and (**h**) Claudin-2 in colon scrapes of control mice on selenium-specific diets was assessed with Western blotting, using β-actin as loading control. N = 1–3 per group, 40 µg protein per lane.

3. Discussion

Previous studies have suggested that the 15 kDa selenoprotein (*Selenof*) is involved in oxidative protein folding, signaling in the cellular misfolded protein response, and may function as a redox quality control for immunoglobulins [15,19–23]. Because functional polymorphisms of *SELENOF* exist in human populations, and have been linked to cancer outcomes, the function of this gene/protein is of great interest. Our previous in vitro and in vivo studies using colon cancer cells with a targeted down-regulation of *SelenoF* or using a systemic Selenof-KO mouse model, saw a reversal of the colon cancer phenotype [18,26] and a decreased number of chemically induced pre-neoplastic lesions [26], respectively. Herein, we evaluated the ability of Selenof-KO mice to develop tumors in an inflammatory colon tumorigenesis model, the impact of dietary selenium, and signaling mechanisms important to colorectal cancer.

Selenof-KO mice and their WT littermate controls were maintained on a selenium-deficient, selenium-adequate, or high selenium diet, were injected with the colon-specific carcinogen AOM, and exposed to the inflammatory tumor promoting agent, DSS. As expected, based on our previous in vivo study, Selenof-KO mice developed dramatically fewer ACF than WT mice, regardless of dietary selenium levels provided. Surprisingly, whereas roughly 60% of the WT mice as expected developed visible tumors in their colons, a similar percentage of Selenof-KO mice also developed tumors. This prompted our investigations to elucidate potential differences in carcinogen metabolism, cell signaling mechanisms relevant to tumorigenesis, and barrier functions specific to intestinal homeostasis.

AOM is well-known to be metabolized by the hepatic CYPE1 machinery. The resulting AOM bioactivation enables this chemical to subsequently form guanine adducts in the colon. Therefore, it was reasonable to hypothesize that Selenof-KO mice may have a lower AOM-bioactivating mechanism, and would be forming fewer ACF, but those fewer ACF might more readily develop into raised polyps and tumors. However, based on our assessment of hepatic CYP2E1 catalytic activity of the enzyme, in addition to the mRNA expression of *Cyp2e1* and other metabolizing enzymes involved in the AOM bioactivation, such a difference that would explain this phenomenon was not detected. Therefore, it appears that overall, the ability to metabolize AOM via the hepatic CYP2E1 pathway only minimally differs between mice with and without functional SELENOF, with an interesting effect of dietary selenium observed. The question remains, why and how Selenof-KO mice with very few ACF would develop a similar number and mass of tumors as WT mice with many more ACF under conditions of inflammation. It should be noted that, while ACF have often been used as biomarkers for intestinal tumorigenesis, some ACF have been shown to regress over time, and most dysplastic ACF do not progress to adenomas [42]. It is furthermore possible that Selenof-KO mice and WT mice use two different mechanisms, develop different types of colorectal tumors, and that ACF are not a good predictor for tumorigenesis in organisms with low or lacking *Selenof* expression. Thus, we continued to investigate potential mechanisms explaining this phenomenon.

Our previous study suggested that Selenof-KO mice had an altered basal inflammation status. We similarly observed in this current study a significantly higher spleen/body mass ratio in KO mice, which was exacerbated in mice exposed to AOM/DSS. We therefore assessed levels of circulating serum cytokines. However, while some serum cytokines, such as IL-6 in AOM/DSS-treated animals, were indeed significantly different in Selenof-KO mice compared to WT littermates, the fact that IL-6 can function both in an inflammatory and anti-inflammatory fashion makes interpretations difficult. We therefore focused on colon cancer-specific signaling pathways, and other factors known to impact intestinal homeostasis and therefore colon cancer.

The primary signaling pathway of interest in colorectal cancer development is the canonical Wnt/β-catenin signaling pathway. Wnt signals can activate gene transcription through nuclear localization of cytosolic β-catenin. Cytosolic levels of β-catenin are controlled by the β-catenin-destruction complex, a multimeric assembly which has been well described [43–45]. Loss of function mutations in the tumor suppressor adenomatous polyposis coli protein (APC) pre-disposes to colorectal adenomas and colorectal cancer, and the vast majority of sporadic colon tumors are found to have mutations in APC [45–50]. Stimulation with the Wnt signal typically leads to the nuclear translocation of β-catenin, and through β-catenin binding to transcriptional activators subsequently to expression of genes important in cell proliferation (e.g., cMYC, cyclins) and cell migration (e.g., matrix metalloproteases, e-cadherin, vascular endothelial growth factor (VEGF)). In addition to its nuclear role in regulating cell proliferation via its downstream targets, membrane-associated stable β-catenin is also involved in regulation and coordination of cell–cell adhesion through its responsibility of the anchoring of cadherins as part of mammalian cell adhesion complexes [51,52], thus impacting barrier integrity in intestinal tissues. Additionally, during colon tumorigenesis, the morphology and synthesis of collagen fibers and other proteins present or active in the extracellular matrix are known to change. Expression of matrix metalloproteinases (MMPs) 2 and MMP9, as well as lysyl oxidase (LOX) are deemed important contributors to tumor invasion and metastasis [53], and are linked to the Wnt/β-catenin pathway. *Selenof* expression did not significantly impact mRNA expression of components of the Wnt/β-catenin pathway or of its downstream targets in colon tumors of mice. Interestingly, dietary selenium modulated mRNA expression of various targets, especially in colon tumors. In mice on deficient selenium levels, a higher mRNA expression of downstream targets was observed in tumors of AOM/DSS-treated animals, which correlated with greater tumor mass. This may, at least in part, explain observations in animal models, where selenium deficiency has been shown to affect the Wnt/β-catenin pathway [54], and epidemiological studies, where selenium status inversely correlated with both cancer incidence and mortality [8].

Several other pathways are known to interact with the Wnt/β-catenin signaling pathway. This includes the NADPH oxidases (NOX), which play roles in cell proliferation via generating ROS [39], and the Notch family of receptors that plays a role in tissue homeostasis and metabolism [55] and in regulation of stem cell properties and cell differentiation [56,57]. Furthermore, the signal transducer and activator of transcription (Stat)-3 is a downstream signaling molecule of IL-6, and acts as a transcription factor with various signaling pathways, including Notch, Nox, and Wnt/β-catenin. Lysyl oxidases (LOX), on the other hand, have been implicated in the inhibition of β-catenin signaling in some cancers [40], whereas the COL1A1 protein appears upregulated in colorectal cancer tissues and promotes metastasis via Wnt signaling [41]. Again, we found some interesting trends where dietary selenium seems to negatively correlate with *Notch* and *Nox* expression, potentially explaining, in part, how selenium deficiency may contribute to increased tumorigenesis. However, much like was observed for the Wnt/β-catenin signaling pathway, neither *Lox*, *Col1a1*, *Tgfβ*, nor *NF-κB* mRNA levels were significantly impacted by *Selenof*-genotype in tumor tissues of AOM/DSS-treated WT and Selenof-KO mice.

Therefore, with the major signaling pathways linked to colorectal tumorigenesis unlikely being significantly modulated by *Selenof* expression, we shifted our focus to the intestinal barrier homeostasis. The single cell layer that forms the intestinal epithelial barrier, is held together by various intercellular junctions that control and regulate permeability and homeostasis in the intestinal epithelium via adherens junctions, desmosomes, and apical tight junctions. Multiple important pathways, including Wnt/β-catenin and Notch/Nox signaling pathways, intersect with the regulation or expression of proteins important in regulation of the intestinal epithelial barrier. Among the many barrier proteins, the members of the families of claudins and occludin localize at and are major constituents of tight junction complexes [58]. Occludin is a cytokine-regulated integral membrane protein that induces adhesion [59], and claudins are involved in selectively controlling paracellular movement of ions [60]. Furthermore, the expression of *CLAUDIN-1* and *CLAUDIN-2* appears elevated in inflammatory bowel diseases and is thought to contribute to tumor progression [61,62]. We hypothesized that mice lacking *Selenof* expressed barrier integrity proteins in intestinal tissues differently than their WT littermate controls, which would result in altered expression of enzymes important to remodeling of colon mucosa and submucosa. This, in turn, could potentially impact the response to a colon carcinogen and/or an inflammatory agent, and possibly explain why Selenof-KO mice develop tumors albeit lacking the development of persistent ACF. In our model, mRNA expression of tight junction proteins claudin-1 and -2, appeared substantially lower in Selenof-KO mice under high-selenium conditions. While such a decrease in claudin-2 would suggest a higher barrier integrity, these observed decreases were not found in tumor tissues from these animals. Furthermore, no differences were found for protein expression for tight junction proteins claudin-1 or -2 in mice. Similarly, mRNA expression of genes relating to adherens junctions, such as the transmembrane glycoprotein epithelial cell adhesion molecule (*Epcam*) which contributes to intercellular adhesion [63], or those relating to general barrier integrity showed interesting trends based on dietary selenium, but not based on *Selenof*-expression of the mice.

However, supporting evidence of altered homeostasis in barrier integrity was observed in tissue sections of untreated (control) WT and Selenof-KO animals, which we prepared with hematoxylin and eosin (H&E) or Masson trichrome stains. Given the possible increased basal systemic inflammation as evidenced by frequent splenomegaly in Selenof-KO mice, we anticipated indications of increased fibrosis, but that was not detected in any of the tissues examined. Instead, the most striking difference was the enormous increase in the size of goblet cells in Selenof-KO mice, though it's unclear whether this was with a concomitant increase in number of goblet cells. Regardless, this finding suggests the potential for increased mucin production in Selenof-KO animals. Mucins are high molecular weight transmembrane glycoproteins that are produced by goblet cells in colonic epithelia, and have been shown to be over-expressed in various cancers, including colorectal cancer. Among these mucins, Mucin-2 (*MUC-2*) is the major secreted form, shown to be expressed by intestinal adenomas and especially by mucinous carcinomas [64,65]. In vivo studies demonstrated that high mucin variant cells injected into nude mice formed twice as large tumors as those of parental cells [65]. In parallel, patients with mucin-producing colorectal cancer appear to have a poor prognosis in terms of outcome. In our study, though we had expected to be able to detect an increase in *Muc-2* transcription in Selenof-KO mice, given the observed increase in goblet cell size, no increase in *Muc-2* mRNA was detected in scrapes of colons nor in colon polyps. We recognize that because colon scrapes constitute a mixture of cell types, it is possible that any changes in gene expression of goblet cells were masked by those of other cell types in the samples. Based on our findings with increased goblet cell sizes in Selenof-KO mice, we furthermore acknowledge the potential role of intestinal microbiota that might contribute to or be the result of potential differences in mucus composition, and therefore impact intestinal barrier integrity. While this is currently beyond the scope of this manuscript, we are looking forward to investigating this in future studies.

Our previous studies suggested that CT-26 mouse colon cancer cells lacking SELENOF displayed limitations in terms of invasion, metastasis, and cell replication [17]. However, these mouse colon cancer cells, though being adenocarcinoma cells, are not considered mucinous carcinoma cells. Mucinous carcinoma cells generally possess a higher degree of invasiveness [66]. Therefore, colorectal cancer cells that are derived from adenocarcinoma cells with low *Selenof* expression may be less aggressive or invasive as we had observed in vitro [17,18]. Whether mucinous carcinoma cells would respond similarly to changes in Selenof-expression remains to be elucidated. Because it has been shown that the predominant mechanisms of tumor progression differ between mucinous carcinoma cells and colorectal adenocarcinoma cells [66], these differences in cell types from which tumors and pre-neoplastic lesions can develop, may explain why Selenof-KO mice appear to be protected initially against ACF formation, but not AOM/DSS-induced tumorigenesis. Therefore, our study showed that Selenof-KO developed tumors in an AOM/DSS-model of colon carcinogenesis, albeit forming dramatically fewer aberrant crypt foci than observed in WT animals. Our main findings showed structural changes in the intestinal tissues of Selenof-KO mice that suggest an altered intestinal barrier integrity.

4. Materials and Methods

4.1. Materials

NuPage® 4–12% polyacrylamide gels, LDS sample buffer, See-Blue Plus2 protein markers, and TRIzol® reagent were purchased from Invitrogen (Carlsbad, CA, USA); iScript™ cDNA synthesis Kit and SYBR™ green supermix from Bio-Rad Laboratories (Hercules, CA, USA), primers for real-time PCR from Integrated DNA Technologies (Coralville, IA, USA). Antibodies against Claudin-1 (which also recognizes Claudin-3) and -2 were purchased from ThermoFisher Scientific (Waltham, MA, USA). Goat polyclonal actin primary antibody, and horseradish peroxidase-conjugated secondary antibody were obtained from

Santa Cruz Biotechnology (Santa Cruz, CA, USA), and SuperSignal West Dura substrate from Pierce (Rockford, IL, USA). A mouse TH1/TH2 9-Plex assay kit was purchased from MesoScale Discovery (Gaithersburg, MD, USA). All other reagents used were commercially available and were of the highest quality available.

4.2. Animal Care Disclosure and Study Organization

All mice used in this experiment were maintained at the National Cancer Institute (National Institutes of Health (NIH)) and were handled and sacrificed in a humane manner in strict accordance with the recommendations in the Guide for the Care and Use of Laboratory Animals of the NIH in Bethesda, Maryland. The Animal Ethics Committee at the NIH previously approved these experiments with proper permit documentation (LCP-011) obtained from the Institutional Animal Care and Use Committee, and documents are on file both at the NIH and at Towson University. Selenof-KO mice (KO) lacking exon 2 of the gene and thus lacking the functional SELENOF protein were generated as described previously [20], and only male Selenof-KO mice and littermate controls (WT) were used to eliminate sex as a variable. Genotypes of the animals were verified by PCR using the following primers: WT allele detection (250 bp): 59-CAGAGTTTGCGTCAGAGGCA-TGCAGAG-39 and 59-CTGAAACTCGTAAAGTCAGAGACTACTGG-39; KO allele detection (312 bp): 59-GGTGTGTTTGCAGATAAGCTAATGC-39 and 59-TACCCGGTAGAATTGACCTGCAG-39.

Weanling mice of both genotypes were weighed, and randomly assigned to be fed a Torula yeast-based customized chow with sodium selenite at 0.02 µg/g diet (selenium-deficient), 0.1 µg/g diet (selenium-adequate), or 2.0 µg/g diet (high-selenium) for the duration of the study (Figure S1). Animals were given free access to deionized water and were monitored closely for any clinical signs of poor health throughout the study. Animals were subcutaneously injected with either azoxymenthane (AOM, 10 mg/kg solubilized in ~100 µL saline) or saline only (controls) at six weeks of age, after having been fed their respective selenium-specific diets for three weeks. At seven weeks of age, AOM-injected mice were given two one-week treatments with 2% dextran sulfate sodium (DSS) via their drinking water separated by a one-week recovery period. All mice were weighed twice weekly for the first 10 weeks in the study, and every other week thereafter. At ten weeks, all mice were maintained on regular drinking water alongside their respective customized selenium diet until the end of the study at 20 weeks (Figure S1). Mice were sacrificed using CO_2 asphyxiation. Animals were weighed, tissues (after determining organ weights) and serum were harvested, flash frozen, and stored at $-80\ ^\circ$C for subsequent analyses.

4.3. Colorectal Tumor and ACF Analyses

Colons from all animals were excised from anus to caecum and rinsed with sterile Dulbecco's phosphate-buffered saline (DPBS). Each colon was measured from anus to caecum in centimeters with a ruler, accurate to one millimeter, opened longitudinally, and stored in 70% ethanol or 10% formalin for subsequent analysis, unless the tissue was used for gene expression analysis. Tumor formation was measured by two independent examiners, counting the total number of tumors formed in each colon using a dissecting microscope. A select number of tumors were excised prior to tissue fixation, the mass of each tumor was determined using a digital scale accurate to 10^{-4} g, and flash frozen for gene expression analyses. To quantitate formation of ACF, ethanol-stored colonic tissues were stained with methylene blue (1 g/L in DPBS) and examined using a dissecting microscope by an examiner blinded to the animal's genotype or treatment to avoid any detection bias. The means were calculated for tumor number, tumor mass, and number of ACF formed in each genotype and treatment group.

4.4. Tissue Staining

Colon tissues of untreated animals were embedded into paraffin and sectioned with a microtome and fixed to glass slides. Subsequently, sections were dewaxed with xylene, washed with ethanol, rinsed with water, and stained with either haemotoxylin and eosin

(H&E) to identify acidic structures like nuclei blue and basic structures such as cytoplasm pink, or Masson's Trichrome (MTC) to stain cytoplasm and muscle fibers red, and collagen with aniline blue. Slides were scanned using Johns Hopkins Medical Institute's Oncology Tissue Services, and images were evaluated by three independent observers.

4.5. Gene Expression Analysis of Mouse Liver and Colon Tissues

For subsequent real-time RT-PCR, total RNA was isolated from liver and colon tissues using the TRIzol (Thermo Fisher Scientific, Carlsbad, CA, USA) reagent following the manufacturer's recommendation, and reverse transcribed using the iScript cDNA synthesis kit (BioRad, Herkules, CA, USA) with 1 µg of total RNA. Gene expression was assessed via real-time RT-PCR using iTaq Universal SYBR Green Supermix (BioRad, Herkules, CA, USA) according to the manufacturer's instructions in 10 µL reactions. mRNA expression was normalized to the expression of glyceraldehyde-3-phosphate dehydrogenase (*Gapdh*).

For Western blotting analyses, colon scrapes were homogenized in lysis buffer with protease inhibitors. Extracted cell lysates were prepared for denaturing gel electrophoresis using NuPAGE LDS 4x sample buffer, heated at 70 °C for 10 min, and 40 µg protein/lane were electrophoresed on NuPAGE 4–12% Bis-Tris polyacrylamide gels. Subsequently, proteins were transferred to polyvinylidene difluoride membranes, and the membranes were blocked in 1% bovine serum albumin in Tris-buffered saline with 0.1% Tween 20 (TBST) for a minimum of 1 h. Membranes were incubated with primary antibodies against Claudin-1 or Claudin-2 for a minimum of one h (1:1000), and then washed in TBST for 10 min three times. Horseradish peroxidase-conjugated secondary antibody (1:10,000) was applied for two h, and the membranes were incubated in Pierce chemiluminescent substrate (ThermoFisher Scientific, Carlsbad, CA, USA) and exposed to X-ray film for detection.

4.6. Cyp2e1 Catalytic Activity Assay

Liver microsomes were isolated following Schenkman and Cinti's protocol [67]. Briefly, liver tissues were homogenized in 0.25 M sucrose in 10 mM Tris-chloride (pH 7.4) and centrifuged at 12,000× g. CaCl$_2$ (8.0 mM final concentration) was added, and microsomes were pelleted via centrifugation at 25,000× g for 15 min and resuspended in 50–75 µL 10 mM KPi/125 mM KCl buffer. CYP2E1 enzyme activity was measured after the modified protocol of Cederbaum [68], using 0.2–0.5 mg microsomal protein and para-nitrophenol to detect formation of para-nitrocatechol at 37 °C. Reactions were initiated by addition of NADPH (1 mM final concentration), and terminated after 10 min by adding trichloroacetic acid (1% final concentration), as described [68]. Proteins were precipitated via centrifugation, and absorbance at 510 nm of the NaOH-treated supernatant was determined with a VersaMax spectrophotometer (ThermoFisher Scientific, Waltham, MA, USA). Para-nitrocatechol concentrations were determined from the extinction coefficient 9.53 mM^{-1} cm^{-1}.

4.7. Serum Cytokine Analysis

Blood was collected from mice by cardiac puncture at sacrifice and centrifuged in heparinized tubes at 3000× g for five min. Serum was then frozen and stored at −80 °C until further analysis. Using the mouse TH1/TH2 7-Plex assay kit, protein levels of interferon-γ, interleukin (IL)-12p70, IL-6, tumor necrosis factor (TNF)-α, KC/GRO (CXCL1, GROα,), IL-1β, and IL-10 were measured in a sandwich immunoassay format using a SECTOR Imager 2400 per manufacturer's protocol (MesoScale Discovery, Rockville, MD, USA). An eight-point standard curve was used to calculate the concentration of cytokines in each murine serum sample, and all samples and standards were analyzed in duplicate (technical replicates).

4.8. Epigenetic Analyses

Genomic DNA was isolated from liver tissues using FitAmp DNA extraction kits (Epigentek, Farmingdale, NY, USA), and global 5-mC DNA methylation was detected using a MethylFlash colorimetric methylated DNA quantification kit (EpiGentek, Farmingdale, NY, USA) following the manufacturer's protocols, with the percentage of methylated DNA proportional to the optical intensity measured with the VersaMax plate reader. Nuclear extracts from mouse livers were isolated using the EpiQuik Nuclear Extraction Kit (EpiGentek, Farmingdale, NY, USA).

4.9. Statistical Analyses

Unless otherwise indicated, data are presented as means +/− SEM, and group means were analyzed with one-way or two-way ANOVA, as appropriate, using GraphPad Prism (v. 9, GraphPad Software, San Diego, CA, USA), followed by Tukey's post hoc analyses. Levels of significance were set to $\alpha = 0.05$.

5. Conclusions

Systemic expression of the 15 kDa selenoprotein, *Selenof*, has been thought to impact cancers in a tissue-specific manner. Whereas effects of *Selenof*-expression in lung cancer cell lines resulted in minimal effects, the effects of *Selenof* in colorectal cancer appeared to be much more substantial [17,18,24]. However, the mechanism behind the reversal of the cancer phenotype in human and mouse colon cancer cells, as well as the dramatic reduction in chemically induced pre-neoplastic lesions in an in vivo Selenof-KO model remained unclear. Our study showed for the first time that the Selenof-KO mouse is capable of developing large tumors in an AOM/DSS-model of colon carcinogenesis albeit forming dramatically fewer aberrant crypt foci than WT animals. Given that the Selenof-KO mouse does not have a strong phenotype other than the early development of cataracts, it may not be surprising that the molecular mechanism remains elusive. Tight junction and other barrier integrity genes appear to have only minor differences in terms of expression, though we recognize the caveat of having to investigate mixtures of cell types present in colon scrapes that may mask any true differences, which will have to be further elucidated. Our main findings point to *Selenof*'s potential role in intestinal barrier integrity and structural changes in glandular and mucin-producing cells in the mucosa and submucosa. Such goblet cells are integral parts of epithelial surfaces in the intestinal barrier but also at the front of the eyes. It would be tempting to speculate that potential changes in intestinal goblet cells would indicate systemic changes that would also affect conjunctival goblet cells, which secrete soluble mucins for the ocular tear film. However, while a protective function of conjunctival goblet cells for regulating surface immune homeostasis is multi-faceted [69], dysregulation of conjunctival mucins generally does not seem to result in cataract development, which is the phenotype observed in Selenof-KO mice [20]. However, our findings of structural changes in intestinal barrier may be of interest to human health, should single nucleotide polymorphisms in the human *SELENOF* gene result in differential expression or activity of SELENOF in the colon. Whether and how this may be further modulated by dietary selenium intake continues to be an area of further studies.

Supplementary Materials: The Supplementary Materials are available online at https://www.mdpi.com/article/10.3390/ijms221910651/s1.

Author Contributions: Conceptualization, P.A.T., D.L.H., B.A.C., C.D.D. and V.N.G.; methodology, P.A.T., B.A.C., C.D.D., Y.Y. and L.C.; formal analysis, P.A.T., D.L.H., B.A.C. and J.A.C.; investigation, P.A.T., D.L.H., B.A.C., K.M.P., J.A.C., S.E.E., N.R.J.T., L.G., P.U., C.D.D., Y.Y. and L.C.; resources, D.L.H., B.A.C., P.A.T. and L.C.; data curation, P.A.T. and B.A.C.; writing—original draft preparation, P.A.T., J.A.C., B.A.C., D.L.H. and C.D.D.; writing—P.A.T., B.A.C., D.L.H. and V.N.G.; funding acquisition, D.L.H., P.A.T. and C.D.D. All authors have read and agreed to the published version of the manuscript.

Funding: This work was funded by intramural support through the National Institutes of Health's Office of Dietary Supplements, and by Towson University's Fisher College of Science & Mathematics and Department of Biological Sciences to Petra Tsuji; by the Office of Graduate studies in support of

Jessica Cancer; by intramural support though the National Cancer Institute to Dolph Hatfield; and by the National Institutes of Health grants to Vadim Gladyshev.

Institutional Review Board Statement: All mice used in this experiment were maintained at the National Cancer Institute (National Institutes of Health (NIH)) and were handled and sacrificed in a humane manner in strict accordance with the recommendations in the Guide for the Care and Use of Laboratory Animals of the NIH in Bethesda, Maryland. The Animal Ethics Committee at the NIH previously approved these experiments with proper permit documentation (LCP-011) obtained from the Institutional Animal Care and Use Committee, and documents are on file both at the NIH and at Towson University.

Informed Consent Statement: Not applicable.

Data Availability Statement: Data generated during the study are contained within the article and Supplementary Materials.

Conflicts of Interest: The authors declare no conflict of interest.

References

1. American Cancer Society. *Cancer Facts & Figures*; American Cancer Society: Atlanta, GA, USA, 2021.
2. Takayama, T.; Katsuki, S.; Takahashi, Y.; Ohi, M.; Nojiri, S.; Sakamaki, S.; Kato, J.; Kogawa, K.; Miyake, H.; Niitsu, Y. Ab-errant crypt foci of the colon as precursors of adenoma and cancer. *N. Engl. J. Med.* **1998**, *339*, 1277–1284. [CrossRef] [PubMed]
3. Fichtner-Feigl, S.; Kesselring, R.; Strober, W. Chronic inflammation and the development of malignancy in the GI tract. *Trends Immunol.* **2015**, *36*, 451–459. [CrossRef] [PubMed]
4. Grivennikov, S.I. Inflammation and colorectal cancer: Colitis-associated neoplasia. *Semin. Immunopath.* **2013**, *35*, 229–244. [CrossRef] [PubMed]
5. Tuomisto, A.E.; Mäkinen, M.J.; Väyrynen, J. Systemic inflammation in colorectal cancer: Underlying factors, effects, and prognostic significance. *World J. Gastroenterol.* **2019**, *25*, 4383–4404. [CrossRef] [PubMed]
6. Arthur, J.C.; Perez-Chanona, E.; Mühlbauer, M.; Tomkovich, S.; Uronis, J.M.; Fan, T.-J.; Campbell, B.J.; Abujamel, T.; Dogan, B.; Rogers, A.B.; et al. Intestinal Inflammation Targets Cancer-Inducing Activity of the Microbiota. *Science* **2012**, *338*, 120–123. [CrossRef]
7. Barrett, C.W.; Short, S.P.; Williams, C.S. Selenoproteins and oxidative stress-induced inflammatory tumorigenesis in the gut. *Cell. Mol. Life Sci.* **2016**, *74*, 607–616. [CrossRef]
8. Barrett, C.W.; Singh, K.; Motley, A.K.; Lintel, M.K.; Matafonova, E.; Bradley, A.M.; Ning, W.; Poindexter, S.V.; Parang, B.; Reddy, V.K.; et al. Dietary selenium deficiency exacerbates DSS-induced epithelial injury and AOM/DSS-induced tumor-igenesis. *PLoS ONE* **2013**, *8*, e67845. [CrossRef]
9. Duntas, L.H. Selenium and Inflammation: Underlying Anti-inflammatory Mechanisms. *Horm. Metab. Res.* **2009**, *41*, 443–447. [CrossRef]
10. Carlson, B.A.; Lee, B.C.; Tsuji, P.A.; Tobe, R.; Park, J.M.; Schweizer, U.; Gladyshev, V.N.; Hatfield, D.L. Selenocysteine tRNA[Ser]Sec: From nonsense suppressor tRNA to the quintessential consituent in selenoprotein biosynthesis. In *Selenium—Its Molecular Biology and Role in Human Health*; Hatfield, D.L., Schweizer, U., Tsuji, P.A., Gladyshev, V.N., Eds.; Springer Science Business Media, LLC: New York, NY, USA, 2016.
11. Tsuji, P.A.; Davis, C.D.; Milner, J.A. Selenium: Dietary sources and human requirements. In *Selenium—Its Molecular Biology and Role in Human Health*, 3rd ed.; Hatfield, D.L., Berry, M.J., Gladyshev, V.N., Eds.; Springer Science Business Media, LLC: New York, NY, USA, 2012.
12. Turanov, A.A.; Lobanov, A.V.; Fomenko, D.E.; Morrison, H.; Sogin, M.L.; Klobutcher, L.; Hatfield, D.L.; Gladyshev, V.N. Genetic Code Supports Targeted Insertion of Two Amino Acids by One Codon. *Science* **2009**, *323*, 259–261. [CrossRef]
13. Hatfield, D.L.; Carlson, B.A.; Tsuji, P.A.; Tobe, R.; Gladyshev, V.N. Selenium and Cancer. In *Molecular, Genetic, and Nutritional Aspects of Major and Trace Minerals*; Collins, J.F., Ed.; Elsevier: New York, NY, USA, 2016; pp. 463–473.
14. Hatfield, D.L.; Tsuji, P.A.; Carlson, B.A.; Gladyshev, V.N. Selenium and selenocysteine: Roles in cancer, health, and development. *Trends Biochem. Sci.* **2014**, *39*, 112–120. [CrossRef]
15. Tsuji, P.A.; Carlson, B.A.; Lee, B.J.; Gladyshev, V.N.; Hatfield, D.L. Interplay of Selenoproteins and Different Antioxidant Systems in Various Cancers. In *Selenium*; Springer Science and Business Media, LLC: New York, NY, USA, 2016; pp. 441–449.
16. Peters, K.M.; Carlson, B.A.; Gladyshev, V.N.; Tsuji, P.A. Selenoproteins in colon cancer. *Free Radic. Biol. Med.* **2018**, *127*, 14–25. [CrossRef]
17. Irons, R.; Tsuji, P.A.; Carlson, B.A.; Ouyang, P.; Yoo, M.-H.; Xu, X.-M.; Hatfield, D.L.; Gladyshev, V.N.; Davis, C.D. Deficiency in the 15-kDa Selenoprotein Inhibits Tumorigenicity and Metastasis of Colon Cancer Cells. *Cancer Prev. Res.* **2010**, *3*, 630–639. [CrossRef]
18. Tsuji, P.A.; Naranjo-Suarez, S.; Carlson, B.A.; Tobe, R.; Yoo, M.-H.; Davis, C.D. Deficiency in the 15 kDa Selenoprotein Inhibits Human Colon Cancer Cell Growth. *Nutrients* **2011**, *3*, 805–817. [CrossRef] [PubMed]
19. Gladyshev, V.N.; Jeang, K.-T.; Wootton, J.C.; Hatfield, D.L. A New Human Selenium-containing Protein. *J. Biol. Chem.* **1998**, *273*, 8910–8915. [CrossRef]

20. Kasaikina, M.V.; Fomenko, D.E.; Labunskyy, V.M.; Lachke, S.A.; Qiu, W.; Moncaster, J.A.; Zhang, J.; Wojnarowicz, M.W.; Natarajan, S.K.; Malinouski, M.; et al. Roles of the 15-kDa selenoprotein (Sep15) in redox homeostasis and cataract devel-opment revealed by the analysis of Sep15 knockout mice. *J. Biol. Chem.* **2011**, *286*, 33203–33212. [CrossRef] [PubMed]
21. Labunskyy, V.M.; Ferguson, A.D.; Fomenko, D.E.; Chelliah, Y.; Hatfield, D.L.; Gladyshev, V.N. A novel cysteine-rich do-main of Sep15 mediates the interaction with UDP-glucose:glycoprotein glucosyltransferase. *J. Biol. Chem.* **2005**, *280*, 37839–37845. [CrossRef] [PubMed]
22. Labunskyy, V.M.; Yoo, M.-H.; Hatfield, D.L.; Gladyshev, V.N. Sep15, a Thioredoxin-like Selenoprotein, Is Involved in the Unfolded Protein Response and Differentially Regulated by Adaptive and Acute ER Stresses. *Biochemistry* **2009**, *48*, 8458–8465. [CrossRef] [PubMed]
23. Yim, S.H.; Everley, R.A.; Schildberg, F.A.; Lee, S.G.; Orsi, A.; Barbati, Z.R.; Karatepe, K.; Fomenko, D.E.; Tsuji, P.A.; Luo, H.R.; et al. Role of Selenof as a gatekeeper of decreted disulfide-tich glycoproteins. *Cell Rep.* **2018**, *23*, 1387–1398. [CrossRef]
24. Tsuji, P.A.; Carlson, B.A.; Yoo, M.-H.; Naranjo-Suarez, S.; Xu, X.-M.; Esther, A.; Asaki, E.; Seifried, H.E.; Reinhold, W.; Davis, C.D.; et al. The 15 kDa Selenoprotein and Thioredoxin Reductase 1 Promote Colon Cancer by Different Pathways. *PLoS ONE* **2015**, *10*, e0124487. [CrossRef]
25. Kasaikina, M.V.; Hatfield, D.L.; Gladyshev, V.N. Understanding selenoprotein function and regulation through the use of rodent models. *Biochim. Biophys. Acta (BBA) Mol. Cell Res.* **2012**, *1823*, 1633–1642. [CrossRef]
26. Tsuji, P.A.; Carlson, B.A.; Naranjo-Suarez, S.; Yoo, M.-H.; Xu, X.-M.; Fomenko, D.E.; Gladyshev, V.N.; Hatfield, D.L.; Davis, C.D. Knockout of the 15 kDa Selenoprotein Protects against Chemically-Induced Aberrant Crypt Formation in Mice. *PLoS ONE* **2012**, *7*, e50574. [CrossRef]
27. Hu, Y.J.; Korotkov, K.; Mehta, R.; Hatfield, D.L.; Rotimi, C.N.; Luke, A.; Prewitt, T.E.; Cooper, R.S.; Stock, W.; Vokes, E.E.; et al. Distribution and functional consequences of nucleotide polymorphisms in the 3'-untranslated region of the human Sep15 gene. *Cancer Res.* **2001**, *61*, 2307–2310.
28. Méplan, C. Association of Single Nucleotide Polymorphisms in Selenoprotein Genes with Cancer Risk. *Adv. Struct. Saf. Stud.* **2017**, *1661*, 313–324. [CrossRef]
29. Mohammaddoust, S.; Salehi, Z.; Saedi, H.S. SEPP1 and SEP15 gene polymorphisms and susceptibility to breast cancer. *Br. J. Biomed. Sci.* **2017**, *75*, 36–39. [CrossRef] [PubMed]
30. Orlando, F.A.; Tan, D.; Baltodano, J.D.; Khoury, T.; Gibbs, J.F.; Hassid, V.J.; Ahmed, B.H.; Alrawi, S.J. Aberrant crypt foci as precursors in colorectal cancer progression. *J. Surg. Oncol.* **2008**, *98*, 207–213. [CrossRef]
31. Bedenne, L.; Faivre, J.; Boutron, M.C.; Piard, F.; Cauvin, J.M.; Hillon, P. Adenoma—Carcinoma sequence or "de novo" car-cinogenesis? A study of adenomatous remnants in a population-based series of large bowel cancers. *Cancer* **1992**, *69*, 883–888. [CrossRef]
32. Gonzalez, F.J. The 2006 Bernard B. Brodie Award lecture. Cyp2e1. *Drug Metab. Dispos.* **2007**, *35*, 1–8. [CrossRef]
33. Sohn, O.S.; Fiala, E.S.; Requeijo, S.P.; Weisburger, J.H.; Gonzalez, F.J. Differential effects of CYP2E1 status on the metabolic activation of the colon carcinogens azoxymethane and methylazoxymethanol. *Cancer Res.* **2001**, *61*, 8435–8440. [PubMed]
34. Feinberg, A.; Zedeck, M.S. Production of a highly reactive alkylating agent from the organospecific carcinogen methyla-zoxymethanol by alcohol dehydrogenase. *Cancer Res.* **1980**, *40*, 4446–4450.
35. Davis, C.D.; Uthus, E.O.; Finley, J.W. Dietary Selenium and Arsenic Affect DNA Methylation In Vitro in Caco-2 Cells and In Vivo in Rat Liver and Colon. *J. Nutr.* **2000**, *130*, 2903–2909. [CrossRef]
36. Davis, C.D.; Uthus, E.O. Dietary Folate and Selenium Affect Dimethylhydrazine-Induced Aberrant Crypt Formation, Global DNA Methylation and One-Carbon Metabolism in Rats. *J. Nutr.* **2003**, *133*, 2907–2914. [CrossRef]
37. Speckman, B.; Schulz, S.; Hiller, F.; Hesse, D.; Schumacher, F.; Kleuser, B.; Geisel, J.; Obeid, R.; Grune, T.; Kipp, A.P. Sele-nium increases hepatic DNA methylation and modulates one-carbon metabolism in the liver of mice. *J. Nutr. Biochem.* **2017**, *48*, 112–119. [CrossRef]
38. Jablonska, E.; Reszka, E. Selenium and Epigenetics in Cancer: Focus on DNA Methylation. In *Advances in Cancer Research*; Elsevier: Amsterdam, The Netherlands, 2017; Volume 136, pp. 193–234.
39. Kajla, S.; Mondol, A.S.; Nagasawa, A.; Zhang, Y.; Kato, M.; Matsuno, K.; Yabe-Nishimura, C.; Kamata, T. A crucial role for Nox 1 in redox-dependent regulation of Wnt-β-catenin signaling. *FASEB J.* **2012**, *26*, 2049–2059. [CrossRef]
40. Liu, F.; Zuo, Z.; Liu, Y.; Deguchi, Y.; Moussali, M.; Chen, W.; Yang, P.; Wei, B.; Tan, L.; Lorenzi, P.L.; et al. Suppression of membranous LRP5 recycling, Wnt/β-catenin signaling, and colon carcinogenesis by 15-LOX-1 oeroxidation of linoleic acid in PI3P. *Cell Rep.* **2020**, *32*, 108049. [CrossRef]
41. Zhang, Z.; Wang, Y.; Zhang, J.; Zhong, J.; Yang, R. COL1A1 promotes metastasis in colorectal cancer by regulating the WNT/PCP pathway. *Mol. Med. Rep.* **2018**, *17*, 5037–5042. [CrossRef]
42. Clapper, M.L.; Chang, W.-C.L.; Cooper, H.S. Dysplastic Aberrant Crypt Foci: Biomarkers of Early Colorectal Neoplasia and Response to Preventive Intervention. *Cancer Prev. Res.* **2020**, *13*, 229–240. [CrossRef]
43. Nakamura, T.; Hamada, F.; Ishidate, T.; Anai, K.; Kawahara, K.; Toyoshima, K.; Akiyama, T. Axin, an inhibitor of the Wnt signalling pathway, interacts with beta-catenin, GSK-3beta and APC and reduces the beta-catenin level. *Genes Cells* **1998**, *3*, 395–403. [CrossRef] [PubMed]
44. Spink, K.E.; Fridman, S.G.; Weis, W.I. Molecular mechanisms of β-catenin recognition by adenomatous polyposis coli re-vealed by the structure of an APC–β-catenin complex. *EMBO J.* **2001**, *20*, 6203–6212. [CrossRef]

45. MacDonald, B.T.; Tamai, K.; He, X. Wnt/β-catenin signaling: Components, mechanisms, and diseases. *Dev. Cell* **2009**, *17*, 9–26. [CrossRef] [PubMed]
46. Kinzler, K.W.; Vogelstein, B. Lessons from Hereditary Colorectal Cancer. *Cell* **1996**, *87*, 159–170. [CrossRef]
47. Segditsas, S.; Tomlinson, I. Colorectal cancer and genetic alterations in the Wnt pathway. *Oncogene* **2006**, *25*, 7531–7537. [CrossRef] [PubMed]
48. Fodde, R. The APC gene in colorectal cancer. *Eur. J. Cancer* **2002**, *38*, 867–871. [CrossRef]
49. Najdi, R.; Holcombe, R.F.; Waterman, M.L. Wnt signaling and colon carcinogenesis: Beyond APC. *J. Carcinog.* **2011**, *10*, 5. [CrossRef] [PubMed]
50. Markowitz, S.D.; Bertagnolli, M.M. Molecular origins of cancer: Molecular basis of colorectal cancer. *N. Engl. J. Med.* **2009**, *361*, 2449–2460. [CrossRef] [PubMed]
51. Bienz, M. β-Catenin: A Pivot between Cell Adhesion and Wnt Signalling. *Curr. Biol.* **2005**, *15*, R64–R67. [CrossRef]
52. Tian, X.; Liu, Z.; Niu, B.; Zhang, J.; Tan, T.K.; Lee, S.R.; Zhao, Y.; Harris, D.C.H.; Zheng, G. E-Cadherin/β-Catenin Complex and the Epithelial Barrier. *J. Biomed. Biotechnol.* **2011**, *2011*, 1–6. [CrossRef]
53. Park, P.-G.; Jo, S.J.; Kim, M.J.; Kim, H.J.; Lee, J.H.; Park, C.K.; Kim, H.; Lee, K.Y.; Kim, H.; Park, J.H.; et al. Role of LOXL2 in the epithelial-mesenchymal transition and colorectal cancer metastasis. *Oncotarget* **2017**, *8*, 80325–80335. [CrossRef]
54. Kipp, A.; Banning, A.; van Schothorst, E.M.; Méplan, C.; Schomburg, L.; Evelo, C.; Coort, S.; Gaj, S.; Keijer, J.; Hesketh, J.; et al. Four selenoproteins, protein synthesis, and Wnt signaling are particularly sensitive to limited selenium intake in mouse colon. *Mol. Nutr. Food Res.* **2009**, *53*, 1561–1572. [CrossRef]
55. Adams, J.M.; Jafar-Nejad, H. The Roles of Notch Signaling in Liver Development and Disease. *Biomolecules* **2019**, *9*, 608. [CrossRef]
56. Miyamoto, S.; Rosenberg, D.W. Role of Notch signaling in colon homeostasis and carcinogenesis. *Cancer Sci.* **2011**, *102*, 1938–1942. [CrossRef]
57. Fazio, C.; Ricciardiello, L. Inflammation and Notch signaling: A crosstalk with opposite effects on tumorigenesis. *Cell Death Dis.* **2016**, *7*, e2515. [CrossRef]
58. Furuse, M.; Fujita, K.; Hiiragi, T.; Fujimoto, K.; Tsukita, S. Claudin-1 and -2: Novel Integral Membrane Proteins Localizing at Tight Junctions with No Sequence Similarity to Occludin. *J. Cell Biol.* **1998**, *141*, 1539–1550. [CrossRef] [PubMed]
59. Van Itallie, C.M.; Fanning, A.S.; Holmes, J.; Anderson, J. Occludin is required for cytokine-induced regulation of tight junction barriers. *J. Cell Sci.* **2010**, *123*, 2844–2852. [CrossRef] [PubMed]
60. Rosenthal, R.; Milatz, S.; Krug, S.M.; Oelrich, B.; Schulzke, J.-D.; Amasheh, S.; Günzel, D.; Fromm, M. Claudin-2, a compo-nent of the tight junction, forms a paracellular water channel. *J. Cell Sci.* **2010**, *123*, 1913–1921. [CrossRef] [PubMed]
61. Garcia-Hernandez, V.; Quiros, M.; Nusrat, A. Intestinal epithelial claudins: Expression and regulation in homeostasis and inflammation. *Ann. N. Y. Acad. Sci.* **2017**, *1397*, 66–79. [CrossRef] [PubMed]
62. Weber, C.R.; Nalle, S.C.; Tretiakova, M.; Rubin, D.T.; Turner, J.R. Claudin-1 and claudin-2 expression is elevated in in-flammatory bowel disease and may contribute to early neoplastic transformation. *Lab. Investig.* **2008**, *88*, 1110–1120. [CrossRef]
63. Huang, L.; Yang, Y.; Yang, F.; Liu, S.; Zhu, Z.; Lei, Z.; Guo, J. Functions of EpCAM in physiological processes and diseases (Review). *Int. J. Mol. Med.* **2018**, *42*, 1771–1785. [CrossRef]
64. Shan, Y.-S.; Hsu, H.-P.; Lai, M.-D.; Yen, M.-C.; Fang, J.-H.; Weng, T.-Y.; Chen, Y.-L. Suppression of mucin 2 promotes interleukin-6 secretion and tumor growth in an orthotopic immune-competent colon cancer animal model. *Oncol. Rep.* **2014**, *32*, 2335–2342. [CrossRef] [PubMed]
65. Kuan, S.F.; Byrd, J.C.; Basbaum, C.B.; Kim, Y.S. Characterization of quantitative mucin variants from a human colon cancer cell line. *Cancer Res.* **1987**, *47*, 5715–5724. [PubMed]
66. Cho, M.; Dahiya, R.; Choi, S.; Siddiki, B.; Yeh, M.; Sleisenger, M.; Kirn, Y. Mucins secreted by cell lines derived from colorectal mucinous carcinoma and adenocarcinoma. *Eur. J. Cancer* **1997**, *33*, 931–941. [CrossRef]
67. Schenkman, J.B.; Cinti, D.L. [6] Preparation of microsomes with calcium. In *Methods in Enzymology*; Elsevier: Amsterdam, The Netherlands, 1978; Volume 52, pp. 83–89. [CrossRef]
68. Cederbaum, A.I. Methodology to assay CYP2E1 mixed function oxidase catalytic activity and its induction. *Redox Biol.* **2014**, *2*, 1048–1054. [CrossRef] [PubMed]
69. Swamynathan, S.K.; Wells, A. Conjunctival goblet cells: Ocular surface functions, disorders that affect them, and the potential for their regeneration. *Ocul. Surf.* **2020**, *18*, 19–26. [CrossRef] [PubMed]

Article

Compensatory Protection of Thioredoxin-Deficient Cells from Etoposide-Induced Cell Death by Selenoprotein W via Interaction with 14-3-3

Hyunwoo Kang, Yeong Ha Jeon †, Minju Ham, Kwanyoung Ko *,‡ and Ick Young Kim *

Laboratory of Cellular and Molecular Biochemistry, Department of Life Sciences, Korea University, Seoul 02841, Korea; khw094@korea.ac.kr (H.K.); yeongha0820@cellabmed.com (Y.H.J.); minjham@hotmail.com (M.H.)
* Correspondence: kko2@mgh.harvard.edu (K.K.); ickkim@korea.ac.kr (I.Y.K.)
† Present address: CellabMED Inc. 1301-1, Seoul 08376, Korea.
‡ Present address: Molecular Neurogenetics Unit, Center for Genomic Medicine, Massachusetts General Hospital, and Department of Neurology, Harvard Medical School, Boston, MA 02114, USA.

Abstract: Selenoprotein W (SELENOW) is a 9.6 kDa protein containing selenocysteine (Sec, U) in a conserved Cys-X-X-Sec (CXXU) motif. Previously, we reported that SELENOW regulates various cellular processes by interacting with 14-3-3β at the U of the CXXU motif. Thioredoxin (Trx) is a small protein that plays a key role in the cellular redox regulatory system. The CXXC motif of Trx is critical for redox regulation. Recently, an interaction between Trx1 and 14-3-3 has been predicted. However, the binding mechanism and its biological effects remain unknown. In this study, we found that Trx1 interacted with 14-3-3β at the Cys32 residue in the CXXC motif, and SELENOW and Trx1 were bound at Cys191 residue of 14-3-3β. In vitro binding assays showed that SELENOW and Trx1 competed for interaction with 14-3-3β. Compared to control cells, Trx1-deficient cells and SELENOW-deficient cells showed increased levels of both the subG1 population and poly (ADP-ribose) polymerase (PARP) cleavage by etoposide treatment. Moreover, Akt phosphorylation of Ser473 was reduced in Trx1-deficient cells and was recovered by overexpression of SELENOW. These results indicate that SELENOW can protect Trx1-deficient cells from etoposide-induced cell death through its interaction with 14-3-3β.

Keywords: selenoprotein W; thioredoxin; 14-3-3; Akt; cell death

Citation: Kang, H.; Jeon, Y.H.; Ham, M.; Ko, K.; Kim, I.Y. Compensatory Protection of Thioredoxin-Deficient Cells from Etoposide-Induced Cell Death by Selenoprotein W via Interaction with 14-3-3. *Int. J. Mol. Sci.* **2021**, *22*, 10338. https://doi.org/10.3390/ijms221910338

Academic Editor: Gautam Sethi

Received: 10 September 2021
Accepted: 22 September 2021
Published: 25 September 2021

Publisher's Note: MDPI stays neutral with regard to jurisdictional claims in published maps and institutional affiliations.

Copyright: © 2021 by the authors. Licensee MDPI, Basel, Switzerland. This article is an open access article distributed under the terms and conditions of the Creative Commons Attribution (CC BY) license (https://creativecommons.org/licenses/by/4.0/).

1. Introduction

Thioredoxin (Trx) is a ubiquitous, antioxidant protein containing a highly conserved Cys-X-X-Cys (CXXC) motif and plays an important role in the redox regulatory system [1]. Trx can reduce the oxidized Cys residues of target proteins, and the oxidized Trx is then reduced by thioredoxin reductase (TrxR) [2]. Trx interacts with various proteins involved in the regulation of cellular signaling pathways, including cell survival and proliferation. For example, Trx binds to phosphatase and tensin homolog (PTEN) and inhibits the lipid phosphatase activity of PTEN, which results in increased Akt activity in the cells [3]. Apoptosis signal-regulating kinase (ASK), which promotes apoptosis, is also a target protein of Trx. Ask1 activity is inhibited by Trx1 binding [4,5]. Moreover, Trx1 is known to be highly expressed in cancer cells [6,7].

Trx1 also plays an important role in cell survival [8–11]. Recently, it was suggested that Akt phosphorylation at Ser473 is upregulated by Trx1 [12,13]. Therefore, in Trx1-deficient cells, phosphorylation at Ser473 of Akt was found to be inhibited [14]. It has also been reported that phosphorylation of Akt at Ser473 can be regulated by the interaction between the 14-3-3 protein and Rictor, a component of the mechanistic target of rapamycin complex 2 (mTORC2) [15,16], and that the binding of 14-3-3β to Rictor is inhibited by the binding of 14-3-3β to selenoprotein W (SELENOW) [17].

SELENOW is the smallest selenoprotein that contains a conserved Cys-X-X-Sec (CXXU) motif, which corresponds to the CXXC redox motif of Trx1 [18]. The intramolecular selenylsulfide bond in the CXXU motif of SELENOW has been recently reported [19,20]. SELENOW is expressed in various tissues; however, it is highly abundant in the brain and muscles of mammals [21]. mRNA and protein expression of SELENOW is upregulated in the early stage of skeletal muscle cell differentiation [22]. Additionally, SELENOW is involved in the protection of cells from oxidative stress-induced cell death in a glutathione-dependent manner [23–27]. SELENOW interacts with the 14-3-3 protein and inhibits its interaction with other target proteins. For example, SELENOW reduced the binding of Rictor to 14-3-3β and increased the phosphorylation of Akt at Ser473 [17]. Downregulation of SELENOW induces cell cycle arrest in cancer cell lines, such as A549 and MCF7 cells, and makes the cells more sensitive to DNA damage by increasing the binding of 14-3-3β to Rictor or CDC25B [17,28]. Etoposide is a cancer chemotherapy drug which induces DNA damage in cells by inhibiting DNA religation activity of topoisomerase II [29]. The subG1 population of DNA damaged cells by etoposide was increased in SELENOW-deficient cells compared to that in control cells [30].

The 14-3-3 protein plays an important role in diverse signaling processes [31]. The function of many cellular proteins is modulated by interactions with the 14-3-3 protein [32–36]. It exists as both homodimers and heterodimers in cells, and dimers can bind to phosphorylated Ser or Thr residues in target proteins [37,38]. However, it has also been reported that the 14-3-3 protein interacts with various target proteins in a phosphorylation-independent manner [39–44]. We previously reported that phosphorylation was not required for the interaction between SELENOW and 14-3-β, and Sec in the CXXU motif of SELENOW was identified as the binding site [30].

Similar to SELENOW, Trx1 is also predicted to bind to 14-3-3 [45,46]. Therefore, in this study, we investigated the interaction of Trx1 with 14-3-3β, the relationship between Trx1 and SELENOW, and the effect of these interactions on cell viability after DNA damage.

2. Results

2.1. Cys32 in the CXXC Motif of Trx1 is Required for Interaction with 14-3-3β

The 14-3-3 protein regulates many signals by interacting with various proteins [32–36]. We previously reported that the interaction between 14-3-3β and SELENOW regulates sensitivity to DNA damage [30]. Sec of the CXXU motif is essential for the interaction. Similar to SELENOW, Trx1 has a conserved CXXC motif, which was also predicted to be a 14-3-3 binding partner [45,46]. To determine the interaction of Trx1 with 14-3-3β, we first performed an immunoprecipitation experiment using HEK293 cells transfected with HA-14-3-3β. Trx1 interacted with 14-3-3β, and this interaction was enhanced by H_2O_2 treatment (Figure 1A,B). To further confirm this interaction, a His-pull-down assay was performed. As Sec13 in the CXXU motif was required for the interaction of SELENOW with 14-3-3β [30], we constructed Trx1-His mutants in which Cys32 or Cys35 in the CXXC motif was changed to Ser. The mutants were designated Trx1(C32S)-His and Trx1(C35S)-His, respectively. The Trx1-His protein and Trx1 mutant proteins were expressed and extracted from *Escherichia coli* BL21(DE3) cells. The purified proteins were confirmed by Coomassie blue staining and Western blot assay (Figure 1C). The purified protein was used as bait in the His-pull-down assay against lysates of HEK293 cells overexpressing HA-14-3-3β. As shown in Figure 1D, Trx1-His interacted with HA-14-3-3β in vitro. Next, we investigated the binding sites of Trx1 for 14-3-3β. Using the purified proteins, a GST-pull-down assay was performed against the purified GST-14-3-3β protein. As shown in Figure 1E, Trx1-His and Trx1(C35S)-His interacted with purified GST-14-3-3β, but not Trx1(C32S)-His. These results indicate that similar to SELENOW, Trx1 also interacts with 14-3-3β and the Cys32 of the CXXC redox motif essential for this interaction.

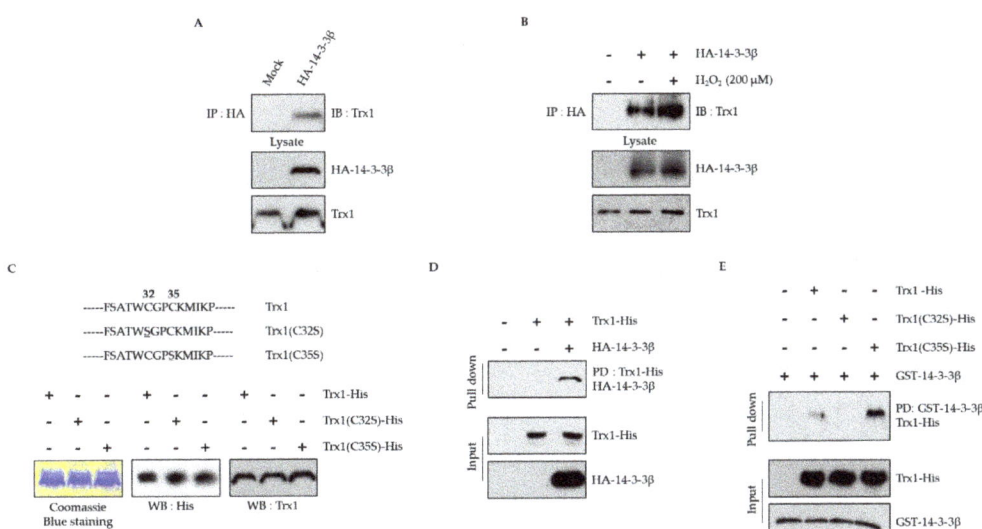

Figure 1. Trx1 interacts with 14-3-3β. (**A**) Human embryonic kidney 293 (HEK293) cells were transfected with HA-14-3-3β, and the cells were harvested 24 h after transfection. The cell extracts were immunoprecipitated with anti-HA antibody and then immunoblotted with anti-Trx1 antibody. (**B**) HEK293 cells were transfected with HA-14-3-3β for 24 h and then treated with H_2O_2 (200 μM) for 30 min. The cell extracts were immunoprecipitated with anti-HA antibody and then immunoblotted with anti-Trx1 antibody. (**C**) Sequences of mutant Trx1. The underlines and numbers indicate the mutation sites. Trx1-His, Trx1(C32S)-His, and Trx1(C35S)-His proteins were purified and verified by Coomassie staining (left) and immunoblotting with anti-His (middle) and anti-Trx1 (right) antibodies. (**D**) Purified Trx1-His protein was incubated with HEK293 cell lysates overexpressing HA-14-3-3β for 1 h. The mixtures were incubated with Ni-NTA beads for 1 h and then immunoblotted with anti-HA or anti-His antibodies. (**E**) Each purified Trx1-His, Trx1(C32S)-His, and Trx1(C35S)-His protein was incubated with purified GST-14-3-3β. The mixtures were incubated with glutathione beads for 1 h and then immunoblotted with anti-His or anti-GST antibodies. The data were obtained from three independent experiments.

2.2. The Cys191 Residue of 14-3-3β Is Identified as the Binding Site of SELENOW and Trx1

Sec13 in the CXXU motif of SELENOW is the binding site for 14-3-3β, and the interaction is redox-regulated [30]. In this study, we determined the binding site of 14-3-3β to SELENOW and Trx1. Human 14-3-3β has two Cys residues at positions 96 and 191 [47]. Mutants for 14-3-3β were constructed in which Cys96 or/and Cys191 were replaced by Ser. The mutants were designated as HA-14-3-3β(C96S), HA-14-3-3β(C191S), and HA-14-3-3β(C96, 191S). The interaction with SELENOW was then investigated via immunoprecipitation using HEK293 cells co-transfected with HA-14-3-3β and His-SELENOW(U13C). As shown in Figure 2A, both HA-14-3-3β(WT) and HA-14-3-3β(C96S) interacted with His-SELENOW(U13C). However, endogenous Trx1 did not interact with either HA-14-3-3β(C191S) or HA-14-3-3β(C96,191S) (Figure 2B). These results indicate that Cys191 of the 14-3-3β is required for interaction with both SELENOW and Trx1.

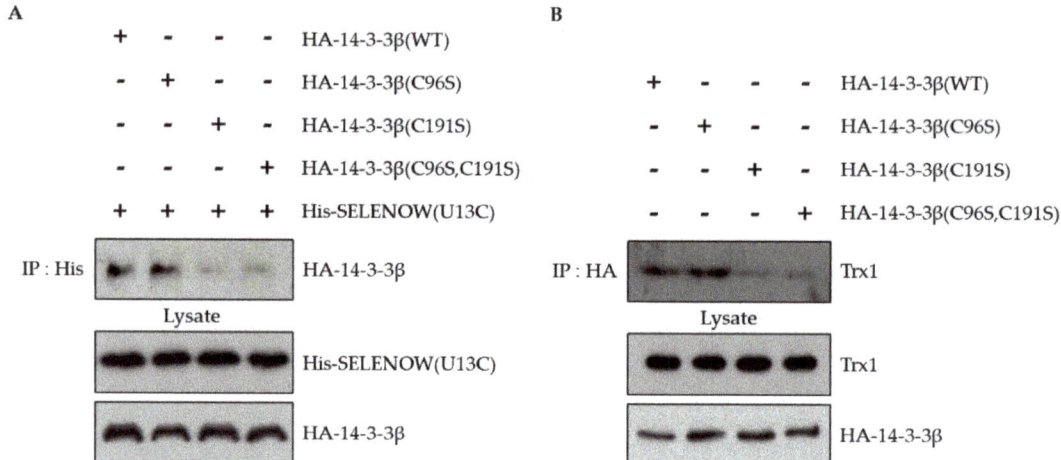

Figure 2. The Cys191 residue of 14-3-3β interacts with SELENOW and Trx1. (**A**) HEK293 cells were co-transfected with His-SELENOW(U13C) and HA-14-3-3β(WT, C96S, C191S or C96,191S). The cell extracts were immunoprecipitated with anti-His antibody, and immunoblotting was performed with anti-HA and anti-His antibodies. (**B**) HEK293 cells were co-transfected with HA-14-3-3β(WT, C96S, C191S or C96,191S). The cell extracts were immunoprecipitated with anti-HA antibody and immunoblotted with anti-Trx1 antibody or anti-HA antibody. The data were obtained from three independent experiments.

2.3. Interaction of 14-3-3β with Trx1 Is Regulated by SELENOW and Vice Versa

Since both SELENOW and Trx1 interacted with the same residue of 14-3-3β, Cys191, we investigated the relationship between these two proteins and their interaction with 14-3-3β. A pull-down assay was performed with purified Trx1-His against the lysates of HEK293 cells overexpressing HA-14-3-3β in different concentrations of purified GST-SELENOW(U13C). As shown in Figure 3A, the binding of Trx1 to 14-3-3β protein decreased with increasing amounts of SELENOW(U13C). A pull-down assay with purified GST-SELENOW(U13C) against the lysates of HEK293 cells overexpressing HA-14-3-3β showed that the interaction between SELENOW and 14-3-3β was also decreased by increasing concentration of Trx1-His in the reaction mixtures (Figure 3B). These results demonstrated that Trx1 and SELENOW may compete to interact with the same site as on the 14-3-3β.

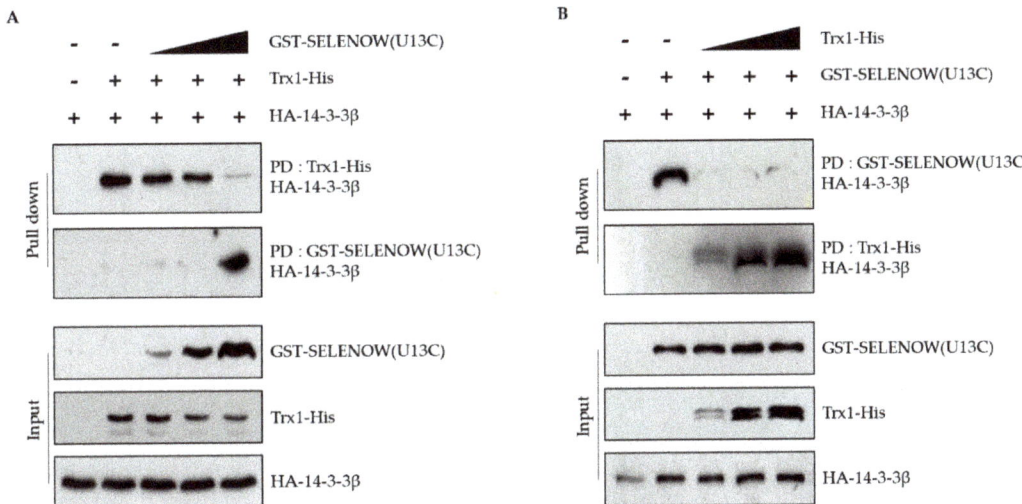

Figure 3. The interaction of Trx1 and SELENOW with 14-3-3β protein is competitive. (**A**) The mixture of purified Trx1-His protein (5 μg) and cell lysates overexpressing HA-14-3-3β (5 μg) was incubated with 0, 1, 2, or 5 μg GST-SELENOW(U13C) for 2 h. The mixtures were then incubated with glutathione or Ni-NTA beads for 1 h. The proteins were separated using non-reducing SDS-PAGE. The separated proteins were immunoblotted with anti-HA antibody. (**B**) The mixture of purified GST-SELENOW(U13C) protein (5 μg) and cell lysates overexpressing HA-14-3-3β (5 μg) was incubated with 0, 1, 2, or 5 μg Trx1-His for 2 h. The mixtures were incubated with glutathione or Ni-NTA beads for 1 h. The proteins were separated using non-reducing SDS-PAGE. The separated proteins were immunoblotted with anti-HA antibody.

2.4. Deficiency of Trx1 and SELENOW Increases the Sensitivity of Cells to Etoposide-Induced Cell Death

A previous report suggested that downregulation of Trx1 increased sensitivity to DNA damage by reducing cyclin D1 expression and phosphorylation of ERK1/2 in A549 and MCF7 cells [48]. It has been reported that the subG1 phase is increased in SELENOW-deficient MCF7 cells. SELENOW-deficient cells were also more sensitive to DNA damage induced by etoposide in A549, T47D, and MCF7 cells. This was observed due to the disruption in the binding of 14-3-3β to Rictor by SELENOW, which regulates phosphorylation of Akt at Ser 473 [17]. Therefore, we next examined the effect of double deficiency of Trx1 and SELENOW on cell survival under etoposide treatment conditions. MCF7 cells were transfected with siSELENOW and/or siTrx1 and then incubated with etoposide for 24 or 48 h to induce DNA damage. The cell cycle distribution was examined using FACS analysis. The subG1 phase population increased in SELENOW-deficient cells as shown in a previous report [17]. In this study, we found that the subG1 population also increased in etoposide treated-Trx1-deficient cells. The subG1 population of SELENOW-deficient cells was higher than that of Trx1-deficient cells (Figure 4A). We have previously shown that poly (ADP-ribose) polymerase (PARP) is rapidly cleaved in SELENOW-deficient cells compared to that in control cells under the DNA damage conditions induced by etoposide [28,30]. In this study, we found that PARP cleavage induced by etoposide also increased in Trx1-deficient cells, and it was further enhanced by double deficiency of SELENOW and Trx1 (Figure 4B). These results suggest that both Trx1 and SELENOW deficiency increase the cell death induced by DNA damage.

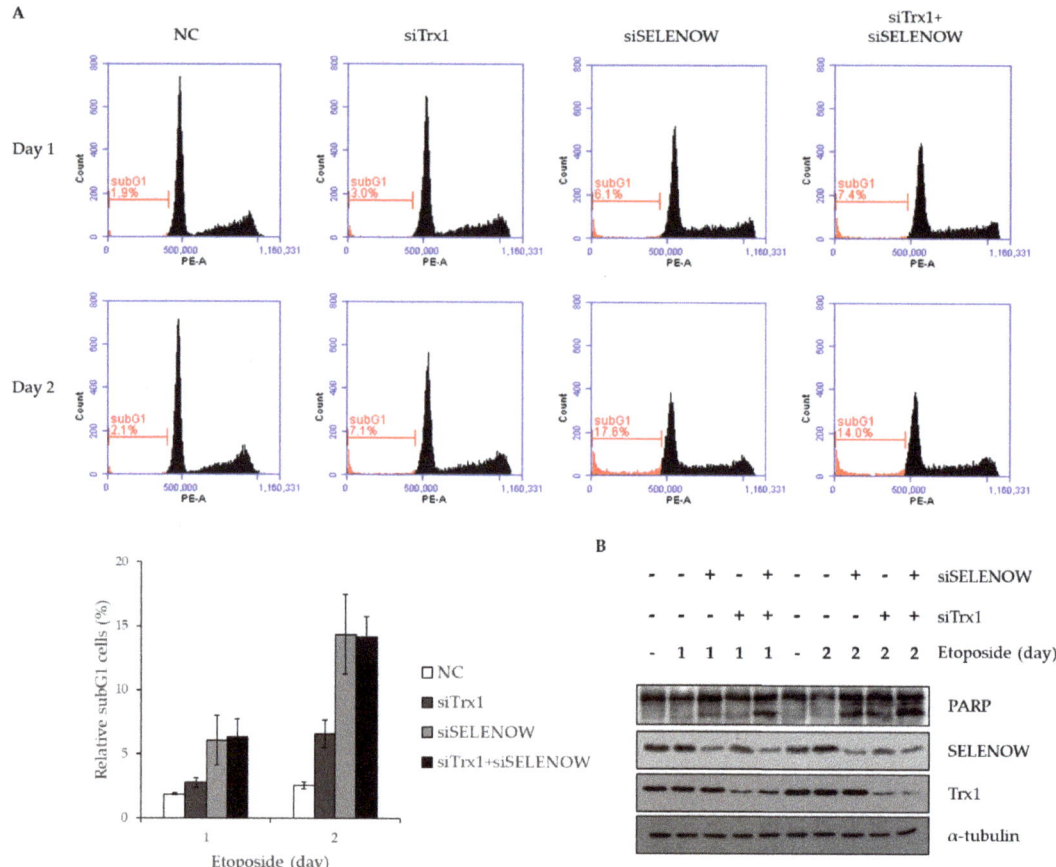

Figure 4. The cell death induced by DNA damage is regulated by the cooperation of SELENOW and Trx1. (**A**) Breast carcinoma MCF7 cells were transfected with siSELENOW or/and siTrx1 and incubated with etoposide (25 μM) for 24 or 48 h to induce DNA damage. Cells were harvested and subjected to FACS analysis (upper panel). The subG1 population is graphically represented (lower panel). The graph indicates the results from three independent experiments. The error bars in the graph indicate ±standard deviations (s.d.). (**B**) MCF7 cells were transiently transfected with siSELENOW or/and siTrx1. After 24 h, the cells were treated with etoposide (25 μM) for 24 or 48 h. The cells lysates were analyzed using Western blot assay using the indicated antibodies.

2.5. Decreased Akt Phosphorylation by Downregulation of Trx1 is Restored by SELENOW Overexpression

Akt activation promotes cell survival [49,50]. The binding of SELENOW to 14-3-3β reduced the interaction between Rictor and 14-3-3β, leading to Akt phosphorylation at Ser473 [51]. As the binding mechanism of Trx1 to 14-3-3β was similar to that of SELENOW, we investigated whether Akt phosphorylation affected by Trx1 depletion can be regulated by SELENOW overexpression. Previous reports showed that SELENOW-deficient cells or Trx1-deficient cells decreased phosphorylation of Akt, and the levels were recovered in SELENOW-deficient cells by ectopic expression of SELENOW(U13C) [14,17,30,51]. In this study, we found that downregulation of Akt phosphorylation by Trx1 depletion was recovered when SELENOW(U13C) or SELENOW(WT) was overexpressed in the cells (Figure 5A,B). These results suggest that SELENOW can compensate for the regulation of Akt phosphorylation by Trx1 via interaction with 14-3-3β in cells.

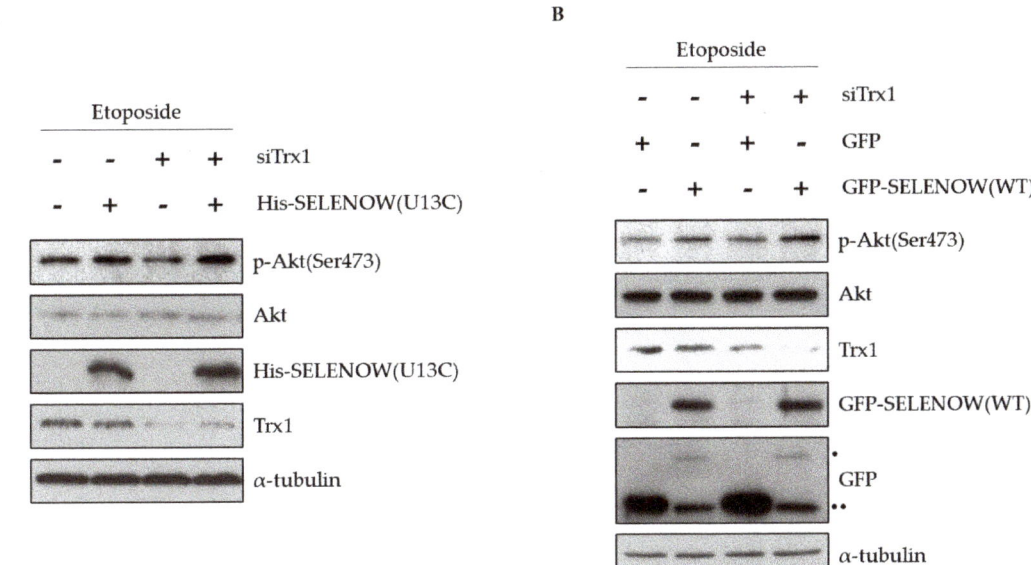

Figure 5. Decreased Akt phosphorylation of Trx1-deficent cells is restored by SELENOW. (**A**) MCF7 cells were transfected with siTrx1 or His-SELENOW(U13C), or co-transfected with siTrx1 and His-SELENOW(U13C) for 24 h and then treated with etoposide (25 μM) for 10 h. The cells lysates were analyzed by immunoblotting with the indicated antibodies. (**B**) MCF7 cells were co-transfected with siTrx1 and GFP-SELENOW(WT) for 24 h and further treated with etoposide for 10 h. The cells lysates were analyzed by immunoblotting with the indicated antibodies. • Full length GFP-SELENOW, •• pre-terminated GFP-SELENOW.

3. Discussion

The 14-3-3 protein interacts with various binding partners to play important roles in diverse signaling processes, including bacterial pathogenesis, cell growth, and development [52,53]. In particular, 14-3-3 functions as a regulator of cell survival under DNA damage [15]. mTORC2 is activated during DNA damage and it phosphorylates Akt at Ser473 [54]. Phosphorylated Akt provides a survival signal to cells from apoptotic stimuli [55]. Akt phosphorylation is inhibited by the interaction between Rictor and 14-3-3 [16]. Therefore, the regulation of the 14-3-3 protein is important for cell survival.

Our previous works showed that the binding of SELENOW to 14-3-3β inhibits the binding of 14-3-3β to its target proteins and regulates various cellular processes, such as cell growth, muscle differentiation, cellular oxidative stress, and cell progression [17,28,56].

SELENOW is a protein containing a Trx-like fold and a CXXU motif, which corresponds to the CXXC motif of Trx [18]. Trx is a small redox protein involved in various processes, such as cell redox homeostasis, cell growth, DNA repair, and cell survival [57–60]. Trx1 binds to proteins such as CDC25, PTEN, and RNR, which are involved in various cellular processes [3,61,62]. Recently, it has been suggested that 14-3-3β might be a target for Trx-like proteins [45,46].

In this study, we found a direct interaction between Trx1 and 14-3-3β by immunoprecipitation and pull-down assays, and the binding was increased during oxidative stress (Figure 1). Trx1 contains two Cys residues, Cys32 and Cys35, in the CXXC motif which are used to reduce the target proteins [63]. Sec13 in the CXXU motif of SELENOW was identified as the binding site for 14-3-3β [30]. Therefore, we constructed Trx1 mutants in which Cys32 or Cys35 was replaced with Ser and Cys32 was identified as the binding site for the 14-3-3β (Figure 1C,E).

Previously, we identified the binding site of SELENOW in the interaction between SELENOW and 14-3-3β, but not the binding site of 14-3-3β. In this study, we determined the binding site of 14-3-3β to SELENOW and investigated whether this binding site is also involved in the interaction with Trx1. Since the interaction of SELENOW with 14-3-3β was redox-regulated, we mutated two Cys residues, Cys96 and Cys191, in 14-3-3β to Ser residues. As shown in Figure 2A,B, the Cys191 of 14-3-3β was required for the interaction with Trx1 or SELENOW, indicating that 14-3-3β Cys191 is a common binding site for Trx1 and SELENOW. Previously, the Cys191 of 14-3-3β was also predicted to bind to SELENOW by computational calculations based on NMR data [47].

Since the binding of SELENOW and Trx1 to 14-3-3β occurs at the same site, Cys191, the relationship between SELENOW and Trx1 in the binding to 14-3-3β was investigated. The binding of Trx1 to 14-3-3β protein decreased with increasing amounts of SELENOW, and the binding of SELENOW to 14-3-3β also decreased with increasing amounts of Trx1 (Figure 3). These results suggest that the 14-3-3β signaling pathway can be cooperatively regulated by Trx1 and SELENOW. There is no interaction between Trx1 and SELENOW [27].

In this study, we found that Trx1 interacted with 14-3-3β. Since sensitivity to DNA damage was increased in SELENOW-deficient cells by etoposide treatment [30], we also determined the effect of Trx1-deficiency on cell viability. Compared to control cells, the subG1 population was increased in SELENOW-deficient cells or Trx1-deficient cells treated with etoposide. Furthermore, PARP was rapidly cleaved in SELENOW-deficient and Trx1-deficient cells compared with that in control cells after treatment with etoposide (Figure 4). This suggests that cell death induced by DNA damage is regulated by both SELENOW and Trx1.

SELENOW has been reported to control cell survival by interacting with 14-3-3β [30]. SELENOW activates the mTORC2/Akt pathway for Akt phosphorylation at Ser473 by interrupting the binding of Rictor to 14-3-3β. In SELENOW-deficient cells, the binding of 14-3-3β to Rictor is enhanced, resulting in the inactivation of Akt by inhibiting phosphorylation at Ser473 [16,17]. Akt inactivation decreases the viability of cells upon treatment with etoposide.

Therefore, we next investigated whether Akt phosphorylation was also regulated by the interaction of Trx1 with 14-3-3β. Phosphorylation of Akt at Ser473 was decreased by depletion of Trx1 in etoposide-treated MCF7 cells and was restored by the expression of SELENOW(U13C) (Figure 5A). Using the wild-type SELENOW containing Sec in the CXXU motif, we confirmed the compensation of SELENOW for Trx1 deficiency (Figure 5B).

Taken together, these results suggest that Trx1 regulates cell survival through interaction with 14-3-3β and that SELENOW compensates Trx1-deficient cells against cell death induced by etoposide treatment.

4. Materials and Methods

4.1. Cell Culture, Transfection, and Reagents

Breast carcinoma MCF7 cells were grown in RPMI 1640 medium containing 10% fetal bovine serum at 37 °C in 5% CO_2. Human embryonic kidney 293 (HEK 293) cells were grown in DMEM containing 10 % FBS at 37 °C in 5% CO_2. The cells were seeded at a density of 5×10^5 cells in 60 mm dishes for transient transfection. Twelve hours after seeding, the cells were transfected using ScreenFect A plus (ScreenFect GmbH, Eggenstein-Leopoldshafen, Germany) and Lipofectamine 2000 (Invitrogen, Carlsbad, CA, USA), according to the manufacturer's instructions. To measure the sensitivity to etoposide, MCF7 cells were treated with 25 μM etoposide (Sigma, St. Louis, MO, USA) after transfection and harvested.

4.2. RNA Interference and Plasmids

The siSELENOW used in this study was designed by Invitrogen. The sequence for human siSELENOW was as follows: siSELENOW 5′-CCA CCG GGU UCU UUG AAG UGA UGG U-3′. Human siTrx1 was purchased from Qiagen (Valencia, CA, USA). GFP-

SELENOW(WT) was generated using the pEGFP-C2 plasmid from wild-type mouse SELENOW (Origene, Rockville, MD, USA). His-SELENOW(U13C) and HA-14-3-3β plasmids have been described previously [28,30].

4.3. Western Blot Assay

Cells were harvested and lysed as previously described [30]. Whole cell lysates were separated using SDS-PAGE and transferred to a PVDF membrane. Membrane was blocked with 5% skim milk for 1h and incubated with specific antibodies at 4 °C overnight with rotation. The antibodies were obtained from the following sources: anti-poly (ADP-ribose) polymerase (PARP) and anti-Akt, anti-phospho-Akt (Ser 473) were from Cell Signaling (Danvers, MA, USA); anti-GST and anti-GFP antibodies were from Santa Cruz Biotechnology (Santa Cruz, CA, USA); anti-His and anti-HA were from ABM (Richmond, BC, Canada); anti-α-tubulin and anti-Trx1 were from AB Frontier (Seoul, Republic of Korea); anti-SELENOW was from Origene (Rockville, MD, USA). After incubating with HRP conjugated secondary antibody for 1 h, immunoreactive bands were visualized using a West Pico Enhanced ECL Detection kit (Pierce, Rockford, IL, USA).

4.4. Immunoprecipitation

Cells were lysed with immunoprecipitation buffer as previously described [30]. The lysates were mixed with antibodies overnight at 4 °C with rotation. Immune complexes were incubated with Protein A or G beads for 1.5 h at 4 °C with rotation. The lysates were washed two times and boiled with SDS sample buffer for 3 min. The samples were separated and detected as described above.

4.5. Protein Purification

The mouse GST-SELENOW protein, GST-14-3-3β and human Trx1-His mutants were expressed and purified in *E. coli*. Briefly, BL21 (DE3) competent cells were transformed with mouse GST-SELENOW and human Trx1-His mutants in pGEX 4T-1 (Amersham Biosciences, Chalfont, UK) and pET26B (Novagen, Madison, WI, USA) plasmids. The proteins were induced by 1 mM IPTG for 16 h at 18 °C and purified using glutathione and Ni-NTA beads as described previously [27,30].

4.6. Pull-Down Assay

Purified proteins were incubated with cell lysate or recombinant proteins in pull-down buffer containing 20 mM HEPES (pH 7.5), 1 mM EDTA and 1 mM for 2 h at 4 °C with rotation, followed by glutathione and NI-NTA beads for 1 h. The beads were washed three times with the wash buffer and then eluted [27,30].

4.7. Flow Cytometry

MCF7 cells were fixed overnight in ice-cold 70% ethanol. Cells were collected by centrifugation and resuspended in PBS. The resuspended cells were treated with RNase A (100 μg/mL) for 30 min at RT, and the cells were then stained with propidium iodide (10 μg/mL). Subsequently, cell cycle distribution was analyzed using a FACS Accuri flow cytometer (BD Bioscience, San Jose, CA, USA).

Author Contributions: H.K. designed and performed most of the experiments and wrote the manuscript. Y.H.J. and M.H. assisted in protein purification, Western blot assay, and DNA cloning experiments. K.K. and I.Y.K. equally contributed to the study design, evaluation of results, and writing of the manuscript. All authors have read and agreed to the published version of the manuscript.

Funding: This work was supported by the National Research Foundation of Korea (NRF) grant by the Korea government (MSIP) (NRF-2019R1A2B5B01069901) and by the Korea University Grant.

Institutional Review Board Statement: Not applicable.

Informed Consent Statement: Not applicable.

Data Availability Statement: The data presented in this study are available upon request from the corresponding author.

Acknowledgments: We would like to thank the College of Life Sciences, Korea University for providing the facilities for cell sorting and FACS analysis.

Conflicts of Interest: The authors declare no conflict of interest.

References

1. Lu, J.; Holmgren, A. Thioredoxin System in Cell Death Progression. *Antioxid. Redox Signal* **2012**, *17*, 1738–1747. [CrossRef]
2. Lee, S.; Kim, S.M.; Lee, R.T. Thioredoxin and Thioredoxin Target Proteins: From Molecular Mechanisms to Functional Significance. *Antioxid. Redox Signal* **2013**, *18*, 1165–1207. [CrossRef]
3. Meuillet, E.J.; Mahadevan, D.; Berggren, M.; Coon, A.; Powis, G. Thioredoxin-1 binds to the C2 domain of PTEN inhibiting PTEN's lipid phosphatase activity and membrane binding: A mechanism for the functional loss of PTEN's tumor suppressor activity. *Arch. Biochem. Biophys.* **2004**, *429*, 123–133. [CrossRef]
4. Saitoh, M.; Nishitoh, H.; Fujii, M.; Takeda, K.; Tobiume, K.; Sawada, Y.; Kawabata, M.; Miyazono, K.; Ichijo, H. Mammalian thioredoxin is a direct inhibitor of apoptosis signal-regulating kinase (ASK). *EMBO J.* **1998**, *17*, 2596–2606. [CrossRef] [PubMed]
5. Karlenius, T.C.; Tonissen, K.F. Thioredoxin and Cancer: A Role for Thioredoxin in all States of Tumor Oxygenation. *Cancers* **2010**, *2*, 209–232. [CrossRef]
6. Bhatia, M.; McGrath, K.L.; Di Trapani, G.; Charoentong, P.; Shah, F.; King, M.M.; Clarke, F.M.; Tonissen, K. The thioredoxin system in breast cancer cell invasion and migration. *Redox Biol.* **2015**, *8*, 68–78. [CrossRef] [PubMed]
7. Berggren, M.; Gallegos, A.; Gasdaska, J.R.; Gasdaska, P.Y.; Warneke, J.; Powis, G. Thioredoxin and thioredoxin reductase gene expression in human tumors and cell lines, and the effects of serum stimulation and hypoxia. *Anticancer Res.* **1996**, *16*, 3459–3466. [PubMed]
8. Chiueh, C.C.; Andoh, T.; Chock, P.B. Induction of Thioredoxin and Mitochondrial Survival Proteins Mediates Preconditioning-Induced Cardioprotection and Neuroprotection. *Ann. N. Y. Acad. Sci.* **2005**, *1042*, 403–418. [CrossRef]
9. Damdimopoulos, A.E.; Miranda-Vizuete, A.; Pelto-Huikko, M.; Gustafsson, J.; Spyrou, G. Human Mitochondrial Thioredoxin. *J. Biol. Chem.* **2002**, *277*, 33249–33257. [CrossRef]
10. Kahlos, K.; Soini, Y.; Säily, M.; Koistinen, P.; Kakko, S.; Pääkkö, P.; Holmgren, A.; Kinnula, V.L. Up-regulation of thioredoxin and thioredoxin reductase in human malignant pleural mesothelioma. *Int. J. Cancer* **2001**, *95*, 198–204. [CrossRef]
11. Mitsui, A.; Hamuro, J.; Nakamura, H.; Kondo, N.; Hirabayashi, Y.; Ishizaki-Koizumi, S.; Hirakawa, T.; Inoue, T.; Yodoi, J. Overexpression of Human Thioredoxin in Transgenic Mice Controls Oxidative Stress and Life Span. *Antioxid. Redox Signal* **2002**, *4*, 693–696. [CrossRef] [PubMed]
12. Zuo, Z.; Zhang, P.; Lin, F.; Shang, W.; Bi, R.; Lu, F.; Wu, J.; Jiang, L. Interplay between Trx-1 and S100P promotes colorectal cancer cell epithelial–mesenchymal transition by up-regulating S100A4 through AKT activation. *J. Cell. Mol. Med.* **2018**, *22*, 2430–2441. [CrossRef] [PubMed]
13. D'annunzio, V.; Perez, V.; Mazo, T.; Muñoz, M.C.; Dominici, F.; Carreras, M.C.; Poderoso, J.J.; Sadoshima, J.; Gelpi, R.J. Loss of myocardial protection against myocardial infarction in middle-aged transgenic mice overexpressing cardiac thioredoxin-1. *Oncotarget* **2016**, *7*, 11889–11898. [CrossRef] [PubMed]
14. González, R.; López-Grueso, M.J.; Muntané, J.; Bárcena, J.A.; Padilla, C.A. Redox regulation of metabolic and signaling pathways by thioredoxin and glutaredoxin in NOS-3 overexpressing hepatoblastoma cells. *Redox Biol.* **2015**, *6*, 122–134. [CrossRef]
15. Yang, H.; Wen, Y.-Y.; Zhao, R.; Lin, Y.-L.; Fournier, K.; Yang, H.-Y.; Qiu, Y.; Diaz, J.; Laronga, C.; Lee, M.-H. DNA Damage–Induced Protein 14-3-3 σ Inhibits Protein Kinase B/Akt Activation and Suppresses Akt-Activated Cancer. *Cancer Res.* **2006**, *66*, 3096–3105. [CrossRef]
16. Dibble, C.C.; Asara, J.M.; Manning, B.D. Characterization of Rictor Phosphorylation Sites Reveals Direct Regulation of mTOR Complex 2 by S6K1. *Mol. Cell. Biol.* **2009**, *29*, 5657–5670. [CrossRef]
17. Jeon, Y.H.; Park, Y.H.; Kwon, J.H.; Lee, J.H.; Kim, I.Y. Inhibition of 14-3-3 binding to Rictor of mTORC2 for Akt phosphorylation at Ser473 is regulated by selenoprotein W. *Biochim. Biophys. Acta (BBA)* **2013**, *1833*, 2135–2142. [CrossRef]
18. Papp, L.V.; Lu, J.; Holmgren, A.; Khanna, K.K. From Selenium to Selenoproteins: Synthesis, Identity, and Their Role in Human Health. *Antioxid. Redox Signal* **2007**, *9*, 775–806. [CrossRef]
19. Liu, J.; Chen, Q.; Rozovsky, S. Utilizing Selenocysteine for Expressed Protein Ligation and Bioconjugations. *J. Am. Chem. Soc.* **2017**, *139*, 3430–3437. [CrossRef]
20. Dery, L.; Reddy, P.S.; Dery, S.; Mousa, R.; Ktorza, O.; Talhami, A.; Metanis, N. Accessing human selenoproteins through chemical protein synthesis. *Chem. Sci.* **2016**, *8*, 1922–1926. [CrossRef]
21. Jeong, D.-W.; Kim, E.H.; Kim, T.S.; Chung, Y.W.; Kim, H.; Kim, I.Y. Different distributions of selenoprotein W and thioredoxin during postnatal brain development and embryogenesis. *Mol. Cells* **2004**, *17*, 156–159. [PubMed]
22. Noh, O.J.; Park, Y.H.; Chung, Y.W.; Kim, I.Y. Transcriptional Regulation of Selenoprotein W by MyoD during Early Skeletal Muscle Differentiation. *J. Biol. Chem.* **2010**, *285*, 40496–40507. [CrossRef] [PubMed]
23. Beilstein, M.; Vendeland, S.; Barofsky, E.; Jensen, O.; Whanger, P. Selenoprotein W of rat muscle binds glutathione and an unknown small molecular weight moiety. *J. Inorg. Biochem.* **1996**, *61*, 117–124. [CrossRef]

24. Gu, Q.-P.; Beilstein, M.A.; Barofsky, E.; Ream, W.; Whanger, P.D. Purification, Characterization, and Glutathione Binding to Selenoprotein W From Monkey Muscle. *Arch. Biochem. Biophys.* **1999**, *361*, 25–33. [CrossRef]
25. Jeong, D.W.; Kim, T.S.; Chung, Y.W.; Lee, B.J.; Kim, I.Y. Selenoprotein W is a glutathione-dependent antioxidant in vivo. *FEBS Lett.* **2002**, *517*, 225–228. [CrossRef]
26. Kim, Y.-J.; Chai, Y.-G.; Ryu, J.-C. Selenoprotein W as molecular target of methylmercury in human neuronal cells is down-regulated by GSH depletion. *Biochem. Biophys. Res. Commun.* **2005**, *330*, 1095–1102. [CrossRef]
27. Ko, K.Y.; Lee, J.H.; Jang, J.K.; Jin, Y.; Kang, H.; Kim, I.Y. S-Glutathionylation of mouse selenoprotein W prevents oxidative stress-induced cell death by blocking the formation of an intramolecular disulfide bond. *Free. Radic. Biol. Med.* **2019**, *141*, 362–371. [CrossRef]
28. Park, Y.H.; Jeon, Y.H.; Kim, I.Y. Selenoprotein W promotes cell cycle recovery from G2 arrest through the activation of CDC25B. *Biochim. Biophys. Acta (BBA)* **2012**, *1823*, 2217–2226. [CrossRef]
29. Pommier, Y.; Leo, E.; Zhang, H.; Marchand, C. DNA Topoisomerases and Their Poisoning by Anticancer and Antibacterial Drugs. *Chem. Biol.* **2010**, *17*, 421–433. [CrossRef]
30. Jeon, Y.H.; Ko, K.Y.; Lee, J.H.; Park, K.J.; Jang, J.K.; Kim, I.Y. Identification of a redox-modulatory interaction between selenoprotein W and 14-3-3 protein. *Biochim. Biophys. Acta (BBA)* **2016**, *1863*, 10–18. [CrossRef] [PubMed]
31. Pennington, K.L.; Chan, T.Y.; Torres, M.; Andersen, J.L. The dynamic and stress-adaptive signaling hub of 14-3-3: Emerging mechanisms of regulation and context-dependent protein–protein interactions. *Oncogene* **2018**, *37*, 5587–5604. [CrossRef] [PubMed]
32. Morrison, D. 14-3-3: Modulators of signaling proteins? *Science* **1994**, *266*, 56–57. [CrossRef]
33. Lin, H.-K.; Wang, G.; Chen, Z.; Teruya-Feldstein, J.; Liu, Y.; Chan, C.-H.; Yang, W.-L.; Erdjument-Bromage, H.; Nakayama, K.I.; Nimer, S.; et al. Phosphorylation-dependent regulation of cytosolic localization and oncogenic function of Skp2 by Akt/PKB. *Nature* **2009**, *11*, 420–432. [CrossRef]
34. Yamagata, K.; Daitoku, H.; Takahashi, Y.; Namiki, K.; Hisatake, K.; Kako, K.; Mukai, H.; Kasuya, Y.; Fukamizu, A. Arginine Methylation of FOXO Transcription Factors Inhibits Their Phosphorylation by Akt. *Mol. Cell* **2008**, *32*, 221–231. [CrossRef]
35. Schlegelmilch, K.; Mohseni, M.; Kirak, O.; Pruszak, J.; Rodriguez, J.R.; Zhou, D.; Kreger, B.T.; Vasioukhin, V.; Avruch, J.; Brummelkamp, T.R.; et al. Yap1 Acts Downstream of α-Catenin to Control Epidermal Proliferation. *Cell* **2011**, *144*, 782–795. [CrossRef]
36. Neal, C.L.; Yu, D. 14-3-3ζ as a prognostic marker and therapeutic target for cancer. *Expert Opin. Ther. Targets* **2010**, *14*, 1343–1354. [CrossRef]
37. Yaffe, M.B.; Rittinger, K.; Volinia, S.; Caron, P.R.; Aitken, A.; Leffers, H.; Gamblin, S.; Smerdon, S.; Cantley, L. The Structural Basis for 14-3-3: Phosphopeptide Binding Specificity. *Cell* **1997**, *91*, 961–971. [CrossRef]
38. Würtele, M.; Jelich-Ottmann, C.; Wittinghofer, A.; Oecking, C. Structural view of a fungal toxin acting on a 14-3-3 regulatory complex. *EMBO J.* **2003**, *22*, 987–994. [CrossRef] [PubMed]
39. Muslin, A.J.; Tanner, J.; Allen, P.M.; Shaw, A.S. Interaction of 14-3-3 with Signaling Proteins Is Mediated by the Recognition of Phosphoserine. *Cell* **1996**, *84*, 889–897. [CrossRef]
40. Seimiya, H.; Sawada, H.; Muramatsu, Y.; Shimizu, M.; Ohko, K.; Yamane, K.; Tsuruo, T. Involvement of 14-3-3 proteins in nuclear localization of telomerase. *EMBO J.* **2000**, *19*, 2652–2661. [CrossRef] [PubMed]
41. Patel, A.; Cummings, N.; Batchelor, M.; Hill, P.J.; Dubois, T.; Mellits, K.H.; Frankel, G.; Connerton, I. Host protein interactions with enteropathogenic Escherichia coli (EPEC): 14-3-3tau binds Tir and has a role in EPEC-induced actin polymerization. *Cell. Microbiol.* **2006**, *8*, 55–71. [CrossRef]
42. Sumioka, A.; Nagaishi, S.; Yoshida, T.; Lin, A.; Miura, M.; Suzuki, T. Role of 14-3-3γ in FE65-dependent Gene Transactivation Mediated by the Amyloid β-Protein Precursor Cytoplasmic Fragment. *J. Biol. Chem.* **2005**, *280*, 42364–42374. [CrossRef]
43. Yasmin, L.; Jansson, A.L.; Panahandeh, T.; Palmer, R.H.; Francis, M.S.; Hallberg, B. Delineation of exoenzyme S residues that mediate the interaction with 14-3-3 and its biological activity. *FEBS J.* **2006**, *273*, 638–646. [CrossRef] [PubMed]
44. Jiang, J.; Balcerek, J.; Rozenova, K.; Cheng, Y.; Bersenev, A.; Wu, C.; Song, Y.; Tong, W. 14-3-3 regulates the LNK/JAK2 pathway in mouse hematopoietic stem and progenitor cells. *J. Clin. Investig.* **2012**, *122*, 2079–2091. [CrossRef] [PubMed]
45. Floen, M.J.; Forred, B.; Bloom, E.J.; Vitiello, P. Thioredoxin-1 redox signaling regulates cell survival in response to hyperoxia. *Free. Radic. Biol. Med.* **2014**, *75*, 167–177. [CrossRef] [PubMed]
46. Nakao, L.; Everley, R.A.; Marino, S.M.; Lo, S.M.; de Souza, L.E.R.; Gygi, S.P.; Gladyshev, V.N. Mechanism-based Proteomic Screening Identifies Targets of Thioredoxin-like Proteins. *J. Biol. Chem.* **2015**, *290*, 5685–5695. [CrossRef]
47. Musiani, F.; Ciurli, S.; Dikiy, A. Interaction of Selenoprotein W with 14-3-3 Proteins: A Computational Approach. *J. Proteome Res.* **2011**, *10*, 968–976. [CrossRef]
48. Mochizuki, M.; Kwon, Y.-W.; Yodoi, J.; Masutani, H. Thioredoxin Regulates Cell Cycleviathe ERK1/2-Cyclin D1 Pathway. *Antioxid. Redox Signal* **2009**, *11*, 2957–2971. [CrossRef]
49. Choudhury, G.G. Akt Serine Threonine Kinase Regulates Platelet-derived Growth Factor-induced DNA Synthesis in Glomerular Mesangial Cells. *J. Biol. Chem.* **2001**, *276*, 35636–35643. [CrossRef]
50. Manning, B.D.; Toker, A. AKT/PKB Signaling: Navigating the Network. *Cell* **2017**, *169*, 381–405. [CrossRef]
51. Lu, T.; Zong, M.; Fan, S.; Lu, Y.; Yu, S.; Fan, L. Thioredoxin 1 is associated with the proliferation and apoptosis of rheumatoid arthritis fibroblast-like synoviocytes. *Clin. Rheumatol.* **2017**, *37*, 117–125. [CrossRef]

52. Fu, H.; Subramanian, R.; Masters, S.C. 14-3-3 Proteins: Structure, Function, and Regulation. *Annu. Rev. Pharmacol. Toxicol.* **2000**, *40*, 617–647. [CrossRef] [PubMed]
53. Satoh, J.-I.; Nanri, Y.; Yamamura, T. Rapid identification of 14-3-3-binding proteins by protein microarray analysis. *J. Neurosci. Methods* **2006**, *152*, 278–288. [CrossRef] [PubMed]
54. Bozulic, L.; Surucu, B.; Hynx, D.; Hemmings, B.A. PKBα/Akt1 Acts Downstream of DNA-PK in the DNA Double-Strand Break Response and Promotes Survival. *Mol. Cell* **2008**, *30*, 203–213. [CrossRef] [PubMed]
55. Song, G.; Ouyang, G.; Bao, S. The activation of Akt/PKB signaling pathway and cell survival. *J. Cell. Mol. Med.* **2005**, *9*, 59–71. [CrossRef]
56. Jeon, Y.H.; Park, Y.H.; Lee, J.H.; Hong, J.-H.; Kim, I.Y. Selenoprotein W enhances skeletal muscle differentiation by inhibiting TAZ binding to 14-3-3 protein. *Biochim. Biophys. Acta (BBA)* **2014**, *1843*, 1356–1364. [CrossRef]
57. Arnér, E.S.; Holmgren, A. The thioredoxin system in cancer. *Semin. Cancer Biol.* **2006**, *16*, 420–426. [CrossRef]
58. Holmgren, A. Thioredoxin. *Annu. Rev. Biochem.* **1985**, *54*, 237–271. [CrossRef]
59. Arnér, E.; Holmgren, A. Physiological functions of thioredoxin and thioredoxin reductase. *JBIC J. Biol. Inorg. Chem.* **2000**, *267*, 6102–6109. [CrossRef]
60. Gromer, S.; Urig, S.; Becker, K. The thioredoxin system? From science to clinic. *Med. Res. Rev.* **2003**, *24*, 40–89. [CrossRef]
61. Sohn, J.; Rudolph, J. Catalytic and Chemical Competence of Regulation of Cdc25 Phosphatase by Oxidation/Reduction. *Biochemistry* **2003**, *42*, 10060–10070. [CrossRef] [PubMed]
62. Rollins, M.F.; van der Heide, D.M.; Weisend, C.M.; Kundert, J.A.; Comstock, K.M.; Suvorova, E.S.; Capecchi, M.R.; Merrill, G.F.; Schmidt, E.E. Hepatocytes lacking thioredoxin reductase 1 have normal replicative potential during development and regeneration. *J. Cell Sci.* **2010**, *123*, 2402–2412. [CrossRef] [PubMed]
63. Watson, W.H.; Pohl, J.; Montfort, W.R.; Stuchlik, O.; Reed, M.S.; Powis, G.; Jones, D.P. Redox Potential of Human Thioredoxin 1 and Identification of a Second Dithiol/Disulfide Motif. *J. Biol. Chem.* **2003**, *278*, 33408–33415. [CrossRef] [PubMed]

Article

Selenoproteome Expression Studied by Non-Radioactive Isotopic Selenium-Labeling in Human Cell Lines

Jordan Sonet [1,†], Anne-Laure Bulteau [2,†], Zahia Touat-Hamici [3,†], Maurine Mosca [1], Katarzyna Bierla [1], Sandra Mounicou [1], Ryszard Lobinski [1,4,5] and Laurent Chavatte [6,7,8,9,10,*]

1 Institut des Sciences Analytiques et de Physico-Chimie Pour l'Environnement et les Matériaux (IPREM), Universite de Pau, CNRS, E2S, UMR 5254, Hélioparc, 64053 Pau, France; jordan.sonet@gmail.com (J.S.); mosca.maurine64@gmail.com (M.M.); katarzyna.bierla@univ-pau.fr (K.B.); sandra.mounicou@univ-pau.fr (S.M.); ryszard.lobinski@univ-pau.fr (R.L.)
2 LVMH Recherche, Life Science Department, 185 Avenue de Verdun, 45800 Saint Jean de Braye, France; abulteau@research.lvmh-pc.com
3 Centre de Génétique Moléculaire, CGM, CNRS, UPR3404, 91198 Gif-sur-Yvette, France; ztouat@gmail.com
4 Laboratory of Molecular Dietetics, I.M. Sechenov First Moscow State Medical University, 19945 Moscow, Russia
5 Chair of Analytical Chemistry, Warsaw University of Technology, Noakowskiego 3, 00-664 Warsaw, Poland
6 Centre International de Recherche en Infectiologie (CIRI), 69007 Lyon, France
7 Institut National de la Santé et de la Recherche Médicale (INSERM), Unité U1111, 69007 Lyon, France
8 Ecole Normale Supérieure de Lyon, 69007 Lyon, France
9 Université Claude Bernard Lyon 1 (UCBL1), 69622 Lyon, France
10 Centre National de la Recherche Scientifique (CNRS), Unité Mixte de Recherche 5308 (UMR5308), 69007 Lyon, France
* Correspondence: laurent.chavatte@ens-lyon.fr; Tel.: +33-4-72-72-86-24
† These authors contributed equally to the work.

Abstract: Selenoproteins, in which the selenium atom is present in the rare amino acid selenocysteine, are vital components of cell homeostasis, antioxidant defense, and cell signaling in mammals. The expression of the selenoproteome, composed of 25 selenoprotein genes, is strongly controlled by the selenium status of the body, which is a corollary of selenium availability in the food diet. Here, we present an alternative strategy for the use of the radioactive ^{75}Se isotope in order to characterize the selenoproteome regulation based on (i) the selective labeling of the cellular selenocompounds with non-radioactive selenium isotopes (^{76}Se, ^{77}Se) and (ii) the detection of the isotopic enrichment of the selenoproteins using size-exclusion chromatography followed by inductively coupled plasma mass spectrometry detection. The reliability of our strategy is further confirmed by western blots with distinct selenoprotein-specific antibodies. Using our strategy, we characterized the hierarchy of the selenoproteome regulation in dose–response and kinetic experiments.

Keywords: selenoproteome; selenoprotein hierarchy; nonradioactive isotopes; SEC-ICP MS; glutathione peroxidase; thioredoxin reductase; SECIS; translation regulation

1. Introduction

The vital role of the essential trace element selenium (Se) in human health has been widely reported. Selenium deficiency is often evoked in the context of cancer, cardiac function, muscular disorders, infections, neurodegenerative disease, and aging [1–8]. Despite the essential nature of selenium, the range of its optimal intake is quite narrow and its toxicity can be quite easily achieved depending on its chemical speciation. Selenium is co-translationally incorporated into a group of proteins, referred to as selenoproteins [9], in the form of a rare amino-acid, selenocysteine (Sec), using a UGA codon, normally used as a stop signal for protein synthesis [10,11]. Twenty-five selenoprotein genes have been identified in the human genome and give rise to selenoproteome expression in tissue and cell-line specific patterns [12]. The selenoproteome is primarily regulated by the bioavailability of selenium from food or from a cell culture medium. This selenium-dependent regulation follows a specific hierarchy that stipulates that "house-keeping" members are kept constant at the expense of "stress-regulated" members, which respond to changes in the selenium level [6,8,13]. Other stimuli, including H_2O_2-induced oxidative stress and replicative senescence, are also able to modulate selenoprotein expression but follow distinct hierarchies [4,14–18].

Selenoprotein biosynthesis follows a non-conventional mechanism which involves the recoding of UGA, normally used as a termination signal, in a Sec insertion codon. To do so, the translation machinery has evolved a sophisticated strategy to cope with using UGA as Sec for selenoprotein mRNAs, while maintaining its use as a stop codon for all other cellular mRNAs. This non-canonical pathway essentially relies on two RNA molecules, namely: (i) the Sec insertion sequence (SECIS) located in the 3' untranslated region (UTR) of selenoprotein mRNAs; (ii) the Sec-tRNA$^{[Ser]Sec}$; and (iii) their interacting partners (as reviewed in [10,11,13,19–21]). Only one Sec residue is present in each selenoprotein (except for SELENOP) [12]. Well-characterized members are glutathione peroxidases (GPX), thioredoxin reductases (TXNRD), iodothyronine deiodinases (DIO), methionine sulfoxide reductase (MSR), and endoplasmic reticulum (ER) selenoproteins (SELENOF, SELENOS, SELENOK, SELENON, SELENOM, and SELENOT). Additionally, selenophosphate synthetase 2 (SEPHS2) is implicated in the synthesis of the selenocysteine precursor, selenophosphate ($SePO_3^{3-}$), and therefore participates in the selenoprotein biosynthesis pathway. The function of about one-third of all selenoproteins remains largely uncharacterized. Human embryonic kidney 293 (HEK293) cells express a wide range of selenoproteins that are highly responsive to selenium levels [15,22]. This cell line offers a validated model for the prioritized selenium-dependent regulation of selenoproteins. Growth media with selenium limiting conditions (3 nM) were engineered and validated [15,16,23]. In particular, the response to selenium levels was reported for the members of the GPX family. It is referred to as the selenoprotein hierarchy and has yet to be characterized for the entire selenoproteome [10].

Selenoprotein detection has long been limited to western immunoblots and radioactivity. Recently, the use of inductively coupled plasma-mass spectrometry (ICP MS) coupled to protein separation methods has proved efficient for the simultaneous detection of abundant selenoproteins [24–28]. First, western immunoblots are limited to the availability of antibodies. Although highly specific and semi-quantitative, this method does not provide information about the relative abundance between selenoproteins. Alternatively, radioactive ^{75}Se (in the form of selenite) has been widely used to label selenoproteins in vivo and in vitro [29], and provides relative abundance data for abundant selenoproteins. Radiolabeled proteins are separated by different biochemical methods, including chromatography or SDS-PAGE, and are revealed by gamma-counting or autoradiography. ^{75}Se is a gamma-emitter with a rather long half-life (120 days) and its use is restricted to a few laboratories. In addition, only abundant selenoproteins are visible by the autoradiography of SDS-PAGE. The presence of selenium in a molecule can be otherwise detected by elemental ion mass spectrometry, such as ICP MS, which is able to differentiate and quantify all natural isotopes of selenium. Indeed, selenium has six stable isotopes with a rather constant natural relative abundance: ^{74}Se (0.89%), ^{76}Se (9.37%), ^{77}Se (7.63%), ^{78}Se (23.77%), ^{80}Se (49.61%), and ^{82}Se (8.73%). Interestingly, each of these selenium isotopes can be enriched to >99% and isotopically labeled selenium compounds such as selenite, selenomethionine, or selenocysteine can readily be synthesized. When selenium is added as pure isotope, its incorporation into selenoproteins can be followed by ICP MS. The efficiency of selenium exchange in proteins is analyzed by size exclusion chromatography (SEC) coupled with ICP MS (SEC-ICP MS).

Here, we developed an innovative strategy to selectively label and trace selenoproteins with non-radioactive selenium-enriched isotopes (^{76}Se, ^{77}Se) in the HEK293 cell line. Cellular extracts were simultaneously analyzed by SEC-ICP MS and by western blots with distinct selenoprotein-specific antibodies. We performed dose–response and kinetic experiments to study the hierarchy of selenoproteome regulation in this particular cell line. Our work further illustrates the potential for non-radioactive selenium-labeling of selenoproteins in the wide field of proteomics.

2. Results

2.1. Fractionation of Intracellular Selenocompounds from HEK293 by Size-Exclusion Chromatography Followed by Inductively Coupled Plasma Mass Spectrometry Detection (SEC-ICP MS)

The fractionation of HEK293 cellular extracts (500 µg of proteins) was reproducibly achieved in native conditions by size exclusion chromatography (SEC). The Superdex 200 used here allowed the separation of protein complexes in a range of molecular weights (MW) comprised between 10 kDa and 600 kDa, with heavier proteins or complexes eluting first (see Figure 1 and Table 1). In our experiment, SEC was linked to an UV^{280nm} monitor followed by an ICP MS tuned for a multi-isotopic selenium detection. Seven fractions were defined according to the selenium signal from SEC-ICP MS chromatograms (Figure 1 and Table 1). F1 to F5 were expected to contain selenoproteins while F6 and F7 should correspond to small MW selenocompounds such as selenite, mono- and di-methyl selenium, selenocyanate and potentially others. When cells were grown in a medium containing 100 nM of selenium, we found that selenoproteins were the major forms of selenocompounds found in cellular extracts. Indeed, the sum of the selenium contained in fractions F1 to F5 reached 75% of the total signal in the chromatogram (Figure 1a and Table 1).

Table 1. Chromatographic data for each fraction collected according to the SEC-ICP MS chromatogram. n.d., not detected.

Fraction N°	Retention Time, s		MW Range, kDa	Selenium Signal Repartition, %				Selenoproteins Detected by Westernblot (Figure 1)
	Start	End		Dpl		Dpl + 100 nM Se		
				Ave	SD	Ave	SD	
F1	600	760	>480	12.0 ±	0.6	17.5 ±	0.9	SelenoS
F2	1000	1120	140–75	4.7 ±	0.2	10.2 ±	0.5	SelenoF, Gpx1, TR1
F3	1120	1300	75–30	25.4 ±	1.3	37.3 ±	1.9	Gpx1, Txnrd1, SelenoP, Sephs2, Txnrd2, SelenoO
F4	1410	1540	8–16	3.5 ±	0.2	5.8 ±	0,3	SelenoM, Gpx4
F5	1540	1650	4–8	2.7 ±	0.1	4.4 ±	0.2	SelenoM, Gpx4, Txnrd2
F6	1660	1870	1.5–4	24.7 ±	1.2	16.5 ±	0.8	Gpx4, Txnrd2
F7	1990	2130	0.38–0.80	27.1 ±	1.4	8.4 ±	0.4	n.d.

The presence of 10 important proteins in the selenoproteome expressed in HEK293 cells was assayed in each fraction by western immunoblotting (Figure 1b). In elution fraction F1, corresponding to the exclusion volume (V_0) of the column, we detected the presence of SELENOS, a rather small protein (21 kDa) that was previously found in a large complex from the endoplasmic reticulum (ER) [30]. Obviously, other selenoproteins other than those tested here may participate in the selenium signal detected in F1. In F2, SELENOF (also known as Sel15 or Sep15) which is also an ER-protein was detected. SELENOF (15 kDa) associates with the UDP-glucose:glycoprotein glucosyltransferase (UGTR) to form a 150–kDa complex [30]. GPX1 and TXNRDs (TXNRD1 and TXNRD2) that form tetramers (~80 kDa) and dimers (~110 kDa), respectively, were observed in F2 and F3. Other selenoproteins studied here (SELENOP, SEPHS2, SELENOO, SELENOM, and GPX4) are believed [6,8] to be predominantly in the monomeric form, and therefore eluted in the expected respective fractions (see Figure 1 and Table 1).

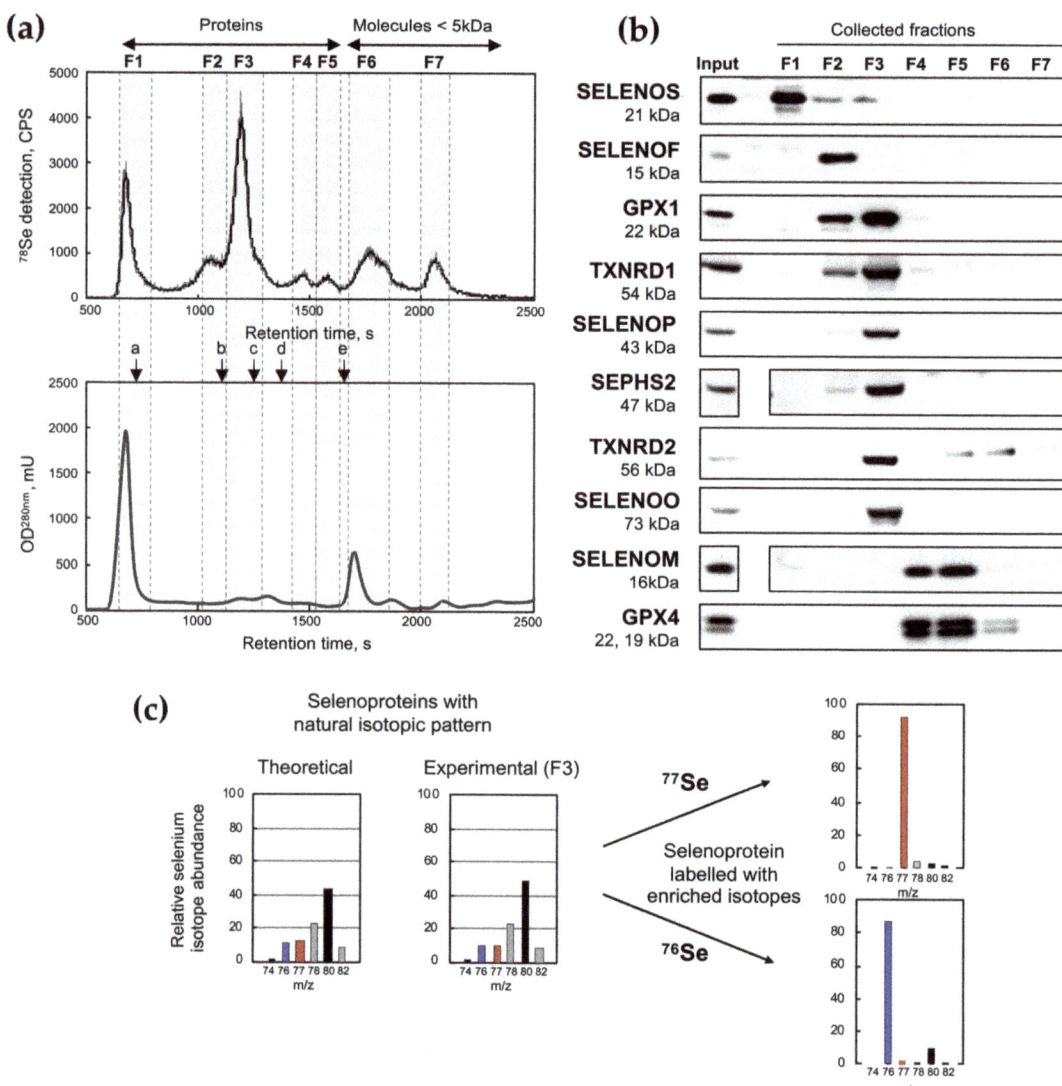

Figure 1. Fractionation of selenoproteins and selenocompounds of HEK293 cells by SEC-ICP MS. (**a**) Cellular extracts of HEK293 cells grown in Unsup + 100 nM of selenium (sodium selenite with natural isotopic pattern) were analyzed by SEC-ICP MS. The chromatographic profiles for ^{78}Se and OD^{280nm} are represented in black and gray lines, respectively. The elution time of the selected calibration standards is shown in the bottom panel (a, thyroglobulin; b, transferrin; c, Mn-SOD; d, myoglobin; e, selenomethionine). The selenium eluted in seven fractions (referred to as F1 to F7) as noted from the ^{78}Se chromatographic profile. Elution time and molecular weight range of these fractions are reported in Table 1. (**b**) The collected fractions F1 to F7 were then analyzed for the presence of 10 selenoproteins (SELENOS, SELENOF, GPX1, TXNRD1, SELENOP, SEPHS2, TXNRD2, SELENOO, SELENOM, and GPX4) using specific antibodies by western immunoblotting in comparison with the raw cellular extract. (**c**) Representation of selenium isotopic pattern in fraction F3 from fractionation SEC-ICP MS performed in panel (**a**). Cellular extracts of HEK293 grown in Unsup +100 nM of selenium (^{76}Se or ^{77}Se enriched selenite) were also analyzed by SEC-ICP MS. The isotopic enrichment of fraction F3 is illustrated in the two right panels.

Interestingly, F3 contained at least six selenoproteins (GPX1, TXNRD1, SELENOP, SEPHS2, TXNRD2, and SELENOO) which accounted for almost 40% of the total selenium content of the extract (Table 1). We knew from previous work that the amount of selenium contained in GPX1 (present here in F3) represented approximately half of the total selenium in HEK293 selenoproteins when grown in a selenium-rich medium [26]. Consequently, F3 appeared as one of the most representative fractions for selenoproteome analyses. When the selenium isotopic pattern of F3 was analyzed, it fitted the theoretical selenium profile (Figure 1c). Then, when natural selenium selenite was replaced by either 100 nM of ^{76}Se- and ^{77}Se-enriched selenite in the culture media, the selenium isotopic profile of F3 was dramatically altered after three days of incubation as illustrated in Figure 1c. Notably, with both ^{76}Se- and ^{77}Se-enriched selenite, almost 90% of the selenoproteins in F3 contained the corresponding isotope, indicating an efficient and rapid exchange of the natural isotope by the enriched one. Our data confirm that isotopically enriched selenium can be used to efficiently label and trace selenoprotein expression in human cultured cell lines.

2.2. Selenium Assimilation in HEK293 Cells as a Function of Selenium Levels

The selenium level in the growth media is known to modulate selenoprotein expression following a prioritized response, also called a hierarchy, but various selenium sources and concentrations have been used in the literature [31]. When 100 nM was added to the culture medium, the selenite induced a selective upregulation of the selenoproteins that was associated with an improved antioxidant defense [15]. Here, we aimed to define the optimal selenium levels (in the form of selenite) for each selenoprotein expression. To do so, increasing concentrations of ^{76}Se-enriched selenite (from 5 to 300 nM) were added to the culture media after a preliminary selenium deprivation of the cells for three days, as illustrated in Figure 2a. In HEK293 cells, the toxic activity of selenite appears at concentrations higher than 1 µM [32]. At these higher levels, an adaptive response to selenium stress may occur. Cell extracts were harvested after three more days of incubation and analyzed by SEC-ICP MS, as mentioned above. The chromatograms (Figure 2b) showed that the ICP MS signal of ^{76}Se increased in all fractions as selenium was added to the cell growth media. Interestingly, the highest selenium concentration used here (i.e., 300 nM) was not sufficient to reach a similar ^{76}Se-isotope enrichment for all the selenocompounds (Figure 2c). Indeed, when a concentration of 300 nM of ^{76}Se selenite was used, a ^{76}Se-enrichment of 80–90% was observed for the protein fractions F1 to F5, while it reached 71% and 45% only in F6 and F7, respectively. This finding indicated a prioritized assimilation of selenium into the selenoprotein pool, with F3 being the most concentrated in terms of the newly synthesized selenoproteins. Then, the isotopic ratio in each fraction was calculated and plotted as a function of the added selenium. Only the most naturally abundant (^{80}Se) and the supplemented isotope (^{76}Se) are represented in Figure 2b. The curve-fit of the ^{76}Se signal with a saturation equation gave us a half-saturation constant ($k_{1/2}$) for each fraction (see Figure 2b). This value expressed the concentration of the added selenium which was necessary to exchange half of the natural selenium with the isotopically enriched element. The comparison of these different values provided the hierarchy of selenium assimilation into the different fractions. It appeared that selenium was primarily incorporated in the selenoprotein pool rather than in the metabolite fractions, with the following hierarchy: F2 and F3 > F1 > F4 and F5 > F6 and F7. Taken together, our data demonstrated that one cell passage of three days in selenium-supplemented conditions was sufficient to efficiently incorporate the isotopically enriched selenite in the newly synthesized selenoproteins.

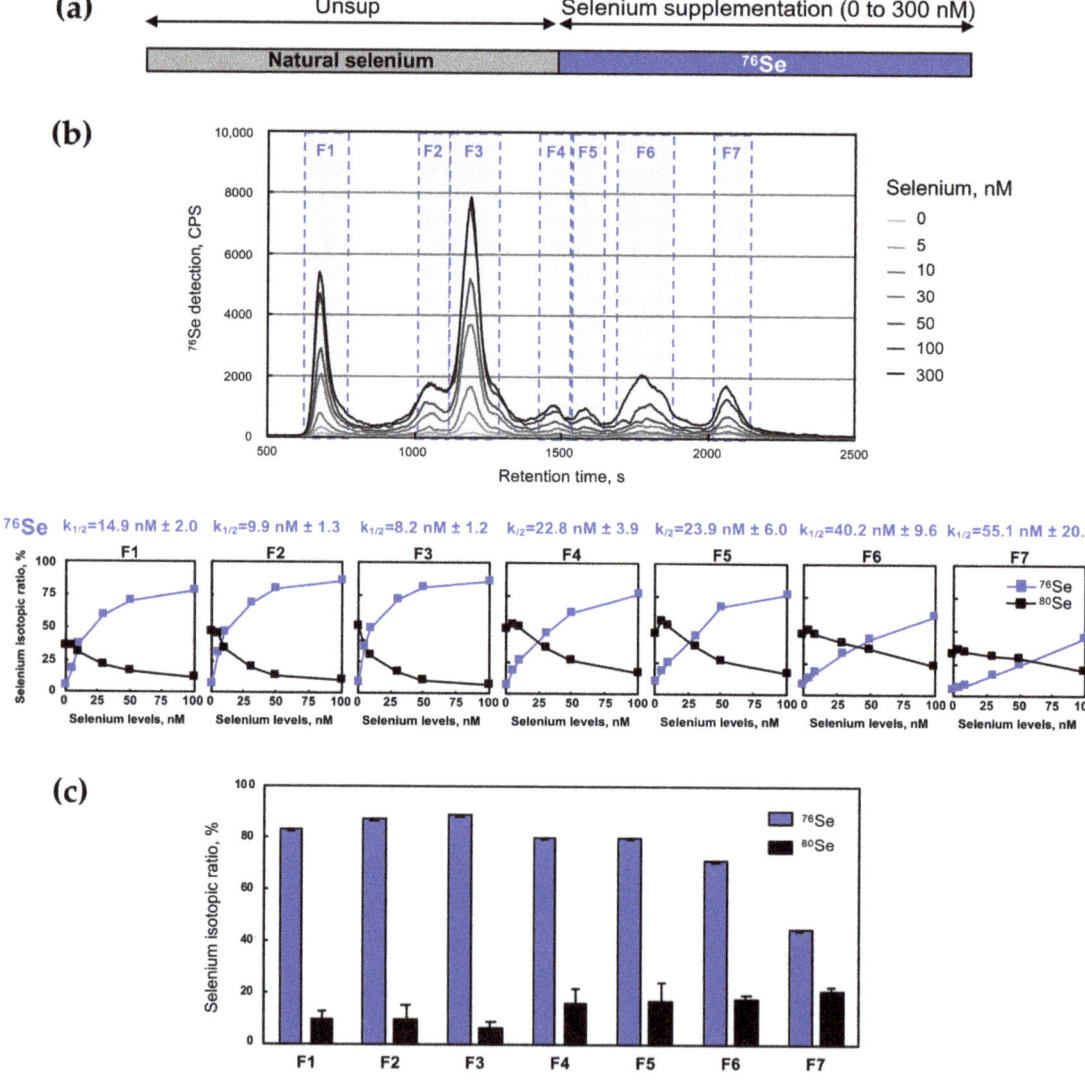

Figure 2. Analysis of HEK193 cells labeled with increasing concentrations of ^{76}Se by SEC ICP MS. (**a**) Cells were grown for three days in a Dpl medium prior to being plated in fresh medium with increasing concentrations of ^{76}Se enriched selenium (from 0 to 300 nM range). (**b**) The different cellular extracts were fractionated by SEC and all selenium isotopes (^{74}Se, ^{76}Se, ^{77}Se, ^{78}Se, ^{80}Se and ^{82}Se) followed by ICP MS detection. The SEC elution profile for ^{76}Se is shown in the bottom panel for the various cellular extracts. The isotopic pattern was analyzed in the seven identified fractions, as indicated by dashed boxes (F1 to F7). The dose–response curves for ^{76}Se abundance in each fraction as a function of selenium levels in the culture media were fitted as described in experimental procedures. The results for $k_{1/2}$ for ^{76}Se labeling are indicated on top of each graph. For clarity, only ^{76}Se and ^{80}Se are represented in blue and black squares, respectively, at a smaller scale of selenium levels (0 to 100 nM). (**c**) The selenium isotopic ratio is represented for the two most abundant isotopes (^{76}Se and ^{80}Se), for each fraction (F1 to F7), in the condition where 300 nM ^{76}Se was used.

2.3. The Selenoproteome Was Differentially Modulated According to the Selenium Level

It has been established that selenoprotein expression is mostly controlled at the translational level in cultured cells in response to the changes in the selenium concentration. In HEK293 cells, selenoprotein mRNAs remained virtually insensitive to the selenium level [14,23]. To grasp this translational regulation, we have previously engineered and validated a set of luciferase-based reporter constructs [15,22,33]. HEK293 cells stably expressing Luc UGA/SECIS GPX1 or Luc UGA/SECIS GPX4 allowed the direct and rapid evaluation of Sec insertion efficiencies in response to various stimuli. Cells were grown with different selenium concentrations as indicated in Figure 3a. Then, cellular extracts were assayed for the luciferase activities, normalized to protein concentrations, and expressed relative to the activity measured in zero-selenium added conditions (unsupplemented, Unsup). Therefore, Figure 3a illustrates the fold stimulation of UGA recoding efficiency in response to the concentration of added selenite. Interestingly, we noticed an important dose–response increase in the Sec insertion with both constructs, with a maximum of a 100- and 50-fold stimulation, respectively, for Luc UGA/SECIS GPX1 and Luc UGA/SECIS GPX4. The Sec insertion efficiency reached a plateau at 50 nM selenium for the Luc UGA/SECIS GPX4 construct, while higher selenium levels seemed to be necessary for the Luc UGA/SECIS GPX1 constructs. These data further confirm the importance of translational control in the selenium-dependent modulation of selenoproteins.

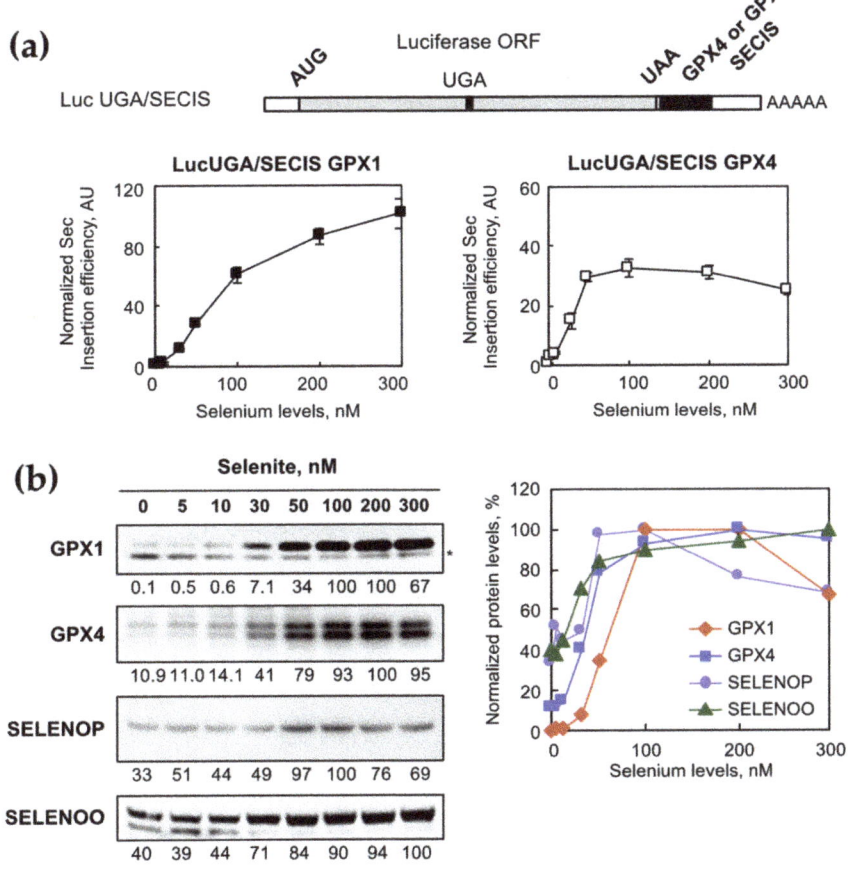

Figure 3. *Cont.*

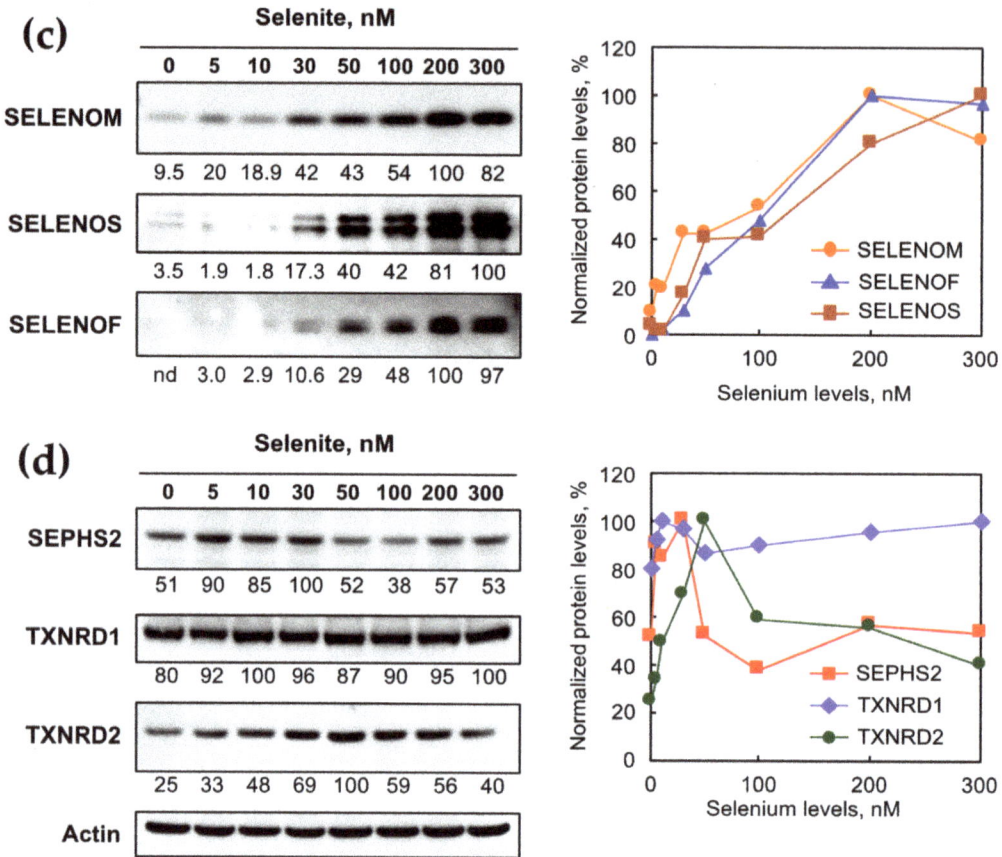

Figure 3. Analysis of selenoprotein expression as a function of selenium levels. (**a**) The response of selenocysteine insertion efficiency to selenium levels depends on the nature of the SECIS element. Two HEK293 cell lines stably expressing luciferase-based reporter constructs varying from the nature of the SECIS element, GPX1 and GPX4, respectively, were grown with increasing concentrations of selenium in the culture media. The structure of the Luc UGA/SECIS constructs was schematized. The Sec insertion efficiency was revealed by the luciferase activity normalized over the amount of proteins in cellular extracts. In both cell lines, the Sec efficiency is expressed relative to the one measured in the Dpl condition, set as 1. The HEK293 cell extracts with increasing amounts of ^{76}Se selenite (similar to Figure 2) were also analyzed by western immunobotting for the presence of selenoproteins GPX1, GPX4, SELENOP, and SELENOO (**b**), SELENOM, SELENOS, and SELENOF (**c**), and SEPHS2, TXNRD1, and TXNRD2 (**d**). Relative quantification of individual selenoprotein over actin is indicated at the bottom of the corresponding immunoblot, with maximum intensity set at 100. The quantification results are also represented as a function of selenium levels in the culture media and sorted into three classes. The optimal selenoprotein expression is either obtained with a low-dose (top-right panel), high-dose (middle-right panel) or in a narrow range (bottom-right panel) of selenium concentration.

To complement this analysis individually at the selenoprotein level, we performed western immunoblots with the cellular extracts from cells grown in the presence of various selenium levels (^{76}Se). Ten selenoproteins that are well expressed and readily detectable with validated antibodies in HEK293 cells were assayed [14,23], and the protein contents in every condition were normalized respective to actin levels. These normalized selenoprotein levels were plotted as a function of the added selenium in the media. It appeared that three main categories of selenium responses emerged from our experiments, as shown in

Figure 3b–d. The first category included GPX1, GPX4, SELENOP, and SELENOO, the levels of which were sensitive to selenium variation and reached an optimal expression level at a 100 nM selenium concentration or lower (Figure 3b). The second category comprised SELENOM, SELENOS, and SELENOF, which reached a maximum expression only at a 200 nM selenium concentration or higher (Figure 3c). The first and second categories were similar and fall into the "stress-related" family of selenoproteins. The last category included "house-keeping" selenoproteins, namely SEPHS2, TXNRD1, and TXNRD2, and had a very distinct response to selenium supplementation. They were either sensitive to the selenium variation with an inverse U-shaped curve (SEPHS2 and TXNRD2) or barely sensitive (TXNRD1), as shown in Figure 3d. Taken together, our data confirmed that the selenoproteome was highly modulated by selenium level variation in HEK293, with a specific hierarchy, mostly controlled at the translation stage.

2.4. Kinetics of Selenium Assimilation by HEK293 Cells as Selenoproteins and as Low-Molecular Selenocompounds

Then, we used the isotopically enriched selenite to analyze the kinetics of the selenium incorporation in selenoproteins. In this experiment, cells were grown in conditions with a limited amount of selenium (Unsup condition) for 3 days and then passaged in a growth medium supplemented with 100 nM of ^{77}Se-selenite, as illustrated in Figure 4a. The HEK293 cells were harvested at different time points (0, 3, 6, 10, 24, 48, and 72 h), and analyzed for the selenium isotopic composition by SEC-ICP MS (see Figure 4b). For each of the seven fractions producing a selenium signal, the isotopic ratio was calculated, and then plotted as a function of time (see the seven histograms above the chromatograms in Figure 4b). As mentioned before, only ^{80}Se and ^{77}Se were represented in these histograms. The ^{77}Se signal was analyzed with an exponential curve-fit to determine the time in the individual fraction when the half replacement of the original isotope was reached, and this time was referred to as $t_{1/2}$ (see in Figure 4a). The $t_{1/2}$ value illustrated the kinetic hierarchy of selenium assimilation by the cell between each fraction, regardless of whether they contained selenoproteins or other selenocompounds. Our experiments indicated that among all the fractions, F6 was the one which assimilated the newly introduced ^{77}Se isotope the fastest. The half replacement of the natural isotopes occurred after a 1.5 h selenium-supplementation. Then, selenoprotein fractions (F1 to F5) came second in the kinetic hierarchy with $t_{1/2}$ between 3.0 and 7.1 h (see Figure 4b). Finally, the selenium metabolite fraction F7 was the one which was the least efficient and the least rapid to incorporate ^{77}Se, with a $t_{1/2}$ longer than one day. Obviously, the selenium exchange in fraction F7 was rather inefficient, even after one passage. Our data suggest that selenium was rapidly assimilated in the selenoprotein fraction, and that one passage was sufficient to reach the plateau. Even if SEC-ICP MS did not allow the differentiation between individual selenoproteins, it provided important information about the hierarchy between selenoproteins and other selenocompounds.

2.5. Selenocysteine Insertion in Selenoprotein Was Timely Controlled in Response to Selenium Supplementation

We took advantage of our luciferase-stable expressing cell lines to further analyze how fast the translational UGA/Sec recoding machinery responded to selenium supplementation (cf. Figure 5a). Clearly, with both constructs, a rapid and massive increase in Sec insertion was observed at the first time point (3 h) to reach a plateau at approximately 6 h after a 100 nM selenium supplementation of the culture media. These data indicate that the translational machinery switched very quickly from an inefficient to an efficient state in response to selenium concentration variation.

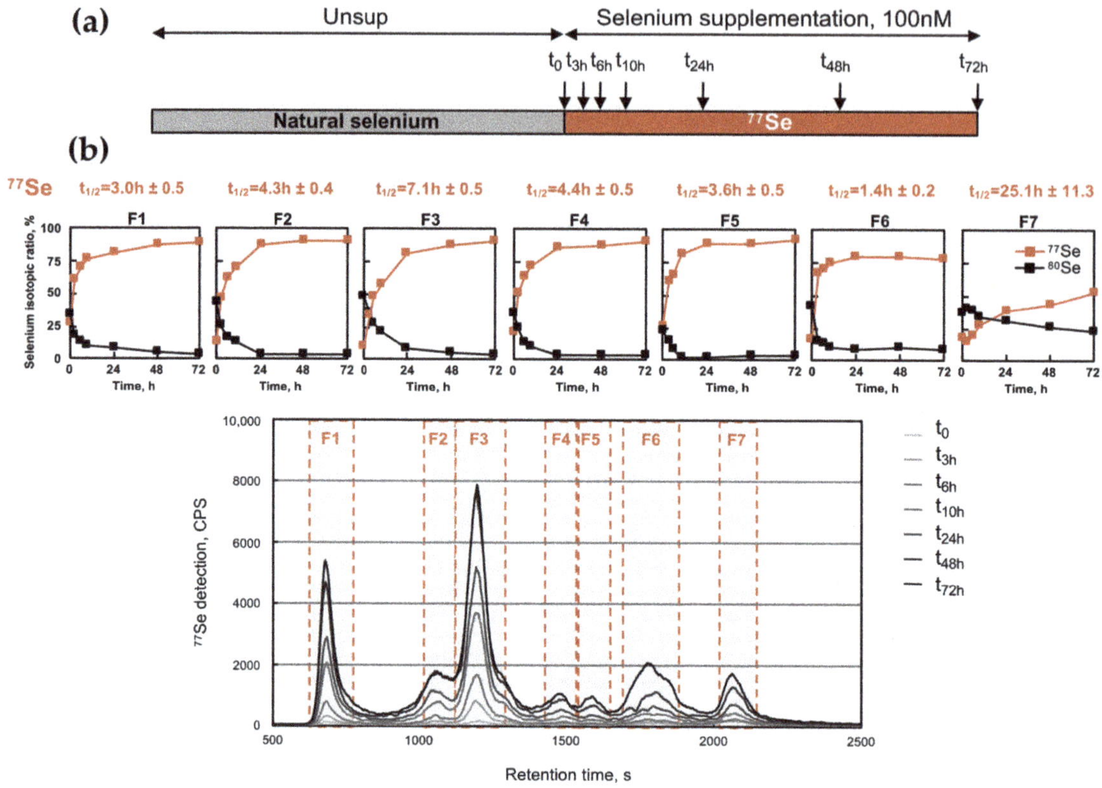

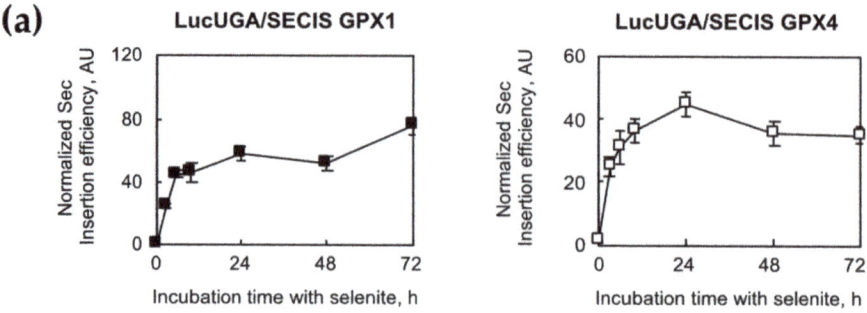

Figure 4. Time-course analysis of HEK193 cells labeled with 100 nM of ^{77}Se by SEC ICP MS. (**a**) Cells were grown for three days in a Dpl medium prior to being plated in fresh medium with 100 nM of ^{76}Se enriched selenium. Then, the cellular extracts were harvested at defined times post treatment (0, 3, 6, 10, 24, 48, and 72 h) and analyzed by SEC-ICP MS as described in Figure 2. (**b**) The chromatographic profile for ^{77}Se is shown in the bottom panel for the various cellular extracts. The isotopic pattern was analyzed in the seven identified fractions, as indicated by dashed boxes (F1 to F7). Only ^{77}Se and ^{80}Se are represented in red and black squares, respectively. The kinetics for ^{77}Se abundance in each fraction as a function of time were fitted as described in experimental procedures. The results of $t_{1/2}$ for ^{76}Se labeling are indicated on top of respective graphs.

Figure 5. *Cont.*

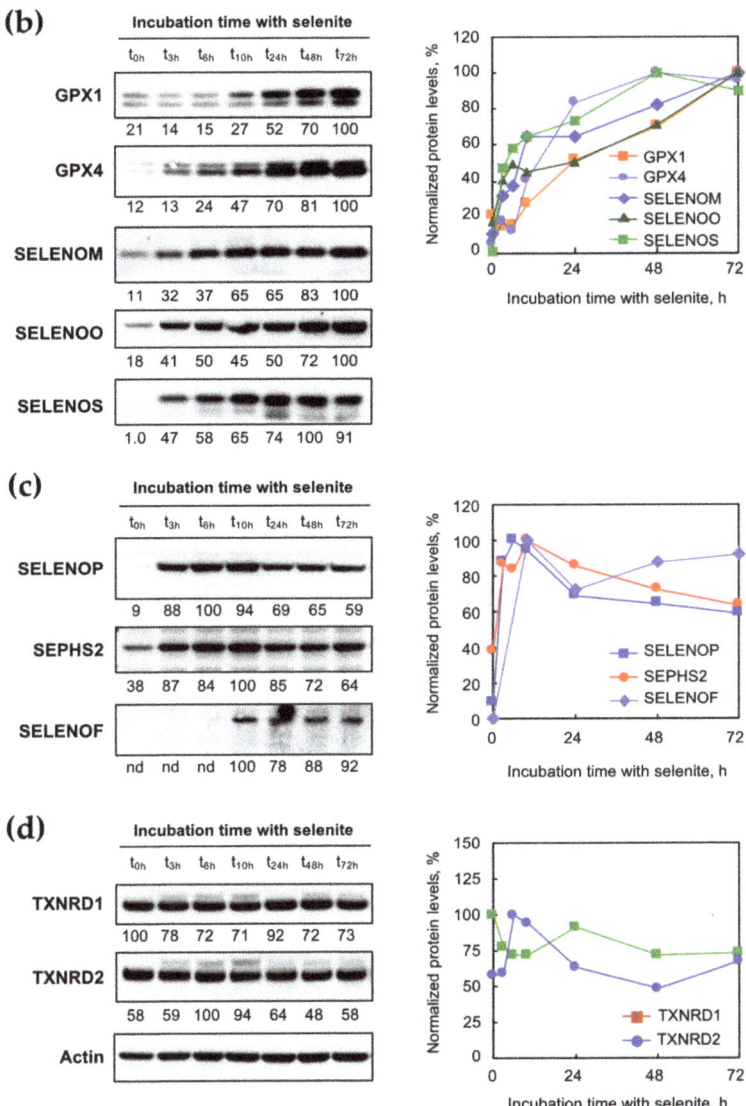

Figure 5. Time-course analysis of selenoprotein expression in medium containing 100 nM ^{77}Se. (**a**) The stimulation of selenocysteine insertion in response to selenium addition is not dependent on the nature of the SECIS element. The two cell lines containing Luc UGA/GPX1 or Luc UGA/GPX4 construct were grown for three days in a Dpl medium prior to being plated in medium containing 100 nM of selenium and harvested at a defined time as described in Figure 4. Cell extracts were analyzed for luciferase activities and Sec efficiency was expressed relative to the one measured at t_0, set as 1. The HEK293 cell extracts labeled with ^{77}Se (similar to Figure 4) were analyzed by western immunoblotting for the presence of selenoproteins GPX1, GPX4, SELENOM, SELENOO, and SELENOS (**b**), SELENOP, SEPHS2, and SELENOF (**c**), and TXNRD1, and TXNRD2 (**d**). Relative quantification of individual selenoprotein over actin is indicated at the bottom of the corresponding immunoblot, with maximum intensity set at 100. The quantification results are also represented as a function of time and sorted into three classes. The optimal selenoprotein expression was reached either after 48 h (top-right panel) or before 10 h (middle-right panel). Alternatively, the expression of selenoprotein can be stable with time (bottom-right panel).

In order to obtain information about the kinetic hierarchy between individual selenoproteins, we performed a western immunoblot with the extracts harvested at different time points. Again, three categories of selenoproteins can be deduced from the kinetic behavior, as shown in Figure 5b–d. The first category may refer to the so-called stress-related selenoproteins, with a rather slow increase in protein expression in response to the selenium supplementation. This category, composed of GPX1, GPX4, SELENOM, SELENOO, and SELENOS (Figure 5b), reached the maximum expression between 48 and 72 h. In contrast to that, the second category of selenoproteins, composed of SELENOP, SEPHS2, and SELENOF (Figure 5c), was characterized by a rapid and massive stimulation of protein expression, with maximal expression at 6 h for SELENOP or after 10 h for SELENOF and SEPHS2. Note that a decrease in the protein expression was observed at 72 h in the case of SELENOP and SEPHS2. This biphasic kinetic may reflect a negative feedback control of gene expression. The third category of selenoproteins, composed of TXNRD1 and TXNRD2 (Figure 5d), was poorly regulated over time by selenium supplementation and can be clearly associated with "housekeeping" selenoproteins. Taken together, our data indicated that the selenoproteome was highly regulated by selenium supplementation in HEK293, with specific individual kinetics.

2.6. Kinetics of Selenium Exchange in HEK293 Grown in Selenium-Supplemented Conditions

After having studied the response to selenium supplementation and having shown the ability to label the selenoproteome and selenocompounds with ^{76}Se and ^{77}Se enriched selenite, we then analyzed the half-lives of selenium in the different fractions detected by SEC-ICP MS, using constant conditions of selenium supplementation in the cell culture. To do so, we initially grew the HEK293 cells in 100 nM of ^{76}Se selenite for three days and then passaged them into a fresh medium containing 100 nM of ^{77}Se selenite, as illustrated in Figure 6. HEK293 cells were harvested at different time points (0, 3, 6, 10, 24, 48, and 72 h), and analyzed for the selenium isotopic composition by SEC-ICP MS as described earlier (see Figure 6b). The isotopic ratio was calculated in each fraction, and then plotted as a function of time. The signals of ^{80}Se, ^{76}Se, and ^{77}Se were represented in the seven histograms above the chromatograms of Figure 6b. The signals of ^{76}Se and ^{77}Se were exponentially curve-fitted to extract $t_{1/2}$, as described previously (see Figure 6a and Table 2). In almost every fraction, the deduced value of $t_{1/2}$ was highly similar for the decrease in ^{76}Se and the increased signal of ^{77}Se, confirming the robustness of our method to label and trace the selenium signal in cultured cells. This $t_{1/2}$ value indicated the rate by which selenium was exchanged in the molecules of each fraction, and was often referred to as the half-life. Indeed, at every time point, this value resulted from the balance between the neo-synthesis or intake (followed by the ^{77}Se signal) and the degradation or secretion (followed by the ^{76}Se signal). From our experiment, it appeared that F6 (metabolites) had the shortest half-life, below 6h. In contrast to that, fractions F2 to F5 containing selenoproteins had an average half-life of approximately one day (24 h). Interestingly, only F1 had a shorter $t_{1/2}$ of 15 h, suggesting a variability within the selenoproteome. Again, a unique behavior was observed for fraction F7. The level of the initial ^{80}Se stayed constant at 25% while ^{76}Se and ^{77}Se exchanged at $t_{1/2}$ of 21 h and 35 h, respectively. These data suggest that only half of the selenium content in F7 was efficiently exchangeable, the other half is likely a long-term storage form of selenium. In summary, our data indicate that our strategy is efficient to label the selenoproteome with different isotopes and to trace the evolution within it. Note that the determination of the individual selenoproteins cannot be performed by western immunoblotting since antibodies are unable to differentiate between ^{76}Se and ^{77}Se.

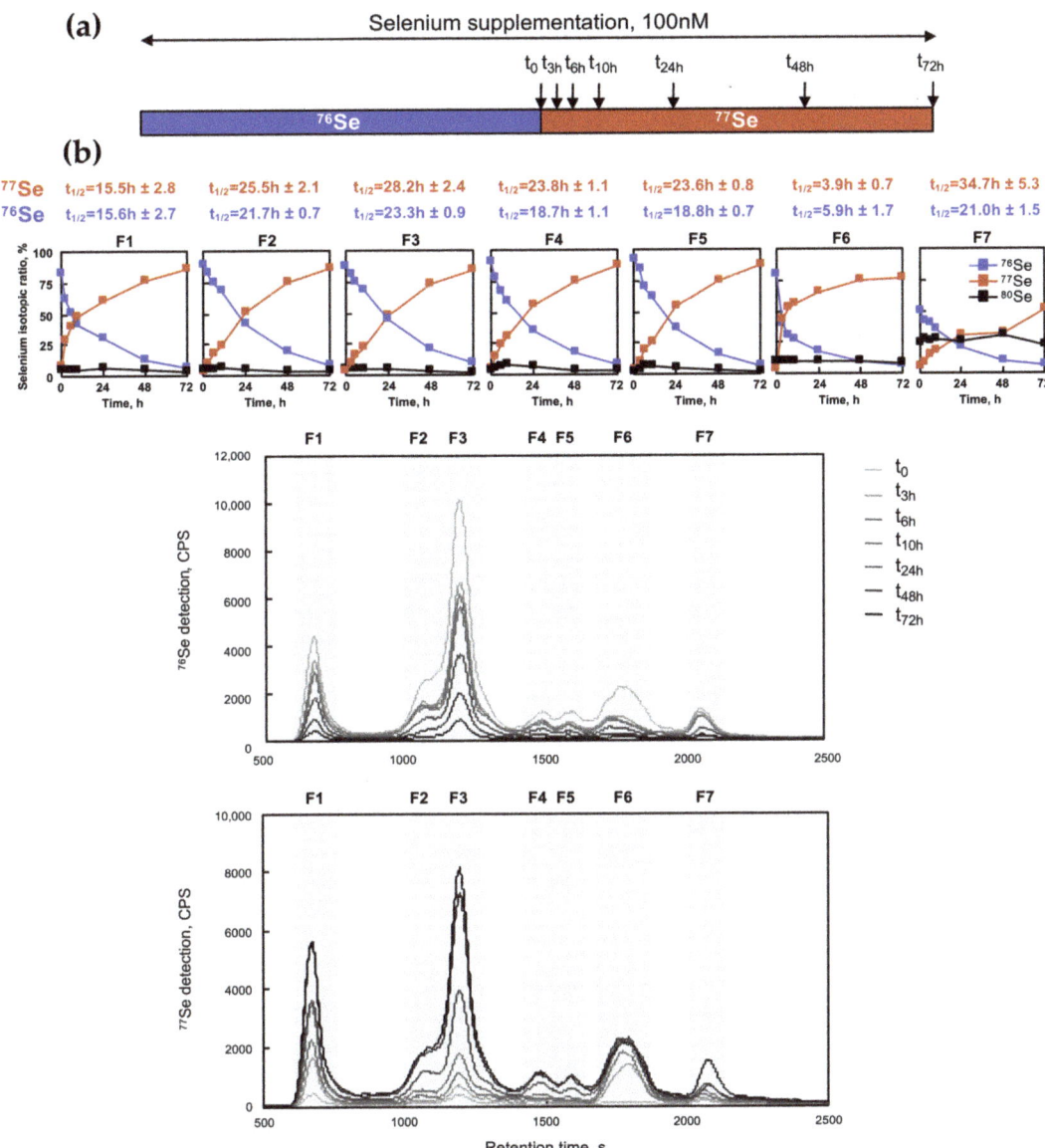

Figure 6. Time-course analysis of selenoprotein expression in medium containing 100 nM ^{77}Se. (**a**) Cells were grown for three days in a medium containing ^{76}Se (100 nM) prior to being plated in a fresh medium with 100 nM of ^{77}Se enriched selenium. Then, the cellular extracts were harvested at defined time post treatment (0, 3, 6, 10, 24, 48, and 72 h) and analyzed by SEC-ICP MS as described in Figure 4. (**b**) The chromatographic profiles for ^{76}Se and ^{77}Se are shown for the various cellular extracts in the bottom two panels. The isotopic pattern was analyzed in the seven identified fractions, as indicated by dashed boxes (F1 to F7). ^{76}Se, ^{77}Se and ^{80}Se are represented in blue, red and black squares, respectively. The kinetics for ^{76}Se and ^{77}Se abundance in each fraction as a function of time were fitted as described in experimental procedures. The results of $t_{1/2}$ for ^{76}Se and ^{77}Se for each fraction are indicated on top of the respective graphs.

Table 2. Summary of the distinct features for each fraction collected according to the SEC-ICP MS chromatogram.

Fraction N°	Dose–Response (Figure 2) $k_{1/2}$ ^{76}Se, nM	Kinetics of Selenium-Labeling (Figure 4) $t_{1/2}$ ^{77}Se, h	Kinetics of Selenium Exchange (Figure 6)	
			$t_{1/2}$ ^{77}Se, h	$t_{1/2}$ ^{76}Se, h
F1	14.9 ± 2.0	3.0 ± 0.5	15.5 ± 2.8	15.6 ± 2.7
F2	9.9 ± 1.3	4.3 ± 0.4	25.5 ± 2.1	21.7 ± 0.7
F3	8.2 ± 1.2	7.1 ± 0.5	28.2 ± 2.4	23.3 ± 0.9
F4	22.8 ± 3.9	4.4 ± 0.5	23.8 ± 1.1	18.7 ± 1.1
F5	23.9 ± 6.0	3.6 ± 0.5	23.6 ± 0.8	18.8 ± 0.7
F6	40.2 ± 9.6	1.4 ± 0.2	3.9 ± 0.7	5.9 ± 1.7
F7	55.1 ± 20.7	25.1 ± 11.3	34.7 ± 5.3	21.0 ± 1.5

3. Discussion

3.1. Hierarchy of Selenoproteins in Response to Selenium Supplementation: Kinetic and Dose–Response Analyses

Selenium is a vital component of selenoproteins and is co-translationally incorporated in the primary structure. As such and due to the fluctuations of its concentration in fluids in the body, there is a tight control of selenoprotein biogenesis by the selenium level. In this context of physiological selenium level variations, a hierarchical expression of the selenoproteome is established which maintains essential selenoproteins at the expense of the others. More importantly, among the stress-related selenoproteins, some are more sensitive than others to the changes in the selenium level and have different optimal expressions. This hierarchy of selenoprotein expression was described for the different glutathione peroxidases in rodents [34,35], but remain uncertain for other members. Yet, it becomes clear that most of this regulation occurs at the translational levels [11], but remains uncertain for other members. This classification between stress-related and housekeeping selenoproteins is not yet clear-cut but is also supported by unique features of Sec-tRNA$^{[Ser]Sec}$ in mammals [36].

Our study shows that the optimal selenium level for a maximum selenoprotein expression varies from one protein to another. We validated this finding in HEK293 cells, and we anticipate that this could also stand true in other cell lines. We observed that several selenoproteins, including GPX1, GPX4, SELENOO, SELENOM, and SELENOS had a similar dose–response and kinetic behavior in response to selenium supplementation. These selenoproteins are among the most responsive to the selenium level variation in terms of time and concentration. On the other side, TXNRD1 and TXNRD2 are poorly sensitive to the selenium level variation and can be considered as housekeeping members. An interesting selenoprotein in terms of selenium regulation is SEPHS2, for which a transient overexpression occurs as a function of selenium concentration and also as a function of time after selenium supplementation. Our data clearly suggest a tight control of this protein level with potential negative feedback. A remarkable characteristic of SEPHS2 is that it is implicated in selenoprotein biogenesis by producing a Sec-tRNA$^{[Ser]Sec}$ precursor, the selenophosphate ($H_2O_3PSe^-$). This process is strictly dependent on the bioavailability of intracellular selenium. Thus, the fact that SEPHS2 also contains a Sec residue in its catalytic site is a way of controlling the level of mature Sec-tRNA$^{[Ser]Sec}$ and, therefore, the selenoproteome. The amplification phase of SEPHS2 expression is easily explained by this mechanism. However, the mechanism for the negative feedback phase remains to be clarified, both temporally and at higher selenium levels.

3.2. Selenium-Enriched Isotope Labeling: An Alternative to Radioactive Labeling and a Novel Multiplexing Strategy for Selenoproteomic Analyses

In this work, we have developed an innovative strategy to label and trace selenoproteins in cultured cell lines, using isotopically enriched forms of selenium. In practice, six naturally present stable isotopes, namely ^{74}Se, ^{76}Se, ^{77}Se, ^{78}Se, ^{80}Se, and ^{82}Se can be used and are commercially available. Interestingly, the signal for these isotopes can be followed by either element (ICP ionization) or molecular (electrospray ionization) mass spectrometry. Here, we validated the potential of this strategy in HEK293 cells, but it can be applied to virtually any cell line which synthesizes selenoproteins. The use of SEC-ICP MS seems to be a robust and quick way to monitor the efficient labeling of a cellular extract. Our data demonstrate that selenite is readily and rapidly available for selenoprotein biogenesis. Our concept is likely to profit from recent advances in terms of proteomics, in terms of detection limits, concentration range, rapidity, and the number of proteins analyzed simultaneously [37]. All selenoproteins (except GPX6) have been detected from human samples in high throughput proteomics (www.proteomicsdb.org, accessed date 1 July 2021). The Sec-containing peptides resulting from tryptic digests have been detected and sequenced (by MS/MS fragmentation) for several human selenoproteins, including GPX1, GPX4, TXNRD1, TXNRD2, and SELENOF [26], but this could definitely be conducted for others. Additionally, the presence of selenium in a molecule such as a peptide can be inferred from its isotopic pattern. In this condition, the difference between protein abundance from one condition to another can be easily measured. Indeed, when using two selenium isotopes, pairs of selenopeptides that differ from selenium isotopes reflect the relative abundance of the corresponding protein, similar to what is conducted in stable isotope labeling with amino acids in cell culture (SILAC). The only, but important, difference with selenium-labeling versus SILAC is related to the availability of six selenium isotopes. Therefore, instead of looking at pairs of peptides, one could follow six different peptides simultaneously. This opens many multiplexing possibilities for selenoprotein labeling and tracing. One could consider that different cellular or tissue extracts originating from samples grown with various selenium isotopes could be quantified for selenoprotein contents using this multiplexing strategy. Non-radioactive multiple selenium-labeling of selenocompounds opens the way to quantitative selenoproteomics in the wide world of proteomics.

Recently, a selenocysteine-specific mass spectrometry-based technique was developed by the selective alkylation of selenocysteine [38]. This method allowed the systematic profiling and quantitative analysis of mouse selenoproteins, but can be applied to other species. In addition to known selenoprotein members described in [12], several novel candidates were proposed in which selenocysteine-containing peptides were characterized by mass spectrometry. In these peptides, the insertion of selenocysteine seemed to result from an inefficient UGA recoding event that occurs without a known SECIS element [38]. These data suggest that at least mammalian selenoproteomes could be more complex than previously expected. This new development in selenoproteome profiling can definitely benefit from selenium isotopic labeling and tracing that we have developed in our present study.

4. Materials and Methods

This manuscript adopts the new systematic nomenclature of selenoprotein names [9].

4.1. Materials

The HEK293 cell line used in this study was purchased from Life technologies. For the HEK293 cells stably expressing Luc UGA/GPX1 or Luc UGA/GPX4, they were generated and validated in [15,22,33]. Fetal calf serum (FCS), cell culture media and supplements were purchased from ThermoFisher Scientific (Waltham, MA, USA). Transferrin, insulin, 3,5,3'-triiodothyronine, hydrocortisone, EDTA, sodium selenite, and DTT were purchased from Merck (Darmstadt, Germany). ^{76}Se (99.8%) and ^{77}Se (99.2%) isotopically enriched selenites were provided by Isoflex (San Francisco, CA, USA). Antibodies were purchased from Abcam (Cambridge, UK) (GPX1, #ab108429; GPX4, #ab125066; SELENOF, #ab124840; SELENOM, #ab133681; SELENOO, #ab172957; SELENOP, #ab109514), ThermoFisher Scientific (TXNRD1, #LF-MA0015) and Merck (SELENOS, #HPA010025; TXNRD2, #HPA003323; SEPHS2, #WH0022928M2; Actin, #A1978). NuPAGE 4–12% bis–Tris polyacrylamide gels, MOPS, MES SDS running buffers and antiprotease inhibitor cocktail were purchased from ThermoFisher Scientific.

4.2. Cell Culture and Incubation with Different Selenium Doses

HEK293 cells were grown and maintained in 75 or 150 cm^2 plates in Dulbecco's Modified Eagle Medium (D-MEM) supplemented with a 2% fetal calf serum (FCS), 100 μg mL^{-1} streptomycin, 100 UI mL^{-1} penicillin, 1 mM sodium pyruvate, 2 mM L-glutamine, 5 mg L^{-1} transferrin, 10 mg L^{-1} insulin, 100 pM 3,5,3′-triiodothyronine (T3), and 50 nM hydrocortisone. This medium is referred to as Unsup and contains 3 nM of selenium as determined in [15,16,22,33]. Cells were cultivated in 5% CO_2 at 37 °C in a humidified atmosphere. The addition of selenium was performed with sodium selenite, either natural (Merck) or isotopically enriched (Isoflex), at the concentration indicated in figure legends for each experiment. After the treatment, cellular extracts were harvested in a 300 μL passive lysis buffer containing 25 mM of Tris phosphate at a pH of 7.8, 2 mM DTT, 2 mM EDTA, 1% Triton X100, and 10% glycerol. Then, protein concentrations were measured using the DC kit protein assay kit (Biorad, Hercules, CA, USA) in microplate assays using the microplate reader FLUOstar OPTIMA (BMG Labtech, Champigny-sur-Marne, France).

4.3. Evaluation of Selenocysteine Insertion Efficiency

To analyze Sec insertion efficiency in HEK293, we used luciferase-based reporter constructs that were validated for UGA/Sec recoding in transfected cells [22,33]. Briefly, the minimal SECIS elements from GPX1 and GPX4 were cloned downstream of a luciferase coding sequence, which was modified to contain an in frame UGA codon at position 258 (Luc UGA/SECIS), as shown in Figure 3a. HEK293 cells stably expressing Luc UGA/GPX1 and Luc UGA/GPX1 SECIS were previously generated and validated [15,22]. After being grown in various concentrations of sodium selenite, cells were harvested and the cellular extracts were assayed for luciferase activities by chemiluminescence (Luciferase assay systems, Promega, Charbonnières, France), in triplicate using a microplate reader FLUOstar OPTIMA (BMG Labtech). We arbitrarily expressed the Sec insertion efficiency relative to the luciferase activity measured in Unsup conditions, which was set at 1.

4.4. Fractionation of Selenium Containing Molecules by Size Exclusion Chromatography (SEC) with ICP MS Detection (SEC-ICP MS)

A precise amount of cellular extract, corresponding to 500 μg of proteins, was fractionated by HPLC (Agilent 1200 series, Santa Clara, CA, USA) using a SEC column (Superdex 200 10/300 GL, GE Healthcare, Chicago, IL, USA). The flow rate was set at 0.7 mL min^{-1} of a mobile phase (ammonium acetate buffered at pH 7.4). The injection volume was 100 μL. The calibration of the SEC column was performed using protein standards (thyroglobulin 670 kDa, transferrin 81 kDa, bovine albumin 66 kDa, chicken albumin 44 kDa, Mn-SOD 39.5 kDa, Cu/Zn SOD 32.5 kDa, carbonate dehydratase 29 kDa, myoglobin 16 kDa, metallothionein 6.8 kDa, and selenomethionine 0.198 kDa). Detection was achieved online by recording the absorbance at 280 nm (Agilent G1365B) and by ICP MS (Agilent 7500) for multi-isotopic detection (^{74}Se, ^{76}Se, ^{77}Se, ^{78}Se, ^{80}Se, and ^{82}Se) with an integration time per element of 0.1 s. UV^{280nm} and ICP MS signals were exported and treated with Microsoft Excel software to present chromatograms and isotopic graphs. Seven chromatographic fractions were defined with the elution times listed in Table 1. To integrate the total count of each selenium isotope ion per fraction, the background levels were determined and subtracted. The dose–response and kinetic curves were fitted using Kaleidragraph software with standard equations. For dose–response curves, it was as follows: $y = a + ((b.x)/(k + x))$, where k was the concentration of selenium allowing half saturation of the signal, and therefore referred to as $k_{1/2}$ throughout the manuscript. Kinetic curves were fitted with $y = a + b.\exp(-k.t)$, where the constant $t_{1/2} = (\ln(2)/k)$, and the represented time allowing an increase or decrease in the signal by 50%.

4.5. Protein Gels and Western Immunoblotting

Equal protein amounts (50 μg) were separated in Bis-Tris NuPAGE Novex Midi Gels and transferred onto nitrocellulose membranes using iBlot® DRy blotting System (ThermoFisher Scientific). Membranes were probed with primary antibodies (as indicated) and HRP-conjugated anti-rabbit or anti-mouse secondary antibodies (Merck). The chemiluminescence signal was detected using the ECL Select Western Blotting Detection Kit (GE Healthcare) and the PXi 4 CCD camera (Ozyme, Saint-Cyr-l'École, France). Image acquisition and data quantifications were performed with the Syngene softwares, GeneSys and Genetools (Ozyme), respectively.

4.6. Large Scale Fraction Collection for Western Immunoblotting Analyses

Larger amounts of cellular extracts, corresponding to 5 mg of proteins, were loaded on an SEC column in the same configuration as before, but fractions were collected as indicated in Table 1 instead of going to an ICP MS detector. The seven different fractions were lyophilized and resuspended in 200 µL of the lysis buffer. An aliquot of 10 µL of each fraction was analyzed by western immunoblotting after migration onto protein gels.

5. Conclusions

The selenoproteome is expressed from 25 selenoprotein genes in humans, where selenocysteine incorporation is genetically encoded. Their expression is finely controlled by the selenium status of the organism or the availability of selenium in the culture medium. Here, we have developed a new analytical strategy to study the expression of the selenoproteome in response to selenium supplementation using non-radioactive selenium isotopes (^{76}Se, ^{77}Se). We characterized the selective regulation of the most abundant selenoproteins in dose–response and kinetic experiments in HEK293 cell line. Our data suggest that the use of other natural isotopes of selenium (^{74}Se, ^{78}Se, ^{80}Se and ^{82}Se) in multiplexing experiments followed by inductively coupled plasma mass spectrometry detection is entirely feasible. Non-radioactive labeling of cellular selenocompounds paves the way for quantitative selenoproteomics and selenometabolomics in the ever-improving world of elemental and molecular mass spectrometry.

Author Contributions: Conceptualization, J.S., A.-L.B., Z.T.-H., R.L. and L.C.; methodology, J.S., A.-L.B., Z.T.-H., K.B., S.M., R.L. and L.C.; validation, J.S., A.-L.B., Z.T.-H. and L.C.; formal analysis, J.S., A.-L.B., Z.T.-H., M.M. and L.C.; investigation, J.S., A.-L.B., Z.T.-H., M.M., K.B., S.M., R.L. and L.C.; writing—original draft preparation, J.S., R.L. and L.C.; writing—review and editing, J.S., A.-L.B., Z.T.-H., M.M., K.B., S.M., R.L. and L.C.; supervision, R.L. and L.C.; project administration, R.L. and L.C.; funding acquisition, R.L. and L.C. All authors have read and agreed to the published version of the manuscript.

Funding: This work was supported by the CNRS (ATIP program to L.C.), the Fondation pour la Recherche Médicale (L.C.), the Ligue Contre le Cancer (Comité de l'Essonne, L.C.), the programme interdisciplinaire de recherche du CNRS longévité et vieillissement (L.C.), the Association pour la recherche sur le cancer [grants numbers 4849, L.C.] and the Agence Nationale de la Recherche [grant number ANR-09-BLAN-0048 to L.C. and R.L.]. J.S. was a Recipient of a fellowship from the Ministère Français de l'Enseignement Supérieur et de la Recherche.

Data Availability Statement: Requests for further information about resources, reagents, and data availability should be directed to the corresponding author.

Conflicts of Interest: The authors declare no conflict of interest.

References

1. Zhang, Z.H.; Song, G.L. Roles of selenoproteins in brain function and the potential mechanism of selenium in Alzheimer's disease. *Front. Neurosci.* **2021**, *15*, 646518. [CrossRef]
2. Guillin, O.M.; Vindry, C.; Ohlmann, T.; Chavatte, L. Selenium, selenoproteins and viral infection. *Nutrients* **2019**, *11*, 2101. [CrossRef] [PubMed]
3. Rocca, C.; Pasqua, T.; Boukhzar, L.; Anouar, Y.; Angelone, T. Progress in the emerging role of selenoproteins in cardiovascular disease: Focus on endoplasmic reticulum-resident selenoproteins. *Cell Mol. Life Sci.* **2019**, *76*, 3969–3985. [CrossRef]
4. Touat-Hamici, Z.; Legrain, Y.; Sonet, J.; Bulteau, A.-L.; Chavatte, L. Alteration of selenoprotein expression during stress and in aging. In *Selenium: Its Molecular Biology and Role in Human Health*, 4th ed.; Hatfield, D.L., Ulrich, S., Tsuji, P.A., Gladyshev, V.N., Eds.; Springer Science & Business Media LLC: New York, NY, USA, 2016; pp. 539–551.
5. Sonet, J.; Bulteau, A.-L.; Chavatte, L. Selenium and selenoproteins in human health and diseases. In *Metallomics: Analytical Techniques and Speciation Methods*; Michalke, B., Ed.; Wiley-VCH Verlag GmbH & Co. KGaA: Hoboken, NJ, USA, 2016; pp. 364–381.

6. Labunskyy, V.M.; Hatfield, D.L.; Gladyshev, V.N. Selenoproteins: Molecular pathways and physiological roles. *Physiol Rev.* **2014**, *94*, 739–777. [CrossRef]
7. Latrèche, L.; Chavatte, L. Selenium incorporation into selenoproteins, implications in human health. *Metal. Ions Biol. Med.* **2008**, *10*, 731–737.
8. Papp, L.V.; Lu, J.; Holmgren, A.; Khanna, K.K. From selenium to selenoproteins: Synthesis, identity, and their role in human health. *Antioxid. Redox. Signal.* **2007**, *9*, 775–806. [CrossRef]
9. Gladyshev, V.N.; Arner, E.S.; Berry, M.J.; Brigelius-Flohe, R.; Bruford, E.A.; Burk, R.F.; Carlson, B.A.; Castellano, S.; Chavatte, L.; Conrad, M.; et al. Selenoprotein gene nomenclature. *J. Biol. Chem.* **2016**, *291*, 24036–24040. [CrossRef] [PubMed]
10. Vindry, C.; Ohlmann, T.; Chavatte, L. Translation regulation of mammalian selenoproteins. *Biochim. Biophys. Acta Gen. Subj.* **2018**, *1862*, 2480–2492. [CrossRef] [PubMed]
11. Bulteau, A.-L.; Chavatte, L. Update on selenoprotein biosynthesis. *Antioxid. Redox. Signal.* **2015**, *23*, 775–794. [CrossRef]
12. Kryukov, G.V.; Castellano, S.; Novoselov, S.V.; Lobanov, A.V.; Zehtab, O.; Guigo, R.; Gladyshev, V.N. Characterization of mammalian selenoproteomes. *Science* **2003**, *300*, 1439–1443. [CrossRef]
13. Driscoll, D.M.; Copeland, P.R. Mechanism and regulation of selenoprotein synthesis. *Annu. Rev. Nutr.* **2003**, *23*, 17–40. [CrossRef]
14. Touat-Hamici, Z.; Bulteau, A.L.; Bianga, J.; Jean-Jacques, H.; Szpunar, J.; Lobinski, R.; Chavatte, L. Selenium-regulated hierarchy of human selenoproteome in cancerous and immortalized cells lines. *Biochim. Biophys. Acta Gen. Subj.* **2018**, *1862*, 2493–2505. [CrossRef]
15. Touat-Hamici, Z.; Legrain, Y.; Bulteau, A.-L.; Chavatte, L. Selective up-regulation of human selenoproteins in response to oxidative stress. *J. Biol. Chem.* **2014**, *289*, 14750–14761. [CrossRef]
16. Legrain, Y.; Touat-Hamici, Z.; Chavatte, L. Interplay between selenium levels, selenoprotein expression, and replicative senescence in WI-38 human fibroblasts. *J. Biol. Chem.* **2014**, *289*, 6299–6310. [CrossRef]
17. Papp, L.V.; Lu, J.; Striebel, F.; Kennedy, D.; Holmgren, A.; Khanna, K.K. The redox state of SECIS binding protein 2 controls its localization and selenocysteine incorporation function. *Mol. Cell Biol.* **2006**, *26*, 4895–4910. [CrossRef]
18. Hammad, G.; Legrain, Y.; Touat-Hamici, Z.; Duhieu, S.; Cornu, D.; Bulteau, A.-L.; Chavatte, L. Interplay between Selenium Levels and Replicative Senescence in WI-38 Human Fibroblasts: A Proteomic Approach. *Antioxidants* **2018**, *7*, 19. [CrossRef] [PubMed]
19. Allmang, C.; Wurth, L.; Krol, A. The selenium to selenoprotein pathway in eukaryotes: More molecular partners than anticipated. *Biochim. Biophys. Acta* **2009**, *1790*, 1415–1423. [CrossRef]
20. Hatfield, D.L.; Gladyshev, V.N. How selenium has altered our understanding of the genetic code. *Mol. Cell Biol.* **2002**, *22*, 3565–3576. [CrossRef] [PubMed]
21. Berry, M.J.; Tujebajeva, R.M.; Copeland, P.R.; Xu, X.M.; Carlson, B.A.; Martin, G.W., 3rd; Low, S.C.; Mansell, J.B.; Grundner-Culemann, E.; Harney, J.W.; et al. Selenocysteine incorporation directed from the 3′UTR: Characterization of eukaryotic EFsec and mechanistic implications. *Biofactors* **2001**, *14*, 17–24. [CrossRef]
22. Latreche, L.; Duhieu, S.; Touat-Hamici, Z.; Jean-Jean, O.; Chavatte, L. The differential expression of glutathione peroxidase 1 and 4 depends on the nature of the SECIS element. *RNA Biol.* **2012**, *9*, 681–690. [CrossRef] [PubMed]
23. Vindry, C.; Guillin, O.; Mangeot, P.E.; Ohlmann, T.; Chavatte, L. A Versatile Strategy to Reduce UGA-Selenocysteine Recoding Efficiency of the Ribosome Using CRISPR-Cas9-Viral-Like-Particles Targeting Selenocysteine-tRNA[Ser]Sec Gene. *Cells* **2019**, *8*, 574. [CrossRef]
24. Sonet, J.; Bierla, K.; Bulteau, A.L.; Lobinski, R.; Chavatte, L. Comparison of analytical methods using enzymatic activity, immunoaffinity and selenium-specific mass spectrometric detection for the quantitation of glutathione peroxidase 1. *Anal. Chim. Acta* **2018**, *1011*, 11–19. [CrossRef]
25. Sonet, J.; Mounicou, S.; Chavatte, L. Detection of selenoproteins by laser ablation inductively coupled plasma mass spectrometry (LA-ICP MS) in immobilized pH gradient (IPG) strips. *Methods Mol. Biol.* **2018**, *1661*, 205–217.
26. Bianga, J.; Touat-Hamici, Z.; Bierla, K.; Mounicou, S.; Szpunar, J.; Chavatte, L.; Lobinski, R. Speciation analysis for trace levels of selenoproteins in cultured human cells. *J. Proteom.* **2014**, *108*, 316–324. [CrossRef] [PubMed]
27. Bianga, J.; Ballihaut, G.; Pecheyran, C.; Touat, Z.; Preud'homme, H.; Mounicou, S.; Chavatte, L.; Lobinski, R.; Szpunar, J. Detection of selenoproteins in human cell extracts by laser ablation-ICP MS after separation by polyacrylamide gel electrophoresis and blotting. *J. Anal. At. Spectrom.* **2012**, *27*, 25–32. [CrossRef]
28. Ballihaut, G.; Mounicou, S.; Lobinski, R. Multitechnique mass-spectrometric approach for the detection of bovine glutathione peroxidase selenoprotein: Focus on the selenopeptide. *Anal. Bioanal. Chem.* **2007**, *388*, 585–591. [CrossRef]
29. Yim, S.H.; Tobe, R.; Turanov, A.A.; Carlson, B.A. Radioactive 75Se labeling and detection of selenoproteins. *Methods Mol. Biol.* **2018**, *1661*, 177–192. [PubMed]
30. Korotkov, K.V.; Kumaraswamy, E.; Zhou, Y.; Hatfield, D.L.; Gladyshev, V.N. Association between the 15-kDa selenoprotein and UDP-glucose:glycoprotein glucosyltransferase in the endoplasmic reticulum of mammalian cells. *J. Biol. Chem.* **2001**, *276*, 15330–15336. [CrossRef]
31. Vindry, C.; Ohlmann, T.; Chavatte, L. Selenium metabolism, regulation, and sex differences in mammals. In *Selenium, Molecular and Integartive Toxicology*; Michalke, B., Ed.; Springer International Publishing AG, Part of Springer Nature: Cham, Switzerland, 2018; pp. 89–107.

32. Sonet, J.; Mosca, M.; Bierla, K.; Modzelewska, K.; Flis-Borsuk, A.; Suchocki, P.; Ksiazek, I.; Anuszewska, E.; Bulteau, A.L.; Szpunar, J.; et al. Selenized plant oil is an efficient source of selenium for selenoprotein biosynthesis in human cell lines. *Nutrients* **2019**, *11*, 1524. [CrossRef]
33. Latreche, L.; Jean-Jean, O.; Driscoll, D.M.; Chavatte, L. Novel structural determinants in human SECIS elements modulate the translational recoding of UGA as selenocysteine. *Nucleic Acids Res.* **2009**, *37*, 5868–5880. [CrossRef] [PubMed]
34. Bermano, G.; Arthur, J.R.; Hesketh, J.E. Selective control of cytosolic glutathione peroxidase and phospholipid hydroperoxide glutathione peroxidase mRNA stability by selenium supply. *FEBS Lett.* **1996**, *387*, 157–160. [CrossRef]
35. Bermano, G.; Nicol, F.; Dyer, J.A.; Sunde, R.A.; Beckett, G.J.; Arthur, J.R.; Hesketh, J.E. Tissue-specific regulation of selenoenzyme gene expression during selenium deficiency in rats. *Biochem. J.* **1995**, *311 Pt 2*, 425–430. [CrossRef]
36. Carlson, B.A.; Lee, B.J.; Tsuji, P.A.; Tobe, R.; Park, J.M.; Schweizer, U.; Gladyshev, V.N.; Hatfield, D.L. Selenocysteine tRNA [Ser]Sec: From nonsense suppressor tRNA to the quintessential constituent in selenoprotein biosynthesis. In *Selenium: Its Molecular Biology and Role in Human Health*, 4th ed.; Hatfield, D.L., Ulrich, S., Tsuji, P.A., Gladyshev, V.N., Eds.; Springer Science & Business Media LLC: New York, NY, USA, 2016; pp. 3–12.
37. Wilhelm, M.; Schlegl, J.; Hahne, H.; Moghaddas Gholami, A.; Lieberenz, M.; Savitski, M.M.; Ziegler, E.; Butzmann, L.; Gessulat, S.; Marx, H.; et al. Mass-spectrometry-based draft of the human proteome. *Nature* **2014**, *509*, 582–587. [CrossRef] [PubMed]
38. Guo, L.; Yang, W.; Huang, Q.; Qiang, J.; Hart, J.R.; Wang, W.; Hu, J.; Zhu, J.; Liu, N.; Zhang, Y. Selenocysteine-specific mass spectrometry reveals tissue-distinct slenoproteomes and candidate selenoproteins. *Cell Chem. Biol.* **2018**, *25*, 1380–1388.e4. [CrossRef] [PubMed]

Review

Role of Selenium in Viral Infections with a Major Focus on SARS-CoV-2

Sabrina Sales Martinez [1], Yongjun Huang [1], Leonardo Acuna [2], Eduardo Laverde [2], David Trujillo [2], Manuel A. Barbieri [2], Javier Tamargo [1], Adriana Campa [1] and Marianna K. Baum [1],*

[1] Robert Stempel College of Public Health and Social Work, Florida International University, Miami, FL 33199, USA; saless@fiu.edu (S.S.M.); hyongjun@fiu.edu (Y.H.); jtamargo@fiu.edu (J.T.); campaa@fiu.edu (A.C.)

[2] College of Arts, Sciences & Education, Florida International University, Miami, FL 33199, USA; lacuna@fiu.edu (L.A.); eelaverd@fiu.edu (E.L.); datrujil@fiu.edu (D.T.); barbieri@fiu.edu (M.A.B.)

* Correspondence: baumm@fiu.edu

Abstract: Viral infections have afflicted human health and despite great advancements in scientific knowledge and technologies, continue to affect our society today. The current coronavirus (COVID-19) pandemic has put a spotlight on the need to review the evidence on the impact of nutritional strategies to maintain a healthy immune system, particularly in instances where there are limited therapeutic treatments. Selenium, an essential trace element in humans, has a long history of lowering the occurrence and severity of viral infections. Much of the benefits derived from selenium are due to its incorporation into selenocysteine, an important component of proteins known as selenoproteins. Viral infections are associated with an increase in reactive oxygen species and may result in oxidative stress. Studies suggest that selenium deficiency alters immune response and viral infection by increasing oxidative stress and the rate of mutations in the viral genome, leading to an increase in pathogenicity and damage to the host. This review examines viral infections, including the novel SARS-CoV-2, in the context of selenium, in order to inform potential nutritional strategies to maintain a healthy immune system.

Keywords: selenium; selenoproteins; virus; viral; infection; reactive oxygen species; antioxidant; HIV; HCV; HBV; coxsackie virus; influenza; glutathione peroxidase; thioredoxin reductase

1. Introduction

Viral infections have afflicted human health despite great advancements in scientific knowledge and technologies [1–3]. Most recently, the novel severe acute respiratory syndrome coronavirus 2 (SARS-CoV-2) has infected over 200 million individuals during 2019–August 2021 and has led to over 4.4 million deaths globally [4]. Selenium (Se), an essential trace element in humans, has a long history of lowering the occurrence and severity of viral infections [5–9]. Se deficiency impacts immune function [10], viral expression [8], selenoprotein expression [11], and alters antioxidant response [12], allowing for greater susceptibility to severe viral and bacterial infections [13]. Supplementing the diet with Se has demonstrated positive effects on enhancing immunity against viral attacks [5]. Much of the benefits derived from Se are due to its incorporation into selenocysteine, an important component of the antioxidant defense systems, including the regulation of glutathione peroxidase (GPXs) and thioredoxin reductase (TXNRD) activities [14]. Low levels of Se can lead to more severe forms of viral infections and adequate selenium levels may provide a protective effect toward the host response by affecting both immune response and oxidative stress [13,15]. Severe pathology in Se deficiency is evidenced by more frequent and graver symptoms, higher viral loads, declining levels of antioxidant enzymes such as GPX, and mutations to the viral genome. Studies conducted by Beck et al. described in this review, demonstrate that Se-deficiency is capable of increasing the virulence of a benign coxsackie

virus through viral mutations and these mutations have led to a reduction in GPX activity, therefore, resulting in oxidative stress [13,15].

The current coronavirus (COVID-19) infection pandemic has put a spotlight on the need to review the evidence on the impact of nutritional strategies to maintain a healthy immune system, as there are limited therapeutic treatments. Therefore, this review principally focuses on Se, in the context of viral infections, including the novel SARS-CoV-2. A review of the most common selenoproteins and their functions will be followed by the evidence on the role and impact of Se on the human host's ability to battle viral infections.

2. Selenoproteins and Functions

Selenoproteins are proteins that have incorporated the 21st amino acid in the genetic code, selenocysteine (Sec) into their polypeptide chain. Selenocysteine is a true proteinogenic amino acid in that it has its own unique codon (UGA), Sec insertion sequence (SECIS), Sec-specific elongation factor (eEFsec), transfer RNA (tRNASec), and is co-translationally inserted [16]. The biological functions of Se are mostly exerted through selenoprotein domains that contain Sec residues [17,18]. Twenty-four selenoprotein genes have been characterized in mice and 25 in humans [19,20]. Some of these selenoproteins demonstrated their essential roles in developmental processes and in disease pathogenesis [21,22]. Selenoproteins have been classified based on their known or suspected cellular functions; for example GPX 1–4 for antioxidation, TXNRD 1–3, methionine sulfoxide reductase B (MSRB)1, selenoproteins (SELENO) H, M, and W for redox regulation, iodothyronine deiodinase (DIO) 1–3 for thyroid hormone metabolism, SELENOP for selenium transport and storage, selenophosphate synthetase (SEPHS) 2 for the synthesis of selenophosphate, SELENOK and T for calcium metabolism, SELENON protein involved in myogenesis, SELENOF, I and S for protein folding, and SELENOO protein with AMPylation activity [21,22].

Only 2 of the 25 selenoproteins identified are extracellular, selenoprotein P (SELENOP), and extracellular glutathione peroxidase (GPX3) [23]. SELENOP is noteworthy in that it carries out the crucial role of distributing Se in plasma from the liver where dietary selenium is metabolized [24,25]. and contains up to 9 Sec residues [23]. SELENOP then binds to apolipoprotein E receptor-2 (apoER2) receptors on various tissues including the brain and testis or lipoprotein receptor megalin (Lrp2) for endocytosis in the kidneys for systemic distribution [18,26]. Different isoforms of SELENOP confer specificity to the various receptors [26]. Once endocytosed, Se can be used for the formation of other selenoproteins.

Among the more well-studied selenoproteins are those involved in maintaining homeostatic redox states, namely GPXs. There are 5 isoforms of GPXs that contain selenocysteine residues and they each occupy distinct regions of the cell. Each GPX isoform catalyzes the reduction of hydrogen peroxides using glutathione (GSH) as a cofactor, and in doing so, maintains cellular homeostasis. In this capacity, GPXs play a vital role not only in the prevention of oxidative stress but also in regulating redox signaling that can have broader effects on cell proliferation, apoptosis, and cytokine expression [27]. This important role of GPX and dietary Se is highlighted by the work of Beck et al., described later in this review, which demonstrated that Se-deficient mice were susceptible to a myocarditic strain of coxsackievirus whereas Se-adequate mice were unperturbed [10,28]. It was hypothesized that diminished activity of GPX was responsible for viral mutations in the Se-deficient mice and the production of more pathogenic virions [10,28].

Thioredoxin reductases (TXNRDs) are a family of selenoproteins, whose main function is to reduce thioredoxins but has broad specificity allowing it to reduce other endogenous and exogenous substrates [18,29]. The reduction of TXNRD's is accomplished by electrons from nicotinamide adenine dinucleotide phosphate (NADPH), which are transferred to the active site of TXNRDs via flavin adenine dinucleotide (FAD), a redox-active coenzyme [18,29]. Thioredoxins themselves reduce a number of small proteins including transcription factors such as nuclear factor kappa beta (NF-κβ), p53, redox factor 1 (REF-1), apurinic/apyrimidinic endonucleases 1 (AP-1), and phosphatase and tensin homologue deleted on chromosome ten (PTEN) thereby controlling the expression of various genes involved in cell growth, proliferation and inflammation [30].

Methionine sulfoxide reductase (MSR) is yet another selenoprotein with enzymatic activity that combats intercellular oxidative damage [18,31–33]. Specifically, MSR reduces the oxidized sulfur of methionine sulfoxide to produce the amino acid methionine [18,31–33]. Methionine sulfoxide alters protein function, may cause misfolding and dysregulates key cellular processes [33]. Lee et al. [32] demonstrated that MSRB1 is involved in cytokine regulation in macrophages by promoting the expression of anti-inflammatory cytokines IL-10 and IL-1RA. Coincidently, MSRB1 is the only methionine sulfoxide reductase that is a selenoprotein [32].

Unlike the aforementioned selenoproteins, SELENOK does not participate directly in redox reactions [34]. Instead SELENOK, a disordered endoplasmic reticulum transmembrane protein is reliant on partner proteins to form complexes and execute various functions [34]. One of the most well-established roles of SELENOK is in the palmitoylation of various substrates when complexed with the acyltransferase DHHC6 [34]. One target of the SELENOK/DHHC6 complex is inositol 1,4,5-trisphosphate receptor, an endoplasmic reticulum (ER) calcium channel protein that is stabilized once acylated [34]. SELENOK, therefore, plays a role in maintaining calcium efflux that is necessary for cell survival and immune cell responses [34].

3. Viral Infections, Reactive Oxygen Species, and Selenium

Viral infections are associated with an increase in reactive oxygen species (ROS), which are known to have both favorable and unfavorable effects on the host's cells and are important for the viral processes to maintain their infectious cycle [35,36]. ROS are a collection of molecules originating from molecular oxygen produced through redox reactions. Radical, having one free electron, and non-radical ROS may be formed by the partial reduction of oxygen [37,38]. Within the host cells, a balance between ROS production and ROS scavengers exists, where viral infections may create an unbalanced situation that develops into oxidative stress [36]. ROS scavengers and antioxidant systems that help to maintain redox homeostasis include catalase (CAT), superoxide dismutases (SODs), GPXs, TXNRDs, peroxidredoxin (PRDXs), and GSH. If oxidative stress remains unchecked, ROS may damage cellular proteins, lipids, and nucleic acids leading to adverse health effects and increasing the risk for several diseases [38,39].

Selenium plays a major role in redox regulation via its incorporation in the form of selenocysteine, into a family of proteins called selenoproteins [6]. Among these proteins, GPXs and TXNRDs play a critical role as antioxidants and confer protection against free radicals released by the immune response as a result of viral infection [8]. TXNRD defense involves the regulation of nuclear factor erythroid 2–related factor 2 (Nrf2) activation, which protects the cell against oxidative stress and inflammation [40], while GPX antioxidant defense involves the reduction of various hydroperoxides and oxidized antioxidants by catalyzing the conversion of GSH to glutathione disulfide [9]. Membrane integrity is also maintained through GPXs [41]. Studies have shown that inadequate Se intake affects GPX and TXNRD levels compromising cell-mediated immunity and humoral immunity linked to an increased inflammatory response by the production of ROS and redox control processes [40,42]. ROS production increases the expression of proinflammatory cytokines such as tumor necrosis factor-alpha (TNF-α) and interleukin (IL)-6, through the upregulation of NF-$\kappa\beta$ activities [42]. Selenium acts as a crucial antioxidant through the modulation of ROS production by inflammatory signaling inhibiting the activation of NF-$\kappa\beta$ cascade and suppressing the production of TNF-α and IL-6 [43]. Low Se levels decrease antioxidant activity thus decreasing free radical neutralization [44]. These studies suggest that Se deficiency alters immune response and viral infection by increasing oxidative stress and the rate of mutations in the viral genome, producing an increase in pathogenicity and damage to the host, as reported on influenza and coxsackie viruses [6].

4. Viral Infections and Selenium

4.1. Coxsackie Virus

Several decades of research have provided sufficient evidence to demonstrate a relationship between Se deficiency and Keshan disease, a grave cardiomyopathy. This cardiomyopathy is believed to be caused by infection with Coxsackie B virus, a nonenveloped single-stranded RNA virus pertaining to the *Picornaviridae* family, and exclusively found within China [10,45]. It was later discovered in the 1970s and 1980s that much of the Se levels in the soil, water, food, and human circulating fluids in areas affected by Keshan disease were deficient compared to other neighboring Chinese providences [46]. Sodium selenite was provided to the population and a prospective study showed that it prevented Keshan disease. Keshan disease was eradicated from endemic areas after the government enacted a Se supplementation policy, therefore, demonstrating that Keshan disease occurred due to two factors, infection with Coxsackie B virus and Se deficiency [47–49].

Animal studies conducted by Beck et al. confirmed the relationship with Se in mice by infection with a non-cardio-virulent strain of Coxsackie B virus (CVB3/0) and a myocarditic strain (CVB3/20). Heart damage was only observed in the mice fed a Se-deficient diet compared to mice fed a Se-sufficient diet for 4 weeks, and the typical human pathology was also observed [28,50]. These studies illustrated that Se deficiency caused a virus that was non-virulent to contribute towards the development of myocarditis in the host, and also increased its pathogenicity as the cardiovirulent strain under Se deficiency produced greater symptoms [28,51].

Additional observations by Beck and colleagues showed higher viral loads in the Se-deficient mice infected with both CVB3/0 and CVB3/20. The Se-deficient mice were found to have reduced T-cell expansion and diminished mRNA levels of cytokines compared to Se-adequate mice [15]. Subsequent studies led to the finding that Se deficiency was responsible for a change in the genotype of the benign coxsackie virus CVB3/0 that caused it to become virulent. Specifically, six nucleotides were modified that mimicked other virulent strains of CVB3 viruses. Due to these mutations, the virus now had the possibility to become pathogenic even in a Se-adequate host [45]. It was then hypothesized that a reduction in GPX activity was responsible for the viral mutations. Therefore, subsequent studies were conducted to demonstrate the protective effect of GPX1 in developing heart damage when infected with a benign strain of Coxsackie B virus (CVB3/0) [52]. Mice with a disrupted *gpx1* gene infected with CVB3/0 compared with wild type mice with an intact *gpx1* gene experienced myocarditis, and sequencing of the viruses from the mice with disrupted *gpx1* gene showed seven nucleotide changes in the Coxsackie virus. Interestingly, six of the seven nucleotide changes in the genome of the virus from the mice with disrupted *gpx1* genes matched the changes found in the Se-deficient mice previously [52]. These classic experiments exhibit how nutritional status as it pertains to Se, and its ability to protect antioxidant systems and immunity may impact the potential evolution of viruses to become more virulent.

4.2. Influenza

Influenza viruses, known to cause the flu, are enveloped, single-stranded RNA viruses within the *Orthomyxoviridae* family. Selenium deficiency has been associated with poor selenoprotein expression [11] and altered antioxidant response in viral influenza A infection [12]. The elegant in vitro [13] and animal experiments conducted by Dr. Beck et al. [28,45,46,50,52,53] were the first to demonstrate the detrimental effects of Se deficiency in influenza A virulence, which occurred due to changes in the viral genome [54]. Se deficiency in mice infected with a highly virulent Influenza A strain (Influenza A/PR/8/34), however, had higher levels of IL-2 expression followed by a higher level of IL-4 expression in the lung, and higher survival compared to Se-adequate mice. These studies demonstrated the essential role of Se in mounting an immune response to influenza A, by changing its virulence and altering the host's immune response [55].

These in vitro and animal studies suggested that in vivo Se supplementation might have a beneficial effect in humans, especially in the elderly, as the immune response is compromised by age. To test this hypothesis, Ivory et al. [56] conducted a 12-week randomized, double-blinded, placebo-controlled clinical trial in six groups of individuals with suboptimal Se status or plasma Se levels < 110 ng/mL to observe the response after the flu vaccine was provided. Four groups were given daily capsules of yeast: 20 participants were given 0 μg Se/day (placebo), 18 participants were given 50 μg Se/day, 21 participants received 100 μg Se/day, and 23 received 200 μg Se/day. Two groups were given onion-containing meals, 17 participants received < 1 μg Se/day (unenriched onions), and 18 participants received 50 μg Se/day (Se enriched onions). After 10 weeks of supplementation, all participants were administered the flu vaccine. Selenium supplementation compared to placebo had beneficial and detrimental effects on the cell immunity response to the flu vaccine that was dependent on the type of Se, and dose administered [56]. Se-yeast dose of 200 μg/day demonstrated enhanced IL-10 secretion and lower granzyme B content, a cytotoxic protease that induces apoptosis of target cells, within a cluster of differentiation 8 (CD8) cells, while 50 μg/day of Se through the enriched onion meal increased granzyme content and perforin in CD8 cells and reduced natural killer T-cells.

The effectiveness of antiviral agents such as amantadine (AM) [57], oseltamivir (OTV) [58], β-thujaplicin (TP) [59], and ribavirin (RBV) [60] to combat viral influence has been limited by the emergence of drug-resistant viruses. Biological Se nanoparticles are increasingly used as an agent to diminish drug resistance by "decorating" the nanoparticles with antiviral drugs to increase effectiveness, such as Se@AM, Se@OTV, Se@TP, Se@RBV. Selenium nanoparticles have been found to decrease oxidative stress, induce apoptosis of infected cells, and reduce lung cell damage during influenza infection, in addition to having low toxicity and increased drug activity in murine [59] and in vitro models [60].

4.3. Human Immunodeficiency Virus (HIV)

It is estimated that over 37 million people globally are living with HIV [61]. HIV is an enveloped, single-stranded RNA virus and without treatment causes a collapse of the immune system. The prevalence of Se deficiency in people living with HIV (PLWH) is reported to be around 7–66% and increases as HIV disease progresses over time [62–65]. Although antiretroviral therapy (ART) has allowed HIV disease to become a chronic disease, the immune system is still not fully reconditioned [66]. The rate of Se deficiency in PLWH in Sub-Saharan Africa is greater than that in the United States (U.S.A.) and the literature shows lower Se soil content in Sub-Saharan Africa [67]. Selenium deficiency in HIV disease is associated with disease progression and mortality, regardless of ART initiation [63,68–71]. Models of simian immunodeficiency virus also corroborate the relationship between Se deficiency and disease progression [72].

The relationship between HIV disease and increased oxidative stress [73–78] was recorded early in the disease, and the development of ROS and its association with HIV disease progression was documented in the very early stages of the emergence of the disease [79,80]. Lower GSH levels were found as HIV advances to acquired immunodeficiency syndrome (AIDS) [81] and alterations in antioxidant defense systems (SOD, CAT, and GPX) have also been observed in PLWH [75,77,82]. Supplementation of 250 μg of L-selenomethionine (100 μg of Se) for one year led to increased GPX activity [83] and adequate dietary Se intake was also associated with lower oxidative stress in PLWH [84].

Studies in children and adults living with HIV have found associations with Se deficiency and adverse health outcomes including mortality. Countries with a high prevalence of HIV such as South Africa, have shown that Se intake in children is not adequate and the overall diet quality is low [85]. In studies conducted in Nigeria, children with HIV had significantly lower Se levels compared to matched HIV-non-infected children in the same region and a high rate (>70%) of Se deficiency [86,87]. In children living with HIV in the U.S.A., Se deficiency was associated with advanced immunodeficiency [88] and mortality [63]. In adult PLWH who were initiating antiretroviral therapy (ART) or were already taking ART, Se deficiency was associated with HIV disease progression and mortality [70,71]. Additionally, Se values have been found to be lower in adult PLWH than in adults without HIV [89,90], as well as in later stages of HIV [90].

Several Se supplementation trials have been conducted within the United States [91,92], Tanzania [93,94], Botswana [95], and Rwanda [96]. These trials have demonstrated that Se supplementation in the dose of 200 µg in PLWH who are ART naïve or on ART may delay HIV disease progression through maintenance of cluster of differentiation 4 (CD4) cell counts. Hurwitz et al. [92] demonstrated that supplementation with Se resulted in significantly suppressed HIV viral load along with improved CD4 cell count. Trials using Se as part of a formula in combination with other micronutrients have not been able to discern the benefits of Se from the other components. We [95] concluded that supplementation with multivitamins and Se was safe and statistically significantly reduced the risk of immune decline and morbidity. Discrepancies between supplementation studies include the ART status of the participants, the baseline CD4 cell counts, and the length of time that the participants were supplemented and followed [97]. A Cochrane review of micronutrient supplementation and HIV concluded that additional trials with single nutrients were needed to build the evidence base for adults and establish long-term benefits [98].

4.4. Hepatitis B and C Viruses (HBV and HCV)

The World Health Organization (WHO) estimates that 257 million people and 71 million people were infected with hepatitis B virus (HBV) and hepatitis C virus (HCV), respectively [99]. Both HBV and HCV can cause acute and chronic hepatitis which can develop into cirrhosis and hepatocellular carcinoma (HCC). In 2015, there were 720,000 and 470,000 deaths from hepatic cirrhosis and HCC, respectively [99]. Though both HBV and HCV are hepatotropic, HCV belongs to the Flaviviridae family, whereas HBV is a member of the Hepadnaviridae family [100]. HBV is a partially double-stranded DNA virus that uses the host RNA polymerase II machinery to produce pre-genomic RNA, which is reverse transcribed into viral DNA [101]. HCV is an enveloped, single-stranded RNA virus, which exhibits extremely high mutation rates—up to one mutation per genome per generation cycle—since proofreading activity is lacking in RNA-dependent RNA polymerases required for its replication [102].

Selenium status determined by GPX3 activity and the concentration of serum/plasma Se and plasma SELENOP have been reported to be influenced in HBV and HCV patients in several studies. Serum Se concentrations are statistically significantly lower in HBV/HCV infected people when compared with the control group [103,104]. Selenium level is also associated with the severity and progression of the HBV/HCV disease [103,105,106]. Increased concentrations of aspartate aminotransferase and alanine aminotransferase (ALT) were independently associated with low Se concentration in chronic HBV patients with more hepatic damage [106]. The Se concentrations in plasma and erythrocytes are significantly lower in HCV-infected people than in controls and have an inversed correlation with HCV viral load [107]. Besides this, plasma Se level is statistically lower in people with HCV-induced cirrhosis with and without HCC when compared with HCV-infected people without liver cirrhosis or HCC [108].

Chronic HBV and HCV infection enhances ROS production and cause elevated oxidative stress and decreased antioxidant activity in liver cells [109–112]. ROS, produced as byproducts during cellular metabolism, have been implicated in several hepatic pathologies to maintain cellular homeostasis, including cell signaling, transcription, apoptosis, and immunomodulation [113–116]. Patients suffering from HBV or HCV infection show significant depletion of GSH and GPX when compared to non-infected participants [117,118]. As part of the antioxidant defense system, Se deficiency may be enhanced by the hepatic viral-induced oxidative stress and the requirement of selenoproteins during viral replication. An in vitro study showed that HCV can inhibit the expression of gastrointestinal-GPX (GPX2), a GPX that is also expressed in the liver, resulting in an increase in viral replication [7,119]. Nonstructural protein 5A (NS5A) of HCV, which is reported to enhance oxidative stress by perturbing Ca^{2+} homeostasis [120], also induces the expression and activity of GPX1 and GPX4 [121]. Besides, the GPX homology region overlaps the highly conserved *NS4* gene in HCV, supporting that the *NS4* gene is a functional GPX module [122]. Although the causes of Se deficiency in HBV and HCV are not fully understood, it is possible that the decreased level of circulating Se is related to the requirement of Se during viral replication.

The demand for Se during HBV and HCV infection causes the systemic deficiency of Se and can be compensated by supplementation. Supplementation of Se has shown to be protective against a wide range of different sources of oxidative stress and optimal immune responses [123,124]. Primary HCC incidence was reduced by 35.1% in Se supplemented people living with HBV as compared with non-supplemented people living with HBV [125]. However, when Se supplementation was stopped, primary HCC incidence began to increase [125]. Selenium also improves the rate and level of antibody response against the HBV vaccine in insulin-dependent diabetes mellitus cases that were on an accelerated vaccination schedule instead of a routine vaccine schedule [126]. A triple antioxidant combination of Se, alpha-lipoic acid, and silymarin supplementation in three chronic HCV-infected patients demonstrated an improvement in ALT [127]. However, a 6-month trial showed that those living with HCV supplemented with vitamins C, vitamin E, and 200 μg Se per day had an increase in antioxidant status with no beneficial effect on ALT, HCV viral load, or liver damage as compared with the non-supplemented individuals living with HCV [128].

As discussed in the previous paragraph, Se deficiency has been involved in the pathogenesis of HBV and HCV infection. In turn, the deficiency of Se leads to elevated oxidative stress, pathological changes, and inflammation in the liver [129]. Histological study shows hepatic sinus expansion, lymphocyte infiltration, and stripe-like hyperplasia in the liver with Se deficiency. Liver inflammation is initiated by Se deficiency as pro-inflammatory factors and molecules, such as IL-1β, IL-6, IL-12, NF-κβ, and NF-κβ p65, were all significantly higher in the Se-deficient group [129]. Hepatic antioxidant capacity is also influenced by Se deficiency as a decrease in both mRNA expression of selenoprotein genes (GPX1 and GPX3), as determined by quantitative real-time PCR and the level of selenoproteins (GPX1, GPX4, and TXNRD1), identified by global proteomics, are observed [129–131]. Interestingly, an in vitro study showed that Se deficiency can result in oxidative stress and apoptosis of non-HBV-infected hepatocytes, whereas HBV-infected hepatocytes gain a survival capacity and escape from the apoptosis consequence [132].

4.5. Poliovirus

Poliovirus is part of the *Picornaviridae* family of RNA viruses that are non-enveloped and may infect vertebrate animals [133]. Infection generates high levels of ROS and reactive nitrogen species as well as antioxidant enzymes being downregulated within cells that have been infected [134]. The supplementation of Se has been shown to improve the response of the vaccine for the poliovirus more in patients that have less optimal immune systems based on Se status, although the impact of supplementation on patients with optimal immune systems based on Se status is unclear [135]. Furthermore, the supplementation of Se did not affect all aspects of an individual's immune response shown in the same trial where a live poliomyelitis vaccine was given to people with low Se status. This resulted in the increase of T cell and IL-10 production but did not affect the natural killer (NK) or B cell count, still resulting in the rapid removal of poliomyelitis from the patients supplemented [136]. Selenium also did not affect the levels of CD4+ T helper (Th) 1 cells to Th2 cells or the humoral immune response [135] in a different trial where patients were given a dose of the poliovirus vaccine and took either a placebo, 50 μg or 100 μg of Se. An increase in the antibody titers within all groups that were relatively equal was shown [135]. Se supplementation prior to the polio vaccine seemed to only enhance the cellular antiviral immune response.

4.6. Severe Acute Respiratory Syndrome Coronavirus 2 (SARS-CoV-2)

The novel COVID-19 is caused by SARS-CoV-2, a single-stranded RNA coronavirus. The severity of the disease has been linked to aging and comorbidities such as hypertension, diabetes, obesity, cardiovascular disease, kidney disease, cancer, and pulmonary diseases [137,138]. Most of the people who test positive for COVID-19 develop mild or no symptoms, while others develop acute respiratory distress syndrome (ARDS), heart failure, blood clots, neurological complications, and elevated inflammatory response [137,139]. SARS-CoV-2 pathology has been associated with an increased immune response, leading to a release of cytokines and chemokines, also known as cytokine storm [140], as well as increased inflammatory markers such as D-dimer and ferritin [141,142]. This hyperactive inflammatory response may also bring about severe pathology in the brain [143]. SARS-CoV-2 may directly impact the central nervous system and enter the brain through various routes [144–147]. Increased systemic inflammation promoted by SARS-CoV-2 has the potential to disturb the blood-brain barrier and co-morbidities associated with severe cases of COVID-19 may enable the attack of the brain by SARS-CoV-2 [143,148].

It has been noted that there is a potential and developing relationship between Se levels and COVID-19 outcomes. Proposed mechanisms by which Se may act upon the SARS-CoV-2 virus based on previous research in RNA viruses include restoration of GPX and TXNRD thus reducing oxidative stress, reduction of viral-induced cell apoptosis, provision of Se for the host's antioxidant needs, protection of endothelial cells, and reduced blood platelet aggregation [149,150]. COVID-19 is associated with a heightened level of oxidative stress and inflammation that are implicated in the pathogenesis of pulmonary disease [151]. GSH provides protection to the epithelial barrier within the lungs, and it has been suggested that improvement of GSH levels would be a strategy that may protect against inflammation and oxidant-related damage in the lungs [151]. A study conducted by Mahmoodpoor et al. [152] supplemented sodium selenite in patients with ARDS, often associated with severe cases of COVID-19, and found that it restored the antioxidant capability of the lungs, reduced inflammation, and improved respiratory mechanics. Lower total lymphocytes and CD4+ T, CD8+ T, B, and NK cells were found in COVID-19 patients and those with severe cases compared to mild cases of COVID-19 had lower lymphocyte subsets [153]. The function and differentiation of B and T cells may be affected by Se status [154]. Deficiency of Se in mice has been associated with lower T cell proliferation, while supplementation increased T cell activity and differentiation [155].

Clinical data investigating Se and COVID-19 are sparse; however, some reports from China and other countries globally have surfaced. In China, where there is a wide range of soil Se levels and thus a variation of Se daily intake, a linear association has been demonstrated between reported cure rates of COVID-19 and Se hair concentration data, dating from 2011 and older [156]. The same research group in China documented higher fatality risk in cities that had selenium-deficient levels in crops and topsoil compared to cities with non-deficient selenium levels in crops and topsoil [157]. Intake of Se varies worldwide, and China is known to be one of the most Se deficient countries in the world, with a wide range of levels that differs from lowest to highest in the world. COVID-19 fatality rate varies across different regions in China, suggesting that Se status may be related to COVID-19 outcomes [156,158]. In the city of Wuhan, where the SARS-CoV-2 virus was first discovered, and in other cities such as Suizhou and Xiaogan, low Se soil status was associated with the highest COVID-19 incidence [156]. In contrast, cities such as Enshi, Yichang, and Xiangyan, where high Se intake occurs, had the lowest COVID-19 incidence [156]. In contrast, in a retrospective study completed in Wuhan, China, with hospitalized COVID-19 patients, the severity of COVID-19 was associated with higher Se levels in urine [159]. The authors hypothesize that liver abnormalities due to the severity of the disease may have impacted the excess urinary Se found in severe COVID-19 patients [159].

Studies conducted in other parts of the world are showing similar relationships to those completed in China. In a study conducted in South Korea on hospitalized COVID-

19 patients, 42% were found to be Se deficient and as the severity of disease increased, Se plasma levels decreased [160]. These patients also experienced additional nutritional deficiencies. COVID-19 patients compared to healthy controls in India, Iran and Russia had significantly lower plasma Se levels [161–163]. A greater rate of low plasma Se levels (<70 ng/mL) was found in COVID-19 patients (43%) compared to controls in India (20) [161]. Lung damage, as assessed by computer tomography, was inversely associated with Se levels in Russia [163].

COVID-19 patients may also experience increases in oxidative stress and increases in Se-related markers and lower Se levels have been documented in these patients. Moghaddam al. [164] observed an association between markers of Se status and COVID-19 outcomes from COVID-19 patients in Germany. Serum Se and SELENOP concentrations were lower in COVID-19 patients compared to a reference European population. A comparison of patients that survived compared to those who died from COVID-19 showed that the deceased had a significantly greater deficiency of serum Se and SELENOP concentrations than those who survived. In addition, those who died had significantly lower serum Se, SELENOP levels, and GPX compared with patients who survived. A study in Belgium using a convenience sample of patients hospitalized with severe COVID-19 pneumonia observed statistically lower GSH levels and higher GPX levels compared with reference intervals among other results showing elevated markers of oxidative stress and lower antioxidant status [165]. Recently, Polonikov [166] hypothesized that GSH deficiency plays a major role in augmenting SARS-CoV-2 oxidative damage, which leads to greater disease progression and mortality. This viewpoint was based on data showing lower GSH and higher ROS levels in COVID-19 patients with mild disease and increasing severity that included higher viral load with GSH deficiency [166] and work completed by Hurwitz et al. [167] that demonstrated improvement in dyspnea with high dose oral and IV GSH in two patients with underlying conditions who tested positive for COVID-19. These conclusions were based on very small samples and therefore require additional larger clinical studies to replicate the findings and eventual intervention studies. The evidence presented above suggests that Se availability contributes to resisting SARS-CoV-2 infection, corresponding with studies that show adequate levels of Se status maintains an appropriate immune response to viral infection [6,134,136].

There are no known published Se supplementation clinical trials in the context of COVID-19 at this time and one study is currently listed on clincaltrials.gov that will examine the efficacy of Se (selenious acid infusion also known as sodium selenite) for the treatment of moderately-ill, severely ill, and critically ill COVID-19 patients (Identifier: NCT04869579). Sodium selenite supplementation has been proposed for the prevention of COVID-19 infections and severe disease [149,168]. Sodium selenite is easily available, short-term toxicity is marginal and may cross the blood-brain barrier [149]. This chemical form may oxidize thiol groups located in the virus protein disulfide isomerase, which would interfere with its ability to infiltrate the cell membrane and produce an infection [168]. TXNRD activity increases quickly after supplementation with sodium selenite in cancer cell lines and critically ill patients [169,170] and has demonstrated reduced ROS production and viral-induced cell apoptosis in cell culture studies [171]. A common feature of COVID-19 is thrombotic complications and altered platelet function is believed to affect the sequelae of this infection [172]. Sodium selenite has also been shown to have an anti-aggregating effect through its reduction of thromboxane A2 formation, an important factor in blood platelet activation and formation [173]. The effectiveness of sodium selenite for the prevention and management of COVID-19 should be tested immediately as the COVID-19 pandemic continues to persist and threaten the health of individuals globally, thus necessitating rapidly accessible treatment strategies.

Since Se has pronounced therapeutic potential for the treatment of viral infections and other conditions such as cancer, Se nanomedicine has received a lot of attention. Se nanoparticles are known to have low toxicity with marked and selective cytotoxic effects with small quantities [174]. Additionally, Se nanoparticles have high effectiveness in the inhibition of oxidative damage [175–177]. Recently published data show that Se nanoparticles activate programmed cell death in target cancer tissue through calcium $(Ca)^{2+}$ signaling pathways [178]. Immune cells also require calcium flux to generate oxidative stress [174]. Through chemical methods, Se nanoparticles may be produced with Se sources that include sodium selenite, selenious acid, and sodium selenosulfate [174]. Due to the developing relationship between Se and COVID-19, Se nanomedicine is being suggested as a tool in the fight against SARS-CoV-2 [179]. Currently, there are tremendous prospects of using nanomedicine in ARDS for the prevention, diagnosis, and treatment, which may have applicability for COVID-19 [180]. Jin et al. [181] discovered that an organic Se compound known as Ebselen, and a promising antioxidant drug, could inhibit SARS-CoV-2 by penetrating the cell membrane and displaying antiviral activity. Ebselen is known to have anti-inflammatory activity, mimic GPX activity, and should be considered for clinical studies [181,182].

5. Nutrition and Recommended Intakes and Supplementation of Selenium

Optimal nutrition is important for regulating inflammatory and oxidative stress processes within the body [183]. These processes are important for the maintenance of the immune system and previous research has shown that nutritional status affects health outcomes in viral infections [13]. The number of chronic conditions globally has increased [184] and seems to have a strong influence on the disease progression of COVID-19 [185]. Therefore, a healthy dietary pattern, a modifiable risk factor, may reduce chronic conditions, the development of infections, and the severity of viral infections [186,187]. Low intakes of Se, zinc, magnesium, copper, vitamins A, B6, B12, C, D, and E, and omega-3 fatty acids have been associated with worse outcomes in viral infection and lower immunity [188–190] and should also be considered for the prevention and management of COVID-19 [191–193]. More research is needed to further define the role of nutrition in COVID-19 infection and disease progression and appropriate doses. However, at a minimum in order to support the functions of the immune system, recommendations for the consumption of nutrients that may impact immunity should be the in amounts directed by the reference nutrient intakes or recommended daily allowance [187].

Intake of Se by humans may vary according to differences in sources of food, accumulation of Se in animals and the content of Se in the soil [8]. Countries with poor Se in the soil include Finland, New Zealand, the United Kingdom, sub-Saharan Africa, and certain areas of China where Keshan is prevalent [194,195]. Consequently, the differences in intake of Se may be quite large, for example, daily Se intake in Europe is estimated to be about 40 µg per day and in the U.S.A. about 90–134 µg per day [196]. Selenium is plentiful in Brazil nuts, seafood, organ meats, muscle meats, cereals, grains, and dairy [197], and the diet in the U.S.A. provides Se mainly from grains, meat, poultry, fish, and eggs [198]. Selenium in vegetables is predominantly found as selenomethionine, selenium-methylselenocysteine or γ-glutamyl-selenium-methylselenocysteine and in meat as selenocysteine [8]. Inorganic Sec compounds including sodium selenite and selenate may be found in dietary supplements [8].

The current recommended intake for Se in the U.S.A. for adults is 55 µg per day. This recommendation is based on the consumption of Se needed to maximize the action of the selenoprotein GPX [199]. The WHO recommended nutrient intakes for Se in adults are 26 µg per day for women and 34 µg per day for men [200]. The Tolerable Upper Intake Level (UL) for Se in adults is 400 µg per day and the limit is based on the increased risk for selenosis [199]. The European Food Safety Authority in the European Union set the daily adequate intake for Se at 70 µg [201]. Selenosis in humans may cause loss of hair, thickened and stratified nails, and a garlic-like odor present in the mouth and skin [202,203]. Toxicity of Se appears to be less common than Se deficiency and has been reported to be caused by over supplementation and accidental consumption of high doses through consumption of foods grown in soil with large amounts of Se present [9,204].

6. Conclusions

Selenium plays an important role in the host during viral infections, assisting in redox homeostasis, antioxidant defense, and minimizing oxidative stress (Table 1). These protective roles are accomplished largely through its incorporation into selenoproteins. Antioxidant defense systems that incorporate selenoproteins, mainly GPXs and TXNRDs, are crucial for reducing oxidative stress created by an imbalance of ROS as a result of viral infections. Selenium deficiency may also have an effect on the viral genome leading to greater pathogenicity. Adequate Se intake is imperative for these systems to be functional and provide full enzymatic activities. The data on the relationship between Se and the novel SARS-CoV-2 are still evolving, however, preliminary results show a link between Se status and severity of COVID-19 outcomes. Therefore, Se status should be reviewed in patients with COVID-19 as a risk factor for graver outcomes. The literature on RNA viruses provides promising mechanisms of action for the use of Se in the prevention and disease management of COVID-19. Sodium selenite has been proposed as a preventive measure and adjuvant therapy for COVID-19 based on its potential ability to restore GPX and TXNRD activity, reduce viral-induced cell apoptosis, protect endothelial cells, and reduce blood platelet aggregation. Se nanoparticles should also be considered as a mechanism to deliver Se to target organs such as the lungs and deliver Se without risks of toxicity. Data available from other viral infections in conjunction with the current COVID-19 data provide sufficient justification for future and timely Se intervention studies.

Table 1. Summary of Selenium Studies.

Topic	Conclusions	References
Viral Infections, Reactive Oxygen Species (ROS), and Selenium (Se)	Viral Infections are associated with ROS. Glutathione peroxidases (GPXs) and thioredoxin reductases (TXNRDs) (family of selenoproteins) play a role as antioxidants and confer protection against free radicals as a result of viral infection. Se intake may affect GPXs and TXNRDs levels.	[8,35,36,40,42]
Coxsackie Virus	Keshan disease responsive to sodium selenite supplementation. Keshan disease due to infection with Coxsackie B virus and Se deficiency. Benign Coxsackie B virus became virulent when mice were Se-deficient and greater pathology in cardiovirulent Coxsackie B virus strain. Se deficiency was responsible for a change in the genotype of the benign coxsackie virus CVB3/0 that caused it to become virulent and decreased the activity of GPX.	[28,45,47–52,54]
Influenza	Se deficiency has been associated with poor selenoprotein expression, altered antioxidant response, and viral genome changes in viral influenza A infection. Se supplementation in healthy older adults yielded beneficial and detrimental effects related to anti-flu immunity.	[11–13,55,56]
Human Immunodeficiency Virus (HIV)	Se deficiency was associated with advanced immunodeficiency and mortality. Se supplementation in HIV has demonstrated benefits on HIV disease progression.	[63,68–71,86–88,91–96]
Hepatitis B and C Viruses	Se levels associated with HBV/HCV infection, severity, and progression of disease. Depletion of GSH and GPX in HBV/HCV. Se supplementation in areas of low intake may prevent HBV and primary liver cancer. Se deficiency associated with inflammation of the liver.	[103–106,117,118,125,129]
Poliovirus	Supplementation of Se to improve the response of polio vaccine remains inconclusive.	[135,136]
Severe Acute Respiratory Syndrome Coronavirus 2 (SARS-CoV-2)	Se soil status may be associated with COVID-19 incidence and severity of COVID-19 outcomes in China. COVID-19 infection and severity associated with lower Se levels, greater oxidative stress, and lower antioxidant status.	[158,159,161–167]

Author Contributions: Conceptualization, S.S.M. and M.K.B.; Methodology, S.S.M.; Resources, M.K.B.; Writing—Original Draft Preparation, S.S.M., L.A., Y.H., A.C., E.L. and D.T.; Writing—Review and Editing, S.S.M., A.C., Y.H., J.T., M.A.B. and M.K.B.; Supervision, M.K.B.; Project Administration, M.K.B. All authors have read and agreed to the published version of the manuscript.

Funding: Research in the lab of MKB is funded by the National Institutes of Health (NIH)/National Institute on Drug Abuse (NIDA), Grant #3U01DA040381-05S1 as a part of the RADx-UP program and Grant #1U01DA040381-01 Cohort Studies on HIV/AIDS and Substance Abuse in Miami.

Institutional Review Board Statement: Not applicable.

Informed Consent Statement: Not Applicable.

Data Availability Statement: Not Applicable.

Conflicts of Interest: The authors declare no conflict of interest.

References

1. Krause, R.M. The Origin of Plagues: Old and New. *Science* **1992**, *257*, 1073–1078. [CrossRef] [PubMed]
2. Piret, J.; Boivin, G. Pandemics Throughout History. *Front. Microbiol.* **2021**, *11*, 3594. [CrossRef] [PubMed]
3. Parvez, M.K.; Parveen, S. Evolution and Emergence of Pathogenic Viruses: Past, Present, and Future. *Intervirology* **2017**, *60*, 1–7. [CrossRef]
4. WHO Coronavirus (COVID-19) Dashboard. WHO Coronavirus (COVID-19) Dashboard with Vaccination Data. Available online: https://covid19.who.int/ (accessed on 18 August 2021).
5. Avery, J.C.; Hoffmann, P.R. Selenium, Selenoproteins, and Immunity. *Nutrients* **2018**, *10*, 1203. [CrossRef]
6. Guillin, O.M.; Vindry, C.; Ohlmann, T.; Chavatte, L. Selenium, Selenoproteins and Viral Infection. *Nutrients* **2019**, *11*, 2101. [CrossRef]
7. Hariharan, S.; Dharmaraj, S. Selenium and Selenoproteins: It's Role in Regulation of Inflammation. *Inflammopharmacology* **2020**, *28*, 1. [CrossRef]
8. Rayman, M.P. Selenium and Human Health. *Lancet* **2012**, *379*, 1256–1268. [CrossRef]
9. Fairweather-Tait, S.J.; Bao, Y.; Broadley, M.R.; Collings, R.; Ford, D.; Hesketh, J.E.; Hurst, R. Selenium in Human Health and Disease. *Antioxid. Redox Signal.* **2011**, *14*, 1337–1383. [CrossRef]
10. Beck, M.A.; Levander, O.A.; Handy, J. Selenium Deficiency and Viral Infection. *J. Nutr.* **2003**, *133*, 1463S–1467S. [CrossRef]
11. Sheridan, D.; Zhong, N.; Carlson, B.; Perella, C.; Hatfield, D.; Beck, M. Decreased Selenoprotein Expression Alters the Immune Response during Influenza Virus Infection in Mice. *J. Nutr.* **2007**, *137*, 1466–1471. [CrossRef]
12. Stýblo, M.; Walton, F.; Harmon, A.; Sheridan, P.; Beck, M. Activation of Superoxide Dismutase in Selenium-Deficient Mice Infected with Influenza Virus. *J. Trace Elem. Med. Biol.* **2007**, *21*, 52–62. [CrossRef]
13. Beck, M.; Handy, J.; Levander, O. Host Nutritional Status: The Neglected Virulence Factor. *Trends Microbiol.* **2004**, *12*, 417–423. [CrossRef]
14. Labunskyy, V.M.; Hatfield, D.L.; Gladyshev, V.N. Selenoproteins: Molecular Pathways and Physiological Roles. *Physiol. Rev.* **2014**, *94*, 739. [CrossRef]
15. Beck, M. Selenium and Host Defence towards Viruses. *Proc. Nutr. Soc.* **1999**, *58*, 707–711. [CrossRef]
16. Vindry, C.; Ohlmann, T.; Chavatte, L. Translation Regulation of Mammalian Selenoproteins. *Biochim. Biophys. Acta Gen. Subj.* **2018**, *1862*, 2480–2492. [CrossRef]
17. Tobe, R.; Mihara, H. Delivery of Selenium to Selenophosphate Synthetase for Selenoprotein Biosynthesis. *Biochim. Biophys. Acta Gen. Subj.* **2018**, *1862*, 2433–2440. [CrossRef]
18. Roman, M.; Jitaru, P.; Barbante, C. Selenium Biochemistry and Its Role for Human Health. *Metallomics* **2013**, *6*, 25–54. [CrossRef]
19. Kryukov, G.V.; Castellano, S.; Novoselov, S.V.; Lobanov, A.V.; Zehtab, O.; Guigó, R.; Gladyshev, V.N. Characterization of Mammalian Selenoproteomes. *Science* **2003**, *300*, 1439–1443. [CrossRef]
20. Ha, H.Y.; Alfulaij, N.; Berry, M.J.; Seale, L.A. From Selenium Absorption to Selenoprotein Degradation. *Biol. Trace Elem. Res.* **2019**, *192*, 26. [CrossRef]
21. Sreelatha, A.; Yee, S.S.; Lopez, V.A.; Park, B.C.; Kinch, L.N.; Pilch, S.; Servage, K.A.; Zhang, J.; Jiou, J.; Karasiewicz-Urbańska, M.; et al. Protein AMPylation by an Evolutionarily Conserved Pseudokinase. *Cell* **2018**, *175*, 809. [CrossRef]
22. Pitts, M.W.; Hoffmann, P.R. Endoplasmic Reticulum-Resident Selenoproteins as Regulators of Calcium Signaling and Homeostasis. *Cell Calcium* **2018**, *70*, 76. [CrossRef]
23. Burk, R.F.; Hill, K.E.; Motley, A.K. Selenoprotein Metabolism and Function: Evidence for More than One Function for Selenoprotein P. *J. Nutr.* **2003**, *133*, 1517S–1520S. [CrossRef]
24. Schweizer, U.; Streckfuß, F.; Pelt, P.; Carlson, B.A.; Hatfield, D.L.; Köhrle, J.; Schomburg, L. Hepatically Derived Selenoprotein P Is a Key Factor for Kidney but Not for Brain Selenium Supply. *Biochem. J.* **2005**, *386*, 221. [CrossRef] [PubMed]

25. Carlson, B.A.; Moustafa, M.E.; Sengupta, A.; Schweizer, U.; Shrimali, R.; Rao, M.; Zhong, N.; Wang, S.; Feigenbaum, L.; Byeong, J.L.; et al. Selective Restoration of the Selenoprotein Population in a Mouse Hepatocyte Selenoproteinless Background with Different Mutant Selenocysteine TRNAs Lacking Um34. *J. Biol. Chem.* **2007**, *282*, 32591–32602. [CrossRef] [PubMed]
26. Burk, R.F.; Hill, K.E. Selenoprotein P-Expression, Functions, and Roles in Mammals. *Biochim. Biophys. Acta* **2009**, *1790*, 1441. [CrossRef]
27. Lubos, E.; Kelly, N.; Oldebeken, S.; Leopold, J.; Zhang, Y.; Loscalzo, J.; Handy, D. Glutathione Peroxidase-1 Deficiency Augments Proinflammatory Cytokine-Induced Redox Signaling and Human Endothelial Cell Activation. *J. Biol. Chem.* **2011**, *286*, 35407–35417. [CrossRef] [PubMed]
28. Beck, M.A.; Kolbeck, P.C.; Rohr, L.H.; Shi, Q.; Morris, V.C.; Levander, O.A. Benign Human Enterovirus Becomes Virulent in Selenium-Deficient Mice. *J. Med. Virol.* **1994**, *43*, 166–170. [CrossRef] [PubMed]
29. Mustacich, D.; Powis, G. Thioredoxin Reductase. *Biochem. J.* **2000**, *346*, 1. [CrossRef]
30. Holmgren, A.; Lu, J. Thioredoxin and Thioredoxin Reductase: Current Research with Special Reference to Human Disease. *Biochem. Biophys. Res. Commun.* **2010**, *396*, 120–124. [CrossRef]
31. Tarrago, L.; Kaya, A.; Weerapana, E.; Marino, S.M.; Gladyshev, V.N. Methionine Sulfoxide Reductases Preferentially Reduce Unfolded Oxidized Proteins and Protect Cells from Oxidative Protein Unfolding. *J. Biol. Chem.* **2012**, *287*, 24448. [CrossRef]
32. Lee, B.C.; Lee, S.-G.; Choo, M.-K.; Kim, J.H.; Lee, H.M.; Kim, S.; Fomenko, D.E.; Kim, H.-Y.; Park, J.M.; Gladyshev, V.N. Selenoprotein MsrB1 Promotes Anti-Inflammatory Cytokine Gene Expression in Macrophages and Controls Immune Response in Vivo. *Sci. Rep.* **2017**, *7*, 5119. [CrossRef]
33. Colombo, G.; Meli, M.; Morra, G.; Gabizon, R.; Gasset, M. Methionine Sulfoxides on Prion Protein Helix-3 Switch on the α-Fold Destabilization Required for Conversion. *PLoS ONE* **2009**, *4*, e4296. [CrossRef]
34. Marciel, M.P.; Hoffmann, P.R. Molecular Mechanisms by Which Selenoprotein K Regulates Immunity and Cancer. *Biol. Trace Elem. Res.* **2019**, *192*, 60. [CrossRef]
35. Khomich, O.A.; Kochetkov, S.N.; Bartosch, B.; Ivanov, A.V. Redox Biology of Respiratory Viral Infections. *Viruses* **2018**, *10*, 392. [CrossRef]
36. Molteni, C.G.; Principi, N.; Esposito, S. Reactive Oxygen and Nitrogen Species during Viral Infections. *Free Radic. Res.* **2014**, *48*, 1163–1169. [CrossRef]
37. Sies, H.; Jones, D.P. Reactive Oxygen Species (ROS) as Pleiotropic Physiological Signalling Agents. *Nat. Rev. Mol. Cell Biol.* **2020**, *21*, 363–383. [CrossRef]
38. Ray, P.D.; Huang, B.W.; Tsuji, Y. Reactive Oxygen Species (ROS) Homeostasis and Redox Regulation in Cellular Signaling. *Cell Signal.* **2012**, *24*, 981–990. [CrossRef]
39. Pizzino, G.; Irrera, N.; Cucinotta, M.; Pallio, G.; Mannino, F.; Arcoraci, V.; Squadrito, F.; Altavilla, D.; Bitto, A. Oxidative Stress: Harms and Benefits for Human Health. *Oxid. Med. Cell. Longev.* **2017**, *2017*, 8416763. [CrossRef] [PubMed]
40. Locy, M.L.; Rogers, L.K.; Prigge, J.R.; Schmidt, E.E.; Arnér, E.S.J.; Tipple, T.E. Thioredoxin Reductase Inhibition Elicits Nrf2-Mediated Responses in Clara Cells: Implications for Oxidant-Induced Lung Injury. *Antioxid. Redox Signal.* **2012**, *17*, 1407. [CrossRef]
41. Ammendolia, D.A.; Bement, W.M.; Brumell, J.H. Plasma Membrane Integrity: Implications for Health and Disease. *BMC Biol.* **2021**, *19*, 71. [CrossRef] [PubMed]
42. Hardy, G.; Hardy, I.; Manzanares, W. Selenium Supplementation in the Critically Ill. *Nutr. Clin. Pract.* **2012**, *27*, 21–33. [CrossRef]
43. Steinbrenner, H.; Sies, H. Protection against Reactive Oxygen Species by Selenoproteins. *Biochim. Biophys. Acta Gen. Subj.* **2009**, *1790*, 1478–1485. [CrossRef] [PubMed]
44. Heyland, D.K.; Dhaliwal, R.; Suchner, U.; Berger, M.M. Antioxidant Nutrients: A Systematic Review of Trace Elements and Vitamins in the Critically Ill Patient. *Intensive Care Med.* **2004**, *31*, 327–337. [CrossRef]
45. Beck, M.A.; Shi, Q.; Morris, V.C.; Levander, O.A. Rapid Genomic Evolution of a Non-Virulent Coxsackievirus B3 in Selenium-Deficient Mice Results in Selection of Identical Virulent Isolates. *Nat. Med.* **1995**, *1*, 433–436. [CrossRef] [PubMed]
46. GQ, Y.; JS, C.; ZM, W.; KY, G.; LZ, Z.; XC, C.; XS, C. The Role of Selenium in Keshan Disease. *Adv. Nutr. Res.* **1984**, *6*, 203–231. [CrossRef]
47. Xu, G.; Wang, S.; Gu, B.; Yang, Y.; Song, H.; Xue, W.; Liang, W.; Zhang, P. Further Investigation on the Role of Selenium Deficiency in the Aetiology and Pathogenesis of Keshan Disease. *Biomed. Environ. Sci.* **1997**, *10*, 316–326.
48. Cheng, Y.Y.; Qian, P.C. The Effect of Selenium-Fortified Table Salt in the Prevention of Keshan Disease on a Population of 1.05 Million. *Biomed. Environ. Sci.* **1990**, *3*, 422–428. [PubMed]
49. Wei-Han, Y. A Study of Nutritional and Bio-Geochemical Factors in the Occurrence and Development of Keshan Disease : The 6th Conference on Prevention for Rheumatic Fever and Rheumatic Heart Disease. *JPN Circ. J.* **1982**, *46*, 1201–1207. [CrossRef]
50. Beck, M.A.; Williams-Toone, D.; Levander, O.A. Coxsackievirus B3-Resistant Mice Become Susceptible in Se/Vitamin E Deficiency. *Free Radic. Biol. Med.* **2003**, *34*, 1263–1270. [CrossRef]
51. Beck, M.A.; Kolbeck, P.C.; Shi, Q.; Rohr, L.H.; Morris, V.C.; Levander, O.A. Increased Virulence of a Human Enterovirus (Coxsackievirus B3) in SeleniumDeficient Mice. *J. Infect. Dis.* **1994**, *170*, 351–357. [CrossRef]
52. Beck, M.A.; Esworthy, R.S.; Ho, Y.-S.; Chu, F.-F. Glutathione Peroxidase Protects Mice from Viral-Induced Myocarditis. *FASEB J.* **1998**, *12*, 1143–1149. [CrossRef] [PubMed]

53. Jaspers, I.; Zhang, W.; Brighton, L.E.; Carson, J.L.; Styblo, M.; Beck, M.A. Selenium Deficiency Alters Epithelial Cell Morphology and Responses to Influenza. *Free Radic. Biol. Med.* **2007**, *42*, 1826. [CrossRef] [PubMed]
54. Beck, M.A.; Nelson, H.K.; Shi, Q.; Dael, P.; van Schiffrin, E.J.; Blum, S.; Barclay, D.; Levander, O.A. Selenium Deficiency Increases the Pathology of an Influenza Virus Infection. *FASEB J.* **2001**, *15*, 1481–1483. [CrossRef] [PubMed]
55. Li, W.; Beck, M.A. Selenium Deficiency Induced an Altered Immune Response and Increased Survival Following Influenza A/Puerto Rico/8/34 Infection. *Exp. Biol. Med.* **2017**, *232*, 412–419. [CrossRef]
56. Ivory, K.; Prieto, E.; Spinks, C.; Armah, C.N.; Goldson, A.J.; Dainty, J.R.; Nicoletti, C. Selenium Supplementation Has Beneficial and Detrimental Effects on Immunity to Influenza Vaccine in Older Adults. *Clin. Nutr.* **2017**, *36*, 407. [CrossRef] [PubMed]
57. Li, Y.; Lin, Z.; Guo, D.; Zhao, M.; Xia, Y.; Wang, C.; Xu, T.; Zhu, B. Inhibition of H1N1 Influenza Virus-Induced Apoptosis by Functionalized Selenium Nanoparticles with Amantadine through ROS-Mediated AKT Signaling Pathways. *Int. J. Nanomed.* **2018**, *13*, 2005. [CrossRef] [PubMed]
58. Li, Y.; Lin, Z.; Guo, M.; Xia, Y.; Zhao, M.; Wang, C.; Xu, T.; Chen, T.; Zhu, B. Inhibitory Activity of Selenium Nanoparticles Functionalized with Oseltamivir on H1N1 Influenza Virus. *Int. Nanomed.* **2017**, *12*, 5733. [CrossRef] [PubMed]
59. Wang, C.; Chen, H.; Chen, D.; Zhao, M.; Lin, Z.; Guo, M.; Xu, T.; Chen, Y.; Hua, L.; Lin, T.; et al. The Inhibition of H1N1 Influenza Virus-Induced Apoptosis by Surface Decoration of Selenium Nanoparticles with β-Thujaplicin through Reactive Oxygen Species-Mediated AKT and P53 Signaling Pathways. *ACS Omega* **2020**, *5*, 30633. [CrossRef]
60. Lin, Z.; Li, Y.; Gong, G.; Xia, Y.; Wang, C.; Chen, Y.; Hua, L.; Zhong, J.; Tang, Y.; Liu, X.; et al. Restriction of H1N1 Influenza Virus Infection by Selenium Nanoparticles Loaded with Ribavirin via Resisting Caspase-3 Apoptotic Pathway. *Int. J. Nanomed.* **2018**, *13*, 5787. [CrossRef]
61. HIV/AIDS. Available online: https://www.who.int/data/gho/data/themes/hiv-aids (accessed on 25 August 2021).
62. Baum, M.K. Role of Micronutrients in HIV-Infected Intravenous Drug Users. *J. Acquir. Immune Defic. Syndr.* **2000**, *25* (Suppl. S1), S49–S52. [CrossRef] [PubMed]
63. Campa, A.; Shor-Posner, G.; Indacochea, F.; Zhang, G.; Lai, H.; Asthana, D.; Scott, G.B.; Baum, M.K. Mortality Risk in Selenium-Deficient HIV-Positive Children. *J. Acquir. Immune Defic. Syndr. Hum. Retrovirol.* **1999**, *20*, 508–513. [CrossRef] [PubMed]
64. Osuna-Padilla, I.A.; Briceño, O.; Aguilar-Vargas, A.; Rodríguez-Moguel, N.C.; Villazon-De la Rosa, A.; Pinto-Cardoso, S.; Flores-Murrieta, F.J.; Perichart-Perera, O.; Tolentino-Dolores, M.; Vargas-Infante, Y.; et al. Zinc and Selenium Indicators and Their Relation to Immunologic and Metabolic Parameters in Male Patients with Human Immunodeficiency Virus. *Nutrition* **2020**, *70*, 110585. [CrossRef]
65. Shivakoti, R.; Gupte, N.; Yang, W.T.; Mwelase, N.; Kanyama, C.; Tang, A.M.; Pillay, S.; Samaneka, W.; Riviere, C.; Berendes, S.; et al. Pre-Antiretroviral Therapy Serum Selenium Concentrations Predict WHO Stages 3, 4 or Death but Not Virologic Failure Post-Antiretroviral Therapy. *Nutrients* **2014**, *6*, 5061. [CrossRef] [PubMed]
66. Bloch, M.; John, M.; Smith, D.; Rasmussen, T.A.; Wright, E. Managing HIV-Associated Inflammation and Ageing in the Era of Modern ART. *HIV Med.* **2020**, *21*, 2–16. [CrossRef]
67. Hurst, R.; Siyame, E.W.P.; Young, S.D.; Chilimba, A.D.C.; Joy, E.J.M.; Black, C.R.; Ander, E.L.; Watts, M.J.; Chilima, B.; Gondwe, J.; et al. Soil-Type Influences Human Selenium Status and Underlies Widespread Selenium Deficiency Risks in Malawi. *Sci. Rep.* **2013**, *3*, 1425. [CrossRef]
68. Stone, C.A.; Kawai, K.; Kupka, R.; Fawzi, W.W. Role of Selenium in HIV Infection. *Nutr. Rev.* **2010**, *68*, 671–681. [CrossRef] [PubMed]
69. Kupka, R.; Msamanga, G.I.; Spiegelman, D.; Rifai, N.; Hunter, D.J.; Fawzi, W.W. Selenium Levels in Relation to Morbidity and Mortality among Children Born to HIV-Infected Mothers. *Eur. J. Clin. Nutr.* **2005**, *59*, 1250–1258. [CrossRef]
70. Baum, M.K.; Shor-Posner, G.; Lai, S.; Zhang, G.; Lai, H.; Fletcher, M.A.; Sauberlich, H.; Page, J.B. High Risk of HIV-Related Mortality Is Associated with Selenium Deficiency. *J. Acquir. Immune Defic. Syndr. Hum. Retrovirol.* **1997**, *15*, 370–374. [CrossRef] [PubMed]
71. Constans, J.; Pellegrin, J.L.; Sergeant, C.; Simonoff, M.; Pellegrin, I.; Fleury, H.; Leng, B.; Conri, C. Serum Selenium Predicts Outcome in HIV Infection. *J. Acquir. Immune Defic. Syndr. Hum. Retrovirol.* **1995**, *10*, 392. [CrossRef] [PubMed]
72. Xu, X.M.; Carlson, B.A.; Grimm, T.A.; Kutza, J.; Berry, M.J.; Arreola, R.; Fields, K.H.; Shanmugam, I.; Jeang, K.T.; Oroszlan, S.; et al. Rhesus Monkey Simian Immunodeficiency Virus Infection as a Model for Assessing the Role of Selenium in AIDS. *J. Acquir. Immune Defic. Syndr.* **2002**, *31*, 453–463. [CrossRef]
73. Repetto, M.; Reides, C.; Gomez Carretero, M.L.; Costa, M.; Griemberg, G.; Llesuy, S. Oxidative Stress in Blood of HIV Infected Patients. *Clin. Chim. Acta* **1996**, *255*, 107–117. [CrossRef]
74. Suresh, D.R.; Annam, V.; Pratibha, K.; Prasad, B.V.M. Total Antioxidant Capacity a Novel Early Bio-Chemical Marker of Oxidative Stress in HIV Infected Individuals. *J. Biomed. Sci.* **2009**, *16*, 61. [CrossRef]
75. Ogunro, P.S.; Ogungbamigbe, T.O.; Elemie, P.O.; Egbewale, B.E.; Adewole, T.A. Plasma Selenium Concentration and Glutathione Peroxidase Activity in HIV-1/AIDS Infected Patients: A Correlation with the Disease Progression. *Niger. Postgrad. Med. J.* **2006**, *13*, 1–5. [PubMed]
76. Pace, G.W.; Leaf, C.D. The Role of Oxidative Stress in HIV Disease. *Free Radic. Biol. Med.* **1995**, *19*, 523–528. [CrossRef]
77. Yano, S.; Colon, M.; Yano, N. An Increase of Acidic Isoform of Catalase in Red Blood Cells from HIV(+) Population. *Mol. Cell. Biochem.* **1996**, *165*, 77–81. [CrossRef]

78. Gil, L.; Martínez, G.; González, I.; Tarinas, A.; Álvarez, A.; Giuliani, A.; Molina, R.; Tápanes, R.; Pérez, J.; León, O.S. Contribution to Characterization of Oxidative Stress in HIV/AIDS Patients. *Pharmacol. Res.* **2003**, *47*, 217–224. [CrossRef]
79. Papadopulos-Eleopulos, E. Reappraisal of Aids—Is the Oxidation Induced by the Risk Factors the Primary Cause? *Med. Hypotheses* **1988**, *25*, 151–162. [CrossRef]
80. Papadopulos-Eleopulos, E.; Hedland-Thomas, B.; Causer, D.A.; Dufty, A.N.P. An Alternative Explanation for the Radiosensitization of AIDS Patients. *Int. J. Radiat. Oncol. Biol. Phys.* **1989**, *17*, 695–697. [CrossRef]
81. Buhl, R.; Holroyd, K.J.; Mastrangeli, A.; Cantin, A.M.; Jaffe, H.A.; Wells, F.B.; Saltini, C.; Crystal, R.G. Systemic Glutathione Deficiency In Symptom-Free Hiv-Seropositive Individuals. *Lancet* **1989**, *334*, 1294–1298. [CrossRef]
82. Skurnick, J.; Bogden, J.; Baker, H.; Kemp, F.; Sheffet, A.; Quattrone, G.; Louria, D. Micronutrient Profiles in HIV-1-Infected Heterosexual Adults. *J. Acquir. Immune Defic. Syndr. Hum. Retrovirol.* **1996**, *12*, 75–83. [CrossRef]
83. Delmas-Beauvieux, M.C.; Peuchant, E.; Couchouron, A.; Constans, J.; Sergeant, C.; Simonoff, M.; Pellegrin, J.L.; Leng, B.; Conri, C.; Clerc, M. The Enzymatic Antioxidant System in Blood and Glutathione Status in Human Immunodeficiency Virus (HIV)-Infected Patients: Effects of Supplementation with Selenium or Beta-Carotene. *Am. J. Clin. Nutr.* **1996**, *64*, 101–107. [CrossRef]
84. McDermid, J.M.; Lalonde, R.G.; Gray-Donald, K.; Baruchel, S.; Kubow, S. Associations between Dietary Antioxidant Intake and Oxidative Stress in HIV-Seropositive and HIV-Seronegative Men and Women. *J. Acquir. Immune Defic. Syndr.* **2002**, *29*, 158–164. [CrossRef]
85. Shiau, S.; Webber, A.; Strehlau, R.; Patel, F.; Coovadia, A.; Kozakowski, S.; Brodlie, S.; Yin, M.T.; Kuhn, L.; Arpadi, S.M. Dietary Inadequacies in HIV-Infected and Uninfected School-Aged Children in Johannesburg, South Africa HHS Public Access. *J. Pediatr. Gastroenterol. Nutr.* **2017**, *65*, 332–337. [CrossRef]
86. Anyabolu, H.C.; Adejuyigbe, E.A.; Adeodu, O.O. Serum Micronutrient Status of Haart-Naïve, HIV Infected Children in South Western Nigeria: A Case Controlled Study. *AIDS Res. Treat* **2014**, *2014*, 351043. [CrossRef]
87. Ubesie, A.C.; Ibe, B.C.; Emodi, I.J.; Iloh, K.K. Serum Selenium Status of HIV-Infected Children on Care and Treatment in Enugu, Nigeria. *SAJCH* **2017**, *11*, 21–25. [CrossRef]
88. Bologna, R.; Indacochea, F.; Shor-Posner, G.; Mantero-Atienza, E.; Grazziutti, M.; Sotomayor, M.C.; Fletcher, M.A.; Cabrejos, C.; Scott, G.B.; Baum, M.K. Selenium and Immunity in HIV-1 Infected Pediatric Patients. *J. Nutr. Immunol.* **1994**, *3*, 41–49. [CrossRef]
89. Allavena, C.; Dousset, B.; May, T.; Dubois, F.; Canton, P.; Belleville, F. Relationship of Trace Element, Immunological Markers, and HIV1 Infection Progression. *Biol. Trace Elem. Res.* **1995**, *47*, 133–138. [CrossRef]
90. Look, M.; Rockstroh, J.; Rao, G.; Kreuzer, K.-A.; Barton, S.; Lemoch, H.; Sudhop, T.; Hoch, J.; Stockinger, K.; Spengler, U.; et al. Serum Selenium, Plasma Glutathione (GSH) and Erythrocyte Glutathione Peroxidase (GSH-Px)-Levels in Asymptomatic versus Symptomatic Human Immunodeficiency Virus-1 (HIV-1)-Infection. *Eur. J. Clin. Nutr.* **1997**, *51*, 266–272. [CrossRef]
91. Burbano, X.; Miguez-Burbano, M.J.; McCollister, K.; Zhang, G.; Rodriguez, A.; Ruiz, P.; Lecusay, R.; Shor-Posner, G. Impact of a Selenium Chemoprevention Clinical Trial on Hospital Admissions of HIV-Infected Participants. *HIV Clin. Trials* **2002**, *3*, 483–491. [CrossRef]
92. Hurwitz, B.E.; Klaus, J.R.; Llabre, M.M.; Gonzalez, A.; Lawrence, P.J.; Maher, K.J.; Greeson, J.M.; Baum, M.K.; Shor-Posner, G.; Skyler, J.S.; et al. Suppression of Human Immunodeficiency Virus Type 1 Viral Load with Selenium Supplementation: A Randomized Controlled Trial. *Arch. Intern. Med.* **2007**, *167*, 148–154. [CrossRef]
93. Kupka, R.; Mugusi, F.; Aboud, S.; Hertzmark, E.; Spiegelman, D.; Fawzi, W.W. Effect of Selenium Supplements on Hemoglobin Concentration and Morbidity among HIV-1-Infected Tanzanian Women. *Clin. Infect. Dis.* **2009**, *48*, 1475–1478. [CrossRef]
94. Kupka, R.; Mugusi, F.; Aboud, S.; Msamanga, G.I.; Finkelstein, J.L.; Spiegelman, D.; Fawzi, W.W. Randomized, Double-Blind, Placebo-Controlled Trial of Selenium Supplements among HIV-Infected Pregnant Women in Tanzania: Effects on Maternal and Child Outcomes. *Am. J. Clin. Nutr.* **2008**, *87*, 1802–1808. [CrossRef]
95. Baum, M.K.; Campa, A.; Lai, S.; Sales Martinez, S.; Tsalaile, L.; Burns, P.; Farahani, M.; Li, Y.; van Widenfelt, E.; Page, J.B.; et al. Effect of Micronutrient Supplementation on Disease Progression in Asymptomatic, Antiretroviral-Naive, HIV-Infected Adults in Botswana: A Randomized Clinical Trial. *JAMA* **2013**, *310*, 2154–2163. [CrossRef]
96. Kamwesiga, J.; Mutabazi, V.; Kayumba, J.; Tayari, J.C.K.; Uwimbabazi, J.C.; Batanage, G.; Uwera, G.; Baziruwiha, M.; Ntizimira, C.; Murebwayire, A.; et al. Effect of Selenium Supplementation on CD4R T-Cell Recovery, Viral Suppression and Morbidity of HIV-Infected Patients in Rwanda: A Randomized Controlled Trial. *AIDS* **2015**, *29*, 1045–1052. [CrossRef]
97. Muzembo, B.A.; Ngatu, N.R.; Januka, K.; Huang, H.L.; Nattadech, C.; Suzuki, T.; Wada, K.; Ikeda, S. Selenium Supplementation in HIV-Infected Individuals: A Systematic Review of Randomized Controlled Trials. *Clin. Nutr. ESPEN* **2019**, *34*, 1–7. [CrossRef]
98. Visser, M.E.; Durao, S.; Sinclair, D.; Irlam, J.H.; Siegfried, N. Micronutrient Supplementation in Adults with HIV Infection. *Cochrane Database Syst. Rev.* **2017**, *2017*, CD003650. [CrossRef]
99. Global Hepatitis Report. 2017. Available online: https://www.who.int/publications/i/item/global-hepatitis-report-2017 (accessed on 16 August 2021).
100. Wieland, S.F.; Chisari, F.V. Stealth and Cunning: Hepatitis B and Hepatitis C Viruses. *J. Virol.* **2005**, *79*, 9369. [CrossRef]
101. Tsukuda, S.; Watashi, K. Hepatitis B Virus Biology and Life Cycle. *Antiviral Res.* **2020**, *182*, 104925. [CrossRef] [PubMed]
102. Lauring, A.S.; Frydman, J.; Andino, R. The Role of Mutational Robustness in RNA Virus. *Nat. Rev. Microbiol.* **2013**, *11*, 327. [CrossRef] [PubMed]
103. Khan, M.S.; Dilawar, S.; Ali, I.; Rauf, N. The Possible Role of Selenium Concentration in Hepatitis B and C Patients. *Saudi J. Gastroenterol.* **2012**, *18*, 106. [CrossRef]

104. Rauf, N.; Tahir, S.S.; Dilawar, S.; Ahmad, I.; Parvez, S. Serum Selenium Concentration in Liver Cirrhotic Patients Suffering from Hepatitis B and C in Pakistan. *Biol. Trace Elem. Res.* **2011**, *145*, 144–150. [CrossRef]
105. Himoto, T.; Yoneyama, H.; Kurokohchi, K.; Inukai, M.; Masugata, H.; Goda, F.; Haba, R.; Watababe, S.; Kubota, S.; Senda, S.; et al. Selenium Deficiency Is Associated with Insulin Resistance in Patients with Hepatitis C Virus–Related Chronic Liver Disease. *Nutr. Res.* **2011**, *31*, 829–835. [CrossRef]
106. Abediankenari, S.; Ghasemi, M.; Nasehi, M.M.; Abedi, S.; Hosseini, V. Determination of Trace Elements in Patients with Chronic Hepatitis B. *Acta Med. Iran.* **2011**, *49*, 667–669.
107. Ko, W.S.; Guo, C.-H.; Yeh, M.-S.; Lin, L.Y.; Hsu, G.S.W.; Chen, P.C.; Luo, M.C.; Lin, C.Y. Blood Micronutrient, Oxidative Stress, and Viral Load in Patients with Chronic Hepatitis C. *World J. Gastroenterol.* **2005**, *11*, 4697. [CrossRef]
108. Bettinger, D.; Schultheiss, M.; Hennecke, N.; Panther, E.; Knüppel, E.; Blum, H.E.; Thimme, R.; Spangenberg, H.C. Selenium Levels in Patients with Hepatitis C Virus-Related Chronic Hepatitis, Liver Cirrhosis, and Hepatocellular Carcinoma: A Pilot Study. *Hepatology* **2013**, *57*, 2543–2544. [CrossRef]
109. Reshi, M.L.; Su, Y.-C.; Hong, J.R. RNA Viruses: ROS-Mediated Cell Death. *Int. J. Cell Biol.* **2014**, *2014*, 467452. [CrossRef]
110. Yu, D.Y. Relevance of Reactive Oxygen Species in Liver Disease Observed in Transgenic Mice Expressing the Hepatitis B Virus X Protein. *Lab. Anim. Res.* **2020**, *36*, 6. [CrossRef]
111. Yuan, K.; Lei, Y.; Chen, H.-N.; Chen, Y.; Zhang, T.; Li, K.; Xie, N.; Wang, K.; Feng, X.; Pu, Q.; et al. HBV-Induced ROS Accumulation Promotes Hepatocarcinogenesis through Snail-Mediated Epigenetic Silencing of SOCS3. *Cell Death Differ.* **2016**, *23*, 616. [CrossRef]
112. Jain, S.K.; Pemberton, P.W.; Smith, A.; McMahon, R.F.T.; Burrows, P.C.; Aboutwerat, A.; Warnes, T.W. Oxidative Stress in Chronic Hepatitis C: Not Just a Feature of Late Stage Disease. *J. Hepatol.* **2002**, *36*, 805–811. [CrossRef]
113. Su, L.J.; Zhang, J.-H.; Gomez, H.; Murugan, R.; Hong, X.; Xu, D.; Jiang, F.; Peng, Z.-Y. Reactive Oxygen Species-Induced Lipid Peroxidation in Apoptosis, Autophagy, and Ferroptosis. *Oxid. Med. Cell. Longev.* **2019**, *2019*, 5080843. [CrossRef]
114. Zhang, J.; Wang, X.; Vikash, V.; Ye, Q.; Wu, D.; Liu, Y.; Dong, W. ROS and ROS-Mediated Cellular Signaling. *Oxid. Med. Cell. Longev.* **2016**, *2016*, 4350965. [CrossRef]
115. Ohl, K.; Tenbrock, K. Reactive Oxygen Species as Regulators of MDSC-Mediated Immune Suppression. *Front. Immunol.* **2018**, *9*, 2499. [CrossRef]
116. Yahfoufi, N.; Alsadi, N.; Jambi, M.; Matar, C. The Immunomodulatory and Anti-Inflammatory Role of Polyphenols. *Nutrients* **2018**, *10*, 1618. [CrossRef] [PubMed]
117. Razzaq, Z.; Malik, A. Viral Load Is Associated with Abnormal Serum Levels of Micronutrients and Glutathione and Glutathione-Dependent Enzymes in Genotype 3 HCV Patients. *BBA Clin.* **2014**, *2*, 72. [CrossRef]
118. Kundu, D.; Roy, A.; Mandal, T.; Bandyopadhyay, U.; Ghosh, E.; Ray, D. Oxidative Stress in Alcoholic and Viral Hepatitis. *N. Am. J. Med. Sci.* **2012**, *4*, 412. [CrossRef] [PubMed]
119. Morbitzer, M.; Herget, T. Expression of Gastrointestinal Glutathione Peroxidase Is Inversely Correlated to the Presence of Hepatitis C Virus Subgenomic RNA in Human Liver Cells. *J. Biol. Chem.* **2005**, *280*, 8831–8841. [CrossRef]
120. Dionisio, N.; Garcia-Mediavilla, M.; Sanchez-Campos, S.; Majano, P.L.; Benedicto, I.; Rosado, J.A.; Salido, G.M.; Gonzalez-Gallego, J. Hepatitis C Virus NS5A and Core Proteins Induce Oxidative Stress-Mediated Calcium Signalling Alterations in Hepatocytes. *J. Hepatol.* **2009**, *50*, 872–882. [CrossRef]
121. Brault, C.; Lévy, P.; Duponchel, S.; Michelet, M.; Sallé, A.; Pécheur, E.-I.; Plissonnier, M.-L.; Parent, R.; Véricel, E.; Ivanov, A.V.; et al. Glutathione Peroxidase 4 Is Reversibly Induced by HCV to Control Lipid Peroxidation and to Increase Virion Infectivity. *Gut* **2016**, *65*, 144–154. [CrossRef] [PubMed]
122. Zhang, W.; Cox, A.G.; Taylor, E.W. Hepatitis C Virus Encodes a Selenium-Dependent Glutathione Peroxidase Gene. *Med. Klin* **1999**, *94*, 2–6. [CrossRef] [PubMed]
123. Kiełczykowska, M.; Kocot, J.; Paździor, M.; Musik, I. Selenium—A Fascinating Antioxidant of Protective Properties. *Adv. Clin. Exp. Med.* **2018**, *27*, 245–255. [CrossRef]
124. Bentley-Hewitt, K.L.; Chen, R.K.-Y.; Lill, R.E.; Hedderley, D.I.; Herath, T.D.; Matich, A.J.; McKenzie, M.J. Consumption of Selenium-Enriched Broccoli Increases Cytokine Production in Human Peripheral Blood Mononuclear Cells Stimulated Ex Vivo, a Preliminary Human Intervention Study. *Mol. Nutr. Food Res.* **2014**, *58*, 2350–2357. [CrossRef]
125. Yu Yu, S.; Zhu, Y.J.; Li, W.G. Protective Role of Selenium against Hepatitis B Virus and Primary Liver Cancer in Qidong. *Biol. Trace Elem. Res.* **1997**, *56*, 117–124. [CrossRef] [PubMed]
126. Janbakhsh, A.; Mansouri, F.; Vaziri, S.; Sayad, B.; Afsharian, M.; Rahimi, M.; Shahebrahimi, K.; Salari, F. Effect of Selenium on Immune Response against Hepatitis B Vaccine with Accelerated Method in Insulin-Dependent Diabetes Mellitus Patients. *Caspian J. Intern. Med.* **2013**, *4*, 603.
127. Berkson, B.M. A Conservative Triple Antioxidant Approach to the Treatment of Hepatitis C. *Med. Klin.* **1999**, *94*, 84–89. [CrossRef] [PubMed]
128. Groenbaek, K.; Friis, H.; Hansen, M.; Ring-Larsen, H.; Krarup, H.B. The Effect of Antioxidant Supplementation on Hepatitis C Viral Load, Transaminases and Oxidative Status: A Randomized Trial among Chronic Hepatitis C Virus-Infected Patients. *Eur. J. Gastroenterol. Hepatol.* **2006**, *18*, 985–989. [CrossRef]
129. Tang, C.; Li, S.; Zhang, K.; Li, J.; Han, Y.; Zhan, T.; Zhao, Q.; Guo, X.; Zhang, J. Selenium Deficiency-Induced Redox Imbalance Leads to Metabolic Reprogramming and Inflammation in the Liver. *Redox Biol.* **2020**, *36*, 101519. [CrossRef] [PubMed]

130. Zhang, Y.; Yu, D.; Zhang, J.; Bao, J.; Tang, C.; Zhang, Z. The Role of Necroptosis and Apoptosis through the Oxidative Stress Pathway in the Liver of Selenium-Deficient Swine. *Metallomics* **2020**, *12*, 607–616. [CrossRef] [PubMed]
131. Burk, R.F.; Hill, K.E.; Nakayama, A.; Mostert, V.; Levander, X.A.; Motley, A.K.; Freeman, M.L.; Austin, L.M. Selenium Deficiency Activates Mouse Liver Nrf2-ARE but Vitamin E Deficiency Does Not. *Free Radic. Biol. Med.* **2008**, *44*, 1617. [CrossRef]
132. Irmak, M.; Ince, G.; Ozturk, M.; Cetin-Atalay, R. Acquired Tolerance of Hepatocellular Carcinoma Cells to Selenium Deficiency: A Selective Survival Mechanism? *Cancer Res.* **2003**, *63*, 6707–6715.
133. Burrill, C.P.; Westesson, O.; Schulte, M.B.; Strings, V.R.; Segal, M.; Andino, R. Global RNA Structure Analysis of Poliovirus Identifies a Conserved RNA Structure Involved in Viral Replication and Infectivity. *J. Virol.* **2013**, *87*, 11670–11683. [CrossRef]
134. Steinbrenner, H.; Al-Quraishy, S.; Dkhil, M.A.; Wunderlich, F.; Sies, H. Dietary Selenium in Adjuvant Therapy of Viral and Bacterial Infections. *Adv. Nutr.* **2015**, *6*, 73–82. [CrossRef] [PubMed]
135. Broome, C.S.; McArdle, F.; Kyle, J.A.M.; Andrews, F.; Lowe, N.M.; Hart, C.A.; Arthur, J.R.; Jackson, M.J. An Increase in Selenium Intake Improves Immune Function and Poliovirus Handling in Adults with Marginal Selenium Status. *Am. J. Clin. Nutr.* **2004**, *80*, 154–162. [CrossRef]
136. Hoffmann, P.R.; Berry, M.J. The Influence of Selenium on Immune Responses. *Mol. Nutr. Food Res.* **2008**, *52*, 1273–1280. [CrossRef] [PubMed]
137. Zhou, F.; Yu, T.; Du, R.; Fan, G.; Liu, Y.; Liu, Z.; Xiang, J.; Wang, Y.; Song, B.; Gu, X.; et al. Clinical Course and Risk Factors for Mortality of Adult Inpatients with COVID-19 in Wuhan, China: A Retrospective Cohort Study. *Lancet* **2020**, *395*, 1054. [CrossRef]
138. Dai, M.; Liu, D.; Liu, M.; Zhou, F.; Li, G.; Chen, Z.; Zhang, Z.; You, H.; Wu, M.; Zheng, Q.; et al. Patients with Cancer Appear More Vulnerable to SARS-CoV-2: A Multicenter Study during the COVID-19 Outbreak. *Cancer Discov.* **2020**, *10*, 783. [CrossRef]
139. Wiersinga, W.J.; Rhodes, A.; Cheng, A.C.; Peacock, S.J.; Prescott, H.C. Pathophysiology, Transmission, Diagnosis, and Treatment of Coronavirus Disease 2019 (COVID-19): A Review. *JAMA* **2020**, *324*, 782–793. [CrossRef] [PubMed]
140. Song, P.; Li, W.; Xie, J.; Hou, Y.; You, C. Cytokine Storm Induced by SARS-CoV-2. *Clin. Chim. Acta* **2020**, *509*, 280. [CrossRef] [PubMed]
141. Mehta, P.; McAuley, D.F.; Brown, M.; Sanchez, E.; Tattersall, R.S.; Manson, J.J.; Collaboration, H.A.S. UK COVID-19: Consider Cytokine Storm Syndromes and Immunosuppression. *Lancet* **2020**, *395*, 1033. [CrossRef]
142. Li, Y.; Zhao, K.; Wei, H.; Chen, W.; Wang, W.; Jia, L.; Liu, Q.; Zhang, J.; Shan, T.; Peng, Z.; et al. Dynamic Relationship between D-dimer and COVID-19 Severity. *Br. J. Haematol.* **2020**, *190*, e24–e27. [CrossRef]
143. Valenza, M.; Steardo, L., Jr.; Steardo, L.; Verkhratsky, A.; Scuderi, C. Systemic Inflammation and Astrocyte Reactivity in the Neuropsychiatric Sequelae of COVID-19: Focus on Autism Spectrum Disorders. *Front. Cell. Neurosci.* **2021**, *15*, 748136. [CrossRef] [PubMed]
144. Zhou, Z.; Kang, H.; Li, S.; Zhao, X. Understanding the Neurotropic Characteristics of SARS-CoV-2: From Neurological Manifestations of COVID-19 to Potential Neurotropic Mechanisms. *J. Neurol.* **2020**, *267*, 1. [CrossRef]
145. Nemoto, W.; Yamagata, R.; Nakagawasai, O.; Nakagawa, K.; Hung, W.Y.; Fujita, M.; Tadano, T.; Tan-No, K. Effect of Spinal Angiotensin-Converting Enzyme 2 Activation on the Formalin-Induced Nociceptive Response in Mice. *Eur. J. Pharmacol.* **2020**, *872*, 172950. [CrossRef]
146. Satarker, S.; Nampoothiri, M. Involvement of the Nervous System in COVID-19: The Bell Should Toll in the Brain. *Life Sci.* **2020**, *262*, 118568. [CrossRef]
147. Merad, M.; Martin, J.C. Pathological Inflammation in Patients with COVID-19: A Key Role for Monocytes and Macrophages. *Nat. Rev. Immunol.* **2020**, *20*, 1. [CrossRef]
148. Erickson, M.A.; Rhea, E.M.; Knopp, R.C.; Banks, W.A. Interactions of SARS-CoV-2 with the Blood–Brain Barrier. *Int. J. Mol. Sci.* **2021**, *22*, 2681. [CrossRef]
149. Hiffler, L.; Rakotoambinina, B. Selenium and RNA Virus Interactions: Potential Implications for SARS-CoV-2 Infection (COVID-19). *Front. Nutr.* **2020**, *7*, 164. [CrossRef]
150. Liu, Q.; Zhao, X.; Ma, J.; Mu, Y.; Wang, Y.; Yang, S.; Wu, Y.; Wu, F.; Zhou, Y. Selenium (Se) Plays a Key Role in the Biological Effects of Some Viruses: Implications for COVID-19. *Environ. Res.* **2021**, *196*, 110984. [CrossRef]
151. Samir, D. Oxidative Stress Associated with SARS-CoV-2 (COVID-19) Increases the Severity of the Lung Disease—A Systematic Review. *J. Infect. Dis. Epidemiol.* **2020**, *6*, 121. [CrossRef]
152. Mahmoodpoor, A.; Hamishehkar, H.; Shadvar, K.; Ostadi, Z.; Sanaie, S.; Saghaleini, S.H.; Nader, N.D. The Effect of Intravenous Selenium on Oxidative Stress in Critically Ill Patients with Acute Respiratory Distress Syndrome. *Immunol. Investig.* **2019**, *48*, 147–159. [CrossRef]
153. Wang, F.; Nie, J.; Wang, H.; Zhao, Q.; Xiong, Y.; Deng, L.; Song, S.; Ma, Z.; Mo, P.; Zhang, Y. Characteristics of Peripheral Lymphocyte Subset Alteration in COVID-19 Pneumonia. *J. Infect. Dis.* **2020**, *221*, 1762–1769. [CrossRef]
154. Huang, Z.; Rose, A.H.; Hoffmann, P.R. The Role of Selenium in Inflammation and Immunity: From Molecular Mechanisms to Therapeutic Opportunities. *Antioxid. Redox Signal.* **2012**, *16*, 705. [CrossRef]
155. Hoffmann, F.K.W.; Hashimoto, A.C.; Shafer, L.A.; Dow, S.; Berry, M.J.; Hoffmann, P.R. Dietary Selenium Modulates Activation and Differentiation of CD4+ T Cells in Mice through a Mechanism Involving Cellular Free Thiols. *J. Nutr.* **2010**, *140*, 1155. [CrossRef]
156. Zhang, J.; Taylor, E.W.; Bennett, K.; Saad, R.; Rayman, M.P. Association between Regional Selenium Status and Reported Outcome of COVID-19 Cases in China. *Am. J. Clin. Nutr.* **2020**, *111*, 1297–1299. [CrossRef]

157. Zhang, H.Y.; Zhang, A.R.; Lu, Q.B.; Zhang, X.A.; Zhang, Z.J.; Guan, X.G.; Che, T.L.; Yang, Y.; Li, H.; Liu, W.; et al. Association between Fatality Rate of COVID-19 and Selenium Deficiency in China. *BMC Infect. Dis.* **2021**, *21*, 452. [CrossRef]
158. Cheng, C.; Chen, S.Y.; Geng, J.; Zhu, P.Y.; Liang, R.N.; Yuan, M.Z.; Wang, B.; Jin, Y.F.; Zhang, R.G.; Zhang, W.D.; et al. Preliminary Analysis on COVID-19 Case Spectrum and Spread Intensity in Different Provinces in China except Hubei Province. *Zhonghua Liu Xing Bing Xue Za Zhi* **2020**, *41*, 1601–1605. [CrossRef]
159. Zeng, H.-L.; Zhang, B.; Wang, X.; Yang, Q.; Cheng, L. Urinary Trace Elements in Association with Disease Severity and Outcome in Patients with COVID-19. *Environ. Res.* **2021**, *194*, 110670. [CrossRef]
160. Im, J.H.; Je, Y.S.; Baek, J.; Chung, M.-H.; Kwon, H.Y.; Lee, J.-S. Nutritional Status of Patients with COVID-19. *Int. J. Infect. Dis.* **2020**, *100*, 390. [CrossRef]
161. Majeed, M.; Nagabhushanam, K.; Gowda, S.; Mundkur, L. An Exploratory Study of Selenium Status in Healthy Individuals and in Patients with COVID-19 in a South Indian Population: The Case for Adequate Selenium Status. *Nutrition* **2021**, *82*, 111053. [CrossRef] [PubMed]
162. Younesian, O.; Khodabakhshi, B.; Abdolahi, N.; Norouzi, A.; Behnampour, N.; Hosseinzadeh, S.; Alarzi, S.S.H.; Joshaghani, H. Decreased Serum Selenium Levels of COVID-19 Patients in Comparison with Healthy Individuals. *Biol. Trace Elem. Res.* **2021**, 1–6. [CrossRef]
163. Skalny, A.; Timashev, P.S.; Aschner, M.; Aaseth, J.; Chernova, L.N.; Belyaev, V.E.; Grabeklis, A.R.; Notova, S.V.; Lobinski, R.; Tsatsakis, A.; et al. Serum Zinc, Copper, and Other Biometals Are Associated with COVID-19 Severity Markers. *Metabolites* **2021**, *11*, 244. [CrossRef]
164. Moghaddam, A.; Heller, R.A.; Sun, Q.; Seelig, J.; Cherkezov, A.; Seibert, L.; Hackler, J.; Seemann, P.; Diegmann, J.; Pilz, M.; et al. Selenium Deficiency Is Associated with Mortality Risk from COVID-19. *Nutrients* **2020**, *12*, 2098. [CrossRef] [PubMed]
165. Pincemail, J.; Cavalier, E.; Charlier, C.; Cheramy–Bien, J.-P.; Brevers, E.; Courtois, A.; Fadeur, M.; Meziane, S.; Goff, C.L.; Misset, B.; et al. Oxidative Stress Status in COVID-19 Patients Hospitalized in Intensive Care Unit for Severe Pneumonia. A Pilot Study. *Antioxidants* **2021**, *10*, 257. [CrossRef]
166. Polonikov, A. Endogenous Deficiency of Glutathione as the Most Likely Cause of Serious and Death in COVID-19 Patients. *ACS Infect. Dis.* **2020**, *6*, 1558–1562. [CrossRef] [PubMed]
167. Horowitz, R.I.; Freeman, P.R.; Bruzzese, J. Efficacy of Glutathione Therapy in Relieving Dyspnea Associated with COVID-19 Pneumonia: A Report of 2 Cases. *Respir. Med. Case Rep.* **2020**, *30*, 101063. [CrossRef] [PubMed]
168. Kieliszek, M.; Lipinski, B. Selenium Supplementation in the Prevention of Coronavirus Infections (COVID-19). *Med. Hypotheses* **2020**, *143*, 109878. [CrossRef]
169. Berggren, M.; Gallegos, A.; Gasdaska, J.; Powis, G. Cellular Thioredoxin Reductase Activity Is Regulated by Selenium. *Anticancer Res.* **1997**, *17*, 3377–3380.
170. Broman, L.M.; Bernardson, A.; Bursell, K.; Wernerman, J.; Fläring, U.; Tjäder, I. Serum Selenium in Critically Ill Patients: Profile and Supplementation in a Depleted Region. *Acta Anaesthesiol. Scand.* **2020**, *64*, 803–809. [CrossRef]
171. Baker, R.D.; Baker, S.S.; Rao, R. Selenium Deficiency in Tissue Culture: Implications for Oxidative Metabolism. *J. Pediatr. Gastroenterol. Nutr.* **1998**, *27*, 387–392. [CrossRef]
172. Kanth Manne, B.; Denorme, F.; Middleton, E.A.; Portier, I.; Rowley, J.W.; Stubben, C.; Petrey, A.C.; Tolley, N.D.; Guo, L.; Cody, M.; et al. Platelet Gene Expression and Function in Patients with COVID-19. *Blood* **2020**, *136*, 1317. [CrossRef] [PubMed]
173. Ersöz, G.; Yakaryilmaz, A.; Turan, B. Effect of Sodium Selenite Treatment on Platelet Aggregation of Streptozotocin-Induced Diabetic Rats. *Thromb Res.* **2003**, *111*, 363–367. [CrossRef]
174. Varlamova, E.G.; Turovsky, E.A.; Blinova, E.V. Therapeutic Potential and Main Methods of Obtaining Selenium Nanoparticles. *Int. J. Mol. Sci.* **2021**, *22*, 10808. [CrossRef]
175. Huang, B.; Zhang, J.; Hou, J.; Chen, C. Free Radical Scavenging Efficiency of Nano-Se in Vitro. *Free Radic. Biol. Med.* **2003**, *35*, 805–813. [CrossRef]
176. Khurana, A.; Tekula, S.; Saifi, M.A.; Venkatesh, P.; Godugu, C. Therapeutic Applications of Selenium Nanoparticles. *Biomed. Pharmacother.* **2019**, *111*, 802–812. [CrossRef]
177. Hosnedlova, B.; Kepinska, M.; Skalickova, S.; Fernandez, C.; Ruttkay-Nedecky, B.; Peng, Q.; Baron, M.; Melcova, M.; Opatrilova, R.; Zidkova, J.; et al. Nano-Selenium and Its Nanomedicine Applications: A Critical Review. *Int. J. Nanomedicine* **2018**, *13*, 2107. [CrossRef]
178. Turovsky, E.A.; Varlamova, E.G. Mechanism of Ca2+-Dependent pro-Apoptotic Action of Selenium Nanoparticles, Mediated by Activation of Cx43 Hemichannels. *Biology* **2021**, *10*, 743. [CrossRef]
179. He, L.; Zhao, J.; Wang, L.; Liu, Q.; Fan, Y.; Li, B.; Yu, Y.L.; Chen, C.; Li, Y.F. Using Nano-Selenium to Combat Coronavirus Disease 2019 (COVID-19)? *Nano Today* **2021**, *36*, 101037. [CrossRef]
180. Qiao, Q.; Liu, X.; Yang, T.; Cui, K.; Kong, L.; Yang, C.; Zhang, Z. Nanomedicine for Acute Respiratory Distress Syndrome: The Latest Application, Targeting Strategy, and Rational Design. *Acta Pharm. Sin. B* **2021**, *11*, 3060. [CrossRef]
181. Jin, Z.; Du, X.; Xu, Y.; Deng, Y.; Liu, M.; Zhao, Y.; Zhang, B.; Li, X.; Zhang, L.; Peng, C.; et al. Structure of Mpro from SARS-CoV-2 and Discovery of Its Inhibitors. *Nature* **2020**, *582*, 289–293. [CrossRef]
182. Azad, G.K.; Tomar, R.S. Ebselen, a Promising Antioxidant Drug: Mechanisms of Action and Targets of Biological Pathways. *Mol. Biol. Rep.* **2014**, *41*, 4865–4879. [CrossRef]

183. Gabriele, M.; Pucci, L. Diet Bioactive Compounds: Implications for Oxidative Stress and Inflammation in the Vascular System. *Endocr. Metab. Immune Disord. Drug Targets* **2017**, *17*, 264–275. [CrossRef]
184. Hajat, C.; Stein, E. The Global Burden of Multiple Chronic Conditions: A Narrative Review. *Prev. Med. Rep.* **2018**, *12*, 284. [CrossRef] [PubMed]
185. Zheng, Z.; Peng, F.; Xu, B.; Zhao, J.; Liu, H.; Peng, J.; Li, Q.; Jiang, C.; Zhou, Y.; Liu, S.; et al. Risk Factors of Critical & Mortal COVID-19 Cases: A Systematic Literature Review and Meta-Analysis. *J. Infect.* **2020**, *81*, e16. [CrossRef]
186. Neuhouser, M.L. The Importance of Healthy Dietary Patterns in Chronic Disease Prevention. *Nutr. Res.* **2019**, *70*, 3. [CrossRef] [PubMed]
187. Zabetakis, I.; Lordan, R.; Norton, C.; Tsoupras, A. COVID-19: The Inflammation Link and the Role of Nutrition in Potential Mitigation. *Nutrients* **2020**, *12*, 1466. [CrossRef]
188. Semba, R.D.; Tang, A.M. Micronutrients and the Pathogenesis of Human Immunodeficiency Virus Infection. *Br. J. Nutr.* **1999**, *81*, 181–189. [CrossRef]
189. Pecora, F.; Persico, F.; Argentiero, A.; Neglia, C.; Esposito, S. The Role of Micronutrients in Support of the Immune Response against Viral Infections. *Nutrients* **2020**, *12*, 3198. [CrossRef]
190. Calder, P.C.; Carr, A.C.; Gombart, A.F.; Eggersdorfer, M. Optimal Nutritional Status for a Well-Functioning Immune System Is an Important Factor to Protect against Viral Infections. *Nutrients* **2020**, *12*, 1181. [CrossRef]
191. Pecoraro, L.; Martini, L.; Salvottini, C.; Carbonare, L.D.; Piacentini, G.; Pietrobelli, A. The Potential Role of Zinc, Magnesium and Selenium against COVID-19: A Pragmatic Review. *Child. Adolesc. Obes.* **2021**, *4*, 127–130. [CrossRef]
192. Shakoor, H.; Feehan, J.; al Dhaheri, A.S.; Ali, H.I.; Platat, C.; Ismail, L.C.; Apostolopoulos, V.; Stojanovska, L. Immune-Boosting Role of Vitamins D, C, E, Zinc, Selenium and Omega-3 Fatty Acids: Could They Help against COVID-19? *Maturitas* **2021**, *143*, 1. [CrossRef]
193. Cámara, M.; Sánchez-Mata, M.C.; Fernández-Ruiz, V.; Cámara, R.M.; Cebadera, E.; Domínguez, L. A Review of the Role of Micronutrients and Bioactive Compounds on Immune System Supporting to Fight against the COVID-19 Disease. *Foods* **2021**, *10*, 1088. [CrossRef]
194. FAO/WHO. Human Vitamin and Mineral Requirments. In Chapter 15, Selenium. 2002. Available online: https://www.fao.org/3/Y2809E/y2809e0l.htm (accessed on 19 December 2021).
195. Harthill, M. Review: Micronutrient Selenium Deficiency Influences Evolution of Some Viral Infectious Diseases. *Biol. Trace Elem. Res.* **2011**, *143*, 1325. [CrossRef]
196. Waegeneers, N.; Thiry, C.; de Temmerman, L.; Ruttens, A. Predicted Dietary Intake of Selenium by the General Adult Population in Belgium. *Food Addit. Contam. Part A Chem. Anal. Control Expo. Risk Assess* **2013**, *30*, 278–285. [CrossRef] [PubMed]
197. Sunde, R. Selenium. In *Modern Nutrition in Health and Disease*; Ross, A., Caballero, B., Cousins, R., Tucker, K., Ziegler, T., Eds.; Lippincott Williams & Williams: Philadelphia, PA, USA, 2012; pp. 225–237.
198. Chun, O.K.; Floegel, A.; Chung, S.-J.; Chung, C.E.; Song, W.O.; Koo, S.I. Estimation of Antioxidant Intakes from Diet and Supplements in U.S. Adults. *J. Nutr.* **2010**, *140*, 317–324. [CrossRef] [PubMed]
199. Institute of Medicine (US) Panel on Dietary Antioxidants and Related Compounds. *Dietary Reference Intakes for Vitamin C, Vitamin E, Selenium, and Carotenoids*; National Academies Press: Washington, DC, USA, 2000.
200. World Health Organization and Food and Agriculture Organization. Food as a Source of Nutrients. *Vitamin and Mineral Requirements in Human Nutrition*, 2nd ed. World Health Organization and Food and Agriculture Organization of the United Nations. 2004. Available online: https://www.who.int/publications/i/item/9241546123 (accessed on 21 December 2021).
201. Scientific Opinion on Dietary Reference Values for Selenium. *EFSA J.* **2014**, *12*, 3846. [CrossRef]
202. MacFarquhar, J.K.; Broussard, D.L.; Melstrom, P.; Hutchinson, R.; Wolkin, A.; Martin, C.; Burk, R.F.; Dunn, J.R.; Green, A.L.; Hammond, R.; et al. Acute Selenium Toxicity Associated with a Dietary Supplement. *Arch. Intern. Med.* **2010**, *170*, 256–261. [CrossRef]
203. Park, Y.C.; Kim, J.B.; Heo, Y.; Park, D.C.; Lee, I.S.; Chung, H.W.; Han, J.H.; Chung, W.G.; Vendeland, S.C.; Whanger, P.D. Metabolism of Subtoxic Level of Selenite by Double-Perfused Small Intestine in Rats. *Biol. Trace Elem. Res.* **2004**, *98*, 143–157. [CrossRef]
204. Rayman, M.P. Food-Chain Selenium and Human Health: Emphasis on Intake. *Br. J. Nutr.* **2008**, *100*, 254–268. [CrossRef]

Review

Historical Roles of Selenium and Selenoproteins in Health and Development: The Good, the Bad and the Ugly

Petra A. Tsuji [1,*], Didac Santesmasses [2], Byeong J. Lee [3], Vadim N. Gladyshev [2] and Dolph L. Hatfield [4]

1. Department of Biological Sciences, Towson University, 8000 York Rd., Towson, MD 21252, USA
2. Brigham and Women's Hospital, Harvard Medical School, Boston, MA 02215, USA; dsantesmassesruiz@bwh.harvard.edu (D.S.); vgladyshev@rics.bwh.harvard.edu (V.N.G.)
3. School of Biological Sciences, College of Natural Sciences, Seoul National University, Seoul 08826, Korea; imbglmg@snu.ac.kr
4. Scientist Emeritus, Mouse Cancer Genetics Program, Center for Cancer Research, National Cancer Institute, National Institutes of Health, Bethesda, MD 20892, USA; hatfielddolph@gmail.com
* Correspondence: ptsuji@towson.edu

Abstract: Selenium is a fascinating element that has a long history, most of which documents it as a deleterious element to health. In more recent years, selenium has been found to be an essential element in the diet of humans, all other mammals, and many other life forms. It has many health benefits that include, for example, roles in preventing heart disease and certain forms of cancer, slowing AIDS progression in HIV patients, supporting male reproduction, inhibiting viral expression, and boosting the immune system, and it also plays essential roles in mammalian development. Elucidating the molecular biology of selenium over the past 40 years generated an entirely new field of science which encompassed the many novel features of selenium. These features were (1) how this element makes its way into protein as the 21st amino acid in the genetic code, selenocysteine (Sec); (2) the vast amount of machinery dedicated to synthesizing Sec uniquely on its tRNA; (3) the incorporation of Sec into protein; and (4) the roles of the resulting Sec-containing proteins (selenoproteins) in health and development. One of the research areas receiving the most attention regarding selenium in health has been its role in cancer prevention, but further research has also exposed the role of this element as a facilitator of various maladies, including cancer.

Keywords: cancer; health; mouse models; selenium; selenocysteine (Sec); tRNA[Ser]^{Sec}; Sec-tRNA[Ser]^{Sec}; selenoproteins

1. Introduction

The element selenium was discovered in 1817 by the Swedish chemist, Jöns Jacob Berzelius [1]. He named selenium after the Greek goddess of the moon, Selene. This fascinating element has a long and unsavory history of use as a dietary component. Its first description as being deleterious for animals to ingest was reported by Marco Polo in the late 13th century [2]. In his travels in Western China, Marco Polo wrote about an illness that his "beasts of burden" acquired wherein their hooves became brittle and fell off after eating certain plants. These plants most likely were seleniferous plants, which absorb large quantities of selenium from the soil and store the selenium in their tissues. Such diseases as Polo described have been found in the 20th century in horses and cattle that grazed on the plains of the Dakota and Nebraska territories of the United States. For example, T.C. Madison, an army physician stationed at Fort Randall in Northern Nebraska in the mid-1850s, described a necrotic hoof disorder that, in addition, involved losses of hair in the mane and tail among the army horses that grazed on the plants around the fort [3].

Subsequently, Franke reported that this malady, which was found to be prevalent in the livestock living in these Great Plains states, resulted from these animals eating seleniferous plants that were rich in selenium absorbed from high levels of this element in the soil [4]. Interestingly, and to further document the harmful effects of selenium on the animals ingesting plants rich in this element, selenium poisoning in horses was thought to have played a role in General George Custer's defeat at the Little Bighorn [5]. The military horses under Custer's command had grazed freely on the plants surrounding the area where he and his men had waited prior to the battle; these plants were later found to be seleniferous plants. At the Battle of the Little Bighorn on 25 June 1876, Custer's horses were reported to have had laminitis, which caused them to be lame, and led to Custer's defeat [5].

Selenium's role as a deleterious dietary element took a major turn for the good in 1957, when Schwarz and Foltz unexpectedly found that it prevented liver necrosis in rats [6]. The Schwartz and Foltz finding was followed by another interesting observation that selenium had an important role in anaerobic growth in *Escherichia coli* when this organism was grown on glucose [7]. It soon became obvious that selenium was an essential element in the diet of mammals and many other life forms when ingested in low levels, but harmful when ingested in high levels (see several chapters in [8] for an in-depth summary on these findings). The window between too little and too much selenium in the diet is rather narrow for most organisms.

Subsequently, the livestock industry found that the inclusion of selenium in the diet of livestock had many health benefits. These included enhanced fertility in male sheep and cattle, and importantly also the alleviation of numerous disorders such as white muscle disease and ill thrift in lambs and calves, pancreatic degeneration and exudative diathesis in fowl, and *hepatosis dietetica* in swine [9]. In many regions of the world where livestock are prevalent, the addition of selenium in the diets of livestock has saved this industry hundreds of millions of dollars [10]. With regard to humans, in certain regions in rural China, where the soil is deficient in selenium and hence the selenium status of the individuals living therein is suboptimal, maladies such as Keshan disease, a cardiomyopathy primarily in children, were found [11]. Similarly, Kashin–Beck disease, a chronic, endemic osteochondropathy, was found primarily in southwestern to northeastern China [12]. Keshan disease has been virtually eradicated in China by supplementing the diets of the populations residing in specific rural areas where the soil is deficient in selenium [13]. In the USA, the recommended daily amount of selenium is set forth by the Food and Nutrition Board of the National Academies of Medicine, and is 55 micrograms per day for men and women above 14 years of age. Women who are pregnant or lactating require 60 or 70 micrograms per day [14,15].

In addition to having roles in preventing heart disease and other muscle disorders, as well as enhancing male fertility, selenium was found to have roles as a chemopreventive agent in certain cancers [16–19], roles in boosting immune function [18,20], in suppressing viral expression [21], in slowing the development of AIDS in HIV positive patients [22] and in Simian Acquired Immunodeficiency Virus (SAIDS)-infected monkeys [23], and possibly in slowing the aging process [24].

Low molecular weight selenium-containing compounds (LMW selenocompounds) also have highly significant roles in providing benefits to mammals. The research carried out in this area constitutes a subfield within the selenium field. There are several excellent reviews on the benefits of LMW selenocompounds in health and numerous other aspects of these selenocompounds [25–27]. This topic will, therefore, not be further discussed herein.

Several seminal studies in the selenium field in the 1970s and 1980s provided the foundation for elucidating the molecular biology of selenium and established it as a separate and highly significant field in science. Initially, selenium was found to be an essential component of glutathione peroxidase 1 (GPX1) in 1973 [28,29], which was subsequently identified in clostridial glycine reductase as selenocysteine (Sec) [30]. Bovine GPX1 was sequenced in 1984, and the position of the Sec moiety was therefore established within the protein [31]. The gene sequences of mammalian *Gpx1* [32] and bacterial glycine formate dehydrogenase [33] were determined, and the Sec residue in the corresponding proteins shown to be encoded by TGA in both genes.

Additional studies that played major roles in defining the molecular biology of selenium as a separate field of science rapidly developed, and encompassed how selenium made its way into protein as the 21st proteinogenic amino acid in the genetic code—the vast machinery dedicated to synthesizing Sec and incorporating it into protein—and the roles of the resulting Sec-containing proteins (selenoproteins) in health and development. One of the research areas receiving much attention has been the role of selenium in cancer prevention, but this finding has also exposed the potential role of this element as a facilitator of various maladies, including cancer. These aspects of the molecular biology of selenium are discussed herein.

2. Selenocysteine (Sec) tRNA[Ser]^{Sec}

This Section describes various aspects of Sec tRNA[Ser]^{Sec} (i.e., the transcription of tRNA[Ser]^{Sec}, primary sequences of the two isoforms, their distributions, the synthesis of Sec on tRNA[Ser]^{Sec}, and the incorporation of Sec into selenoproteins as the 21st amino acid in the genetic code). The reason Sec tRNA is designated tRNA[Ser]^{Sec} is that it is initially aminoacylated with serine (Ser) by Ser-tRNA synthetase (SARS), and the Ser moiety is then uniquely converted to Sec on the tRNA (see Section 2.4 below).

2.1. Transcription of the tRNA[Ser]^{Sec} Gene (Trsp)

Trsp is a single-copy gene in most genomes, but several organisms, including zebrafish, have more than one copy [34,35]. Transfer RNA[Ser]^{Sec} is transcribed, like all canonical tRNAs, by RNA polymerase (Pol) III, except in *Trypanosoma brucei* which was reported to be transcribed by Pol II [36]. However, the promoter structure and TATA-box-binding protein utilization of tRNA[Ser]^{Sec} are distinct from those of other tRNA genes [37,38]. While the transcription of canonical tRNA genes is dependent on the internal promoters, so called A- and B-boxes, the upstream promoters including TATA-boxes govern the transcription of tRNA[Ser]^{Sec} and other TATA-less Pol III genes such as snU6 and 7SK. *Trsp* transcription is initiated at the first nucleotide within the gene, whereas all other tRNAs are transcribed with a leader sequence that must be removed by processing from the resulting transcript [39]. The tRNA[Ser]^{Sec} transcript has a trailer sequence, and like all other tRNAs, the trailer must be processed to yield the primary sequence, wherein the ubiquitous CCA terminus is then added to prepare the completed transcript which is now ready for modification.

2.2. Primary Sequence of Sec tRNA[Ser]^{Sec}

The primary sequence of Sec tRNA[Ser]^{Sec}, which is the longest tRNA described to date, contains 90 or more nucleotides, depending on the species that encodes Trsp. In higher animals (e.g., *Xenopus*, birds, and mammals), four bases are modified on the 90 nucleotide primary transcript, and a portion of the Sec tRNA population is modified on the 2′-O-ribosyl moiety forming the only nucleoside modification (reviewed in [40]). The four base modifications are 5-methoxycarbonylmethyluracil (mcm^{5}U) at position 34 (the wobble position in the anticodon), isopentenyladenosine (i^{6}A) at position 37 (the base immediately 5′ to the anticodon), pseudouridine (Ψ) at position 55, and N1-methyladenosine (m1A) at position 58 (wherein the last two modifications occur within the TΨC loop). The single nucleoside modification occurs when a portion of the mcm^{5}U isoform is converted to 5-methoxycarbonylmethyluracil-2′-O-methylribose (mcm^{5}Um). The methylation of mcm^{5}U to mcm^{5}Um requires other modifications such as i^{6}A and m^{1}A, and the correct tertiary structure [41]. Interestingly, methylation of mcm^{5}U is influenced by the selenium levels in the cell [41]. The primary structures of tRNA[Ser]^{Sec}$_{mm^5U}$ and tRNA[Ser]^{Sec}$_{mm^5Um}$ are shown in a cloverleaf form in Figure 1.

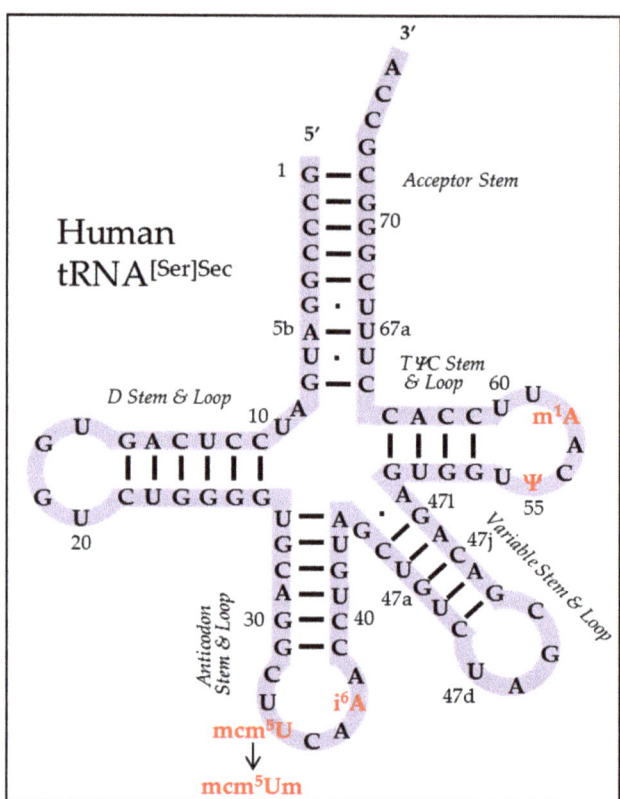

Figure 1. Cloverleaf model of human tRNA[Ser]^{Sec}. The image shows the 90 bases in human tRNA[Ser]^{Sec}. The paired 5′ and 3′ terminal bases constitute the acceptor stem, and on the left portion of the tRNA, the D stem and loop constitute six paired and four unpaired bases. On the lower portion of the tRNA, the anticodon stem and loop constitute six paired and seven unpaired bases, and the variable stem and loop constitute five paired and four unpaired bases. On the right portion of the tRNA, the TΨC stem and loop constitute four paired and seven unpaired bases. Human tRNA[Ser]^{Sec} contains base modifications at the following positions: 34 (mcm^{5}U), 37 (i^{6}A), 55 (Ψ), and 58 (m^{1}A). The two isoforms differ from one another at position 34 by a single methyl group on the 2′-O-ribosyl moiety.

2.3. The Sec-tRNA[Ser]^{Sec} Population

The Sec-tRNA[Ser]^{Sec} populations in bacteria and archaea consist of a single tRNA that is aminoacylated with Ser by seryl-tRNA synthetase (SARS). The tRNA[Ser]^{Sec} populations in mammals, birds, and Xenopus consist of two isoforms: tRNA[Ser]$^{Sec}_{mcm^5U}$ and tRNA[Ser]$^{Sec}_{mm^5Um}$, both of which are aminoacylated with Ser by their corresponding SARS [18]. The levels of the two isoforms are not limiting; however, reducing the tRNA[Ser]^{Sec} population by about half or increasing it several-fold does not generally appear to affect overall selenoprotein expression in various mammalian cells and mouse tissues [42]—albeit some minor differences in selenoprotein expression have been observed [43].

The levels of mcm^{5}U are enriched and those of mcm^{5}Um diminished under conditions of selenium deficiency in mammalian cells and organs, while the reverse is true under conditions of selenium sufficiency, i.e., the levels of mcm^{5}Um are enriched, and the levels of mcm^{5}U diminished. Interestingly, the two Sec-tRNA[Ser]^{Sec} isoforms are involved in the synthesis of different classes of selenoproteins. Housekeeping selenoproteins, such as GPX4 and thioredoxin reductase 1 (TXNRD1), are essential to cellular function and are expressed even during selenium-deficient conditions. Housekeeping selenoproteins are expressed by the Sec-tRNA[Ser]$^{Sec}_{mcm^5U}$ isoform. Stress-related selenoproteins, such as GPX1, are expressed in higher amounts in response to enriched selenium levels. This class of selenoproteins is expressed by the Sec-tRNA[Ser]$^{Sec}_{mcm5Um}$ isoform [43,44].

Detailed examinations of Ser-tRNA[Ser]$^{Sec}_{mcm^5U}$ and Ser-tRNA[Ser]$^{Sec}_{mcm^5Um}$ levels were carried out in various mammalian cell lines by growing human leukemia (HL-60) cells, Chinese hamster ovary (CHO) cells, and rat mammary tumor (RMT) cells in media with or without selenium supplementation (Table 1), and in various mammalian organs by subjecting mice to diets with or without selenium supplementation (Table 2). The respective tRNA populations were isolated from each cell line and from each mouse organ, labeled with ^{3}H-serine, and the distributions of the two Sec isoforms were determined. The total amount of the two Ser-tRNA[Ser]^{Sec} isoforms varied considerably in the different cell lines and mammalian organs (Tables 1 and 2), respectively. The consistent observation was that in every case the tRNA[Ser]$^{Sec}_{mcm^5U}$ isoform was enriched in selenium-deficient conditions and the Ser-tRNA[Ser]$^{Sec}_{mcm^5Um}$ isoform was enriched in selenium-sufficient conditions.

Table 1. Sec-tRNA[Ser]^{Sec} isoforms in cultured mammalian cells.

			Sec tRNA[Ser]^{Sec}				
				mcm^{5}U		mcm^{5}Um	
Cell Line	Selenium Supplementation [a]	% of Total [b]	%	% of Total [c]	%	% of Total [d]	mcm^{5}Um/ mcm^{5}U [e]
HL-60	+(chem. defined media)	9.6	38.5	3.70	61.5	5.90	1.60
	−(chem. defined media)	7.5	61.3	4.60	38.7	2.90	0.63
HL-60	+(FBS)	9.4	55.3	5.20	44.7	4.20	0.81
	−(FBS)	7.4	77.0	5.70	23.0	1.70	0.30
CHO	+(FBS)	1.01	45.1	0.46	54.9	0.55	1.22
	−(FBS)	0.86	56.2	0.48	43.8	0.38	0.78
RMT	+(chem. defined media)	1.7	11.8	0.20	88.2	1.50	7.47
	−(chem. defined media)	1.4	35.7	0.50	64.3	0.90	1.80

[a] FBS: fetal bovine serum; [b] percentage of tRNA[Ser]^{Sec} population within total Ser-tRNA population; [c] percentages of mcm^{5}U and mcm^{5}Um isoforms within total tRNA[Ser]^{Sec} population; [d] percentages of mcm^{5}U or mcm^{5}Um isoforms within total Ser-tRNA population; [e] amount of mcm^{5}Um/amount of mcm^{5}U isoforms.

Table 2. Sec-tRNA[Ser]^{Sec} isoforms in murine tissues.

			Sec tRNA[Ser]^{Sec}				
				mcm^{5}U		mcm^{5}Um	
Organ	Dietary Selenium Supplementation	% of Total [a]	%	% of Total [b]	%	% of Total [c]	mcm^{5}Um/ mcm^{5}U [d]
Heart	+	4.3	38.1	1.64	61.9	2.66	1.62
	−	3.2	66.4	2.12	33.6	1.08	0.51
Kidney	+	7.5	33.7	2.52	66.3	4.97	1.97
	−	3.7	59.2	2.19	40.8	1.51	0.69
Liver	+	4.5	33.3	1.50	66.7	3.00	2.00
	−	2.8	57.7	1.62	42.3	1.18	0.73
Muscle	+	1.9	38.6	0.73	61.4	1.17	1.59
	−	1.5	73.3	1.10	26.7	0.40	0.35

[a] Percentage of tRNA[Ser]^{Sec} population within total Ser-tRNA population; [b] percentages of mcm^{5}U and mcm^{5}Um isoforms within total tRNA[Ser]^{Sec} population; [c] percentages of mcm^{5}U or mcm^{5}Um isoforms within total Ser-tRNA population; [d] amount of mcm^{5}Um/amount of mcm^{5}U isoforms.

2.4. Biosynthesis of Sec on Sec tRNA[Ser]^{Sec}

The incorporation of selenium into protein occurs as the amino acid Sec. This amino acid is biosynthesized in a unique manner on its tRNA, named Sec-tRNA[Ser]^{Sec} [45,46]. The pathway of Sec biosynthesis is different in archaea and eukaryotes (Figure 2a), and in bacteria (Figure 2b). The unacylated tRNA[Ser]^{Sec} is initially aminoacylated with Ser by SARS to form Ser-tRNA[Ser]^{Sec} in all three groups. The Ser moiety on Ser-tRNA[Ser]^{Sec} in archaea and eukaryotes (Figure 2a) is transferred to an intermediate, phosphorseryl-tRNA[Ser]^{Sec}, by phosphoseryl-tRNA kinase (PSTK). The intermediate is then converted to Sec-tRNA[Ser]^{Sec} in the presence of selenophosphate 2 (SEPHS2) (see [18] and references therein). On the other hand, bacteria use an enzyme, Sec synthetase (SecS or SepSecS), to convert Ser-tRNA[Ser]^{Sec} to Sec-tRNA[Ser]^{Sec} (Figure 2b). There is an abundance of complex machinery dedicated to incorporating Sec into protein in response to the UGA Sec codon in selenoprotein mRNA. This topic has been reviewed elsewhere in this Special Issue by Copeland and Howard [47], and, therefore, will not be further discussed herein. The insertion of Sec into protein to generate selenoproteins occurs in organisms within all three of the taxonomic domains, eukaryotes, archaea, and bacteria. Selenoproteins are found in only about 15% of archaea, about 25% of bacteria, and about half of eukaryotes [48].

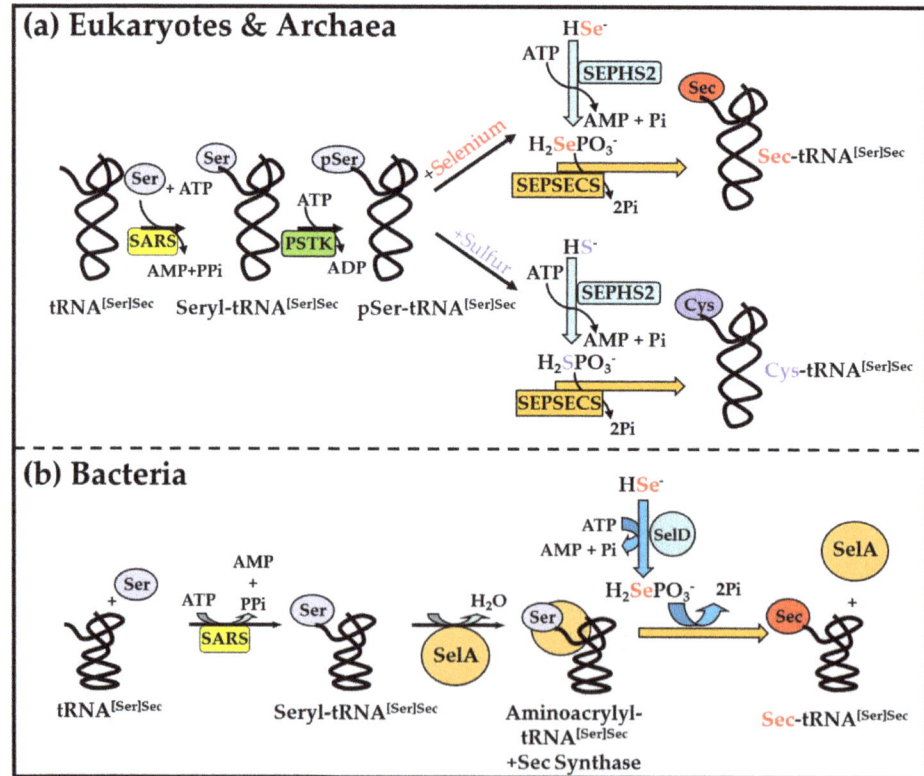

Figure 2. Pathways of Selenocysteine (Sec) biosynthesis. The biosynthetic pathways of Sec in: (**a**) eukaryotes and archaea; (**b**) bacteria. Abbreviations are: Pi—inorganic phosphate; PPi—inorganic pyrophosphate; SARS—Ser-tRNA synthetase; SelA—selenocysteine synthase; SelD—selenophosphate synthetase; H$_2$SePO$_3^-$—selenophosphate; SEPHS2—selenophosphate synthetase 2.

2.5. Sec, the 21st Amino Acid in the Genetic Code

Sec constitutes the 21st proteinogenic amino acid in the genetic code, and Sec is encoded in selenoprotein mRNA, as noted above, by the genetic codeword UGA. UGA is a shared codon in the genetic code in organisms containing selenoproteins, and it designates either Sec or the cessation of protein synthesis depending on its location within the mRNA. A specific sequence called the Sec insertion sequence (SECIS) element, which is located immediately downstream of the Sec UGA codon in bacteria and located much further downstream of the Sec UGA codon in archaea and eukaryotes, is responsible for designating the upstream codon as Sec [49]. The classes of SECIS elements and their roles have been reviewed in detail elsewhere (see [18,45] and references therein) and will not be further considered herein.

3. Selenoproteins

3.1. Mammalian Selenoproteins

There are 25 selenoprotein genes in the human genome. They are highly conserved across mammals, with the only two known exceptions being GPX6 and SELENOV. GPX6 contains Cys instead of Sec in some species, including mice and rats, and SELENOV was lost in gorillas [50]. Some of the mammalian selenoproteins are shared with non-mammalian eukaryotes, including glutathione peroxidases (GPXs), thioredoxin reductases (TXNRDs), and selenophosphate synthetase (SEPHS), indicating an early evolutionary origin for these proteins [51].

3.1.1. Glutathione Peroxidases (GPX)

GPXs comprise a large superfamily that is widespread across all kingdoms of life [52]. They use glutathione or thiol oxidoreductases as major reductants, and their functions include detoxification of hydroperoxides, regulation of ferroptosis, and hydrogen hydroperoxide signaling, among others [51]. There are eight GPXs in mammals, five of which are selenoproteins (GPX1-4, GPX6), and three are Cys-containing homologs (GPX5, GPX7, and GPX8). GPX1 is the most abundant mammalian selenoprotein found in the cytosol of most cells. It was the first mammalian selenoprotein described, and it was instrumental in developing methods for the insertion of selenium in the form of Sec into proteins. GPX2 and GPX3 have a more localized expression. GPX2 is often termed gastrointestinal based on its initial detection in gastrointestinal tissues, and GPX3 is expressed in the kidney at very high levels and is secreted into the plasma. GPX4 is unique among the GPXs because it reduces phospholipid hydroperoxides and has a broad substrate specificity. It has received much attention recently due to its essentiality for embryonic development in mice [53,54] and its role as a master regulator of ferroptosis [55,56]. GPX6 is the most recently evolved Sec-containing GPX, present only in mammals. In mice and in a few other species, Sec was then replaced by Cys [50].

3.1.2. Thioredoxin Reductases (TXNRD)

There are three TXNRDs in mammals, all of which contain Sec, and their functions are selenium-dependent. Sec is located in the penultimate C-terminal position, as part of a characteristic GCUG motif [57]. TXNRD1 is primarily localized in the cytosol and nucleus, and uses thioredoxin 1 (TXN1) as a major substrate. Its main physiological role is the NADPH-dependent reduction of TXN1, which in turn is involved in many physiological processes. TXNRD2 is localized in the mitochondria, and it is involved in the reduction of mitochondrial thioredoxin (TXN2) and glutaredoxin 2 (GLRX2). Both TXNRD1 and TXNRD2 are present in all vertebrates and are essential in mice [58,59]. TXNRD3 contains an additional N-terminal GLRX domain, and displays glutaredoxin activity [60].

3.1.3. Iodothyronine Deiodinases (DIO)

The thyroid hormone deiodinases (DIO) consist of three selenoproteins (DIO1, DIO2, and DIO3) that are involved in the metabolism of thyroid hormones by iodothyronine deiodination [61]. Like most selenoproteins, they are thioredoxin-like proteins. Thyroid hormones regulate a variety of processes, including growth, development, and metabolic rate. Thyroxine (T4) is the main thyroid hormone in circulation, produced in the thyroid gland, and is the precursor of triiodothyronine (T3), which has a higher affinity for thyroid hormone receptors [62]. DIO1 and DIO2 catalyze the activation of T4 to T3. Conversely, DIO3, and in some conditions DIO1, can inactivate T3 by producing the inactive metabolites T2 and reverse T3 (rT3), respectively. Studies in deiodinase-deficient mice have confirmed the function of deiodinases for T3 formation [63–65]. Distantly related homologs have been identified in single-celled eukaryotes; however, their function is not known, but it must be different from that of mammalian deiodinases.

3.1.4. Methionine-R-Sulfoxide Reductase 1 (MSRB1)

MSRB1 is a zinc-containing selenoprotein that was initially identified as selenoprotein R [66] and selenoprotein X [67] using bioinformatic tools. It was later found to be methionine-R-sulfoxide reductase based on its repair activity on the R enantiomer of oxidized methionine residues in proteins. Methionine and cysteine are the two sulfur-containing amino acids that are the most susceptible to oxidation, which may lead to a significant alteration of their structure and the disruption of protein function. MSRB1 may protect proteins against oxidative damage by catalyzing the reduction of methionine sulfoxide back to methionine. Though structurally different to MSRA (methionine-S-sulfoxide reductase), both proteins perform complementary functions by acting in only one of the two stereoisomers. MSRA is also a selenoprotein in some unicellular eukaryotes. Two additional homologs, MSRB2 and MSRB3, which contain catalytic Cys instead of Sec, are present in mammals. MSRB2 is localized in the mitochondria, whereas MSRB3 is targeted to the endoplasmic reticulum [68].

3.1.5. Selenophosphate Synthetase 2 (SEPHS2)

As discussed above, SEPHS2 provides the active Se donor for the synthesis of Sec. Selenophosphate is synthesized from selenide and ATP [69]. SEPHS2 is a widespread protein found in all Sec-containing prokaryotes and eukaryotes. In prokaryotes, in addition to Sec, SEPHS2 also supports other forms of selenium utilization. Selenium is used in the form of selenouridine in certain tRNAs [70,71], and as a cofactor in some molybdenum-containing proteins [72,73]. In eukaryotes, Sec is the only known selenium trait.

A SEPHS2 paralog called SEPHS1 is found in some animals [74,75]. SEPHS1 is not a selenoprotein, it does not support selenophosphate and selenoprotein synthesis [76], and never carries Sec or Cys at the catalytic site. Instead, other amino acids have been observed at that position (arginine, threonine, glycine, and leucine). SEPHS1 is an essential gene in fruit flies [77] and mice [78], but its function remains unknown. The fact that SEPHS1 is present in selenoprotein-less animals [79,80] suggests that its function is not related to selenium. Nonetheless, it is believed to participate in redox homeostasis [78]. Interestingly, SEPHS1 genes in different animal lineages, e.g., insects and vertebrates, are believed to be functional homologs, but they originated independently [81].

3.1.6. Selenoprotein P (SELENOP)

SELENOP is the only selenoprotein with multiple Sec residues in mammals. It is a major selenoprotein in plasma and is synthesized primarily in the liver [82]. Its sequence contains a Sec-containing thioredoxin-like domain in its N-terminal region, and a Sec-rich domain in the C-terminus [83]. Its main function is to provide selenium to several tissues, especially the testis and brain [84,85]. SELENOP is present across metazoans, but its Sec content is highly variable. Among vertebrates, the number ranges from five in mole rats to up to 37 in amberjack fish. Human and mouse SELENOP contains 10 Sec residues. A remarkable diversity is observed in invertebrates. SELENOP was lost in most nematodes, most insects, tunicates, and Platyhelminthes, whereas in other lineages, SELENOP is particularly Sec-rich, including in annelids, ribbon worms (Nemertea), and mollusks. Topping the list is the bivalve *Elliptio complanata* with 133 Sec residues [86].

3.1.7. Selenoprotein N (SELENON)

SELENON (formerly SEPN1) is an endoplasmic reticulum (ER) glycoprotein that contains a calcium-binding EF-hand domain and a Sec-containing domain with a possible oxidoreductase function. The specific function of the protein remains unknown. One possible function that has been suggested is a calcium sensor through the EF-hand domain, which activates the sarcoplasmic/ER calcium ATPase 2 (SERCA2)-mediated calcium uptake into the ER in a redox-dependent manner [87]. SELENON was first identified using computational methods [67], and shortly after was associated with congenital rigid spine muscular dystrophy [88], becoming the first selenoprotein to be associated with a genetic disease. Mutations in SELENON cause a group of recessive neuromuscular disorders collectively known as SELENON-related myopathies [89]. Many mutations have been identified in homozygous or heterozygous compound patients, including mutations in the UGA codon and the SECIS element that prevent the incorporation of Sec [88,90,91].

3.1.8. Selenoprotein O (SELENOO)

SELENOO is a widespread selenoprotein present in both prokaryotes and eukaryotes. The mammalian homologs carry a Sec residue at their C-terminal penultimate position [92], while in bacteria and in many other eukaryotes, including yeast and plants, homologs have a Cys instead. Its sequence contains a protein kinase-like domain [93] but its function remained elusive [92]. A recent study [94] uncovered a novel activity for the protein kinase superfamily, wherein SELENOO transfers AMP from ATP to Ser, Thr, and Tyr residues on protein substrates, an activity termed AMPylation. The protein is localized in the mitochondria [95] and AMPylates proteins involved in redox homeostasis [94].

3.1.9. Selenoprotein I (SELENOI)

SELENOI is essential for embryonic development in mice [96]. It is a recently evolved selenoprotein only found in vertebrates [50]. Its sequence contains a CDP-alcohol phosphatidyltransferase domain, characteristic of phospholipid synthases. It was suggested that SELENOI may be an ethanolamine phosphotransferase that catalyzes the last step of the Kennedy pathway for the synthesis of phosphatidylethanolamine [97] and is localized in the Golgi apparatus [98]. However, this activity was reported for a protein truncated at Sec, so more studies are needed to establish the function of this selenoprotein. Mutations in SELENOI have been identified in patients with a form of hereditary spastic paraplegia [99,100].

3.1.10. Other Selenoproteins

Other selenoproteins have no known functions, albeit some suggested ones. SELENOW, SELENOT, SELENOH, and SELENOV belong to the redox family of selenoproteins [101]. They possess a CXXU motif within a thioredoxin-like fold domain. Based on this observation, they are proposed to be oxidoreductases of unknown functions. SELENOV is the most recently evolved selenoprotein, only present in placental mammals. It appeared by duplication of SELENOW; the two genes share the same exonic structure, but SELENOV contains a long highly variable N-terminal extension [50]. SELENOV is exclusively expressed in testis [92].

SELENOF and SELENOM are thioredoxin-like fold ER-resident selenoproteins. These proteins share ~30% of sequence identity in mammals and are distantly related homologs with a common evolutionary origin [51]. Their function, however, is not well understood. SELENOF may be involved in the regulation of protein folding, and its deletion promotes nuclear cataracts in mice [102]. Several studies examined its possible role in cancer [103,104]. Altered expression, both high and low, has been linked to a higher cancer risk in different tissues, including lung, breast, prostate, and liver [105]. Similarly, common genetic variants in SELENOF have been studied for their relationship with higher cancer risk [105]. SELENOM is expressed in the brain and shows neuroprotective properties. Altered levels of SELENOM have been linked to the early onset of Alzheimer's disease and hepatocellular carcinoma [106]. In addition, it has also been implicated in calcium release from the ER in response to hydrogen peroxide [107]. Its deletion in mice leads to increased body weight but does not affect neuronal and cognitive function [108].

SELENOK and SELENOS share a few features that set them apart from other selenoproteins. They are ER-resident transmembrane proteins with a single transmembrane domain in the N-terminal sequence, contain a Gly-rich segment with positively charged residues, and their Sec residues are characteristically located near the C-terminus. They have been implicated in the ER-associated degradation (ERAD) of misfolded proteins [51]. SELENOK has also been proposed to link selenium levels and immunity through association with a partner enzyme, DHHC6, for protein palmitoylation [109].

3.2. Phylogenetic Distribution of Selenoproteins

Selenoproteins are present across the three domains of life: bacteria, archaea, and eukaryotes. The evidence supports that the Sec trait evolved only once: prokaryotes and eukaryotes use analogous Sec biosynthesis and insertion pathways, and some selenoprotein families are shared among bacteria, archaea, and eukaryotes. The use of Sec is not universal, however. In selenoprotein-less organisms, the functions of selenoproteins are typically replaced by Cys homologs and the Sec synthesis machinery genes are lost.

Among prokaryotes, it is estimated that 20–25% of bacteria use selenoproteins [81,110], and the proportion is even lower in archaea, with a narrow distribution of Sec-containing genomes [111]. Nonetheless, the closest relatives to eukaryotes, the archaeal Asgard lineage, was identified as the intermediate form between the prokaryotic and eukaryotic Sec insertion systems [60,112,113].

Sec is much more common among eukaryotes, and selenoproteins show a highly dynamic evolution in terms of Sec to Cys conversions and gene losses. A scattered pattern of the presence/absence of selenoproteins was already evident from the analysis of the early sequenced genomes [114,115]. Since then, thousands of genomes have been sequenced, and the use of bioinformatic tools for large-scale analyses has provided a much more detailed picture. No selenoproteins have been identified in land plants so far, although they are abundant in other Archaeplastida (plantae) lineages. Recent works have explored the diversity of selenoproteins in plantae and especially algae lineages, reporting novel eukaryotic selenoprotein families [116,117]. Fungi were traditionally thought to have lost all selenoproteins at the root of the lineage. This was recently challenged by the discovery of several Sec-containing fungi genomes, outlining multiple independent Sec- loss events, including in the yeast lineage [118]. Among animals, Sec losses have been reported in all major insect lineages [119,120], in mites [34], and in nematodes [121].

4. Mouse Models

In the early 2000s, various mouse models were developed to elucidate the role of selenoproteins in health and development [122–125]. These mouse models took advantage of the fact that selenoprotein expression is dependent on the presence of a single tRNA, Sec tRNA[Ser]^{Sec}, and the fact that this tRNA is encoded as a single copy gene, designated Trsp [126]. Thus, by manipulating Trsp in numerous ways, several models were developed that identified the presence of two selenoprotein classes, housekeeping selenoproteins and stress-related selenoproteins, their cellular roles, and also the roles of various individual selenoproteins within these two classes.

4.1. Trsp Transgenic Mouse Models

The first mouse models that examined the role of Sec-tRNA[Ser]^{Sec} in selenoprotein synthesis were created in 2001 and involved *Trsp* wild-type or mutant transgenes [92]. Three different constructs encoding either the wild-type or two different mutant transgenes were prepared. By substituting one of the bases within the anticodon loop of *Trsp*, the role of the mutant Sec-tRNA[Ser]^{Sec} in selenoprotein expression could be assessed. Changing either the T at position 34 to A, or the A at position 37 to G, prevented the synthesis of the 2′-0-methyluridine at position 34 [124,127]. Therefore, these mutant mice carrying either mutant transgene permitted the evaluation of the methylribose in selenoprotein expression. Synthesis of stress-related selenoproteins was virtually abolished, while housekeeping selenoprotein expression was virtually unchanged. Stress-related selenoprotein synthesis, therefore, is dependent on the 2′-0-hydroxymethyl group, while housekeeping selenoprotein expression is carried out by Sec-tRNA[Ser]$^{Sec}_{mcm5U}$. These studies did not resolve the question of whether Sec-tRNA[Ser]$^{Sec}_{mcm5Um}$ also supports housekeeping selenoprotein synthesis. However, it seems very plausible that this isoform can also synthesize the essential class of selenoproteins.

Various aspects of the effects of stress-related selenoprotein loss on health were also examined. Interestingly, reduced stress-related selenoprotein expression in G37 mutant mice was tissue-specific, wherein the loss was highly significant in the kidney and liver but not in the testes [124]. These mice were further studied regarding health issues and were found to be more susceptible to colon cancer [128], viral infection [129], and X-ray damage [130]. Crossing these mice with C3/TAg mice yielded offspring with accelerated rates of prostatic epithelial neoplasia, suggesting a protective role of selenoproteins in prostate cancer development [131]. Glucose intolerance was also observed in these G37 mice, which led to a diabetic-like phenotype [132].

4.2. Trsp Conditional Knockout Mouse Models

Although the total removal of *Trsp* in mice is embryonic lethal [122], the targeted removal of *Trsp* in specific tissues and organs provides an alternative model for examining the role of selenoproteins in health and development [123]. Highly significant functions of selenoproteins were elucidated in numerous organs and tissues such as skeletal muscle; heart and endothelial cells [133]; bone [134]; skin [135]; immune cells, including macrophages, hematopoietic tissues, and T cells [136–138]; neurons [139]; liver [82,140]; podocytes [141]; osteochondroprogenitor [134]; thyroid [142]; prostate [143]; kidney; and mammary glands [123]. For convenience to the reader, and to keep this review within the allotted length, we have summarized the major findings in each of the above in vivo *Trsp* conditional knockout studies in Table 3. It should also be noted that more in-depth, recent studies of endothelial cells in cell death have been carried out (see [144] in this Special Issue and references therein).

Table 3. *Trsp* conditional knockout mouse models.

Targeted Tissue or Organ [1]	Main Findings Regarding Role of Selenoproteins in Genetically-Altered Mice, Relative to Control Mice in the Study	*Cre* Promoter
Endothelial cells	Endothelial cell development/function: embryonic lethal. 14.5 d.p.c. embryos were smaller, more fragile, had poorly or under-developed vascular systems, limbs, head, and tail [133].	*TieTek2-Cre*
Heart & Skeletal Muscle	Heart disease prevention: mice died from acute myocardial failure 12 days after birth.	*MCK-Cre*
Kidney	No increase in oxidative stress or nephropathy found in podocytes of selenoprotein-deficient mice [141].	*NPHS2-Cre*
Liver	Liver function: severe hepatocellular degeneration—mice died between 1 and 3 months of age [82]. SELENOP and GPX3 were reduced in serum and kidney, supporting a selenium-transport role for liver-derived SELENOP [140]. Enhanced expression of phase II response genes compensated for loss of hepatic *Trsp* [145]. Mice used as controls to monitor selenium pools in kidney due to reduction of GPX3 imported from liver [146]. *Secisbp2* gene inactivation was less detrimental than *Trsp* inactivation [147].	*Alb-Cre*

Table 3. Cont.

Targeted Tissue or Organ [1]	Main Findings Regarding Role of Selenoproteins in Genetically-Altered Mice, Relative to Control Mice in the Study	*Cre* Promoter
Macrophages	Immune function: increased oxidative stress and expression of cytoprotective antioxidant and detoxification genes, accumulation of ROS levels, and impaired invasiveness. Altered expression of ECM and fibrosis-associated genes [148]. Balance of pro- and anti-inflammatory oxylipids during inflammation [149]. Selenoproteins protect mice from chemically-induced colitis by alleviating inflammation [150]. Role in epigenetic modulation of pro-inflammatory genes [151]. When infected with *N. brasiliensis*, selenium-supplemented KO mice showed a complete abrogation in M2-marker expression with a significant increase in intestinal worms and fecal eggs [152].	LysM-Cre
Mammary glands	First *Trsp* conditional KO mouse, providing an important tool for elucidating the role of selenoproteins in health and development [123]. MMTV-Cre mice treated with DMBA had significantly more tumors, suggesting that selenoproteins protect against carcinogen-induced mammary cancer [153].	MMTV-Cre; Wap-Cre
Neurons	Neuronal function: enhanced neuronal excitation followed by neurodegeneration of hippocampus. Cerebellar hypoplasia associated with degeneration of Purkinje and granule cells. Cerebellar interneurons essentially absent [139]. Selenoproteins required in post-mitotic neurons of the developing cerebellum [154].	Tal-Cre; CamK-Cre
Osteo-chondroprogenitor	Kashin–Beck disease model: mice had post-natal growth retardation, chondrodysplasia, chondronecrosis, and delayed skeletal ossification characteristic of Kashin–Beck disease [134].	Col2a1-Cre
Prostate	Mice developed PIN-like lesions and microinvasive carcinoma by 24 weeks, which were associated with loss of basement membrane, increased cell cycle, and apoptotic activity [143].	PB-Cre4
Skin	Role in skin and hair follicle development: runt phenotype, premature death, alopecia with flaky and fragile skin, epidermal hyperplasia with disturbed hair cycle, and an early regression of hair follicles [135].	K14-Cre
T-cells	Immune function: reduction of mature T cells and a defect in T-cell-dependent antibody response. Antioxidant hyperproduction and suppression of T cell proliferation in response to T cell receptor stimulation [137].	LCK-Cre
Thyroid	Mice lacking selenoproteins in thyrocytes showed increased oxidative stress in thyroid. Gross morphology remained intact for at least 6 months. Thyroid hormone levels remained normal in knockout mice; thyrotropin levels moderately elevated [142].	Pax8-Cre; Tg-Cre^{ER}

[1] Target organs/tissues in alphabetical order. Abbreviations: days-post-coitum (d.p.c.); 7,12-dimethylbenz[a]anthracene (DMBA); extracellular matrix (ECM); mouse mammary tumor virus (MMTV); prostatic intraepithelial neoplasia (PIN).

4.3. *Trsp* Knockout/Transgenic and *Trsp* Conditional Knockout/Transgenic Mouse Models

Models involving *Trsp* knockout and *Trsp* conditional knockout mice that were rescued with the *Trsp* wild-type transgene, or the G37 or A34 mutant transgenes, were constructed [125] to examine the ability of these transgenes to restore selenoprotein synthesis. As expected, the wild-type transgene completely restored selenoprotein biosynthesis, while the G37 mutant transgene restored housekeeping selenoprotein synthesis but virtually did not refurbish stress-related selenoprotein synthesis [125]. Importantly, these mice demonstrated unequivocally that stress-related selenoproteins are not essential to the livelihood of the animal, although these mice were found to be very susceptible to selenium status (see Table 4). Mice carrying the G37 mutant transgene appeared phenotypically normal, but male mice produced sperm with an abnormal morphology which accounted for their reduced fertility, while female mice produced smaller-sized litters than the corresponding wild-type mice.

The A in the wobble position of the anticodon in tRNA is converted to inosine (I) which decodes A, U, and C in the 3′ position of the corresponding codewords. Hence, this tRNA decodes UGA and the cysteine codons, UCU and UCC, and likely promotes misreading in protein synthesis, which most certainly accounts for the reason why *Trsp* negative mice could not be rescued with the A34 mutant transgene [155].

Additional mouse models lacking *Trsp* specifically in the liver and rescued with the wild-type transgene or G37 or A34 mutant transgenes were generated [82]. A mouse model involving the loss of *Trsp*, and rescued with a transgene carrying a deletion within the activator element, was also prepared [25]. The activator element is required for the binding of the transcription factor, STAF, to transcribe Sec tRNA[Ser]^{Sec} [156]. The major findings of these studies are summarized in Table 4.

Table 4. Mouse models involving Trsp knockout (KO) or Trsp conditional KO mice rescued with wild-type (WT), G37 mutant, or A34 transgenes.

Target Site	Model Description	Major Findings Observed in Genetically Altered Mice in Comparison to Control Mice
Whole Mouse	*Trsp* KO rescued with WT *Trsp* transgene	Selenoprotein synthesis was completely recovered [125].
	Trsp KO rescued with G37 *Trsp* transgene	Proper base modification in the anticodon is essential, as mutant mice synthesize stress-related selenoproteins very poorly. Male mutant mice show abnormal sperm and reduced fertility; females produced reduced litter size [43]. Trsp KO could not be rescued with A34 mutant transgene most likely due to misreading (see Text).
Whole Mouse	*Trsp* KO rescued with promoter mutant *Trsp* transgene	Mice expressed tissue- and organ-specific amounts of tRNA[Ser]^{Sec}. Lower levels of the mcm^{5}Um isoform were observed in promoter mutant *Trsp* mice. Mice developed a similar neurological phenotype as SELENOP-KO mice and a reduced life span [157].
Liver *Alb-Cre*	*Trsp* liver KO rescued with *Trsp* WT transgene	Selenoprotein synthesis was completely recovered [82].
	Trsp liver KO rescued with G37 mutant *Trsp* transgene	Housekeeping selenoprotein synthesis was recovered while stress-related selenoprotein synthesis was poorly recovered [82].
	Trsp liver KO rescued with A34 mutant *Trsp* transgene	Housekeeping selenoprotein synthesis was recovered while stress-related selenoprotein synthesis was poorly recovered. Replacement of selenoprotein synthesis in conditional Trsp mutants resulted in normal gene expression of Phase II response enzymes [127,145].

5. Conclusions

So much of the molecular biology of selenium has been resolved in the past 30 to 40 years (see chapters in [8]) that the selenium field has tapered off considerably. There is still much to be done, primarily in providing more detailed functions of numerous individual selenoproteins and in understanding the roles of LMW selenocompounds in health and development. Herein, we have assembled a variety of topics in the selenium field that involve the unique characteristics of transcribing *Trsp*, determining the primary sequences of Sec-tRNA[Ser]^{Sec}$_{mcm5U}$ and tRNA[Ser]^{Sec}$_{mcm5Um}$ isoforms and their roles in translating housekeeping and stress-related selenoproteins, discussing numerous aspects of the historical roles of selenium in health and disease, and the molecular biology of selenium and selenoproteins in health and development. We borrowed the title of the classic Italian Western "The Good, the Bad and the Ugly" to include as part of our title as it seemed to perfectly describe the historical roles of selenium in health. This element has its "good" (an essential element in the diet of mammals and many other life forms), its "bad" (the consequences of too little or too much selenium in the diet), and its "ugly" aspects (in some cases, selenium may promote cancer and other health disorders, and extreme selenium deficiency may be lethal). Likewise, selenoproteins also have their "good" (these proteins are responsible in large part for the many health benefits of selenium), their "bad" characteristics (they can promote many health disorders including cancer), and can be "ugly" (loss of several selenoproteins is lethal). It will be of considerable interest to witness the many discoveries in the selenium/selenoprotein field as they continue to unfold in the years to come.

Author Contributions: All authors contributed in all aspects of the preparation and writing of this review. All authors have read and agreed to the published version of the manuscript.

Funding: This work was funded by Towson University's Fisher College of Science and Mathematics and the Department of Biological Sciences, awarded to Petra Tsuji, and by grants from the National Institutes of Health, awarded to Vadim Gladyshev.

Institutional Review Board Statement: Not applicable.

Informed Consent Statement: Not applicable.

Data Availability Statement: Not applicable.

Acknowledgments: We gratefully acknowledge Bradley A. Carlson for the many contributions he made to this review, and for his vast accomplishments in the selenium/selenoprotein research field that resulted in the more than 170 papers he has published, many of which are referenced herein.

Conflicts of Interest: The authors declare that there are on conflict of interest.

References

1. Berzelius, J.J. Undersökning af en ny Mineral-kropp, funnen i de orenare sorterna af det I Falun tillverkade svafl et. *Afhandlingar Fysik Kemi Och Mineral.* **1818**, *6*, 42.
2. Marsden, W. *The travels of Marco Polo, the Venetian: The translation of Marsden Revised, with A Selection of His Notes*; Wright, T., Ed.; Franklin Classics: London, UK, 1854.
3. Madison, T.C. Sanitary report—Fort Randall. In *Statistical Report on the Sickness and Mortality in the Army of the United States*; 36th Congress Senate Executive Document; Coolidge, R.H., Ed.; United States, Surgeon General's Office: Washington, DC, USA, 1856; pp. 37–41.
4. Franke, K.W. A new toxicant occurring naturally in certain samples of plant foodstuffs. *J. Nutr.* **1934**, *8*, 597. [CrossRef]
5. Hintz, H.F.; Thompson, L.J. Custer, selenium and swainsonine. *Veter Hum. Toxicol.* **2000**, *42*, 242–243.
6. Schwarz, K.; Foltz, C.M. Factor 3 Activity of Selenium Compounds. *J. Biol. Chem.* **1958**, *233*, 245–251. [CrossRef]
7. Enoch, H.G.; Lester, R.L. Effects of Molybdate, Tungstate, and Selenium Compounds on Formate Dehydrogenase and Other Enzyme Systems in Escherichia coli. *J. Bacteriol.* **1972**, *110*, 1032–1040. [CrossRef] [PubMed]
8. Hatfield, D.L.; Schweizer, U.; Tsuji, P.A.; Gladyshev, V.N. (Eds.) *Selenium Its Molecular Biology and Role in Human Health*, 4th ed.; Springer: New York, NY, USA, 2016.
9. Oldfield, J.E. Selenium: A historical perspective. In *Selenium—Its Molecular Biology and Role in Human Health*, 2nd ed.; Hatfield, D.L., Berry, M.J., Gladyshev, V.N., Eds.; Springer: New York, NY, USA, 2006; pp. 1–6.
10. Combs, G.F.; Yan, L. Status of dietary selenium in cancer prevention. In *Selenium—Its Molecular Biology and Role in Human Health*, 4th ed.; Hatfield, D.L., Schweizer, U., Tsuji, P.A., Gladyshev, V.N., Eds.; Springer: New York, NY, USA, 2016; pp. 321–332.
11. Ge, K.; Xue, A.; Bai, J.; Wang, S. Keshan disease-an endemic cardiomyopathy in China. *Virchows Archiv A Pathol. Anat. Histopathol.* **1983**, *401*, 1–15. [CrossRef]
12. Yu, F.F.; Qi, Z.; Shang, Y.-N.; Ping, Z.-G.; Guo, X. Prevention and control strategies for children Kashin-Beck disease in China: A systematic review and meta-analysis. *Medicine* **2019**, *98*, e16823. [CrossRef]
13. Zhou, H.; Wang, T.; Li, Q.; Li, D. Prevention of Keshan Disease by Selenium Supplementation: A Systematic Review and Meta-analysis. *Biol. Trace Element Res.* **2018**, *186*, 98–105. [CrossRef]
14. Institute of Medicine. *Dietary Reference Intakes: Vitamin C, Vitamin E, Selenium, and Carotenoids*; Food and Nutrition Board: Washington, DC, USA, 2000.
15. Peters, K.M.; Galinn, S.E.; Tsuji, P.A. Selenium: Dietary Sources, Human Nutritional Requirements and Intake Across Populations. In *Selenium—Its Molecular Biology and Role in Human Health*, 4th ed.; Hatfield, D.L., Schweizer, U., Tsuji, P.A., Gladyshev, V.N., Eds.; Springer: New York, NY, USA, 2016; Volume 1, pp. 295–305.
16. Diwadkar-Navsariwala, V.; Diamond, A.M. The Link between Selenium and Chemoprevention: A Case for Selenoproteins. *J. Nutr.* **2004**, *134*, 2899–2902. [CrossRef] [PubMed]
17. Fairweather-Tait, S.J.; Bao, Y.; Broadley, M.; Collings, R.; Ford, D.; Hesketh, J.E.; Hurst, R. Selenium in Human Health and Disease. *Antioxid. Redox Signal.* **2011**, *14*, 1337–1383. [CrossRef]

18. Hatfield, D.L.; Tsuji, P.A.; Carlson, B.A.; Gladyshev, V.N. Selenium and selenocysteine: Roles in cancer, health, and development. *Trends Biochem. Sci.* **2014**, *39*, 112–120. [CrossRef]
19. Rayman, M.P. Selenium in cancer prevention: A review of the evidence and mechanism of action. In Proceedings of the Nutrition Society; CABI Publishing: Oxfordshire, UK, 2005; Volume 64, pp. 527–542.
20. Hoffmann, P.R.; Berry, M.J. The influence of selenium on immune responses. *Mol. Nutr. Food Res.* **2008**, *52*, 1273–1280. [CrossRef] [PubMed]
21. Guillin, O.M.; Vindry, C.; Ohlmann, T.; Chavatte, L. Selenium, Selenoproteins and Viral Infection. *Nutrients* **2019**, *11*, 2101. [CrossRef] [PubMed]
22. Campa, A.; Shor-Posner, G.; Indachochea, F.; Zhang, G.; Lai, H.; Asthana, D.; Scott, G.B.; Baum, M.K. Mortality risk in selenium-deficient HIV-positive children. *J. Acquir. Immune Defic. Syndr. Hum. Retrovirol.* **1999**, *20*, 508–513. [CrossRef]
23. Xu, X.-M.; Carlson, B.A.; Grimm, T.A.; Kutza, J.; Berry, M.J.; Arreola, R.; Fields, K.H.; Shanmugam, I.; Jeang, K.-T.; Oroszlan, S.; et al. Rhesus Monkey Simian Immunodeficiency Virus Infection as a Model for Assessing the Role of Selenium in AIDS. *JAIDS J. Acquir. Immune Defic. Syndr.* **2002**, *31*, 453–463. [CrossRef]
24. Cai, Z.; Zhang, J.; Li, H. Selenium, aging and aging-related diseases. *Aging Clin. Exp. Res.* **2018**, *31*, 1035–1047. [CrossRef]
25. Bartolini, D.; Sancineto, L.; Fabro de Bem, A.; Tew, K.D.; Santi, C.; Radi, R.; Toquato, P.; Galli, F. Selenocompounds in cancer therapy: An overview. *Adv. Cancer Res.* **2017**, *136*, 259–302. [PubMed]
26. Whanger, P.D. Selenocompounds in Plants and Animals and their Biological Significance. *J. Am. Coll. Nutr.* **2002**, *21*, 223–232. [CrossRef]
27. Ferreira, R.L.U.; Sena-Evangelista, K.C.M.; de Azevedo, E.P.; Pinheiro, F.I.; Cobucci, R.N.; Pedrosa, L.F.C. Selenium in Human Health and Gut Microflora: Bioavailability of Selenocompounds and Relationship with Diseases. *Front. Nutr.* **2021**, *8*, 685317. [CrossRef] [PubMed]
28. Flohe, L.; Günzler, W.; Schock, H. Glutathione peroxidase: A selenoenzyme. *FEBS Lett.* **1973**, *32*, 132–134. [CrossRef]
29. Rotruck, J.T.; Pope, A.L.; Ganther, H.E.; Swanson, A.B.; Hafeman, D.G.; Hoekstra, W.G. Selenium: Biochemical Role as a Component of Glutathione Peroxidase. *Science* **1973**, *179*, 588–590. [CrossRef]
30. Cone, J.E.; Del Rio, R.M.; Davis, J.N.; Stadtman, T.C. Chemical characterization of the selenoprotein component of clostridial glycine reductase: Identification of selenocysteine as the organoselenium moiety. *Proc. Natl. Acad. Sci. USA* **1976**, *73*, 2659–2663. [CrossRef]
31. Günzler, W.A.; Steffens, G.J.; Grossmann, A.; Kim, S.-M.A.; Ötting, F.; Wendel, A.; Flohé, L. The Amino-Acid Sequence of Bovine Glutathione Peroxidase. *Hoppe-Seyler's Z Physiol. Chem.* **1984**, *365*, 195–212. [CrossRef] [PubMed]
32. Chambers, I.; Frampton, J.; Goldfarb, P.; Affara, N.; McBain, W.; Harrison, P. The structure of the mouse glutathione peroxidase gene: The selenocysteine in the active site is encoded by the 'termination' codon, TGA. *EMBO J.* **1986**, *5*, 1221–1227. [CrossRef] [PubMed]
33. Zinoni, F.; Birkmann, A.; Stadtman, T.C.; Bock, A. Nucleotide sequence and expression of the selenocysteine-containing polypeptide of formate dehydrogenase (formate-hydrogen-lyase-linked) from Escherichia coli. *Proc. Natl. Acad. Sci. USA* **1986**, *83*, 4650–4654. [CrossRef] [PubMed]
34. Santesmasses, D.; Mariotti, M.; Guigó, R. Computational identification of the selenocysteine tRNA (tRNASec) in genomes. *PLoS Comput. Biol.* **2017**, *13*, e1005383. [CrossRef]
35. Xu, X.M.; Zhou, X.; Carlson, B.A.; Kim, L.K.; Huh, T.L.; Lee, B.J.; Hatfield, D.L. The zebrafish genome contains two distinct selenocysteine tRNA[Ser]sec genes. *FEBS Lett.* **1999**, *454*, 16–20. [CrossRef]
36. Aeby, E.; Ullu, E.; Yepiskoposyan, H.; Schimanski, B.; Roditi, I.; Mühlemann, O.; Schneider, A. tRNASec is transcribed by RNA polymerase II in Trypanosoma brucei but not in humans. *Nucleic Acids Res.* **2010**, *38*, 5833–5843. [CrossRef] [PubMed]
37. Park, J.M.; Lee, J.Y.; Hatfield, D.L.; Lee, B.J. Differential mode of TBP utilization in transcription of the tRNA[Ser]Sec gene and TATA-less class III genes. *Gene* **1997**, *196*, 99–103. [CrossRef]
38. Park, J.M.; Yang, E.S.; Hatfield, L.L.; Lee, B.J. Analysis of the Selenocysteine tRNA[SER]SEC Gene Transcription in vitro Using Xenopus Oocyte Extracts. *Biochem. Biophys. Res. Commun.* **1996**, *226*, 231–236. [CrossRef] [PubMed]
39. Lee, B.J.; De-La-Pena-Cortines, P.; Tobian, J.A.; Zasloff, M.; Hatfield, D. Unique pathway of expression of an opal suppressor phosphoserine tRNA. *Proc. Natl. Acad. Sci. USA* **1987**, *84*, 6384–6388. [CrossRef]
40. Carlson, B.A.; Lee, B.C.; Tsuji, P.A.; Tobe, R.; Park, J.M.; Schweizer, U.; Gladyshev, V.N.; Hatfield, D.L. Selenocysteine tRNA[Ser]Sec: From nonsense suppressor tRNA to the quintessential constituent in selenoprotein biosynthesis. In *Selenium—Its Molecular Biology and Role in Human Health*; Hatfield, D.L., Schweizer, U., Tsuji, P.A., Gladyshev, V.N., Eds.; Springer: New York, NY, USA, 2016.
41. Kim, L.K.; Matsufuji, T.; Matsufuji, S.; Carlson, B.A.; Kim, S.S.; Hatfield, D.L.; Lee, B.J. Methylation of the ribosyl moiety at position 34 of selenocysteine tRNA[Ser]Sec is governed by both primary and tertiary structure. *RNA* **2000**, *6*, 1306–1315. [CrossRef]
42. Chittum, H.S.; Baek, H.J.; Diamond, A.M.; Fernandez-Salguero, P.; Gonzalez, F.; Ohama, T.; Hatfield, D.L.; Kuehn, M.; Lee, B.J. Selenocysteine tRNA[Ser]Sec Levels and Selenium-Dependent Glutathione Peroxidase Activity in Mouse Embryonic Stem Cells Heterozygous for a Targeted Mutation in the tRNA[Ser]Sec Gene. *Biochemistry* **1997**, *36*, 8634–8639. [CrossRef]
43. Carlson, B.A.; Xu, X.-M.; Gladyshev, V.N.; Hatfield, D.L. Selective Rescue of Selenoprotein Expression in Mice Lacking a Highly Specialized Methyl Group in Selenocysteine tRNA. *J. Biol. Chem.* **2005**, *280*, 5542–5548. [CrossRef] [PubMed]

44. Carlson, B.A. Um34 in selenocysteine tRNA is required for the expression of stress-related selenoproteins in mammals. In *Fine-tuning of RNA Functions by Modification and Editing*; Grosjean, H., Ed.; Topis in Current Genetics; Springer: Berlin/Heidelberg, Germany, 2005; Volume 12, pp. 431–438.
45. Xu, X.-M.; Carlson, B.A.; Mix, H.; Zhang, Y.; Saira, K.; Glass, R.S.; Berry, M.J.; Gladyshev, V.N.; Hatfield, D.L. Biosynthesis of Selenocysteine on Its tRNA in Eukaryotes. *PLoS Biol.* **2006**, *5*, e4. [CrossRef]
46. Yuan, J.; Palioura, S.; Salazar, J.C.; Su, D.; O'Donoghue, P.; Hohn, M.J.; Cardoso, A.; Whitman, W.; Söll, D. RNA-dependent conversion of phosphoserine forms selenocysteine in eukaryotes and archaea. *Proc. Natl. Acad. Sci. USA* **2006**, *103*, 18923–18927. [CrossRef] [PubMed]
47. Copeland, P.R.; Howard, M.T. Ribosome Fate during Decoding of UGA-Sec Codons. *Int. J. Mol. Sci.* **2021**, *22*, 13204. [CrossRef]
48. Zhang, Y.; Gladyshev, V.N. Comparative Genomics of Trace Element Dependence in Biology. *J. Biol. Chem.* **2011**, *286*, 23623–23629. [CrossRef] [PubMed]
49. Berry, M.J.; Banu, L.; Chen, Y.; Mandel, S.J.; Kieffer, J.D.; Harney, J.W.; Larsen, P.R. Recognition of UGA as a selenocysteine codon in Type I deiodinase requires sequences in the 3' untranslated region. *Nat. Cell Biol.* **1991**, *353*, 273–276. [CrossRef] [PubMed]
50. Mariotti, M.; Ridge, P.G.; Zhang, Y.; Lobanov, A.V.; Pringle, T.H.; Guigo, R.; Hatfield, D.L.; Gladyshev, V.N. Composition and Evolution of the Vertebrate and Mammalian Selenoproteomes. *PLoS ONE* **2012**, *7*, e33066. [CrossRef]
51. Labunskyy, V.M.; Hatfield, D.L.; Gladyshev, V.N. Selenoproteins: Molecular Pathways and Physiological Roles. *Physiol. Rev.* **2014**, *94*, 739–777. [CrossRef] [PubMed]
52. Toppo, S.; Vanin, S.; Bosello, V.; Tosatto, S.C. Evolutionary and Structural Insights into the Multifaceted Glutathione Peroxidase (Gpx) Superfamily. *Antioxid. Redox Signal.* **2008**, *10*, 1501–1514. [CrossRef] [PubMed]
53. Imai, H.; Hirao, F.; Sakamoto, T.; Sekine, K.; Mizukura, Y.; Saito, M.; Kitamoto, T.; Hayasaka, M.; Hanaoka, K.; Nakagawa, Y. Early embryonic lethality caused by targeted disruption of the mouse PHGPx gene. *Biochem. Biophys. Res. Commun.* **2003**, *305*, 278–286. [CrossRef]
54. Yant, L.; Ran, Q.; Rao, L.; Van Remmen, H.; Shibatani, T.; Belter, J.G.; Motta, L.; Richardson, A.; Prolla, T.A. The selenoprotein GPX4 is essential for mouse development and protects from radiation and oxidative damage insults. *Free. Radic. Biol. Med.* **2003**, *34*, 496–502. [CrossRef]
55. Ingold, I.; Berndt, C.; Schmitt, S.; Doll, S.; Poschmann, G.; Buday, K.; Roveri, A.; Peng, X.; Porto Freitas, F.P.; Seibt, T.; et al. Selenium Utilization by GPX4 Is Required to Prevent Hydroperoxide-Induced Ferroptosis. *Cell* **2017**, *172*, 409–422.e21. [CrossRef]
56. Stockwell, B.R.; Angeli, J.P.F.; Bayir, H.; Bush, A.; Conrad, M.; Dixon, S.J.; Fulda, S.; Gascón, S.; Hatzios, S.K.; Kagan, V.E.; et al. Ferroptosis: A Regulated Cell Death Nexus Linking Metabolism, Redox Biology, and Disease. *Cell* **2017**, *171*, 273–285. [CrossRef] [PubMed]
57. Arnér, E.S. Focus on mammalian thioredoxin reductases—Important selenoproteins with versatile functions. *Biochim. Biophys. Acta BBA Gen. Subj.* **2009**, *1790*, 495–526. [CrossRef]
58. Conrad, M.; Jakupoglu, C.; Moreno, S.; Lippl, S.; Banjac, A.; Schneider, M.; Beck, H.; Hatzopoulos, A.K.; Just, U.; Sinowatz, F.; et al. Essential Role for Mitochondrial Thioredoxin Reductase in Hematopoiesis, Heart Development, and Heart Function. *Mol. Cell. Biol.* **2004**, *24*, 9414–9423. [CrossRef]
59. Jakupoglu, C.; Przemeck, G.K.H.; Schneider, M.; Moreno, S.; Mayr, N.; Hatzopoulos, A.K.; de Angelis, M.H.; Wurst, W.; Bornkamm, G.W.; Brielmeier, M.; et al. Cytoplasmic Thioredoxin Reductase Is Essential for Embryogenesis but Dispensable for Cardiac Development. *Mol. Cell. Biol.* **2005**, *25*, 1980–1988. [CrossRef] [PubMed]
60. Sun, Q.-A.; Kirnarsky, L.; Sherman, S.; Gladyshev, V.N. Selenoprotein oxidoreductase with specificity for thioredoxin and glutathione systems. *Proc. Natl. Acad. Sci. USA* **2001**, *98*, 3673–3678. [CrossRef] [PubMed]
61. Bianco, A.C.; Salvatore, D.; Gereben, B.; Berry, M.J.; Larsen, P.R. Biochemistry, Cellular and Molecular Biology, and Physiological Roles of the Iodothyronine Selenodeiodinases. *Endocr. Rev.* **2002**, *23*, 38–89. [CrossRef] [PubMed]
62. Larsen, P.R.; Dick, T.E.; Markovitz, B.P.; Kaplan, M.M.; Gard, T.G. Inhibition of intrapituitary thyroxine to 3.5.3'-triiodothyronine conversion prevents the acute suppression of thyrotropin release by thyroxine in hypothyroid rats. *J. Clin. Investig.* **1979**, *64*, 117–128. [CrossRef] [PubMed]
63. Martinez, M.E.; Karaczyn, A.; Stohn, J.P.; Donnelly, W.T.; Croteau, W.; Peeters, R.P.; Galton, V.A.; Forrest, D.; St Germain, D.; Hernandez, A. The Type 3 Deiodinase Is a Critical Determinant of Appropriate Thyroid Hormone Action in the Developing Testis. *Endocrinology* **2016**, *157*, 1276–1288. [CrossRef] [PubMed]
64. Schneider, M.J.; Fiering, S.N.; Pallud, S.E.; Parlow, A.F.; Germain, D.L.S.; Galton, V.A. Targeted Disruption of the Type 2 Selenodeiodinase Gene (DIO2) Results in a Phenotype of Pituitary Resistance to T4. *Mol. Endocrinol.* **2001**, *15*, 2137–2148. [CrossRef] [PubMed]
65. Schneider, M.J.; Fiering, S.N.; Thai, B.; Wu, S.-Y.; Germain, E.S.; Parlow, A.F.; Germain, D.L.S.; Galton, V.A. Targeted Disruption of the Type 1 Selenodeiodinase Gene (Dio1) Results in Marked Changes in Thyroid Hormone Economy in Mice. *Endocrinology* **2006**, *147*, 580–589. [CrossRef] [PubMed]
66. Kryukov, G.; Kryukov, V.M.; Gladyshev, V.N. New Mammalian Selenocysteine-containing Proteins Identified with an Algorithm That Searches for Selenocysteine Insertion Sequence Elements. *J. Biol. Chem.* **1999**, *274*, 33888–33897. [CrossRef] [PubMed]
67. Lescure, A.; Gautheret, D.; Carbon, P.; Krol, A. Novel Selenoproteins Identified in Silico and in Vivo by Using a Conserved RNA Structural Motif. *J. Biol. Chem.* **1999**, *274*, 38147–38154. [CrossRef]

68. Kim, H.-Y.; Gladyshev, V.N. Methionine Sulfoxide Reduction in Mammals: Characterization of Methionine-R-Sulfoxide Reductases. *Mol. Biol. Cell* **2004**, *15*, 1055–1064. [CrossRef] [PubMed]
69. Veres, Z.; Kim, I.; Scholz, T.; Stadtman, T.; Veres, Z.; Kim, I.; Scholz, T.; Stadtman, T. Selenophosphate synthetase. Enzyme properties and catalytic reaction. *J. Biol. Chem.* **1994**, *269*, 10597–10603. [CrossRef]
70. Ching, W.-M.; Wittwer, A.J.; Tsai, L.; Stadtman, T.C. Distribution of two selenonucleosides among the selenium-containing tRNAs from Methanococcus vannielii. *Proc. Natl. Acad. Sci. USA* **1984**, *81*, 57–60. [CrossRef]
71. Payne, N.C.; Geissler, A.; Button, A.; Sasuclark, A.R.; Schroll, A.L.; Ruggles, E.L.; Gladyshev, V.N.; Hondal, R.J. Comparison of the redox chemistry of sulfur- and selenium-containing analogs of uracil. *Free. Radic. Biol. Med.* **2017**, *104*, 249–261. [CrossRef] [PubMed]
72. Haft, D.H.; Self, W.T. Orphan SelD proteins and selenium-dependent molybdenum hydroxylases. *Biol. Direct* **2008**, *3*, 1–6. [CrossRef]
73. Zhang, Y.; Turanov, A.A.; Hatfield, D.L.; Gladyshev, V.N. In silico identification of genes involved in selenium metabolism: Evidence for a third selenium utilization trait. *BMC Genom.* **2008**, *9*, 251. [CrossRef]
74. Ma, C. Animal models of disease. *Mod. Drug Discov.* **2004**, *7*, 30–36.
75. Mariotti, M.; Santesmasses, D.; Guigó, R. Evolution of selenophosphate synthetase. In *Selenium—Its Molecular Biology and Role in Human Health*, 4th ed.; Hatfield, D.L., Schweizer, U., Tsuji, P.A., Gladyshev, V.N., Eds.; Springer: New York, NY, USA, 2016; pp. 85–99.
76. Xu, X.-M.; Carlson, B.A.; Irons, R.; Mix, H.; Zhong, N.; Gladyshev, V.N.; Hatfield, D.L. Selenophosphate synthetase 2 is essential for selenoprotein biosynthesis. *Biochem. J.* **2007**, *404*, 115–120. [CrossRef] [PubMed]
77. Alsina, B.; Corominas, M.; Berry, M.; Baguna, J.; Serras, F. Disruption of selenoprotein biosynthesis affects cell proliferation in the imaginal discs and brain of Drosophila melanogaster. *J. Cell Sci.* **1999**, *112*, 2875–2884. [CrossRef] [PubMed]
78. Na, J.; Jung, J.; Bang, J.; Lu, Q.; Carlson, B.A.; Guo, X.; Gladyshev, V.N.; Kim, J.-H.; Hatfield, D.L.; Lee, B.J. Selenophosphate synthetase 1 and its role in redox homeostasis, defense and proliferation. *Free. Radic. Biol. Med.* **2018**, *127*, 190–197. [CrossRef]
79. Chapple, C.E.; Guigó, R. Relaxation of Selective Constraints Causes Independent Selenoprotein Extinction in Insect Genomes. *PLoS ONE* **2008**, *3*, e2968. [CrossRef]
80. Lobanov, A.V.; Hatfield, D.L.; Gladyshev, V.N. Selenoproteinless animals: Selenophosphate synthetase SPS1 functions in a pathway unrelated to selenocysteine biosynthesis. *Protein Sci.* **2007**, *17*, 176–182. [CrossRef]
81. Mariotti, M.; Santesmasses, D.; Capella-Gutierrez, S.; Mateo, A.; Arnan, C.; Johnson, R.; D'Aniello, S.; Yim, S.H.; Gladyshev, V.N.; Serras, F.; et al. Evolution of selenophosphate synthetases: Emergence and relocation of function through independent duplications and recurrent subfunctionalization. *Genome Res.* **2015**, *25*, 1256–1267. [CrossRef]
82. Carlson, B.A.; Novoselov, S.V.; Kumaraswamy, E.; Lee, B.J.; Anver, M.R.; Gladyshev, V.N.; Hatfield, D.L. Specific excision of the selenocysteine tRNA[Ser]Sec (Trsp) gene in mouse liver demonstrates an essential role of selenoproteins in liver function. *J Biol. Chem.* **2004**, *279*, 8011–8017. [CrossRef]
83. Schweizer, U.; Schomburg, L.; Köhrle, J. Selenoprotein P and selenium distribution in mammals. In *Selenium—Its Molecular Biology and Role in Human Health*, 4th ed.; Hatfield, D.L., Schweizer, U., Tsuji, P.A., Gladyshev, V.N., Eds.; Springer: New York, NY, USA, 2016; pp. 261–274.
84. Hill, K.E.; Zhou, J.; McMahan, W.J.; Motley, A.K.; Atkins, J.; Gesteland, R.F.; Burk, R.F. Deletion of Selenoprotein P Alters Distribution of Selenium in the Mouse. *J. Biol. Chem.* **2003**, *278*, 13640–13646. [CrossRef]
85. Motsenbocker, M.A.; Tappel, A. A selenocysteine-containing selenium-transport protein in rat plasma. *Biochim. Biophys. Acta BBA Gen. Subj.* **1982**, *719*, 147–153. [CrossRef]
86. Baclaocos, J.; Santesmasses, D.; Mariotti, M.; Bierła, K.; Vetick, M.B.; Lynch, S.; McAllen, R.; Mackrill, J.J.; Loughran, G.; Guigó, R.; et al. Processive Recoding and Metazoan Evolution of Selenoprotein P: Up to 132 UGAs in Molluscs. *J. Mol. Biol.* **2019**, *431*, 4381–4407. [CrossRef]
87. Chernorudskiy, A.; Varone, E.; Colombo, S.F.; Fumagalli, S.; Cagnotto, A.; Cattaneo, A.; Briens, M.; Baltzinger, M.; Kuhn, L.; Bachi, A.; et al. Selenoprotein N is an endoplasmic reticulum calcium sensor that links luminal calcium levels to a redox activity. *Proc. Natl. Acad. Sci. USA* **2020**, *117*, 21288–21298. [CrossRef] [PubMed]
88. Moghadaszadeh, B.; Petit, N.; Jaillard, C.; Brockington, M.; Roy, S.Q.; Merlini, L.; Romero, N.; Estournet, B.; Desguerre, I.; Chaigne, D.; et al. Mutations in SEPN1 cause congenital muscular dystrophy with spinal rigidity and restrictive respiratory syndrome. *Nat. Genet.* **2001**, *29*, 17–18. [CrossRef]
89. Villar-Quiles, R.N.; von der Hagen, M.; Métay, C.; Gonzalez, V.; Donkervoort, S.; Bertini, E.; Castiglioni, C.; Chaigne, D.; Colomer, J.; Cuadrado, M.L.; et al. The clinical, histologic, and genotypic spectrum of SEPN1-related myopathy. *Neurology* **2020**, *95*, e1512–e1527. [CrossRef]
90. Allamand, V.; Richard, P.; Lescure, A.; Ledeuil, C.; Desjardin, D.; Petit, N.; Gartioux, C.; Ferreiro, A.; Krol, A.; Pellegrini, N.; et al. A single homozygous point mutation in a 3′untranslated region motif of selenoprotein N mRNA causes SEPN1-related myopathy. *EMBO Rep.* **2006**, *7*, 450–454. [CrossRef] [PubMed]
91. Maiti, B.; Arbogast, S.; Moyle, M.W.; Anderson, C.B.; Richard, P.; Guicheney, P.; Ferreiro, A.; Flanigan, K.; Howard, M.T. A mutation in the SEPN1 selenocysteine redefinition element (SRE) reduces selenocysteine incorporation and leads toSEPN1-related myopathy. *Hum. Mutat.* **2009**, *30*, 411–416. [CrossRef] [PubMed]

92. Kryukov, G.V.; Castellano, S.; Novoselov, S.V.; Lobanov, A.V.; Zehtab, O.; Guigó, R.; Gladyshev, V.N. Characterization of Mammalian Selenoproteomes. *Science* **2003**, *300*, 1439–1443. [CrossRef]
93. Dudkiewicz, M.; Szczepińska, T.; Grynberg, M.; Pawłowski, K. A Novel Protein Kinase-Like Domain in a Selenoprotein, Widespread in the Tree of Life. *PLoS ONE* **2012**, *7*, e32138. [CrossRef] [PubMed]
94. Sreelatha, A.; Yee, S.S.; Lopez, V.A.; Park, B.C.; Kinch, L.N.; Pilch, S.; Servage, K.; Zhang, J.; Jiou, J.; Karasiewicz-Urbańska, M.; et al. Protein AMPylation by an Evolutionarily Conserved Pseudokinase. *Cell* **2018**, *175*, 809–821.e19. [CrossRef]
95. Han, S.-J.; Lee, B.C.; Yim, S.H.; Gladyshev, V.N.; Lee, S.-R. Characterization of Mammalian Selenoprotein O: A Redox-Active Mitochondrial Protein. *PLoS ONE* **2014**, *9*, e95518. [CrossRef]
96. Avery, J.C.; Yamazaki, Y.; Hoffmann, F.W.; Folgelgren, B.; Hoffmann, P.R. Selenoprotein I is essential for murine embryogenesis. *Arch. Biochem. Biophys.* **2020**, *689*, 108444. [CrossRef]
97. Horibata, Y.; Hirabayashi, Y. Identification and characterization of human ethanolaminephosphotransferase. *J. Lipid Res.* **2007**, *48*, 503–508. [CrossRef]
98. Horibata, Y.; Ando, H.; Sugimoto, H. Locations and contributions of the phosphotransferases EPT1 and CEPT1 to the biosynthesis of ethanolamine phospholipids. *J. Lipid Res.* **2020**, *61*, 1221–1231. [CrossRef] [PubMed]
99. Horibata, Y.; Elpeleg, O.; Eran, A.; Hirabayashi, Y.; Savitzki, D.; Tal, G.; Mandel, H.; Sugimoto, H. EPT1 (selenoprotein I) is critical for the neural development and maintenance of plasmalogen in humans. *J. Lipid Res.* **2018**, *59*, 1015–1026. [CrossRef] [PubMed]
100. Ahmed, M.Y.; Al-Khayat, A.; Al-Murshedi, F.; Al-Futaisi, A.; Chioza, B.A.; Fernandez-Murray, J.P.; Self, J.E.; Salter, C.G.; Harlalka, G.V.; Rawlins, L.E.; et al. A mutation ofEPT1 (SELENOI)underlies a new disorder of Kennedy pathway phospholipid biosynthesis. *Brain* **2017**, *140*, 547–554. [CrossRef]
101. Dikiy, A.; Novoselov, S.V.; Fomenko, D.E.; Sengupta, A.; Carlson, B.A.; Cerny, R.L.; Ginalski, K.; Grishin, N.V.; Hatfield, D.L.; Gladyshev, V.N. SelT, SelW, SelH, and Rdx12: Genomics and Molecular Insights into the Functions of Selenoproteins of a Novel Thioredoxin-like Family. *Biochemistry* **2007**, *46*, 6871–6882. [CrossRef]
102. Kasaikina, M.V.; Fomenko, D.E.; Labunskyy, V.M.; Lachke, S.A.; Qiu, W.; Moncaster, J.A.; Zhang, J.; Wojnarowicz, M.W.; Natarajan, S.K.; Malinouski, M.; et al. Roles of the 15-kDa Selenoprotein (Sep15) in Redox Homeostasis and Cataract Development Revealed by the Analysis of Sep 15 Knockout Mice. *J. Biol. Chem.* **2011**, *286*, 33203–33212. [CrossRef]
103. Canter, J.A.; Ernst, S.E.; Peters, K.M.; Carlson, B.A.; Thielman, N.R.J.; Grysczyk, L.; Udofe, P.; Yu, Y.; Cao, L.; Davis, C.D.; et al. Selenium and the 15kDa Selenoprotein Impact Colorectal Tumorigenesis by Modulating Intestinal Barrier Integrity. *Int. J. Mol. Sci.* **2021**, *22*, 10651. [CrossRef] [PubMed]
104. Tsuji, P.A.; Carlson, B.A.; Naranjo-Suarez, S.; Yoo, M.-H.; Xu, X.-M.; Fomenko, D.E.; Gladyshev, V.N.; Hatfield, D.L.; Davis, C.D. Knockout of the 15 kDa Selenoprotein Protects against Chemically-Induced Aberrant Crypt Formation in Mice. *PLoS ONE* **2012**, *7*, e50574. [CrossRef]
105. Carlson, B.A.; Hartman, J.M.; Tsuji, P.A. The 15 kDa Selenoprotein: Insights into Its Regulation and Function. In *Selenium—Its Molecular Biology and Role in Human Health*, 4th ed.; Hatfield, D.L., Schweizer, U., Tsuji, P.A., Gladyshev, V.N., Eds.; Springer: New York, NY, USA, 2016; pp. 235–243.
106. Gong, T.; Berry, M.J.; Pitts, M.W. Selenoprotein M: Structure, Expression and Functional Relevance. In *Selenium—Its Molecular Biology and Role in Human Health*, 4th ed.; Hatfield, D.L., Schweizer, U., Tsuji, P.A., Gladyshev, V.N., Eds.; Springer: New York, NY, USA, 2016; pp. 253–260.
107. Reeves, M.A.; Bellinger, F.P.; Berry, M.J. The Neuroprotective Functions of Selenoprotein M and its Role in Cytosolic Calcium Regulation. *Antioxid. Redox Signal.* **2010**, *12*, 809–818. [CrossRef]
108. Pitts, M.W.; Reeves, M.A.; Hashimoto, A.C.; Ogawa, A.; Kremer, P.; Seale, L.A.; Berry, M.J. Deletion of Selenoprotein M Leads to Obesity without Cognitive Deficits. *J. Biol. Chem.* **2013**, *288*, 26121–26134. [CrossRef]
109. Fredericks, G.J.; Hoffmann, F.W.; Hondal, R.J.; Rozovsky, S.; Urschitz, J.; Hoffmann, P.R. Selenoprotein K Increases Efficiency of DHHC6 Catalyzed Protein Palmitoylation by Stabilizing the Acyl-DHHC6 Intermediate. *Antioxidants* **2017**, *7*, 4. [CrossRef] [PubMed]
110. Peng, T.; Lin, J.; Xu, Y.-Z.; Zhang, Y. Comparative genomics reveals new evolutionary and ecological patterns of selenium utilization in bacteria. *ISME J.* **2016**, *10*, 2048–2059. [CrossRef]
111. Lin, J.; Peng, T.; Jiang, L.; Ni, J.-Z.; Liu, Q.; Chen, L.; Zhang, Y. Comparative genomics reveals new candidate genes involved in selenium metabolism in prokaryotes. *Genome Biol. Evol.* **2015**, *7*, 664–676. [CrossRef] [PubMed]
112. Liu, Y.; Makarova, K.S.; Huang, W.-C.; Wolf, Y.I.; Nikolskaya, A.N.; Zhang, X.; Cai, M.; Zhang, C.-J.; Xu, W.; Luo, Z.; et al. Expanded diversity of Asgard archaea and their relationships with eukaryotes. *Nat. Cell Biol.* **2021**, *593*, 553–557. [CrossRef] [PubMed]
113. Mariotti, M.; Lobanov, A.V.; Manta, B.; Santesmasses, D.; Bofill, A.; Guigó, R.; Gabaldón, T.; Gladyshev, V.N. Lokiarchaeota Marks the Transition between the Archaeal and Eukaryotic Selenocysteine Encoding Systems. *Mol. Biol. Evol.* **2016**, *33*, 2441–2453. [CrossRef]
114. Lobanov, A.V.; Fomenko, D.E.; Zhang, Y.; Sengupta, A.; Hatfield, D.L.; Gladyshev, V.N. Evolutionary dynamics of eukaryotic selenoproteomes: Large selenoproteomes may associate with aquatic life and small with terrestrial life. *Genome Biol.* **2007**, *8*, 1–16. [CrossRef]
115. Lobanov, A.V.; Hatfield, D.L.; Gladyshev, V.N. Eukaryotic selenoproteins and selenoproteomes. *Biochim. Biophys. Acta BBA Gen. Subj.* **2009**, *1790*, 1424–1428. [CrossRef]

116. Jiang, L.; Lu, Y.; Zheng, L.; Li, G.; Chen, L.; Zhang, M.; Ni, J.; Liu, Q.; Zhang, Y. The algal selenoproteomes. *BMC Genom.* **2020**, *21*, 1–16. [CrossRef]
117. Liang, H.; Wei, T.; Xu, Y.; Li, L.; Sahu, S.K.; Wang, H.; Li, H.; Fu, X.; Zhang, G.; Melkonian, M.; et al. Phylogenomics Provides New Insights into Gains and Losses of Selenoproteins among Archaeplastida. *Int. J. Mol. Sci.* **2019**, *20*, 3020. [CrossRef]
118. Mariotti, M.; Salinas, G.; Gabaldón, T.; Gladyshev, V.N. Utilization of selenocysteine in early-branching fungal phyla. *Nat. Microbiol.* **2019**, *4*, 759–765. [CrossRef]
119. Mariotti, M. Selenocysteine extinction in insect. In *Short Views on Insect Genomics and Proteomics*; Springer: Cham, Switzerland, 2016; Volume 4, pp. 113–140.
120. Rispe, C.; Legeai, F.; Nabity, P.D.; Fernández, R.; Arora, A.K.; Baa-Puyoulet, P.; Banfill, C.R.; Bao, L.; Barberà, M.; Bouallègue, M.; et al. The genome sequence of the grape phylloxera provides insights into the evolution, adaptation, and invasion routes of an iconic pest. *BMC Biol.* **2020**, *18*, 90. [CrossRef]
121. Otero, L.; Romanelli-Cedrez, L.; Turanov, A.A.; Gladyshev, V.N.; Miranda-Vizuete, A.; Salinas, G. Adjustments, extinction, and remains of selenocysteine incorporation machinery in the nematode lineage. *RNA* **2014**, *20*, 1023–1034. [CrossRef]
122. Bösl, M.R.; Takaku, K.; Oshima, M.; Nishimura, S.; Taketo, M.M. Early embryonic lethality caused by targeted disruption of the mouse selenocysteine tRNA gene (Trsp). *Proc. Natl. Acad. Sci. USA* **1997**, *94*, 5531–5534. [CrossRef]
123. Kumaraswamy, E.; Carlson, B.A.; Morgan, F.; Miyoshi, K.; Robinson, G.W.; Su, D.; Wang, S.; Southon, E.; Tessarollo, L.; Lee, B.J.; et al. Selective removal of the selenocysteine tRNA [Ser]Sec gene (Trsp) in mouse mammary epithelium. *Mol. Cell Biol.* **2003**, *23*, 1477–1488. [CrossRef]
124. Moustafa, M.E.; Carlson, B.A.; El-Saadani, M.A.; Kryukov, G.V.; Sun, Q.-A.; Harney, J.W.; Hill, K.E.; Combs, G.F.; Feigenbaum, L.; Mansur, D.B.; et al. Selective Inhibition of Selenocysteine tRNA Maturation and Selenoprotein Synthesis in Transgenic Mice Expressing Isopentenyladenosine-Deficient Selenocysteine tRNA. *Mol. Cell. Biol.* **2001**, *21*, 3840–3852. [CrossRef]
125. Moustafa, M.E.; Kumaraswamy, E.; Zhong, N.; Rao, M.; Carlson, B.A.; Hatfield, D.L. Models for assessing the role of selenoproteins in health. *J. Nutr.* **2003**, *133*, 2494S–2496S. [CrossRef]
126. Hatfield, D.L.; Gladyshev, V.; Park, J.; Park, S.; Chittum, H.; Huh, J.; Carlson, B.; Kim, M.; Moustafa, M.; Lee, B.J. Biosynthesis of selenocysteine and its incorporation into protein as the 21st amino acid. In *Comprehensive Natural Products Chemistry*; Barton, D., Nakanishi, K., Meth-Cohn, O., Kelly, J.W., Eds.; Amino Acids, Peptides, Porphyrins, and Alkaloids; Pergamon; Elsevier Science: Oxford, UK, 1999; Volume 4, pp. 353–380.
127. Carlson, B.A.; Moustafa, M.; Sengupta, A.; Schweizer, U.; Shrimali, R.; Rao, M.; Zhong, N.; Wang, S.; Feigenbaum, L.; Lee, B.J.; et al. Selective Restoration of the Selenoprotein Population in a Mouse Hepatocyte Selenoproteinless Background with Different Mutant Selenocysteine tRNAs Lacking Um. *J. Biol. Chem.* **2007**, *282*, 32591–32602. [CrossRef]
128. Irons, R.; Carlson, B.A.; Hatfield, D.L.; Davis, C.D. Both selenoproteins and low molecular weight selenocompounds reduce colon cancer risk in mice with genetically impaired selenoprotein expression. *J. Nutr.* **2006**, *135*, 1311–1317. [CrossRef]
129. Sheridan, P.A.; Zhong, N.; Carlson, B.A.; Perella, C.M.; Hatfield, D.L.; Beck, M.A. Decreased selenoprotein expression alters the immune response during influenzavirus infection in mice. *J. Nutr.* **2007**, *137*, 1466–1471. [CrossRef]
130. Baliga, M.S.; Diwadkar-Navsariwala, V.; Koh, T.; Fayad, R.; Fantuzzi, G.; Diamond, A.M. Selenoprotein deficiency enhances radiation-induced micronucleiformation. *Mol. Nutr. Food Res.* **2008**, *52*, 1300–1304. [CrossRef] [PubMed]
131. Diwadkar-Navsariwala, V.; Prins, G.S.; Swanson, S.M.; Birch, L.A.; Ray, V.H.; Hedayat, S.; Lantvit, D.L.; Diamond, A.M. Selenoprotein deficiency accelerates pros-tate carcinogenesis in a transgenic model. *Proc. Natl. Acad. Sci. USA* **2006**, *103*, 8179–8184. [CrossRef] [PubMed]
132. Labunskyy, V.; Lee, B.C.; Handy, D.; Loscalzo, J.; Hatfield, D.L.; Gladyshev, V.N. Both Maximal Expression of Selenoproteins and Selenoprotein Deficiency Can Promote Development of Type 2 Diabetes-Like Phenotype in Mice. *Antioxid. Redox Signal* **2011**, *14*, 2327–2336. [CrossRef]
133. Shrimali, R.K.; Weaver, J.A.; Miller, G.F.; Starost, M.F.; Carlson, B.A.; Novoselov, S.V.; Kumaraswamy, E.; Gladyshev, V.N.; Hatfield, D.L. Selenoprotein expression is essential in endothelial cell development and cardiac muscle function. *Neuromuscul. Disord.* **2007**, *17*, 135–142. [CrossRef]
134. Downey, C.M.; Horton, C.R.; Carlson, B.A.; Parsons, T.E.; Hatfield, D.L.; Hallgrímsson, B.; Jirik, F.R. Osteo-Chondroprogenitor-Specific Deletion of the Selenocysteine tRNA Gene, Trsp, Leads to Chondronecrosis and Abnormal Skeletal Development: A Putative Model for Kashin-Beck Disease. *PLoS Genet.* **2009**, *5*, e1000616. [CrossRef] [PubMed]
135. Sengupta, A.; Lichti, U.F.; Carlson, B.A.; Ryscavage, A.O.; Gladyshev, V.N.; Yuspa, S.H.; Hatfield, D.L. Selenoproteins are essential for proper keratinocyte function andskin development. *PLoS ONE* **2010**, *5*, e12249. [CrossRef]
136. Kawatani, Y.; Suzuki, T.; Shimizu, R.; Kelly, T.K.; Yamamoto, M. Nrf2 and selenoproteins are essential for maintaining oxidative homeostasis in erythrocytes andprotecting against hemolytic anemia. *Blood* **2011**, *117*, 986–996. [CrossRef]
137. Shrimali, R.K.; Irons, R.D.; Carlson, B.A.; Sano, Y.; Gladyshev, V.N.; Park, J.M.; Hatfield, D.L. Selenoproteins Mediate T Cell Immunity through an Antioxidant Mechanism. *J. Biol. Chem.* **2008**, *283*, 20181–20185. [CrossRef]
138. Suzuki, T.; Kelly, V.P.; Motohashi, H.; Nakajima, O.; Takahashi, S.; Nishimura, S.; Yamamoto, M. Deletion of the selenocysteine tRNA gene in macrophages and liver results in compensatory gene induction of cytoprotective enzymes by Nrf2. *J. Biol. Chem.* **2008**, *283*, 2021–2030. [CrossRef] [PubMed]

139. Wirth, E.K.; Conrad, M.; Winterer, J.; Wozny, C.; Carlson, B.A.; Roth, S.; Schmitz, D.; Bornkamm, G.W.; Coppola, V.; Tessarollo, L.; et al. Neuronal selenoprotein expression is required for interneuron development and prevents seizures and neurodegeneration. *FASEB J.* **2010**, *24*, 844–852. [CrossRef]
140. Schweizer, U.; Streckfuß, F.; Pelt, P.; Carlson, B.A.; Hatfield, D.L.; Köhrle, J.; Schomburg, L. Hepatically derived selenoprotein P is a key factor for kidney but not for brain selenium supply. *Biochem. J.* **2005**, *386*, 221–226. [CrossRef]
141. Blauwkamp, M.N.; Yu, J.; Schin, M.A.; Burke, K.A.; Berry, M.J.; Carlson, B.A.; Brosius III, F.C.; Koenig, R.J. Podocyte specific knock out of selenoproteins does not enhance nephropathy in streptozotocin diabetic C57BL/6 mice. *BMC Nephrol.* **2008**, *9*, 7. [CrossRef]
142. Chiu-Ugalde, J.; Wirth, E.K.; Klein, M.O.; Sapin, R.; Fradejas-Villar, N.; Renko, K.; Schomburg, L.; Köhrle, J.; Schweizer, U. Thyroid Function Is Maintained Despite Increased Oxidative Stress in Mice Lacking Selenoprotein Biosynthesis in Thyroid Epithelial Cells. *Antioxid. Redox Signal.* **2012**, *17*, 902–913. [CrossRef] [PubMed]
143. Luchman, H.A.; Villemaire, M.L.; Bismar, T.A.; Carlson, B.A.; Jirik, F.R. Prostate epithelium-specific deletion of the selenocysteine tRNA gene Trsp leads to early onset intraepithelial neoplasia. *Am. J. Pathol.* **2014**, *184*, 871–877. [CrossRef] [PubMed]
144. Jung, J.; Kim, Y.; Na, J.; Qiao, L.; Bang, J.; Kwon, D.; Yoo, T.-J.; Kang, D.; Kim, L.K.; Carlson, B.A.; et al. Constitutive Oxidative Stress by SEPHS1 Deficiency Induces Endothelial Cell Dysfunction. *Int. J. Mol. Sci.* **2021**, *22*, 11646. [CrossRef] [PubMed]
145. Sengupta, A.; Carlson, B.A.; Weaver, J.A.; Novoselov, S.V.; Fomenko, D.E.; Gladyshev, V.N.; Hatfield, D.L. A functional link between housekeeping selenoproteins and phase II enzymes. *Biochem. J.* **2008**, *413*, 151–161. [CrossRef]
146. Malinouski, M.; Kehr, S.; Finney, L.; Vogt, S.; Carlson, B.A.; Seravalli, J.; Jin, R.; Handy, D.E.; Park, T.J.; Loscalzo, J.; et al. High-Resolution Imaging of Selenium in Kidneys: A Localized Selenium Pool Associated with Glutathione Peroxidase *Antioxid. Redox Signal.* **2012**, *16*, 185–192. [CrossRef]
147. Seeher, S.; Atassi, T.; Mahdi, Y.; Carlson, B.A.; Braun, D.; Wirth, E.K.; Klein, M.O.; Reix, N.; Miniard, A.C.; Schomburg, L.; et al. Secisbp2 Is Essential for Embryonic Development and Enhances Selenoprotein Expression. *Antioxid. Redox Signal.* **2014**, *21*, 835–849. [CrossRef] [PubMed]
148. Carlson, B.A.; Yoo, M.-H.; Sano, Y.; Sengupta, A.; Kim, J.Y.; Irons, R.; Gladyshev, V.N.; Hatfield, D.L.; Park, J.M. Selenoproteins regulate macrophage invasiveness and extracellular matrix-related gene expression. *BMC Immunol.* **2009**, *10*, 57. [CrossRef]
149. Mattmiller, S.A.; Carlson, B.A.; Gandy, J.C.; Sordillo, L.M. Reduced macrophage selenoprotein expression alters oxidized lipid metabolite biosynthesis from arachidonic and linoleic acid. *J. Nutr. Biochem.* **2014**, *25*, 647–654. [CrossRef] [PubMed]
150. Kaushal, N.; Kudva, A.K.; Patterson, A.D.; Chiaro, C.; Kennett, M.J.; Desai, D.; Amin, S.; Carlson, B.A.; Cantorna, M.T.; Prabhu, K.S. Crucial Role of Macrophage Selenoproteins in Experimental Colitis. *J. Immunol.* **2014**, *193*, 3683–3692. [CrossRef] [PubMed]
151. Narayan, V.; Ravindra, K.C.; Liao, C.; Kaushal, N.; Carlson, B.A.; Prabhu, K.S. Epigenetic regulation of inflammatory gene expression in macrophages by selenium. *J. Nutr. Biochem.* **2015**, *26*, 138–145. [CrossRef]
152. Nelson, S.M.; Shay, A.E.; James, J.L.; Carlson, B.A.; Urban, J.; Prabhu, K.S. Selenoprotein Expression in Macrophages Is Critical for Optimal Clearance of Parasitic Helminth Nippostrongylus brasiliensis. *J. Biol. Chem.* **2016**, *291*, 2787–2798. [CrossRef]
153. Hudson, T.S.; Carlson, B.A.; Hoeneroff, M.J.; Young, H.A.; Sordillo, L.; Muller, W.J.; Hatfield, D.L.; Green, J.E. Selenoproteins reduce susceptibility to DMBA-induced mammary carcinogenesis. *Carcinogenesis* **2012**, *33*, 1225–1230. [CrossRef]
154. Wirth, E.K.; Bharathi, B.S.; Hatfield, D.; Conrad, M.; Brielmeier, M.; Schweizer, U. Cerebellar Hypoplasia in Mice Lacking Selenoprotein Biosynthesis in Neurons. *Biol. Trace Element Res.* **2014**, *158*, 203–210. [CrossRef]
155. Carlson, B.A. Selenocysteine tRNA[Ser]Sec mouse models for elucidating roles of selenoproteins in health and development. In *Selenium—Its Molecular Biology and Role in Human Health*; Hatfield, D.L., Schweizer, U., Tsuji, P.A., Gladyshev, V.N., Eds.; Springer: New York, NY, USA, 2016.
156. Park, J.M.; Choi, I.S.; Kang, S.G.; Lee, J.Y.; Hatfield, D.L.; Lee, B.J. Upstream promoter elements are sufficient for selenocysteine tRNA[Ser]Sec gene transcription and to determine the transcription start point. *Gene* **1995**, *162*, 13–19. [CrossRef]
157. Carlson, B.A.; Schweizer, U.; Perella, C.; Shrimali, R.K.; Feigenbaum, L.; Shen, L.; Spersansky, S.; Floss, T.; Jeong, S.-J.; Watts, J.; et al. The selenocysteine tRNA STAF-binding region is essential for adequate selenocysteine tRNA status, selenoprotein expression and early age survival of mice. *Biochem. J.* **2009**, *418*, 61–71. [CrossRef] [PubMed]

Review

Ribosome Fate during Decoding of UGA-Sec Codons

Paul R. Copeland [1,*] and Michael T. Howard [2,*]

1. Department of Biochemistry and Molecular Biology, Rutgers-Robert Wood Johnson Medical School, Piscataway, NJ 08854, USA
2. Department of Human Genetics, University of Utah, Salt Lake City, UT 84112, USA
* Correspondence: copelapr@rwjms.rutgers.edu (P.R.C.); mhoward@genetics.utah.edu (M.T.H.)

Abstract: Decoding of genetic information into polypeptides occurs during translation, generally following the codon assignment rules of the organism's genetic code. However, recoding signals in certain mRNAs can overwrite the normal rules of translation. An exquisite example of this occurs during translation of selenoprotein mRNAs, wherein UGA codons are reassigned to encode for the 21st proteogenic amino acid, selenocysteine. In this review, we will examine what is known about the mechanisms of UGA recoding and discuss the fate of ribosomes that fail to incorporate selenocysteine.

Keywords: selenocysteine; selenoprotein; SECIS; recoding; SECIS-binding protein; translation termination; nonsense-mediated decay; ribosome rescue

1. Introduction

The genetic code was well established by the mid-1960s and initially thought to be fixed and unalterable. This dogma was challenged in the 1970s and 1980s as a number of studies demonstrated that some mRNAs contain signals in addition to the 64 codons that could modify readout of the genetic code in various ways. The signals may involve intramolecular RNA structures such as hairpins or pseudoknots, combinations of specific nucleotides in tandem, recognition sites for RNA-binding proteins, and inter-molecular base pairing between mRNA and RNA within the ribosome [1]. These signals can induce the ribosome to shift to the −1 or +1 reading frame, skip over bases in the mRNA, or even redefine the meaning of a codon. While many of these mechanisms are quite elegant, the process of redefining a UGA codon from a termination codon to one that encodes selenocysteine (Sec) during translation of selenoprotein mRNAs is arguably one of the most complex. The UGA recoding mechanism involves cis-acting RNA structures, several enzymes required for synthesis of Sec-on-Sec tRNA, a specialized elongation factor, and an impressive list of RNA-binding proteins. Details of Sec synthesis on the Sec tRNA are discussed elsewhere in this issue, and the basic features of the Sec incorporation process have been recently reviewed in detail [2,3]. Here, we present a focused review that delves into the cis and trans-acting recoding signals that promote UGA recoding and the resulting competing outcomes for ribosomes translating UGA-Sec.

2. Co-Translational Sec-Incorporation

2.1. Cis-Acting Elements

2.1.1. Required Selenocysteine Insertion Sequences

The potential for UGA recoding from termination to Sec incorporation occurs in a subset of organisms from all three domains of life. In all instances examined, an RNA secondary structure called the Sec insertion sequence (SECIS) marks the mRNA for recoding of UGA codons. The product of recoding is the co-translational insertion of Sec and the production of a selenoprotein. Most selenoproteins are oxidoreductases requiring Sec for their function [4,5].

In prokaryotes, the SECIS element is composed of a stem-loop located just downstream from the UGA codon (Figure 1A). The sequence of prokaryotic SECIS elements is highly variable to accommodate the amino acid constraints of the protein being produced. The conserved features are a bulged U in the upper part of the stem-loop and unpaired adjacent G and U nucleotides in the loop (Figure 1A) [6,7]. Sec-incorporation is facilitated by a bifunctional protein, SelB, which consists of SECIS-binding and Sec-tRNA-specific elongation factor domains [8]. The co-location of the SECIS element adjacent to the UGA-Sec codon allows for both positional and temporal delivery of the Sec-tRNA^{Sec} to the ribosome during UGA recoding. Consequently, it came as a surprise when it was shown that SECIS elements in eukaryotes are located in the 3′ UTR of selenoprotein mRNAs [9], necessitating delivery of a signal for UGA recoding from a distance.

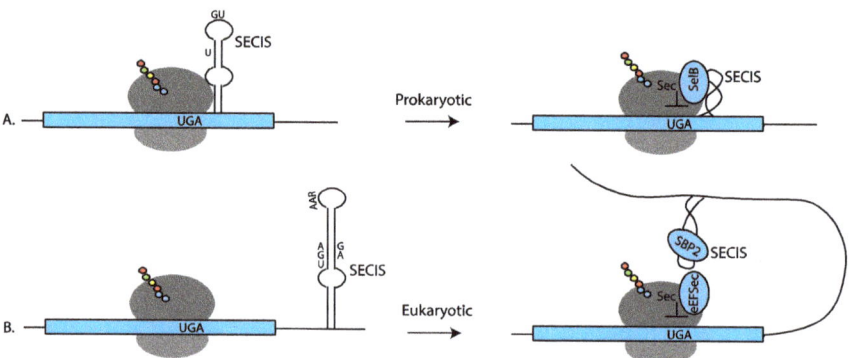

Figure 1. Prokaryotic versus eukaryotic Sec insertion showing the SECIS elements and core components required for Sec-tRNA^{Sec} delivery to the ribosome. (**A**) In prokaryotes, the SECIS element is located just downstream from the UGA codon within the coding sequence. The bulged U and G U nucleotides required for SelB binding are shown. SelB binds to the SECIS element and delivers the Sec-tRNA^{Sec} to the ribosome for UGA recoding. (**B**) In eukaryotes, the SECIS element is located in the 3′ UTR. The SECIS-binding protein, SECISBP2 (SBP2), recruits the Sec-specific elongation factor EEFSEC (eEFSec) to deliver Sec-tRNA^{Sec} to the ribosome for UGA recoding. On the right-hand side of the figure, the SECIS elements are depicted as a helix.

One consequence of eukaryotic SECIS action from a distance is that UGA recoding may occur at any in-frame UGA in the message [10–12]. Not only does this allow for greater flexibility in the coding region around the UGA codon, but it also allows for recoding of multiple UGA codons during translation of a single mRNA, as is observed in the eukaryotic selenoprotein (SELENOP) mRNA. In eukaryotes, UGA recoding is generally independent of where the UGA occurs in the mRNA; however, as with most phenomena in biology, there are exceptions, and it has been shown that some eukaryotic SECIS elements have evolved the ability to limit Sec-incorporation to UGAs found only in certain regions of the mRNA [13,14].

The conserved features of eukaryotic SECIS elements include an AAR motif in the loop and an AUGA:GA within the stem (Figure 1B) [9,10]. The latter motif is positioned such that the G.A nucleotides form a non-Watson/Crick quartet of sheared G.A/A.G pairs [15,16] that induce a "kink-turn" in the RNA structure. This motif interacts with the RNA-binding protein SECISBP2 (SBP2) to recruit the EEFSEC ternary complex carrying Sec-tRNA^{Sec} (Figure 1B). Although variations in these two motifs exist [4,17,18], these features are sufficient to allow for SECIS search programs to be generated that can predict the entire selenoproteome from genomic sequence alone [4,19]. In addition to facilitating UGA recoding, several studies have demonstrated that some SECIS elements have additional unique molecular functions, such as the ability to direct processive Sec-incorporation at multiple UGA codons in the same transcript [20,21], the ability to alter Sec-incorporation efficiency in response to variations in available selenium [22–24], and the ability to discriminate UGA codons depending on their location within the mRNA [13,14]. The remainder of this review will focus on the eukaryotic mechanism of Sec insertion.

2.1.2. Accessory Cis-Acting Selenocysteine Insertion Elements

In cases of stop codon readthrough, where standard near-cognate tRNAs are used to decode the stop codon rather than Sec-tRNA^{Sec}, RNA structures and sequences located adjacent to the termination codon have been shown to be stimulatory [25–30]. Nucleotides adjacent to UGA-Sec codons were likewise shown to impact UGA-Sec recoding efficiency [31–33]. Phylogenetic analyses identified several potential RNA structures downstream from UGA-Sec codons [22,34] that have been designated stop codon/selenocysteine redefinition elements, or SREs. Mutations of the sequences comprising the SRE structure of *selenoprotein N* (*SELENON*) demonstrated the importance of the stem, and of the sequences within the loop and the spacer separating it from the UGA-Sec codon to inhibit termination and stimulate Sec-incorporation [34–36]. The signals for two of these structures in SELENON and *selenoprotein T* (*SELENOT*) were also independently identified in a genome-wide search for deeply conserved RNA structures [37]. In addition to the SREs, structural and sequence motifs located close to a UGA-Sec codon in selenoprotein S (*SELENOS*) mRNA have been shown to impact Sec-incorporation. The sequence motif referred to as the *SELENOS* positive UGA recoding element, or SPUR, has been shown to enhance Sec-incorporation efficiency [38,39]. The authors propose that the SPUR element interacts with the SECIS element itself to facilitate Sec incorporation.

2.2. Trans-Acting Factors

2.2.1. The Core SECIS-Binding Proteins

Just over 20 years ago, two eukaryotic trans-acting factors essential for Sec-incorporation were identified: the SECIS RNA-binding protein (SECISBP2), and the Sec-tRNA^{Sec} specific elongation factor (EEFSEC) [40–42]. Unlike the bacterial SelB, SECISBP2 lacks the ability to bind the Sec-tRNA^{Sec}. This function is provided by EEFSEC.

EEFSEC differs from the standard eukaryotic elongation factor (EEF1A) in that it has the ability to interact with SECISBP2, has a higher affinity for GTP than GDP [41,43], and has an extended C-terminal extension that, unlike EEF1A and EF-Tu, undergoes a conformational change upon guanine nucleotide exchange [44]. The latter observation suggests a non-canonical mechanism for release of Sec-tRNA^{Sec} to the ribosome.

SECISBP2 contains a canonical L7Ae RNA-binding domain [45–47] that is responsible for interactions with the kink-turn motif of the SECIS element, a central Sec-incorporation domain (SID) that is required for Sec-incorporation but not SECIS binding, and a poorly conserved N-terminal domain of unknown function. A crucial insight came from the observation that SECISBP2 and EEFSEC interact in a manner that is stimulated by the presence of Sec-tRNA^{Sec} [48], suggesting a mechanism whereby SECISBP2 would only recruit EEFSEC carrying Sec-tRNA^{Sec} and would dissociate from EEFSEC upon delivery of Sec-tRNA^{Sec} to the ribosome.

While SECISBP2 is clearly a central player in Sec-incorporation, studies of SECISBP2 deletions in mice [49] and cultured mammalian cells [50] revealed, surprisingly, that SECISBP2 is not absolutely required for Sec-incorporation, although the efficiency of UGA recoding is greatly reduced in its absence. One possible explanation is that SECISBP2L, a paralogue of SECISBP2, may compensate for SECISBP2 in its absence or under certain physiological conditions [51]. An alternative explanation involves direct binding of EEFSEC to the SECIS element. Initial studies of SECISBP2- and EEFSEC-binding to SECIS RNA in vitro illustrated direct binding of EEFSEC in the absence of SECISBP2 [42]. SECISBP2 may not be absolutely required, but rather strongly enhances efficient EEFSEC ternary complex delivery to the ribosome or facilitates exchange following UGA-Sec recoding, perhaps through direct interactions with the 28S ribosomal RNA [52,53].

Emerging from these early studies is a picture in which the SECIS elements act as a scaffold for RNA-binding proteins and this ribonucleoprotein complex recruits EEFSEC to accommodate UGA recoding by Sec-tRNA^{Sec}. In the following section, we will discuss additional SECIS-binding proteins that have been proposed as part of the SECIS ribonucleoprotein complex, and how these may further modulate the efficiency of UGA-Sec recoding.

2.2.2. Accessory Trans-Acting Factors

Levels of complexity were added to the mechanism of Sec-incorporation with the discovery of several additional SECIS-binding proteins. The first of these, nucleolin (NCL), best known for its role in ribosome biogenesis, was shown to bind specifically to the SECIS element [54] around the same time as the discovery of SECISBP2. Subsequent studies showed that NCL binds to the upper part of the SECIS stem for a subset of selenoprotein mRNAs, and thus is unlikely to compete with SECISBP2-binding [55]. Knockdowns of NCL cause a decrease in levels of selenoproteins without changing mRNA levels or localization, suggesting that it is a positive regulator of selenoprotein translation.

In contrast to NCL, the SECIS-binding ribosomal protein L30 contains an L7Ae domain and competes directly with SECISBP2-binding at the kink-turn motif of SECIS elements [56]. The canonical role of L30 includes interaction with the 60S rRNA subunit [57] mediated through binding to a kink-turn motif in helix 58 of the 28S rRNA, as well as binding the 5′ UTR of its own mRNA to auto-regulate expression [58]. Increasing levels of L30 in cultured cells were shown to enhance Sec-incorporation [59], whereas addition of free L30 to in vitro translation reaction was shown to decrease UGA-Sec recoding [56]. A possible explanation for these disparate results comes from the finding that ribosome-associated L30 has a higher affinity for the SECIS element than free protein. One model proposes that within the context of the ribosome, L30 may have transient interactions with the SECIS element that displace SECISBP2 and stabilize a SECIS conformation that is required for EEFSEC delivery.

Several years after the identification of L30 as a SECIS-binding protein, EIF4A3, a member of the DEAD-box family of RNA-dependent ATPases, was also shown to bind SECIS elements contained in selenoprotein mRNAs that are known to be sensitive to degradation and have reduced UGA recoding efficiency when selenium is limiting [60]. Mapping of the EIF4A3-binding site demonstrated that it overlaps with the SECISBP2-binding site, and it was further shown that when selenium is limiting, there is an increase in cellular EIF4A3 protein expression. Importantly, EIF4A3 is also known to be a key component of the exon junction complex (EJC) involved in nonsense-mediated decay (NMD) [61–63], suggesting a possible direct link between inhibition of Sec-incorporation and degradation of selenoprotein mRNAs.

2.2.3. Selenoprotein mRNA 5′ Cap Modifications and Recruitment of the SMN Complex

Most mRNAs contain a 5′ m7G cap that plays important roles in RNA processing and stability and is bound by the translation initiation factor EIF4E as a key step in the process of translation initiation [64]. Several of the selenoprotein mRNAs are inefficiently recognized by EIF4E because the cap is hypermethylated by the trimethyl-guanosine synthase (TGS1) via a pathway related to that which processes small nuclear RNAs and snoRNAs [65,66]. The tri-methyl guanosine (TMG) capped selenoprotein mRNAs are localized to the cytoplasm and actively associate with ribosomes; at least one selenoprotein mRNA, GPX1, appears to require TMG to support efficient translation.

TGS1 is recruited to selenoprotein mRNAs by interactions between SECISBP2 and the survival of motor neuron protein complex (SMN). The SMN protein is a component of a ribonucleoprotein assembly chaperone pathway first described as being essential for assembly of small nuclear RNPs involved in splicing [67]. In addition to the selective 5′ TMG modification of selenoprotein mRNAs, perhaps the recruitment of the SMN complex to select selenoprotein mRNAs helps chaperone the formation of functional ribonucleoprotein complexes involved in UGA recoding.

Most recently, the RNA-binding protein PTBP1 has been shown by RNA affinity chromatography to interact with 3′ UTR sequences of SELENOP in the U-rich region separating the two 3′ UTR SECIS elements [68]. Deletion of this region inhibited the regulation of translation that normally occurs during oxidative stress in cultured human liver cells, indicating that there may yet be undiscovered regulatory elements in the 3′ UTRs of individual selenoproteins.

Although many of the key components of the Sec-incorporation machinery have been identified, we have an incomplete picture of the dynamic nature of the Sec ribonucleoprotein complex, how it delivers EEFSEC and the Sec-tRNA^{Sec} to the ribosome during UGA recoding, and how this process competes with the standard decoding process of translation termination. It has been clearly established through in vitro reporter assays and in vivo ribosome profiling experiments that UGA recoding is an inherently inefficient process (with the notable exception of Sec-incorporation at the C-terminal UGA-Sec codons of SELENOP mRNAs), such that the majority of ribosomes that initiate translation on selenoprotein mRNAs fail to incorporate Sec and never reach the natural termination codon.

Several outcomes can be envisioned for ribosomes that fail to incorporate Sec: (1) ribosomes may decode the UGA-Sec codon as a stop codon and terminate translation (Figure 2A, and in some cases this event will be recognized as premature termination leading to mRNA degradation through the nonsense-mediated decay (NMD) pathway; (2) ribosomes may stall at or near the UGA-Sec codon. Stalled ribosomes may either be removed by ribosome rescue or be rescued and continue translation (Figure 2C). RNAs that escape NMD or those on which ribosomes have been rescued may resume translation and go on to incorporate Sec (Figure 2B). What is known about these competing events will be discussed in Section 3.

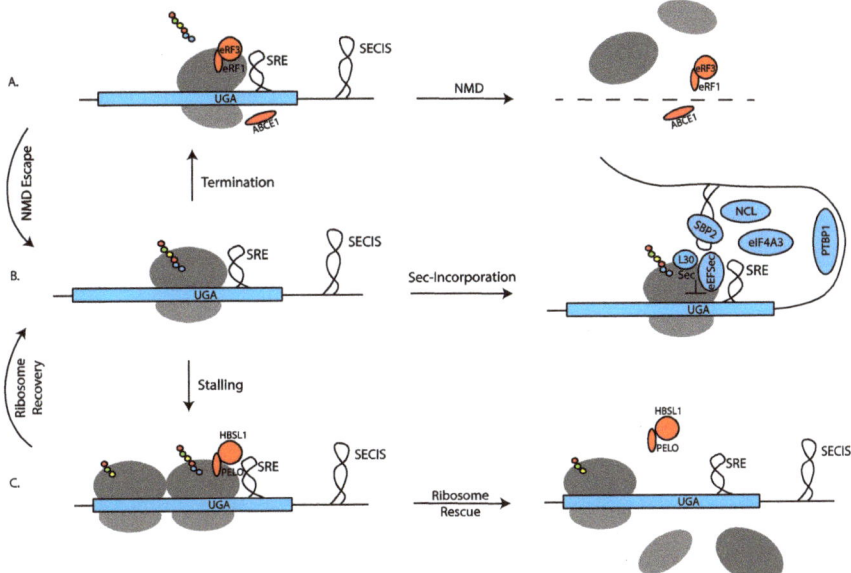

Figure 2. Possible fates of ribosomes encountering UGA-Sec codons. (**A**) Ribosomes failing to incorporate Sec may prematurely terminate translation, leading to nonsense-mediated decay (NMD) or NMD escape and continued translation by ribosomes located upstream of the UGA codon. (**B**) Sec-incorporation mediated by Sec insertion machinery. (**C**) Ribosomes that fail to incorporate Sec or terminate may be subject to various translational quality control pathways leading to ribosome release or recovery.

3. Competing Ribosome Fates at UGA-Sec Codons

Every step of the information-flow pathway in a living cell is monitored and kept accurate by quality control mechanisms. Replication has DNA repair, transcription has proofreading Pol II, tRNA aminoacylation has noncognate hydrolysis, and ribosomes both execute and are subjected to multiple control mechanisms during translation. These quality control pathways are broad and arguably nonspecific, so naturally there are a multitude of exceptions that must either evade or modify specific components of quality control. As a case in point, the mechanism by which mRNAs containing a misplaced in-frame stop (also known as nonsense) codon are sensed and shunted to a degradation pathway was discovered in 1979 [69], and a vast literature provides substantial mechanistic insight into this process termed nonsense-mediated decay [70]. The prevailing model posits that a stop codon is considered to be in the "wrong" position if it is not in the last exon. Thus, if the ribosome encounters a premature stop codon, it will "sense" the existence of a downstream exon/exon boundary, which is bound by a host of factors collectively referred to as the exon junction complex (EJC). Essentially the EJC is thought to cause inefficient translation termination which is sufficient to recruit the lynchpin NMD factor UPF1, which in turn recruits other factors required to initiate mRNA decay. It is important to stress that inefficient termination is sufficient to induce NMD, thus explaining the considerable evidence that EJC-independent NMD does occur and represents a significant fraction of total NMD events, particularly in organisms that have very few introns (e.g., *Saccharomyces cerevisiae*).

Selenoprotein mRNAs, with in-frame stop codons that are reprogrammed to allow Sec incorporation, stand as interesting case studies in the regulation of NMD. When selenium is replete, it is plausible to assume that efficient use of UGA as a sense codon would be sufficient to prevent NMD activation. However, it has long been reported that the efficiency of Sec incorporation is low, in the order of 10–25% [20,32,35,71]. Since the vast majority of Sec codons lie upstream from exon/exon boundaries [72], nearly every case of failed Sec incorporation might initiate NMD. Although the half-lives of all selenoprotein mRNAs have not been directly measured under the array of conditions required to precisely answer this question, the reality is that steady state levels of selenoprotein mRNAs are sufficient to provide adequate selenoprotein production under normal conditions. This may be due to the fact that selenoprotein mRNAs are shielded from NMD factors during incorporation, so the termination events that do occur are not able to signal the NMD machinery. This hypothesis was generally supported by a study where liver-specific knockout of SECISBP2 or tRNA^{Sec} resulted in an overall 60–70% reduction of selenoprotein mRNA levels [49]. However, a deeper analysis of the relative contributions of translation efficiency and mRNA abundance revealed a complex array of differential regulation, depending on the identity of the selenoprotein mRNA. For example, their findings for GPX1 were consistent with prior work that showed it to be a strong NMD target during selenium deficiency. This stands in contrast to GPX4, whose mRNA levels are dependent on SECISBP2, but not on selenium or Sec-tRNA^{Sec} levels. In addition, there appeared a third case, SELENOT, where mRNA levels were reduced but the translational efficiency remained the same. Part of this complexity is undoubtedly due to the existence of an SECISBP2 orthologue, SECISBP2L, which is very likely responsible for supporting selenoprotein production when SECISBP2 is absent. Overall, it is clear that active Sec incorporation is essential for maintaining selenoprotein mRNA levels, albeit to varying extents, but the question remains as to how much higher they would be if there were no in-frame UGA codon. One study did tangentially address this question by stably transfecting cDNAs encoding the zebrafish and human versions of SELENOP into the human hepatocyte cell line HepG2. One version of the cDNA was wild-type, which contains multiple Sec codons, and the other had replaced those with Cys codons. Quantitative RT-PCR revealed that the steady state mRNA level for the Cys codon-containing construct was about twice that of the Sec codon counterpart [73]. Interpreting this result is confounded by the fact that SELENOP is an exception among selenoproteins in possessing multiple UGA codons. Overall, while there is no doubt that NMD and Sec incorporation are deeply intertwined, there are still mechanistic questions that remain unanswered.

As shown in Figure 2, a byproduct of inefficient Sec incorporation is stalled ribosomes. In general, ribosome-stalling is an undesirable event that can result from a multitude of aberrant or regulated translation elongation reactions, so a surveillance and quality control pathway termed no-go decay (NGD) evolved to prevent ribosomes from accumulating on inefficiently translated mRNAs [74,75]. As such, the components that make up the NGD pathway may also play a role in regulating the fate of selenoprotein mRNAs. Perhaps not surprisingly, the two key factors in signaling NGD (PELO and HBS1) are evolutionarily related to the termination factors eRF1 and eRF3. One of the key signals for the NGD pathway event is an empty ribosomal A site, which would occur during inefficient translation elongation. Similar to a translation termination reaction, where an eRF1/eRF3 complex recognizes the ribosomal A site at stop codons, the PELO/HBS1 complex accesses any A site that is not occupied by an elongation complex and recruits factors that release the ribosome and degrade the mRNA. So, in the case of a Sec codon, three different complexes are competing for the same site: EEFSec/Sec-tRNA^{Sec}, eRF1/eRF3, and PELO/HBS1. Although some attention has been paid to the mechanism of the interplay between termination and Sec incorporation [31,32], very little has been studied about the role of the NGD pathway as a potential regulator. It is likely, therefore, that Sec incorporation efficiency is directly related to the ability of Sec-tRNA^{Sec} to outcompete the access of eRF1 or PELO. The multitude of factors that would impact this competition include local RNA structure, codon sequence context, relative concentrations of the A-site binders, and the nascent peptide chain sequence that is known to regulate the processivity of ribosome transit during translation elongation [76]. The major mechanistic question for the field centers around the extent to which Sec incorporation "actively" competes with these processes. For example, specific occlusion of PELO or eRF1 by the SECISBP2/SECIS complex, even in the absence of an EEFSec/Sec-tRNA^{Sec} complex, would represent an active competition. On the other hand, termination and NGD might be passively out-competed purely as a function of the relative concentrations of factors. Intriguingly, it has been reported that increasing the amount of eRF1 (but not eRF3) caused an increase rather than the expected decrease in Sec incorporation in transfected cells [31]. Similarly, the addition of eRF1 to an in vitro translation system also failed to inhibit Sec incorporation [32]. Taken at face value, these results may favor the "active" model of Sec incorporation where specific events prevent termination regardless of eRF1 concentration. While it is likely that a combination of active and passive processes is at play, a thorough investigation of the role that NGD factors might play in regulating the efficiency of Sec incorporation will be required to shed light on any role that PELO (or PELO exclusion) may play.

In the context of considerable complexity regarding processes that monitor A-site occupancy during the translation elongation reaction, another quality control mechanism monitors the stalling of ribosomes. The ribosome quality control (RQC) system induces a general cellular stress pathway that is coordinated by activation of the heatshock protein Hsf1 when nascent chain peptides are stalled [77]. The outcome of activating this pathway is ribosomal subunit ubiquitination and nascent chain degradation [78]. More recent work has shown that the signal is actually conveyed by "colliding" ribosomes [79]. Again, since ribosomes stall on many selenoprotein mRNAs, the interplay between the RQC pathway and Sec incorporation stands as yet another potential contributor to the overall efficiency. Indeed, the question of whether ribosome-stalling on selenoprotein mRNAs is sufficient to signal the stress pathway raises an intriguing possibility. While it is unlikely to be the case under normal conditions, the RQC pathway could, however, be an important signaling mechanism for the cell to detect limiting selenium concentrations, because selenoprotein mRNAs that are not turned over by NMD would likely accumulate collided ribosomes.

4. Conclusions

In this focused review, we have highlighted the fact that Sec incorporation must coexist with multiple quality control pathways. In theory, each of these pathways should effectively reduce Sec incorporation, but the Sec incorporation machinery co-evolved to deploy mechanisms that both subvert and exploit these pathways to optimize efficiency and signal stress conditions. The key for future exploration will be to decipher the mechanistic bases of the workarounds that organisms were required to evolve in order to effectively utilize Sec.

Author Contributions: P.R.C. and M.T.H. co-wrote this review. All authors have read and agreed to the published version of the manuscript.

Funding: This research was funded by the National Institutes of Health, grant number R01GM114291 (M.T.H.) and GM077073 (P.R.C.).

Institutional Review Board Statement: Not applicable.

Informed Consent Statement: Not applicable.

Data Availability Statement: Not applicable.

Conflicts of Interest: The authors declare no conflict of interest.

References

1. Atkins, J.F.; Gesteland, R.F. Recoding: Expansion of decoding rules enriches gene expression. In *Nucleic Acids and Molecular Biology*; Springer: New York, NY, USA, 2010.
2. Vindry, C.; Ohlmann, T.; Chavatte, L. Translation regulation of mammalian selenoproteins. *Biochim. Biophys. Acta Gen. Subj.* **2018**, *1862*, 2480–2492. [CrossRef] [PubMed]
3. Howard, M.T.; Copeland, P.R. New directions for understanding the codon redefinition required for selenocysteine incorporation. *Biol. Trace Elem. Res.* **2019**, *192*, 18–25. [CrossRef] [PubMed]
4. Kryukov, G.V.; Castellano, S.; Novoselov, S.V.; Lobanov, A.V.; Zehtab, O.; Guigo, R.; Gladyshev, V.N. Characterization of mammalian selenoproteomes. *Science* **2003**, *300*, 1439–1443. [CrossRef]
5. Kryukov, G.V.; Gladyshev, V.N. The prokaryotic selenoproteome. *EMBO Rep.* **2004**, *5*, 538–543. [CrossRef] [PubMed]
6. Heider, J.; Baron, C.; Bock, A. Coding from a distance: Dissection of the mRNA determinants required for the incorporation of selenocysteine into protein. *EMBO J.* **1992**, *11*, 3759–3766. [CrossRef]
7. Liu, Z.; Reches, M.; Groisman, I.; Engelberg-Kulka, H. The nature of the minimal 'selenocysteine insertion sequence' (secis) in *Escherichia coli*. *Nucleic Acids Res.* **1998**, *26*, 896–902. [CrossRef]
8. Kromayer, M.; Wilting, R.; Tormay, P.; Bock, A. Domain structure of the prokaryotic selenocysteine-specific elongation factor Selb. *J. Mol. Biol.* **1996**, *262*, 413–420. [CrossRef]
9. Berry, M.J.; Banu, L.; Chen, Y.Y.; Mandel, S.J.; Kieffer, J.D.; Harney, J.W.; Larsen, P.R. Recognition of uga as a selenocysteine codon in type i deiodinase requires sequences in the 3′ untranslated region. *Nature* **1991**, *353*, 273–276. [CrossRef]
10. Berry, M.J.; Banu, L.; Harney, J.W.; Larsen, P.R. Functional characterization of the eukaryotic secis elements which direct selenocysteine insertion at uga codons. *EMBO J.* **1993**, *12*, 3315–3322. [CrossRef]
11. Hill, K.E.; Lloyd, R.S.; Burk, R.F. Conserved nucleotide sequences in the open reading frame and 3′ untranslated region of selenoprotein p mRNA. *Proc. Natl. Acad. Sci. USA* **1993**, *90*, 537–541. [CrossRef]
12. Shen, Q.; Chu, F.F.; Newburger, P.E. Sequences in the 3′-untranslated region of the human cellular glutathione peroxidase gene are necessary and sufficient for selenocysteine incorporation at the uga codon. *J. Biol. Chem.* **1993**, *268*, 11463–11469. [CrossRef]
13. Turanov, A.A.; Lobanov, A.V.; Fomenko, D.E.; Morrison, H.G.; Sogin, M.L.; Klobutcher, L.A.; Hatfield, D.L.; Gladyshev, V.N. Genetic code supports targeted insertion of two amino acids by one codon. *Science* **2009**, *323*, 259–261. [CrossRef] [PubMed]
14. Turanov, A.A.; Lobanov, A.V.; Hatfield, D.L.; Gladyshev, V.N. Uga codon position-dependent incorporation of selenocysteine into mammalian selenoproteins. *Nucleic. Acids. Res.* **2013**, *41*, 6952–6959. [CrossRef] [PubMed]
15. Walczak, R.; Carbon, P.; Krol, A. An essential non-watson-crick base pair motif in 3′utr to mediate selenoprotein translation. *RNA* **1998**, *4*, 74–84.
16. Walczak, R.; Westhof, E.; Carbon, P.; Krol, A. A novel rna structural motif in the selenocysteine insertion element of eukaryotic selenoprotein mrnas. *RNA* **1996**, *2*, 367–379. [PubMed]
17. Buettner, C.; Harney, J.W.; Berry, M.J. The Caenorhabditis elegans homologue of thioredoxin reductase contains a selenocysteine insertion sequence (secis) element that differs from mammalian secis elements but directs selenocysteine incorporation. *J. Biol. Chem.* **1999**, *274*, 21598–21602. [CrossRef]
18. Grundner-Culemann, E.; Martin, G.W., 3rd; Harney, J.W.; Berry, M.J. Two distinct secis structures capable of directing selenocysteine incorporation in eukaryotes. *RNA* **1999**, *5*, 625–635. [CrossRef]
19. Mariotti, M. Secisearch3 and seblastian: In-silico tools to predict secis elements and selenoproteins. *Methods Mol. Biol.* **2018**, *1661*, 3–16.
20. Fixsen, S.M.; Howard, M.T. Processive selenocysteine incorporation during synthesis of eukaryotic selenoproteins. *J. Mol. Biol.* **2010**, *399*, 385–396. [CrossRef]
21. Shetty, S.P.; Sturts, R.; Vetick, M.; Copeland, P.R. Processive incorporation of multiple selenocysteine residues is driven by a novel feature of the selenocysteine insertion sequence. *J. Biol. Chem.* **2018**, *293*, 19377–19386. [CrossRef]
22. Mariotti, M.; Shetty, S.; Baird, L.; Wu, S.; Loughran, G.; Copeland, P.R.; Atkins, J.F.; Howard, M.T. Multiple RNA structures affect translation initiation and uga redefinition efficiency during synthesis of selenoprotein p. *Nucleic Acids Res.* **2017**, *45*, 13004–13015. [CrossRef]

23. Stoytcheva, Z.; Tujebajeva, R.M.; Harney, J.W.; Berry, M.J. Efficient incorporation of multiple selenocysteines involves an inefficient decoding step serving as a potential translational checkpoint and ribosome bottleneck. *Mol. Cell Biol.* **2006**, *26*, 9177–9184. [CrossRef]
24. Wu, S.; Mariotti, M.; Santesmasses, D.; Hill, K.E.; Baclaocos, J.; Aparicio-Prat, E.; Li, S.; Mackrill, J.; Wu, Y.; Howard, M.T.; et al. Human selenoprotein p and s variant mrnas with different numbers of secis elements and inferences from mutant mice of the roles of multiple secis elements. *Open Biol.* **2016**, *6*, 160241. [CrossRef]
25. Feng, Y.X.; Yuan, H.; Rein, A.; Levin, J.G. Bipartite signal for read-through suppression in murine leukemia virus mRNA: An eight-nucleotide purine-rich sequence immediately downstream of the gag termination codon followed by an RNA pseudoknot. *J. Virol.* **1992**, *66*, 5127–5132. [CrossRef]
26. Li, G.; Rice, C.M. The signal for translational readthrough of a uga codon in sindbis virus rna involves a single cytidine residue immediately downstream of the termination codon. *J. Virol.* **1993**, *67*, 5062–5067. [CrossRef]
27. Martin, R.; Phillips-Jones, M.K.; Watson, F.J.; Hill, L.S. Codon context effects on nonsense suppression in human cells. *Biochem. Soc. Trans.* **1993**, *21*, 846–851. [CrossRef]
28. Mottagui-Tabar, S.; Bjornsson, A.; Isaksson, L.A. The second to last amino acid in the nascent peptide as a codon context determinant. *EMBO J.* **1994**, *13*, 249–257. [CrossRef]
29. Mottagui-Tabar, S.; Tuite, M.F.; Isaksson, L.A. The influence of 5′ codon context on translation termination in Saccharomyces cerevisiae. *Eur. J. Biochem.* **1998**, *257*, 249–254. [CrossRef]
30. Wills, N.M.; Gesteland, R.F.; Atkins, J.F. Evidence that a downstream pseudoknot is required for translational read-through of the moloney murine leukemia virus gag stop codon. *Proc. Natl. Acad. Sci. USA* **1991**, *88*, 6991–6995. [CrossRef] [PubMed]
31. Grundner-Culemann, E.; Martin, G.W., 3rd; Tujebajeva, R.; Harney, J.W.; Berry, M.J. Interplay between termination and translation machinery in eukaryotic selenoprotein synthesis. *J. Mol. Biol.* **2001**, *310*, 699–707. [CrossRef] [PubMed]
32. Gupta, M.; Copeland, P.R. Functional analysis of the interplay between translation termination, selenocysteine codon context, and selenocysteine insertion sequence-binding protein 2. *J. Biol. Chem.* **2007**, *282*, 36797–36807. [CrossRef]
33. McCaughan, K.K.; Brown, C.M.; Dalphin, M.E.; Berry, M.J.; Tate, W.P. Translational termination efficiency in mammals is influenced by the base following the stop codon. *Proc. Natl. Acad. Sci. USA* **1995**, *92*, 5431–5435. [CrossRef]
34. Howard, M.T.; Aggarwal, G.; Anderson, C.B.; Khatri, S.; Flanigan, K.M.; Atkins, J.F. Recoding elements located adjacent to a subset of eukaryal selenocysteine-specifying uga codons. *EMBO J.* **2005**, *24*, 1596–1607. [CrossRef] [PubMed]
35. Howard, M.T.; Moyle, M.W.; Aggarwal, G.; Carlson, B.A.; Anderson, C.B. A recoding element that stimulates decoding of uga codons by sec trna[ser]sec. *RNA* **2007**, *13*, 912–920. [CrossRef]
36. Maiti, B.; Arbogast, S.; Allamand, V.; Moyle, M.W.; Anderson, C.B.; Richard, P.; Guicheney, P.; Ferreiro, A.; Flanigan, K.M.; Howard, M.T. A mutation in the sepn1 selenocysteine redefinition element (SRE) reduces selenocysteine incorporation and leads to Sepn1-related myopathy. *Hum. Mutat.* **2009**, *30*, 411–416. [CrossRef] [PubMed]
37. Pedersen, J.S.; Bejerano, G.; Siepel, A.; Rosenbloom, K.; Lindblad-Toh, K.; Lander, E.S.; Kent, J.; Miller, W.; Haussler, D. Identification and classification of conserved rna secondary structures in the human genome. *PLoS Comput. Biol.* **2006**, *2*, e33. [CrossRef]
38. Bubenik, J.L.; Miniard, A.C.; Driscoll, D.M. Alternative transcripts and 3′utr elements govern the incorporation of selenocysteine into selenoprotein s. *PLoS ONE* **2013**, *8*, e62102. [CrossRef]
39. Cockman, E.M.; Narayan, V.; Willard, B.; Shetty, S.P.; Copeland, P.R.; Driscoll, D.M. Identification of the selenoprotein s positive uga recoding (spur) element and its position-dependent activity. *RNA Biol.* **2019**, *16*, 1682–1696. [CrossRef]
40. Copeland, P.R.; Fletcher, J.E.; Carlson, B.A.; Hatfield, D.L.; Driscoll, D.M. A novel RNA binding protein, SBP2, is required for the translation of mammalian selenoprotein mrnas. *EMBO J.* **2000**, *19*, 306–314. [CrossRef] [PubMed]
41. Fagegaltier, D.; Hubert, N.; Yamada, K.; Mizutani, T.; Carbon, P.; Krol, A. Characterization of mselb, a novel mammalian elongation factor for selenoprotein translation. *EMBO J.* **2000**, *19*, 4796–4805. [CrossRef]
42. Tujebajeva, R.M.; Copeland, P.R.; Xu, X.M.; Carlson, B.A.; Harney, J.W.; Driscoll, D.M.; Hatfield, D.L.; Berry, M.J. Decoding apparatus for eukaryotic selenocysteine insertion. *EMBO Rep.* **2000**, *1*, 158–163. [CrossRef] [PubMed]
43. Hilgenfeld, R.; Bock, A.; Wilting, R. Structural model for the selenocysteine-specific elongation factor selb. *Biochimie* **1996**, *78*, 971–978. [CrossRef]
44. Dobosz-Bartoszek, M.; Pinkerton, M.H.; Otwinowski, Z.; Chakravarthy, S.; Soll, D.; Copeland, P.R.; Simonovic, M. Crystal structures of the human elongation factor EEFSec suggest a non-canonical mechanism for selenocysteine incorporation. *Nat. Commun.* **2016**, *7*, 12941. [CrossRef] [PubMed]
45. Allmang, C.; Carbon, P.; Krol, A. The SBP2 and 15.5 kd/snu13p proteins share the same RNA binding domain: Identification of SBP2 amino acids important to secisRNA binding. *RNA* **2002**, *8*, 1308–1318. [CrossRef]
46. Caban, K.; Kinzy, S.A.; Copeland, P.R. The l7ae rna binding motif is a multifunctional domain required for the ribosome-dependent sec incorporation activity of sec insertion sequence binding protein 2. *Mol. Cell Biol.* **2007**, *27*, 6350–6360. [CrossRef] [PubMed]
47. Fletcher, J.E.; Copeland, P.R.; Driscoll, D.M.; Krol, A. The selenocysteine incorporation machinery: Interactions between the secis RNA and the secis-binding protein SBP2. *RNA* **2001**, *7*, 1442–1453.
48. Zavacki, A.M.; Mansell, J.B.; Chung, M.; Klimovitsky, B.; Harney, J.W.; Berry, M.J. Coupled tRNA(sec)-dependent assembly of the selenocysteine decoding apparatus. *Mol. Cell* **2003**, *11*, 773–781. [CrossRef]

49. Fradejas-Villar, N.; Seeher, S.; Anderson, C.B.; Doengi, M.; Carlson, B.A.; Hatfield, D.L.; Schweizer, U.; Howard, M.T. The RNA-binding protein SECISBP2 differentially modulates uga codon reassignment and RNA decay. *Nucleic Acids Res.* **2017**, *45*, 4094–4107. [CrossRef]
50. Dubey, A.; Copeland, P.R. The selenocysteine-specific elongation factor contains unique sequences that are required for both nuclear export and selenocysteine incorporation. *PLoS ONE* **2016**, *11*, e0165642. [CrossRef]
51. Donovan, J.; Copeland, P.R. Selenocysteine Insertion Sequence Binding Protein 2L is implicated as a novel post-transcriptional regulator of selenoprotein expression. *PLoS ONE* **2012**, *7*, e35581. [CrossRef]
52. Copeland, P.R.; Stepanik, V.A.; Driscoll, D.M. Insight into mammalian selenocysteine insertion: Domain structure and ribosome binding properties of sec insertion sequence binding protein 2. *Mol. Cell Biol.* **2001**, *21*, 1491–1498. [CrossRef] [PubMed]
53. Kossinova, O.; Malygin, A.; Krol, A.; Karpova, G. The SBP2 protein central to selenoprotein synthesis contacts the human ribosome at expansion segment 7l of the 28s rRNA. *RNA* **2014**, *20*, 1046–1056. [CrossRef]
54. Wu, R.; Shen, Q.; Newburger, P.E. Recognition and binding of the human selenocysteine insertion sequence by nucleolin. *J. Cell Biochem.* **2000**, *77*, 507–516. [CrossRef]
55. Miniard, A.C.; Middleton, L.M.; Budiman, M.E.; Gerber, C.A.; Driscoll, D.M. Nucleolin binds to a subset of selenoprotein mRNAs and regulates their expression. *Nucleic Acids Res.* **2010**, *38*, 4807–4820. [CrossRef]
56. Bifano, A.L.; Atassi, T.; Ferrara, T.; Driscoll, D.M. Identification of nucleotides and amino acids that mediate the interaction between ribosomal protein L30 and the secis element. *BMC Mol. Biol.* **2013**, *14*, 12. [CrossRef]
57. Halic, M.; Becker, T.; Frank, J.; Spahn, C.M.; Beckmann, R. Localization and dynamic behavior of ribosomal protein L30e. *Nat. Struct. Mol. Biol.* **2005**, *12*, 467–468. [CrossRef]
58. Macias, S.; Bragulat, M.; Tardiff, D.F.; Vilardell, J. L30 binds the nascent rpl30 transcript to repress u2 snrnp recruitment. *Mol. Cell* **2008**, *30*, 732–742. [CrossRef]
59. Chavatte, L.; Brown, B.A.; Driscoll, D.M. Ribosomal protein L30 is a component of the uga-selenocysteine recoding machinery in eukaryotes. *Nat. Struct. Mol. Biol.* **2005**, *12*, 408–416. [CrossRef] [PubMed]
60. Budiman, M.E.; Bubenik, J.L.; Driscoll, D.M. Identification of a signature motif for the eIF4a3-secis interaction. *Nucleic Acids Res.* **2011**, *39*, 7730–7739. [CrossRef] [PubMed]
61. Ferraiuolo, M.A.; Lee, C.-S.; Ler, L.W.; Hsu, J.L.; Costa-Mattioli, M.; Luo, M.J.; Reed, R.; Sonenberg, N. A nuclear translation-like factor eIF4aiii is recruited to the mRNA during splicing and functions in nonsense-mediated decay. *Proc. Natl. Acad. Sci. USA* **2004**, *101*, 4118–4123. [CrossRef] [PubMed]
62. Palacios, I.M.; Gatfield, D.; St Johnston, D.; Izaurralde, E. An eif4aiii-containing complex required for mRNA localization and nonsense-mediated mRNA decay. *Nature* **2004**, *427*, 753–757. [CrossRef] [PubMed]
63. Shibuya, T.; Tange, T.O.; Sonenberg, N.; Moore, M.J. Eif4aiii binds spliced mRNA in the exon junction complex and is essential for nonsense-mediated decay. *Nat. Struct. Mol. Biol.* **2004**, *11*, 346–351. [CrossRef]
64. Sonenberg, N. EIF4e, the mRNA cap-binding protein: From basic discovery to translational research. *Biochem. Cell Biol.* **2008**, *86*, 178–183. [CrossRef] [PubMed]
65. Gribling-Burrer, A.S.; Leichter, M.; Wurth, L.; Huttin, A.; Schlotter, F.; Troffer-Charlier, N.; Cura, V.; Barkats, M.; Cavarelli, J.; Massenet, S.; et al. Secis-binding protein 2 interacts with the smn complex and the methylosome for selenoprotein mrnp assembly and translation. *Nucleic Acids Res.* **2017**, *45*, 5399–5413. [CrossRef] [PubMed]
66. Wurth, L.; Gribling-Burrer, A.S.; Verheggen, C.; Leichter, M.; Takeuchi, A.; Baudrey, S.; Martin, F.; Krol, A.; Bertrand, E.; Allmang, C. Hypermethylated-capped selenoprotein mRNAs in mammals. *Nucleic Acids Res.* **2014**, *42*, 8663–8677. [CrossRef] [PubMed]
67. Li, D.K.; Tisdale, S.; Lotti, F.; Pellizzoni, L. Smn control of rnp assembly: From post-transcriptional gene regulation to motor neuron disease. *Semin. Cell Dev. Biol.* **2014**, *32*, 22–29. [CrossRef]
68. Shetty, S.P.; Kiledjian, N.T.; Copeland, P.R. The polypyrimidine tract binding protein, PTBP1, regulates selenium homeostasis via the selenoprotein p 3′ untranslated region. *bioRxiv* **2020**. [CrossRef]
69. Losson, R.; Lacroute, F. Interference of nonsense mutations with eukaryotic messenger RNA stability. *Proc. Natl. Acad. Sci. USA* **1979**, *76*, 5134–5137. [CrossRef]
70. Kurosaki, T.; Popp, M.W.; Maquat, L.E. Quality and quantity control of gene expression by nonsense-mediated mRNA decay. *Nat. Rev. Mol. Cell Biol.* **2019**, *20*, 406–420. [CrossRef]
71. Novoselov, S.V.; Lobanov, A.V.; Hua, D.; Kasaikina, M.V.; Hatfield, D.L.; Gladyshev, V.N. A highly efficient form of the selenocysteine insertion sequence element in protozoan parasites and its use in mammalian cells. *Proc. Natl. Acad. Sci. USA* **2007**, *104*, 7857–7862. [CrossRef]
72. Shetty, S.P.; Copeland, P.R. Selenocysteine incorporation: A trump card in the game of mRNA decay. *Biochimie* **2015**, *114*, 97–101. [CrossRef] [PubMed]
73. Shetty, S.P.; Copeland, P.R. The selenium transport protein, selenoprotein p, requires coding sequence determinants to promote efficient selenocysteine incorporation. *J. Mol. Biol.* **2018**, *430*, 5217–5232. [CrossRef]
74. Doma, M.K.; Parker, R. Endonucleolytic cleavage of eukaryotic mRNAs with stalls in translation elongation. *Nature* **2006**, *440*, 561–564. [CrossRef]
75. Morris, C.; Cluet, D.; Ricci, E.P. Ribosome dynamics and mRNA turnover, a complex relationship under constant cellular scrutiny. *Wiley Interdiscip. Rev. RNA* **2021**, *12*, e1658. [CrossRef] [PubMed]

76. Choi, J.; Grosely, R.; Prabhakar, A.; Lapointe, C.P.; Wang, J.; Puglisi, J.D. How messenger RNA and nascent chain sequences regulate translation elongation. *Annu. Rev. Biochem.* **2018**, *87*, 421–449. [CrossRef]
77. Brandman, O.; Stewart-Ornstein, J.; Wong, D.; Larson, A.; Williams, C.C.; Li, G.W.; Zhou, S.; King, D.; Shen, P.S.; Weibezahn, J.; et al. A ribosome-bound quality control complex triggers degradation of nascent peptides and signals translation stress. *Cell* **2012**, *151*, 1042–1054. [CrossRef]
78. Matsuo, Y.; Ikeuchi, K.; Saeki, Y.; Iwasaki, S.; Schmidt, C.; Udagawa, T.; Sato, F.; Tsuchiya, H.; Becker, T.; Tanaka, K.; et al. Ubiquitination of stalled ribosome triggers ribosome-associated quality control. *Nat. Commun.* **2017**, *8*, 159. [CrossRef] [PubMed]
79. Ikeuchi, K.; Tesina, P.; Matsuo, Y.; Sugiyama, T.; Cheng, J.; Saeki, Y.; Tanaka, K.; Becker, T.; Beckmann, R.; Inada, T. Collided ribosomes form a unique structural interface to induce hel2-driven quality control pathways. *EMBO J.* **2019**, *38*, e100276. [CrossRef]

Review

Human Genetic Disorders Resulting in Systemic Selenoprotein Deficiency

Erik Schoenmakers and Krishna Chatterjee *

Metabolic Research Laboratories, Wellcome Trust-MRC Institute of Metabolic Science, Addenbrooke's Hospital, University of Cambridge, Cambridge CB2 0QQ, UK; es308@cam.ac.uk
* Correspondence: kkc1@medschl.cam.ac.uk; Tel.: +44-1223-336842

Abstract: Selenium, a trace element fundamental to human health, is incorporated as the amino acid selenocysteine (Sec) into more than 25 proteins, referred to as selenoproteins. Human mutations in *SECISBP2*, *SEPSECS* and *TRU-TCA1-1*, three genes essential in the selenocysteine incorporation pathway, affect the expression of most if not all selenoproteins. Systemic selenoprotein deficiency results in a complex, multifactorial disorder, reflecting loss of selenoprotein function in specific tissues and/or long-term impaired selenoenzyme-mediated defence against oxidative and endoplasmic reticulum stress. *SEPSECS* mutations are associated with a predominantly neurological phenotype with progressive cerebello-cerebral atrophy. Selenoprotein deficiency due to *SECISBP2* and *TRU-TCA1-1* defects are characterized by abnormal circulating thyroid hormones due to lack of Sec-containing deiodinases, low serum selenium levels (low SELENOP, GPX3), with additional features (myopathy due to low SELENON; photosensitivity, hearing loss, increased adipose mass and function due to reduced antioxidant and endoplasmic reticulum stress defence) in *SECISBP2* cases. Antioxidant therapy ameliorates oxidative damage in cells and tissues of patients, but its longer term benefits remain undefined. Ongoing surveillance of patients enables ascertainment of additional phenotypes which may provide further insights into the role of selenoproteins in human biological processes.

Keywords: selenoprotein deficiency; SECISBP2; Sec-tRNA[Ser]Sec; SEPSECS; selenium

1. Introduction

Dietary selenium (Se) is absorbed as inorganic Se (e.g., selenate; selenite) or organic Se (e.g., Se-methionine; selenocysteine) and metabolized to hydrogen selenide (H_2Se) before incorporation into the amino acid selenocysteine (Sec) [1]. Selenocysteine is different from other amino acids in that, uniquely, it is synthesized on its own tRNA, (Sec-tRNA[Ser]Sec; encoded by *TRU-TCA1-1*), via a well described pathway including O-phosphoserine-tRNA:selenocysteine tRNA synthase (SEPSECS) (Figure 1) [2,3]. Selenocysteine is incorporated into selenoproteins, at the position of a UGA codon in the mRNA, which ordinarily encodes a termination codon that dictates the cessation of protein synthesis. Unique Sec-insertion machinery, involving a cis-acting SEleniumCysteine Insertion Sequence (SECIS) stem-loop structure located in the 3'-UTR of all selenoprotein mRNAs and the UGA codon, interacting with trans-acting factors (SECIS binding protein 2 (SECISBP2), Sec-tRNA specific eukaryotic elongation factor (EEFSEC) and Sec-tRNA[Ser]Sec) (Figure 1), recodes UGA as a codon mediating Sec incorporation rather than termination of protein translation [3–5].

At least 25 human selenoproteins are described and recognized functions include maintenance of redox potential, regulating redox sensitive biochemical pathways, protection of genetic material, proteins and membranes from oxidative damage, metabolism of thyroid hormones, regulation of gene expression and control of protein folding (Table 1) [3,6]. The importance of selenoproteins is illustrated by the embryonic lethal phenotype of *Trsp* (mouse Sec-tRNA[Ser]Sec) and *Secisbp2* knockout mice [7,8]. It is well known that dietary

Se intake affects systemic Se-status and selenoprotein expression, but not all selenoproteins are affected to the same extent. Thus, expression of housekeeping selenoproteins (e.g., TXNRD1, TXNRD3, GPX4) is less affected by low circulating Se-levels compared to stress-related selenoproteins (e.g., GPX1, GPX3, SELENOW). Such differential preservation of selenoprotein expression is attributed to the existence of a "hierarchy of selenoprotein synthesis", whose underlying molecular basis is unclear [3,9]. With this knowledge, it is no surprise that mutations in key components of the Sec-insertion pathway (*SEPSECS, SECISBP2, TRU-TCA1-1*) result in generalized deficiency of selenoproteins associated with a complex, multisystem phenotypes. Here, we describe the clinical consequences of mutations in these three human genes and suggest possible links with loss-of-function of known selenoproteins.

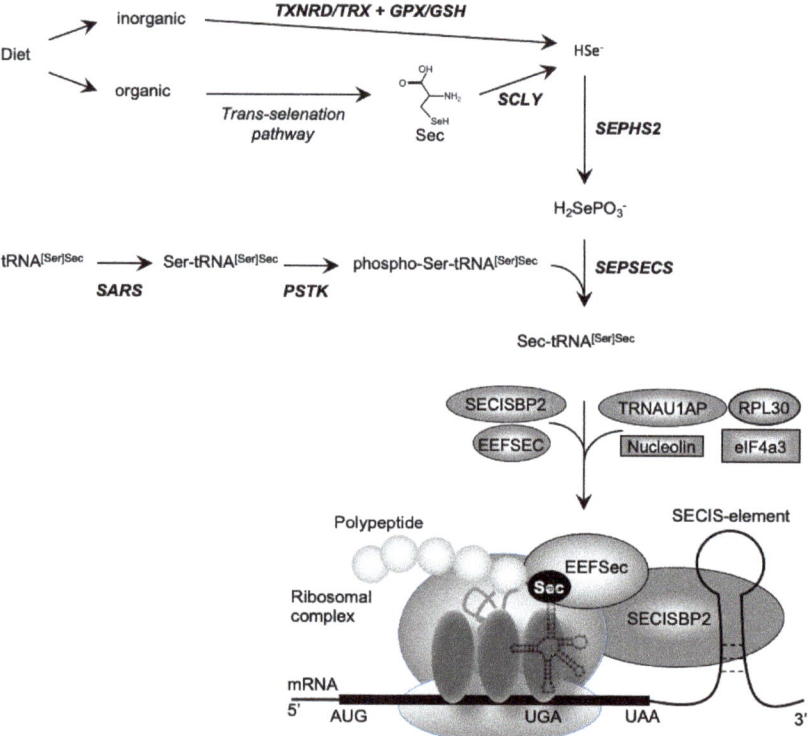

Figure 1. Biosynthesis of selenocysteine (Sec) and selenoproteins. Dietary sources of selenium exist in inorganic form (e.g., selenate, selenite) and organic form (e.g., Sec, SeMet). Inorganic selenium is reduced to selenide by TXNRD/TRX or GPX/GSH systems and organic selenium is metabolized to Sec, used by SCLY to generate selenide. De novo Sec synthesis takes place on its own tRNA (tRNA[Ser]Sec), which undergoes maturation through sequential modifications (SARS-mediated addition of Ser, PSTK-mediated phosphorylation of Ser), with acceptance of a selenophosphate (generated from selenide by SEPHS2) catalysed by SEPSECS as final step. Expression of selenoproteins requires recoding of an UGA codon as the amino acid Sec instead of a premature stop. The incorporation of Sec is mediated by a multiprotein complex containing SECISBP2, bound to the SECIS element situated in the 3'-untranslated region of selenoproteins, the Sec elongation factor EEFSEC, together with Sec-tRNA[Ser]Sec at the ribosomal acceptor site. The other factors (e.g., ribosomal protein L30, eukaryotic initiation factor eIF4a3, nucleolin) have regulatory roles.

Table 1. Human selenoproteins.

Selenoprotein	Function	Expression Subcellular Localization
GPX1 glutathione peroxidase 1	Oxidoreductase	most tissues cytoplasmic
GPX2 glutathione peroxidase 2	Oxidoreductase	limited number of tissues Nucleus and cytoplasmic
GPX3 glutathione peroxidase 3	Oxidoreductase	most tissues secreted
GPX4 glutathione peroxidase 4	Oxidoreductase	most tissues Nucleus and mitochondria
GPX6 glutathione peroxidase 6	Oxidoreductase	testis, epididymis, olfactory system predicted secreted
TXNRD1 thioredoxin reductase 1	Oxidoreductase	Ubiquitous Nucleus and cytoplasmic
TXNRD2 Thioredoxin reductase 2	Oxidoreductase	Ubiquitous cytoplasmic and mitochondria
TXNRD3 Thioredoxin reductase 3	Oxidoreductase	most tissues, high in testis Intracellular
DIO1 Iodothyronine deiodinase 1	Thyroid hormone metabolism	kidney, liver, thyroid gland Intracellular membrane-associated
DIO2 Iodothyronine deiodinase 2	Thyroid hormone metabolism	central nervous system, pituitary Intracellular membrane-associated
DIO3 Iodothyronine deiodinase 3	Thyroid hormone metabolism	several tissues Intracellular membrane-associated
MSRB1 methionine sulfoxide reductase B1	Met sulfoxide reduction	Ubiquitous Nucleus and cytoplasmic
SELENOF Selenoprotein F	Protein folding control	Ubiquitous endoplasmic reticulum
SELENOH Selenoprotein H	Unknown oxidoreductase	Ubiquitous Nucleus
SELENOI Selenoprotein I	Phospholipid biosynthesis	Ubiquitous transmembrane
SELENOK Selenoprotein K	Protein folding control	Ubiquitous ER, plasma membrane
SELENOM Selenoprotein M	Unknown	Ubiquitous Nuclear and perinuclear
SELENON Selenoprotein N	Redox-calcium homeostasis	Ubiquitous endoplasmic reticulum
SELENOO Selenoprotein O	Protein AMPylation activity	Ubiquitous mitochondria
SELENOP Selenoprotein P	Transport/oxidoreductase	most tissues secreted, cytoplasmic
SELENOS Selenoprotein S	Protein folding control	Ubiquitous endoplasmic reticulum
SELENOT Selenoprotein T	Unknown oxidoreductase	Ubiquitous endoplasmic reticulum
SELENOV Selenoprotein V	Unknown	thyroid, parathyroid, testis, brain Intracellular
SELENOW Selenoprotein W	Oxidoreductase	Ubiquitous Intracellular
SEPHS2 Selenophosphate synthetase 2	Selenophosphate synthesis	Ubiquitous, high in liver and kidney Intracellular

2. SECISBP2 Mutations

SECISBP2 is an essential and limiting factor for biosynthesis of selenoproteins [4,10] and functions as a scaffold, recruiting ribosomes, EEFSEC, and Sec-tRNA[Ser]Sec to the UGA codon by binding to SECIS-elements in selenoprotein mRNAs, generating a dynamic ribosome-Sec-incorporation complex (Figure 1). The first 400 amino (N-)terminal residues of SECISBP2 have no clear function; in contrast the carboxy (C-)terminal region (amino acids 399–784) is both necessary and sufficient for Sec-incorporation (Sec incorporation domain: SID) and binding to the SECIS element (RNA-binding domain: RBD) in vitro (Figure 2). The RBD, contains a L7Ae-type RNA interaction module and a lysine-rich domain, mediating specific recognition of "stem-loop" structures adopted by SECIS elements and other regulatory RNA motifs [11–13]. The C-terminal region also contains motifs (nuclear localization signal; nuclear export signal) involved in cellular localization of SECISBP2 and a cysteine rich domain (Figure 2) [14]. In the N-terminal region, alternative splicing events and ATG start codons have been described, generating different SECISBP2 isoforms [14], but all containing the essential C-terminal region. These events, together with regulatory domains in the C-terminal region, are thought to control SECISBP2-dependent Sec incorporation and the hierarchy of selenoprotein expression in vivo.

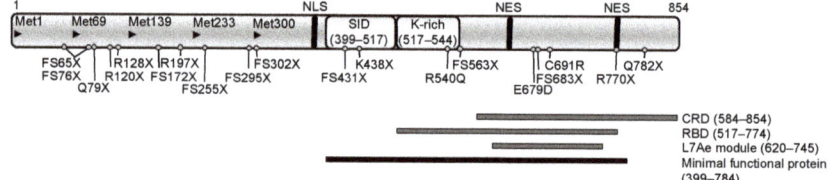

Figure 2. Functional domains of human SECISBP2 with the position of mutations described hitherto. Arrowheads denote the location of ATG codons; NLS: nuclear localisation signal (380–390); NES: nuclear export signals (634–657 and 756–770); SID: Sec incorporation domain; CRD: cysteine rich domain; RBD: minimal RNA-binding domain with the Lysine-rich domain (K-rich) and the L7Ae RNA-binding module; the black bar denotes the minimal protein region required for full functional activity in vitro.

Homozygous or compound heterozygous mutations in SECISBP2 have been described in individuals from 11 families from diverse ethnic backgrounds [15] (Table 2, Figure 2); hitherto no phenotypes have been described in heterozygous individuals. Most SECISBP2 mutations identified to date result in premature stops in the N-terminal region upstream of an alternative start codon (Met 300), permitting synthesis of the shorter, C-terminal, minimal functional domain of SECISBP2 (Figure 2) [14,16–23]. Conversely, stop mutations (e.g., R770X, Q782X) [18,23], distal to the minimal functional domain might generate C-terminally truncated proteins with residual but altered function. In one patient with an intronic mutation (IVS8ds + 29G > A) leading to a stop in the SID-domain, correct mRNA splicing was only reduced by 50% [16], a mechanism preserving some SECISBP2 synthesis that may operate in other splice site mutation contexts.

Table 2. Human SECISBP2 mutations.

Age in Years (Gender)	Mutation	Protein Change	Alleles Affected	Ethnicity	Reference
26 (M [1]); 19 (M); 19 (F [2])	c.1619 G > A	R540Q	homozygous	Saudi Arabian	[16]
25 (M)	c.1312 A > T c.IVS8ds + 29 G > A	K438 * fs431 *	compound heterozygous	Irish	[16]

Table 2. Cont.

Age in Years (Gender)	Mutation	Protein Change	Alleles Affected	Ethnicity	Reference
19 (M)	c.382 C > T	R128 *	homozygous	Ghanaian	[17]
18 (F)	c.358 C > T c.2308 C > T	R120 * R770 *	compound heterozygous	Brazilian	[18]
44 (M)	c.668delT c.IVS7 -155, T > A	F223fs255 * fs295 * + fs302 *	compound heterozygous	British	[19]
13 (M)	c. 2017 T > C 1–5 intronic SNP's	C691R fs65 * + fs76 *	compound heterozygous	British	[19]
15 (M)	c.1529_1541dup CCAGCGCCCCACT c.235 C > T	M515fs563 * Q79 *	compound heterozygous	Japanese	[20]
10 (M)	c.800_801insA	K267Kfs * 2	homozygous	Turkish	[21]
3.5 (M)	c.283delT c.589 C > T	T95Ifs31 * R197 *	compound heterozygous	N/A [3]	[22]
11 (F)	c.2344 C > T c.2045–2048 delAACA	Q782 * K682fs683 *	compound heterozygous	Turkish	[23]
5 (F)	c.589 C > T c.2108 G > T or C	R197 * E679D	compound heterozygous	Argentinian	[23]

[1] M: Male; [2] F: Female; [3] N/A: Not available.

Only three missense *SECISBP2* mutations, situated in the RBD (R540Q, E679D and C691R) are described. The R540Q mutation, in the lysine-rich domain, fails to bind a specific subset of SECIS-elements in vitro and a mouse model revealed an abnormal pattern of Secisbp2 and selenoprotein expression in tissues [16,24,25]. The E679D and C691R mutations are situated in the L7Ae RNA-binding module and part of the CRD. C691R mutant SECISBP2 undergoes enhanced proteasomal degradation, with loss of RNA-binding [19,25]. The E679D mutation has not been investigated but is predicted to be deleterious (PolyPhen-2 algorithm score of 0.998), possibly affecting RNA-binding [23].

Complete knockout of *Secisbp2* in mice is embryonic lethal [8], suggesting some functional protein, or an alternative rescue mechanism, is present in humans with *SECISBP2* mutations. Studies suggest that most combinations of *SECISBP2* mutations in patients hitherto are hypomorphic, with at least one allele directing synthesis of protein at either reduced levels or that is partially functional (Table 2). Because it is rate limiting for Sec incorporation, decreased SECISBP2 function will affect most if not all selenoprotein synthesis, as confirmed by available selenoprotein expression data in the patients [16,19].

Hitherto, only a small number of patients are described, from different ethnic and geographical backgrounds, often with compound heterozygous mutations and with limited information of their phenotypes. Some clinical phenotypes are attributable to deficiencies of particular selenoproteins in specific tissues whilst other features have a complex, multifactorial, basis possible linked to unbalanced antioxidant defence or protein folding pathways or loss of selenoproteins of unknown function. Increased cellular oxidative stress, readily measurable in most cells and tissues from patients, results in cumulative membrane and DNA damage. A common biochemical signature in all patients consists of low circulating selenium (reflecting low plasma SELENOP and GPX3) and abnormal thyroid hormone levels due to diminished activity of deiodinases resulting in raised FT4, normal to low FT3, raised reverse T3 and normal or high TSH concentrations [15,16,19]. Most cases were diagnosed in childhood with growth retardation (e.g., failure to thrive, short stature, delayed bone age) and developmental delay (e.g., delayed speech, intellectual- and motor coordination deficits) as common features, due not only to abnormal thyroid hormone metabolism [26,27] but also effects of specific selenoproteins deficiency in tissues (e.g., neuronal [8] or skeletal [28]). Fatigue and muscle weakness is a recognized feature in several patients and is attributable at least in part to a progressive muscular dystrophy affecting axial and proximal limb muscles, and very similar to the phenotype of myopathy due to selenoprotein N-deficiency [29]. Mild, bilateral, high-frequency sensorineural hearing loss is observed in some patients and is possibly due to ROS-mediated damage in the auditory system [30,31] as it is progressive in nature, worsening in older patients. An adult male patient was azoospermic, with marked deficiency of testis-expressed selenoproteins (GPX4, TXNRD3, SELENOV) [32–36]. Several other recorded phenotypes (increased whole body, subcutaneous fat mass, increased systemic insulin sensitivity, cutaneous photosensitivity) probably have a multifactorial basis which includes loss of antioxidant and endoplasmic reticulum stress defence. Studies of mouse models and in humans provide a substantial body of evidence to suggest a link between selenoproteins and most of these phenotypes [19,37–41].

Clinical management of these patients is mostly limited to correcting abnormal thyroid hormone metabolism with liothyronine supplementation if necessary. No specific therapies exist for other phenotypes (e.g., myopathy), but their progressive nature can require supportive intervention (e.g., nocturnal assisted ventilation for respiratory muscle weakness, aid for hearing loss). Oral selenium supplementation did raise total serum Se levels in some SECISBP2-deficient patients, but without clinical [17,18,21] or biochemical (circulating GPX's, SELENOP, thyroid hormone metabolism) effect [42]. Antioxidant (alpha tocopherol) treatment was beneficial in one patient, reducing circulating levels of products of lipid peroxidation with reversal of these changes after treatment withdrawal [40]. These observations suggest that treatment with antioxidants is a rational therapeutic approach, but the longterm consequences in this multisystem disorder are unpredictable.

3. *TRU-TCA1-1* Mutations

Selenocysteine is different from other amino acids in that it is synthesized uniquely on its own tRNA, encoded by *TRU-TCA1-1*, via a well described pathway including SEPSECS (Figure 1) [2,3]. Two major isoforms of the mature Sec-tRNA$^{[Ser]Sec}$ have been identified, containing either 5-methoxycarbonyl-methyluridine (mcm^{5}U) or its methylated form 5-methoxycarbonylmethyl-2'-O-methyluridine (mcm^{5}Um) at position 34, with the level of methylation being dependent on selenium status (Figure 3). The methylation state of uridine 34, located in the anticodon loop, is thought to contribute to stabilization of the codon–anticodon interaction and to play a role in mediating the hierarchy of selenoprotein expression. Thus, expression of essential, cellular housekeeping selenoproteins (e.g., TXNRDs, GPX4) is dependent on the mcm^{5}U isoform, whilst synthesis of cellular, stress-related selenoproteins (e.g., GPX1, GPX3) synthesis require the mcm^{5}Um isoform [43,44].

The first patient with a homozygous nucleotide change at position 65 (C > G) in *TRU-TCA1-1* (Figure 3) [45], presented with a similar clinical and biochemical phenotype (fatigue and muscle weakness, raised FT4, normal T3, raised rT3 and TSH, low plasma selenium concentrations) to that seen in patients with *SECISBP2* deficiency. However, the pattern of selenoprotein expression in his cells differed, with preservation housekeeping selenoproteins (e.g., TXNRDs, GPX4), but not stress-related selenoproteins (e.g., GPX1, GPX3) in cells from the *TRU-TCA1-1* mutation patient. This pattern is similar to the differential preservation of selenoprotein synthesis described in murine Sec-tRNA$^{[Ser]Sec}$ mutant models [3,44]. Recently, a second, unrelated patient with the same, homozygous *TRU-TCA1-1* mutation (C65G) with raised FT4 and low plasma GPX3 levels has been described [46].

The mechanism for such differential selenoprotein expression is unresolved, but a possible explanation is the observation that the *TRU-TCA1-1* C65G mutation results in lower total Sec-tRNA$^{[Ser]Sec}$ expression in patients cells, with disproportionately greater diminution in Sec-tRNA$^{[Ser]Sec}$ mcm^{5}Um levels. This suggest that the low Sec-tRNA$^{[Ser]Sec}$ levels in the proband are still sufficient for normal synthesis of housekeeping selenoproteins, whereas diminution of Sec-tRNA$^{[Ser]Sec}$ mcm^{5}Um levels accounts for reduced synthesis of stress-related selenoproteins.

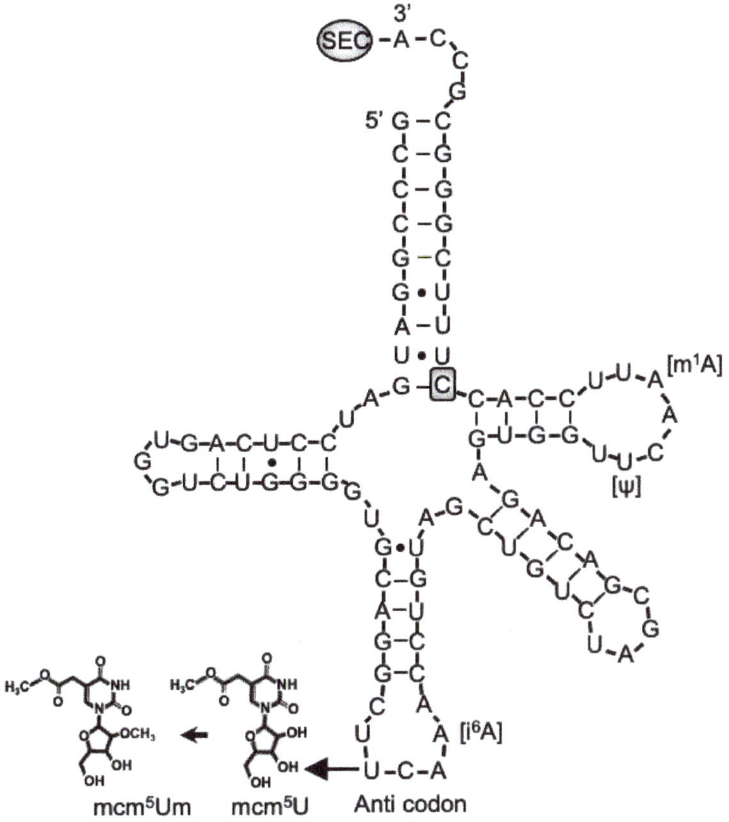

Figure 3. Sec-tRNA$^{[Ser]Sec}$ showing the position of human mutation. The primary structure of human Sec-tRNA$^{[Ser]Sec}$ is shown in a cloverleaf model, with the location of C65G mutation and posttranscriptional modification at positions U34 (mcm^{5}U or mcm^{5}Um, in the anticodon), A37 (i^{6}A), U55 (Ψ) and A58 (m^{1}A).

Clinical management of the patient is limited to alleviating clinical symptoms. However, with the knowledge that changing systemic selenium status can alter the relative proportions of the Sec-RNA$^{[Ser]Sec}$ isoforms [44,47,48], selenium supplementation of this patient, aiming to restore specific selenoprotein deficiencies, may be a rational therapeutic approach.

4. SEPSECS Mutations

The human SEPSECS protein was first characterized as an autoantigen (soluble liver antigen/liver pancreas, SLA) in autoimmune hepatitis [49]. The observation that it was present in a ribonucleoprotein complex with Sec-tRNA$^{[Ser]Sec}$, led to its identification as the enzyme that catalyzes the final step of Sec formation by converting O-phosphoserine-tRNA$^{[Ser]Sec}$ into Sec-tRNA$^{[Ser]Sec}$ using selenophosphate as substrate donor [50,51] (Figure 1).

Homozygous and compound heterozygous mutations in SEPSECS have been identified in 20 patients (Table 3, Figure 4). The availability of the crystal structures of the archaeal and murine SEPSECS apo-enzymes as well as human wild type and mutant SECSEPS (A239T, Y334C, T325S and Y429X) complexed with Sec-tRNA$^{[Ser]Sec}$ provides functional information [52–55]. The four premature stop mutants are predicted to be insoluble and inactive, as documented for the Y429X mutant. Mutants at Tyrosine 334 are predicted to fold like wild type SEPSECS and retain binding to Sec-tRNA[Ser]Sec, but with reduced enzyme activity. The A239T mutant failed to form stable tetramers, possible as result of a steric clash destabilizing the enzyme's core, rendering it inactive [55]. The other mutants for which no structure is available have been analyzed in silico and are predicted to be deleterious to varying degrees [15].

Table 3. Human SEPSECS mutations.

Age in Year (Gender)	Mutation	Protein Change	Alleles Affected	Ethnicity	Reference
6 (F [1]); 7.5 ([2])	c.1001 A > G	Y334C	homozygous	Jewish/Iraq	[56]
4 (F); 2.5 (M)	c.715 G > A c.1001 A > G	A239T Y334C	compound heterozygous	Iraqi/Moroccan	[56]
7 (F); 4 (F); 2 (F)	c.1466 A > T	D489V	homozygous	Jordan	[57]
0 (M); 0 (F); 0 (F); 0 (F)	c.974 C > G c.1287 C > A	T325S Y429X	compound heterozygous	Finnish	[58]
14 (F)	c.1 A > G c.388 + 3 A > G	M1V G130Vfs * 5	compound heterozygous	N/A [3]	[59]
N/A	c.1027–1120del	E343Lfs * 2	Homozygous	N/A [3]	[60]
9 (M)	c.1001 A > C	Y334H	homozygous	Arabian	[61]
10 (F)	c.77delG c.356 A > G	R26Pfs * 42 N119S	compound heterozygous	Japanese	[62]
21 (F)	c.356 A > G c.467 G > A	N119S R156Q	compound heterozygous	Japanese	[62]
1 (M)	c.176 C > T	A59V	Homozygous	N/A [3]	[63]
23 (F)	c.1321 G > A	G441R	Homozygous	N/A [3]	[64]
4 (F)	c.114 + 3 A > G	N/A [3]	Homozygous	Moroccan	[65]
N/A [1]	c.877 G > A	A293T	Homozygous	N/A [3]	[66]

[1] F: Female; [2] M: Male; [3] N/A: Not available.

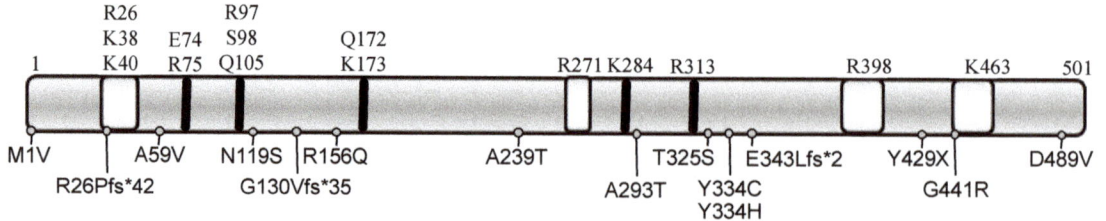

Figure 4. Functional domains of SEPSECS with the positions of the human mutations. Schematic of the human SEPSECS protein with key amino acids (above) that are part of the active domain (black bars) or interact with tRNA[ser]sec] (white shaded boxes) and mutations described hitherto below.

Patients with mutations in *SEPSECS* have profound intellectual disability, global developmental delay, spasticity, epilepsy, axonal neuropathy, optic atrophy and hypotonia with progressive microcephaly due to cortical and cerebellar atrophy [56,58,61,63]. The disorder is classified as autosomal recessive pontocerebellar hypoplasia type 2D (PCH2D, OMIM#613811), also referred to as progressive cerebellocerebral atrophy (PCCA) [56,67]. SEPSECS is required for generation of Sec-tRNA[Ser]Sec, which is essential for survival as demonstrated by the *Trsp* (mouse tRNA[Ser]Sec) knockout mouse model [44]. The Y334C-

Sepsecs mouse model exhibits a phenotype similar to features described in patients [68]. However, there is some variation in impact of *SEPSECS* mutations and specific phenotypes, with three patients (homozygous for G441R; compound heterozygous for R26Pfs*42/N119S or N119S/R156Q), presenting with late-onset PCH2D and progressive but milder degree of CNS atrophy [62,64]. In silico analyses suggest that these mutations have a less deleterious effect on SEPSECS function [15], although environmental factors or patients' genetic background modulating phenotype cannot be excluded.

The young age and severity of neurological problems in *SEPSECS* patients has precluded detailed investigation of selenoprotein expression and associated phenotypes. Studies of brain tissue from some patients showed decreased selenoprotein expression, correlating with increased cellular oxidative stress, but selenoprotein expression in other cell types (fibroblasts, muscle cells) was not significantly affected [58]. Serum selenium concentrations and thyroid status has been partially investigated in five patients, documenting either normal levels [61,65] or normal T4 but elevated TSH levels [58]. This suggests that the biochemical hallmarks of selenoprotein deficiency in *SECISBP2* and *TRU-TCA1-1* disorders (low circulating selenium and abnormal thyroid hormone levels) are not a significant feature in patients with *SEPSECS* mutations. Myopathic features with raised CK levels, abnormal mitochondria, cytoplasmic bodies and increased lipid accumulation in muscle are documented in one *SEPSECS* mutation case [61], with broad-based gait and postural instability suggesting muscle weakness in another patient [64]. These findings are similar to observations in selenoprotein N-deficient patients with *SECISBP2* mutations [19]. Overall, limited studies to date suggest that *SEPSECS* patients exhibit phenotypes associated with selenoprotein deficiency, but that these features can be mutation and tissue dependent.

5. Conclusions

In humans, 25 genes, encoding different selenoproteins containing the amino acid selenocysteine (Sec), have been identified. In selenoprotein mRNAs the amino acid Sec is encoded by the triplet UGA which usually constitutes a stop codon, requiring its recoding by a complex, multiprotein mechanism. Failure of selenoprotein synthesis due to *SECISBP2*, *TRU-TCA1-1* or *SEPSECS* defects, essential components of the selenoprotein biosynthesis pathway, results in complex disorders.

Individuals with *SECISBP2* defects exhibit a multisystem phenotype including growth retardation, fatigue and muscle weakness, sensorineural hearing loss, increased whole body fat mass, azoospermia and cutaneous photosensitivity. Most patients were identified due to a characteristic biochemical signature with raised FT4, normal to low FT3, raised rT3 and normal/slightly high TSH and low plasma selenium levels. A similar biochemical phenotype and clinical features are described in one individual with a *TRU-TCA1-1* mutation, although with relative preservation of essential housekeeping versus stress-related selenoprotein expression in his cells. Individuals with *SEPSECS* defects, essential for Sec-tRNA$^{[Ser]Sec}$ synthesis, present with a disorder characterized by cerebello-cerebral atrophy. Due to the young age and severe phenotype of patients, the effect of *SEPSECS* mutations on selenoprotein expression has not been studied in detail. In contrast, it is noteworthy that an early-onset central nervous system phenotype is not a feature in patients with *SECISBP2* or *TRU-TCA1-1* mutations.

As the function of many selenoproteins is unknown, or simultaneous deficiency of several selenoproteins exerts additive, synergistic or antagonistic effects culminating in complex dysregulation, linking disease phenotypes with altered expression of specific selenoproteins is challenging. Nevertheless, some causal links between specific selenoprotein deficiencies and phenotypes (e.g., abnormal thyroid function and deiodinase enzymes; low plasma Se and SELENOP, GPX3; azoospermia and SELENOV, GPX4, TXRND3; myopathy and SELENON) can be made. Other, progressive, phenotypes (e.g., photosensitivity, age-dependent hearing loss, neurodegeneration) may reflect absence of selenoenzymes mediating defence against oxidative and endoplasmic reticulum stress, resulting in cumulative tissue damage.

Triiodothyronine supplementation can correct abnormal thyroid hormone metabolism, with other medical intervention being mainly supportive. Selenium supplementation is of no proven benefit in *SECISBP2* mutation patients, but needs evaluation in the *TRU-TCA1-1* mutation case. Antioxidants, targeting the imbalance in oxidoredox and protein folding control pathways, could be beneficial in many selenoprotein deficient patients, but due to the complex interplay between different selenoproteins and their role in diverse biological processes, such treatment will require careful evaluation.

Author Contributions: E.S., writing—original draft preparation; K.C., writing—review and editing. All authors have read and agreed to the published version of the manuscript.

Funding: This research was funded by Wellcome Trust Investigator Award, grant number 210755/Z/18/Z and the NIHR Cambridge Biomedical Research Centre.

Institutional Review Board Statement: Not applicable.

Informed Consent Statement: Not applicable.

Data Availability Statement: Not applicable.

Conflicts of Interest: The authors declare no conflict of interest. The funders had no role in the design of the study; in the collection, analyses, or interpretation of data; in the writing of the manuscript, or in the decision to publish the results.

References

1. Ferreira, R.; Sena-Evangelista, K.; de Azevedo, E.P.; Pinheiro, F.I.; Cobucci, R.N.; Pedrosa, L. Selenium in Human Health and Gut Microflora: Bioavailability of Selenocompounds and Relationship With Diseases. *Frontiers in nutrition* **2021**, *8*, 685317. [CrossRef]
2. Turanov, A.A.; Xu, X.M.; Carlson, B.A.; Yoo, M.H.; Gladyshev, V.N.; Hatfield, D.L. Biosynthesis of selenocysteine, the 21st amino acid in the genetic code, and a novel pathway for cysteine biosynthesis. *Ad. Nutr.* **2011**, *2*, 122–128. [CrossRef]
3. Labunskyy, V.M.; Hatfield, D.L.; Gladyshev, V.N. Selenoproteins: Molecular pathways and physiological roles. *Physiol. Rev.* **2014**, *94*, 739–777. [CrossRef]
4. Copeland, P.R.; Fletcher, J.E.; Carlson, B.A.; Hatfield, D.L.; Driscoll, D.M. A novel RNA binding protein, SBP2, is required for the translation of mammalian selenoprotein mRNAs. *EMBO J.* **2000**, *19*, 306–314. [CrossRef]
5. Martin, G.W.; Harney, J.W.; Berry, M.J. Selenocysteine incorporation in eukaryotes: Insights into mechanism and efficiency from sequence, structure, and spacing proximity studies of the type 1 deiodinase SECIS element. *RNA* **1996**, *2*, 171–182.
6. Zoidis, E.; Seremelis, I.; Kontopoulos, N.; Danezis, G.P. Selenium-Dependent Antioxidant Enzymes: Actions and Properties of Selenoproteins. *Antioxidants* **2018**, *7*, 66–91. [CrossRef]
7. Bösl, M.R.; Takaku, K.; Oshima, M.; Nishimura, S.; Taketo, M.M. Early embryonic lethality caused by targeted disruption of the mouse selenocysteine tRNA gene (*Trsp*). *PNAS* **1997**, *94*, 5531–5534. [CrossRef]
8. Seeher, S.; Atassi, T.; Mahdi, Y.; Carlson, B.A.; Braun, D.; Wirth, E.K.; Klein, M.O.; Reix, N.; Miniard, A.C.; Schomburg, L.; et al. Secisbp2 is essential for embryonic development and enhances selenoprotein expression. *Antioxid. Redox Signal* **2014**, *21*, 835–849. [CrossRef]
9. Sunde, R.A.; Raines, A.M. Selenium regulation of the selenoprotein and nonselenoprotein transcriptomes in rodents. *Ad. Nutr.* **2011**, *2*, 138–150. [CrossRef] [PubMed]
10. Copeland, P.R.; Driscoll, D.M. Purification, redox sensitivity, and RNA binding properties of SECIS-binding protein 2, a protein involved in selenoprotein biosynthesis. *J. Biol. Chem.* **1999**, *274*, 25447–25454. [CrossRef]
11. Caban, K.; Kinzy, S.A.; Copeland, P.R. The L7Ae RNA binding motif is a multifunctional domain required for the ribosome-dependent Sec incorporation activity of Sec insertion sequence binding protein 2. *Mol. Cell. Biol.* **2007**, *27*, 6350–6360. [CrossRef]
12. Donovan, J.; Caban, K.; Ranaweera, R.; Gonzalez-Flores, J.N.; Copeland, P.R. A novel protein domain induces high affinity selenocysteine insertion sequence binding and elongation factor recruitment. *J. Biol. Chem.* **2008**, *283*, 35129–35139. [CrossRef]
13. Takeuchi, A.; Schmitt, D.; Chapple, C.; Babaylova, E.; Karpova, G.; Guigo, R.; Krol, A.; Allmang, C. A short motif in Drosophila SECIS Binding Protein 2 provides differential binding affinity to SECIS RNA hairpins. *Nucleic acids res.* **2009**, *37*, 2126–2141. [CrossRef]
14. Papp, L.V.; Lu, J.; Holmgren, A.; Khanna, K.K. From selenium to selenoproteins: Synthesis, identity, and their role in human health. *Antioxid. Redox Signal* **2007**, *9*, 775–806. [CrossRef]
15. Schoenmakers, E.; Chatterjee, K. Human Disorders Affecting the Selenocysteine Incorporation Pathway Cause Systemic Selenoprotein Deficiency. *Antioxid Redox Signal.* **2020**, *33*, 481–497. [CrossRef]
16. Dumitrescu, A.M.; Liao, X.H.; Abdullah, M.S.; Lado-Abeal, J.; Majed, F.A.; Moeller, L.C.; Boran, G.; Schomburg, L.; Weiss, R.E.; Refetoff, S. Mutations in SECISBP2 result in abnormal thyroid hormone metabolism. *Nat. Genet.* **2005**, *37*, 1247–1252. [CrossRef] [PubMed]

17. Di Cosmo, C.; McLellan, N.; Liao, X.H.; Khanna, K.K.; Weiss, R.E.; Papp, L.; Refetoff, S. Clinical and molecular characterization of a novel selenocysteine insertion sequence-binding protein 2 (SBP2) gene mutation (R128X). *J. Clin. Endocrinol. Metab.* **2009**, *94*, 4003–4009. [CrossRef]
18. Azevedo, M.F.; Barra, G.B.; Naves, L.A.; Ribeiro Velasco, L.F.; Godoy Garcia Castro, P.; de Castro, L.C.; Amato, A.A.; Miniard, A.; Driscoll, D.; Schomburg, L.; et al. Selenoprotein-related disease in a young girl caused by nonsense mutations in the SBP2 gene. *J. Clin. Endocrinol. Metab.* **2010**, *95*, 4066–4071. [CrossRef]
19. Schoenmakers, E.; Agostini, M.; Mitchell, C.; Schoenmakers, N.; Papp, L.; Rajanayagam, O.; Padidela, R.; Ceron-Gutierrez, L.; Doffinger, R.; Prevosto, C.; et al. Mutations in the selenocysteine insertion sequence-binding protein 2 gene lead to a multisystem selenoprotein deficiency disorder in humans. *J. Clin. Invest.* **2010**, *120*, 4220–4235. [CrossRef]
20. Hamajima, T.; Mushimoto, Y.; Kobayashi, H.; Saito, Y.; Onigata, K. Novel compound heterozygous mutations in the SBP2 gene: Characteristic clinical manifestations and the implications of GH and triiodothyronine in longitudinal bone growth and maturation. *Eur. J. Endocrinol.* **2012**, *166*, 757–764. [CrossRef]
21. Çatli, G.; Fujisawa, H.; Kirbiyik, Ö.; Mimoto, M.S.; Gençpinar, P.; Özdemir, T.R.; Dündar, B.N.; Dumitrescu, A.M. A Novel Homozygous Selenocysteine Insertion Sequence Binding Protein 2 (SECISBP2, SBP2) Gene Mutation in a Turkish Boy. *Thyroid* **2018**, *28*, 1221–1223. [CrossRef]
22. Korwutthikulrangsri, M.; Raimondi, C.; Dumitrescu, A.M. *Novel Compound Heterozygous SBP2 Gene Mutations in a Boy with Developmental Delay and Failure to Thrive*; 13th IWRTH: Doorn, The Netherlands, 2018; p. 22, Abstract Book.
23. Fu, J.; Korwutthikulrangsri, M.; Gönç, E.N.; Sillers, L.; Liao, X.H.; Alikaşifoğlu, A.; Kandemir, N.; Menucci, M.B.; Burman, K.D.; Weiss, R.E.; et al. Clinical and Molecular Analysis in 2 Families With Novel Compound Heterozygous SBP2 (SECISBP2) Mutations. *J. Clin. Endocrinol. Metab.* **2020**, *105*, e6–e11. [CrossRef]
24. Bubenik, J.L.; Driscoll, D.M. Altered RNA binding activity underlies abnormal thyroid hormone metabolism linked to a mutation in selenocysteine insertion sequence-binding protein 2. *J. Biol. Chem.* **2007**, *282*, 34653–34662. [CrossRef]
25. Zhao, W.; Bohleber, S.; Schmidt, H.; Seeher, S.; Howard, M.T.; Braun, D.; Arndt, S.; Reuter, U.; Wende, H.; Birchmeier, C.; et al. Ribosome profiling of selenoproteins in vivo reveals consequences of pathogenic *Secisbp2* missense mutations. *J. Biol. Chem.* **2019**, *294*, 14185–14200. [CrossRef]
26. Ng, L.; Goodyear, R.J.; Woods, C.A.; Schneider, M.J.; Diamond, E.; Richardson, G.P.; Kelley, M.W.; Germain, D.L.; Galton, V.A.; Forrest, D. Hearing loss and retarded cochlear development in mice lacking type 2 iodothyronine deiodinase. *Proc. Natl. Acad. Sci. USA* **2004**, *101*, 3474–3479. [CrossRef]
27. Hernandez, A.; Martinez, M.E.; Fiering, S.; Galton, V.A.; St Germain, D. Type 3 deiodinase is critical for the maturation and function of the thyroid axis. *J. Clin. Invest.* **2006**, *116*, 476–484. [CrossRef] [PubMed]
28. Downey, C.M.; Horton, C.R.; Carlson, B.A.; Parsons, T.E.; Hatfield, D.L.; Hallgrímsson, B.; Jirik, F.R. Osteo-chondroprogenitor-specific deletion of the selenocysteine tRNA gene, Trsp, leads to chondronecrosis and abnormal skeletal development: A putative model for Kashin-Beck disease. *PLoS Genet.* **2009**, *5*, e1000616. [CrossRef]
29. Silwal, A.; Sarkozy, A.; Scoto, M.; Ridout, D.; Schmidt, A.; Laverty, A.; Henriques, M.; D'Argenzio, L.; Main, M.; Mein, R.; et al. Selenoprotein N-related myopathy: A retrospective natural history study to guide clinical trials. *Ann. Clin. Transl. neurol.* **2020**, *7*, 2288–2296. [CrossRef] [PubMed]
30. McFadden, S.L.; Ohlemiller, K.K.; Ding, D.; Shero, M.; Salvi, R.J. The Influence of Superoxide Dismutase and Glutathione Peroxidase Deficiencies on Noise-Induced Hearing Loss in Mice. *Noise health* **2001**, *3*, 49–64.
31. Riva, C.; Donadieu, E.; Magnan, J.; Lavieille, J.P. Age-related hearing loss in CD/1 mice is associated to ROS formation and HIF target proteins up-regulation in the cochlea. *Exp. Geront.* **2007**, *42*, 327–336. [CrossRef]
32. Ursini, F.; Heim, S.; Kiess, M.; Maiorino, M.; Roveri, A.; Wissing, J.; Flohé, L. Dual function of the selenoprotein PHGPx during sperm maturation. *Science* **1999**, *285*, 1393–1396. [CrossRef]
33. Foresta, C.; Flohé, L.; Garolla, A.; Roveri, A.; Ursini, F.; Maiorino, M. Male fertility is linked to the selenoprotein phospholipid hydroperoxide glutathione peroxidase. *Biol. reprod.* **2002**, *67*, 967–971. [CrossRef]
34. Kryukov, G.V.; Castellano, S.; Novoselov, S.V.; Lobanov, A.V.; Zehtab, O.; Guigó, R.; Gladyshev, V.N. Characterization of mammalian selenoproteomes. *Science* **2003**, *300*, 1439–1443. [CrossRef]
35. Su, D.; Novoselov, S.V.; Sun, Q.A.; Moustafa, M.E.; Zhou, Y.; Oko, R.; Hatfield, D.L.; Gladyshev, V.N. Mammalian selenoprotein thioredoxin-glutathione reductase. Roles in disulfide bond formation and sperm maturation. *J. Biol. Chem.* **2005**, *280*, 26491–26498. [CrossRef]
36. Schneider, M.; Förster, H.; Boersma, A.; Seiler, A.; Wehnes, H.; Sinowatz, F.; Neumüller, C.; Deutsch, M.J.; Walch, A.; Hrabé de Angelis, M.; et al. Mitochondrial glutathione peroxidase 4 disruption causes male infertility. *FASEB J.* **2009**, *23*, 3233–3242. [CrossRef]
37. Schweikert, K.; Gafner, F.; Dell'Acqua, G. A bioactive complex to protect proteins from UV-induced oxidation in human epidermis. *Int. J. Cosm. Sci.* **2010**, *32*, 29–34. [CrossRef]
38. Sengupta, A.; Lichti, U.F.; Carlson, B.A.; Ryscavage, A.O.; Gladyshev, V.N.; Yuspa, S.H.; Hatfield, D.L. Selenoproteins are essential for proper keratinocyte function and skin development. *PLoS ONE* **2010**, *5*, e12249. [CrossRef] [PubMed]
39. Verma, S.; Hoffmann, F.W.; Kumar, M.; Huang, Z.; Roe, K.; Nguyen-Wu, E.; Hashimoto, A.S.; Hoffmann, P.R. Selenoprotein K knockout mice exhibit deficient calcium flux in immune cells and impaired immune responses. *J. Immunol.* **2011**, *186*, 2127–2137. [CrossRef]

40. Saito, Y.; Shichiri, M.; Hamajima, T.; Ishida, N.; Mita, Y.; Nakao, S.; Hagihara, Y.; Yoshida, Y.; Takahashi, K.; Niki, E.; et al. Enhancement of lipid peroxidation and its amelioration by vitamin E in a subject with mutations in the SBP2 gene. *Lipid Res.* **2015**, *56*, 2172–2182. [CrossRef]
41. Liao, C.; Carlson, B.A.; Paulson, R.F.; Prabhu, K.S. The intricate role of selenium and selenoproteins in erythropoiesis. *Free Radic. Biol. Med.* **2018**, *127*, 165–171. [CrossRef]
42. Schomburg, L.; Dumitrescu, A.M.; Liao, X.H.; Bin-Abbas, B.; Hoeflich, J.; Köhrle, J.; Refetoff, S. Selenium supplementation fails to correct the selenoprotein synthesis defect in subjects with SBP2 gene mutations. *Thyroid* **2009**, *19*, 277–281. [CrossRef]
43. Shetty, S.P.; Copeland, P.R. Selenocysteine incorporation: A trump card in the game of mRNA decay. *Biochimie* **2015**, *114*, 97–101. [CrossRef]
44. Carlson, B.A.; Yoo, M.H.; Tsuji, P.A.; Gladyshev, V.N.; Hatfield, D.L. Mouse models targeting selenocysteine tRNA expression for elucidating the role of selenoproteins in health and development. *Molecules* **2009**, *14*, 3509–3527. [CrossRef]
45. Schoenmakers, E.; Carlson, B.; Agostini, M.; Moran, C.; Rajanayagam, O.; Bochukova, E.; Tobe, R.; Peat, R.; Gevers, E.; Muntoni, F.; et al. Mutation in human selenocysteine transfer RNA selectively disrupts selenoprotein synthesis. *J. Clin. Invest.* **2016**, *126*, 992–996. [CrossRef]
46. Geslot, A.; Savagner, F.; Caron, P. Inherited selenocysteine transfer RNA mutation: Clinical and hormonal evaluation of 2 patients. *Eur. Thyroid, J.* **2021**, *10*, 542–547. [CrossRef]
47. Hatfield, D.; Lee, B.J.; Hampton, L.; Diamond, A.M. Selenium induces changes in the selenocysteine tRNA$^{[Ser]Sec}$ population in mammalian cells. *Nucleic acids res.* **1991**, *19*, 939–943. [CrossRef]
48. Diamond, A.M.; Choi, I.S.; Crain, P.F.; Hashizume, T.; Pomerantz, S.C.; Cruz, R.; Steer, C.J.; Hill, K.E.; Burk, R.F.; McCloskey, J.A.; et al. Dietary selenium affects methylation of the wobble nucleoside in the anticodon of selenocysteine tRNA$^{([Ser]Sec)}$. *J. Biol. Chem.* **1993**, *268*, 14215–14223. [CrossRef]
49. Kernebeck, T.; Lohse, A.W.; Grötzinger, J. A bioinformatical approach suggests the function of the autoimmune hepatitis target antigen soluble liver antigen/liver pancreas. *Hepatology* **2001**, *34*, 230–233. [CrossRef]
50. Small-Howard, A.; Morozova, N.; Stoytcheva, Z.; Forry, E.P.; Mansell, J.B.; Harney, J.W.; Carlson, B.A.; Xu, X.M.; Hatfield, D.L.; Berry, M.J. Supramolecular complexes mediate selenocysteine incorporation in vivo. *Mol. Cell. Biol.* **2006**, *26*, 2337–2346. [CrossRef]
51. Xu, X.M.; Mix, H.; Carlson, B.A.; Grabowski, P.J.; Gladyshev, V.N.; Berry, M.J.; Hatfield, D.L. Evidence for direct roles of two additional factors, SECp43 and soluble liver antigen, in the selenoprotein synthesis machinery. *J. Biol. Chem.* **2005**, *280*, 41568–41575. [CrossRef] [PubMed]
52. Araiso, Y.; Palioura, S.; Ishitani, R.; Sherrer, R.L.; O'Donoghue, P.; Yuan, J.; Oshikane, H.; Domae, N.; Defranco, J.; Söll, D.; et al. Structural insights into RNA-dependent eukaryal and archaeal selenocysteine formation. *Nucleic Acids Res.* **2008**, *36*, 1187–1199. [CrossRef]
53. Ganichkin, O.M.; Xu, X.M.; Carlson, B.A.; Mix, H.; Hatfield, D.L.; Gladyshev, V.N.; Wahl, M.C. Structure and catalytic mechanism of eukaryotic selenocysteine synthase. *J. Biol. Chem.* **2008**, *283*, 5849–5865. [CrossRef]
54. Palioura, S.; Sherrer, R.L.; Steitz, T.A.; Söll, D.; Simonovic, M. The human SepSecS-tRNASec complex reveals the mechanism of selenocysteine formation. *Science* **2009**, *325*, 321–325. [CrossRef]
55. Puppala, A.K.; French, R.L.; Matthies, D.; Baxa, U.; Subramaniam, S.; Simonović, M. Structural basis for early-onset neurological disorders caused by mutations in human selenocysteine synthase. *Sci. Rep.* **2016**, *6*, 32563. [CrossRef]
56. Agamy, O.; Ben Zeev, B.; Lev, D.; Marcus, B.; Fine, D.; Su, D.; Narkis, G.; Ofir, R.; Hoffmann, C.; Leshinsky-Silver, E.; et al. Mutations disrupting selenocysteine formation cause progressive cerebello-cerebral atrophy. *Am. J. Hum. Genet.* **2010**, *87*, 538–544. [CrossRef] [PubMed]
57. Makrythanasis, P.; Nelis, M.; Santoni, F.A.; Guipponi, M.; Vannier, A.; Béna, F.; Gimelli, S.; Stathaki, E.; Temtamy, S.; Mégarbané, A.; et al. Diagnostic exome sequencing to elucidate the genetic basis of likely recessive disorders in consanguineous families. *Hum. Mutat.* **2014**, *35*, 1203–1210. [CrossRef] [PubMed]
58. Anttonen, A.K.; Hilander, T.; Linnankivi, T.; Isohanni, P.; French, R.L.; Liu, Y.; Simonović, M.; Söll, D.; Somer, M.; Muth-Pawlak, D.; et al. Selenoprotein biosynthesis defect causes progressive encephalopathy with elevated lactate. *Neurology* **2015**, *85*, 306–315. [CrossRef]
59. Zhu, X.; Petrovski, S.; Xie, P.; Ruzzo, E.K.; Lu, Y.F.; McSweeney, K.M.; Ben-Zeev, B.; Nissenkorn, A.; Anikster, Y.; Oz-Levi, D.; et al. Whole-exome sequencing in undiagnosed genetic diseases: Interpreting 119 trios. *Genet. Med.* **2015**, *17*, 774–781. [CrossRef] [PubMed]
60. Alazami, A.M.; Patel, N.; Shamseldin, H.E.; Anazi, S.; Al-Dosari, M.S.; Alzahrani, F.; Hijazi, H.; Alshammari, M.; Aldahmesh, M.A.; Salih, M.A.; et al. Accelerating novel candidate gene discovery in neurogenetic disorders via whole-exome sequencing of prescreened multiplex consanguineous families. *Cell Rep.* **2015**, *10*, 148–161. [CrossRef] [PubMed]
61. Pavlidou, E.; Salpietro, V.; Phadke, R.; Hargreaves, I.P.; Batten, L.; McElreavy, K.; Pitt, M.; Mankad, K.; Wilson, C.; Cutrupi, M.C.; et al. Pontocerebellar hypoplasia type 2D and optic nerve atrophy further expand the spectrum associated with selenoprotein biosynthesis deficiency. *Eur. J. Paediatr. Neurol.* **2016**, *20*, 483–488. [CrossRef] [PubMed]
62. Iwama, K.; Sasaki, M.; Hirabayashi, S.; Ohba, C.; Iwabuchi, E.; Miyatake, S.; Nakashima, M.; Miyake, N.; Ito, S.; Saitsu, H.; et al. Milder progressive cerebellar atrophy caused by biallelic SEPSECS mutations. *J. Hum. Genet.* **2016**, *61*, 527–531. [CrossRef]

63. Olson, H.E.; Kelly, M.; LaCoursiere, C.M.; Pinsky, R.; Tambunan, D.; Shain, C.; Ramgopal, S.; Takeoka, M.; Libenson, M.H.; Julich, K.; et al. Genetics and genotype-phenotype correlations in early onset epileptic encephalopathy with burst suppression. *Ann. Neurol.* **2017**, *81*, 419–429. [CrossRef] [PubMed]
64. Van Dijk, T.; Vermeij, J.D.; van Koningsbruggen, S.; Lakeman, P.; Baas, F.; Poll-The, B.T. A SEPSECS mutation in a 23-year-old woman with microcephaly and progressive cerebellar ataxia. *J. Inherit. Metab. Dis* **2018**, *41*, 897–898. [CrossRef] [PubMed]
65. Arrudi-Moreno, M.; Fernández-Gómez, A.; Peña-Segura, J.L. A new mutation in the SEPSECS gene related to pontocerebellar hypoplasia type 2D. *Med. Clin.* **2021**, *156*, 94–95. [CrossRef] [PubMed]
66. Nejabat, M.; Inaloo, S.; Sheshdeh, A.T.; Bahramjahan, S.; Sarvestani, F.M.; Katibeh, P.; Nemati, H.; Tabei, S.; Faghihi, M.A. Genetic Testing in Various Neurodevelopmental Disorders Which Manifest as Cerebral Palsy: A Case Study From Iran. *Frontiers in Pediatrics* **2021**, *9*, 734946. [CrossRef] [PubMed]
67. Ben-Zeev, B.; Hoffman, C.; Lev, D.; Watemberg, N.; Malinger, G.; Brand, N.; Lerman-Sagie, T. Progressive cerebellocerebral atrophy: A new syndrome with microcephaly, mental retardation, and spastic quadriplegia. *J. Med. Genet.* **2003**, *40*, e96. [CrossRef]
68. Fradejas-Villar, N.; Zhao, W.; Reuter, U.; Doengi, M.; Ingold, I.; Bohleber, S.; Conrad, M.; Schweizer, U. Missense mutation in selenocysteine synthase causes cardio-respiratory failure and perinatal death in mice which can be compensated by selenium-independent GPX4. *Redox Biol.* **2021**, *48*, 102188. [CrossRef]

International Journal of Molecular Sciences

Review

Pathogenic Variants in Selenoproteins and Selenocysteine Biosynthesis Machinery

Didac Santesmasses and Vadim N. Gladyshev *

Division of Genetics, Department of Medicine, Brigham and Women's Hospital, Harvard Medical School, Boston, MA 02115, USA; dsantesmassesruiz@bwh.harvard.edu
* Correspondence: vgladyshev@rics.bwh.harvard.edu

Abstract: Selenium is incorporated into selenoproteins as the 21st amino acid selenocysteine (Sec). There are 25 selenoproteins encoded in the human genome, and their synthesis requires a dedicated machinery. Most selenoproteins are oxidoreductases with important functions in human health. A number of disorders have been associated with deficiency of selenoproteins, caused by mutations in selenoprotein genes or Sec machinery genes. We discuss mutations that are known to cause disease in humans and report their allele frequencies in the general population. The occurrence of protein-truncating variants in the same genes is also presented. We provide an overview of pathogenic variants in selenoproteins genes from a population genomics perspective.

Keywords: selenium; selenoprotein; selenocysteine; genetic variance; human disease

Citation: Santesmasses, D.; Gladyshev, V.N. Pathogenic Variants in Selenoproteins and Selenocysteine Biosynthesis Machinery. *Int. J. Mol. Sci.* **2021**, *22*, 11593. https://doi.org/10.3390/ijms222111593

Academic Editor: Antonella Roveri

Received: 27 September 2021
Accepted: 22 October 2021
Published: 27 October 2021

Publisher's Note: MDPI stays neutral with regard to jurisdictional claims in published maps and institutional affiliations.

Copyright: © 2021 by the authors. Licensee MDPI, Basel, Switzerland. This article is an open access article distributed under the terms and conditions of the Creative Commons Attribution (CC BY) license (https://creativecommons.org/licenses/by/4.0/).

1. Introduction

Selenium is an essential trace element in mammals. Its main biological functions are mediated by selenoproteins, which contain selenium in the form of the 21st amino acid selenocysteine (Sec). Selenoproteins are important oxidoreductase enzymes widely conserved across mammals [1]. They are involved in diverse molecular pathways, with Sec typically found at the catalytic site [2]. Sec is incorporated into selenoproteins in response to a UGA codon, normally a stop codon, through a recoding mechanism that is selenium-dependent and involves a dedicated machinery [3]. There are 25 known selenoprotein genes in humans, and 24 in mice [4]. Mouse models have been particularly instrumental in interrogating the functions of selenoproteins and assessing their gene essentiality [5]. Five selenoproteins have been reported to be essential in mice: Gpx4 [6,7], Txnrd1 [8], Txnrd2 [9], Selenot [10], and Selenoi [11]. In humans, the significance of selenium and selenoproteins for health is manifested by inherited congenital disorders caused by mutations that disrupt selenoprotein synthesis or affect individual selenoproteins.

Here we provide an overview of clinically relevant genetic variants based on the analysis of the literature and specialized databases. We used ClinVar, a public archive of reports of the relationships of genetic variation and phenotypes with assessment of clinical relevance of variants submitted by researchers and genetic testing labs [12], and gnomAD, an aggregate of exome and genome sequencing data of unrelated individuals sequenced as part of various disease-specific and population genetic studies [13].

2. Pathogenic Variants in Selenoproteins

Selenoproteins associated with human syndromes thus far include 5 proteins: SELENON, GPX4, TXNRD1, TXNRD2, and SELENOI. Disruption of the selenoprotein synthesis pathway, which causes generalized selenoprotein deficiency, has been associated with mutations in SEPSECS, SECISBP2, and Sec tRNA (*TRU-TCA1-1*). The consequences of selenoprotein deficiency and causal mutations have been reviewed recently [14–17]. ClinVar currently lists a total of 72 pathogenic or likely pathogenic variants in selenoprotein genes, and 35 in the Sec synthesis machinery factors. 44 of those variants are observed in at least one individual in gnomAD.

2.1. SELENON

Selenoprotein N (SELENON, also known as SEPN1) is a transmembrane protein located in the endoplasmic reticulum (ER) [18]. Initially described by the in silico identification of its SECIS element [19], SELENON was shortly after linked to a congenital rigid spine muscular dystrophy [20], becoming the first selenoprotein associated with human disease. Its protein sequence contains an EF-hand domain and a Sec residue as part of a SCUG motif, where U corresponds to Sec, reminiscent of the Sec-containing motif in thioredoxin reductases. Since its association with disease, there has been a lot of interest in the characterization of its functions in the muscle [21,22]. A recent study showed how SELENON functions as a calcium sensor through its calcium-binding EF-hand domain and activates sarco/endoplasmic reticulum calcium ATPase (SERCA2)-mediated calcium uptake into the ER in a redox-dependent manner [23]. Knockout mice were also developed [24,25].

SELENON-related myopathy (SELENON-RM, formerly SEPN1-RM) is a congenital disorder caused by loss-of-function variants in the SELENON gene. SELENON-RM comprises four neuromuscular disorders initially described separately as rigid spine muscular dystrophy [20,26], multi-minicore disease [27], congenital fiber type disproportion [28], and desmin-related myopathy with Mallory body-like inclusions [29]. The clinical phenotype is characterized by early-onset axial muscle weakness, spinal rigidity, and scoliosis, with respiratory failure. Transmission is autosomal recessive, and patients are homozygous or compound heterozygous. To date, many genetic variants that disrupt SELENON, or prevent Sec insertion, have been identified in SELENO-RM patients [30]. ClinVar currently lists 65 variants in SELENON as pathogenic/likely pathogenic. Notably, those variants include a change in the Sec codon TGA to TAA [20], which produces a premature stop codon (currently erroneously classified as synonymous in ClinVar); a variant that affects the conserved quartet core of the SECIS [31]; and several variants in the Selenocysteine Redefinition Element (SRE), a small RNA hairpin loop adjacent to the UGA codon [20,32]. In those cases where Sec synthesis efficiency is reduced, transcript levels have been shown to be decreased, suggesting that the mRNA is targeted by nonsense-mediated decay [31,32]. The remaining pathogenic variants are missense changes in conserved residues, small insertions and deletions that lead to frameshift, and splice donor/acceptor variants.

The gnomAD database contains 30 of the pathogenic or likely pathogenic variants, which include both missense and protein-truncating variants (PTV: frameshift, stop gain, and splice variants) (Figure 1). Though they can be considered rare, those included in gnomAD are, presumably, the most common SELENON-RM associated variants in the general population. Their allele frequencies range from 0.000004 to 0.0002, and carriers are all heterozygotes. Some of the variants are more frequent in certain populations. For example, the frameshift variant p.Asn204LysfsTer63 (g. 26135244C>CA) had an allele frequency of 0.001353 in the Ashkenazi Jewish population. In addition to those listed in ClinVar, other PTVs are present in gnomAD (Figure 1), which have not been assessed for clinical significance. Given that all but one of the PTVs that have been submitted to ClinVar are pathogenic (19 in total), it raises the question whether the additional 23 truncating variants in gnomAD could potentially cause SELENON-RM.

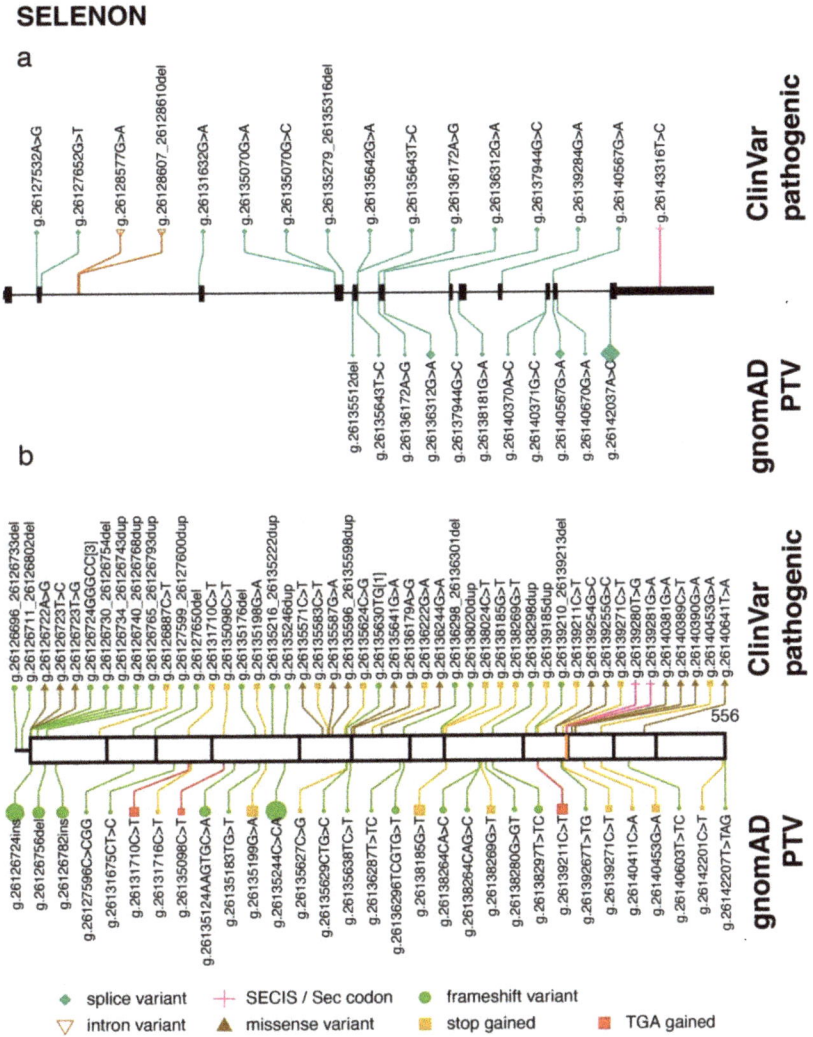

Figure 1. Pathogenic variants and protein-truncating variants in *SELENON*. (**a**) Genomic organization of *SELENON* exons, and location of ClinVar pathogenic variants (above) and gnomAD protein-truncating variants (PTV) (below). The shape and color of each variant corresponds to its predicted consequence (see legend). For gnomAD variants only, the size of the symbol is proportional to the allele frequency. The genomic notation is used to describe each variant. (**b**) Location of variants along the SELENON protein sequence. The length of the protein is indicated on the right. Vertical black lines correspond to exon boundaries, and the vertical orange line correspond to the Sec residue. The same aesthetics for variants as (**a**) are used. The genomic coordinates correspond to genome build GRCh37/hg19; the *SELENON* gene structure and protein sequence correspond to transcript ENST00000374315.

2.2. GPX4

GPX4 is unique among the five glutathione peroxidases that depend on selenium in humans due to its ability to reduce lipid hydroperoxides and to use protein thiols as donors of electrons in addition to glutathione [33,34]. GPX4 also functions as a major regulator of ferroptosis, a form of regulated cell death characterized by the accumulation of lipid hydroperoxides [35,36]. Gpx4 is essential for embryonic mouse development [6,7], and its deficiency in mice leads to neuronal degeneration, ataxia, and seizures [37].

GPX4 is associated with Sedaghatian-type spondylometaphyseal dysplasia (SSDM). The syndrome was described in 1980 as a congenital autosomal recessive disorder [38], and more recently, inactivating variants in GPX4 were identified in SSDM patients [39]. Patients show skeletal disorder and brain atrophy and die shortly after birth due to respiratory failure [40].

There are five pathogenic or likely pathogenic variants in GPX4 associated with SSDM in ClinVar (Figure 2). Three of them were reported in [39], while for the other two (p.Gly51_His52insTer and p.Ile170fs) there is no study citation in ClinVar. Only one of the pathogenic variants is observed in gnomAD, c.476 + 5G>A (g. 1105813G>A), with allele frequency 0.00003586. Additional protein-truncating variants in GPX4 are observed in gnomAD, which are not present in ClinVar. Their allele frequencies range from 0.000004 to 0.000065.

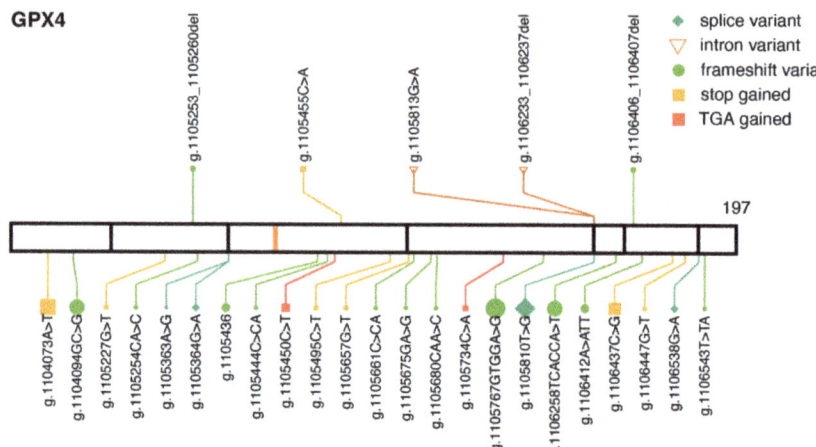

Figure 2. Pathogenic variants and protein-truncating variants in *GPX4*. Location of ClinVar pathogenic variants (above) and gnomAD protein-truncating variants (below), along the GPX4 protein sequence. The same aesthetics as in Figure 1b are used. GPX4 transcript: ENST00000354171.

2.3. Thioredoxin Reductases

Thioredoxin reductases (TXNRD) are oxidoreductases that play a major role in the disulfide reduction system of the cell by converting thioredoxins to their reduced state. There are three TXNRDs in mammals, with different cellular or tissue localization. TXNRD1 is localized mainly in the cytosol and nucleus throughout different tissues, TXNRD2 is localized in the mitochondria also throughout tissues, and TXNRD3 is highly expressed in the testis [2]. All three proteins have the carboxy-terminal motif GCUG that contains the Sec residue [1]. Both Txnrd1 and Txnrd2 are essential for embryonic development in mice [8,9]. TXNRD2 has been associated with two different, unrelated diseases: dilated cardiomyopathy [41] and familial glucocorticoid deficiency [42]. TXNRD1 has been associated with generalized epilepsy [43].

Two variants in TXNRD2, p.Ala59Thr and p.Gly375Arg, were identified in patients that suffered from dilated cardiomyopathy [41]. The clinical significance of p.Gly375Arg is currently uncertain in ClinVar, although it is predicted to be deleterious by the algorithms Polyphen and SIFT, used in gnomAD. The missense variant p.Gly375Arg (c.1123G>A, allele frequency (AF) = 0.000016; and c.1123G>C, AF = 0.00003) is observed in five heterozygous subjects in gnomAD. Only one of them is part of the control set, therefore it is not possible to assume that they are all healthy. A change to glutamic acid in this position is also present in gnomAD, p.Gly375Glu (AF = 0.000004). The other variant associated with dilated cardiomyopathy, p.Ala59Thr, is not present in ClinVar, but is also observed in gnomAD (AF = 0.000004). In this position, a change to proline is also observed, p.Ala59Pro, with an allele frequency of 0.00001. All missense variants in position 59 are heterozygous.

The variant p.Y447Ter (g.19865895A>C) in TXNRD2 introduces a premature UAG stop codon leading to a truncated protein that lacks the Sec residue. The homozygous form of this variant was identified in several members of a consanguineous Kashmiri family affected with familial glucocorticoid deficiency [42]. The absence of cardiomyopathy in the family was surprising because it implied clinical heterogeneity associated with TXNRD2. The variant was also observed in a heterozygous patient with dilated cardiomyopathy [44]. The global allele frequency in gnomAD is 0.0005 (Figure 3), observed in 141 heterozygotes, but the variant appears to be much more common in the South Asian population, with an allele frequency of 0.0035.

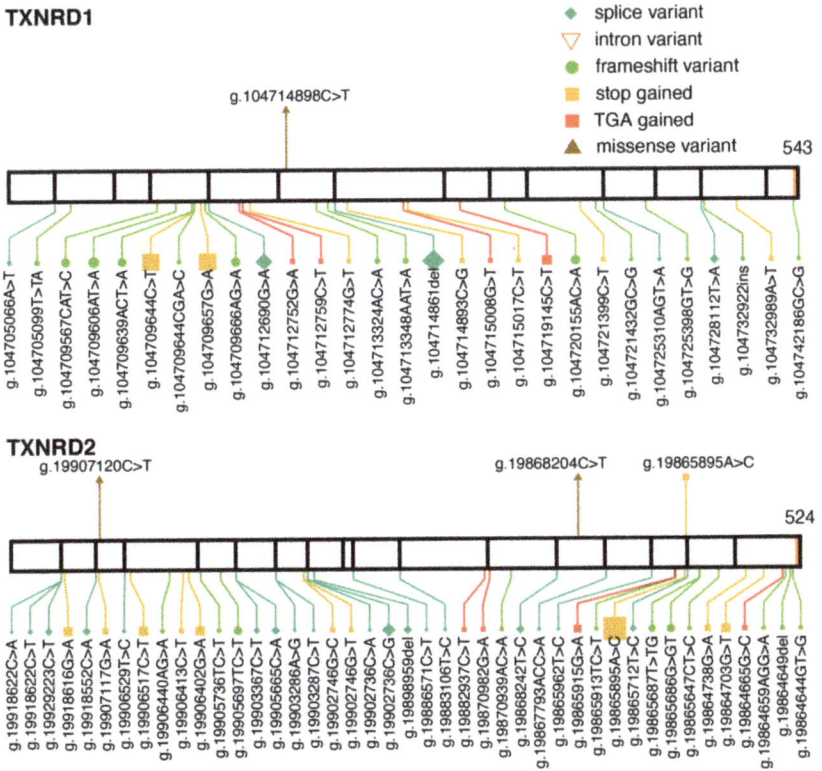

Figure 3. Pathogenic variants and protein-truncating variants in TXNRD1 and TXNRD2. Location of pathogenic variants (above) and gnomAD protein-truncating variants (below) along the TXNRD1 and TXNRD2 protein sequence. The same aesthetics as Figure 1b are used. TXNRD1 transcript: ENST00000526390; TXNRD2 transcript: ENST00000400521.

TXNRD1 has been associated with genetic generalized epilepsy. The homozygous variant p.Pro190Leu (g.104714898C>T) was shown to cause decreased TXNRD1 protein levels and turnover rate, and segregated with the disease in a family [43]. The variant is found in one heterozygous subject in gnomAD (AF = 0.000004).

2.4. SELENOI

Selenoprotein I (SELENOI; also known as EPT1, Ethanolamine phosphotransferase 1) has been recently added to the list of essential selenoproteins in mice [11]. In humans, two pathogenic variants are currently known. They cause spastic paraplegia, described in two different families [45,46]. The missense variant p.Arg112Pro (g.26596259G>C) [45] hits a highly conserved arginine residue within the CDP-alcohol phosphatidyltransferase (Pfam PF01066). The splice variant g.26607825A>G leads to aberrantly spliced transcripts with exons 6 and 8 affected [46]. The mode of inheritance is autosomal recessive, all patients were homozygous, and unaffected direct relatives were heterozygous. Those two variants are not present in gnomAD (Figure 4), and presumably, are very rare in the population. In concordance with the importance of SELENOI, a strong selection against protein-truncating variants was observed in humans [47].

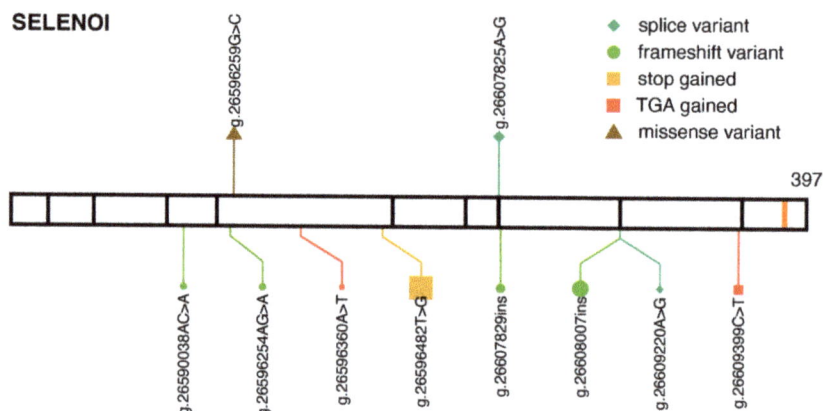

Figure 4. Pathogenic variants and protein-truncating variants in *SELENOI*. Location of ClinVar pathogenic variants (above) and gnomAD protein-truncating variants (below), along the SELENOI protein sequence. The same aesthetics as Figure 1b are used. *SELENOI* transcript: ENST00000260585.

3. Pathogenic Variants Disrupting Selenoprotein Synthesis
3.1. TRU-TCA1-1

The Sec-specific tRNA (tRNA[Ser]Sec), encoded by *TRU-TCA1-1*, plays a central role in the synthesis of selenoproteins. The Sec tRNA provides the backbone for the biosynthesis of Sec [48], and its anticodon recognizes context-dependent UGA codons to specify Sec insertion [49]. Its deletion in mice (encoded by *Trsp*) is embryonic lethal [50]. Several mouse models have been developed to study its role in health [51]. The Sec tRNA has unique features that distinguish it from other tRNAs: it is the longest tRNA with 90 nucleotides compared to ~75 in other tRNAs; it has a unique structure, with a long acceptor 9-base pairs (bp) stem, a long 6-bp D stem, and an unusually long variable arm [52]. It is transcribed by RNA Pol III, like other tRNAs, but has unique promoter elements [53]. The mature tRNA contains a few modified bases [49], and the tRNA pool is composed of two major isoforms, containing either 5-methoxycarbonylmethyluridine (mcm^{5}U) or 5-methoxycarbonylmethyl-2'-O-methyluridine (mcm^{5}Um) at position 34 [54]. The presence of mcm^{5}Um governs the expression of stress-related selenoproteins [55].

The single nucleotide change C65G was identified in patients [56,57]. The first patient described [56] exhibited a similar clinical phenotype to that observed in SECISBP2 mutation patients. Primary cells from the proband showed low Sec tRNA[Ser[Sec]] expression with a reduction in mcm^{5}Um levels and decreased i6A modification in position 37. This suggests that the tRNA post-transcriptional maturation was impaired. The selenoprotein expression profile showed a deficiency of stress-related selenoprotein levels, but preserved the levels of housekeeping selenoproteins. The position C65 is located in the acceptor arm, adjacent to C64 in the TΨC arm, which interacts with SEPSECS [58]. The precise mechanism leading to the imbalance of Sec tRNA$^{\text{Ser[Sec]}}$ isoforms is unclear, with impaired post-transcriptional maturation or unstable interaction with SEPSECS being possibilities [16]. The variant C65G was not observed in gnomAD, but in that same position, the variant C65T was observed in two subjects, both heterozygotes. In total, 83 variants in 55 sites of *TRU-TCA1-1* are currently present in gnomAD, with allele frequencies ranging from 0.000007 to 0.0008. All variants are heterozygous, except for one: the change C28T, in the anticodon arm is observed in two homozygous subjects. Given that many heterozygous variants exist in the general population, it would be reasonable to assume that a single wild type *TRU-TCA1-1* allele is enough to maintain adequate expression of selenoproteins, as observed in heterozygotes for the variant C65G [56]. Remarkably, two subjects have a variant in position 35 of the anticodon triplet. One variant is C35T, which produces a TTA anticodon that is complementary to the stop codon UAA. The other one is C35G, producing a TGA anticodon, which is a Ser anticodon. Both subjects are heterozygous, but the change in the anticodon could have consequences not only on the synthesis of selenoproteins, but potentially on the entire proteome.

3.2. SECISBP2

The SECIS binding protein 2 (SECISBP2) binds the SECIS element in the 3′UTR of selenoprotein mRNAs, and interacts with EEFSEC, tRNA$^{\text{[Ser]Sec}}$, and the ribosome, to incorporate Sec into the growing peptide. It is an obligate limiting factor for selenoprotein synthesis [59], and it is an essential gene in mice [60]. Its deficiency disrupts the synthesis of selenoproteins, which, in humans, is manifested by a multisystem disorder characterized by low circulating selenium and abnormal thyroid hormone levels [16]. A total of 18 pathogenic variants have been identified in 13 individuals from 11 families [16]. Three of the subjects are homozygous, and the rest are compound heterozygous. Most variants produce a truncated protein, either by stop gain or frameshift, and three are missense variants.

The first 400 N-terminal amino acids of the human SECISBP2 protein are dispensable for Sec incorporation [59,61]. The C-terminal region (positions 399 to 784) comprises two domains, the Sec incorporation domain (SID), and the RNA binding domain (RBD) [62]. The RBD domain contains an L7Ae RNA-binding domain that interacts with the SECIS and the 28S ribosomal RNA [63]. Mouse models carrying human mutations have been developed to study the effect of specific pathogenic variants that affect either the RBD or SID regions [64]. The results showed that the variant in the RBD domain abrogates Sec insertion, while in the SID domain, the particular variant tested results in residual SECISBP2 activity in the brain, but it rendered the protein unstable in a tissue-specific manner, being completely degraded in mouse liver.

ClinVar lists seven variants classified as pathogenic/likely pathogenic (Figure 5). Four of them are observed in gnomAD, all with an allele frequency below 0.00003. In addition, 93 protein-truncating variants are observed in gnomAD, which are not listed in ClinVar. They are all heterozygous and their allele frequencies are below 0.00009.

A paralog of SECISBP2, named SECISBP2L, is found in vertebrates [65]. Its function has not been elucidated, but it was shown to lack Sec incorporation activity [66]. Based on gnomAD, SECISBP2L has strong selective constraints against truncating variants [47]. Mice lacking Secisbp2l have been phenotyped and deposited on the International Mouse Phenotyping Consortium (IMPC). Deletion of Secisbp2l has, apparently, no effect on viability, but it shows significant phenotypes in body size, metabolism/adipose tissue, cardiovascular system, skeleton, hearing, and vision (https://www.mousephenotype.org/data/genes/MGI:1917604, accessed on 17 October 2021). However, deletion of Secisbp2 is also reported to have no effect on viability (https://www.mousephenotype.org/data/genes/MGI:1922670, accessed on 17 October 2021), which is in contradiction with the previous body of work that reported Secisbp2 as essential [60].

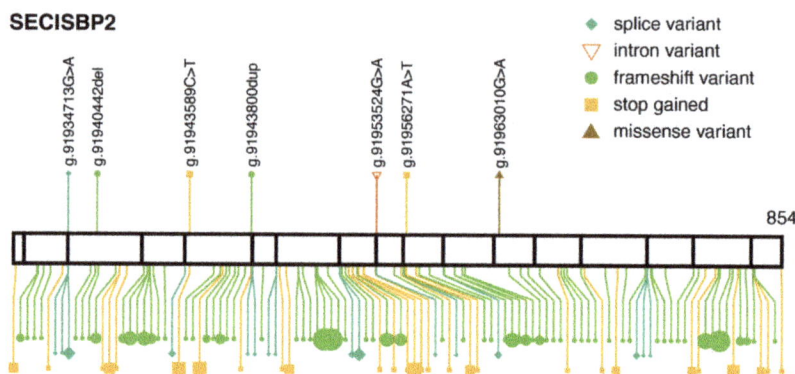

Figure 5. Pathogenic variants and protein-truncating variants in *SECISBP2*. Location of ClinVar pathogenic variants (above) and gnomAD protein-truncating variants (below), along the SECISBP2 protein sequence. The same aesthetics as Figure 1b are used. *SECISBP2* transcript: ENST00000375807.

3.3. SEPSECS

SEPSECS catalyzed the conversion of Ser-tRNA$^{[Ser]Sec}$ to Sec-tRNA$^{[Ser]Sec}$, the final step in the biosynthesis of selenocysteine [67]. The crystal structure of human SEPSECS in complex with tRNA$^{[Ser]Sec}$ has been solved [58], providing substantial information on its function. The interaction of SEPSECS and tRNA$^{[Ser]Sec}$ occurs through the tRNA long acceptor-TΨC arms. There is no published study on mice *Sepsecs* knockout, but its deletion is reported as embryonic lethal, with significant phenotypes in heterozygotes including cardiovascular, pigmentation, and vision (https://www.mousephenotype.org/data/genes/MGI:1098791, accessed on 17 October 2021).

Deficiency of SEPSECS in humans causes pontocerebellar hypoplasia 2D (PCH2D), a neurological condition characterized by neurodegeneration and epilepsy. Multiple families from diverse ethnic backgrounds carry homozygous or compound heterozygous mutations that show similar clinical characteristics [16,68].

ClinVar currently lists 27 pathogenic or likely pathogenic variants in SEPSECS (Figure 6). Most are protein-truncating variants, and three of them are missense variants. Eleven of those are present in gnomAD and their global allele frequencies are below 0.00004. There are many observed protein-truncating variants observed in gnomAD, which are not present in ClinVar, and have no clinical significance category. The most common is p.Tyr429Ter (g.25125772G>T), which has been observed in three unrelated patients in Finland [69]. All three were compound heterozygous for the same two variants, p.Tyr429Ter and p.Thr325Ser. p.Tyr429Ter is reportedly enriched in the population in Finland [69]. In gnomAD, both variants are observed exclusively within the Finnish population, where the allele frequencies are 0.002791 for p.Tyr429Ter, and 0.0002792 for p.Thr325Ser. The second most common truncating SEPSECS variant in gnomAD is g.25155152C>T, which disrupts the canonical splice acceptor site in intron 4. Interestingly, this variant appears to be also enriched in the Finnish population, with allele frequency 0.002192. Among the 58 subjects in gnomAD, only three are non-Finnish. ClinVar lists its clinical significance as uncertain as it has not been reported as pathogenic or benign. These observations raise the question whether the Finnish population is enriched with pathogenic variants in SEPSECS.

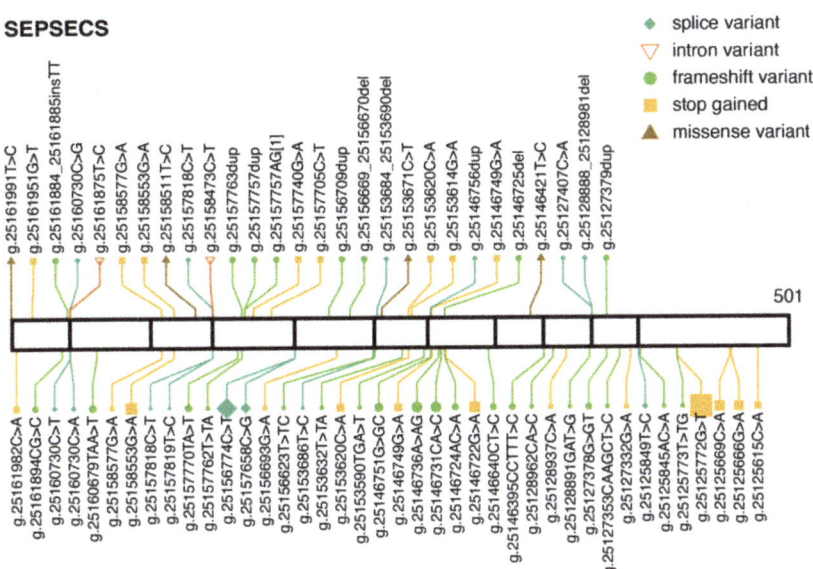

Figure 6. Pathogenic variants and protein-truncating variants in *SEPSECS*. Location of ClinVar pathogenic variants (above) and gnomAD protein-truncating variants (below) along the SEPSECS protein sequence. The same aesthetics as Figure 1b are used. *SEPSECS* transcript: ENST00000382103.

4. Common Genetic Variance

Genetic variation is often shared among many individuals in a population. This common variation reflects coinheritance of haplotypes. Functional annotation of haplotypes (groups of single nucleotide variants) is needed to understand hereditary factors linked to complex disease. Genome-wide association studies (GWAS) are designed to find associations between common genetic variants and particular traits.

The NHGRI-EBI GWAS catalog (https://www.ebi.ac.uk/gwas, accessed on 17 October 2021) is a repository of genetic variant-trait associations from published studies [70]. The catalog currently lists 24 associations involving variants in selenoprotein genes. Most of those variants fall in introns and it is not clear what impact in protein function they might produce, or even whether they are the causal variants. Three of them are missense and produce an amino acid change in three selenoproteins. The variant rs225014 (p.Thr92Ala) in DIO2 has been shown to be involved in insulin resistance [71], thyroid functionality [72,73] and the onset and progression of osteoarthritis [74] in humans. DIO2 activates the prohormone thyroxine (T4) into the active thyroid hormone 3,3′,5-triiodothyronine (T3) [2]. The variant p.Thr92Ala is particularly common, with an allele frequency of 0.41. The variant rs5771225 (p.Val3Ala) in SELENOO is associated with late onset Alzheimer's disease [75] and has an allele frequency of 0.22. SELENOO is a pseudokinase that transfers AMP from ATP to Ser, Thr, and Tyr residues (AMPylation) [76]. Lastly, the variant rs1050450 (p.Pro200Leu) in GPX1, with an allele frequency of 0.28, is associated with decreased hemoglobin levels. Its clinical significance is currently classified as benign according to ClinVar.

Four GWAS on circulating selenium concentrations have been published in recent years [77–80]. Perhaps not surprisingly, individual variants in selenoproteins did not reach genome-wide significance levels. The authors discuss that though the selenium measurements used (circulating and toenail selenium) are an accepted selenium level biomarker, the selenium concentration might not reflect its functional significance in selenium-sufficient populations [79]. Nonetheless, a locus overlapping genes involved in metabolism of sulfur-containing amino acids reached significance level in two independent cohorts [77,79].

The region overlaps with genes dimethylglycine dehydrogenase (DMGDH) and betaine-homocysteine S-methyltransferase (BHMT), both involved in conversion of homocysteine to methionine. A prospective GWAS also found strong association with greater increase of circulating selenium after supplementation on the same locus in chromosome 5 [80]. These studies revealed a link between selenium exposure and homocysteine metabolism.

The main source of selenium and other essential trace elements in humans is through the diet. Studies on bioavailability in plants and animals suggest that selenium levels vary widely across world regions, with several countries containing relatively low selenium, notably in some parts of China [81,82]. To explore how human populations have adapted to varying levels of dietary selenium levels during history, the signatures of positive selection were assessed by Castellano and collaborators through a survey using genetic polymorphisms in selenoproteins, Cys-containing homologs, and Sec synthesis machinery factors, in 50 human populations [83]. The strongest signals were observed in populations from China. The genes with largest contributions included selenoproteins DIO2, SELENOS, GPX1, SELENOM and SELENOF, and the Sec machinery factors SEPHS2 and SEPSECS. Several single nucleotide polymorphism with known functional consequences showed high levels of population differentiation, including the missense substitution p.Thr92Ala in DIO2 [84], and in GPX1, the missense variant p.Pro200Leu, and a noncoding change A to G (rs3811699) in its promoter region. Positive selection in GPX1 in human populations has been reported also in other studies [85,86].

5. Conclusions

Selenocysteine is a genetic trait shared by bacteria, archaea, and eukaryotes. Its origin maps at the root of the tree of life. Throughout evolution, some organisms lost the ability to synthesize selenoproteins and replaced selenocysteine in essential proteins with cysteine, which contains sulfur instead of selenium. Non-selenoprotein cysteine-containing homologs exist for almost all known selenoproteins; they are less costly for the cell to synthesize, but are preferred when it comes to catalysis. It is not fully understood what the advantage of Sec over Cys is. Nevertheless, the presence of 25 selenoproteins in the human genome should convey the requirement for the unique properties of Sec that cannot be compensated by the use of Cys.

Elucidation of the molecular biology behind the insertion of Sec into proteins in response to UGA, paved the way for the identification of a genetic association between selenoproteins and human disease twenty years ago. Since then, several disorders caused by selenoprotein deficiency have been described. Their clinical phenotypes are diverse and different systems are affected, which is not surprising, given that selenoproteins carry out diverse functions and are expressed in different tissues. Moreover, global deficiency of selenoproteins caused by defects in the synthesis machinery result in more complex phenotypes. More surprising, however, is the fact that there are the differences observed between patients with defects in different components of the Sec machinery, SECISBP2 and *TRU-TCA1-1* affect mainly thyroid function, muscle, and growth, while SEPSECS affects the brain and causes a more severe phenotype.

A subset of the observed stop gain variants introduce a UGA codon. This is a special type of genetic variation when it occurs in selenoprotein genes because selenoprotein mRNAs carrying additional UGA codons may produce proteins with extra Sec residues. Sec insertion can occur in multiple sites [87,88], which may be partially functional or have a negative gain of function. mRNAs with early termination codons can be targeted by nonsense-mediated decay (NMD) to prevent its translation. But we previously showed that, in case of selenoproteins, NMD is less efficient if the stop gained is UGA, compared to the other two stop codons [47], which may increase insertion of extra Sec residues. TGA gain is observed in multiple sites in all selenoproteins included in this review and are indicated in the corresponding figures.

The advent of genome sequencing, particularly exome sequencing for clinical diagnosis, has led to the identification of dozens of variants that cause disease through deficiency of selenoproteins, either by inactivating individual selenoprotein genes or by disrupting the selenoprotein synthesis pathway. Initiatives like ClinVar and gnomAD have become instrumental for researchers and clinicians by giving access to a vast amount of genomic data, and will help accelerate our understanding of the associations between genetic variation and health. Undoubtedly, more pathogenic variants in selenoproteins will be discovered, which will provide insights into the function of selenoproteins and opportunities to better understand the role of selenium in human health and disease.

Author Contributions: Conceptualization, D.S. and V.N.G.; writing—original draft preparation, D.S.; writing—review and editing, V.N.G.; funding acquisition, V.N.G. All authors have read and agreed to the published version of the manuscript.

Funding: This research was funded by the National Institutes of Health grants to V.N.G.

Institutional Review Board Statement: Not applicable.

Informed Consent Statement: Not applicable.

Conflicts of Interest: The authors declare no conflict of interest.

References

1. Mariotti, M.; Ridge, P.G.; Zhang, Y.; Lobanov, A.V.; Pringle, T.H.; Guigo, R.; Hatfield, D.L.; Gladyshev, V.N. Composition and Evolution of the Vertebrate and Mammalian Selenoproteomes. *PLoS ONE* **2012**, *7*, e33066. [CrossRef]
2. Labunskyy, V.M.; Hatfield, D.L.; Gladyshev, V.N. Selenoproteins: Molecular Pathways and Physiological Roles. *Physiol. Rev.* **2014**, *94*, 739–777. [CrossRef] [PubMed]
3. Allmang, C.; Wurth, L.; Krol, A. The Selenium to Selenoprotein Pathway in Eukaryotes: More Molecular Partners than Anticipated. *Biochim. Biophys. Acta* **2009**, *1790*, 1415–1423. [CrossRef]
4. Kryukov, G.V.; Castellano, S.; Novoselov, S.V.; Lobanov, A.V.; Zehtab, O.; Guigó, R.; Gladyshev, V.N. Characterization of Mammalian Selenoproteomes. *Science* **2003**, *300*, 1439–1443. [CrossRef]
5. Conrad, M.; Schweizer, U. Mouse Models That Target Individual Selenoproteins. In *Selenium*; Hatfield, D.L., Schweizer, U., Tsuji, P.A., Gladyshev, V.N., Eds.; Springer International Publishing: Cham, Switzerland, 2016; pp. 567–578. [CrossRef]
6. Yant, L.J.; Ran, Q.; Rao, L.; Van Remmen, H.; Shibatani, T.; Belter, J.G.; Motta, L.; Richardson, A.; Prolla, T.A. The Selenoprotein GPX4 Is Essential for Mouse Development and Protects from Radiation and Oxidative Damage Insults. *Free Radic. Biol. Med.* **2003**, *34*, 496–502. [CrossRef]
7. Imai, H.; Hirao, F.; Sakamoto, T.; Sekine, K.; Mizukura, Y.; Saito, M.; Kitamoto, T.; Hayasaka, M.; Hanaoka, K.; Nakagawa, Y. Early Embryonic Lethality Caused by Targeted Disruption of the Mouse PHGPx Gene. *Biochem. Biophys. Res. Commun.* **2003**, *305*, 278–286. [CrossRef]
8. Jakupoglu, C.; Przemeck, G.K.H.; Schneider, M.; Moreno, S.G.; Mayr, N.; Hatzopoulos, A.K.; de Angelis, M.H.; Wurst, W.; Bornkamm, G.W.; Brielmeier, M.; et al. Cytoplasmic Thioredoxin Reductase Is Essential for Embryogenesis but Dispensable for Cardiac Development. *Mol. Cell. Biol.* **2005**, *25*, 1980–1988. [CrossRef] [PubMed]
9. Conrad, M.; Jakupoglu, C.; Moreno, S.G.; Lippl, S.; Banjac, A.; Schneider, M.; Beck, H.; Hatzopoulos, A.K.; Just, U.; Sinowatz, F.; et al. Essential Role for Mitochondrial Thioredoxin Reductase in Hematopoiesis, Heart Development, and Heart Function. *Mol. Cell. Biol.* **2004**, *24*, 9414–9423. [CrossRef] [PubMed]
10. Boukhzar, L.; Hamieh, A.; Cartier, D.; Tanguy, Y.; Alsharif, I.; Castex, M.; Arabo, A.; El Hajji, S.; Bonnet, J.-J.; Errami, M.; et al. Selenoprotein T Exerts an Essential Oxidoreductase Activity That Protects Dopaminergic Neurons in Mouse Models of Parkinson's Disease. *Antioxid. Redox Signal.* **2016**, *24*, 557–574. [CrossRef]
11. Avery, J.C.; Yamazaki, Y.; Hoffmann, F.W.; Folgelgren, B.; Hoffmann, P.R. Selenoprotein I Is Essential for Murine Embryogenesis. *Arch. Biochem. Biophys.* **2020**, *689*, 108444. [CrossRef]
12. Landrum, M.J.; Chitipiralla, S.; Brown, G.R.; Chen, C.; Gu, B.; Hart, J.; Hoffman, D.; Jang, W.; Kaur, K.; Liu, C.; et al. ClinVar: Improvements to Accessing Data. *Nucleic Acids Res.* **2020**, *48*, D835–D844. [CrossRef] [PubMed]
13. Karczewski, K.J.; Francioli, L.C.; Tiao, G.; Cummings, B.B.; Alföldi, J.; Wang, Q.; Collins, R.L.; Laricchia, K.M.; Ganna, A.; Birnbaum, D.P.; et al. The Mutational Constraint Spectrum Quantified from Variation in 141,456 Humans. *Nature* **2020**, *581*, 434–443. [CrossRef]
14. Schweizer, U.; Fradejas-Villar, N. Why 21? The Significance of Selenoproteins for Human Health Revealed by Inborn Errors of Metabolism. *FASEB J.* **2016**, *30*, 3669–3681. [CrossRef] [PubMed]
15. Fradejas-Villar, N. Consequences of Mutations and Inborn Errors of Selenoprotein Biosynthesis and Functions. *Free Radic. Biol. Med.* **2018**, *127*, 206–214. [CrossRef]
16. Schoenmakers, E.; Chatterjee, K. Human Disorders Affecting the Selenocysteine Incorporation Pathway Cause Systemic Selenoprotein Deficiency. *Antioxid. Redox Signal.* **2020**, *33*, 481–497. [CrossRef]

17. Schweizer, U.; Bohleber, S.; Zhao, W.; Fradejas-Villar, N. The Neurobiology of Selenium: Looking Back and to the Future. *Front. Neurosci.* **2021**, *15*, 652099. [CrossRef]
18. Castets, P.; Lescure, A.; Guicheney, P.; Allamand, V. Selenoprotein N in Skeletal Muscle: From Diseases to Function. *J. Mol. Med.* **2012**, *90*, 1095–1107. [CrossRef] [PubMed]
19. Lescure, A.; Gautheret, D.; Carbon, P.; Krol, A. Novel Selenoproteins Identified in Silico and in Vivo by Using a Conserved RNA Structural Motif. *J. Biol. Chem.* **1999**, *274*, 38147–38154. [CrossRef]
20. Moghadaszadeh, B.; Petit, N.; Jaillard, C.; Brockington, M.; Quijano Roy, S.; Merlini, L.; Romero, N.; Estournet, B.; Desguerre, I.; Chaigne, D.; et al. Mutations in SEPN1 Cause Congenital Muscular Dystrophy with Spinal Rigidity and Restrictive Respiratory Syndrome. *Nat. Genet.* **2001**, *29*, 17–18. [CrossRef]
21. Jurynec, M.J.; Xia, R.; Mackrill, J.J.; Gunther, D.; Crawford, T.; Flanigan, K.M.; Abramson, J.J.; Howard, M.T.; Grunwald, D.J. Selenoprotein N Is Required for Ryanodine Receptor Calcium Release Channel Activity in Human and Zebrafish Muscle. *Proc. Natl. Acad. Sci. USA* **2008**, *105*, 12485–12490. [CrossRef] [PubMed]
22. Petit, N.; Lescure, A.; Rederstorff, M.; Krol, A.; Moghadaszadeh, B.; Wewer, U.M.; Guicheney, P. Selenoprotein N: An Endoplasmic Reticulum Glycoprotein with an Early Developmental Expression Pattern. *Hum. Mol. Genet.* **2003**, *12*, 1045–1053. [CrossRef]
23. Chernorudskiy, A.; Varone, E.; Colombo, S.F.; Fumagalli, S.; Cagnotto, A.; Cattaneo, A.; Briens, M.; Baltzinger, M.; Kuhn, L.; Bachi, A.; et al. Selenoprotein N Is an Endoplasmic Reticulum Calcium Sensor That Links Luminal Calcium Levels to a Redox Activity. *Proc. Natl. Acad. Sci. USA* **2020**, *117*, 21288–21298. [CrossRef]
24. Rederstorff, M.; Castets, P.; Arbogast, S.; Lainé, J.; Vassilopoulos, S.; Beuvin, M.; Dubourg, O.; Vignaud, A.; Ferry, A.; Krol, A.; et al. Increased Muscle Stress-Sensitivity Induced by Selenoprotein N Inactivation in Mouse: A Mammalian Model for SEPN1-Related Myopathy. *PLoS ONE* **2011**, *6*, e23094. [CrossRef] [PubMed]
25. Castets, P.; Bertrand, A.T.; Beuvin, M.; Ferry, A.; Le Grand, F.; Castets, M.; Chazot, G.; Rederstorff, M.; Krol, A.; Lescure, A.; et al. Satellite Cell Loss and Impaired Muscle Regeneration in Selenoprotein N Deficiency. *Hum. Mol. Genet.* **2011**, *20*, 694–704. [CrossRef] [PubMed]
26. Moghadaszadeh, B.; Desguerre, I.; Topaloglu, H.; Muntoni, F.; Pavek, S.; Sewry, C.; Mayer, M.; Fardeau, M.; Tomé, F.M.; Guicheney, P. Identification of a New Locus for a Peculiar Form of Congenital Muscular Dystrophy with Early Rigidity of the Spine, on Chromosome 1p35-36. *Am. J. Hum. Genet.* **1998**, *62*, 1439–1445. [CrossRef] [PubMed]
27. Ferreiro, A.; Quijano-Roy, S.; Pichereau, C.; Moghadaszadeh, B.; Goemans, N.; Bönnemann, C.; Jungbluth, H.; Straub, V.; Villanova, M.; Leroy, J.-P.; et al. Mutations of the Selenoprotein N Gene, Which Is Implicated in Rigid Spine Muscular Dystrophy, Cause the Classical Phenotype of Multiminicore Disease: Reassessing the Nosology of Early-Onset Myopathies. *Am. J. Hum. Genet.* **2002**, *71*, 739–749. [CrossRef]
28. Clarke, N.F.; Kidson, W.; Quijano-Roy, S.; Estournet, B.; Ferreiro, A.; Guicheney, P.; Manson, J.I.; Kornberg, A.J.; Shield, L.K.; North, K.N. SEPN1: Associated with Congenital Fiber-Type Disproportion and Insulin Resistance. *Ann. Neurol.* **2006**, *59*, 546–552. [CrossRef]
29. Ferreiro, A.; Ceuterick-de Groote, C.; Marks, J.J.; Goemans, N.; Schreiber, G.; Hanefeld, F.; Fardeau, M.; Martin, J.-J.; Goebel, H.H.; Richard, P.; et al. Desmin-Related Myopathy with Mallory Body-like Inclusions Is Caused by Mutations of the Selenoprotein, N. Gene. *Ann. Neurol.* **2004**, *55*, 676–686. [CrossRef]
30. Villar-Quiles, R.N.; von der Hagen, M.; Métay, C.; Gonzalez, V.; Donkervoort, S.; Bertini, E.; Castiglioni, C.; Chaigne, D.; Colomer, J.; Cuadrado, M.L.; et al. The Clinical, Histologic, and Genotypic Spectrum of SEPN1-Related Myopathy: A Case Series. *Neurology* **2020**, *95*, e1512–e1527. [CrossRef]
31. Allamand, V.; Richard, P.; Lescure, A.; Ledeuil, C.; Desjardin, D.; Petit, N.; Gartioux, C.; Ferreiro, A.; Krol, A.; Pellegrini, N.; et al. A Single Homozygous Point Mutation in a 3′untranslated Region Motif of Selenoprotein N MRNA Causes SEPN1-Related Myopathy. *EMBO Rep.* **2006**, *7*, 450–454. [CrossRef]
32. Maiti, B.; Arbogast, S.; Allamand, V.; Moyle, M.W.; Anderson, C.B.; Richard, P.; Guicheney, P.; Ferreiro, A.; Flanigan, K.M.; Howard, M.T. A Mutation in the SEPN1 Selenocysteine Redefinition Element (SRE) Reduces Selenocysteine Incorporation and Leads to SEPN1-Related Myopathy. *Hum. Mutat.* **2009**, *30*, 411–416. [CrossRef]
33. Ursini, F.; Maiorino, M.; Brigelius-Flohé, R.; Aumann, K.D.; Roveri, A.; Schomburg, D.; Flohé, L. Diversity of Glutathione Peroxidases. *Methods Enzymol.* **1995**, *252*, 38–53. [CrossRef]
34. Ursini, F.; Heim, S.; Kiess, M.; Maiorino, M.; Roveri, A.; Wissing, J.; Flohé, L. Dual Function of the Selenoprotein PHGPx during Sperm Maturation. *Science* **1999**, *285*, 1393–1396. [CrossRef] [PubMed]
35. Ingold, I.; Berndt, C.; Schmitt, S.; Doll, S.; Poschmann, G.; Buday, K.; Roveri, A.; Peng, X.; Porto Freitas, F.; Seibt, T.; et al. Selenium Utilization by GPX4 Is Required to Prevent Hydroperoxide-Induced Ferroptosis. *Cell* **2018**, *172*, 409–422.e21. [CrossRef] [PubMed]
36. Stockwell, B.R.; Friedmann Angeli, J.P.; Bayir, H.; Bush, A.I.; Conrad, M.; Dixon, S.J.; Fulda, S.; Gascón, S.; Hatzios, S.K.; Kagan, V.E.; et al. Ferroptosis: A Regulated Cell Death Nexus Linking Metabolism, Redox Biology, and Disease. *Cell* **2017**, *171*, 273–285. [CrossRef]
37. Hambright, W.S.; Fonseca, R.S.; Chen, L.; Na, R.; Ran, Q. Ablation of Ferroptosis Regulator Glutathione Peroxidase 4 in Forebrain Neurons Promotes Cognitive Impairment and Neurodegeneration. *Redox Biol.* **2017**, *12*, 8–17. [CrossRef]
38. Sedaghatian, M.R. Congenital Lethal Metaphyseal Chondrodysplasia: A Newly Recognized Complex Autosomal Recessive Disorder. *Am. J. Med. Genet.* **1980**, *6*, 269–274. [CrossRef]

39. Smith, A.C.; Mears, A.J.; Bunker, R.; Ahmed, A.; MacKenzie, M.; Schwartzentruber, J.A.; Beaulieu, C.L.; Ferretti, E.; FORGE Canada Consortium; Majewski, J.; et al. Mutations in the Enzyme Glutathione Peroxidase 4 Cause Sedaghatian-Type Spondylometaphyseal Dysplasia. *J. Med. Genet.* **2014**, *51*, 470–474. [CrossRef] [PubMed]
40. Aygun, C.; Celik, F.C.; Nural, M.S.; Azak, E.; Kucukoduk, S.; Ogur, G.; Incesu, L. Simplified Gyral Pattern with Cerebellar Hypoplasia in Sedaghatian Type Spondylometaphyseal Dysplasia: A Clinical Report and Review of the Literature. *Am. J. Med. Genet. A* **2012**, *158A*, 1400–1405. [CrossRef] [PubMed]
41. Sibbing, D.; Pfeufer, A.; Perisic, T.; Mannes, A.M.; Fritz-Wolf, K.; Unwin, S.; Sinner, M.F.; Gieger, C.; Gloeckner, C.J.; Wichmann, H.-E.; et al. Mutations in the Mitochondrial Thioredoxin Reductase Gene TXNRD2 Cause Dilated Cardiomyopathy. *Eur. Heart J.* **2011**, *32*, 1121–1133. [CrossRef]
42. Prasad, R.; Chan, L.F.; Hughes, C.R.; Kaski, J.P.; Kowalczyk, J.C.; Savage, M.O.; Peters, C.J.; Nathwani, N.; Clark, A.J.L.; Storr, H.L.; et al. Thioredoxin Reductase 2 (TXNRD2) Mutation Associated with Familial Glucocorticoid Deficiency (FGD). *J. Clin. Endocrinol. Metab.* **2014**, *99*, E1556–E1563. [CrossRef]
43. Kudin, A.P.; Baron, G.; Zsurka, G.; Hampel, K.G.; Elger, C.E.; Grote, A.; Weber, Y.; Lerche, H.; Thiele, H.; Nürnberg, P.; et al. Homozygous Mutation in TXNRD1 Is Associated with Genetic Generalized Epilepsy. *Free Radic. Biol. Med.* **2017**, *106*, 270–277. [CrossRef]
44. Rojnueangnit, K.; Sirichongkolthong, B.; Wongwandee, R.; Khetkham, T.; Noojarern, S.; Khongkraparn, A.; Wattanasirichaigoon, D. Identification of Gene Mutations in Primary Pediatric Cardiomyopathy by Whole Exome Sequencing. *Pediatr. Cardiol.* **2020**, *41*, 165–174. [CrossRef]
45. Ahmed, M.Y.; Al-Khayat, A.; Al-Murshedi, F.; Al-Futaisi, A.; Chioza, B.A.; Fernandez-Murray, J.P.; Self, J.E.; Salter, C.G.; Harlalka, G.V.; Rawlins, L.E.; et al. A Mutation of EPT1 (SELENOI) Underlies a New Disorder of Kennedy Pathway Phospholipid Biosynthesis. *Brain* **2017**, *140*, 547–554. [CrossRef]
46. Horibata, Y.; Elpeleg, O.; Eran, A.; Hirabayashi, Y.; Savitzki, D.; Tal, G.; Mandel, H.; Sugimoto, H. EPT1 (Selenoprotein I) Is Critical for the Neural Development and Maintenance of Plasmalogen in Humans. *J. Lipid Res.* **2018**, *59*, 1015–1026. [CrossRef]
47. Santesmasses, D.; Mariotti, M.; Gladyshev, V.N. Tolerance to Selenoprotein Loss Differs between Human and Mouse. *Mol. Biol. Evol.* **2020**, *37*, 341–354. [CrossRef]
48. Turanov, A.A.; Xu, X.-M.; Carlson, B.A.; Yoo, M.-H.; Gladyshev, V.N.; Hatfield, D.L. Biosynthesis of Selenocysteine, the 21st Amino Acid in the Genetic Code, and a Novel Pathway for Cysteine Biosynthesis. *Adv. Nutr.* **2011**, *2*, 122–128. [CrossRef]
49. Diamond, A.; Dudock, B.; Hatfield, D. Structure and Properties of a Bovine Liver UGA Suppressor Serine TRNA with a Tryptophan Anticodon. *Cell* **1981**, *25*, 497–506. [CrossRef]
50. Bösl, M.R.; Takaku, K.; Oshima, M.; Nishimura, S.; Taketo, M.M. Early Embryonic Lethality Caused by Targeted Disruption of the Mouse Selenocysteine TRNA Gene (Trsp). *Proc. Natl. Acad. Sci. USA* **1997**, *94*, 5531–5534. [CrossRef]
51. Carlson, B.A. Selenocysteine TRNA[Ser]Sec Mouse Models for Elucidating Roles of Selenoproteins in Health and Development. In *Selenium*; Springer International Publishing: Cham, Switzerland, 2016; pp. 555–566. [CrossRef]
52. Itoh, Y.; Chiba, S.; Sekine, S.-I.; Yokoyama, S. Crystal Structure of Human Selenocysteine TRNA. *Nucleic Acids Res.* **2009**, *37*, 6259–6268. [CrossRef]
53. Carlson, B.A.; Lee, B.J.; Tsuji, P.A.; Tobe, R.; Park, J.M.; Schweizer, U.; Gladyshev, V.N.; Hatfield, D.L. Selenocysteine TRNA[Ser]Sec: From Nonsense Suppressor TRNA to the Quintessential Constituent in Selenoprotein Biosynthesis. In *Selenium*; Springer International Publishing: Cham, Switzerland, 2016; pp. 3–12. [CrossRef]
54. Diamond, A.M.; Choi, I.S.; Crain, P.F.; Hashizume, T.; Pomerantz, S.C.; Cruz, R.; Steer, C.J.; Hill, K.E.; Burk, R.F.; McCloskey, J.A.; et al. Dietary Selenium Affects Methylation of the Wobble Nucleoside in the Anticodon of Selenocysteine TRNA([Ser]Sec). *J. Biol. Chem.* **1993**, *268*, 14215–14223. [CrossRef]
55. Carlson, B.A.; Moustafa, M.E.; Sengupta, A.; Schweizer, U.; Shrimali, R.; Rao, M.; Zhong, N.; Wang, S.; Feigenbaum, L.; Lee, B.J.; et al. Selective Restoration of the Selenoprotein Population in a Mouse Hepatocyte Selenoproteinless Background with Different Mutant Selenocysteine TRNAs Lacking Um34. *J. Biol. Chem.* **2007**, *282*, 32591–32602. [CrossRef]
56. Schoenmakers, E.; Carlson, B.; Agostini, M.; Moran, C.; Rajanayagam, O.; Bochukova, E.; Tobe, R.; Peat, R.; Gevers, E.; Muntoni, F.; et al. Mutation in Human Selenocysteine Transfer RNA Selectively Disrupts Selenoprotein Synthesis. *J. Clin. Investig.* **2016**, *126*, 992–996. [CrossRef]
57. Geslot, A.; Savagner, F.; Caron, P. Inherited Selenocysteine Transfer RNA Mutation: Clinical and Hormonal Evaluation of 2 Patients. *Eur. Thyroid J.* **2021**, 1–6. [CrossRef]
58. Palioura, S.; Sherrer, R.L.; Steitz, T.A.; Söll, D.; Simonovic, M. The Human SepSecS-TRNASec Complex Reveals the Mechanism of Selenocysteine Formation. *Science* **2009**, *325*, 321–325. [CrossRef]
59. Copeland, P.R.; Fletcher, J.E.; Carlson, B.A.; Hatfield, D.L.; Driscoll, D.M. A Novel RNA Binding Protein, SBP2, Is Required for the Translation of Mammalian Selenoprotein MRNAs. *EMBO J.* **2000**, *19*, 306–314. [CrossRef]
60. Seeher, S.; Atassi, T.; Mahdi, Y.; Carlson, B.A.; Braun, D.; Wirth, E.K.; Klein, M.O.; Reix, N.; Miniard, A.C.; Schomburg, L.; et al. Secisbp2 Is Essential for Embryonic Development and Enhances Selenoprotein Expression. *Antioxid. Redox Signal.* **2014**, *21*, 835–849. [CrossRef]
61. Mehta, A.; Rebsch, C.M.; Kinzy, S.A.; Fletcher, J.E.; Copeland, P.R. Efficiency of Mammalian Selenocysteine Incorporation. *J. Biol. Chem.* **2004**, *279*, 37852–37859. [CrossRef] [PubMed]

62. Copeland, P.R.; Stepanik, V.A.; Driscoll, D.M. Insight into Mammalian Selenocysteine Insertion: Domain Structure and Ribosome Binding Properties of Sec Insertion Sequence Binding Protein 2. *Mol. Cell. Biol.* **2001**, *21*, 1491–1498. [CrossRef]
63. Allmang, C.; Carbon, P.; Krol, A. The SBP2 and 15.5 KD/Snu13p Proteins Share the Same RNA Binding Domain: Identification of SBP2 Amino Acids Important to SECIS RNA Binding. *RNA* **2002**, *8*, 1308–1318. [CrossRef]
64. Zhao, W.; Bohleber, S.; Schmidt, H.; Seeher, S.; Howard, M.T.; Braun, D.; Arndt, S.; Reuter, U.; Wende, H.; Birchmeier, C.; et al. Ribosome Profiling of Selenoproteins in Vivo Reveals Consequences of Pathogenic Secisbp2 Missense Mutations. *J. Biol. Chem.* **2019**, jbc.RA119.009369. [CrossRef]
65. Donovan, J.; Copeland, P.R. Selenocysteine Insertion Sequence Binding Protein 2L Is Implicated as a Novel Post-Transcriptional Regulator of Selenoprotein Expression. *PLoS ONE* **2012**, *7*, e35581. [CrossRef]
66. Donovan, J.; Copeland, P.R. Evolutionary History of Selenocysteine Incorporation from the Perspective of SECIS Binding Proteins. *BMC Evol. Biol.* **2009**, *9*, 229. [CrossRef]
67. Small-Howard, A.; Morozova, N.; Stoytcheva, Z.; Forry, E.P.; Mansell, J.B.; Harney, J.W.; Carlson, B.A.; Xu, X.-M.; Hatfield, D.L.; Berry, M.J. Supramolecular Complexes Mediate Selenocysteine Incorporation in Vivo. *Mol. Cell. Biol.* **2006**, *26*, 2337–2346. [CrossRef]
68. Schoenmakers, E.; Schoenmakers, N.; Chatterjee, K. Mutations in Humans That Adversely Affect the Selenoprotein Synthesis Pathway. In *Selenium. Its Molecular Biology and Role in Human Health*; Hatfield, D.L., Berry, M.J., Gladyshev, N.V., Eds.; Springer International Publishing: Cham, Switzerland, 2016; pp. 523–538. [CrossRef]
69. Anttonen, A.-K.; Hilander, T.; Linnankivi, T.; Isohanni, P.; French, R.L.; Liu, Y.; Simonović, M.; Söll, D.; Somer, M.; Muth-Pawlak, D.; et al. Selenoprotein Biosynthesis Defect Causes Progressive Encephalopathy with Elevated Lactate. *Neurology* **2015**, *85*, 306–315. [CrossRef]
70. Buniello, A.; MacArthur, J.A.L.; Cerezo, M.; Harris, L.W.; Hayhurst, J.; Malangone, C.; McMahon, A.; Morales, J.; Mountjoy, E.; Sollis, E.; et al. The NHGRI-EBI GWAS Catalog of Published Genome-Wide Association Studies, Targeted Arrays and Summary Statistics 2019. *Nucleic Acids Res.* **2019**, *47*, D1005–D1012. [CrossRef]
71. Mentuccia, D.; Proietti-Pannunzi, L.; Tanner, K.; Bacci, V.; Pollin, T.I.; Poehlman, E.T.; Shuldiner, A.R.; Celi, F.S. Association between a Novel Variant of the Human Type 2 Deiodinase Gene Thr92Ala and Insulin Resistance: Evidence of Interaction with the Trp64Arg Variant of the Beta-3-Adrenergic Receptor. *Diabetes* **2002**, *51*, 880–883. [CrossRef] [PubMed]
72. Castagna, M.G.; Dentice, M.; Cantara, S.; Ambrosio, R.; Maino, F.; Porcelli, T.; Marzocchi, C.; Garbi, C.; Pacini, F.; Salvatore, D. DIO2 Thr92Ala Reduces Deiodinase-2 Activity and Serum-T3 Levels in Thyroid-Deficient Patients. *J. Clin. Endocrinol. Metab.* **2017**, *102*, 1623–1630. [CrossRef] [PubMed]
73. Park, E.; Jung, J.; Araki, O.; Tsunekawa, K.; Park, S.Y.; Kim, J.; Murakami, M.; Jeong, S.-Y.; Lee, S. Concurrent TSHR Mutations and DIO2 T92A Polymorphism Result in Abnormal Thyroid Hormone Metabolism. *Sci. Rep.* **2018**, *8*, 10090. [CrossRef]
74. Meulenbelt, I.; Min, J.L.; Bos, S.; Riyazi, N.; Houwing-Duistermaat, J.J.; van der Wijk, H.-J.; Kroon, H.M.; Nakajima, M.; Ikegawa, S.; Uitterlinden, A.G.; et al. Identification of DIO2 as a New Susceptibility Locus for Symptomatic Osteoarthritis. *Hum. Mol. Genet.* **2008**, *17*, 1867–1875. [CrossRef]
75. Mez, J.; Chung, J.; Jun, G.; Kriegel, J.; Bourlas, A.P.; Sherva, R.; Logue, M.W.; Barnes, L.L.; Bennett, D.A.; Buxbaum, J.D.; et al. Two Novel Loci, COBL and SLC10A2, for Alzheimer's Disease in African Americans. *Alzheimers. Dement.* **2017**, *13*, 119–129. [CrossRef]
76. Sreelatha, A.; Yee, S.S.; Lopez, V.A.; Park, B.C.; Kinch, L.N.; Pilch, S.; Servage, K.A.; Zhang, J.; Jiou, J.; Karasiewicz-Urbańska, M.; et al. Protein AMPylation by an Evolutionarily Conserved Pseudokinase. *Cell* **2018**, *175*, 809–821.e19. [CrossRef] [PubMed]
77. Evans, D.M.; Zhu, G.; Dy, V.; Heath, A.C.; Madden, P.A.F.; Kemp, J.P.; McMahon, G.; St Pourcain, B.; Timpson, N.J.; Golding, J.; et al. Genome-Wide Association Study Identifies Loci Affecting Blood Copper, Selenium and Zinc. *Hum. Mol. Genet.* **2013**, *22*, 3998–4006. [CrossRef]
78. Gong, J.; Hsu, L.; Harrison, T.; King, I.B.; Stürup, S.; Song, X.; Duggan, D.; Liu, Y.; Hutter, C.; Chanock, S.J.; et al. Genome-Wide Association Study of Serum Selenium Concentrations. *Nutrients* **2013**, *5*, 1706–1718. [CrossRef] [PubMed]
79. Cornelis, M.C.; Fornage, M.; Foy, M.; Xun, P.; Gladyshev, V.N.; Morris, S.; Chasman, D.I.; Hu, F.B.; Rimm, E.B.; Kraft, P.; et al. Genome-Wide Association Study of Selenium Concentrations. *Hum. Mol. Genet.* **2015**, *24*, 1469–1477. [CrossRef] [PubMed]
80. Batai, K.; Trejo, M.J.; Chen, Y.; Kohler, L.N.; Lance, P.; Ellis, N.A.; Cornelis, M.C.; Chow, H.-H.S.; Hsu, C.-H.; Jacobs, E.T. Genome-Wide Association Study of Response to Selenium Supplementation and Circulating Selenium Concentrations in Adults of European Descent. *J. Nutr.* **2021**, *151*, 293–302. [CrossRef]
81. Thomson, C.D.; Burton, C.E.; Robinson, M.F. On Supplementing the Selenium Intake of New Zealanders. 1. Short Experiments with Large Doses of Selenite or Selenomethionine. *Br. J. Nutr.* **1978**, *39*, 579–587. [CrossRef]
82. Xia, Y.; Hill, K.E.; Byrne, D.W.; Xu, J.; Burk, R.F. Effectiveness of Selenium Supplements in a Low-Selenium Area of China. *Am. J. Clin. Nutr.* **2005**, *81*, 829–834. [CrossRef]
83. White, L.; Romagné, F.; Müller, E.; Erlebach, E.; Weihmann, A.; Parra, G.; Andrés, A.M.; Castellano, S. Genetic Adaptation to Levels of Dietary Selenium in Recent Human History. *Mol. Biol. Evol.* **2015**, *32*, 1507–1518. [CrossRef]
84. Canani, L.H.; Capp, C.; Dora, J.M.; Meyer, E.L.S.; Wagner, M.S.; Harney, J.W.; Larsen, P.R.; Gross, J.L.; Bianco, A.C.; Maia, A.L. The Type 2 Deiodinase A/G (Thr92Ala) Polymorphism Is Associated with Decreased Enzyme Velocity and Increased Insulin Resistance in Patients with Type 2 Diabetes Mellitus. *J. Clin. Endocrinol. Metab.* **2005**, *90*, 3472–3478. [CrossRef]

85. Foster, C.B.; Aswath, K.; Chanock, S.J.; McKay, H.F.; Peters, U. Polymorphism Analysis of Six Selenoprotein Genes: Support for a Selective Sweep at the Glutathione Peroxidase 1 Locus (3p21) in Asian Populations. *BMC Genet.* **2006**, *7*, 56. [CrossRef] [PubMed]
86. Engelken, J.; Espadas, G.; Mancuso, F.M.; Bonet, N.; Scherr, A.-L.; Jímenez-Álvarez, V.; Codina-Solà, M.; Medina-Stacey, D.; Spataro, N.; Stoneking, M.; et al. Signatures of Evolutionary Adaptation in Quantitative Trait Loci Influencing Trace Element Homeostasis in Liver. *Mol. Biol. Evol.* **2016**, *33*, 738–754. [CrossRef] [PubMed]
87. Turanov, A.A.; Lobanov, A.V.; Hatfield, D.L.; Gladyshev, V.N. UGA Codon Position-Dependent Incorporation of Selenocysteine into Mammalian Selenoproteins. *Nucleic Acids Res.* **2013**, *41*, 6952–6959. [CrossRef] [PubMed]
88. Wen, W.; Weiss, S.L.; Sunde, R.A. UGA Codon Position Affects the Efficiency of Selenocysteine Incorporation into Glutathione Peroxidase-1. *J. Biol. Chem.* **1998**, *273*, 28533–28541. [CrossRef] [PubMed]

International Journal of Molecular Sciences

Review

Roles for Selenoprotein I and Ethanolamine Phospholipid Synthesis in T Cell Activation

Chi Ma, Verena Martinez-Rodriguez and Peter R. Hoffmann *

Department of Cell and Molecular Biology, John A. Burns School of Medicine, University of Hawaii, Honolulu, HI 96813, USA; chima@hawaii.edu (C.M.); verenam@hawaii.edu (V.M.-R.)
* Correspondence: peterrh@hawaii.edu

Citation: Ma, C.; Martinez-Rodriguez, V.; Hoffmann, P.R. Roles for Selenoprotein I and Ethanolamine Phospholipid Synthesis in T Cell Activation. *Int. J. Mol. Sci.* **2021**, *22*, 11174. https://doi.org/10.3390/ijms222011174

Academic Editors: Petra A. Tsuji and Dolph L. Hatfield

Received: 21 September 2021
Accepted: 14 October 2021
Published: 16 October 2021

Publisher's Note: MDPI stays neutral with regard to jurisdictional claims in published maps and institutional affiliations.

Copyright: © 2021 by the authors. Licensee MDPI, Basel, Switzerland. This article is an open access article distributed under the terms and conditions of the Creative Commons Attribution (CC BY) license (https://creativecommons.org/licenses/by/4.0/).

Abstract: The selenoprotein family includes 25 members, many of which are antioxidant or redox regulating enzymes. A unique member of this family is Selenoprotein I (SELENOI), which does not catalyze redox reactions, but instead is an ethanolamine phosphotransferase (Ept). In fact, the characteristic selenocysteine residue that defines selenoproteins lies far outside of the catalytic domain of SELENOI. Furthermore, data using recombinant SELENOI lacking the selenocysteine residue have suggested that the selenocysteine amino acid is not directly involved in the Ept reaction. SELENOI is involved in two different pathways for the synthesis of phosphatidylethanolamine (PE) and plasmenyl PE, which are constituents of cellular membranes. Ethanolamine phospholipid synthesis has emerged as an important process for metabolic reprogramming that occurs in pluripotent stem cells and proliferating tumor cells, and this review discusses roles for upregulation of SELENOI during T cell activation, proliferation, and differentiation. SELENOI deficiency lowers but does not completely diminish de novo synthesis of PE and plasmenyl PE during T cell activation. Interestingly, metabolic reprogramming in activated SELENOI deficient T cells is impaired and this reduces proliferative capacity while favoring tolerogenic to pathogenic phenotypes that arise from differentiation. The implications of these findings are discussed related to vaccine responses, autoimmunity, and cell-based therapeutic approaches.

Keywords: selenium; selenoprotein; seleocysteine; autoimmunity; lymphocyte

1. Introduction

Selenium (Se) is an essential dietary trace mineral that is important for various aspects of human health, including optimal immunity [1]. The biological effects of Se are mainly exerted through its incorporation into selenoproteins as the 21st amino acid, selenocysteine (Sec) [2]. The 25 members of the selenoprotein family exhibit a wide variety of functions including the control of reactive oxygen species and cellular redox tone, regulating thyroid hormone metabolism, facilitating sperm maturation and protection, and promoting optimal immunity [3]. Under conditions of low Se status, the translation of selenoproteins stalls at the Sec-encoding UGA codon and both the mRNA and truncated protein become degraded through nonsense-mediated decay and destruction via C-end degrons, respectively [4,5]. In a Se deficient individual, the brain, muscle, and testes receive 'priority' for bioavailable Se at a cost to other tissues, such as those comprising the immune system [6]. Insufficient Se intake or other factors (e.g., defects in selenoprotein gene expression or some chronic infections that deplete Se) can impair adaptive immunity, especially T cell responses that are critical for producing effective vaccine responses and fighting infections [7,8]. Interestingly, not all types of immune responses are equivalently affected by Se deficiency or by Se supplementation [9,10]. Although the reasons for this are unclear due to an inadequate understanding of the mechanisms by which Se affects the immune system, data have recently emerged regarding roles for individual selenoproteins in immunity. This is particularly the case for T cell immunity, and a better understanding of how selenoproteins regulate T cell immunity may provide new targets for therapeutic intervention for immune based disease.

2. T Cell Immunity and Selenoprotein I Expression

CD4+ T cells constitute the topmost regulatory layer of the adaptive immune system, providing cytokine 'help' to CD8+ T cells (effector cells of cell mediated immunity) and B cells (effector cells of humoral immunity), thus coordinating acquired immune responses [11]. CD4+ T cells are activated through the T cell receptor (TCR), proliferate and differentiate into helper subsets, which forms the foundation for their ability to shape immune response and mediate host protection [12]. CD8+ T cells are also activated through their TCR to produce effector and memory cells required for optimal immunity [13]. Levels of Se and selenoproteins regulate T cell functions that drive both cell-mediated and humoral immunity [14–17]. However, it remains unclear which selenoproteins are involved in the different steps of T cell activation, proliferation, and differentiation. To gain insight into the roles that different selenoproteins play in T cell activation, we used real-time PCR to evaluate the selenoprotein transcriptome in naïve vs. activated T cells purified from C57BL/6 mice [18]. Results showed that a subset of selenoprotein mRNAs increased during T cell activation. These included the thioredoxin reductases (TXNRD1-3) enzymatically regenerate reduced thioredoxin and promote reducing capacity within cells, and we previously published that mRNA levels for these enzymes were increased as a mechanism for regulating redox tone in during T cell activation [7]. An interesting result found during T cell activation was an increase in the mRNA encoding selenoprotein I (SELENOI), which was upregulated nearly 3-fold [18]. Protein levels and enzyme activity for SELENOI were similarly increased in activated T cells. This raised the question: Why do SELENOI levels increase during T cell activation and what is its role in regulating T cell immunity?

To understand how SELENOI is involved in T cell activation, a bit more background on this selenoprotein is necessary. After the initial identification of the SELENOI gene in 2003 [19], the sequence was subsequently characterized through a homology search that found the cytidine diphosphate (CDP) alcohol phosphatidyltransferase signature, a common motif conserved in phospholipid synthases [20]. This study went on to demonstrate that SELENOI exhibits ethanolamine phosphotransferase (Ept) activity in vitro, providing the first data showing that SELENOI (which this group called EPT1) is an enzyme that transfers phosphoethanolamine from CDP-ethanolamine to 1,2-DAG acceptors to produce phosphatidylethanolamine (PE). SELENOI is only found in vertebrates, and in humans is expressed in a wide variety of cells [19,20]. A more recent study showed that fibroblasts from a patient with a mutation (exon skipping) leading to nonfunctional SELENOI had impaired Ept activity and reduced levels of several PE species, especially plasmenyl PE [21]. This study focused on the neurodevelopment defects exhibited by the patient, and another earlier clinical report substantiated the effects of a Arg112Pro mutation in the gene encoding SELENOI on the central nervous system (CNS) [22]. Understandably, these patients with severe CNS impairments were not evaluated in terms of immune cell function. Furthermore, there has been a paucity of data published regarding dietary Se regulating less overt changes in SELENOI. This is significant because the central nervous system retains Se under conditions Se deficiency, at a cost to the immune system [6]. Thus, individuals deficient in Se may maintain higher SELENOI levels in the brain compared to the immune system, and Se deficient T cells expressing lower levels of SELENOI may impact immunity. SELENOI protein or enzyme activity have not been measured in individuals with different Se status, but these data would be useful in determining how Se intake is related to function for this selenoprotein.

3. Structure of SELENOI Related to the Synthesis of PE and Plasmenyl PE

Plasma membrane phospholipids are distributed asymmetrically between the outer and inner leaflets of viable mammalian cells. The outer leaflet of the plasma membrane is composed primarily of sphingomyelin (SM) and phosphatidylcholine (PC), whereas the inner leaflet contains phosphatidylserine (PS) and phosphatidylethanolamine (PE) and phosphatidylinositol (PI), with cholesterol distributed equally [23,24]. In the plasma membrane and organelle membranes of mammalian cells, PE comprises 15–25% of total phospholipids [25]. PE is comprised of 1,2-diacylglycerol (DAG) and ethanolamine phosphate (Figure 1A). A structurally related ethanolamine phospholipid, plasmenyl PE, is an ether-linked lipid comprised of 1-alkyl-2-acylglycerol (AAG) and ethanolamine phosphate (Figure 1B), and is found at a lower abundance in cellular membranes compared to PE. SELENOI is involved in de novo synthesis of both PE and plasmenyl PE, carrying out the transfer of phosphoethanolamine from cytidine diphosphate (CDP)-ethanolamine to DAG to generate PE or to AAG to generate plasmenyl PE. In particular, SELENOI catalyzes the final step of the Kennedy pathway in the interface of the cytosol and endoplasmic reticulum (ER) that is responsible for synthesizing PE [26]. Although an alternative pathway operates in the mitochondria for synthesizing PE from PS, the importance of SELENOI in the synthesis of PE for cellular membranes through the Kennedy pathway has been clearly demonstrated in primary human fibroblasts and HeLa cells [21]. Plasmenyl PE is synthesized through a separate pathway beginning in peroxisomes (Reactions 1–3) and finishing in the ER membrane (Reactions 4–7). SELENOI catalyzes the sixth reaction to generate plasmenyl PE [27]. The vinyl ether bond at the sn-1 position of glycerol backbone contributes to a difference in biophysical properties compared to PE, and plasmenyl PE species are enriched in lipid rafts [28–30]. Thus, synthesis of PE and plasmenyl PE species are largely dependent on SELENOI and serve structural functions in cellular membranes, but possible roles in signaling pathways and in metabolic regulation are beginning to emerge [31,32]. Related to SELENOI (aka ethanolamine phosphotransferase 1 or EPT1) is the enzyme choline/ethanolamine phosphotransferase (CEPT1). CEPT1 is not a selenoprotein and can use both CDP-ethanolamine and CDP-choline as substrates, while SELENOI only uses CDP-ethanolamine [6]. This seems to suggest that SELENOI uses the selenocysteine residue to select cytidine diphosphate (CDP)-ethanolamine instead of CDP-choline as a substrate, although there are no data to include or exclude this possibility. Experimental evidence strongly suggests that CEPT1 can partially compensate for a lack of SELENOI expression [33], and PE may be generated by converting other phospholipids like phosphatidylserine to PE [34]. This other pathway occurs within the mitochondrial inner membrane, involving the conversion of the serine base to ethanolamine by PS decarboxylase (PSD) [26]. The PS decarboxylation pathway generates different PE species from CDP-ethanolamine pathway (i.e., the Kennedy pathway) [35]. This implies that phospholipid synthesis may occur by different routes and highlights the priority that cells place on maintaining balanced membrane composition.

As discussed above, SELENOI belongs to the family of selenoproteins and also to the family of phospholipid transferases. The latter enzymes are identified by a conserved CDP-alcohol phosphotransferase motif $D(X)_2DG(X)_2AR(X)_{7-12}G(X)_3D(X)_3D$ (X represents any amino acid and subscript numbers represent number of residues) commonly shared by all enzymes catalyzing the biosynthesis of phospholipids [20,36]. SELENOI is a predicted transmembrane protein that is fully embedded into the lipid bilayer, with its transmembrane helices traversing the membrane multiple times, as shown in Figure 2 [37]. Since most phospholipid transferases reside primarily in the (ER) membranes [38], SELENOI was initially thought to share the same localization. Recent studies showed that SELENOI is mainly localized in the Golgi apparatus [33], although these data involved overexpressed, tagged SELENOI in a cell line. Similar to other phospholipid transferases, the catalytic domain of the enzyme is predicted to be located within a pocket of the lipid bilayer that is accessible from the cytoplasm [37]. In redox selenoenzymes, selenocysteine residue is located within the catalytic domain in a prototypical C-X-X-U motif (X represents any amino acid), and replacement of selenocysteine with cysteine has been shown to reduce the catalytic activity of some of these enzymes [39,40]. In contrast, SELENOI does not contain a C-X-X-U motif and SELENOI's selenocysteine is located near the C-terminus at amino acid position 387 that is predicted to reside in the cytosol, apart from active site in within the membrane pocket [37]. This raises the question if selenocysteine is necessary for the catalytic function of this enzyme? Overexpression of SELENOI cDNA in a bacterial expression system led to a truncated form of the protein due to the lack of recognition of eukaryotic SECIS elements, and this Sec-deficient protein retained in vitro Ept activity [20]. This suggests the selenocysteine residue is not directly involved in the catalytic function of SELENOI, which is further supported by the observation described above that CEPT1 (that lacks a selenocysteine residue) may exert Ept activity when compensating for a lack of SELENOI.

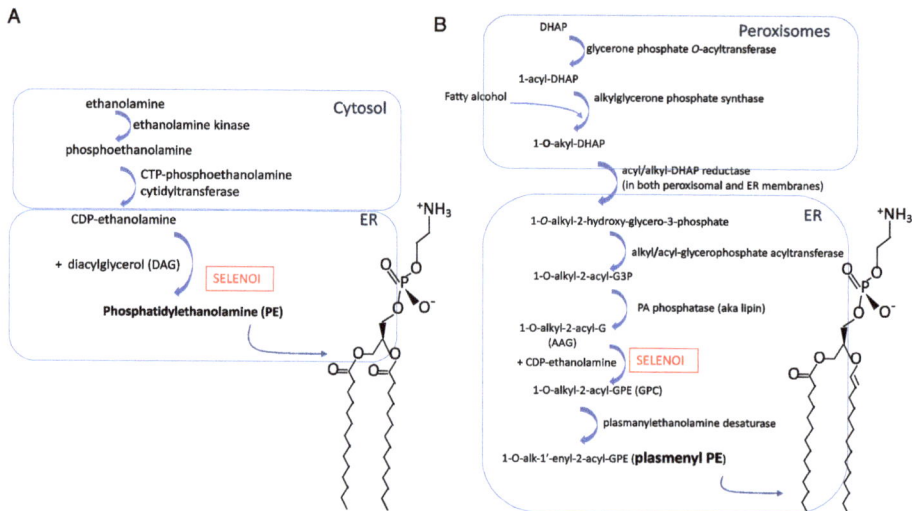

Figure 1. Synthesis pathways and molecular structures for (**A**) PE and (**B**) plasmenyl PE.

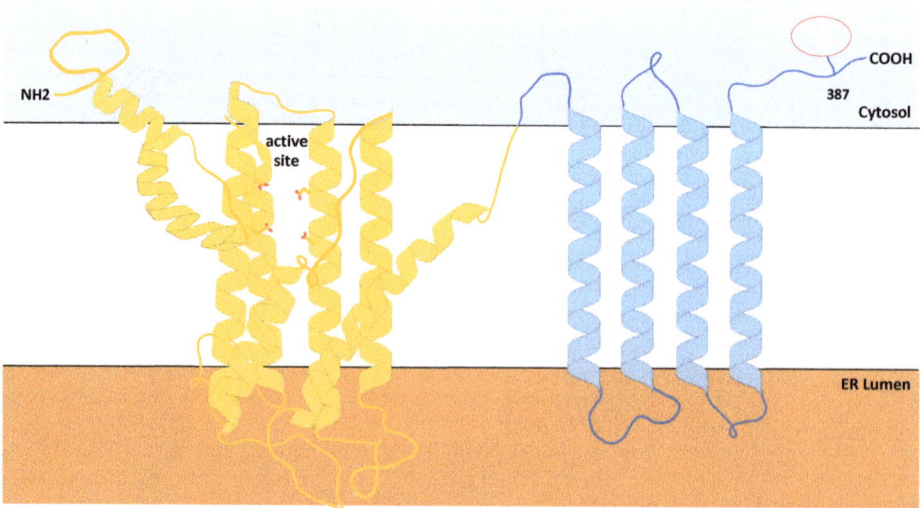

Figure 2. Predicted structure of Selenoprotein I. Results from Phyre Alarm and other online prediction programs show that, similar to other phosphotransferases, SELENOI is predominantly comprised of hydrophobic amino acids (~90%) with the catalytic domain residing within a membrane pocket. Note the C-terminal Sec residue is located outside of the catalytic domain.

4. Cellular Membranes and SELENOI KO

Published data show that SELENOI KO does not affect TCR induced signaling [18], which is somewhat surprising given that SELENOI deficiency decreases phospholipids involved in cellular membrane structure. In particular, the ratio of PE levels relative to PC levels plays a key role in membrane fluidity or rigidity [28,41]. Sufficient membrane rigidity is crucial in TCR signaling strength required to drive proliferation and differentiation [42]. Thus, one may expect a disruption of TCR signaling or, at a minimum, weaker TCR signaling with the lowered PE and plasmenyl PE levels in T cells that accompanies SELENOI KO. The lack of an effect of SELENOI KO on TCR signaling may suggest a compensatory process serves to maintain cellular membrane integrity in the absence of SELENOI. There is also a question regarding lipid rafts, which are fluctuating nanoscale assemblies of sphingolipid, cholesterol, and proteins that can be stabilized to coalesce, forming platforms that function in membrane signaling and trafficking [43]. Lipid rafts are enriched for plasmenyl PE [44], so it would reason that SELENOI deficiency in T cells that reduces plasmenyl PE should affect lipid raft organization. However, membrane raft organization does not appear differ between SELENOI KO versus WT control T cells as determined by fluorescent staining (Figure 3). This is consistent with the lack of effect of SELENOI KO on TCR signaling described above. Overall, there is insufficient evidence to date suggesting that membrane integrity or structure is affected by SELENOI KO in T cells.

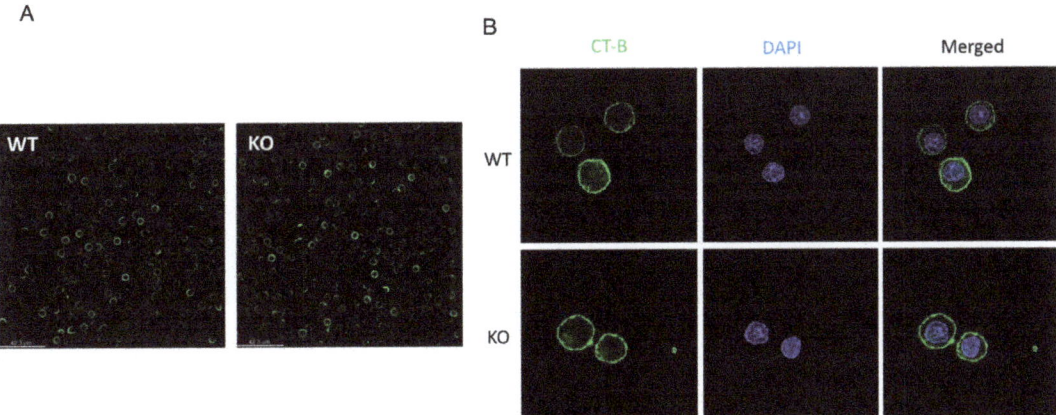

Figure 3. Lipid raft staining is similar between SELENOI KO T cells and WT controls. Mouse T cells isolated from mice and activated through the TCR for 24 h were stained for lipid rafts using standard fluorescent cholera toxin subunit B protocols. (**A**) Fluorescent microscopy images from live cells at 20× and (**B**) confocal microscopy images of paraformaldehyde fixed cells at 63× reveal similar patterns of staining between KO and WT T cells.

5. Metabolic Reprogramming During T Cell Activation Requires SELENOI for Proliferation

In the non-activated or quiescent state, T cells mainly exhibit catabolic activity involving the break-down of nutrients to fuel cell survival. Upon T cell receptor (TCR) triggered activation, T cells transition to a state of anabolism in which nutrients are used to construct the molecular building blocks that are incorporated into cellular biomass to support proliferation [45,46]. In addition to small molecule precursors, energy is also needed for proliferation. This requirement is met during TCR-induced activation through increased glucose uptake via upregulated glucose transporters, which is accompanied by induced aerobic metabolism [47]. These changes are akin to shifts in cancer cell metabolism known as the Warburg effect. TCR-induced metabolic reprogramming is critical for generating sufficient energy and precursor molecules for subsequent rounds of mitosis, which promotes the T cell expansion phase that is so crucial for immune clearance of pathogens. The balance of catabolic and anabolic pathways in a cell determines how much adenosine triphosphate (ATP) is generated versus consumed, the availability of biosynthetic precursors, and the redox status of the cell [48]. Redox status may be controlled by antioxidant and redox regulating selenoproteins, with free thiols playing a key role [16]. However, SELENOI is an unconventional selenoprotein that is not involved in redox reactions. Since SELENOI is directly involved in two different anabolic pathways, one for PE synthesis and the other for plasmenyl PE synthesis, it follows that this selenoprotein is likely an integral part of the metabolic reprogramming during T cell activation.

To understand the role of SELENOI in T cell proliferation, our research group conducted loss-of-function studies in T cells isolated from different transgenic mouse models. In particular, an inducible knockdown (KD) mouse model along with a T cell specific knockout (KO) mouse model were compared to wild-type (WT) controls for TCR-induced proliferative capacity. SELENOI KD led to a ~22% decrease and KO to a ~56% decrease in proliferation compared to WT controls [18]. In vivo T cell expansion to vaccination was also decreased in T cell specific SELENOI KO mice compared to WT controls. Surprisingly, the TCR signaling was not affected by SELENOI deficiency and levels of PE and plasmenyl PE were only partially decreased. The latter observation may be explained by the fact that enzyme activity of a related phospholipid transferase may compensate for PE and plasmenyl PE when SELENOI is absent [33]. The most impressive result of SELENOI deficiency in activated T cells was a progressive accumulation of ATP, which was detected by the metabolic sensor AMPK and led to lower activation of this kinase. In fact, treating WT T cells with the AMPK inhibitor, dorsomorphin, reduced proliferation by similar levels as SELENOI KO. These data collectively suggest that SELENOI serves a critical function during T cell activation to maintain ethanolamine phospholipid synthesis and thereby keep a balanced metabolism within the cells as they undergo metabolic reprogramming. Removing SELENOI causes a ripple effect-these synthesis pathways are disrupted and this causes ATP to accumulate, which is sensed within the cells by AMPK and eventually proliferation is reduced.

6. SELENOI and T Helper Cell Differentiation

As proliferation takes place, the fates of the daughter T cells differ through asymmetric cell division and differentiation [49]. In particular, naïve $CD4^+$ T cells differentiate into one of several T helper cell lineages depending on signals from antigen presenting cells and cytokines present in the surrounding environment. In addition to these external stimuli, internal factors, such as cellular metabolism, can influence differentiation [50]. $CD4^+$ T cell subsets express unique sets of cell surface markers and transcription factors, while secreting a defined array of cytokines that determines their functional properties [51]. It was recently demonstrated that $CD4^+$ T cell differentiation was dependent on ethanolamine kinase 1, CTP-phosphoethanolamine cytidyltransferase, and SELENOI enzymes that comprise the CDP-ethanolamine pathway for de novo synthesis of PE and plasmenyl PE [52]. In particular, this pathway was important for T follicular helper (T_{FH}) cell differentiation by promoting the surface expression and functional effects of the chemokine receptor, CXCR5. Unlike our SELENOI KO T cells that were decreased, but not lacking, PE and plasmenyl PE as described above, these studies largely focused on T cells lacking these phospholipids. A state of complete loss of PE and plasmenyl PE in T cells is unlikely given the essential roles these phospholipids play in development [53,54], but the importance of ethanolamine phospholipid synthesis in T_{FH} cell formation does highlight how certain metabolic pathways promote T cell differentiation outcomes over others.

T helper type 17 (Th17) cells represent another subtype of T cells differentiating from naïve Th cells that promote inflammation. Th17 cells are required for immune responses to specific extracellular bacteria and fungi [55], but dysregulated Th17 responses may also contribute to pathogenesis in autoimmune diseases, such as rheumatoid arthritis and multiple sclerosis (MS) [56,57]. In contrast, $CD4^+$ regulatory T (Treg) cells contribute to the suppression of immune responses and immune homeostasis, countering Th17 cells to protect against autoimmune disorders. Treg differentiation relies on the upregulation of the transcription factor FoxP3, and Treg cells function to secrete immunosuppressive cytokines TGF-β and IL-10 [58]. A major factor influencing Th17/Treg fates during T cell activation is the type of metabolic reprogramming that occurs after TCR engagement, which serves to meet increased demand for energy and metabolites [59].

Th17 cells take on a distinct metabolic signature that promotes differentiation into this subtype, heavily relying on finely tuned glycolysis, glutamine metabolism, and fatty acid synthesis for their differentiation and pro-inflammatory function [60–62]. Sustained mitochondrial oxidative phosphorylation has also been shown to regulate the fate decision between pathogenic Th17 and Treg cells [63]. These coordinated shifts in specific metabolic pathways that occur during T helper cell differentiation may be disrupted in a number of ways, and we recently identified SELENOI as a metabolic enzyme involved in regulating T helper cell differentiation. SELENOI mRNA and protein levels increased at early stages of Th17 differentiation, similar to levels of the master regulator that drives Th17 phenotypes, RORγt (manuscript submitted). Using T cell specific SELENOI KO mice, our studies have found that naïve CD4$^+$ T cells were directed toward Treg and away from Th1 and Th17 subtypes when activated through the TCR. Moreover, T cell specific KO of SELENOI protected mice from a mouse model of MS, experimental autoimmune encephalitis (EAE), showing that SELENOI deficiency reduces Th17 pathology. These data may suggest that targeting SELENOI in T cells may represent a potential therapeutic approach to treating MS. Indeed, cell-based therapies have been proposed for restoring homeostasis in MS patients such as tolerogenic dendritic cells, Tregs, mesenchymal stem cells, and vaccination with T cells [64]. Interfering with SELENOI activity to skew Th17/Treg cells toward a tolerogenic phenotype certainly presents its challenges, but may provide a new cell-based therapeutic approach for this or other autoimmune disorders.

7. Conclusions

Overall, insights into roles for SELENOI in T cell functions have been made using SELENOI loss-of-function models (Table 1). Upregulated SELENOI triggered by TCR engagement contributes to both proliferation and differentiation, which may be crucial for a variety of immune responses (Figure 4). The emerging field of metabolic reprogramming during T cell activation is providing new insights into factors and pathways that regulate optimal responses to pathogens and tumors. SELENOI has recently been identified as an anabolic enzyme involved in metabolic reprogramming, and disruption of its activity in mouse models has been shown to effectively decrease EAE. Ethanolamine phospholipid synthesis is particularly important for metabolic reprogramming in pluripotent stem cells and proliferating tumor cells [31,65], and our recent work showed a crucial role for ethanolamine phospholipid synthesis during T cell activation. How this may be translated into new therapies remains uncertain, but targeting metabolism may be incorporated into cell-based therapies involving T cells. For example, chimeric antigen receptor (CAR) expressing T cells have become an effective approach for treating some cancers, and advances have been made to improve the metabolic fitness and efficacy of CAR T cells [66]. This may provide a framework for developing new strategies in treating diseases, including autoimmune disorders, focused on regulating T cell metabolism and providing optimal immunity.

Table 1. A summary of roles for SELENOI in T cell functions.

T Cell Function	Role for SELENOI	References
TCR signaling	SELENOI KO has minimal effect on signaling pathways downstream of TCR engagement	[18]
Membrane raft distribution	SELENOI KO has minimal effect on raft distribution or membrane fluidity	Data presented herein
Metabolic reprogramming	SELENOI KO disrupts ATP generation/consumption during TCR activation	[18]
Metabolic sensing	SELENOI KO impairs sensing by AMPK during TCR activation	[18]
T cell proliferation	SELENOI KO decreases in vivo and ex vivo TCR induced proliferation	[18]
T cell differentiation	T helper cell differentiation is affected by SELENOI KO; Tfh and Th17 cells are reduced, while Treg cells are increased	[52]

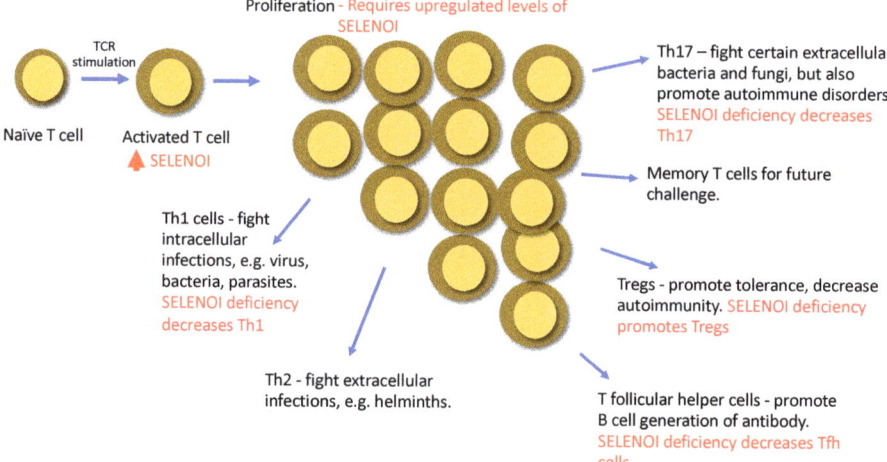

Figure 4. A summary of roles of SELENOI in T cells.

Funding: This research was supported by NIH grant R01AI147496.

Institutional Review Board Statement: No human studies were included. Animal protocols were approved by the University of Hawaii Institutional Animal Care and Use Committee.

Informed Consent Statement: Not applicable.

Data Availability Statement: Not applicable.

Conflicts of Interest: The authors declare no conflict of interest.

References

1. Rayman, M.P. Selenium and human health. *Lancet* **2012**, *379*, 1256–1268. [CrossRef]
2. Reeves, M.A.; Hoffmann, P.R. The human selenoproteome: Recent insights into functions and regulation. *Cell. Mol. Life Sci.* **2009**, *66*, 2457–2478. [CrossRef]
3. Schweizer, U.; Fradejas-Villar, N. Why 21? The significance of selenoproteins for human health revealed by inborn errors of metabolism. *FASEB J.* **2016**, *30*, 3669–3681. [CrossRef] [PubMed]
4. Seyedali, A.; Berry, M.J. Nonsense-mediated decay factors are involved in the regulation of selenoprotein mRNA levels during selenium deficiency. *RNA* **2014**, *20*, 1248–1256. [CrossRef] [PubMed]

4. Lin, H.-C.; Yeh, C.-W.; Chen, Y.-F.; Lee, T.-T.; Hsieh, P.-Y.; Rusnac, D.V.; Lin, S.-Y.; Elledge, S.J.; Zheng, N.; Yen, H.-C.S. C-Terminal End-Directed Protein Elimination by CRL2 Ubiquitin Ligases. *Mol. Cell* **2018**, *70*, 602e3–613e3. [CrossRef] [PubMed]
5. Burk, R.F.; Hill, K.E. Regulation of Selenium Metabolism and Transport. *Annu. Rev. Nutr.* **2015**, *35*, 109–134. [CrossRef] [PubMed]
6. Hoffmann, F.W.; Hashimoto, A.C.; Shafer, L.A.; Dow, S.; Berry, M.J.; Hoffmann, P.R. Dietary Selenium Modulates Activation and Differentiation of CD4+ T Cells in Mice through a Mechanism Involving Cellular Free Thiols. *J. Nutr.* **2010**, *140*, 1155–1161. [CrossRef] [PubMed]
7. Broome, C.S.; McArdle, F.; Kyle, J.A.M.; Andrews, F.; Lowe, N.; Hart, C.A.; Arthur, J.R.; Jackson, M. An increase in selenium intake improves immune function and poliovirus handling in adults with marginal selenium status. *Am. J. Clin. Nutr.* **2004**, *80*, 154–162. [CrossRef] [PubMed]
8. Hoffmann, P.R.; Berry, M.J. The influence of selenium on immune responses. *Mol. Nutr. Food Res.* **2008**, *52*, 1273–1280. [CrossRef]
9. Ivory, K.; Prieto, E.; Spinks, C.; Armah, C.N.; Goldson, A.J.; Dainty, J.R.; Nicoletti, C. Selenium supplementation has beneficial and detrimental effects on immunity to influenza vaccine in older adults. *Clin. Nutr.* **2015**, *36*, 407–415. [CrossRef] [PubMed]
10. Kaiko, G.E.; Horvat, J.C.; Beagley, K.; Hansbro, P.M. Immunological decision-making: How does the immune system decide to mount a helper T-cell response? *Immunology* **2008**, *123*, 326–338. [CrossRef]
11. Zhu, J.; Yamane, H.; Paul, W.E. Differentiation of Effector CD4 T Cell Populations. *Annu. Rev. Immunol.* **2010**, *28*, 445–489. [CrossRef]
12. Cui, W.; Kaech, S.M. Generation of effector CD8+ T cells and their conversion to memory T cells. *Immunol. Rev.* **2010**, *236*, 151–166. [CrossRef]
13. Fredericks, G.J.; Hoffmann, F.W.; Rose, A.H.; Osterheld, H.J.; Hess, F.M.; Mercier, F.; Hoffmann, P.R. Stable expression and function of the inositol 1,4,5-triphosphate receptor requires palmitoylation by a DHHC6/selenoprotein K complex. *Proc. Natl. Acad. Sci. USA* **2014**, *111*, 16478–16483. [CrossRef]
14. Carlson, B.A.; Yoo, M.-H.; Shrimali, R.K.; Irons, R.; Gladyshev, V.N.; Hatfield, D.L.; Park, J.M. Role of selenium-containing proteins in T-cell and macrophage function. *Proc. Nutr. Soc.* **2010**, *69*, 300–310. [CrossRef]
15. Shrimali, R.K.; Irons, R.D.; Carlson, B.A.; Sano, Y.; Gladyshev, V.N.; Park, J.M.; Hatfield, D.L. Selenoproteins Mediate T Cell Immunity through an Antioxidant Mechanism. *J. Biol. Chem.* **2008**, *283*, 20181–20185. [CrossRef] [PubMed]
16. Gladyshev, V.N.; Stadtman, T.C.; Hatfield, D.L.; Jeang, K.-T. Levels of major selenoproteins in T cells decrease during HIV infection and low molecular mass selenium compounds increase. *Proc. Natl. Acad. Sci. USA* **1999**, *96*, 835–839. [CrossRef] [PubMed]
17. Ma, C.; Hoffmann, F.W.; Marciel, M.P.; Page, K.E.; Williams-Aduja, M.A.; Akana, E.N.; Gojanovich, G.S.; Gerschenson, M.; Urschitz, J.; Moisyadi, S.; et al. Upregulated ethanolamine phospholipid synthesis via selenoprotein I is required for effective metabolic reprogramming during T cell activation. *Mol. Metab.* **2021**, *47*, 101170. [CrossRef]
18. Kryukov, G.V.; Castellano, S.; Novoselov, S.V.; Lobanov, A.V.; Zehtab, O.; Guigó, R.; Gladyshev, V.N. Characterization of Mammalian Selenoproteomes. *Science* **2003**, *300*, 1439–1443. [CrossRef] [PubMed]
19. Horibata, Y.; Hirabayashi, Y. Identification and characterization of human ethanolaminephosphotransferase1. *J. Lipid Res.* **2007**, *48*, 503–508. [CrossRef]
20. Horibata, Y.; Elpeleg, O.; Eran, A.; Hirabayashi, Y.; Savitzki, D.; Tal, G.; Mandel, H.; Sugimoto, H. Ethanolamine phosphotransferase 1 (selenoprotein I) is critical for the neural development and maintenance of plasmalogen in human. *J. Lipid Res.* **2018**, *59*, 1015–1026. [CrossRef]
21. Ahmed, M.Y.; Al-Khayat, A.; Al-Murshedi, F.; Al-Futaisi, A.; Chioza, B.A.; Fernandez-Murray, J.P.; Self, J.E.; Salter, C.G.; Harlalka, G.V.; Rawlins, L.E.; et al. A mutation of EPT1 (SELENOI) underlies a new disorder of Kennedy pathway phospholipid biosynthesis. *Brain* **2017**, *140*, 547–554. [CrossRef]
22. Devaux, P.F. Static and dynamic lipid asymmetry in cell membranes. *Biochemistry* **1991**, *30*, 1163–1173. [CrossRef]
23. Murate, M.; Abe, M.; Kasahara, K.; Iwabuchi, K.; Umeda, M.; Kobayashi, T. Transbilayer distribution of lipids at nano scale. *J. Cell Sci.* **2015**, *128*, 1627–1638. [CrossRef]
24. Vance, J.E. Phospholipid Synthesis and Transport in Mammalian Cells. *Traffic* **2014**, *16*, 1–18. [CrossRef]
25. Gibellini, F.; Smith, T.K. The Kennedy pathway-De novo synthesis of phosphatidylethanolamine and phosphatidylcholine. *IUBMB Life* **2010**, *62*, 414–428. [CrossRef]
26. Braverman, N.E.; Moser, A.B. Functions of plasmalogen lipids in health and disease. *Biochim. Biophys.* **2012**, *1822*, 1442–1452. [CrossRef]
27. Dawaliby, R.; Trubbia, C.; Delporte, C.; Noyon, C.; Ruysschaert, J.-M.; Van Antwerpen, P.; Govaerts, C. Phosphatidylethanolamine Is a Key Regulator of Membrane Fluidity in Eukaryotic Cells. *J. Biol. Chem.* **2016**, *291*, 3658–3667. [CrossRef]
28. Silin, V.I.; Hoogerheide, D.P. pH dependent electrical properties of the inner- and outer- leaflets of biomimetic cell membranes. *J. Colloid Interface Sci.* **2021**, *594*, 279–289. [CrossRef] [PubMed]
29. Rubio, J.; Astudillo, A.M.; Casas, J.; Balboa, M.A.; Balsinde, J. Regulation of Phagocytosis in Macrophages by Membrane Ethanolamine Plasmalogens. *Front. Immunol.* **2018**, *9*, 1723. [CrossRef] [PubMed]
30. Wu, Y.; Chen, K.; Xing, G.; Li, L.; Ma, B.; Hu, Z.; Duan, L.; Liu, X. Phospholipid remodeling is critical for stem cell pluripotency by facilitating mesenchymal-to-epithelial transition. *Sci. Adv.* **2019**, *5*, eaax7525. [CrossRef]

32. Lebrero, P.; Astudillo, A.M.; Rubio, J.M.; Fernandez-Caballero, L.; Kokotos, G.; Balboa, M.A.; Balsinde, J. Cellular Plasmalogen Content Does Not Influence Arachidonic Acid Levels or Distribution in Macrophages: A Role for Cytosolic Phospholipase A2gamma in Phospholipid Remodeling. *Cells* **2019**, *8*, 799. [CrossRef]
33. Horibata, Y.; Ando, H.; Sugimoto, H. Locations and contributions of the phosphotransferases EPT1 and CEPT1 to the biosynthesis of ethanolamine phospholipids. *J. Lipid Res.* **2020**, *61*, 1221–1231. [CrossRef]
34. Vance, J.E. Historical perspective: Phosphatidylserine and phosphatidylethanolamine from the 1800s to the present. *J. Lipid Res.* **2018**, *59*, 923–944. [CrossRef]
35. Bleijerveld, O.B.; Brouwers, J.F.; Vaandrager, A.B.; Helms, J.B.; Houweling, M. The CDP-ethanolamine Pathway and Phosphatidylserine Decarboxylation Generate Different Phosphatidylethanolamine Molecular Species. *J. Biol. Chem.* **2007**, *282*, 28362–28372. [CrossRef] [PubMed]
36. McMaster, C.R.; Bell, R.M. CDP-ethanolamine:1,2-diacylglycerol ethanolaminephosphotransferase. *Biochim. Biophys. Acta* **1997**, *1348*, 117–123. [CrossRef]
37. Liu, J.; Rozovsky, S. Membrane-Bound Selenoproteins. *Antioxidants Redox Signal.* **2015**, *23*, 795–813. [CrossRef] [PubMed]
38. Vance, J.E.; Vance, D.E. Does rat liver Golgi have the capacity to synthesize phospholipids for lipoprotein secretion? *J Biol Chem* **1988**, *263*, 5898–5909. [CrossRef]
39. Kuiper, G.G.J.M.; Klootwijk, W.; Visser, T.J. Substitution of Cysteine for Selenocysteine in the Catalytic Center of Type III Iodothyronine Deiodinase Reduces Catalytic Efficiency and Alters Substrate Preference. *Endocrinology* **2003**, *144*, 2505–2513. [CrossRef]
40. Kim, H.-Y.; Fomenko, D.E.; Yoon, Y.-E.; Gladyshev, V.N. Catalytic Advantages Provided by Selenocysteine in Methionine-S-Sulfoxide Reductases†. *Biochemistry* **2006**, *45*, 13697–13704. [CrossRef]
41. Vance, J.E.; Tasseva, G. Formation and function of phosphatidylserine and phosphatidylethanolamine in mammalian cells. *Biochim. Biophys. Acta* **2013**, *1831*, 543–554. [CrossRef]
42. He, H.-T.; Bongrand, P. Membrane dynamics shape TCR-generated signaling. *Front. Immunol.* **2012**, *3*, 90. [CrossRef] [PubMed]
43. Lingwood, D.; Simons, K. Lipid Rafts as a Membrane-Organizing Principle. *Science* **2010**, *327*, 46–50. [CrossRef] [PubMed]
44. Pike, L.J.; Han, X.; Chung, K.-N.; Gross, R.W. Lipid Rafts Are Enriched in Arachidonic Acid and Plasmenylethanolamine and Their Composition Is Independent of Caveolin-1 Expression: A Quantitative Electrospray Ionization/Mass Spectrometric Analysis. *Biochemistry* **2002**, *41*, 2075–2088. [CrossRef]
45. Fox, C.J.; Hammerman, P.S.; Thompson, C.B. Fuel feeds function: Energy metabolism and the T-cell response. *Nat. Rev. Immunol.* **2005**, *5*, 844–852. [CrossRef]
46. MacIver, N.; Michalek, R.D.; Rathmell, J.C. Metabolic Regulation of T Lymphocytes. *Annu. Rev. Immunol.* **2013**, *31*, 259–283. [CrossRef] [PubMed]
47. Jacobs, S.R.; Herman, C.E.; MacIver, N.; Wofford, J.A.; Wieman, H.L.; Hammen, J.J.; Rathmell, J.C. Glucose Uptake Is Limiting in T Cell Activation and Requires CD28-Mediated Akt-Dependent and Independent Pathways. *J. Immunol.* **2008**, *180*, 4476–4486. [CrossRef]
48. Van der Windt, G.J.; Pearce, E.L. Metabolic switching and fuel choice during T-cell differentiation and memory development. *Immunol. Rev.* **2012**, *249*, 27–42. [CrossRef]
49. Chang, J.T.; Palanivel, V.R.; Kinjyo, I.; Schambach, F.; Intlekofer, A.M.; Banerjee, A.; Longworth, S.A.; Vinup, K.E.; Mrass, P.; Oliaro, J.; et al. Asymmetric T Lymphocyte Division in the Initiation of Adaptive Immune Responses. *Science* **2007**, *315*, 1687–1691. [CrossRef]
50. Chisolm, D.A.; Weinmann, A.S. Connections Between Metabolism and Epigenetics in Programming Cellular Differentiation. *Annu. Rev. Immunol.* **2018**, *36*, 221–246. [CrossRef]
51. Pawlak, M.; Ho, A.W.; Kuchroo, V.K. Cytokines and transcription factors in the differentiation of CD4+ T helper cell subsets and induction of tissue inflammation and autoimmunity. *Curr. Opin. Immunol.* **2020**, *67*, 57–67. [CrossRef] [PubMed]
52. Fu, G.; Guy, C.S.; Chapman, N.M.; Palacios, G.; Wei, J.; Zhou, P.; Long, L.; Wang, Y.-D.; Qian, C.; Dhungana, Y.; et al. Metabolic control of TFH cells and humoral immunity by phosphatidylethanolamine. *Nature* **2021**, *595*, 724–729. [CrossRef]
53. Wang, L.; Magdaleno, S.; Tabas, I.; Jackowski, S. Early Embryonic Lethality in Mice with Targeted Deletion of the CTP:Phosphocholine Cytidylyltransferase α Gene (Pcyt1a). *Mol. Cell. Biol.* **2005**, *25*, 3357–3363. [CrossRef]
54. Avery, J.C.; Yamazaki, Y.; Hoffmann, F.W.; Folgelgren, B.; Hoffmann, P.R. Selenoprotein I is essential for murine embryogenesis. *Arch. Biochem. Biophys.* **2020**, *689*, 108444. [CrossRef]
55. E Harrington, L.; Hatton, R.; Mangan, P.R.; Turner, H.; Murphy, T.L.; Murphy, K.M.; Weaver, C. Interleukin 17–producing CD4+ effector T cells develop via a lineage distinct from the T helper type 1 and 2 lineages. *Nat. Immunol.* **2005**, *6*, 1123–1132. [CrossRef] [PubMed]
56. Gaffen, S.L.; Jain, R.; Garg, A.V.; Cua, D.J. The IL-23–IL-17 immune axis: From mechanisms to therapeutic testing. *Nat. Rev. Immunol.* **2014**, *14*, 585–600. [CrossRef]
57. Ghoreschi, K.; Laurence, A.; Yang, X.-P.; Hirahara, K.; O'Shea, J.J. T helper 17 cell heterogeneity and pathogenicity in autoimmune disease. *Trends Immunol.* **2011**, *32*, 395–401. [CrossRef]
58. Sakaguchi, S.; Sakaguchi, N.; Asano, M.; Itoh, M.; Toda, M. Immunologic self-tolerance maintained by activated T cells expressing IL-2 receptor alpha-chains (CD25). Breakdown of a single mechanism of self-tolerance causes various autoimmune diseases. *J. Immunol.* **1995**, *155*, 1151–1164. [PubMed]

59. Shen, H.; Shi, L.Z. Metabolic regulation of TH17 cells. *Mol. Immunol.* **2019**, *109*, 81–87. [CrossRef] [PubMed]
60. Araujo, L.; Khim, P.; Mkhikian, H.; Mortales, C.-L.; Demetriou, M. Glycolysis and glutaminolysis cooperatively control T cell function by limiting metabolite supply to N-glycosylation. *eLife* **2017**, *6*, e21330. [CrossRef]
61. Berod, L.; Friedrich, C.; Nandan, A.; Freitag, J.; Hagemann, S.; Harmrolfs, K.; Sandouk, A.; Hesse, C.; Castro, C.N.; Bahre, H.; et al. De novo fatty acid synthesis controls the fate between regulatory T and T helper 17 cells. *Nat. Med.* **2014**, *20*, 1327–1333. [CrossRef] [PubMed]
62. Delgoffe, G.M.; Kole, T.P.; Zheng, Y.; Zarek, P.E.; Matthews, K.L.; Xiao, B.; Worley, P.F.; Kozma, S.C.; Powell, J.D. The mTOR Kinase Differentially Regulates Effector and Regulatory T Cell Lineage Commitment. *Immunity* **2009**, *30*, 832–844. [CrossRef] [PubMed]
63. Shin, B.; Benavides, G.A.; Geng, J.; Koralov, S.; Hu, H.; Darley-Usmar, V.M.; Harrington, L.E. Mitochondrial Oxidative Phosphorylation Regulates the Fate Decision between Pathogenic Th17 and Regulatory T Cells. *Cell Rep.* **2020**, *30*, 1898–1909. [CrossRef] [PubMed]
64. Mansilla, M.J.; Presas-Rodríguez, S.; Teniente-Serra, A.; González-Larreategui, I.; Quirant-Sánchez, B.; Fondelli, F.; Djedovic, N.; Iwaszkiewicz-Grześ, D.; Chwojnicki, K.; Miljković, D.; et al. Paving the way towards an effective treatment for multiple sclerosis: Advances in cell therapy. *Cell. Mol. Immunol.* **2021**, *18*, 1353–1374. [CrossRef] [PubMed]
65. Wang, B.; Rong, X.; Palladino, E.N.; Wang, J.; Fogelman, A.M.; Martín, M.G.; Alrefai, W.A.; Ford, D.A.; Tontonoz, P. Phospholipid Remodeling and Cholesterol Availability Regulate Intestinal Stemness and Tumorigenesis. *Cell Stem Cell* **2018**, *22*, 206–220. [CrossRef] [PubMed]
66. Pellegrino, M.; del Bufalo, F.; de Angelis, B.; Quintarelli, C.; Caruana, I.; de Billy, E. Manipulating the Metabolism to Improve the Efficacy of CAR T-Cell Immunotherapy. *Cells* **2020**, *10*, 14. [CrossRef]

Review

The Impact of Selenium Deficiency on Cardiovascular Function

Briana K. Shimada, Naghum Alfulaij and Lucia A. Seale *

Pacific Biosciences Research Center, University of Hawaii at Manoa, Honolulu, HI 96822, USA; bkshimad@hawaii.edu (B.K.S.); alfulaij@hawaii.edu (N.A.)
* Correspondence: lseale@hawaii.edu; Tel.: +1-808-956-8296

Abstract: Selenium (Se) is an essential trace element that is necessary for various metabolic processes, including protection against oxidative stress, and proper cardiovascular function. The role of Se in cardiovascular health is generally agreed upon to be essential yet not much has been defined in terms of specific functions. Se deficiency was first associated with Keshan's Disease, an endemic disease characterized by cardiomyopathy and heart failure. Since then, Se deficiency has been associated with multiple cardiovascular diseases, including myocardial infarction, heart failure, coronary heart disease, and atherosclerosis. Se, through its incorporation into selenoproteins, is vital to maintain optimal cardiovascular health, as selenoproteins are involved in numerous crucial processes, including oxidative stress, redox regulation, thyroid hormone metabolism, and calcium flux, and inadequate Se may disrupt these processes. The present review aims to highlight the importance of Se in cardiovascular health, provide updated information on specific selenoproteins that are prominent for proper cardiovascular function, including how these proteins interact with microRNAs, and discuss the possibility of Se as a potential complemental therapy for prevention or treatment of cardiovascular disease.

Keywords: selenium; cardiovascular; heart; selenoproteins; Keshan's Disease

1. Introduction

Selenium (Se) is an essential trace element necessary for a variety of biological functions in animals, including cardiovascular function. Se deficiency has been linked to different cardiovascular diseases, including cardiomyopathies such as Keshan's disease (KD) [1,2], heart failure [3,4], and myocardial infarction [5]. Although there has been increasing evidence of the importance of Se for optimal cardiovascular function, the role of Se in cardiovascular syndromes, particularly under dietary Se deficiency, remains only partially understood.

The main effects of Se are manifested when it is part of the catalytic center of selenoproteins in the form of the amino acid selenocysteine. Se compounds obtained in the diet are rapidly metabolized via the trans-selenation pathway or reduced in the presence of glutathione (GSH), to allow for the production of the common intermediate, selenocysteine (Sec) [6], an amino acid that is then incorporated into peptide chains to form selenoproteins. Interestingly, for Sec to be used in selenoprotein translation, it needs to be decomposed into selenide by an enzyme called selenocysteine lyase (SCLY) [7], which will then allow for its synthesis attached to a specific tRNA, a process thoroughly reviewed elsewhere [8]. Briefly, the selenophosphate synthetase enzyme (SEPHS2) utilizes the selenide released from SCLY to produce monoselenophosphate for Sec biosynthesis. Sec is then attached to its tRNA and utilized in selenoprotein translation.

Overall, 25 genes encoding for selenoproteins have been found in humans, and 24 in rodents [9]. Selenoproteins are expressed in a wide range of tissues and have a variety of functions, including curbing free radicals, controlling calcium flux, and maintaining thyroid hormone levels [10]. Several of these selenoproteins have vital roles in the heart, including glutathione peroxidases (GPXs) [11,12], iodothyronine deiodinases (DIOs) [13],

thioredoxin reductases (TXNRDs) [14], selenoprotein W (SELENOW) [15], selenoprotein P (SELENOP) [16], MsrB1 (previously known as selenoprotein R) [17], selenoprotein T (SELENOT) [18], and selenoprotein K (SELENOK) [19]. Low Se levels disrupt the synthesis of a subgroup of stress-induced selenoproteins, including GPX1, leading to the shortage of one or more of these crucial proteins in the heart [3], with a potential impact on overall cardiovascular health [20]. In fact, removal of selenoproteins by deleting Sec tRNA in the heart and skeletal muscle resulted in sudden cardiac arrest due to increased oxidative stress and inflammation [21]. Nevertheless, Se metabolism and the molecular mechanism by which selenoproteins participate in and regulate heart function has not been completely determined, particularly under Se deficiency. Therefore, a better understanding of the molecular mechanisms behind Se and selenoproteins involvement in cardiac function is needed to explain how Se deficiency may contribute to the development of cardiovascular disease. As selenoproteins involvement in the heart have been reviewed previously [3,10,22], we aim in this review to briefly cover select key selenoproteins and focus mainly on recent studies (within the past five years), particularly the current knowledge about Se metabolism and cardiac selenoproteins in the context of cardiovascular diseases, updating what has been previously discussed in the literature [10,22]. We focus on the role of Se metabolism and some selenoproteins in the heart function, its involvement in the pathogenesis of cardiovascular diseases, and finally debate Se as a potential therapeutic or preventative supplement for cardiovascular syndromes.

2. Methods

The following criteria were used to select references for this review:
Databases searched: Pubmed (NLM).
Keywords searched: selenium, selenium metabolism, selenium deficiency, selenoproteins, cardiovascular diseases, myocardial infarction, ischemia reperfusion injury, atherosclerosis, oxidative stress, Keshan's Disease, heart, cardiovascular, glutathione peroxidase, thioredoxin reductase, iodothyronine deiodinases, microRNAs, Selenoprotein P, Selenoprotein T, Selenoprotein K, Selenoprotein W, selenite, selenomethionine, Selenoprotein MsrB1, Selenoprotein S, and Selenoprotein P.

3. The relationship between Se Deficiency and Cardiovascular Health

The first evidence of Se involvement in cardiovascular function came from the discovery that Se deficiency was involved in KD, a severe form of cardiomyopathy that is sometimes fatal. It is an endemic disease, first found in Keshan County in northeast China. KD is characterized by myocardial necrosis, calcification, and fibrosis, leading to various clinical manifestations such as cardiogenic shock, cardiac arrhythmias, and congestive heart failure [23]. Observations that KD shared similar morphology to white muscle disease, a degenerative cardiac and skeletal muscle disease found in foals from Se-poor areas [24], provided a clue that Se deficiency may play a role in KD [2]. Supporting this hypothesis, administration of oral Se reversed Se deficiency and improved outcomes of KD [25,26]. Although there is some debate about whether Se deficiency is the primary cause of KD as there are several additional underlying causes, it remains until now the most convincing possibility [1,26].

Further strengthening the argument that Se is important in cardiovascular function is that Se deficiency also contributes to the pathogenesis of other cardiomyopathies and heart failure. A recently reported rare case of dilated cardiomyopathy in a young boy was attributed to chronic starvation and a severe Se deficiency [27]. Notably, nutritional support with Se supplementation resulted in the reversal of the disease. Another case recently reported involved a malnourished woman with Se deficiency that resulted in cardiomyopathy, with the condition being reversed by Se treatment [28].

Se deficiency also is associated with heart failure [4,29–31]. Serum Se concentrations measured in a large European cohort of patients with worsening heart failure were low in 70% of these patients. Low Se levels in these patients were associated with a poorer quality of life, impaired exercise capacity, and an inferior prognosis with worsening heart failure [4]. Moreover, this same study demonstrated human cardiomyocytes generated from human pluripotent stem cells (hPSCs) cultured in low Se resulted in reduced mitochondrial function and increased reactive oxygen species (ROS), thereby showing impairment of key metabolic processes. Despite the increasing evidence of Se involvement in the pathogenesis of cardiovascular diseases, molecular mechanisms linking Se deficiency to cardiovascular diseases remain to be elucidated.

Potential Mechanisms behind Se Deficiency and Cardiovascular Disease

With the likely linkage between Se and cardiovascular health, studies have been conducted to examine the mechanism behind this association. Se has been suggested to participate in cell survival [32–34], a role reviewed recently [35]. The key findings in the studies connecting Se to cardiomyocyte survival is that, predominantly, Se deficiency increased apoptosis while inhibiting autophagy in the heart. Pro-apoptotic proteins, such as cleaved caspases-3, -8, and -9 and Bax, were upregulated, while anti-apoptotic ones, such as BCL-2, were reduced [32–34]. Combined, these findings suggest that Se regulates cardiomyocyte apoptosis.

Additional reports suggest a role of microRNAs (miRNAs) during Se deficiency in sustaining cardiovascular function. MiRNAs are an emerging topic in cardiovascular research as they have been shown to play an important role in a variety of cardiovascular diseases. These roles include cell survival, inflammation, and curbing of oxidative stress. Therefore, it has been hypothesized that miRNAs could also be regulating or contributing in cardiomyopathies related to Se deficiency. Studies using miRNA array or miRNA specific omics identified potential miRNAs that could be involved in such regulation. One study examined miRNA expression in Se-deficient rats using an miRNA array and found the cardiac dysfunction of the Se-deficient rat to be associated with five upregulated microRNAs, namely, miR-374, miR-16, miR-199a-5p, miR-195, and miR-30e, as well as three downregulated miRNAs, namely, miR-3571, miR-675, and miR-450a [36]. Another report used microRNAome analysis to examine miRNAs in the myocardium of Se-deficient chickens and found miR-2954 to have increased expression [37]. Bioinformatic analysis predicted phosphoinositide 3-kinase (PI3K), a key protein regulating cell apoptosis and autophagy, as the target gene of miR-2954. Overexpression of this miR led to autophagy and apoptosis of myocardial cells through the regulation of the PI3K pathway. This same microRNAome analysis also identified miR-200a-5p, and its target gene for ring finger protein 11 (RNF11), as triggering necroptosis in cardiomyocytes [38]. RNF11 is involved in the necroptosis pathway, a form of programmed necrosis. Knockdown of miR-200a-5p in Se-deficient cardiomyocytes resulted in enhanced cell survival after treatment with z-VAD-fmk, a necroptosis inducer. It is interesting to note that the miRNAs found by these microRNAomics are mainly involved in cell death pathways. This is consistent with the previous findings that Se deficiency leads to the increase in pro-cell death pathways. Further investigation into the interplay between miRNAs and Se in the heart is needed to determine how both contribute to molecular mechanisms leading to cardiomyopathy and other cardiovascular diseases.

4. Se Metabolism in the Heart

Se is an essential trace element that supports heart function [39]. It is vital to maintain adequate Se levels in the body as both low and excess levels of Se can have detrimental effects to cardiovascular health. As we previously mentioned, Se deficiency has been linked to cardiomyopathies, including KD and heart failure [3,29,40,41]. Excess Se intake, on the other hand, may result in severe toxicity and cardiac symptoms that may be fatal [42,43]. Although it is clear that adequate Se levels maintain normal cardiovascular function, a comprehensive understanding of the molecular mechanisms linking how Se metabolism contributes to cardiovascular disease is still lacking. Se exists in both inorganic and organic chemical forms, entering the food chain through plants and microorganisms via uptake by sulfate transporters. These forms include selenomethionine (SeMet), Sec, selenite, selenate, and selenious acid, among others. Depending on the chemical form of Se ingested, the Se compound may need to be either reduced (inorganic forms) or metabolized using the trans-selenation pathway (most organic forms) before it can be incorporated into a selenoprotein as Sec [7]. These mechanisms of Se metabolism have been demonstrated to occur in cells that are highly dependent on Se, such as hepatocytes [44–46]; however, they have not been determined in cardiomyocytes. Below we summarize how Se might be metabolized in the heart and how a diet deficient in Se may contribute to the development of several cardiovascular diseases (Figure 1). This will be discussed in greater detail in subsequent sections.

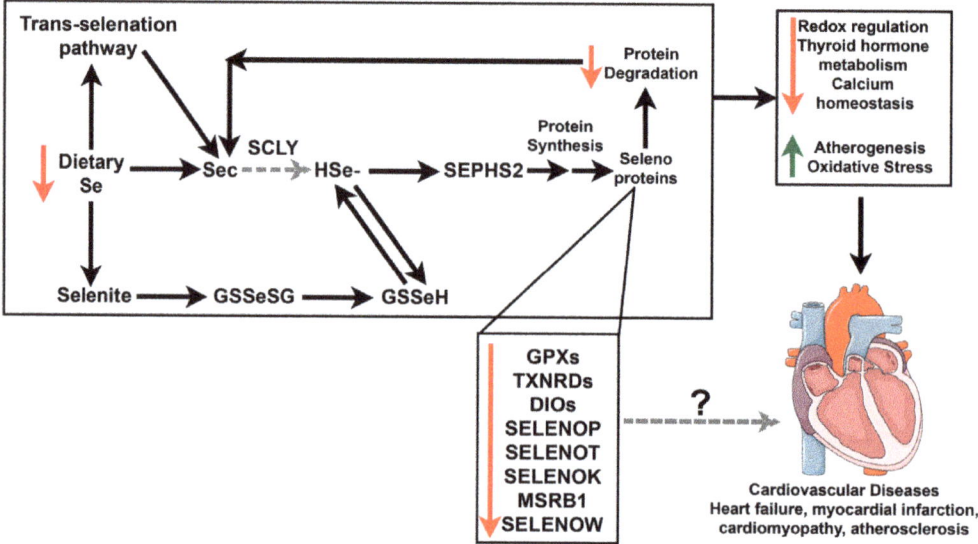

Figure 1. Potential mechanism of how Se deficiency impacts Se metabolism and contributes to cardiovascular diseases. Se acquired through the diet can go through the trans-selenation pathway and be directly converted to Sec, becoming hydrogen selenide. In cases of low Se, stress-responsive selenoprotein synthesis may be affected. Those selenoproteins with known function in the heart are shown above but it is still unknown how some of these proteins contribute to the development of cardiovascular disease in cases of Se deficiency. Decreased levels of Se may negatively impact redox regulation, thyroid hormone metabolism, and calcium flux while increasing atherogenesis and oxidative stress. This, in turn, may lead to several cardiovascular diseases, including heart failure, myocardial infarction, cardiomyopathy, and atherosclerosis. Red arrows indicate processes and proteins that are decreased while green arrows indicate processes that are increased during Se deficiency. Black arrows point to known relationships, and dashed gray lines indicate the relationships that have not been established yet in the heart. The "?" indicates mechanisms that have not been determined. Se, selenium; Sec, selenocysteine; HSe–, hydrogen selenide; SEPHS2, selenophosphate synthetase 2; GPX, glutathione peroxidase; TXNRDs, thioredoxin reductases; DIO, iodothyronine deiodinases; SELENOP, selenoprotein P; SELENOT, selenoprotein T; SELENOK, selenoprotein K; MSRB1, methionine sulfoxide reductase B1; SELENOW, selenoprotein W. Heart was used from Servier Medical Art (smart.servier.com).

4.1. Trans-Selenation Enzymes in the Heart

An essential component of Se metabolism is the trans-selenation pathway that produces Sec as a byproduct of selenomethionine (SeMet) metabolism [44]. There are three main enzymes involved in this process: cystathionine beta-synthase (CBS), cystathionine gamma-lyase (CGL), and SCLY. CGL and CBS sequentially convert selenohomocysteine to Sec while SCLY decomposes L-Sec into L-alanine and hydrogen selenide (H_2Se) in various mammalian tissues [7]. It remains unknown whether SCLY has a similar function in the heart; however, CGL and CBS have both been purported to protect the heart against ischemic damage [47,48]. We will discuss the function of these enzymes and their potential role in the heart below.

4.1.1. Cystathionine Beta-Synthase (CBS)

CBS is the first enzyme in the transsulfuration pathway that catalyzes the conversion of serine and homocysteine to cystathionine and H_2O [49]. Functional knowledge about this enzyme's action in the heart is poor; however, a recent study, using a combination of sodium thiosulfate and a CGL inhibitor, propargylglycine (PAG), to treat the rat cardiac myoblast cell line H9C2 cells undergoing hypoxia and reoxygenation, or rat hearts undergoing ischemia/reperfusion (I/R), revealed a partial recovery from these states [48], indicating that CBS may play a protective role as the function of CGL was inhibited by PAG, therefore being ruled out as a player in this protective mechanism in the heart.

4.1.2. Cystathionine Gamma-Lyase (CGL)

CGL is a sulfide (H_2S)-producing enzyme with L-cysteine as its main substrate, and therefore a member of the transsulfuration pathway that follows the methionine cycle. In the heart, specific overexpression of CGL protected against heart failure induced by permanent coronary ligation and significantly improved survival in mice [47]. Exogenous H_2S administration at the time of reperfusion in mice also protected against detrimental left ventricular remodeling that can lead to heart failure, and preserved cardiac function by attenuating oxidative stress and mitochondrial dysfunction [47]. H_2S administration also led to isoproterenol–caffeine-induced left ventricular hypertrophy in rats through upregulation of myocardial CGL [50].

4.1.3. Selenocysteine Lyase (SCLY)

SCLY was first identified as an enzyme that decomposes L-Sec into L-alanine and H_2Se in eukaryotes [51]. Sec for SCLY decomposition is typically acquired in the diet, produced via the trans-selenation pathway or released from selenoprotein degradation [52]. This enzyme was first purified from pig liver [51], and has been detected in several human tissues, including the heart; however, it is found at modest levels [53], and has a relatively unexplored role in the heart. Interestingly, one study revealed that after treating H9C2 cells with the H_2Se homologue, hydrogen sulfide (H_2S), oxidative stress was significantly attenuated [54]. Moreover, H_2S treatment increased SCLY/H_2Se signaling and resulted in the increased activity and expression of multiple selenoproteins, including GPX1 and TXNRD2. Therefore, it is likely that SCLY plays a role in the heart, but whether it responds also to Se deficiency, as it does in other tissues, and whether it is involved in the pathogenesis of cardiovascular diseases, remains to be explored.

5. Selenoproteins in the Heart

Selenoprotein involvement in heart function has been well documented and reviewed previously [3,10,22]. Therefore, as abovementioned, we will briefly cover select key selenoproteins and focus mainly on recent studies (within the past five years). These proteins are discussed below:

5.1. Glutathione Peroxidases (GPXs)

GPXs neutralize reactive oxygen and nitrogen species by catalyzing the reduction of hydrogen peroxides (H_2O_2) to water. There are five Sec-containing GPX enzymes in humans: GPX1, GPX2, GPX3, GPX4, and GPX6 [55]. Of the five, GPX1 is the most well-documented in cardiac tissues and known to be cardioprotective against I/R injury in mice [11]. Mice lacking GPX1 exhibited increased susceptibility to I/R injury due to increased apoptosis. GPX1 also regulates oxidative stress; hypoxia inducible factor (HIF)-2a knockout (KO) mice exhibited increased oxidative stress that was associated with low levels of GPX1 [56]. Interestingly, GPX1 is a stress-responsive selenoprotein, i.e., it is tightly regulated by Se levels. It is possible that Se deficiency may contribute to downregulate its expression, deteriorating the heart capacity to respond to oxidative stress, contributing to poor cell survival.

GPX3 and GPX4 are also highly expressed in the heart although their role in cardiac tissue is not as clear as GPX1. GPX3, a mostly plasma GPX enzyme, scavenges ROS in extracellular spaces and in the vasculature. It has been shown to protect against stroke by regulating the bioavailability of nitric oxide [57] and was also found to be upregulated during myocardial hypertrophy [17]. GPX4 protects cellular lipids from oxidative damage. Although GPX4 has a crucial role in the cellular antioxidant defense, it is still unclear what the specific mechanism in which GPX4 participates is being affected in times of cardiac stress. In other organs, GPX4 is resistant to Se deficiency, although it is still unknown if this is true in the heart. GPX4 overexpression in the mitochondria protected neonatal rat cardiomyocytes against ischemic damage [58]. It has also been shown that GPX4 protects cellular lipids from oxidative damage during hypertrophy [17]. More recently, a proteomic study revealed that the downregulation of GPX4 exacerbates ferroptosis, a form of iron-dependent nonapoptotic cell death, during acute myocardial infarction (MI) [59]. This finding is supported by other studies that have revealed the ferroptosis inducer erastin to downregulate GPX4 expression in H9C2 rat cardiac myoblasts [60]. Moreover, inhibition of ferroptosis using another ferroptosis inhibitor, liproxstatin-1, protected hearts against ischemic damage and restored GPX4 expression [61]. These interesting studies suggest that increasing GPX4 expression in the heart may help prevent ferroptosis, a contributor to myocardial infarction (MI). Inhibition of ferroptosis has been repeatedly demonstrated to protect the heart against ischemic damage and may suggest a new role for GPX4 as regulating cardiovascular function [62].

5.2. Iodothyronine Deiodinases (DIOs)

The main function of the DIOs is to regulate thyroid hormone levels [63]. There are three DIO isoforms: DIO1, DIO2, and DIO3. DIO1 and DIO2 catalyze the release of outer ring iodine from thyronine hormones to convert the prohormone thyroxine (T4) to the active 3-3'-5-triiodothyronine (T3) form, while DIO3 inactivates both T3 and T4 by removing an inner ring iodine from the molecule. DIO2 is highly expressed in the heart as well as brown adipose tissue and the pituitary. DIO1 is predominantly expressed in the liver, kidneys, and thyroid, as well as many other tissues, while DIO3 is expressed in the placenta, uterus, brain, and fetal tissues [64]. Dysregulation of thyroid hormones mainly impacts myocardial development and differentiation, although there are indications that these hormones may also be involved in cardiac hypertrophy and I/R injury [65,66]. Overexpression of DIO2 in mice enhanced myocardial contractility in a pressure-overload hypertrophy model that was accompanied by increased expression of SERCA2a and improved contractility, likely due to increased sarcoplasmic reticulum (SR) Ca^{2+} uptake [65]. As thyroid hormones are known to regulate SERCA expression in skeletal muscle [67] and heart [68], and deiodinases control thyroid hormones, it is not surprising that upregulation of DIO2 also impacts SERCA expression.

Thyroid hormones also attenuate cardiac remodeling after MI [66,69]. Reduced plasma levels of these hormones, particularly T3, is associated with ventricular failure and increased expression of DIO3 [70,71]. Consistent with these earlier findings, a recent report used an infusion of 6 µg/kg/day of T3 prior to subjecting rats to I/R injury, and the T3 treatment improved cardiac function following the injury [13]. Interestingly, DIO1 and DIO2 expression were significantly increased in the area at risk (the area around the ischemic injury) in rats given T3 before I/R. In the remote zone (the region distant to the ischemic injury), DIO1 and DIO3 were downregulated in sham and rats not given T3 prior to I/R [13]. This suggests novel roles for all three isoforms of the DIOs in I/R injury. Additionally, an interesting study that examined the regulation of an miRNAs on DIO3 revealed that miR-214 was found to be co-expressed with DIO3 in cardiomyocytes post-MI. Fascinatingly, DIO3 expression preceded miR-214 expression in the left ventricle and locally suppressed the known cardioprotective effect of hormone T3. High expression of T3 significantly reduced miR-214, suggesting a possible negative feedback loop where miR-214 controls DIO3 expression [72]. This is a potential novel area for the field to expand upon, understanding how various miRNAs may fine-tune DIO expression and, consequently, T3 availability, in the context of cardiovascular diseases such as myocardial infarction.

5.3. Thioredoxin Reductases (TXNRDs)

In mammals, TXNRDs mainly regulate intracellular redox reactions for functions as varied as DNA synthesis, redox status of transcription factors (e.g., NF-kB and AP-1), immunomodulation, and regulation of apoptosis [73]. Thioredoxins are utilized by TXNRD enzymes that use NADPH/H+ as a reducing agent to regenerate reduced thioredoxins, which are used to reduce oxidized cysteine residues in cellular proteins. There are three TXNRD enzymes, TXNRD1, TXNRD2, and TXNRD3, and all have well-documented roles in cardiovascular function, such as mitigating oxidative stress in response to pressure overload-induced hypertrophy [14] and ameliorating left ventricular remodeling [74].

Recent studies have continued to examine the impact of the thioredoxins (Trxs) and TXNRDs in multiple cardiovascular diseases and during oxidative stress in rodent and cell models with one study exploring what happens to selenoprotein expression after knockdown of Trx in a Se-deficient chicken cardiomyocyte model [75]. Chicken cardiomyocytes were given low Se (0.033 mg/kg) and treated with siRNA to knockdown thioredoxins. By qRT-PCR, mRNA expression of several selenoproteins, including all three TXNRDs, were significantly reduced [75], indicating that the thioredoxin levels may be important in cardiomyocytes to regulate TXNRDs. However, no functional analysis was performed in this study; therefore, it is unknown whether this was a maladaptive or adaptive response. In another study, BALB/C mice were provided with excess iron to determine how iron affects the thioredoxin system in the mouse heart [76]. It was discovered that iron overload altered protein expression of the thioredoxin-interacting protein (TXNIP) and TXNRD1 but that gene expression of these proteins remained unchanged. It would be interesting if the field focused on how TXNRDs and Trxs interact when there is inadequate Se in the heart as there are few studies focusing on how Se deficiency impacts the function of TXNRDs and Trxs.

5.4. Selenoprotein P (SELENOP)

SELENOP is of particular interest as Se is largely bound to circulating SELENOP in the plasma, providing Se to several tissues. Nevertheless, there are conflicting reports in the literature as to whether SELENOP may be beneficial or detrimental to the heart. Low SELENOP levels are associated with increased risk of mortality in acute heart failure patients and all-cause cardiovascular mortality [77,78]. A decrease in circulating levels of SELENOP was also associated with a greater risk of metabolic syndrome in patients with documented cardiovascular disease [79]. Despite this clinical association, it is unclear if SELENOP is indeed necessary to prevent cardiovascular disease, as SELENOP KO mice subjected to I/R injury exhibited significantly reduced infarct sizes, indicating that less tissue was damaged, and cardiac apoptosis compared to WT mice. This reduction in tissue injury corresponded to an increase in phosphorylation of several proteins involved in the reperfusion injury salvage kinase (RISK) pathway, such as Akt and Erk [16]. SELENOP levels were also higher in patients with cardiogenic shock and complicating acute MI, and thirty-day mortality was significantly higher in patients with SELENOP levels above the 75th percentile 3 days post-MI [80].

5.5. Selenoprotein T (SELENOT)

SELENOT has been shown to be involved in cardiac development, as it has showed a dramatically increased expression following an ex vivo Langendorff I/R model. This upregulation suggests that SELENOT may not be required when heart function is normal, but can be activated in times of stress [18]. It was suggested that SELENOT regulates calcium homeostasis, and this upregulation effect may be a compensation by the heart to protect against calcium overload-mediated cell death [81]. It should be noted that studies analyzing SELENOT are very limited, and thus any current conclusions regarding the role of SELENOT in I/R injury should be taken with caution.

5.6. Selenoprotein K (SELENOK)

In the heart, SELENOK was first identified to have an antioxidant effect in cardiomyocytes [19]. Overexpression of SELENOK in cardiomyocytes using a recombinant adenovirus system attenuated ROS and protected against oxidative stress induced by H_2O_2 [19]. In addition to this protein's antioxidant capabilities, SELENOK has also been identified as a regulator of calcium flux in immune cells [82]. In the heart, immune cells may exacerbate tissue damage after MI due to increased inflammation and these cells contribute to the development of atherosclerosis. Interestingly, SELENOK itself may also play a role in atherogenesis. SELENOK expression was detected in aortic plaques, particularly in macrophages, and SELENOK KO animals exhibited reduced atherosclerosis as indicated by lesion formation, which may contribute to foam cell formation and atherogenesis [83].

5.7. Selenoprotein MsrB1

The role of selenoprotein MsrB1 in cardiac health has been reviewed previously [10]; therefore, we will only provide an update here. MsrB1 reduces methionine sulfoxide, a byproduct of ROS oxidation of methionine residues. MsrB1 was identified as a selenoprotein highly expressed in the heart during cardiac stress after T3 and isoproterenol treatment [17]. It is thought that MsrB1 may be upregulated as a compensatory response to prevent oxidative damage or induced to regulate actin remodeling during myocardial hypertrophy. Additional studies are required to assess the molecular mechanism behind MsrB1 upregulation during hypertrophy.

5.8. Selenoprotein W (SELENOW)

Like other selenoproteins, SELENOW serves as an antioxidant [84], acting in calcium regulation and redox regulation [85]. It may also be involved in muscle growth and differentiation as this protein was shown to be highly expressed in proliferating myoblasts but not differentiated ones [86]. SELENOW levels are highest in the heart, muscle, and brain in sheep and primates, but is low in rodent hearts [87], suggesting that its role in the heart is species-specific, potentially only in non-rodent mammals. There are few studies to confirm this, however, SELENOW expression was increased by Se treatment in myocardial chicken cells [15], and is therefore a cardiac selenoprotein sensitive to Se levels. Future studies can determine if Se deficiency modulates SELENOW actions and what role this enzyme might play in heart function.

5.9. Selenoprotein S (SELENOS)

Another selenoprotein that may be involved in cardiovascular function is SELENOS. In Finnish and Chinese cohorts, single nucleotide polymorphisms (SNPs) were associated with increased risk for coronary heart disease and ischemic stroke [88], as well as type 2 diabetes, which carries an increased risk of cardiovascular disease [89]. Nevertheless, despite the association between certain SNPs and cardiovascular disease, there have not been follow-up studies and not much is known about the role of this protein in the heart.

6. Se Supplementation for Mitigation of Cardiovascular Diseases

As mentioned before, Se plays a role in the pathogenesis and responses of several cardiovascular diseases, including cardiomyopathy, myocardial infarction, cardiovascular stress response, and hypertrophy, particularly under deficient intake. From early on, Se deficiency has been implicated in KD and heart failure, and over time it was shown that a subset of selenoproteins play vital roles during I/R injury, cardiomyopathy, and heart failure. Excess Se has been connected to decreased cardiac output in pigs [90] and increased incidence of type 2 diabetes [91], a syndrome that greatly increases the risk of cardiovascular disease [92]. Below, we will explore recent studies that investigate the effects of Se supplementation in the context of several cardiovascular diseases.

Clinically, low Se levels are associated with higher rates of all-cause mortality due to heart failure [4], increased rates of myocardial infarction [5], and cardiomyopathies [27,93]. It is therefore unsurprising that several studies have focused on Se supplementation to mitigate cardiovascular disease despite mixed results, likely due to a host of factors, including the different pharmacokinetics of various forms of Se, the range and duration of treatment, and whether Se was used alone or in combination with another therapy. Among the cardiovascular conditions that have employed Se supplementation in an attempt to mitigate the disease pathogenesis in animal/cell models or clinically, include all-cause cardiovascular mortality [94], peripartum cardiomyopathy [95], I/R injury [96,97], atherosclerosis [98], and coronary heart disease [99]. More rigorous studies are needed to determine solely the effects of Se supplementation on specific cardiovascular mechanisms and prognostic outcomes. Moreover, studies comparing the pharmacokinetics of multiple chemical forms of Se supplementation in heart health are lacking. We will discuss the studies so far that have utilized Se as a supplement to mitigate various cardiovascular diseases and discuss why their results have varied significantly.

6.1. Se Supplementation in Myocardial Infarction and I/R Injury

Se plays vital roles during MI, such as the reduction of ischemic injury and left ventricular remodeling, likely due to decreased oxidative stress. It was recently reported that Se deficiency was found in the majority of MI patients [5,100]. Moreover, patients with high Se levels exhibited the lowest prevalence of cardiovascular outcomes, including MI in an Inuit cohort in Canada [101]. This suggests that strategies aiming at Se adequacy, such as Se supplementation in deficient individuals or populations, may mitigate ischemic damage to the heart. Nevertheless, there have been limited mechanistic studies addressing this possibility. Pretreatment of rats with Se prior to in vivo I/R injury led to better cardiovascular outcomes [96]. Another study found that radioactive Se 75 in the form of selenide targets damaged tissue after myocardial I/R injury [97]. Based on markers of cell damage, such as neutrophil accumulation, cardiac troponin levels, and measurements of cardiac function, it was concluded that selenide treatment in solution reduced damage to the heart [97]. Intriguingly, this was the only inorganic form of Se that was capable of doing so as treatment with reduced Se, oxidized Se, and selenite had no effect. As different forms of Se have varying pharmacokinetics, besides also being absorbed differentially by the gut microbiome [102], these differences serve as reminders when investigating whether a particular form of Se may be suitable to use as a supplement to protect the heart from ischemic damage.

The different pharmacokinetics of Se compounds may be one reason why Se supplementation as a strategy to prevent or mitigate I/R injury has had mixed results, although studies in this area are limited. A recent report demonstrated SeMet was effective at preventing necrotic cell death in a rat H9C2 myoblasts [103]. However, SeMet only modestly improved cardiac function and was incapable of preventing remodeling following I/R injury in Wistar rats. Another chemical form of Se, the inorganic selenite, was used in rats undergoing I/R injury during cardiovascular surgery. Selenite suppressed tissue damage as indicated by markers of cardiac injury such as lactate dehydrogenase and cardiac troponin but did not prevent the production of inflammatory cytokines such as IL-6 and TNF-α [104]. There have been combination therapies as well, although the added complication of mixing compounds compromises the ability to distinguish if, and which, Se form is indeed the primary effector for improving I/R injury outcomes. One of these tested therapies utilized aloe vera biomacromolecules conjugated with Se, finding that pretreatment of the compound decreased infarct sizes, cardiac injury markers, and cardiomyocyte apoptosis [105]. Clinically, a combination therapy of Se and coenzyme Q10 was purported to have beneficial effects in preventing all-cause mortality from cardiovascular disease, including in those individuals with ischemic heart disease. Moreover, this effect continued for 12 years after supplementation was stopped [94,106]. While intriguing, this study is problematic in that the protection from Se is unclear, since the subjects were also treated with coenzyme Q10. Therefore, studies that focus solely on the effect of Se as a therapeutic treatment to determine if Se supplementation is cardioprotective in MI and I/R injury are needed.

6.2. Se Supplementation in Atherosclerosis

Atherosclerosis is a disease of the arteries that develops slowly over many years and is characterized by fat and cholesterol deposition in large and medium sized arteries. Mechanistically, atherosclerosis develops through the increased transcytosis of low-density lipoprotein (LDL), increased inflammation, endothelial dysfunction, and leukocyte migration into the arterial intima where mononuclear phagocytes proliferate and engulf lipids. After enough lipids have been engulfed, these phagocytes eventually transform into foam cells. Excessive foam cell formation results in the accumulation of cholesterol esters, and is termed a fatty streak. In the fatty streak, lymphocytes and macrophages secrete inflammatory cytokines and enhance vascular smooth muscle cell (VSMC) migration into the intima. This, in turn, thickens the arterial wall, and the fatty streak evolves into a stable plaque. Eventually, in more advanced stages, these VSMCs proliferate, secrete extracellular matrix proteins, and generate a fibrous cap, while the apoptotic and necrotic cell death of foam cells produces what is termed a necrotic core [107]. Increased oxidative stress is considered a major contributor to the development of atherosclerosis; ROS are key mediators of inflammatory signaling pathways that lead to atherogenesis and the formation of the fatty streak [108]. As selenoproteins are heavily involved in the protection against oxidative stress, it is likely that Se plays a role in atherogenesis. Indeed, several studies have previously shown in various animal models that Se supplementation prevents atherosclerosis [109–112]. More recently, SeMet supplementation was tested in a mouse model of atherosclerosis [98]. Apolipoprotein E-deficient (ApoE$^{-/-}$) mice were placed on a high-fat diet without supplemental Se or were given a high fat diet with 2 mg/kg of SeMet for 6 or 12 weeks. After either 6 or 12 weeks of SeMet treatment, the study demonstrated a significant decrease in lesion burden, which was accompanied by the formation of a more stabilized plaque. To concur with a role of selenoproteins in this protective effect, GPX1 expression and activity were increased. Interestingly, an insertion and deletion (INDEL) SNP in SCLY, the enzyme that catalyzes a key step in SeMet metabolism, has also been linked with increased risk to develop atherosclerosis in Mexican-American subjects [113].

More recently, studies have focused on using engineered forms of Se such as Se nanoparticles [114,115] and Se quantum dots [116] to determine if they prevent atherosclerosis in murine and rat models of atherosclerosis. Animals treated with 50 mg/kg/day of Se nanoparticles (SeNPs) for 8 weeks significantly attenuated vascular injury [114]. This was associated with significantly lower levels of triglycerides, total cholesterol, and LDL-cholesterol in the serum of mice given the SeNPs. Oxidative stress, as measured by serum malondialdehyde (MDA) levels, was significantly lower compared with the control, and corresponded with an increase in the activity of antioxidant enzymes, GPX, and super oxide dismutase (SOD), suggesting SeNPs may mitigate the development of atherosclerosis, at least in a murine model. Indeed, the other study using SeNPs by the same group demonstrated that SeNP administration for 12 weeks significantly decreased atherosclerotic lesions in ApoE$^{-/-}$ mice [115]. This was again associated with a decrease in hyperlipidemia and oxidative stress. Finally, rats or ApoE$^{-/-}$ mice on a high-fat diet were treated with Se quantum dots (SeQDs) at a dose of 0.1 mg/kg/day, or a combination of lithium chloride (LiCl) and quantum dots [116]. SeQDs protected against endothelial dysfunction in rats, an effect associated with the inhibition of NHE1 by LiCl, a pathway known to protect against endothelial dysfunction. In ApoE$^{-/-}$ mice, SeQDs decreased the serum nitric oxide levels, attenuated endothelial dysfunction, and inhibited the formation of atherosclerotic plaques. It is evident that, at least in rodent models, Se supplementation is beneficial to protect against atherosclerosis; however, it remains to be seen if this is also true in humans, as clinical studies are currently limited. Moreover, the mechanism by which both SeQDs and SeNPs are being processed inside the cells to exert these effects is still undefined.

6.3. Se Supplementation in Other Cardiovascular Diseases

Aside from MI and I/R injury, Se supplementation for the prevention of other cardiovascular diseases, such as peripartum cardiomyopathy and coronary heart disease, has been reported, again with mixed results. Patients with peripartum cardiomyopathy (left ventricular ejection fraction < 45%) and Se deficiency (< 70 mg/L) were randomly assigned to receive oral SeMet (200 µg/day) for 3 months [95]. While symptoms of heart failure were reduced in the patients given SeMet, the percentage of left ventricular ejection fraction (LVEF) remained similar between the control group and the Se-supplemented group. It is unknown whether a longer course of SeMet treatment would eventually improve the LVEF; however, the results suggest that heart failure was somewhat mitigated in the Se-supplemented group. Similarly, there have been multiple randomized controlled trials using different forms of Se to mitigate coronary heart disease with mixed results [99]. Meta-analysis of these studies revealed that Se supplementation decreased serum C-reactive protein levels and elevated GPX1, but it had no effect on mortality from coronary heart disease, nor did it alter any lipid profiles.

7. Discussion

Since the discovery of Se involvement in Keshan's Disease more than fifty years ago [2,23,93], the knowledge of the involvement of Se in cardiovascular diseases has expanded significantly. Se is now known to participate in a host of different cardiovascular disorders, including myocardial infarction, heart failure, cardiomyopathies, atherosclerosis, and coronary heart disease, as Se deficiency is associated with increased risk of these cardiovascular diseases [4,5,30,99,117]. Therefore, it is unsurprising that Se supplementation has been explored as a potential therapeutic to treat several of these cardiovascular disorders when they are linked to a deficient intake status for Se. However, assessing the benefits of Se supplementation has proven challenging as the duration of Se treatment, dose, and the varying pharmokinetics of different chemical forms of Se have made it difficult to assess the therapeutic potential of Se supplementation, at least in clinical studies. Combination studies have also played a role as many clinical studies have combined Se with another nutritional supplement, making it problematic to determine the effects of Se alone. Experimental animal models using Se have been more successful, likely due to the ability to control more factors, such as the environment and diet, among other factors. Due to these complicating factors, it is still unknown whether Se supplementation can be useful as a nutritional supplement for patients with cardiovascular disease. More rigorous studies that focus solely on Se are needed to assess the true therapeutic potential of Se to mitigate cardiovascular diseases, or new targets such as selenoproteins themselves.

Like Se, significant progress has been made elucidating the roles of different selenoproteins in the heart. Selenoproteins serve as antioxidants, regulators of oxidative stress, controllers of calcium flux, and mediators of thyroid hormones. They may also play a role in immune cell migration, contributing to atherogenesis [83]. However, many selenoproteins are still understudied in the heart, particularly in disease situations, and it remains unknown what these proteins are doing in these contexts. There is some evidence that selenoproteins regulate/are regulated by miRNAs [38,72]—post-transcriptional regulators of gene expression that are involved in numerous signaling pathways. The miRNAs and their selenoprotein targets potentially provide hundreds of novel targets for study and it is essential that we learn more about the mechanisms behind selenoprotein and miRNA involvement in cardiovascular diseases, as they could eventually open new avenues for therapies.

8. Conclusions

Much has been learned about the role of selenium and selenoproteins in the heart; however, there remains substantial work to be done, particularly in studying how selenium deficiency impacts selenoproteins during disease conditions. These studies could significantly improve our understanding of how selenium and selenoproteins work on a molecular basis during the pathogenesis of cardiovascular diseases, and potentially lay the foundation for the development of novel therapeutics or improved nutritional guidance.

Author Contributions: Writing—original draft preparation, B.K.S., N.A. and L.A.S.; writing—review and editing, B.K.S., N.A. and L.A.S. All authors have read and agreed to the published version of the manuscript.

Funding: Work in our laboratory is funded by grants R01 DK128390 from the National Institute of Diabetes and Digestive and Kidney Diseases and 20ADVC-102166 from the Hawaii Community Foundation (L.A.S.), and start-up funds.

Institutional Review Board Statement: Not applicable.

Informed Consent Statement: Not applicable.

Data Availability Statement: Not applicable.

Conflicts of Interest: The authors declare no conflict of interest.

References

1. Shi, Y.; Yang, W.; Tang, X.; Yan, Q.; Cai, X.; Wu, F. Keshan Disease: A Potentially Fatal Endemic Cardiomyopathy in Remote Mountains of China. *Front. Pediatr.* **2021**, *9*, 576916. [CrossRef]
2. Yu, W.H. A study of nutritional and bio-geochemical factors in the occurrence and development of Keshan disease. *Jpn. Circ. J.* **1982**, *46*, 1201–1207.
3. Al-Mubarak, A.A.; van der Meer, P.; Bomer, N. Selenium, Selenoproteins, and Heart Failure: Current Knowledge and Future Perspective. *Curr. Heart Fail. Rep.* **2021**, *18*, 122–131. [CrossRef] [PubMed]
4. Bomer, N.; Grote Beverborg, N.; Hoes, M.F.; Streng, K.W.; Vermeer, M.; Dokter, M.M.; Ijmker, J.; Anker, S.D.; Cleland, J.G.F.; Hillege, H.L.; et al. Selenium and outcome in heart failure. *Eur. J. Heart Fail.* **2020**, *22*, 1415–1423. [CrossRef] [PubMed]
5. Fraczek-Jucha, M.; Szlosarczyk, B.; Kabat, M.; Czubek, U.; Nessler, J.; Gackowski, A. Low triiodothyronine syndrome and serum selenium status in the course of acute myocardial infarction. *Pol. Merkur. Lekarski* **2019**, *47*, 45–51. [PubMed]
6. Evenson, J.K.; Sunde, R.A. Metabolism of Tracer (75)Se Selenium From Inorganic and Organic Selenocompounds Into Selenoproteins in Rats, and the Missing (75)Se Metabolites. *Front. Nutr.* **2021**, *8*, 699652. [CrossRef]
7. Seale, L.A. Selenocysteine beta-Lyase: Biochemistry, Regulation and Physiological Role of the Selenocysteine Decomposition Enzyme. *Antioxidants* **2019**, *8*, 357. [CrossRef]
8. Labunskyy, V.M.; Hatfield, D.L.; Gladyshev, V.N. Selenoproteins: Molecular pathways and physiological roles. *Physiol. Rev.* **2014**, *94*, 739–777. [CrossRef] [PubMed]
9. Kryukov, G.V.; Castellano, S.; Novoselov, S.V.; Lobanov, A.V.; Zehtab, O.; Guigo, R.; Gladyshev, V.N. Characterization of mammalian selenoproteomes. *Science* **2003**, *300*, 1439–1443. [CrossRef]
10. Rose, A.H.; Hoffmann, P.R. Selenoproteins and cardiovascular stress. *Thromb. Haemost.* **2015**, *113*, 494–504. [CrossRef]
11. Yoshida, T.; Watanabe, M.; Engelman, D.T.; Engelman, R.M.; Schley, J.A.; Maulik, N.; Ho, Y.S.; Oberley, T.D.; Das, D.K. Transgenic mice overexpressing glutathione peroxidase are resistant to myocardial ischemia reperfusion injury. *J. Mol. Cell Cardiol.* **1996**, *28*, 1759–1767. [CrossRef]
12. Yoshida, T.; Maulik, N.; Engelman, R.M.; Ho, Y.S.; Magnenat, J.L.; Rousou, J.A.; Flack, J.E., 3rd; Deaton, D.; Das, D.K. Glutathione peroxidase knockout mice are susceptible to myocardial ischemia reperfusion injury. *Circulation* **1997**, *96*, II-216–II-220.
13. Sabatino, L.; Kusmic, C.; Iervasi, G. Modification of cardiac thyroid hormone deiodinases expression in an ischemia/reperfusion rat model after T3 infusion. *Mol. Cell. Biochem.* **2020**, *475*, 205–214. [CrossRef]
14. Yamamoto, M.; Yang, G.; Hong, C.; Liu, J.; Holle, E.; Yu, X.; Wagner, T.; Vatner, S.F.; Sadoshima, J. Inhibition of endogenous thioredoxin in the heart increases oxidative stress and cardiac hypertrophy. *J. Clin. Investig.* **2003**, *112*, 1395–1406. [CrossRef]
15. Liu, W.; Yao, H.; Zhao, W.; Shi, Y.; Zhang, Z.; Xu, S. Selenoprotein W was Correlated with the Protective Effect of Selenium on Chicken Myocardial Cells from Oxidative Damage. *Biol. Trace Elem. Res.* **2016**, *171*, 419–426. [CrossRef] [PubMed]
16. Chadani, H.; Usui, S.; Inoue, O.; Kusayama, T.; Takashima, S.I.; Kato, T.; Murai, H.; Furusho, H.; Nomura, A.; Misu, H.; et al. Endogenous Selenoprotein P, a Liver-Derived Secretory Protein, Mediates Myocardial Ischemia/Reperfusion Injury in Mice. *Int. J. Mol. Sci.* **2018**, *19*, 878. [CrossRef] [PubMed]
17. Hoffmann, F.W.; Hashimoto, A.S.; Lee, B.C.; Rose, A.H.; Shohet, R.V.; Hoffmann, P.R. Specific antioxidant selenoproteins are induced in the heart during hypertrophy. *Arch. Biochem. Biophys.* **2011**, *512*, 38–44. [CrossRef] [PubMed]

18. Rocca, C.; Boukhzar, L.; Granieri, M.C.; Alsharif, I.; Mazza, R.; Lefranc, B.; Tota, B.; Leprince, J.; Cerra, M.C.; Anouar, Y.; et al. A selenoprotein T-derived peptide protects the heart against ischaemia/reperfusion injury through inhibition of apoptosis and oxidative stress. *Acta Physiol.* **2018**, *223*, e13067. [CrossRef] [PubMed]
19. Lu, C.; Qiu, F.; Zhou, H.; Peng, Y.; Hao, W.; Xu, J.; Yuan, J.; Wang, S.; Qiang, B.; Xu, C.; et al. Identification and characterization of selenoprotein K: An antioxidant in cardiomyocytes. *FEBS Lett.* **2006**, *580*, 5189–5197. [CrossRef] [PubMed]
20. Al-Mubarak, A.A.; Grote Beverborg, N.; Anker, S.D.; Samani, N.J.; Dickstein, K.; Filippatos, G.; van Veldhuisen, D.J.; Voors, A.A.; Bomer, N.; van der Meer, P. A Clinical Tool to Predict Low Serum Selenium in Patients with Worsening Heart Failure. *Nutrients* **2020**, *12*, 2541. [CrossRef]
21. Benstoem, C.; Goetzenich, A.; Kraemer, S.; Borosch, S.; Manzanares, W.; Hardy, G.; Stoppe, C. Selenium and its supplementation in cardiovascular disease–what do we know? *Nutrients* **2015**, *7*, 3094–3118. [CrossRef]
22. Xu, G.L.; Wang, S.C.; Gu, B.Q.; Yang, Y.X.; Song, H.B.; Xue, W.L.; Liang, W.S.; Zhang, P.Y. Further investigation on the role of selenium deficiency in the aetiology and pathogenesis of Keshan disease. *Biomed. Environ. Sci.* **1997**, *10*, 316–326.
23. Lofstedt, J. White muscle disease of foals. *Vet. Clin. N. Am. Equine Pract.* **1997**, *13*, 169–185. [CrossRef]
24. Loscalzo, J. Keshan disease, selenium deficiency, and the selenoproteome. *N. Engl. J. Med.* **2014**, *370*, 1756–1760. [CrossRef]
25. Zhu, Y.H.; Wang, X.F.; Yang, G.; Wei, J.; Tan, W.H.; Wang, L.X.; Guo, X.; Lammi, M.J.; Xu, J.H. Efficacy of Long-term Selenium Supplementation in the Treatment of Chronic Keshan Disease with Congestive Heart Failure. *Curr. Med. Sci.* **2019**, *39*, 237–242. [CrossRef]
26. Dasgupta, S.; Aly, A.M. Dilated Cardiomyopathy Induced by Chronic Starvation and Selenium Deficiency. *Case Rep. Pediatr.* **2016**, *2016*, 8305895. [CrossRef] [PubMed]
27. Munguti, C.M.; Al Rifai, M.; Shaheen, W. A Rare Cause of Cardiomyopathy: A Case of Selenium Deficiency Causing Severe Cardiomyopathy that Improved on Supplementation. *Cureus* **2017**, *9*, e1627. [CrossRef]
28. Burke, M.P.; Opeskin, K. Fulminant heart failure due to selenium deficiency cardiomyopathy (Keshan disease). *Med. Sci. Law* **2002**, *42*, 10–13. [CrossRef] [PubMed]
29. Mirdamadi, A.; Rafiei, R.; Kahazaipour, G.; Fouladi, L. Selenium Level in Patients with Heart Failure versus Normal Individuals. *Int J. Prev. Med.* **2019**, *10*, 210. [CrossRef] [PubMed]
30. Zhang, Z.; Chang, C.; Zhang, Y.; Chai, Z.; Li, J.; Qiu, C. The association between serum selenium concentration and prognosis in patients with heart failure in a Chinese population. *Sci. Rep.* **2021**, *11*, 14533. [CrossRef]
31. Shan, H.; Yan, R.; Diao, J.; Lin, L.; Wang, S.; Zhang, M.; Zhang, R.; Wei, J. Involvement of caspases and their upstream regulators in myocardial apoptosis in a rat model of selenium deficiency-induced dilated cardiomyopathy. *J. Trace Elem. Med. Biol.* **2015**, *31*, 85–91. [CrossRef]
32. Zhang, L.; Gao, Y.; Feng, H.; Zou, N.; Wang, K.; Sun, D. Effects of selenium deficiency and low protein intake on the apoptosis through a mitochondria-dependent pathway. *J. Trace Elem. Med. Biol.* **2019**, *56*, 21–30. [CrossRef]
33. Yang, J.; Zhang, Y.; Hamid, S.; Cai, J.; Liu, Q.; Li, H.; Zhao, R.; Wang, H.; Xu, S.; Zhang, Z. Interplay between autophagy and apoptosis in selenium deficient cardiomyocytes in chicken. *J. Inorg. Biochem.* **2017**, *170*, 17–25. [CrossRef] [PubMed]
34. Shalihat, A.; Hasanah, A.N.; Mutakin; Lesmana, R.; Budiman, A.; Gozali, D. The role of selenium in cell survival and its correlation with protective effects against cardiovascular disease: A literature review. *Biomed. Pharmacother.* **2021**, *134*, 111125. [CrossRef]
35. Xing, Y.; Liu, Z.; Yang, G.; Gao, D.; Niu, X. MicroRNA expression profiles in rats with selenium deficiency and the possible role of the Wnt/beta-catenin signaling pathway in cardiac dysfunction. *Int. J. Mol. Med.* **2015**, *35*, 143–152. [CrossRef] [PubMed]
36. Liu, Q.; Cai, J.; Gao, Y.; Yang, J.; Gong, Y.; Zhang, Z. miR-2954 Inhibits PI3K Signaling and Induces Autophagy and Apoptosis in Myocardium Selenium Deficiency. *Cell Physiol. Biochem.* **2018**, *51*, 778–792. [CrossRef] [PubMed]
37. Yang, T.; Cao, C.; Yang, J.; Liu, T.; Lei, X.G.; Zhang, Z.; Xu, S. miR-200a-5p regulates myocardial necroptosis induced by Se deficiency via targeting RNF11. *Redox Biol.* **2018**, *15*, 159–169. [CrossRef] [PubMed]
38. Deagen, J.T.; Butler, J.A.; Beilstein, M.A.; Whanger, P.D. Effects of dietary selenite, selenocystine and selenomethionine on selenocysteine lyase and glutathione peroxidase activities and on selenium levels in rat tissues. *J. Nutr.* **1987**, *117*, 91–98. [CrossRef]
39. Fairweather-Tait, S.J.; Bao, Y.; Broadley, M.R.; Collings, R.; Ford, D.; Hesketh, J.E.; Hurst, R. Selenium in human health and disease. *Antioxid. Redox Signal.* **2011**, *14*, 1337–1383. [CrossRef]
40. Oropeza-Moe, M.; Wisloff, H.; Bernhoft, A. Selenium deficiency associated porcine and human cardiomyopathies. *J. Trace Elem. Med. Biol.* **2015**, *31*, 148–156. [CrossRef]
41. Hadrup, N.; Ravn-Haren, G. Acute human toxicity and mortality after selenium ingestion: A review. *J. Trace Elem. Med. Biol.* **2020**, *58*, 126435. [CrossRef]
42. Spiller, H.A.; Pfiefer, E. Two fatal cases of selenium toxicity. *Forensic Sci. Int.* **2007**, *171*, 67–72. [CrossRef] [PubMed]
43. Seale, L.A.; Ha, H.Y.; Hashimoto, A.C.; Berry, M.J. Relationship between selenoprotein P and selenocysteine lyase: Insights into selenium metabolism. *Free Radic. Biol. Med.* **2018**, *127*, 182–189. [CrossRef]
44. Geillinger, K.E.; Rathmann, D.; Kohrle, J.; Fiamoncini, J.; Daniel, H.; Kipp, A.P. Hepatic metabolite profiles in mice with a suboptimal selenium status. *J. Nutr. Biochem.* **2014**, *25*, 914–922. [CrossRef] [PubMed]
45. Meredith, M.J. Cystathionase activity and glutathione metabolism in redifferentiating rat hepatocyte primary cultures. *Cell Biol. Toxicol.* **1987**, *3*, 361–377. [CrossRef] [PubMed]

46. Calvert, J.W.; Elston, M.; Nicholson, C.K.; Gundewar, S.; Jha, S.; Elrod, J.W.; Ramachandran, A.; Lefer, D.J. Genetic and pharmacologic hydrogen sulfide therapy attenuates ischemia-induced heart failure in mice. *Circulation* **2010**, *122*, 11–19. [CrossRef] [PubMed]
47. Kannan, S.; Boovarahan, S.R.; Rengaraju, J.; Prem, P.; Kurian, G.A. Attenuation of cardiac ischemia-reperfusion injury by sodium thiosulfate is partially dependent on the effect of cystathione beta synthase in the myocardium. *Cell Biochem. Biophys.* **2019**, *77*, 261–272. [CrossRef]
48. Jhee, K.H.; Kruger, W.D. The role of cystathionine beta-synthase in homocysteine metabolism. *Antioxid. Redox Signal.* **2005**, *7*, 813–822. [CrossRef]
49. Ahmad, A.; Sattar, M.A.; Rathore, H.A.; Abdulla, M.H.; Khan, S.A.; Azam, M.; Abdullah, N.A.; Johns, E.J. Up Regulation of cystathione gamma lyase and Hydrogen Sulphide in the Myocardium Inhibits the Progression of Isoproterenol-Caffeine Induced Left Ventricular Hypertrophy in Wistar Kyoto Rats. *PLoS ONE* **2016**, *11*, e0150137. [CrossRef]
50. Esaki, N.; Nakamura, T.; Tanaka, H.; Soda, K. Selenocysteine lyase, a novel enzyme that specifically acts on selenocysteine. Mammalian distribution and purification and properties of pig liver enzyme. *J. Biol. Chem.* **1982**, *257*, 4386–4391. [CrossRef]
51. Kurokawa, S.; Takehashi, M.; Tanaka, H.; Mihara, H.; Kurihara, T.; Tanaka, S.; Hill, K.; Burk, R.; Esaki, N. Mammalian selenocysteine lyase is involved in selenoprotein biosynthesis. *J. Nutr. Sci. Vitaminol.* **2011**, *57*, 298–305. [CrossRef]
52. Daher, R.; Van Lente, F. Characterization of selenocysteine lyase in human tissues and its relationship to tissue selenium concentrations. *J. Trace Elem. Electrolytes Health Dis.* **1992**, *6*, 189–194.
53. Greasley, A.; Zhang, Y.; Wu, B.; Pei, Y.; Belzile, N.; Yang, G. H2S Protects against Cardiac Cell Hypertrophy through Regulation of Selenoproteins. *Oxid. Med. Cell. Longev.* **2019**, *2019*, 6494306. [CrossRef]
54. Brigelius-Flohe, R.; Maiorino, M. Glutathione peroxidases. *Biochim. Biophys. Acta* **2013**, *1830*, 3289–3303. [CrossRef] [PubMed]
55. Scortegagna, M.; Ding, K.; Oktay, Y.; Gaur, A.; Thurmond, F.; Yan, L.J.; Marck, B.T.; Matsumoto, A.M.; Shelton, J.M.; Richardson, J.A.; et al. Multiple organ pathology, metabolic abnormalities and impaired homeostasis of reactive oxygen species in Epas1-/- mice. *Nat. Genet.* **2003**, *35*, 331–340. [CrossRef] [PubMed]
56. Freedman, J.E.; Frei, B.; Welch, G.N.; Loscalzo, J. Glutathione peroxidase potentiates the inhibition of platelet function by S-nitrosothiols. *J. Clin. Investig.* **1995**, *96*, 394–400. [CrossRef] [PubMed]
57. Hollander, J.M.; Lin, K.M.; Scott, B.T.; Dillmann, W.H. Overexpression of PHGPx and HSP60/10 protects against ischemia/reoxygenation injury. *Free Radic. Biol. Med.* **2003**, *35*, 742–751. [CrossRef]
58. Park, T.J.; Park, J.H.; Lee, G.S.; Lee, J.Y.; Shin, J.H.; Kim, M.W.; Kim, Y.S.; Kim, J.Y.; Oh, K.J.; Han, B.S.; et al. Quantitative proteomic analyses reveal that GPX4 downregulation during myocardial infarction contributes to ferroptosis in cardiomyocytes. *Cell Death Dis.* **2019**, *10*, 835. [CrossRef] [PubMed]
59. Bai, Y.T.; Chang, R.; Wang, H.; Xiao, F.J.; Ge, R.L.; Wang, L.S. ENPP2 protects cardiomyocytes from erastin-induced ferroptosis. *Biochem. Biophys. Res. Commun.* **2018**, *499*, 44–51. [CrossRef]
60. Feng, Y.; Madungwe, N.B.; Imam Aliagan, A.D.; Tombo, N.; Bopassa, J.C. Liproxstatin-1 protects the mouse myocardium against ischemia/reperfusion injury by decreasing VDAC1 levels and restoring GPX4 levels. *Biochem. Biophys. Res. Commun.* **2019**, *520*, 606–611. [CrossRef]
61. Baba, Y.; Higa, J.K.; Shimada, B.K.; Horiuchi, K.M.; Suhara, T.; Kobayashi, M.; Woo, J.D.; Aoyagi, H.; Marh, K.S.; Kitaoka, H.; et al. Protective effects of the mechanistic target of rapamycin against excess iron and ferroptosis in cardiomyocytes. *Am. J. Physiol. Heart Circ. Physiol.* **2018**, *314*, H659–H668. [CrossRef] [PubMed]
62. Schomburg, L.; Kohrle, J. On the importance of selenium and iodine metabolism for thyroid hormone biosynthesis and human health. *Mol. Nutr. Food Res.* **2008**, *52*, 1235–1246. [CrossRef] [PubMed]
63. Dentice, M.; Marsili, A.; Zavacki, A.; Larsen, P.R.; Salvatore, D. The deiodinases and the control of intracellular thyroid hormone signaling during cellular differentiation. *Biochim. Biophys. Acta* **2013**, *1830*, 3937–3945. [CrossRef] [PubMed]
64. Trivieri, M.G.; Oudit, G.Y.; Sah, R.; Kerfant, B.G.; Sun, H.; Gramolini, A.O.; Pan, Y.; Wickenden, A.D.; Croteau, W.; Morreale de Escobar, G.; et al. Cardiac-specific elevations in thyroid hormone enhance contractility and prevent pressure overload-induced cardiac dysfunction. *Proc. Natl. Acad. Sci. USA* **2006**, *103*, 6043–6048. [CrossRef]
65. Pantos, C.; Mourouzis, I.; Markakis, K.; Dimopoulos, A.; Xinaris, C.; Kokkinos, A.D.; Panagiotou, M.; Cokkinos, D.V. Thyroid hormone attenuates cardiac remodeling and improves hemodynamics early after acute myocardial infarction in rats. *Eur. J. Cardiothorac. Surg.* **2007**, *32*, 333–339. [CrossRef]
66. Simonides, W.S.; Thelen, M.H.; van der Linden, C.G.; Muller, A.; van Hardeveld, C. Mechanism of thyroid-hormone regulated expression of the SERCA genes in skeletal muscle: Implications for thermogenesis. *Biosci. Rep.* **2001**, *21*, 139–154. [CrossRef]
67. Wu, P.S.; Moriscot, A.S.; Knowlton, K.U.; Hilal-Dandan, R.; He, H.; Dillmann, W.H. Alpha 1-adrenergic stimulation inhibits 3,5,3'-triiodothyronine-induced expression of the rat heart sarcoplasmic reticulum Ca2+ adenosine triphosphatase gene. *Endocrinology* **1997**, *138*, 114–120. [CrossRef]
68. Pantos, C.; Mourouzis, I.; Markakis, K.; Tsagoulis, N.; Panagiotou, M.; Cokkinos, D.V. Long-term thyroid hormone administration reshapes left ventricular chamber and improves cardiac function after myocardial infarction in rats. *Basic Res. Cardiol.* **2008**, *103*, 308–318. [CrossRef]
69. Pol, C.J.; Muller, A.; Zuidwijk, M.J.; van Deel, E.D.; Kaptein, E.; Saba, A.; Marchini, M.; Zucchi, R.; Visser, T.J.; Paulus, W.J.; et al. Left-ventricular remodeling after myocardial infarction is associated with a cardiomyocyte-specific hypothyroid condition. *Endocrinology* **2011**, *152*, 669–679. [CrossRef]

70. Simonides, W.S.; Mulcahey, M.A.; Redout, E.M.; Muller, A.; Zuidwijk, M.J.; Visser, T.J.; Wassen, F.W.; Crescenzi, A.; da-Silva, W.S.; Harney, J.; et al. Hypoxia-inducible factor induces local thyroid hormone inactivation during hypoxic-ischemic disease in rats. *J. Clin. Investig.* **2008**, *118*, 975–983. [CrossRef]
71. Janssen, R.; Zuidwijk, M.J.; Muller, A.; van Mil, A.; Dirkx, E.; Oudejans, C.B.; Paulus, W.J.; Simonides, W.S. MicroRNA 214 Is a Potential Regulator of Thyroid Hormone Levels in the Mouse Heart Following Myocardial Infarction, by Targeting the Thyroid-Hormone-Inactivating Enzyme Deiodinase Type III. *Front. Endocrinol.* **2016**, *7*, 22. [CrossRef]
72. Arner, E.S.; Holmgren, A. Physiological functions of thioredoxin and thioredoxin reductase. *Eur. J. Biochem.* **2000**, *267*, 6102–6109. [CrossRef]
73. Ago, T.; Sadoshima, J. Thioredoxin and ventricular remodeling. *J. Mol. Cell. Cardiol.* **2006**, *41*, 762–773. [CrossRef]
74. Yang, J.; Hamid, S.; Liu, Q.; Cai, J.; Xu, S.; Zhang, Z. Gene expression of selenoproteins can be regulated by thioredoxin(Txn) silence in chicken cardiomyocytes. *J. Inorg. Biochem.* **2017**, *177*, 118–126. [CrossRef]
75. Altun, S.; Budak, H. The protective effect of the cardiac thioredoxin system on the heart in the case of iron overload in mice. *J. Trace Elem. Med. Biol.* **2021**, *64*, 126704. [CrossRef] [PubMed]
76. Jujic, A.; Melander, O.; Bergmann, A.; Hartmann, O.; Nilsson, P.M.; Bachus, E.; Struck, J.; Magnusson, M. Selenoprotein P Deficiency and Risk of Mortality and Rehospitalization in Acute Heart Failure. *J. Am. Coll Cardiol.* **2019**, *74*, 1009–1011. [CrossRef] [PubMed]
77. Schomburg, L.; Orho-Melander, M.; Struck, J.; Bergmann, A.; Melander, O. Selenoprotein-P Deficiency Predicts Cardiovascular Disease and Death. *Nutrients* **2019**, *11*, 1852. [CrossRef] [PubMed]
78. Gharipour, M.; Sadeghi, M.; Salehi, M.; Behmanesh, M.; Khosravi, E.; Dianatkhah, M.; Haghjoo Javanmard, S.; Razavi, R.; Gharipour, A. Association of expression of selenoprotein P in mRNA and protein levels with metabolic syndrome in subjects with cardiovascular disease: Results of the Selenegene study. *J. Gene Med.* **2017**, *19*. [CrossRef]
79. Buttner, P.; Obradovic, D.; Wunderlich, S.; Feistritzer, H.J.; Holzwirth, E.; Lauten, P.; Fuernau, G.; de Waha-Thiele, S.; Desch, S.; Thiele, H. Selenoprotein P in Myocardial Infarction With Cardiogenic Shock. *Shock* **2020**, *53*, 58–62. [CrossRef]
80. Hawkins, C.L. A therapeutic role for selenoprotein T in reducing ischaemia/reperfusion injury in the heart? *Acta Physiol.* **2018**, *223*, e13106. [CrossRef]
81. Verma, S.; Hoffmann, F.W.; Kumar, M.; Huang, Z.; Roe, K.; Nguyen-Wu, E.; Hashimoto, A.S.; Hoffmann, P.R. Selenoprotein K knockout mice exhibit deficient calcium flux in immune cells and impaired immune responses. *J. Immunol.* **2011**, *186*, 2127–2137. [CrossRef]
82. Meiler, S.; Baumer, Y.; Huang, Z.; Hoffmann, F.W.; Fredericks, G.J.; Rose, A.H.; Norton, R.L.; Hoffmann, P.R.; Boisvert, W.A. Selenoprotein K is required for palmitoylation of CD36 in macrophages: Implications in foam cell formation and atherogenesis. *J. Leukoc. Biol.* **2013**, *93*, 771–780. [CrossRef]
83. Whanger, P.D. Selenoprotein expression and function-selenoprotein W. *Biochim. Biophys. Acta* **2009**, *1790*, 1448–1452. [CrossRef]
84. Yao, H.; Fan, R.; Zhao, X.; Zhao, W.; Liu, W.; Yang, J.; Sattar, H.; Zhao, J.; Zhang, Z.; Xu, S. Selenoprotein W redox-regulated Ca2+ channels correlate with selenium deficiency-induced muscles Ca2+ leak. *Oncotarget* **2016**, *7*, 57618–57632. [CrossRef] [PubMed]
85. Loflin, J.; Lopez, N.; Whanger, P.D.; Kioussi, C. Selenoprotein W during development and oxidative stress. *J. Inorg. Biochem.* **2006**, *100*, 1679–1684. [CrossRef] [PubMed]
86. Whanger, P.D. Selenoprotein W: A review. *Cell Mol. Life Sci.* **2000**, *57*, 1846–1852. [CrossRef] [PubMed]
87. Alanne, M.; Kristiansson, K.; Auro, K.; Silander, K.; Kuulasmaa, K.; Peltonen, L.; Salomaa, V.; Perola, M. Variation in the selenoprotein S gene locus is associated with coronary heart disease and ischemic stroke in two independent Finnish cohorts. *Hum. Genet.* **2007**, *122*, 355–365. [CrossRef] [PubMed]
88. Zhao, L.; Zheng, Y.Y.; Chen, Y.; Ma, Y.T.; Yang, Y.N.; Li, X.M.; Ma, X.; Xie, X. Association of genetic polymorphisms of SelS with Type 2 diabetes in a Chinese population. *Biosci. Rep.* **2018**, *38*, BSR20181696. [CrossRef]
89. Nebbia, C.; Soffietti, M.G.; Zittlau, E.; Fink-Gremmels, J. Pathogenesis of sodium selenite and dimethylselenide acute toxicosis in pigs: Cardiovascular changes. *Res. Vet. Sci.* **1991**, *50*, 269–272. [CrossRef]
90. Stranges, S.; Marshall, J.R.; Natarajan, R.; Donahue, R.P.; Trevisan, M.; Combs, G.F.; Cappuccio, F.P.; Ceriello, A.; Reid, M.E. Effects of long-term selenium supplementation on the incidence of type 2 diabetes: A randomized trial. *Ann. Intern. Med.* **2007**, *147*, 217–223. [CrossRef]
91. Johansen, O.E. Cardiovascular disease and type 2 diabetes mellitus: A multifaceted symbiosis. *Scand. J. Clin. Lab. Investig.* **2007**, *67*, 786–800. [CrossRef]
92. Chen, J. An original discovery: Selenium deficiency and Keshan disease (an endemic heart disease). *Asia Pac. J. Clin. Nutr.* **2012**, *21*, 320–326.
93. Alehagen, U.; Aaseth, J.; Alexander, J.; Johansson, P. Still reduced cardiovascular mortality 12 years after supplementation with selenium and coenzyme Q10 for four years: A validation of previous 10-year follow-up results of a prospective randomized double-blind placebo-controlled trial in elderly. *PLoS ONE* **2018**, *13*, e0193120. [CrossRef]
94. Karaye, K.M.; Sa'idu, H.; Balarabe, S.A.; Ishaq, N.A.; Sanni, B.; Abubakar, H.; Mohammed, B.L.; Abdulsalam, T.; Tukur, J.; Mohammed, I.Y. Selenium supplementation in patients with peripartum cardiomyopathy: A proof-of-concept trial. *BMC Cardiovasc. Disord.* **2020**, *20*, 457. [CrossRef]

95. Tanguy, S.; Rakotovao, A.; Jouan, M.G.; Ghezzi, C.; de Leiris, J.; Boucher, F. Dietary selenium intake influences Cx43 dephosphorylation, TNF-alpha expression and cardiac remodeling after reperfused infarction. *Mol. Nutr. Food Res.* **2011**, *55*, 522–529. [CrossRef]
96. Iwata, A.; Morrison, M.L.; Blackwood, J.E.; Roth, M.B. Selenide Targets to Reperfusing Tissue and Protects It From Injury. *Crit. Care Med.* **2015**, *43*, 1361–1367. [CrossRef]
97. Zhang, Y.; Cartland, S.P.; Henriquez, R.; Patel, S.; Gammelgaard, B.; Flouda, K.; Hawkins, C.L.; Rayner, B.S. Selenomethionine supplementation reduces lesion burden, improves vessel function and modulates the inflammatory response within the setting of atherosclerosis. *Redox Biol.* **2020**, *29*, 101409. [CrossRef] [PubMed]
98. Ju, W.; Li, X.; Li, Z.; Wu, G.R.; Fu, X.F.; Yang, X.M.; Zhang, X.Q.; Gao, X.B. The effect of selenium supplementation on coronary heart disease: A systematic review and meta-analysis of randomized controlled trials. *J. Trace Elem. Med. Biol.* **2017**, *44*, 8–16. [CrossRef] [PubMed]
99. Fraczek-Jucha, M.; Kabat, M.; Szlosarczyk, B.; Czubek, U.; Nessler, J.; Gackowski, A. Selenium deficiency and the dynamics of changes of thyroid profile in patients with acute myocardial infarction and chronic heart failure. *Kardiol. Pol.* **2019**, *77*, 674–682. [CrossRef]
100. Hu, X.F.; Eccles, K.M.; Chan, H.M. High selenium exposure lowers the odds ratios for hypertension, stroke, and myocardial infarction associated with mercury exposure among Inuit in Canada. *Environ. Int.* **2017**, *102*, 200–206. [CrossRef] [PubMed]
101. Ferreira, R.L.U.; Sena-Evangelista, K.C.M.; de Azevedo, E.P.; Pinheiro, F.I.; Cobucci, R.N.; Pedrosa, L.F.C. Selenium in Human Health and Gut Microflora: Bioavailability of Selenocompounds and Relationship With Diseases. *Front. Nutr.* **2021**, *8*, 685317. [CrossRef]
102. Reyes, L.; Bishop, D.P.; Hawkins, C.L.; Rayner, B.S. Assessing the Efficacy of Dietary Selenomethionine Supplementation in the Setting of Cardiac Ischemia/Reperfusion Injury. *Antioxidants* **2019**, *8*, 546. [CrossRef]
103. Steinbrenner, H.; Bilgic, E.; Pinto, A.; Engels, M.; Wollschlager, L.; Dohrn, L.; Kellermann, K.; Boeken, U.; Akhyari, P.; Lichtenberg, A. Selenium Pretreatment for Mitigation of Ischemia/Reperfusion Injury in Cardiovascular Surgery: Influence on Acute Organ Damage and Inflammatory Response. *Inflammation* **2016**, *39*, 1363–1376. [CrossRef]
104. Yang, Y.; Yang, M.; Ai, F.; Huang, C. Cardioprotective Effect of Aloe vera Biomacromolecules Conjugated with Selenium Trace Element on Myocardial Ischemia-Reperfusion Injury in Rats. *Biol. Trace Elem. Res.* **2017**, *177*, 345–352. [CrossRef]
105. Alehagen, U.; Aaseth, J.; Johansson, P. Reduced Cardiovascular Mortality 10 Years after Supplementation with Selenium and Coenzyme Q10 for Four Years: Follow-Up Results of a Prospective Randomized Double-Blind Placebo-Controlled Trial in Elderly Citizens. *PLoS ONE* **2015**, *10*, e0141641. [CrossRef] [PubMed]
106. Libby, P. The changing landscape of atherosclerosis. *Nature* **2021**, *592*, 524–533. [CrossRef] [PubMed]
107. Bonomini, F.; Tengattini, S.; Fabiano, A.; Bianchi, R.; Rezzani, R. Atherosclerosis and oxidative stress. *Histol. Histopathol.* **2008**, *23*, 381–390. [CrossRef] [PubMed]
108. Wojcicki, J.; Rozewicka, L.; Barcew-Wiszniewska, B.; Samochowiec, L.; Juzwiak, S.; Kadlubowska, D.; Tustanowski, S.; Juzyszyn, Z. Effect of selenium and vitamin E on the development of experimental atherosclerosis in rabbits. *Atherosclerosis* **1991**, *87*, 9–16. [CrossRef]
109. Schwenke, D.C.; Behr, S.R. Vitamin E combined with selenium inhibits atherosclerosis in hypercholesterolemic rabbits independently of effects on plasma cholesterol concentrations. *Circ. Res.* **1998**, *83*, 366–377. [CrossRef]
110. Mehta, U.; Kang, B.P.; Kukreja, R.S.; Bansal, M.P. Ultrastructural examination of rabbit aortic wall following high-fat diet feeding and selenium supplementation: A transmission electron microscopy study. *J. Appl. Toxicol.* **2002**, *22*, 405–413. [CrossRef]
111. Krohn, R.M.; Lemaire, M.; Negro Silva, L.F.; Lemarie, C.; Bolt, A.; Mann, K.K.; Smits, J.E. High-selenium lentil diet protects against arsenic-induced atherosclerosis in a mouse model. *J. Nutr. Biochem.* **2016**, *27*, 9–15. [CrossRef] [PubMed]
112. Gao, C.; Hsu, F.C.; Dimitrov, L.M.; Okut, H.; Chen, Y.I.; Taylor, K.D.; Rotter, J.I.; Langefeld, C.D.; Bowden, D.W.; Palmer, N.D. A genome-wide linkage and association analysis of imputed insertions and deletions with cardiometabolic phenotypes in Mexican Americans: The Insulin Resistance Atherosclerosis Family Study. *Genet. Epidemiol.* **2017**, *41*, 353–362. [CrossRef] [PubMed]
113. Guo, L.; Xiao, J.; Liu, H.; Liu, H. Selenium nanoparticles alleviate hyperlipidemia and vascular injury in ApoE-deficient mice by regulating cholesterol metabolism and reducing oxidative stress. *Metallomics* **2020**, *12*, 204–217. [CrossRef] [PubMed]
114. Xiao, S.; Mao, L.; Xiao, J.; Wu, Y.; Liu, H. Selenium nanoparticles inhibit the formation of atherosclerosis in apolipoprotein E deficient mice by alleviating hyperlipidemia and oxidative stress. *Eur. J. Pharmacol.* **2021**, *902*, 174120. [CrossRef] [PubMed]
115. Zhu, M.L.; Wang, G.; Wang, H.; Guo, Y.M.; Song, P.; Xu, J.; Li, P.; Wang, S.; Yang, L. Amorphous nano-selenium quantum dots improve endothelial dysfunction in rats and prevent atherosclerosis in mice through Na(+)/H(+) exchanger 1 inhibition. *Vascul. Pharmacol.* **2019**, *115*, 26–32. [CrossRef]
116. Flores-Mateo, G.; Navas-Acien, A.; Pastor-Barriuso, R.; Guallar, E. Selenium and coronary heart disease: A meta-analysis. *Am. J. Clin. Nutr.* **2006**, *84*, 762–773. [CrossRef] [PubMed]
117. Yang, T.; Liu, T.; Cao, C.; Xu, S. miR-200a-5p augments cardiomyocyte hypertrophy induced by glucose metabolism disorder via the regulation of selenoproteins. *J. Cell Physiol.* **2019**, *234*, 4095–4103. [CrossRef] [PubMed]

Review

Selenium Deficiency Due to Diet, Pregnancy, Severe Illness, or COVID-19—A Preventable Trigger for Autoimmune Disease

Lutz Schomburg

Charité–Universitätsmedizin Berlin, Corporate Member of Freie Universität Berlin and Humboldt Universität zu Berlin, Institut für Experimentelle Endokrinologie, Cardiovascular–Metabolic–Renal (CMR)-Research Center, Hessische Straße 3-4, Charitéplatz 1, 10117 Berlin, Germany; lutz.schomburg@charite.de; Tel.: +49-30-450524289

Abstract: The trace element selenium (Se) is an essential part of the human diet; moreover, increased health risks have been observed with Se deficiency. A sufficiently high Se status is a prerequisite for adequate immune response, and preventable endemic diseases are known from areas with Se deficiency. Biomarkers of Se status decline strongly in pregnancy, severe illness, or COVID-19, reaching critically low concentrations. Notably, these conditions are associated with an increased risk for autoimmune disease (AID). Positive effects on the immune system are observed with Se supplementation in pregnancy, autoimmune thyroid disease, and recovery from severe illness. However, some studies reported null results; the database is small, and randomized trials are sparse. The current need for research on the link between AID and Se deficiency is particularly obvious for rheumatoid arthritis and type 1 diabetes mellitus. Despite these gaps in knowledge, it seems timely to realize that severe Se deficiency may trigger AID in susceptible subjects. Improved dietary choices or supplemental Se are efficient ways to avoid severe Se deficiency, thereby decreasing AID risk and improving disease course. A personalized approach is needed in clinics and during therapy, while population-wide measures should be considered for areas with habitual low Se intake. Finland has been adding Se to its food chain for more than 35 years—a wise and commendable decision, according to today's knowledge. It is unfortunate that the health risks of Se deficiency are often neglected, while possible side effects of Se supplementation are exaggerated, leading to disregard for this safe and promising preventive and adjuvant treatment options. This is especially true in the follow-up situations of pregnancy, severe illness, or COVID-19, where massive Se deficiencies have developed and are associated with AID risk, long-lasting health impairments, and slow recovery.

Keywords: autoimmune thyroid disease; diabetes mellitus; Graves' disease; Hashimoto thyroiditis; infection; inflammation; long-COVID; rheumatoid arthritis; selenoprotein P; sepsis

1. Selenium and Selenoproteins

The trace element selenium (Se) is a micronutrient that is notable for several reasons, e.g., its essentiality, high medical relevance, and unique biochemistry [1]. It is part of the proteinogenic amino acid selenocysteine (Sec), for which an elaborate biochemical pathway has evolved and been preserved in many species [2]. Sec is synthesized on a seryl-loaded tRNA as a template, and it is used via one of two specific tRNA$^{[Ser]Sec}$ isoforms [3,4], before being inserted during ribosomal translation into a small family of selenoproteins [5]. The distribution of Se to different organs and tissues, as well as to the different gene transcripts for selenoprotein biosynthesis, is hierarchically regulated and involves the two tRNA$^{[Ser]Sec}$ isoforms [6–8]. Importantly, some selenoproteins are essential for life and are preferentially synthesized; inactivation of the respective genes in mouse models turned out embryonically lethal [9,10]. There is some overlap, but also certain distinctive differences to their roles and essentiality in humans [11]. A growing number of inherited diseases provide evidence for the distinctive roles played by selenoproteins in human health [12]. Hereby, the essentiality of Se is impressively underlined, as originally highlighted in 1957 by Schwarz and Foltz studying liver necrosis associated with vitamin E deficiency [13]. The tissue degeneration observed was efficiently prevented by supplemental Se [14], via increased biosynthesis of selenoenzymes [15]. It was only recently that the tight interplay among vitamin E, selenoproteins, and the accumulation of lipid peroxides inducing tissue degeneration and damage was identified as a conserved and tightly coordinated mechanism of cell death, referred to as ferroptosis [16]. These insights contribute to a better understanding of the molecular interrelation of Se and selenoproteins with energy metabolism and damage, as well as aging and decay, as underlying bases for the protective effects of Se in acute and chronic diseases [17].

1.1. Model Systems, Inherited Diseases, and Hierarchical Selenium Supply

Besides being essential for embryonic development, the identified inherited diseases with mutations in genes encoding selenoproteins or in essential factors involved in Se metabolism provide a more refined view on the essential role of selenoproteins [18–20]. The first identified example was a central component of the biosynthesis machinery, i.e., the RNA-binding protein SECISBP2 involved in correctly decoding the codon UGA for directing co-translational Sec insertion [21]. The affected children presented with an altered thyroid hormone pattern, reduced selenoprotein levels, and delayed growth that was not correctable by Se supplementation [22]. A more comprehensive characterization of additional young and adult subjects with mutation in *SECISBP2* indicated a spectrum of phenotypes, many of which resemble findings known from specific mouse models, e.g., delayed bone development, male infertility, metabolic dysregulation, elevated stress sensitivity and, importantly, disrupting effects on the immune system [23]. A detailed overview on the immune-relevant selenoproteins and their role in lymphocyte development, immune response, and effects on cytokine release and signaling can be found in excellent pieces of work published elsewhere [24–30].

The strong effects observed in the human patients with inherited defects indicate the extreme end of a spectrum of phenotypes that may result from insufficient selenoprotein expression. Certain single nucleotide polymorphisms (SNPs) in selenoprotein genes are associated with more subtle effects that may still be of relevance for normal development and human health, as seen in population-wide studies [31]. This line of research can efficiently study the full set of selenoprotein encoding genes along with additional components involved in Se metabolism and biosynthesis [20,31–33]. Notably, evolutionary selection of certain genotypes may have enabled a better adaptation to nutritional restrictions, and render subjects less sensitive to poor supply in a Se-deficient environment [34]. A full overview of the potential interactions between variations in Se-related genes and health issues can be found elsewhere [35–37]. Notably, certain genotypes associate with autoimmune disease (AID) risk and autoantibody titers, e.g., in selenoprotein S (*SELENOS*) [38] or selenoprotein P (*SELENOP*) [39].

Apart from genetics, the undisputed greatest influence on selenoprotein expression is the dietary supply of the essential trace element, which varies widely around the world [40]. Under insufficient supply, the hierarchical principles that govern the targeted Se distribution within the organism protect essential tissues and biochemical pathways from deficiency [6–8]. A poor Se status of a given subject therefore does not cause an obvious phenotype or detectable health symptoms, and usually goes unnoticed if no laboratory analysis is conducted. Nevertheless, large analytical studies indicate that a Se supply below the recommended intake levels is associated with increased disease risks, especially for subjects with chronic disease, inflammation, or other predispositions [40,41]. It is time to take this knowledge more seriously and to realize that Se deficiency is difficult to detect, but easily avoidable. The essentiality is similar to, e.g., iodine, where population-wide nutritional measures and suitable supplements yield remarkable positive health effects [42]. In the case of iodine, deficiencies are visible and better known. Until now, meaningful measures to correct a poor Se supply have rarely been taken by the authorities. Finland is a remarkable exception and decided more than three decades ago (1984) to take action, and to add supplemental Se systematically to the fertilizers used in agriculture [43]. This countrywide supplementation program was successful in increasing the average Se intake of the Finish population to a likely health-supporting level [44]. Unfortunately, a control population is not at hand and direct health-supporting effects are difficult to be deduced, while positive effects on animals as well as plants are recorded [45]. A sufficiently high Se intake will improve general selenoprotein expression levels and raise the Se status in particular in those tissues that are stringently dependent on the Se supply, i.e., those that are low in the hierarchical supply, including liver, muscle, the gastrointestinal tract, and the immune system [24,46]. A sufficiently high Se status supports their integrity, metabolic activities, and responsiveness to regulatory signals, while Se deficiency compromises the adequate functioning of these organ systems and reactive cell types.

1.2. The General Role of Se Status and Supplemental Intake for Human Health

A Se deficit alone does not lead to an obvious phenotype, but it predisposes to certain diseases. The health risks of a relatively low Se intake become apparent in observational or intervention studies when biomarkers of Se status are analyzed in relation to disease prevalence and incidence. Using dietary Se intake data for deducing Se status has proven most difficult, as the same food items vary strongly in Se content in relation to the origin of production [40,47–51]. Among the most suitable biomarkers of Se status are total Se concentrations in serum or plasma and circulating selenoproteins [52–56]. These biomarkers are applied to epidemiological or intervention studies where only small amounts of biosamples are available. Unfortunately, only a few studies have determined the Se status by assessing more than one biomarker, or tried to closely monitor changes in Se status prior, during and after an intervention [41]. This constitutes a shortcoming in many fields of nutrition research, and unfortunately also in the studies on Se in AID.

The number of well-controlled and sufficiently powered randomized control trials (RCT) with supplemental Se is small, and the largest and most comprehensive ones have been conducted in the USA, where a large fraction of the population exhibits a replete Se status [57]. This is in contrast to the majority of subjects residing in other parts of the world [58]. The fortunate situation in North America is mainly due to the high Se content in the soils used for agriculture. The scientific strength of the large US trials for the role of Se in human health is therefore largely restricted to the issue of pharmacological Se effects, i.e., questions of side effects and toxicity in sufficiently supplied subjects [41,48]. This notion is best exemplified in the field of chemoprevention, where two large RCT with supplemental Se have been conducted over long time-periods. In the Nutritional Prevention of Cancer (NPC) study, roughly two thirds of the participants had a high baseline Se status [59], whereas in the Selenium and Vitamin E Cancer Prevention Trial (SELECT) follow-up study, almost all of the enrolled subjects were sufficiently supplied with Se [60]. Accordingly, chemopreventive effects were observed in the small fraction of Se-deficient participants in the lowest tertile of baseline Se in NPC only [61].

When summarizing both studies, it is inappropriate to combine the trials into one meta-analysis, as subjects with suboptimal Se status were under-represented, i.e., around 1–2% only when the studies are aggregated (Figure 1). From today's viewpoint, correcting a Se deficit is a meaningful health-supporting measure with biochemical effects, whereas supplementing well-supplied subjects with additional Se appears unnecessary [41,62,63]. Consequently, meta-analyses on supplemental Se by combining studies that differ profoundly in baseline Se status fail [64]. Supplementing subjects beyond their needs may be conducted as experimental attempts of targeted therapy, e.g., in oncology, where exceedingly high pharmacological Se dosages are studied for death-inducing effects on tumor cells [65–67]. However, no meaningful insights into the role of Se deficiency for the immune system and AID risk can be expected from these approaches.

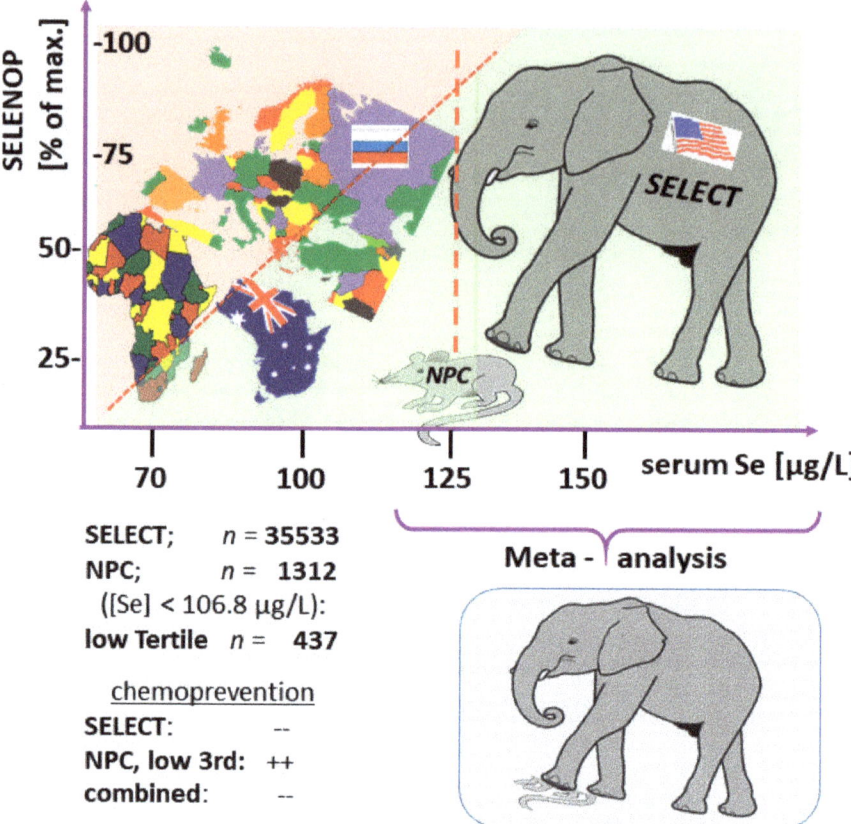

Figure 1. Importance of baseline Se status for the interpretation of RCT results. Dietary intake of Se varies widely in different parts of the world, and large parts of, e.g., the USA, are well supplied. This is in contrast to many other countries. A saturated expression of the Se transporter selenoprotein P (SELENOP, (% of max.)) indicates a sufficiently high Se status and corresponds to serum Se of 120–130 µg/L (broken line). As only one third of participants in the NPC study were below this threshold and showed positive health benefits from supplemental Se, their contribution to the overall effects is diluted and lost in meta-analyses. Positive health effects of supplemental Se can only be expected if selenoprotein expression is affected, and studies performing substitution versus supplementation need to be distinguished.

1.3. Uneven Worldwide Clinical Research on the Role of Se for Human Health

The excursion on supplemental Se and well-controlled large RCT may seem of little relevance to Se in AID, while in fact the knowledge and understanding of this background is of central importance for the interpretation of the current database and the general appreciation of Se in health and disease. It is almost impossible not to mention these two contradictory RCT when discussing the perspectives of Se in the ambulant and clinical care of patients. While the large and expensive SELECT study failed to achieve its primary goal of prostate cancer prevention, its scientific quality and technical rigor was undoubtedly of the highest order and provided insights into the human organism's tolerance to high Se intake. However, its relevance for the risks from Se deficiency and the importance of correcting a pre-existent Se deficit for preserving health and reducing disease risks is marginal. Despite these limits, these particular RCT are constantly referred to when Se supplementation trials are planned, discussed, and neglected, analytical research is considered and then dismissed, or when funding is applied for and then finally denied. The failure of SELECT has not served the basic and clinical research on Se well. A thorough understanding of the reason for failure of this RCT also helps to understand why smaller analytical and intervention studies can still be of profound scientific value, when conducted in populations of low or moderate Se status under high scientific standards. This is another peculiarity for Se in medicine; many important and meaningful studies are conducted in areas with Se deficiency [68,69], which explains the uneven distribution of collaborations on Se in AID, highlighting in particular central Europe, China, Turkey, and Iran as exceptionally active regions (Figure 2).

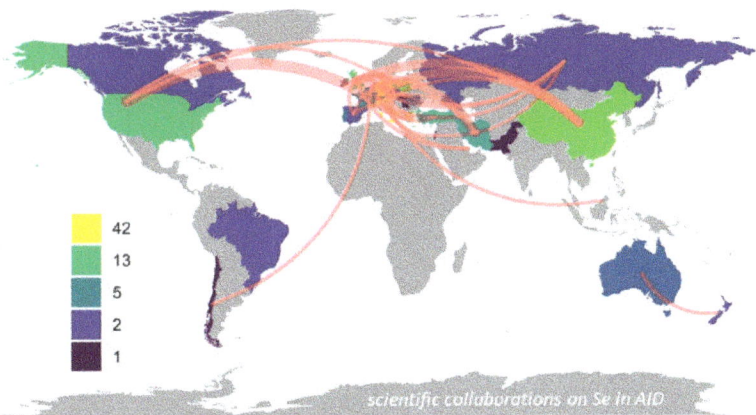

Figure 2. Overview on international collaborations on a potential role of Se in autoimmune disease. Countries with prevalent Se deficiency are particularly active in this line of research. The red lines indicate collaborations as published in the Web of Science (accessed on 4 April, 2021), with line thickness corresponding to the number of entries. The countries are color-coded for the number of contributions, with Denmark, Germany, and Italy yielding highest marks (yellow). Analysis was conducted for references matching the term "ALL = (selenium autoimmune thyroid)", using the 'bibliometrix' and 'ggplot2' packages in R (Version 2.1.0, R: A Language and Environment for Statistical Computing; The R Foundation for Statistical Computing: Vienna, Austria, 2020).

The concept of "substitution" versus "supplementation" may be introduced here, with the former indicating a targeted supply in order to correct a nutritional deficit, while the latter indicates pharmacological supply on top of a sufficient baseline status [41,70]. The chemopreventive effects of Se in the NPC study were restricted to the intake as substitution, since the responsive subjects were residing in the lowest tertile of baseline Se status, where full expression of selenoproteins was not yet achieved [71,72]. This setting is in contrast to the unsuccessful SELECT intervention, where Se was applied on top of a sufficiently high baseline Se status as a pharmacological supplementation [48,62]. Consequently, when we follow the same hypothesis and assume that a Se deficit constitutes a risk factor for AID, we need to focus on trials with Se-deficient subjects residing in areas of general poor Se supply. The published evidence for a potential role of Se in AID needs to be interpreted with this distinction in mind, and null results from areas with high baseline intake are unlikely of value for this issue. Given the inconsistent Se status around the world, no coherent and uniform guidelines from the various national expert committees on Se substitution can be expected, and recommendations on supplemental Se will differ with good cause between areas with low and high habitual Se intake.

2. Clinical Studies Linking Se Status and Autoimmune Diseases

Clinical studies on Se are either observational and compare Se status between diseased and healthy subjects or in relation to disease severity, or trials are interventional, supplying extra dosages like in the aforementioned RCT that used a daily supplement of 200 µg Se, either as Se-rich yeast (NPC) or selenomethionine (SELECT). As mentioned above, analytical studies relying on dietary Se intake are difficult to conduct; therefore, biomarkers of Se status are analyzed [51]. To this end, three biomarkers have achieved high acceptance in the field, i.e., the enzymatic activity of the extracellular glutathione peroxidase-3 (GPX3), total serum or plasma Se concentration (total Se), and the circulating transporter selenoprotein P (SELENOP) [54,71,73]. Some analytical studies have combined more than one biomarker (or determined Se content in hair or nails in parallel to blood), and yielded some congruent and meaningful results [53,74]. The published large RCT mentioned have unfortunately analyzed only one Se status biomarker.

2.1. Selenium Status and Hashimoto's Thyroiditis

Hashimoto's thyroiditis (HT) is a highly prevalent, renowned, and dynamic AID, characterized by autoreactive lymphocytes invading the thyroid gland. The immune process causes swelling, initial signs of hyperthyroidism and malaise, followed by a long and often chronic phase of progressive gland destruction, hypothyroidism, and increasing need for thyroid hormone replacement therapy [75]. The incidence is high and on the rise, exhibits individual disease courses and it is more prevalent in women than in men [76]. Observational studies in the EU have indicated that Se deficiency is positively associated with thyroid gland volume and nodule formation [77,78]. Conclusive evidence for a causal role of Se deficiency in promoting hypothyroidism and autoimmune-related thyroid gland damage (i.e., chronic stages of Hashimoto's thyroiditis) has come from a large cross-sectional study in China enrolling >6000 subjects. The participants were residing either in a Se-deficient area (Ningshan) or in an area with moderate Se status (Ziyang) [79]. Median plasma Se concentrations differed two-fold (104 vs. 57 µg/L), i.e., from sufficient for a saturated expression of most selenoproteins to a moderate deficiency. Accordingly, thyroid disease prevalence was almost twice as high in the Se-deficient area of Ningshan as compared to Ziyang (Figure 3A,B). The findings highlight that chronic Se-deficiency enhances the risk for AID of the thyroid, despite the notion that the populations were accustomed to their general Se supply [34]. Summarizing the different observational studies at hand, it becomes apparent that Se deficiency is a risk factor for increased thyroid gland volume, hypothyroidism, HT, thyroid nodules and associated health problems, and a population-wide approach as taken in Finland would likely reduce AID incidence and disease load in many areas of the world. Sex-specific differences in relation to the role of Se in goiter or HT incidence were reported from a large study in Europe [77], but were not observed in the large Chinese study [79]. Baseline iodine supply may modify the interaction of Se and thyroid disease [80], as Se status and goiter development are particularly interrelated under iodine deficiency [78,81]. Besides differences in the dietary Se intake, certain acute conditions are also associated with higher Se demands or increased Se loss, including severe disease, infection, or pregnancy. During pregnancy, increasing amounts of Se are transferred from the mother to the growing fetus, leading to an increasing deterioration of the pregnant woman's Se status if the increased need is not counteracted by additional Se intake [82,83]. The deficit may even worsen thereafter due to lactation (Figure 3C), potentially eliciting negative health effects on both the child and mother under limiting Se supply [84]. After pregnancy, postpartum thyroid disease (PPTD) constitutes a frequently found AID [85], in particular for women with positive thyroid autoantibodies (aAb). In view of the developing Se deficit in pregnancy, a highly interesting and instructive RCT from Italy tested supplemental Se to prevent PPTD and permanent hypothyroidism (PHT) [86]. To this end, pregnant women with or without positive TPO-aAb were enrolled and monitored. The intervention was conducted with 200 µg of selenomethionine per day, and it was successful [86]. The incidence of both PPTD and PHT was reduced by almost twofold, and no side effects were noted (Figure 3D).

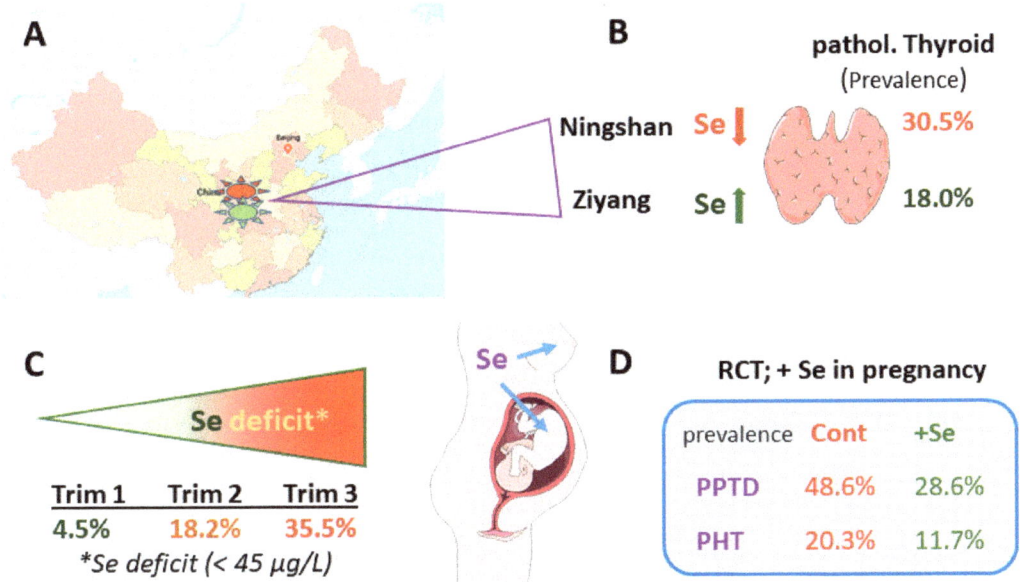

Figure 3. Importance of Se deficiency as druggable risk factor for AID. (**A**) High thyroid disease prevalence (pathol. Thyroid; goiter, (sub-) clinical hypothyroidism or autoimmune thyroiditis) is observed under low baseline Se intake. (**B**) The population on low Se intake (Ningshan) had an almost twice-higher incidence of thyroid disease as compared to the population with higher Se intake (Ziyang). (**C**) During pregnancy, Se status in the pregnant mothers decline from trimester 1 (Trim 1) to Trim 3, causing a severe Se deficit (<45 μg/L) in one-third of pregnancies in a European observational study. (**D**) Supplemental Se during pregnancy was capable of suppressing the high incidence of postpartum thyroid dysfunction (PPTD) and permanent hypothyroidism (PHT) in predisposed women ca. two-fold.

2.2. Selenium Status and Graves' Disease

Graves' Disease (GD), the second major AID of the thyroid gland, occurs when the aAb mimicking thyroid-stimulating hormone (TSH) directly activates the TSH-receptor, resulting in an overactive endocrine gland that is not controlled by negative feedback regulation and, thus, causes clinical symptoms of hyperthyroidism [87]. Like HT, incidence and prevalence of GD show a strong female-biased sex-specific difference [88]. The relation between Se status and GD is more complex and less-well understood in comparison to Se and HT [89,90], probably due to the positive effects of thyroid hormone on Se status and hepatic SELENOP biosynthesis [91]. Interestingly, no general interrelation between habitual Se intake and prevalence of hyperthyroidism was observed in the aforementioned large cross-sectional study in the two Chinese provinces with different Se intake and status [79]. However, a remarkable difference in GD prevalence was apparent when the data were analyzed for males and females separately; male subjects only seemed to react sensitively to Se deficits, and exhibited a higher prevalence of GD under low Se supply as compared to men with higher baseline Se intake, while disease prevalence of women remained largely unaffected [92]. Accordingly, the female-to-male ratio of hyperthyroidism (mainly due to GD) was 1.6 in the Se-poor area and 4.2 in the area with higher baseline Se intake (Figure 3B). This finding was statistically significant and completely unexpected, as it is widely assumed that sex-specific differences in AID incidence and prevalence are rather related to (epi-)genetic [93,94] or hormonal effects [95,96], like dysregulated X-chromosome inactivation [97] or estrogens and estradiol receptors [98,99], but not to micronutrient intake, again highlighting a peculiar sex-specific oddity of Se for the immune system. The surprising notion that a particular micronutrient, such as Se, which is associated with a number of sexual dimorphic effects in medicine and biology [7], is also strongly modifying the sex-specific risk for a prominent AID, opens a new and challenging perspective on the interplay of nutrition, genetics, and AID risk. Besides a number of analytical studies that provide a partly overlapping (but largely inconclusive) overall picture on the importance of Se for GD, one notable RCT reported remarkable positive health effects in GD-associated eye disease, i.e., Graves' orbitopathy (GO). Proptosis of the eye ball and inflammatory markers were successfully suppressed and eye motility and quality of life were strongly improved by daily Se supplementation over six months in patients with mild GO [100]. The positive health effects lasted beyond study termination, but the underlying mechanisms are unknown, and the promising findings have not been replicated yet. Still, the results are taken seriously and Se is recommended in mild GO by leading experts in the field [101,102].

2.3. Selenium Status and Type 1 Diabetes Mellitus

Type 1 diabetes mellitus (T1DM) is also known as "insulin dependent diabetes mellitus" or "juvenile diabetes", and describes the autoimmune form of diabetes that mainly develops at a young age [103], but also occurs as a slowly developing latent autoimmune diabetes in adults (LADA) or as classical and abrupt adult-onset T1DM [104]. Disease course involves a progressive destruction of insulin-producing pancreatic beta cells, sooner or later necessitating insulin replacement therapy due to gland insufficiency, eventually along with other medication [105]. It is estimated that around one in ten cases of diabetes mellitus are due to the destructive AID, with certain population-specific differences, while the majority are associated with insulin insensitivity of target tissue, being overweight, and often older age, typically denoted as type 2 diabetes mellitus (T2DM) [106]. Recent research indicates that diabetes mellitus should be categorized more precisely, as overlapping and distinct phenotypes and disease courses require personalized analyses and treatment regimens [107–111]. Despite the high prevalence and rising incidence of diabetes mellitus along with the essential role of Se for the endocrine and immune system, research activities on Se in the AID form of diabetes (T1DM) are sparse.

This is in contrast to some knowledge on the interrelationship of Se with T2DM, where an increased serum or plasma Se concentration is observed in many patients [112–114]. The dysregulation involves insulin resistance of liver, and its positive effect on hepatic SELENOP biosynthesis, causing higher serum Se concentrations [41,115,116]. Conversely, Se deficits are associated with hypoglycemia, i.e., a dysregulation of blood glucose levels in the opposite direction [117]. In contrast to initial reports, the results from large and well-controlled RCT indicate that Se supplementation is not causally related to T2DM incidence, not even in Se replete populations [41,118,119]. In comparison to T2DM, the data on Se and T1DM risk and course are few. In children, a small case-control study indicated a slightly increased serum Se status in the pediatric patients with T1DM as compared to controls (74 ± 8 vs. 65 ± 8 µg/L) [120]. In a larger study, an inverse relationship was observed, which was especially pronounced in children with T1DM and poorly controlled blood glucose, who displayed reduced serum Se and relatively low erythrocyte GPX1 activity [121]. This finding accords with a report on relatively low Se in erythrocytes of children with T1DM [122]. A case-control analysis using siblings of diseased children observed no difference in relation to GPX activity between T1DM and controls [123], in agreement with a small study on adults from the UK, where plasma Se was comparable between patients with T1DM, T2DM, and controls [124].

From these few studies, a picture emerges with two opposing forces apparently influencing the relationship of Se with diabetes mellitus; both the autoimmune nature of T1DM and the increased fat mass in many patients with T2DM are associated with an activated immune system, inflammation, and increased cytokines/adipokines including IL-1 and IL-6 [125,126]. This condition is known to impair hepatic selenoprotein biosynthesis and SELENOP expression [127–129]. At the same time, insulin resistance and increased glucose availability exert positive metabolic effects on hepatic SELENOP biosynthesis, consistently in T2DM and with an age- and disease-stage dependence in T1DM, compatible with a generally increased Se status in T2DM, and an unpredictable relationship in T1DM.

In Finland, it was observed that the diet could influence the interrelation between Se status and disease in T1DM. Children with T1DM had a higher Se status as compared to age-matched controls before the nation-wide Se supplementation was in place. This difference may have been due to a better food quality higher in Se content selected by the patients and their well-caring parents. However, the difference in serum Se disappeared between T1DM and control children with the population-wide supplementation and a general increase in Se intake [130], indicating that the potential advantage of choosing high quality food items in T1DM was crucial and decisive at times of poor Se supply, but became dispensable in face of the universal enrichment of Finish food with Se. This notion underlines again the particular relevance of population-wide measures for improving Se status under poor habitual Se availability.

2.4. Selenium Status and Rheumatoid Arthritis

Rheumatoid arthritis (RA) describes a third major group of inflammatory and chronic AID affecting primarily the joints, but eventually also spreading to remote sites including skin, kidneys, nerve tissue, lung, heart, and other organ systems [131]. The disease activity shows some circadian rhythm, and it can precipitate in sporadic longer-lasting waves of increased or decreased activity in a very personalized manner. In general, RA constitutes a constant and increasing threat to joint function, movement, quality of life, and overall health [132]. In agreement with other AID, its prevalence is higher in women than in men, with a disease peak in late adulthood [133]. Serum Se concentrations were found decreased in a set of 101 patients with seropositive RA compared to controls, even among US patients with high baseline Se concentrations exceeding the levels needed for full expression of circulating selenoproteins (148 ± 42 vs. 160 ± 25 µg/L) [134]. In a study on patients suffering from a rare and severe form of RA, i.e., systemic sclerosis, all three biomarkers of Se status (GPx3, total Se, and SELENOP) were significantly decreased as compared to controls [135]. This tendency was also reported in a meta-analysis from 2016, highlighting a relatively low Se status in patients with RA [136]. The deficiency is obviously not restricted to the trace element Se, but can also be observed for a second immune relevant micronutrient, i.e., zinc (Zn) [137]. In parallel, the trace element Cu seems to be slightly elevated in serum of RA patients, consistent with the positive acute phase reaction of liver-derived ceruloplasmin as systemic Cu transporter [138,139].

The imbalance in trace element concentrations underlines the chronic inflammatory nature of RA and an elevation of pro-inflammatory cytokines, with IL-6 and IL-1ß likely taking center stage for the observed effects on the micronutrient status [140,141]. Notably, the interaction between Se deficiency and enhanced inflammation is not necessarily unidirectional and self-limiting, but rather constitutes a vicious cycle with self-amplifying characteristics, similar to the situation in sepsis [142]. The Se status is known to affect NF-kB activity, and a reduced Se status will cause an up-regulation of a whole set of inflammation-relevant genes [143]. During NF-kB activation, and in response to other noxae, a number of intracellular stress-regulated selenoproteins are induced including essential components affecting ER quality control mechanisms like SELENOF, SELENON, SELENOS, SELENOT, or SELENOV [144–148]. Notably, SNPs in the promoter of *SELENOS* have directly been associated with IL-6, IL1ß, and tumor necrosis factor-alpha expression in human subjects [149]. Accordingly, an interaction of *SELENOS* genotype and IL-1 was identified for RA risk [150]. The potential molecular interactions are supported by recent clinical data. Anti-rheumatic treatment showed a sustained positive effect on Se concentrations [151]. The increase was parallel to a reduction in inflammatory activity and suppression of pro-inflammatory signaling [151]. Collectively, the chronic inflammatory nature of RA seems to directly impair Se metabolism, with a negative effect on hepatic SELENOP biosynthesis and consequently a reduced serum Se status in patients as compared to healthy controls, which closes a self-sustaining but druggable feedforward pathway [152].

Some pilot intervention studies with supplemental Se have been conducted, albeit with small groups of RA patients only, short duration, and a limited monitoring of changes in the important biomarkers of Se status. An increase in GPX activity to control levels was observed in a six months Se supplementation trial in a group of six patients in relation to six control subjects as early as in 1987 [153]. An RCT using 200 µg Se per day as selenized yeast for 90 days reported an increase in serum Se concentrations and an improvement in several parameters of RA, including quality of life [154]. However, the positive effects were observed in both study arms and a specific health impact of the trace element exceeding the placebo effect was not detected with statistical significance. Collectively, the clinical experience with Se in RA is very rudimentary in view of the importance and high prevalence of this potentially devastating AID. Supplemental Se may protect from severe disease course, in view that RA is associated with progressive Se loss, and supplemental Se may counteract efficiently the decline. It is not known why this cost-efficient and safe adjuvant treatment option is not explored in sufficiently sized and well-controlled RCT, as the chronic, severe, and progressive nature of RA serves as a paradigm for an inflammatory, most prevalent and relevant AID, which affects millions of people, in particular young and adult women, and is responsible for many lost years of quality of life.

3. Correcting Se Deficits in Prevention and Treatment of Autoimmune Diseases

The majority of analytical studies indicate a profound Se deficit in patients with an AID, likely linked to disease activity, ongoing inflammation, and an elevated activity of an autoreactive immune system. It is also known that a sufficient Se supply is needed to support and moderate the activity of immune cells and to avoid an overshooting immune response [27,39,129,155]. Consequently, it appears plausible that a Se deficit aggravates AID course, and supplemental Se may elicit positive health effects, contribute to the normalization of the immune response, reduce autoreactive disease activity, and support other therapeutic measures in an adjuvant mode.

Several supplementation studies have been conducted, albeit mostly with small numbers of patients and for short periods of time only. Comparing the three types of AID mentioned above, the majority of intervention studies with supplemental Se have addressed the two autoimmune thyroid diseases, i.e., HT and GD [89,90]. Despite an initial enthusiasm, the overall picture is ambiguous as several trials have reported positive effects, but a number of similar studies reported no health improvements [89,90]. Negative side effects have not been observed. At present, the major reason for these equivocal results is unknown, as neither the nature of the most suitable selenocompound nor the optimal treatment duration, nor best dosage have been specified. Similarly, the ideal baseline Se status and best time for treatment initiation are not identified. The paucity of knowledge is partly due to the small groups studied, the heterogeneous groups of patients, and a lack of detailed monitoring of Se status at baseline, during supplementation and at study end. In order to resolve some of these inconsistencies, two well-controlled intervention trials with Se in patients with HT and GD, respectively, are under way and will hopefully shed some more light on the potential efficacy of supplemental Se on disease severity and course, as well as on the underlying reasons for the contradictory experiences. The chronic autoimmune thyroiditis quality of life Se trial (CATALYST) has enrolled almost 500 patients, and the effects of supplemental Se (200 µg per day) or placebo on the disease-related quality of life along with changes in thyroid aAb concentrations during an intervention period of one year are evaluated [156]. The GRAves' disease Selenium Supplementation (GRASS) trial is of similar size and design, and primarily studies the effects of Se on anti-thyroid drug treatment failure [157]. Importantly, both studies are conducted in Denmark, i.e., with European patients of insufficient baseline Se status for full expression of selenoproteins. Until the results from these well-designed and promising RCT are at hand, the data collectively indicate that dosages up to 200 µg Se per day, as also used in the aforementioned large RCT in the USA, do not harm and may reduce inflammation and autoantibody concentrations in AID.

4. Immune Dysfunction in Se Deficiency–Possible Molecular Mechanisms

Lymphocytes and the immune system show certain functional alterations in Se deficiency and upon Se substitution, indicating that their position within the hierarchical order of supply with the essential trace element is not at the top, where Se status is maintained even in low supply, e.g., the endocrine or central nervous system [7,24,29]. However, Se supply beyond the needs for saturated expression of selenoproteins shows little if any effects on the immune system, underlining the essential role of selenoproteins and their adequate expression for a regular functioning of lymphocytes [158]. Declining Se status in disease or pregnancy will consequently impair regular selenoprotein biosynthesis and regular immune system function. Accordingly, the members of the selenoprotein family with proven relevance for lymphocyte activity along with those having a general role in antigen processing and MHC-dependent presentation are the prime targets of Se substitution, and constitute the main culprit for immune dysfunction in Se deficiency [29,159]. The biochemical and physiological pathways affected by suboptimal expression of particular selenoproteins range from a poorly controlled intracellular peroxide and redox tone, failing quality control of newly synthesized proteins in the ER, impaired Ca signaling to an excessive oxidation of cellular components and death by ferroptosis (Table 1).

Table 1. Selenoproteins with particular functions in the immune system.

Glutathione peroxidases 1, 2 (GPX1, 2)	control intracellular peroxide tone	[160]
Glutathione peroxidase 4 (GPX4)	reduces lipid hydroperoxides, prevents ferroptosis	[161]
Iodothyronine deiodinases (DIO2, 3)	regulate activity of innate immune cells	[162]
Methionine-R-sulfoxide reductase (MSRB1)	reduces oxidized Met, affects F-actin formation	[163]
Selenoprotein H (SELENOH)	controls redox-sensitive transcription and damage	[164]
Selenoprotein I (SELENOI)	contributes to the biosynthesis of phospholipids	[165]
Selenoprotein K (SELENOK)	affects Ca-signaling and activity of lymphocytes	[166]
Selenoprotein P (SELENOP)	mediates hierarchical Se supply, indicates Se status	[8]
Selenoprotein S (SELENOS)	controls quality of proteins synthesized in ER	[149]
Selenoprotein T (SELENOT)	controls redox state and protein quality in ER	[144]
Selenophosphate Synthetase 2 (SEPHS2)	controls selenoprotein biosynthesis rate	[167]
Selenoprotein 15 (SEP15, SELENOF)	gatekeeper for ER exit of immunoglobulins	[168]
Thioredoxin reductases 1, 2 (TXNRD1, 2)	regenerate thioredoxin, balance mitochondrial ROS	[169]

It is at present unknown which of these immune-relevant selenoproteins respond most sensitively to Se deficiency and constitute the rate-limiting factors for a dysfunctional immune system under low Se supply. Besides the selenoproteins that affect lymphocyte activity, migration, proliferation, survival, and interaction directly, a large number of selenoproteins is involved in quality control of newly synthesized proteins in the ER affecting correct protein folding and retrotranslocation of misfolded proteins into the cytosol for degradation [170]. Consequently, a disease- or pregnancy-induced decline of the Se status into a critical zone where these selenoproteins are not synthesized any more to the required expression levels will increase the risk for lymphocytes loosing self-tolerance and at the same time for a widespread presentation of wrongly processed novel autoantigens due to failing quality control in the ER. This hypothesis is compatible with many preclinical and clinical studies on autoimmune thyroid disease [89], but lacks solid and more comprehensive clinical data from the other prevalent autoimmune diseases, such as RA or T1DM, as mentioned above.

5. Potential Toxicity of Supplemental Se

A critical evaluation of the role of Se in AID needs to consider the potential toxicity of supplemental Se. The biological activity of Se as an essential trace element is mainly exerted via its incorporation into Sec-containing selenoproteins. High Se supply in excess of the required amounts can be toxic due to poorly characterized molecular effects [171], and in relation to the molecular form of the selenocompound ingested and its metabolic fate [67,172,173]. An incidence of endemic Se-related poisoning ("selenosis") was observed in the naturally Se-rich Enshi county in Hubei Province, China, where the Se intake peaked to about 5000 µg per day due to the usage of stony coal of extremely high Se content [174]. In comparison, the recommended intake of Se for healthy adults in Austria, Germany, or Switzerland ranges at 60–70 µg per day, i.e., around one percent of these amounts [175].

A second relevant risk for selenosis is encountered when using Se-containing supplements that have been prepared with a lack of appropriate quality control. The consequences from this type of error were impressively documented in veterinary medicine, where wrongly formulated preparations caused several incidents of animal poisoning, e.g., sudden death of polo ponies upon receiving excessive amounts of supplemental Se causing serum concentrations exceeding 1000 µg/L [176]. Similar incidences are also known from human subjects consuming misformulated dietary supplements [177]. However, humans seem to be relatively robust to acute Se intoxication, as most of the symptoms described from a daily intake of pills containing 22–32 mg of Se per serving were reversible, and luckily, no fatal course occurred [177]. Even more surprising are sporadic case reports, e.g., from a pregnant mother taking 200 mg of Se per day (i.e., 1000-fold higher than the commonly accepted maximal dosage used in clinical trials) during gestational weeks 7 to 12 [178]. The women lost hair and fingernails, as expected from severe selenosis, but surprisingly gave birth at term to a healthy child [178]. These examples indicate that supplemental Se can be toxic and even fatal, irrespective of the underlying motivation for health-support or suicidal intention [179]. However, the dosages causing acute selenosis are far higher than the recommended amounts applied in clinical studies or provided by high-quality supplements, where up to 200 µg Se per serving per day proved safe, irrespective of the selenocompound used. The health risks from chronic oversupply are however poorly characterized, potentially due to some adaptation of the population [34].

6. Se-Deficiency as Potential Trigger of Autoimmune Disease

The hypothesis that severe Se deficiency contributes to AID risk as a trigger factor is supported by a limited body of evidence, but appears reasonable and scientifically congruent. The immune system is constantly balancing its activity for reliably recognizing self, responding to non-self and identifying variants resulting from malignant transformation, chemical modification, infection or mimicry [180]. A number of highly specialized immune cells help maintain this delicate balance and prevent AID. In order to work reliably, essential vitamins and trace elements are needed, and deficiencies are established risk factors for infections and inappropriate immune responses [181]. Several selenoproteins are known to contribute to ER quality control, intracellular calcium signaling, and antigen presentation, as well as to immune cell activation, suppression, proliferation, and differentiation [182]. Insufficient supply with trace elements needed interferes with these pathways, in particular in disease and under certain demanding conditions.

Bacterial and viral infections, acute and chronic inflammation, trauma, burn injury, childbirth, and surgical interventions are clinical conditions causing a declining Se status, as shown both in model systems and clinical studies alike (Figure 4A). On a molecular level, the downregulation of hepatic SELENOP biosynthesis by pro-inflammatory cytokines seems central for the observed decline in serum Se concentrations [183–186]. In parallel, the specific Se transport via SELENOP to target tissues becomes disrupted, and a systemic Se deficit may develop, in particular when baseline Se status is already low [8]. Notably, declining Se status and increasing cytokine concentrations are closing a feed-forward regulation, i.e., a self-amplifying loop [129,142,158]. The clinical conditions causing a decline in Se status show a considerable overlap to potential triggers for AID development, and both processes may be causally linked, i.e., acute Se deficiency may constitute an as-yet poorly appreciated trigger for AID development (Table 1). A decline below a certain threshold may tip the balance from regular lymphocyte function to disruption of self-tolerance, triggering autoreactive processes (Figure 4B).

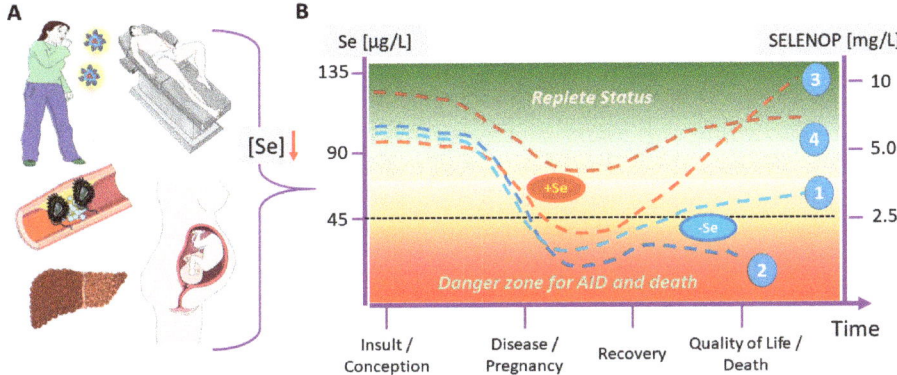

Figure 4. Hypothesis on Se decline into a critical zone as trigger for immune system failure, AID development, or even death. (**A**) Bacterial or viral infections, acute or chronic illness, AID, surgery, liver disease, or pregnancy are associated with a vicious cycle of inflammation, increasing cytokine levels and decreasing Se status. During and following to these conditions, AID may develop. (**B**) A disease-related drop in Se status below a certain threshold into a critical concentration range ("danger zone") impairs regular immune system function and potentially disrupts self-tolerance, leading to AID. (1) Under regular conditions, disease-associated Se decline is transient, recovering with time. (2) Fatal disease course is associated with strong Se status decline and lack of its recovery. (3) Supplemental Se (+Se) reduces both the Se trough and time spent in severe deficiency, thereby likely improving odds of convalescence. (4) The risk for dropping into the danger zone of severe Se deficiency and immune system failure can be reduced by early Se supplementation, thereby starting on a sufficiently high status, ideally in combination with adequate monitoring in order to provide the amounts necessary and avoid side effects, i.e., to substitute what is needed without supplementing beyond requirement.

Among the major modifiers of this process are the baseline Se status, before and during the early stages of disease, the severity, and course of the condition that is causing the Se decline along with the minimal Se concentrations that are reached during the disease. Supportive adjuvant therapy including supplemental Se will counteract the drop into the dangerous zone of severe Se deficiency (Figure 4B). Besides potentially triggering AID development, a severe Se deficit has been identified as mortality risk factor in severe illness, e.g., sepsis, COVID-19, liver disease, along with other critical conditions [187–190]. Notably, a sufficiently high baseline Se status is capable of preventing an uncontrolled and exaggerated immune response [149,191], with some sex-specific characteristics [7,129], and high potential in the prevention of severe COVID-19 [192]. The principle of positively affecting the immune system and decreasing inflammation by Se has just been verified in COVID-19 with patients displaying severe acute respiratory distress syndrome, where early supplementation was capable of restoring Se status [193].

In view that severe COVID-19 causes very strongly decreasing Se status, high mortality risk, and newly developing autoimmunity, it may also be hypothesized that severe Se deficiency may be related to post-acute sequelae and long-COVID symptoms [194,195]. It will be of high importance to monitor Se status during the current pandemic in both mildly and severely affected patients, and to delineate Se status decline and recovery to long-term health issues including autoimmune reactions to peripheral and central antigens. The author is convinced that supplemental Se to subjects with proven or predicted low Se status is eliciting protective and immune supportive health benefits, both in prevention and adjuvant therapy of the different AID and COVID-19, in particular for patients residing in areas of low habitual Se intake. From all we know today, there is no guarantee that supplemental Se will provide measurable health benefits to all patients, but the probability that a subset will profit from Se when provided as substitution is high. The avoidance of severe Se deficiency will not only reduce AID risk, but also confer some protection from other relevant diseases, including cancer, and cardiovascular and infectious diseases. There are no indications that Se substitution is associated with side effects if not applied as a supplement or as a drug of excessively high dosage, in particular when well-supplied subjects are excluded. Accordingly, the potential health benefits of supplemental Se are not of equal relevance and importance around the world, but largely restricted to areas with sub-optimal baseline supply, as reflected in the map shown above (Figure 2). However, even in areas of ample supply, severe diseases, such as sepsis or COVID-19, may cause immune-relevant Se deficiency irrespective of geography, and monitoring under such conditions is recommended (Table 2). First, this recommendation applies to critically ill patients in intensive care, irrespective of underlying disease. According to the aforementioned definition of substitution versus supplementation, there is at present no scientific rationale for Se supplementation, but substitution is mandatory in case of suspected or diagnosed Se deficiency, both for avoiding health risks (e.g., AID) and for supporting the immune system of an already or not-yet diseased organism.

Table 2. Inflammatory conditions associated with decreasing Se status and increasing AID risk.

Condition	Effect on Se Status	Reference *	Effect on AID Risk	Reference *
bacterial infection	suppression	[142]	enhancement	[196]
viral infection	suppression	[187]	enhancement	[197]
cancer	suppression	[198]	enhancement	[199]
(poly-)trauma	suppression	[200]	enhancement	[201]
burn injury	suppression	[202]	enhancement	[203]
smoking	suppression	[204]	enhancement	[205]
surgery	suppression	[206]	enhancement	[207]
transplantation	suppression	[190]	enhancement	[208]
pregnancy	suppression	[82]	enhancement	[85]

* out of many suitable references, one is chosen for reasons of space and clarity, with an apology to the authors of relevant studies not listed here. Admittedly, this overview is a biased selection intended to motivate and stimulate reflection.

7. Conclusions

The trace element Se is essential for a normal functioning of the immune system, and severe Se deficits impair immune responses and predispose to AID, as characterized best for AID of the thyroid gland. An active immune system with elevated cytokine levels exerts a suppressive effect on hepatic selenoprotein biosynthesis and SELENOP secretion into the circulation, causing reduced Se metabolism and an overall suppressed Se status. An acute or chronic severe Se deficit perturbs the immune system, irrespective of the underlying reason (e.g., infection, pregnancy, trauma, cancer, poor nutrition, surgery). Notably, the conditions causing low Se status overlap with known triggers for AID, and indicate a potential direct causal interrelation (Table 2). This interrelationship and the benefit of supplemental Se are firmly established for the thyroid, but not yet for the other prevalent AID, such as RA or T1DM. Consequently, avoiding severe Se deficits may reduce AID risks and alleviate disease symptoms, both in prevention and during therapy. Should this theory be verified by suitable RCT, the insight would be of broad medical relevance and contribute to a better understanding, control, and reduction of the elevated postpartum, post-infection, and post-injury AID risk. Accordingly, Se substitution is strongly recommended for chronic or acute deficiency, whereas supplementation to healthy subjects with sufficiently high baseline levels does not appear to be warranted.

Funding: Research in the lab of L.S. is funded by the Deutsche Forschungsgemeinschaft (DFG), research unit FOR-2558 "TraceAge" (Scho 849/6-2), and CRC/TR 296 "Local control of TH action" (LocoTact, P17). Financial support was provided by the Open Access Publication Fund of Charité–Universitätsmedizin Berlin.

Acknowledgments: Essential intellectual support from my research team and inspiring colleagues from the International Society for Selenium Research (ISSR) is gratefully acknowledged, especially the help from Raban A. Heller in the construction of Figure 2. Some elements of the figures were taken from Servier Medical Art (licensed under a Creative Commons Attribution 3.0 Unported License).

Conflicts of Interest: L.S. holds shares of selenOmed GmbH, a company involved in Se status assessment and supplementation.

References

1. Hatfield, D.L.; Gladyshev, V.N. How selenium has altered our understanding of the genetic code. *Mol. Cell. Biol.* **2002**, *22*, 3565–3576. [CrossRef]
2. Xu, X.M.; Carlson, B.A.; Mix, H.; Zhang, Y.; Saira, K.; Glass, R.S.; Berry, M.J.; Gladyshev, V.N.; Hatfield, D.L. Biosynthesis of selenocysteine on its tRNA in eukaryotes. *PLoS Biol.* **2007**, *5*, e4. [CrossRef]
3. Commans, S.; Bock, A. Selenocysteine inserting tRNAs: An overview. *FEMS Microbiol. Rev.* **1999**, *23*, 335–351. [CrossRef] [PubMed]
4. Songe-Moller, L.; van den Born, E.; Leihne, V.; Vagbo, C.B.; Kristoffersen, T.; Krokan, H.E.; Kirpekar, F.; Falnes, P.O.; Klungland, A. Mammalian ALKBH8 possesses tRNA methyltransferase activity required for the biogenesis of multiple wobble uridine modifications implicated in translational decoding. *Mol. Cell. Biol.* **2010**, *30*, 1814–1827. [CrossRef] [PubMed]
5. Lacourciere, G.M.; Stadtman, T.C. Catalytic properties of selenophosphate synthetases: Comparison of the selenocysteine-containing enzyme from Haemophilus influenzae with the corresponding cysteine-containing enzyme from Escherichia coli. *Proc. Natl. Acad. Sci. USA* **1999**, *96*, 44–48. [CrossRef]
6. Motsenbocker, M.A.; Tappel, A.L. A selenocysteine-containing selenium-transport protein in rat plasma. *Biochim. Biophys. Acta Gen. Subj.* **1982**, *719*, 147–153. [CrossRef]
7. Schomburg, L.; Schweizer, U. Hierarchical regulation of selenoprotein expression and sex-specific effects of selenium. *Biochim. Biophys. Acta Gen. Subj.* **2009**, *1790*, 1453–1462. [CrossRef]
8. Burk, R.F.; Hill, K.E. Regulation of Selenium Metabolism and Transport. *Annu. Rev. Nutr.* **2015**, *35*, 109–134. [CrossRef]
9. Carlson, B.A.; Yoo, M.H.; Tsuji, P.A.; Gladyshev, V.N.; Hatfield, D.L. Mouse models targeting selenocysteine tRNA expression for elucidating the role of selenoproteins in health and development. *Molecules* **2009**, *14*, 3509–3527. [CrossRef]
10. Conrad, M. Transgenic mouse models for the vital selenoenzymes cytosolic thioredoxin reductase, mitochondrial thioredoxin reductase and glutathione peroxidase 4. *Biochim. Biophys. Acta Gen. Subj.* **2009**, *1790*, 1575–1585. [CrossRef] [PubMed]
11. Santesmasses, D.; Mariotti, M.; Gladyshev, V.N. Tolerance to Selenoprotein Loss Differs between Human and Mouse. *Mol. Biol. Evol.* **2020**, *37*, 341–354. [CrossRef] [PubMed]
12. Schweizer, U.; Fradejas-Villar, N. Why 21? The significance of selenoproteins for human health revealed by inborn errors of metabolism. *FASEB J.* **2016**, *30*, 3669–3681. [CrossRef] [PubMed]
13. Schwarz, K.; Foltz, C.M. Selenium as an integral part of Factor 3 against dietary necrotic liver degeneration. *J. Am. Chem. Soc.* **1957**, *79*, 3292. [CrossRef]
14. Schwarz, K. Factors protecting against dietary necrotic liver degeneration. *Ann. N. Y. Acad. Sci.* **1954**, *57*, 878–888. [CrossRef]
15. Flohé, L. The labour pains of biochemical selenology: The history of selenoprotein biosynthesis. *Biochim. Biophys. Acta Gen. Subj.* **2009**, *1790*, 1389–1403. [CrossRef]
16. Ingold, I.; Berndt, C.; Schmitt, S.; Doll, S.; Poschmann, G.; Buday, K.; Roveri, A.; Peng, X.; Porto Freitas, F.; Seibt, T.; et al. Selenium Utilization by GPX4 Is Required to Prevent Hydroperoxide-Induced Ferroptosis. *Cell* **2018**, *172*, 409–422. [CrossRef] [PubMed]
17. Bock, F.J.; Tait, S.W.G. Mitochondria as multifaceted regulators of cell death. *Nat. Rev. Mol. Cell Bio.* **2020**, *21*, 85–100. [CrossRef]
18. Papp, L.V.; Holmgren, A.; Khanna, K.K. Selenium and selenoproteins in health and disease. *Antioxid. Redox Signal.* **2010**, *12*, 793–795. [CrossRef] [PubMed]

19. Fradejas-Villar, N. Consequences of mutations and inborn errors of selenoprotein biosynthesis and functions. *Free Radic. Biol. Med.* **2018**, *127*, 206–214. [CrossRef]
20. Schoenmakers, E.; Chatterjee, K. Human Disorders Affecting the Selenocysteine Incorporation Pathway Cause Systemic Selenoprotein Deficiency. *Antioxid. Redox Signal.* **2020**, *33*, 481–497. [CrossRef]
21. Dumitrescu, A.M.; Liao, X.H.; Abdullah, M.S.; Lado-Abeal, J.; Majed, F.A.; Moeller, L.C.; Boran, G.; Schomburg, L.; Weiss, R.E.; Refetoff, S. Mutations in SECISBP2 result in abnormal thyroid hormone metabolism. *Nat. Genet.* **2005**, *37*, 1247–1252. [CrossRef]
22. Schomburg, L.; Dumitrescu, A.M.; Liao, X.H.; Bin-Abbas, B.; Hoeflich, J.; Köhrle, J.; Refetoff, S. Selenium supplementation fails to correct the selenoprotein synthesis defect in subjects with SBP2 gene mutations. *Thyroid* **2009**, *19*, 277–281. [CrossRef]
23. Schoenmakers, E.; Agostini, M.; Mitchell, C.; Schoenmakers, N.; Papp, L.; Rajanayagam, O.; Padidela, R.; Ceron-Gutierrez, L.; Doffinger, R.; Prevosto, C.; et al. Mutations in the selenocysteine insertion sequence-binding protein 2 gene lead to a multisystem selenoprotein deficiency disorder in humans. *J. Clin. Investig.* **2010**, *120*, 4220–4235. [CrossRef] [PubMed]
24. Arthur, J.R.; McKenzie, R.C.; Beckett, G.J. Selenium in the immune system. *J. Nutr.* **2003**, *133*, 1457S–1459S. [CrossRef]
25. Broome, C.S.; McArdle, F.; Kyle, J.A.; Andrews, F.; Lowe, N.M.; Hart, C.A.; Arthur, J.R.; Jackson, M.J. An increase in selenium intake improves immune function and poliovirus handling in adults with marginal selenium status. *Am. J. Clin. Nutr.* **2004**, *80*, 154–162. [CrossRef]
26. Hawkes, W.C.; Kelley, D.S.; Taylor, P.C. The effects of dietary selenium on the immune system in healthy men. *Biol. Trace Elem. Res.* **2001**, *81*, 189–213. [CrossRef]
27. Hoffmann, P.R.; Berry, M.J. The influence of selenium on immune responses. *Mol. Nutr. Food Res.* **2008**, *52*, 1273–1280. [CrossRef]
28. Spallholz, J.E.; Boylan, L.M.; Larsen, H.S. Advances in understanding selenium's role in the immune system. *Ann. NY Acad. Sci.* **1990**, *587*, 123–139. [CrossRef]
29. Avery, J.C.; Hoffmann, P.R. Selenium, Selenoproteins, and Immunity. *Nutrients* **2018**, *10*, 1203. [CrossRef] [PubMed]
30. Bermano, G.; Meplan, C.; Mercer, D.K.; Hesketh, J.E. Selenium and viral infection: Are there lessons for COVID-19? *Br. J. Nutr.* **2020**, *125*, 618–627. [CrossRef] [PubMed]
31. Meplan, C.; Crosley, L.K.; Nicol, F.; Beckett, G.J.; Howie, A.F.; Hill, K.E.; Horgan, G.; Mathers, J.C.; Arthur, J.R.; Hesketh, J.E. Genetic polymorphisms in the human selenoprotein P gene determine the response of selenoprotein markers to selenium supplementation in a gender-specific manner (the SELGEN study). *FASEB J.* **2007**, *21*, 3063–3074. [CrossRef]
32. Fedirko, V.; Jenab, M.; Meplan, C.; Jones, J.S.; Zhu, W.; Schomburg, L.; Siddiq, A.; Hybsier, S.; Overvad, K.; Tjonneland, A.; et al. Association of Selenoprotein and Selenium Pathway Genotypes with Risk of Colorectal Cancer and Interaction with Selenium Status. *Nutrients* **2019**, *11*, 935. [CrossRef] [PubMed]
33. Shibata, T.; Arisawa, T.; Tahara, T.; Ohkubo, M.; Yoshioka, D.; Maruyama, N.; Fujita, H.; Kamiya, Y.; Nakamura, M.; Nagasaka, M.; et al. Selenoprotein S (SEPS1) gene -105G>A promoter polymorphism influences the susceptibility to gastric cancer in the Japanese population. *BMC Gastroenterol.* **2009**, *9*, 2. [CrossRef]
34. White, L.; Romagne, F.; Muller, E.; Erlebach, E.; Weihmann, A.; Parra, G.; Andres, A.M.; Castellano, S. Genetic adaptation to levels of dietary selenium in recent human history. *Mol. Biol. Evol.* **2015**, *32*, 1507–1518. [CrossRef] [PubMed]
35. Meplan, C. Selenium and chronic diseases: A nutritional genomics perspective. *Nutrients* **2015**, *7*, 3621–3651. [CrossRef] [PubMed]
36. Kadkol, S.; Diamond, A.M. The Interaction between Dietary Selenium Intake and Genetics in Determining Cancer Risk and Outcome. *Nutrients* **2020**, *12*, 2424. [CrossRef] [PubMed]
37. Mathers, J.C.; Meplan, C.; Hesketh, J.E. Polymorphisms affecting trace element bioavailability. *Int. J. Vitam. Nutr. Res.* **2010**, *80*, 314–318. [CrossRef]
38. Santos, L.R.; Duraes, C.; Mendes, A.; Prazeres, H.; Alvelos, M.I.; Moreira, C.S.; Canedo, P.; Esteves, C.; Neves, C.; Carvalho, D.; et al. A Polymorphism in the Promoter Region of the Selenoprotein S Gene (SEPS1) Contributes to Hashimoto's Thyroiditis Susceptibility. *J. Clin. Endocrinol. Metab.* **2014**, *99*, E719–E723. [CrossRef] [PubMed]
39. Wang, W.; Mao, J.; Zhao, J.; Lu, J.; Yan, L.; Du, J.; Lu, Z.; Wang, H.; Xu, M.; Bai, X.; et al. Decreased Thyroid Peroxidase Antibody Titer in Response to Selenium Supplementation in Autoimmune Thyroiditis and the Influence of a Selenoprotein P Gene Polymorphism: A Prospective, Multicenter Study in China. *Thyroid* **2018**, *28*, 1674–1681. [CrossRef]
40. Rayman, M.P. Food-chain selenium and human health: Emphasis on intake. *Br. J. Nutr.* **2008**, *100*, 254–268. [CrossRef]
41. Schomburg, L. The other view: The trace element selenium as a micronutrient in thyroid disease, diabetes, and beyond. *Hormones* **2020**, *19*, 15–24. [CrossRef]
42. Vanderpump, M.P. Epidemiology of iodine deficiency. *Minerva Med.* **2017**, *108*, 116–123. [CrossRef]
43. Pietinen, P.; Mannisto, S.; Valsta, L.M.; Sarlio-Lahteenkorva, S. Nutrition policy in Finland. *Public Health Nutr.* **2010**, *13*, 901–906. [CrossRef] [PubMed]
44. Alfthan, G.; Eurola, M.; Ekholm, P.; Venalainen, E.R.; Root, T.; Korkalainen, K.; Hartikainen, H.; Salminen, P.; Hietaniemi, V.; Aspila, P.; et al. Effects of nationwide addition of selenium to fertilizers on foods, and animal and human health in Finland: From deficiency to optimal selenium status of the population. *J. Trace Elem. Med. Biol.* **2015**, *31*, 142–147. [CrossRef] [PubMed]
45. Hartikainen, H. Biogeochemistry of selenium and its impact on food chain quality and human health. *J. Trace Elem. Med. Biol.* **2005**, *18*, 309–318. [CrossRef] [PubMed]
46. Hoffmann, P.R. Mechanisms by which selenium influences immune responses. *Arch. Immunol. Ther. Exp.* **2007**, *55*, 289–297. [CrossRef] [PubMed]
47. Combs, G.F., Jr. Selenium in global food systems. *Br. J. Nutr.* **2001**, *85*, 517–547. [CrossRef]

48. Nicastro, H.L.; Dunn, B.K. Selenium and prostate cancer prevention: Insights from the selenium and vitamin E cancer prevention trial (SELECT). *Nutrients* **2013**, *5*, 1122–1148. [CrossRef]
49. Spadoni, M.; Voltaggio, M.; Carcea, M.; Coni, E.; Raggi, A.; Cubadda, F. Bioaccessible selenium in Italian agricultural soils: Comparison of the biogeochemical approach with a regression model based on geochemical and pedoclimatic variables. *Sci. Total Environ.* **2007**, *376*, 160–177. [CrossRef]
50. Navarro-Alarcon, M.; Cabrera-Vique, C. Selenium in food and the human body: A review. *Sci. Total Environ.* **2008**, *400*, 115–141. [CrossRef] [PubMed]
51. Belhadj, M.; Kazi Tani, L.S.; Dennouni Medjati, N.; Harek, Y.; Dali Sahi, M.; Sun, Q.; Heller, R.; Behar, A.; Charlet, L.; Schomburg, L. Se Status Prediction by Food Intake as Compared to Circulating Biomarkers in a West Algerian Population. *Nutrients* **2020**, *12*, 3599. [CrossRef]
52. Burk, R.F.; Norsworthy, B.K.; Hill, K.E.; Motley, A.K.; Byrne, D.W. Effects of chemical form of selenium on plasma biomarkers in a high-dose human supplementation trial. *Cancer Epidemiol. Prev. Biomark.* **2006**, *15*, 804–810. [CrossRef]
53. Combs, G.F., Jr. Biomarkers of selenium status. *Nutrients* **2015**, *7*, 2209–2236. [CrossRef] [PubMed]
54. Hoeflich, J.; Hollenbach, B.; Behrends, T.; Hoeg, A.; Stosnach, H.; Schomburg, L. The choice of biomarkers determines the selenium status in young German vegans and vegetarians. *Br. J. Nutr.* **2010**, *104*, 1601–1604. [CrossRef]
55. Satia, J.A.; King, I.B.; Morris, J.S.; Stratton, K.; White, E. Toenail and plasma levels as biomarkers of selenium exposure. *Ann. Epidemiol.* **2006**, *16*, 53–58. [CrossRef] [PubMed]
56. Taylor, R.M.; Sunde, R.A. Selenoprotein Transcript Level and Enzyme Activity as Biomarkers for Selenium Status and Selenium Requirements in the Turkey (Meleagris gallopavo). *PLoS ONE* **2016**, *11*, e0151665. [CrossRef]
57. Laclaustra, M.; Stranges, S.; Navas-Acien, A.; Ordovas, J.M.; Guallar, E. Serum selenium and serum lipids in US adults: National Health and Nutrition Examination Survey (NHANES) 2003–2004. *Atherosclerosis* **2010**, *210*, 643–648. [CrossRef]
58. Stoffaneller, R.; Morse, N.L. A review of dietary selenium intake and selenium status in Europe and the Middle East. *Nutrients* **2015**, *7*, 1494–1537. [CrossRef] [PubMed]
59. Clark, L.C.; Combs, G.F., Jr.; Turnbull, B.W.; Slate, E.H.; Chalker, D.K.; Chow, J.; Davis, L.S.; Glover, R.A.; Graham, G.F.; Gross, E.G.; et al. Effects of selenium supplementation for cancer prevention in patients with carcinoma of the skin. A randomized controlled trial. Nutritional Prevention of Cancer Study Group. *JAMA* **1996**, *276*, 1957–1963. [CrossRef] [PubMed]
60. Klein, E.A.; Thompson, I.M., Jr.; Tangen, C.M.; Crowley, J.J.; Lucia, M.S.; Goodman, P.J.; Minasian, L.M.; Ford, L.G.; Parnes, H.L.; Gaziano, J.M.; et al. Vitamin E and the risk of prostate cancer: The Selenium and Vitamin E Cancer Prevention Trial (SELECT). *JAMA* **2011**, *306*, 1549–1556. [CrossRef] [PubMed]
61. Duffield-Lillico, A.J.; Dalkin, B.L.; Reid, M.E.; Turnbull, B.W.; Slate, E.H.; Jacobs, E.T.; Marshall, J.R.; Clark, L.C. Selenium supplementation, baseline plasma selenium status and incidence of prostate cancer: An analysis of the complete treatment period of the Nutritional Prevention of Cancer Trial. *BJU Int.* **2003**, *91*, 608–612. [CrossRef]
62. Hatfield, D.L.; Gladyshev, V.N. The Outcome of Selenium and Vitamin E Cancer Prevention Trial (SELECT) reveals the need for better understanding of selenium biology. *Mol. Interv.* **2009**, *9*, 18–21. [CrossRef]
63. Dunn, B.K.; Richmond, E.S.; Minasian, L.M.; Ryan, A.M.; Ford, L.G. A nutrient approach to prostate cancer prevention: The Selenium and Vitamin E Cancer Prevention Trial (SELECT). *Nutr. Cancer* **2010**, *62*, 896–918. [CrossRef]
64. Dennert, G.; Zwahlen, M.; Brinkman, M.; Vinceti, M.; Zeegers, M.P.; Horneber, M. Selenium for preventing cancer. *Cochrane Database Syst. Rev.* **2011**, *5*, CD005195. [CrossRef]
65. Brodin, O.; Eksborg, S.; Wallenberg, M.; Asker-Hagelberg, C.; Larsen, E.H.; Mohlkert, D.; Lenneby-Helleday, C.; Jacobsson, H.; Linder, S.; Misra, S.; et al. Pharmacokinetics and Toxicity of Sodium Selenite in the Treatment of Patients with Carcinoma in a Phase I Clinical Trial: The SECAR Study. *Nutrients* **2015**, *7*, 4978–4994. [CrossRef]
66. Subburayan, K.; Thayyullathil, F.; Pallichankandy, S.; Cheratta, A.R.; Galadari, S. Superoxide-mediated ferroptosis in human cancer cells induced by sodium selenite. *Transl. Oncol.* **2020**, *13*, 100843. [CrossRef] [PubMed]
67. Carlisle, A.E.; Lee, N.; Matthew-Onabanjo, A.N.; Spears, M.E.; Park, S.J.; Youkana, D.; Doshi, M.B.; Peppers, A.; Li, R.; Joseph, A.B.; et al. Selenium detoxification is required for cancer-cell survival. *Nat. Metab.* **2020**, *2*, 603–611. [CrossRef]
68. Winkel, L.H.; Vriens, B.; Jones, G.D.; Schneider, L.S.; Pilon-Smits, E.; Banuelos, G.S. Selenium cycling across soil-plant-atmosphere interfaces: A critical review. *Nutrients* **2015**, *7*, 4199–4239. [CrossRef]
69. Jones, G.D.; Droz, B.; Greve, P.; Gottschalk, P.; Poffet, D.; McGrath, S.P.; Seneviratne, S.I.; Smith, P.; Winkel, L.H. Selenium deficiency risk predicted to increase under future climate change. *Proc. Natl. Acad. Sci. USA* **2017**, *114*, 2848–2853. [CrossRef] [PubMed]
70. Schomburg, L. Selenium in sepsis–substitution, supplementation or pro-oxidative bolus? *Crit. Care* **2014**, *18*, 444. [CrossRef] [PubMed]
71. Hurst, R.; Armah, C.N.; Dainty, J.R.; Hart, D.J.; Teucher, B.; Goldson, A.J.; Broadley, M.R.; Motley, A.K.; Fairweather-Tait, S.J. Establishing optimal selenium status: Results of a randomized, double-blind, placebo-controlled trial. *Am. J. Clin. Nutr.* **2010**, *91*, 923–931. [CrossRef]
72. Brodin, O.; Hackler, J.; Misra, S.; Wendt, S.; Sun, Q.; Laaf, E.; Stoppe, C.; Bjornstedt, M.; Schomburg, L. Selenoprotein P as Biomarker of Selenium Status in Clinical Trials with Therapeutic Dosages of Selenite. *Nutrients* **2020**, *12*, 1067. [CrossRef]

73. Combs, G.F., Jr.; Jackson, M.I.; Watts, J.C.; Johnson, L.K.; Zeng, H.; Idso, J.; Schomburg, L.; Hoeg, A.; Hoefig, C.S.; Chiang, E.C.; et al. Differential responses to selenomethionine supplementation by sex and genotype in healthy adults. *Br. J. Nutr.* **2012**, *107*, 1514–1525. [CrossRef]
74. Ashton, K.; Hooper, L.; Harvey, L.J.; Hurst, R.; Casgrain, A.; Fairweather-Tait, S.J. Methods of assessment of selenium status in humans: A systematic review. *Am. J. Clin. Nutr.* **2009**, *89*, 2025S–2039S. [CrossRef]
75. Ralli, M.; Angeletti, D.; Fiore, M.; D'Aguanno, V.; Lambiase, A.; Artico, M.; de Vincentiis, M.; Greco, A. Hashimoto's thyroiditis: An update on pathogenic mechanisms, diagnostic protocols, therapeutic strategies, and potential malignant transformation. *Autoimmun. Rev.* **2020**, *19*, 102649. [CrossRef] [PubMed]
76. Pyzik, A.; Grywalska, E.; Matyjaszek-Matuszek, B.; Rolinski, J. Immune disorders in Hashimoto's thyroiditis: What do we know so far? *J. Immunol. Res.* **2015**, *2015*, 979167. [CrossRef]
77. Derumeaux, H.; Valeix, P.; Castetbon, K.; Bensimon, M.; Boutron-Ruault, M.C.; Arnaud, J.; Hercberg, S. Association of selenium with thyroid volume and echostructure in 35- to 60-year-old French adults. *Eur. J. Endocrinol.* **2003**, *148*, 309–315. [CrossRef] [PubMed]
78. Rasmussen, L.; Schomburg, L.; Köhrle, J.; Pedersen, I.B.; Hollenbach, B.; Hog, A.; Ovesen, L.; Perrild, H.; Laurberg, P. Selenium status, thyroid volume and multiple nodule formation in an area with mild iodine deficiency. *Eur. J. Endocrinol.* **2011**, *164*, 585–590. [CrossRef] [PubMed]
79. Wu, Q.; Rayman, M.P.; Lv, H.; Schomburg, L.; Cui, B.; Gao, C.; Chen, P.; Zhuang, G.; Zhang, Z.; Peng, X.; et al. Low Population Selenium Status Is Associated with Increased Prevalence of Thyroid Disease. *J. Clin. Endocrinol. Metab.* **2015**, *100*, 4037–4047. [CrossRef] [PubMed]
80. Schomburg, L.; Kohrle, J. On the importance of selenium and iodine metabolism for thyroid hormone biosynthesis and human health. *Mol. Nutr. Food Res.* **2008**, *52*, 1235–1246. [CrossRef]
81. Liu, Y.; Huang, H.; Zeng, J.; Sun, C. Thyroid volume, goiter prevalence, and selenium levels in an iodine-sufficient area: A cross-sectional study. *BMC Public Health* **2013**, *13*, 1153. [CrossRef]
82. Ambroziak, U.; Hybsier, S.; Shahnazaryan, U.; Krasnodebska-Kiljanska, M.; Rijntjes, E.; Bartoszewicz, Z.; Bednarczuk, T.; Schomburg, L. Severe selenium deficits in pregnant women irrespective of autoimmune thyroid disease in an area with marginal selenium intake. *J. Trace Elem. Med. Biol.* **2017**, *44*, 186–191. [CrossRef]
83. Mantovani, G.; Isidori, A.M.; Moretti, C.; Di Dato, C.; Greco, E.; Ciolli, P.; Bonomi, M.; Petrone, L.; Fumarola, A.; Campagna, G.; et al. Selenium supplementation in the management of thyroid autoimmunity during pregnancy: Results of the "SERENA study", a randomized, double-blind, placebo-controlled trial. *Endocrine* **2019**, *66*, 542–550. [CrossRef] [PubMed]
84. Varsi, K.; Bolann, B.; Torsvik, I.; Rosvold Eik, T.C.; Hol, P.J.; Bjorke-Monsen, A.L. Impact of Maternal Selenium Status on Infant Outcome during the First 6 Months of Life. *Nutrients* **2017**, *9*, 486. [CrossRef] [PubMed]
85. Di Bari, F.; Granese, R.; Le Donne, M.; Vita, R.; Benvenga, S. Autoimmune Abnormalities of Postpartum Thyroid Diseases. *Front. Endocrinol.* **2017**, *8*, 8. [CrossRef]
86. Negro, R.; Greco, G.; Mangieri, T.; Pezzarossa, A.; Dazzi, D.; Hassan, H. The influence of selenium supplementation on postpartum thyroid status in pregnant women with thyroid peroxidase autoantibodies. *J. Clin. Endocrinol. Metab.* **2007**, *92*, 1263–1268. [CrossRef] [PubMed]
87. Smith, T.J.; Hegedus, L. Graves' Disease. *N. Engl. J. Med.* **2017**, *376*, 185. [CrossRef] [PubMed]
88. Amur, S.; Parekh, A.; Mummaneni, P. Sex differences and genomics in autoimmune diseases. *J. Autoimmun.* **2012**, *38*, J254–J265. [CrossRef] [PubMed]
89. Schomburg, L. Selenium, selenoproteins and the thyroid gland: Interactions in health and disease. *Nat. Rev. Endocrinol.* **2011**, *8*, 160–171. [CrossRef] [PubMed]
90. Winther, K.H.; Rayman, M.P.; Bonnema, S.J.; Hegedus, L. Selenium in thyroid disorders—Essential knowledge for clinicians. *Nat. Rev. Endocrinol.* **2020**, *16*, 165–176. [CrossRef]
91. Mittag, J.; Behrends, T.; Hoefig, C.S.; Vennstrom, B.; Schomburg, L. Thyroid hormones regulate selenoprotein expression and selenium status in mice. *PLoS ONE* **2010**, *5*, e12931. [CrossRef]
92. Wang, Y.; Zhao, F.; Rijntjes, E.; Wu, L.; Wu, Q.; Sui, J.; Liu, Y.; Zhang, M.; He, M.; Chen, P.; et al. Role of Selenium Intake for Risk and Development of Hyperthyroidism. *J. Clin. Endocrinol. Metab.* **2019**, *104*, 568–580. [CrossRef]
93. Murphy, T.M.; Mill, J. Epigenetics in health and disease: Heralding the EWAS era. *Lancet* **2014**, *383*, 1952–1954. [CrossRef]
94. Zhang, L.; Lu, Q.; Chang, C. Epigenetics in Health and Disease. *Adv. Exp. Med. Biol.* **2020**, *1253*, 3–55. [CrossRef] [PubMed]
95. Moulton, V.R. Sex Hormones in Acquired Immunity and Autoimmune Disease. *Front. Immunol.* **2018**, *9*, 2279. [CrossRef] [PubMed]
96. Angum, F.; Khan, T.; Kaler, J.; Siddiqui, L.; Hussain, A. The Prevalence of Autoimmune Disorders in Women: A Narrative Review. *Cureus* **2020**, *12*, e8094. [CrossRef] [PubMed]
97. Yuen, G.J. Autoimmunity in women: An eXamination of eXisting models. *Clin. Immunol.* **2020**, *210*, 108270. [CrossRef]
98. Jones, B.G.; Penkert, R.R.; Surman, S.L.; Sealy, R.E.; Pelletier, S.; Xu, B.; Neale, G.; Maul, R.W.; Gearhart, P.J.; Hurwitz, J.L. Matters of life and death: How estrogen and estrogen receptor binding to the immunoglobulin heavy chain locus may influence outcomes of infection, allergy, and autoimmune disease. *Cell. Immunol.* **2019**, *346*, 103996. [CrossRef] [PubMed]
99. Ortona, E.; Pierdominici, M.; Maselli, A.; Veroni, C.; Aloisi, F.; Shoenfeld, Y. Sex-based differences in autoimmune diseases. *Ann. Ist Super. Sanita* **2016**, *52*, 205–212. [CrossRef] [PubMed]

100. Marcocci, C.; Kahaly, G.J.; Krassas, G.E.; Bartalena, L.; Prummel, M.; Stahl, M.; Altea, M.A.; Nardi, M.; Pitz, S.; Boboridis, K.; et al. Selenium and the course of mild Graves' orbitopathy. *N. Engl. J. Med.* **2011**, *364*, 1920–1931. [CrossRef]
101. Lanzolla, G.; Marino, M.; Marcocci, C. Selenium in the Treatment of Graves' Hyperthyroidism and Eye Disease. *Front. Endocrinol.* **2020**, *11*, 608428. [CrossRef]
102. Duntas, L.H. The evolving role of selenium in the treatment of graves' disease and ophthalmopathy. *J. Thyroid Res.* **2012**, *2012*, 736161. [CrossRef]
103. Craig, M.E.; Kim, K.W.; Isaacs, S.R.; Penno, M.A.; Hamilton-Williams, E.E.; Couper, J.J.; Rawlinson, W.D. Early-life factors contributing to type 1 diabetes. *Diabetologia* **2019**, *62*, 1823–1834. [CrossRef]
104. Buzzetti, R.; Zampetti, S.; Maddaloni, E. Adult-onset autoimmune diabetes: Current knowledge and implications for management. *Nat. Rev. Endocrinol.* **2017**, *13*, 674–686. [CrossRef]
105. Warnes, H.; Helliwell, R.; Pearson, S.M.; Ajjan, R.A. Metabolic Control in Type 1 Diabetes: Is Adjunctive Therapy the Way Forward? *Diabetes Ther.* **2018**, *9*, 1831–1851. [CrossRef]
106. Carstensen, B.; Ronn, P.F.; Jorgensen, M.E. Prevalence, incidence and mortality of type 1 and type 2 diabetes in Denmark 1996–2016. *BMJ Open Diabetes Res. Care* **2020**, *8*, e001071. [CrossRef] [PubMed]
107. Alkorta-Aranburu, G.; Carmody, D.; Cheng, Y.W.; Nelakuditi, V.; Ma, L.; Dickens, J.T.; Das, S.; Greeley, S.A.W.; Del Gaudio, D. Phenotypic heterogeneity in monogenic diabetes: The clinical and diagnostic utility of a gene panel-based next-generation sequencing approach. *Mol. Genet. Metab.* **2014**, *113*, 315–320. [CrossRef]
108. Bollyky, J.B.; Xu, P.; Butte, A.J.; Wilson, D.M.; Beam, C.A.; Greenbaum, C.J.; Type 1 Diabetes TrialNet Study Group. Heterogeneity in recent-onset type 1 diabetes—A clinical trial perspective. *Diabetes Metab. Res. Rev.* **2015**, *31*, 588–594. [CrossRef] [PubMed]
109. Park, Y.; Wintergerst, K.A.; Zhou, Z. Clinical heterogeneity of type 1 diabetes (T1D) found in Asia. *Diabetes Metab. Res. Rev.* **2017**, *33*, e2907. [CrossRef] [PubMed]
110. Rubio-Cabezas, O.; Ellard, S. Diabetes mellitus in neonates and infants: Genetic heterogeneity, clinical approach to diagnosis, and therapeutic options. *Horm. Res. Paediatr.* **2013**, *80*, 137–146. [CrossRef]
111. Ahlqvist, E.; Storm, P.; Karajamaki, A.; Martinell, M.; Dorkhan, M.; Carlsson, A.; Vikman, P.; Prasad, R.B.; Aly, D.M.; Almgren, P.; et al. Novel subgroups of adult-onset diabetes and their association with outcomes: A data-driven cluster analysis of six variables. *Lancet Diabetes Endocrinol.* **2018**, *6*, 361–369. [CrossRef]
112. Kohler, L.N.; Florea, A.; Kelley, C.P.; Chow, S.; Hsu, P.; Batai, K.; Saboda, K.; Lance, P.; Jacobs, E.T. Higher Plasma Selenium Concentrations Are Associated with Increased Odds of Prevalent Type 2 Diabetes. *J. Nutr.* **2018**, *148*, 1333–1340. [CrossRef] [PubMed]
113. Moon, S.; Chung, H.S.; Yu, J.M.; Yoo, H.J.; Park, J.H.; Kim, D.S.; Park, Y.K.; Yoon, S.N. Association between serum selenium level and the prevalence of diabetes mellitus in U.S. population. *J. Trace Elem. Med. Biol.* **2019**, *52*, 83–88. [CrossRef]
114. Xi, J.; Zhang, Q.; Wang, J.; Guo, R.; Wang, L. Factors Influencing Selenium Concentration in Community-Dwelling Patients with Type 2 Diabetes Mellitus. *Biol. Trace Elem. Res.* **2020**, *199*, 1657–1663. [CrossRef]
115. Mao, J.; Teng, W. The relationship between selenoprotein P and glucose metabolism in experimental studies. *Nutrients* **2013**, *5*, 1937–1948. [CrossRef]
116. Speckmann, B.; Sies, H.; Steinbrenner, H. Attenuation of hepatic expression and secretion of selenoprotein P by metformin. *Biochem. Biophys. Res. Commun.* **2009**, *387*, 158–163. [CrossRef] [PubMed]
117. Wang, Y.; Rijntjes, E.; Wu, Q.; Lv, H.; Gao, C.; Shi, B.; Schomburg, L. Selenium deficiency is linearly associated with hypoglycemia in healthy adults. *Redox Biol.* **2020**, *37*, 101709. [CrossRef]
118. Kohler, L.N.; Foote, J.; Kelley, C.P.; Florea, A.; Shelly, C.; Chow, H.S.; Hsu, P.; Batai, K.; Ellis, N.; Saboda, K.; et al. Selenium and Type 2 Diabetes: Systematic Review. *Nutrients* **2018**, *10*, 1924. [CrossRef] [PubMed]
119. Strozyk, A.; Osica, Z.; Przybylak, J.D.; Kolodziej, M.; Zalewski, B.M.; Mrozikiewicz-Rakowska, B.; Szajewska, H. Effectiveness and safety of selenium supplementation for type 2 diabetes mellitus in adults: A systematic review of randomised controlled trials. *J. Hum. Nutr. Diet.* **2019**, *32*, 635–645. [CrossRef] [PubMed]
120. Gebre-Medhin, M.; Ewald, U.; Plantin, L.O.; Tuvemo, T. Elevated serum selenium in diabetic children. *Acta Paediatr. Scand.* **1984**, *73*, 109–114. [CrossRef] [PubMed]
121. Alghobashy, A.A.; Alkholy, U.M.; Talat, M.A.; Abdalmonem, N.; Zaki, A.; Ahmed, I.A.; Mohamed, R.H. Trace elements and oxidative stress in children with type 1 diabetes mellitus. *Diabetes Metab. Syndr. Obes. Targets Ther.* **2018**, *11*, 85–92. [CrossRef]
122. Kruse-Jarres, J.D. Limited usefulness of essential trace element analyses in hair. *Am. Clin. Lab.* **2000**, *19*, 8–10.
123. Salmonowicz, B.; Krzystek-Korpacka, M.; Noczynska, A. Trace elements, magnesium, and the efficacy of antioxidant systems in children with type 1 diabetes mellitus and in their siblings. *Adv. Clin. Exp. Med.* **2014**, *23*, 259–268. [CrossRef]
124. Sobczak, A.I.S.; Stefanowicz, F.; Pitt, S.J.; Ajjan, R.A.; Stewart, A.J. Total plasma magnesium, zinc, copper and selenium concentrations in type-I and type-II diabetes. *Biometals* **2019**, *32*, 123–138. [CrossRef]
125. Fandrich, F.; Ungefroren, H. Customized cell-based treatment options to combat autoimmunity and restore beta-cell function in type 1 diabetes mellitus: Current protocols and future perspectives. *Adv. Exp. Med. Biol.* **2010**, *654*, 641–665. [CrossRef] [PubMed]
126. Donath, M.Y.; Dinarello, C.A.; Mandrup-Poulsen, T. Targeting innate immune mediators in type 1 and type 2 diabetes. *Nat. Rev. Immunol.* **2019**, *19*, 734–746. [CrossRef]
127. Renko, K.; Hoefig, C.S.; Hiller, F.; Schomburg, L.; Köhrle, J. Identification of iopanoic acid as substrate of type 1 deiodinase by a novel nonradioactive iodide-release assay. *Endocrinology* **2012**, *153*, 2506–2513. [CrossRef] [PubMed]

128. Mostert, V. Selenoprotein P: Properties, functions, and regulation. *Arch. Biochem. Biophys.* **2000**, *376*, 433–438. [CrossRef] [PubMed]
129. Stoedter, M.; Renko, K.; Hog, A.; Schomburg, L. Selenium controls the sex-specific immune response and selenoprotein expression during the acute-phase response in mice. *Biochem.J.* **2010**, *429*, 43–51. [CrossRef]
130. Wang, W.C.; Makela, A.L.; Nanto, V.; Makela, P. Serum selenium levels in diabetic children. A followup study during selenium-enriched agricultural fertilization in Finland. *Biol. Trace Elem. Res.* **1995**, *47*, 355–364. [CrossRef]
131. Smolen, J.S.; Aletaha, D.; Barton, A.; Burmester, G.R.; Emery, P.; Firestein, G.S.; Kavanaugh, A.; McInnes, I.B.; Solomon, D.H.; Strand, V.; et al. Rheumatoid arthritis. *Nat. Rev. Dis. Primers* **2018**, *4*, 18001. [CrossRef]
132. Cutolo, M. Rheumatoid arthritis: Circadian and circannual rhythms in RA. *Nat. Rev. Rheumatol.* **2011**, *7*, 500–502. [CrossRef]
133. Lambert, N.C. Nonendocrine mechanisms of sex bias in rheumatic diseases. *Nat. Rev. Rheumatol.* **2019**, *15*, 673–686. [CrossRef] [PubMed]
134. O'Dell, J.R.; Lemley-Gillespie, S.; Palmer, W.R.; Weaver, A.L.; Moore, G.F.; Klassen, L.W. Serum selenium concentrations in rheumatoid arthritis. *Ann. Rheum. Dis.* **1991**, *50*, 376–378. [CrossRef] [PubMed]
135. Sun, Q.; Hackler, J.; Hilger, J.; Gluschke, H.; Muric, A.; Simmons, S.; Schomburg, L.; Siegert, E. Selenium and Copper as Biomarkers for Pulmonary Arterial Hypertension in Systemic Sclerosis. *Nutrients* **2020**, *12*, 1894. [CrossRef]
136. Yu, N.; Han, F.; Lin, X.J.; Tang, C.; Ye, J.H.; Cai, X.Y. The Association Between Serum Selenium Levels with Rheumatoid Arthritis. *Biol. Trace Elem. Res.* **2016**, *172*, 46–52. [CrossRef]
137. Ma, Y.; Zhang, X.; Fan, D.; Xia, Q.; Wang, M.; Pan, F. Common trace metals in rheumatoid arthritis: A systematic review and meta-analysis. *J. Trace Elem. Med. Biol.* **2019**, *56*, 81–89. [CrossRef] [PubMed]
138. Hellman, N.E.; Gitlin, J.D. Ceruloplasmin metabolism and function. *Annu. Rev. Nutr.* **2002**, *22*, 439–458. [CrossRef]
139. Hackler, J.; Wisniewska, M.; Greifenstein-Wiehe, L.; Minich, W.B.; Cremer, M.; Buhrer, C.; Schomburg, L. Copper and selenium status as biomarkers of neonatal infections. *J. Trace Elem. Med. Biol.* **2020**, *58*, 126437. [CrossRef] [PubMed]
140. Franco, P.; Laura, F.; Valentina, C.; Simona, A.; Gloria, A.; Eleonora, N. Interleukin-6 in Rheumatoid Arthritis. *Int. J. Mol. Sci.* **2020**, *21*, 5238. [CrossRef]
141. Dinarello, C.A. The IL-1 family of cytokines and receptors in rheumatic diseases. *Nat. Rev. Rheumatol.* **2019**, *15*, 612–632. [CrossRef] [PubMed]
142. Forceville, X.; Van Antwerpen, P.; Preiser, J.C. Selenocompounds and Sepsis: Redox Bypass Hypothesis for Early Diagnosis and Treatment: Part A-Early Acute Phase of Sepsis: An Extraordinary Redox Situation (Leukocyte/Endothelium Interaction Leading to Endothelial Damage). *Antioxid. Redox Signal.* **2021**, *35*, 113–138. [CrossRef]
143. Kretz-Remy, C.; Arrigo, A.P. Selenium: A key element that controls NF-kappa B activation and I kappa B alpha half life. *Biofactors* **2001**, *14*, 117–125. [CrossRef]
144. Pothion, H.; Jehan, C.; Tostivint, H.; Cartier, D.; Bucharles, C.; Falluel-Morel, A.; Boukhzar, L.; Anouar, Y.; Lihrmann, I. Selenoprotein T: An Essential Oxidoreductase Serving as a Guardian of Endoplasmic Reticulum Homeostasis. *Antioxid. Redox Signal.* **2020**, *33*, 1257–1275. [CrossRef] [PubMed]
145. Addinsall, A.B.; Wright, C.R.; Andrikopoulos, S.; van der Poel, C.; Stupka, N. Emerging roles of endoplasmic reticulum-resident selenoproteins in the regulation of cellular stress responses and the implications for metabolic disease. *Biochem. J.* **2018**, *475*, 1037–1057. [CrossRef] [PubMed]
146. Chernorudskiy, A.; Varone, E.; Colombo, S.F.; Fumagalli, S.; Cagnotto, A.; Cattaneo, A.; Briens, M.; Baltzinger, M.; Kuhn, L.; Bachi, A.; et al. Selenoprotein N is an endoplasmic reticulum calcium sensor that links luminal calcium levels to a redox activity. *Proc. Natl. Acad. Sci. USA* **2020**, *117*, 21288–21298. [CrossRef]
147. Ren, B.; Liu, M.; Ni, J.; Tian, J. Role of Selenoprotein F in Protein Folding and Secretion: Potential Involvement in Human Disease. *Nutrients* **2018**, *10*, 1619. [CrossRef]
148. Zhang, X.; Xiong, W.; Chen, L.L.; Huang, J.Q.; Lei, X.G. Selenoprotein V protects against endoplasmic reticulum stress and oxidative injury induced by pro-oxidants. *Free Radic. Biol. Med.* **2020**, *160*, 670–679. [CrossRef] [PubMed]
149. Curran, J.E.; Jowett, J.B.; Elliott, K.S.; Gao, Y.; Gluschenko, K.; Wang, J.; Abel Azim, D.M.; Cai, G.; Mahaney, M.C.; Comuzzie, A.G.; et al. Genetic variation in selenoprotein S influences inflammatory response. *Nat. Genet.* **2005**, *37*, 1234–1241. [CrossRef]
150. Marinou, I.; Walters, K.; Dickson, M.C.; Binks, M.H.; Bax, D.E.; Wilson, A.G. Evidence of epistasis between interleukin 1 and selenoprotein-S with susceptibility to rheumatoid arthritis. *Ann. Rheum. Dis.* **2009**, *68*, 1494–1497. [CrossRef]
151. Deyab, G.; Hokstad, I.; Aaseth, J.; Smastuen, M.C.; Whist, J.E.; Agewall, S.; Lyberg, T.; Tveiten, D.; Hjeltnes, G.; Zibara, K.; et al. Effect of anti-rheumatic treatment on selenium levels in inflammatory arthritis. *J. Trace Elem. Med. Biol.* **2018**, *49*, 91–97. [CrossRef] [PubMed]
152. Sanmartin, C.; Plano, D.; Font, M.; Palop, J.A. Selenium and clinical trials: New therapeutic evidence for multiple diseases. *Curr. Med. Chem.* **2011**, *18*, 4635–4650. [CrossRef] [PubMed]
153. Tarp, U.; Hansen, J.C.; Overvad, K.; Thorling, E.B.; Tarp, B.D.; Graudal, H. Glutathione peroxidase activity in patients with rheumatoid arthritis and in normal subjects: Effects of long-term selenium supplementation. *Arthritis Rheum.* **1987**, *30*, 1162–1166. [CrossRef] [PubMed]
154. Peretz, A.; Siderova, V.; Neve, J. Selenium supplementation in rheumatoid arthritis investigated in a double blind, placebo-controlled trial. *Scand. J. Rheumatol.* **2001**, *30*, 208–212. [CrossRef]

155. Duntas, L.H. Selenium and inflammation: Underlying anti-inflammatory mechanisms. *Horm. Metab. Res.* **2009**, *41*, 443–447. [CrossRef]
156. Winther, K.H.; Watt, T.; Bjorner, J.B.; Cramon, P.; Feldt-Rasmussen, U.; Gluud, C.; Gram, J.; Groenvold, M.; Hegedus, L.; Knudsen, N.; et al. The chronic autoimmune thyroiditis quality of life selenium trial (CATALYST): Study protocol for a randomized controlled trial. *Trials* **2014**, *15*, 115. [CrossRef] [PubMed]
157. Watt, T.; Cramon, P.; Bjorner, J.B.; Bonnema, S.J.; Feldt-Rasmussen, U.; Gluud, C.; Gram, J.; Hansen, J.L.; Hegedus, L.; Knudsen, N.; et al. Selenium supplementation for patients with Graves' hyperthyroidism (the GRASS trial): Study protocol for a randomized controlled trial. *Trials* **2013**, *14*, 119. [CrossRef] [PubMed]
158. Huang, Z.; Rose, A.H.; Hoffmann, P.R. The role of selenium in inflammation and immunity: From molecular mechanisms to therapeutic opportunities. *Antioxid. Redox Signal.* **2012**, *16*, 705–743. [CrossRef]
159. Qian, F.; Misra, S.; Prabhu, K.S. Selenium and selenoproteins in prostanoid metabolism and immunity. *Crit. Rev. Biochem. Mol. Biol.* **2019**, *54*, 484–516. [CrossRef]
160. Ren, Z.; Fan, Y.; Zhang, Z.; Chen, C.; Chen, C.; Wang, X.; Deng, J.; Peng, G.; Hu, Y.; Cao, S.; et al. Sodium selenite inhibits deoxynivalenol-induced injury in GPX1-knockdown porcine splenic lymphocytes in culture. *Sci. Rep.* **2018**, *8*, 17676. [CrossRef]
161. Matsushita, M.; Freigang, S.; Schneider, C.; Conrad, M.; Bornkamm, G.W.; Kopf, M. T cell lipid peroxidation induces ferroptosis and prevents immunity to infection. *J. Exp. Med.* **2015**, *212*, 555–568. [CrossRef]
162. Van der Spek, A.H.; Fliers, E.; Boelen, A. Thyroid Hormone and Deiodination in Innate Immune Cells. *Endocrinology* **2021**, *162*, bqaa200. [CrossRef]
163. Lee, B.C.; Lee, S.G.; Choo, M.K.; Kim, J.H.; Lee, H.M.; Kim, S.; Fomenko, D.E.; Kim, H.Y.; Park, J.M.; Gladyshev, V.N. Selenoprotein MsrB1 promotes anti-inflammatory cytokine gene expression in macrophages and controls immune response in vivo. *Sci. Rep.* **2017**, *7*, 5119. [CrossRef] [PubMed]
164. Cox, A.G.; Tsomides, A.; Kim, A.J.; Saunders, D.; Hwang, K.L.; Evason, K.J.; Heidel, J.; Brown, K.K.; Yuan, M.; Lien, E.C.; et al. Selenoprotein H is an essential regulator of redox homeostasis that cooperates with p53 in development and tumorigenesis. *Proc. Natl. Acad. Sci. USA* **2016**, *113*, E5562–E5571. [CrossRef]
165. Ma, C.; Hoffmann, F.W.; Marciel, M.P.; Page, K.E.; Williams-Aduja, M.A.; Akana, E.N.L.; Gojanovich, G.S.; Gerschenson, M.; Urschitz, J.; Moisyadi, S.; et al. Upregulated ethanolamine phospholipid synthesis via selenoprotein I is required for effective metabolic reprogramming during T cell activation. *Mol. Metab.* **2021**, *47*, 101170. [CrossRef] [PubMed]
166. Verma, S.; Hoffmann, F.W.; Kumar, M.; Huang, Z.; Roe, K.; Nguyen-Wu, E.; Hashimoto, A.S.; Hoffmann, P.R. Selenoprotein K knockout mice exhibit deficient calcium flux in immune cells and impaired immune responses. *J. Immunol.* **2011**, *186*, 2127–2137. [CrossRef] [PubMed]
167. Xu, X.M.; Carlson, B.A.; Irons, R.; Mix, H.; Zhong, N.X.; Gladyshev, V.N.; Hatfield, D.L. Selenophosphate synthetase 2 is essential for selenoprotein biosynthesis. *Biochem. J.* **2007**, *404*, 115–120. [CrossRef] [PubMed]
168. Yim, S.H.; Everley, R.A.; Schildberg, F.A.; Lee, S.G.; Orsi, A.; Barbati, Z.R.; Karatepe, K.; Fomenko, D.E.; Tsuji, P.A.; Luo, H.R.; et al. Role of Selenof as a Gatekeeper of Secreted Disulfide-Rich Glycoproteins. *Cell Rep.* **2018**, *23*, 1387–1398. [CrossRef] [PubMed]
169. Kirsch, J.; Schneider, H.; Pagel, J.I.; Rehberg, M.; Singer, M.; Hellfritsch, J.; Chillo, O.; Schubert, K.M.; Qiu, J.; Pogoda, K.; et al. Endothelial Dysfunction, and A Prothrombotic, Proinflammatory Phenotype Is Caused by Loss of Mitochondrial Thioredoxin Reductase in Endothelium. *Arterioscler. Thromb. Vasc. Biol.* **2016**, *36*, 1891–1899. [CrossRef]
170. Shchedrina, V.A.; Zhang, Y.; Labunskyy, V.M.; Hatfield, D.L.; Gladyshev, V.N. Structure-function relations, physiological roles, and evolution of mammalian ER-resident selenoproteins. *Antioxid. Redox Signal.* **2010**, *12*, 839–849. [CrossRef]
171. Cui, Z.; Huang, J.; Peng, Q.; Yu, D.; Wang, S.; Liang, D. Risk assessment for human health in a seleniferous area, Shuang'an, China. *Environ. Sci Pollut. Res. Int* **2017**, *24*, 17701–17710. [CrossRef]
172. Hoefig, C.S.; Renko, K.; Kohrle, J.; Birringer, M.; Schomburg, L. Comparison of different selenocompounds with respect to nutritional value vs. toxicity using liver cells in culture. *J. Nutr. Biochem.* **2011**, *22*, 945–955. [CrossRef] [PubMed]
173. Suzuki, K.T.; Ogra, Y. Metabolic pathway for selenium in the body: Speciation by HPLC-ICP MS with enriched Se. *Food Addit. Contam.* **2002**, *19*, 974–983. [CrossRef] [PubMed]
174. Yang, G.Q.; Wang, S.Z.; Zhou, R.H.; Sun, S.Z. Endemic selenium intoxication of humans in China. *Am. J. Clin. Nutr.* **1983**, *37*, 872–881. [CrossRef]
175. Kipp, A.P.; Strohm, D.; Brigelius-Flohe, R.; Schomburg, L.; Bechthold, A.; Leschik-Bonnet, E.; Heseker, H.; German Nutrition, S. Revised reference values for selenium intake. *J. Trace Elem. Med. Biol.* **2015**, *32*, 195–199. [CrossRef]
176. Desta, B.; Maldonado, G.; Reid, H.; Puschner, B.; Maxwell, J.; Agasan, A.; Humphreys, L.; Holt, T. Acute selenium toxicosis in polo ponies. *J. Vet. Diagn Investig.* **2011**, *23*, 623–628. [CrossRef]
177. Morris, J.S.; Crane, S.B. Selenium toxicity from a misformulated dietary supplement, adverse health effects, and the temporal response in the nail biologic monitor. *Nutrients* **2013**, *5*, 1024–1057. [CrossRef]
178. D'Oria, L.; Apicella, M.; De Luca, C.; Licameli, A.; Neri, C.; Pellegrino, M.; Simeone, D.; De Santis, M. Chronic exposure to high doses of selenium in the first trimester of pregnancy: Case report and brief literature review. *Birth Defects Res.* **2018**, *110*, 372–375. [CrossRef]
179. Hadrup, N.; Ravn-Haren, G. Acute human toxicity and mortality after selenium ingestion: A review. *J. Trace Elem. Med. Biol.* **2020**, *58*, 126435. [CrossRef] [PubMed]
180. Geenen, V. The thymus and the science of self. *Semin Immunopathol.* **2021**, *43*, 5–14. [CrossRef]

181. Wintergerst, E.S.; Maggini, S.; Hornig, D.H. Contribution of selected vitamins and trace elements to immune function. *Ann. Nutr. Metab.* **2007**, *51*, 301–323. [CrossRef]
182. Ma, C.; Hoffmann, P.R. Selenoproteins as regulators of T cell proliferation, differentiation, and metabolism. *Sem Cell Dev. Biol* **2021**, *115*, 54–61. [CrossRef] [PubMed]
183. Dreher, I.; Jakobs, T.C.; Kohrle, J. Cloning and characterization of the human selenoprotein P promoter. Response of selenoprotein P expression to cytokines in liver cells. *J. Biol. Chem.* **1997**, *272*, 29364–29371. [CrossRef] [PubMed]
184. Nichol, C.; Herdman, J.; Sattar, N.; O'Dwyer, P.J.; O'Reilly, D.S.J.; Littlejohn, D.; Fell, G. Changes in the concentrations of plasma selenium and selenoproteins after minor elective surgery: Further evidence for a negative acute phase response? *Clin. Chem.* **1998**, *44*, 1764–1766. [CrossRef] [PubMed]
185. Renko, K.; Hofmann, P.J.; Stoedter, M.; Hollenbach, B.; Behrends, T.; Kohrle, J.; Schweizer, U.; Schomburg, L. Down-regulation of the hepatic selenoprotein biosynthesis machinery impairs selenium metabolism during the acute phase response in mice. *FASEB J.* **2009**, *23*, 1758–1765. [CrossRef]
186. Sherlock, L.G.; Sjostrom, K.; Lei, S.A.; Delaney, C.; Tipple, T.E.; Krebs, N.F.; Nozik-Grayck, E.; Wright, C.J. Hepatic-Specific Decrease in the Expression of Selenoenzymes and Factors Essential for Selenium Processing After Endotoxemia. *Front. Immunol.* **2020**, *11*, 595282. [CrossRef]
187. Moghaddam, A.; Heller, R.A.; Sun, Q.; Seelig, J.; Cherkezov, A.; Seibert, L.; Hackler, J.; Seemann, P.; Diegmann, J.; Pilz, M.; et al. Selenium Deficiency Is Associated with Mortality Risk from COVID-19. *Nutrients* **2020**, *12*, 2098. [CrossRef] [PubMed]
188. Forceville, X.; Vitoux, D.; Gauzit, R.; Combes, A.; Lahilaire, P.; Chappuis, P. Selenium, systemic immune response syndrome, sepsis, and outcome in critically ill patients. *Crit. Care Med.* **1998**, *26*, 1536–1544. [CrossRef]
189. Meyer, H.A.; Endermann, T.; Stephan, C.; Stoedter, M.; Behrends, T.; Wolff, I.; Jung, K.; Schomburg, L. Selenoprotein P Status Correlates to Cancer-Specific Mortality in Renal Cancer Patients. *PLoS ONE* **2012**, *7*, e46644. [CrossRef]
190. Gul-Klein, S.; Haxhiraj, D.; Seelig, J.; Kastner, A.; Hackler, J.; Sun, Q.; Heller, R.A.; Lachmann, N.; Pratschke, J.; Schmelzle, M.; et al. Serum Selenium Status as a Diagnostic Marker for the Prognosis of Liver Transplantation. *Nutrients* **2021**, *13*, 619. [CrossRef]
191. Capelle, C.M.; Zeng, N.; Danileviciute, E.; Rodrigues, S.F.; Ollert, M.; Balling, R.; He, F.Q. Identification of VIMP as a gene inhibiting cytokine production in human CD4+ effector T cells. *Iscience* **2021**, *24*, 102289. [CrossRef] [PubMed]
192. Kieliszek, M.; Lipinski, B. Selenium supplementation in the prevention of coronavirus infections (COVID-19). *Med. Hypotheses* **2020**, *143*, 109878. [CrossRef]
193. Notz, Q.; Herrmann, J.; Schlesinger, T.; Helmer, P.; Sudowe, S.; Sun, Q.; Hackler, J.; Roeder, D.; Lotz, C.; Meybohm, P.; et al. Clinical Significance of Micronutrient Supplementation in Critically Ill COVID-19 Patients with Severe ARDS. *Nutrients* **2021**, *13*, 2113. [CrossRef] [PubMed]
194. Candan, S.A.; Elibol, N.; Abdullahi, A. Consideration of prevention and management of long-term consequences of post-acute respiratory distress syndrome in patients with COVID-19. *Physiother. Theory Pract.* **2020**, *36*, 663–668. [CrossRef]
195. Nalbandian, A.; Sehgal, K.; Gupta, A.; Madhavan, M.V.; McGroder, C.; Stevens, J.S.; Cook, J.R.; Nordvig, A.S.; Shalev, D.; Sehrawat, T.S.; et al. Post-acute COVID-19 syndrome. *Nat. Med.* **2021**, *27*, 601–615. [CrossRef] [PubMed]
196. Qiu, C.C.; Caricchio, R.; Gallucci, S. Triggers of Autoimmunity: The Role of Bacterial Infections in the Extracellular Exposure of Lupus Nuclear Autoantigens. *Front. Immunol.* **2019**, *10*, 2608. [CrossRef]
197. Bornstein, S.R.; Voit-Bak, K.; Donate, T.; Rodionov, R.N.; Gainetdinov, R.R.; Tselmin, S.; Kanczkowski, W.; Muller, G.M.; Achleitner, M.; Wang, J.; et al. Chronic post-COVID-19 syndrome and chronic fatigue syndrome: Is there a role for extracorporeal apheresis? *Mol. Psychiatry* **2021**, in press. [CrossRef]
198. Muecke, R.; Micke, O.; Schomburg, L.; Buentzel, J.; Kisters, K.; Adamietz, I.A. Selenium in Radiation Oncology-15 Years of Experiences in Germany. *Nutrients* **2018**, *10*, 483. [CrossRef] [PubMed]
199. Yshii, L.; Bost, C.; Liblau, R. Immunological Bases of Paraneoplastic Cerebellar Degeneration and Therapeutic Implications. *Front. Immunol.* **2020**, *11*, 991. [CrossRef]
200. Braunstein, M.; Kusmenkov, T.; Zuck, C.; Angstwurm, M.; Becker, N.P.; Bocker, W.; Schomburg, L.; Bogner-Flatz, V. Selenium and Selenoprotein P Deficiency Correlates With Complications and Adverse Outcome After Major Trauma. *Shock* **2020**, *53*, 63–70. [CrossRef] [PubMed]
201. Asherson, R.A. The catastrophic antiphospholipid syndrome, 1998. A review of the clinical features, possible pathogenesis and treatment. *Lupus* **1998**, *7* (Suppl. 2), S55–S62. [CrossRef]
202. Ben-Hamouda, N.; Charriere, M.; Voirol, P.; Berger, M.M. Massive copper and selenium losses cause life-threatening deficiencies during prolonged continuous renal replacement. *Nutrition* **2017**, *34*, 71–75. [CrossRef] [PubMed]
203. Mai, Y.; Nishie, W.; Sato, K.; Hotta, M.; Izumi, K.; Ito, K.; Hosokawa, K.; Shimizu, H. Bullous Pemphigoid Triggered by Thermal Burn Under Medication with a Dipeptidyl Peptidase-IV Inhibitor: A Case Report and Review of the Literature. *Front. Immunol.* **2018**, *9*, 542. [CrossRef]
204. Arnaud, J.; Bertrais, S.; Roussel, A.M.; Arnault, N.; Ruffieux, D.; Favier, A.; Berthelin, S.; Estaquio, C.; Galan, P.; Czernichow, S.; et al. Serum selenium determinants in French adults: The SU.VI.M.AX study. *Br. J. Nutr.* **2006**, *95*, 313–320. [CrossRef] [PubMed]
205. Perricone, C.; Versini, M.; Ben-Ami, D.; Gertel, S.; Watad, A.; Segel, M.J.; Ceccarelli, F.; Conti, F.; Cantarini, L.; Bogdanos, D.P.; et al. Smoke and autoimmunity: The fire behind the disease. *Autoimmun. Rev.* **2016**, *15*, 354–374. [CrossRef]